ACIDIFICATION RESEARCH IN THE NETHERLANDS: FINAL REPORT

ACIDIFICATION RESEARCH IN THE NETHERLANDS

FINAL REPORT OF THE DUTCH PRIORITY PROGRAMME ON ACIDIFICATION

Studies in Environmental Science 46

ACIDIFICATION RESEARCH IN THE NETHERLANDS

FINAL REPORT OF THE DUTCH PRIORITY PROGRAMME ON ACIDIFICATION

Edited by

G.J. Heij and **T. Schneider**

Rijksinstituut voor Volksgezondheid en Milieuhygiëne (RIVM),
3720 BA Bilthoven, The Netherlands

ELSEVIER
Amsterdam — London — New York — Tokyo 1991

ELSEVIER SCIENCE PUBLISHERS B.V.
Molenwerf 1
P.O. Box 211, 1000 AE Amsterdam, The Netherlands

Distributors for the United States and Canada:

ELSEVIER SCIENCE PUBLISHING COMPANY INC.
655, Avenue of the Americas
New York, NY 10010, U.S.A.

ISBN 0-444-88831-4

This book is printed on acid-free paper.

Printed in The Netherlands

Studies in Environmental Science
Other volumes in this series

1 **Atmospheric Pollution 1978** edited by M.M. Benarie
2 **Air Pollution Reference Measurement Methods and Systems** edited by T. Schneider, H.W. de Koning and L.J. Brasser
3 **Biogeochemical Cycling of Mineral-Forming Elements** edited by P.A. Trudinger and D.J. Swaine
4 **Potential Industrial Carcinogens and Mutagens** by L. Fishbein
5 **Industrial Waste Management** by S.E. Jørgensen
6 **Trade and Environment: A Theoretical Enquiry** by H. Siebert, J. Eichberger, R. Gronych and R. Pethig
7 **Field Worker Exposure during Pesticide Application** edited by W.F. Tordoir and E.A.H. van Heemstra-Lequin
8 **Atmospheric Pollution 1980** edited by M.M. Benarie
9 **Energetics and Technology of Biological Elimination of Wastes** edited by G. Milazzo
10 **Bioengineering, Thermal Physiology and Comfort** edited by K. Cena and J.A. Clark
11 **Atmospheric Chemistry. Fundamental Aspects** by E. Mészáros
12 **Water Supply and Health** edited by H. van Lelyveld and B.C.J. Zoeteman
13 **Man under Vibration. Suffering and Protection** edited by G. Bianchi, K.V. Frolov and A. Oledzki
14 **Principles of Environmental Science and Technology** by S.E. Jørgensen and I. Johnsen
15 **Disposal of Radioactive Wastes** by Z. Dlouhý
16 **Mankind and Energy** edited by A. Blanc-Lapierre
17 **Quality of Groundwater** edited by W. van Duijvenbooden, P. Glasbergen and H. van Lelyveld
18 **Education and Safe Handling in Pesticide Application** edited by E.A.H. van Heemstra-Lequin and W.F. Tordoir
19 **Physicochemical Methods for Water and Wastewater Treatment** edited by L. Pawlowski
20 **Atmospheric Pollution 1982** edited by M.M. Benarie
21 **Air Pollution by Nitrogen Oxides** edited by T. Schneider and L. Grant
22 **Environmental Radioanalysis** by H.A. Das, A. Faanhof and H.A. van der Sloot
23 **Chemistry for Protection of the Environment** edited by L. Pawlowski, A.J. Verdier and W.J. Lacy
24 **Determination and Assessment of Pesticide Exposure** edited by M. Siewierski
25 **The Biosphere: Problems and Solutions** edited by T.N. Veziroğlu
26 **Chemical Events in the Atmosphere and their Impact on the Environment** edited by G.B. Marini-Bettòlo
27 **Fluoride Research 1985** edited by H. Tsunoda and Ming-Ho Yu
28 **Algal Biofouling** edited by L.V. Evans and K.D. Hoagland
29 **Chemistry for Protection of the Environment 1985** edited by L. Pawlowski, G. Alaerts and W.J. Lacy
30 **Acidification and its Policy Implications** edited by T. Schneider
31 **Teratogens: Chemicals which Cause Birth Defects** edited by V. Kolb Meyers
32 **Pesticide Chemistry** by G. Matolcsy, M. Nádasy and V. Andriska
33 **Principles of Environmental Science and Technology (second revised edition)** by S.E. Jørgensen and I. Johnsen
34 **Chemistry for Protection of the Environment 1987** edited by L. Pawlowski, E. Mentasti, C. Sarzanini and W.J. Lacy

35 **Atmospheric Ozone Research and its Policy Implications** edited by T. Schneider, S.D. Lee, G.J.R. Wolters and L.D. Grant
36 **Valuation Methods and Policy Making in Environmental Economics** edited by H. Folmer and E. van Ierland
37 **Asbestos in the Natural Environment** by H. Schreier
38 **How to Conquer Air Pollution. A Japanese Experience** edited by H. Nishimura
39 **Aquatic Bioenvironmental Studies: The Hanford Experience, 1944–1984** by C.D. Becker
40 **Radon in the Environment** by M. Wilkening
41 **Evaluation of Environmental Data for Regulatory and Impact Assessment** by S. Ramamoorthy and E. Baddaloo
42 **Environmental Biotechnology** edited by A. Blazej and V. Prívarová
43 **Applied Isotope Hydrogeology** by F.J. Pearson, Jr., W. Balderer, H.H. Loosli, B.E. Lehmann, A. Matter, Tj. Peters, H. Schmassmann and A. Gautschi
44 **Highway Pollution** edited by R.S. Hamilton and R.M. Harrison
45 **Freight Transport and the Environment** edited by M. Kroon, R. Smit and J. van Ham

CONTENTS

ACKNOWLEDGEMENTS

This report contains the results and conclusions of extensive research (the second phase of the Dutch Priority Programme on Acidification) on the causes and effects of acidification of forests/forest soils and heathland in the Netherlands. This research was financed by the Ministeries of Housing, Physical Planning and the Environment, of Agriculture, Nature Management and Fisheries, of Economic Affairs, and of Transport and Public Works, and also by the refineries and producers of electricity.

The research itself, which formed the basis for the final report, has been described in the reports on individual projects. A summary of the scientific results and conclusions is given in the thematic reports. This report gives results and conclusions relevant for policy-making. The Project Group Acidification Research is responsible for all these reports.

Apart from the Project Group, many people have contributed to this final report by supplying additional information, writing part of the text, or commenting upon it. We especially wish to mention those who, together with people from the Project Group, worked on the final report: W.de Vries (Winand Staring Centre), J.W.Erisman (RIVM), Mrs.H.Marseille, also on behalf of V.G.Keizer (VROM), P.Evers and G.M.J.Mohren (De Dorschkamp), J.G.M.Roelofs (Catholic University of Nijmegen), and A.H.M.Bresser (RIVM). The following people contributed to the individual chapters:

Chapter 2	K.de Winkel and H.Hannessen (VROM), R.Thomas, J.van Til and K.W.van der Hoek (RIVM)
Chapter 3	F.G.Römer (KEMA), P.A.van den Tweel (LNV), F.A.A.M.de Leeuw, and J.A.van Jaarsveld (RIVM)
Chapter 4	M.Kropff, W.L.M.Smeets, L.W.A.van Hove, Mrs.A.Jansen (Agricultural University of Wageningen), L.J.M.van der Eerden and F.Tonneijck (IPO), P.A.van den Tweel and G.van Tol (LNV), H.Visser (KEMA) and P.P.Th.M. Maessen (De Dorschkamp), J.J.M.van Grinsven and A.J.Schouten (RIVM)
Chapter 5	A.H.Bakema (RIVM)
Chapter 6	P.A.van den Tweel
Chapter 7	L.J.M.van der Eerden, P.A.van den Tweel, and J.J.M.van Grinsven

Mrs.O.van Steenis, assisted by Mrs.T.Buijtendijk-Olij, has put an enormous amount of work into the realization of the report and played a key role here. Contributions were also made by M.J.C.Middelburg and A.J. Berends (graphic design) and J.J.Winters (reproduction). The translation was done by Mrs.A.J.Blommesteijn, Mrs.S.Dickerson,

T.C.Tinkler and J.Burn.
We would like to thank them all for their contributions to this final report.

T. Schneider, Programme Director G.J. Heij, Secretary

SUMMARY

1. INTRODUCTION

In 1985 the Dutch Priority Programme on Acidification (Additioneel Programma Verzuringsonderzoek, APV) was started, in order to give a more concrete form to the increasing interest of policy-makers in the effects of air pollution on ecosystems in particular. One of the most important reasons for this growing interest was probably the decline of the forests in Germany and the Netherlands.

Since 1985 almost all the research on acidification in the Netherlands has been co-ordinated within the Dutch Priority Programme on Acidification. The first phase of this research programme (1985 - 1988) aimed to answer the following questions:

- Which substances are responsible for the damage caused by "acid rain" and to what extent?
- How (by what means and in what way) is this damage inflicted?
- How effective are abatement measures?

To a certain extent the answers to these questions were given in the Evaluation Report on Acidification, which was published in 1988.

On the basis of an interim evaluation, carried out in 1987, the Steering Committee for Acidification Research (which commissioned the programme) decided to start a Second Phase Programme, which was carried out during the period 1988 -1990.

This Second Phase Programme was focused on further quantifying of the findings of the First Phase Programme, so that effective policy measures could be recommended.

Research in this second phase was focused on obtaining a more accurate estimate of the emission of ammonia and the deposition of SO_x, NO_y and NH_x, and also on quantifying effects on forest and heathland ecosystems. This quantification of effects included model analyses, and the derivation of critical loads and levels for forest and heathland ecosystems. Furthermore, scenario analyses were made with the Dutch Acidification Systems Model (DAS) in order to evaluate the effectiveness of policy measures. Below a summary will be given of the results and conclusions of the research programme. The numbers of the different sections correspond with the numbers of the chapters of the final report.

In June 1990, an independent Review Team of eight researchers from abroad has been convened, to provide a critical assessment of the second phase programme and also to indicate whether there is justification for future work in this area. The Review Report has been addes as Annex 2.

2. AMMONIA EMISSIONS AND ABATEMENT

The livestock industry is the main source of ammonia emissions in the Netherlands. The research on ammonia emissions was focused on the development of methods to measure and quantify ammonia emissions from different sources. Together with the Research Programme on Animal Manure and Ammonia, this led, for a number of source categories, to better insights into the extent of ammonia emissions and the possibilities of reducing them. So far it has not been possible, however, to make a sufficiently accurate estimate of total ammonia emissions from the livestock industry in the Netherlands on the basis of measurements alone. Therefore, in order to obtain an emission figure for the Netherlands as a whole and an overview of the spatial distribution over the Netherlands, emission factors were used for the different sources. These factors had been published before, for the benefit of policy makers, by an interdepartmental working group (1988). On the basis of these emission factors, the most recent (1988) figure of total annual emission in the Netherlands is about 250 kton NH_3. Since 1980 the emission figure for the Netherlands as a whole has hardly changed. The total uncertainty in the estimate of spatial distribution is at least 40%.
The estimated emission figures in this report for the total emission in the Netherlands and its spatial distribution are consistent with measurements of concentrations and with results of model computations for the pathway from emission via air transport to concentration/deposition, and also with deposition and throughfall data. Differences are only found for heavily loaded areas, where measured concentrations are systematically higher than the estimates. Meanwhile a monitoring programme was started which is aimed at obtaining more data on local concentrations.
The major source of ammonia emission through volatilization is the surface application of manure. Most ammonia is emitted during the first hours after spreading the manure. Emissions can be reduced considerably if the manure is quickly worked into the soil, through ploughing in or injection. Application of more manure than can be taken up by the crop will cause problems with regard to the quality of the soil and groundwater. If manure is injected, the total amount of fertilizer and manure can be adjusted.

3. DEPOSITION OF ACIDIFYING COMPOUNDS

Since 1980 the annual average total potential acid deposition has shown a downward trend: from about 6800 mol H^+ per ha in 1980 (estimate from first phase: 5800 mol) to about 4800 mol H^+ per ha in 1989. About one third of this 4800 mol consists of wet deposition and two thirds consists of dry deposition. One of the major causes of this downward trend is the decrease in SO_2 emissions in Western Europe, including the Netherlands. Apart from this, meteorological conditions are of importance. The deposition from natural sources amounts to about 300 mol H^+ $ha^{-1}yr^{-1}$.

Particularly the calculated contribution of NH_x to total potential acid deposition differs from the figure presented in 1988, in the first phase of the Dutch Priority Programme on Acidification. The NH_x deposition for 1989, as based on the latest insights, is about 50% higher that the figure based on the method used during the first phase. This is mainly caused by higher emission factors and a higher rate of deposition being used[1)]. Through that,the regional differences are also greater than assumed earlier. Confidence in the latest deposition figures is based on a comparison between model calculations and concentration, deposition and throughfall measurements.

With regard to NO_y depositions, the results also vary from the figures presented in the first phase of the Dutch Priority Programme on Acidification, although the differences are smaller. These differences are caused by lower HNO_3 concentration estimates (based on measurements in Petten and Speuld) and higher surface resistance for NO and NO_2. As a result, NO_y deposition estimates are lower on the whole.

The contributions of the various compounds to total potential acid deposition in 1989 were: NH_x about 46%, NO_y about 24%, and SO_x about 28%. In 1989, almost 55% of the total potential acid deposition in the Netherlands was caused by emissions in the Netherlands itself. The origin of the potential acid deposition in the Netherlands in 1989 and the contribution per sector to the Dutch portion are shown in Table 1.

Unlike in the first phase, the influence of large-scale differences in roughness have now been accounted for in the estimated figures for regional deposition. These differences lead to a higher dry deposition rate for forests, heathland and heathland lakes than for an average Dutch landscape. As a result of this influence, combined with the location of forests in relation to sources areas, the average dry deposition on large stretches of forest and heathland in the Netherlands is estimated to be higher by about 20 % and 10% respectively. Changes in roughness and other features of the vegetation have not been taken into account in these estimates. At the local level, for instance in the case of forest edges, inclusion of these aspects can lead to a higher deposition figure than would be the case for a large stretch of forest. The Dutch forest area is rather fragmented; about 40 % of the forest plots are smaller than 5 ha.

With regard to throughfall (the rainwater which falls through the canopy on to the forest floor), measurements adjusted for sea salt and neutral aerosols were compared with deposition estimates. At two research sites throughfall measurements were compared with deposition values derived from micrometeorological measurements. Both comparisions showed that there are still considerable differences between throughfall and atmospheric deposition, in spite of adjustments being made. However, the differences are smaller than those reported in the first phase. Further process-oriented research is required to find a direct causal relationship between throughfall and atmospheric deposition.

Table 1. Origin of potential acid deposition in the Netherlands in 1989 (per compound and total), and the contributions from the various sectors to the Netherlands fraction (%)

	total acid	SO_x	NO_y	NH_x
UK + Ireland	9	17	11	2
France	8	9	13	5
Belgium	10	17	8	5
FRG	10	10	19	5
the Netherlands	54	28	41	81
Eastern Europe	8	16	5	2
Rest	2	2	3	1
TOTAL	100%	100%	100%	100%
Refineries	5	24	2	0
Power stations	3	13	6	0
Road traffic	18	27	77	0
Industry	9	35	9	2
Agriculture	62	1	2	94
Households	3	1	4	4
TOTAL NL	100%	100%	100%	100%

1) In the Netherlands Acidification Abatement Plan (1989) the estimated total emission (about 250 kton) is based on the new emission factors. The deposition figures from the first phase are used, however.

The uncertainty in the total potential acid deposition figure for 1988 amounts to 45 - 80% for values per grid (5 x 5 km^2) and to 15 - 50 % for the Netherlands as a whole. The largest factor in the total uncertainty is the uncertainty in NH_x and NO_y depositions on the 5 x 5 km^2 grid and in the NH_x deposition for the Netherlands as a whole. The rates of uncertainty for 1989 are much the same as those estimated for 1988, because the total acid deposition for both years is more or less the same.

4. EFFECTS ON FORESTS AND FOREST FLOORS

4.1 Introduction

SO_x, NO_y and NH_x can cause damage to trees through direct effects on the parts of the tree above ground and through indirect effects via the soil solution on the parts of the tree below ground. Effects on the soil solution can be expressed as changes in the concentrations and budgets of S, N and Al in particular. For Dutch forests these changes have been assessed at the national level by measurements and calculations. With regard to nitrogen the acidification problem is closely linked with the problem of N eutrophication. The research into effects on forests has mainly been confined to Douglas fir forests on dry, poor sandy soils.

In the next section a description is given of the effects in the event of unchanged atmospheric deposition.

4.2 Direct effects

With regard to the direct effects of air pollution in the Netherlands, the short-term effects are relatively unimportant apart from the visible damage near local sources (NH_x, at low temperatures) and during episodes of high concentration O_3, and probably also a reduced activity of photosynthesis at low temperatures and high humidity (SO_2). At current daily O_3 concentration values, a slight reduction of the photosynthesis activity was also observed in mature Douglas trees in the field. In general, such a change in photosynthesis should not have any effects on annual growth.

So far, there are insufficient data on possible long-term effects of air pollution (during a period of more than one year), because there have been few long-term experiments and not enough monitoring has been done. There are, however, strong indications that long-term effects are important. The effects of long-term accumulation are unknown. The same is true of the risk of direct damage from combined stress. Effects of O_3 can lead to accelerated leaf/needle ageing. The interaction of direct effects (from SO_2, NH_3 and NO_2) and indirect effects (from NH_4+/NO_3- ratio in the soil solution, and from K and Mg deficiencies) can lead to internal acidification of the plant, resulting in leaf/needle damage and leaf/needle loss. The wax morphology and amount of wax on needles of Douglas fir has been studied in the field and in fumigation experiments with young trees. These experiments showed that the needle wax morphology cannot be considered to be an indicator of damage from air pollution.

4.3 Indirect effects

Effects on the soil

Results from seventeen intensive monitoring studies in forests showed that in almost all these cases there was sulphate saturation, which means that sulphate leaching is equal to the sulphate load. Uptake and retention of sulphate is therefore usually negligible. This means that the actual acid load from sulphur is usually equal to the potential acid load from sulphur. The actual acid load from nitrogen, however, is only 50% of the potential load, as was shown by a national survey. In the event of unchanged deposition, however, the leaching of N from forest ecosystems on dry sandy soils is expected to increase and N saturation is also

expected to occur more often. The results of the above-mentioned national survey show that the NO_3/SO_4 ratio is higher than or equal to 1.0 in about 15% of the cases, which indicates the percentage of N saturated forest ecosystems.
Given the large-scale occurrence of considerable nitrification, it is not right to make an adjustment for partial nitrification for the total soil profile in the calculation of the potential acid load. The difference between potential acid load and actual acidification of the soil (in the sense of a reduced buffer capacity) is mainly the result of the removal or fixation of N through uptake by plants, denitrification, or immobilization of N in the soil organic matter. In the long term, immobilization can be considered to be zero owing to N saturation. The current contribution of N to the actual soil acidification is estimated to be about 35%, the contribution of S being about 65 %.
The acidification of the soil has led to increasing Al concentrations in the root zone, which has reduced the availability of cations such as K and Mg to the roots. The current composition of the soil solution in the topsoil of Dutch forests is characterized by Al/Ca ratios just above the critical value of 1.0 and Al concentrations far above the critical value of 2 $mg.l^{-1}$. (The critical value of 1 for the Al/ Ca ratio has been derived from laboratory experiments. The field situation is somewhat more complicated, but does not conflict with the value of 1. In other countries this is an usual value too.)In forests with low vitality or no vitality (in terms of loss and discolouration of needles), the NH_4/K ratio in the top 10 cm of the soil profile is often above the critical value of 5. In forests within the vitality range of reasonably vital to vital, the ratio is often below the critical value. In the top 30 cm of the soil, the average NH_4/K ratio is usually below the value of 5.0; this is connected with the high nitrification rate in the topsoil, which also accounts for the average NH_4/NO_3 ratio being well below 1.0. In the top 10 cm of the soil the NH_4/K ratio is higher.
Mobilization of Al is currently the main buffer in the sandy forest soils in the Netherlands. This Al buffer in the soil is limited, however. Model results show that, at current deposition levels, the Al buffer in the top 10 cm of most sandy forest soils will be depleted by between 10 and 100 years. When there is no longer an Al buffer, the pH in the soil solution may decline even further to a value of 2.8 to 2.9. Moreover, there may be increasing P deficiency owing to the formation of iron phosphates and phosphorus-aluminium compounds.
These developments can be reversed if the deposition of acidifying substances is reduced considerably in the coming years. In this connection, scenarios have been developed (partly extrapolations of the current situation with regard to acid deposition, and partly already implemented policy). These scenarios have been evaluated with the DAS acidification model. The results are presented in section 6 of this summary.
Corrective measures such as liming and fertilization involve the danger of increased mineralization and N leaching.

Indirect effects on forests

- In forest ecosystems in which N is the growth-limiting factor, input of extra N will lead primarily to increased production. In the Netherlands, however, this is hardly ever the case (nowadays). Here, increased availability of N strongly stimulates the growth of nitrogen-loving grasses and herbs, particularly in forests where the canopy is opening as a result of a decline in the number of needles. Grass species such as Molinia caerulea and Deschampsia flexuosa are not only crowding out heathland vegetation in the Netherlands, but are also increasingly found in forests. It should be noted here, however, that dominance of grasses can be a natural phase in the succession of young forest ecosystems on sandy soils. On the basis of long-running tests with high N doses, it is to be expected that the growth of nitrogen-loving species will increase in the event of unchanged N deposition.

  A distinct relationship has moreover been found between increased N contents and the occurrence of Sphaeropsis affection. The increased N contents also lead to relative K, Mg and P deficiencies, as a result of disturbed nutrient ratios. These nutrient deficiences are further increased by deposition of NH_4 ions on the needles, which stimulates the leaching of nutrients from the needles.

  Most of the forested areas in the Netherlands are located on poor, dry, sandy soils. As a result the forests are more sensitive to water stress and nutrient deficiency. In these forests the nitrogen supply, together with the deposition of SO_x, has led to considerable acidification of the soil too. The interaction between the increased availability of N and soil acidification has resulted in a decreased availability of cations (nutrient deficiency). Thus the forest is becoming increasingly vulnerable to damage resulting from diseases, pests, frost and drought.

- In forests, ectomycorrhizas are very important for the uptake of nutrients and probably also for trees' resistance to disease. It is possible that mycorrhiza infestations protect the roots and the tree against aluminium. In the field a negative correlation has been observed between air pollution and mycorrhiza parameters such as number of species, number of fruit bodies and degree of infestation. The mechanism involved, however, is still not completely clear. Decrease of pH in combination with strong nitrogen enrichment would seem to be the most important cause. Quantified general statements concerning the effects of pH, aluminium and nitrogen on the occurrence and functioning of mycorrhizas in forests are not (yet) possible. The effects of an impoverished mycorrhiza population and of a shift in species on the nutrient supply of trees can not yet be quantified either.

- The role of the soil (micro)fauna in (forest)ecosystems is still not very clear. Microfauna probably play a key role in decomposition and mineralization processes. In the Dutch Priority Programme on Acidification no research has been carried out on soil fauna. From other research, carried out on Scots pine forests in the Netherlands, it appears that in soils which are sensitive to acidification there is a shift in species, and that, in general, the diversity of soil organisms is declining. With regard to the direct effects of pH and Al on free-living nematodes (eelworms), laboratory experiments have shown that a continuing acidification of the soil will strongly affect the survival probability of nematodes in acid forest soils. Especially the effects of H^+ are acute and occur within a relatively narrow concentration range. It seems that many nematode species have already reached the limit of their tolerance capacity.

Effects on the groundwater under forests

- Research on the composition of shallow groundwater at 150 different locations (forests and heathland) has shown that the nitrate content was higher than the drinking water standard in almost 30% of the coniferous forest sites investigated. In the case of deciduous forests this was the case in 13% of the sites. The standard for nitrate in drinking water is 50 mg/l (0.8 mmol/l). It is true that in the subsoil (on the way to the wells) denitrification can occur, but where (under what conditions) and to what extent that occurs is still not known.

  In addition, there are signs that Al will become a problem for the drinking water supply, particularly in the case of shallow private wells. The Al content of the shallow groundwater at the above-mentioned 150 sites is often above the drinking water standard (7 $\mu mol.l^{-1}$). This is true for almost 90% of the coniferous forest sites investigated and for 70% of the deciduous forest sites. It should be mentioned, however, that these data provide an indication of the quality trend of the shallow groundwater under forest areas. The water that is withdrawn at a pumping site comes from a particular area around the well field, called the recharge area, where, in general, various forms of soil use are to be found, as a result of which there will be mixing of various types of water. Furthermore, Al retention in the deeper subsoil has an important effect here.

  If the acidification of the soil continues as described above, the groundwater quality will further deteriorate owing to increasing nitrate and aluminium contents. Finally, there is the danger that the increased amount of nitrogen fixed in the soil will be released as NO_3 through increasing mineralization and nitrification (in the event of a temperature rise) or through a temporary decrease in the N uptake (in the event of the disappearance of forests as a result of diseases or large-scale felling).

4.4 Combined direct and indirect effects

Research into the combined effects of direct and indirect influences has focused on sap flow velocity, the growth of individual trees and forest stand growth.

Field experiments have not revealed any clear relationship between sap flow velocity and growth of individual trees, on the one hand, and air pollution, on the other. Modelling has been used to extrapolate from physiological processes within individual trees to consequences for growth of whole forest stands. This has shown that growth reduction caused by direct effects of air pollution is insignificant with an atmospheric load such as occurred at the two research sites at Speuld and Kootwijk in 1988 and 1989. It is not yet possible to determine long-term direct effects with the applied forest stand model. But first impressions would indicate that these effects could be relevant.

At the Speuld and Kootwijk sites no growth reductions occurred in 1988 and 1989 as a result of nutrient deficiencies. However, root density at both locations is low, and any further reduction through a shift in the root/shoot ratio as a result of an increased nitrogen availability or through soil acidification and aluminium toxicity, will reinforce the effect of what may be a very limited availability of water and other nutrients apart from N. Most of the forests in the Netherlands are on poor soils, which makes them sensitive to water stress and nutrient deficiencies. The average acid load (throughfall) on Dutch forests is not known. The forest locations where throughfall has been measured are mostly in heavily loaded areas. In these areas the acid load is much higher than at Speuld and Kootwijk.

The integration of direct and indirect influences has not yet been modelled. This linkage is necessary in order to make a more realistic risk evaluation.

4.5 Main conclusions concerning the effects on forests and forest soils

Usually there is not a monocausal relationship between the acid load on forests and their health, in terms of needle density, needle discolouration (observed under field conditions as "vitality") and growth. There is rather a combined and complex influence of various biotic and abiotic factors. Acidifying deposition generally reinforces the impact of traditional stress factors (frost, drought, disease and pests) on forest health (greater risk).

Increasing nitrogen deposition over a period of several decades has led, first of all, to a removal of N deficiencies and increased growth and, secondly, to nutrient imbalances as a result of Mg, K and P deficiencies. More and more forest ecosystems are moving from a situation of nitrogen deficiency to a situation of nitrogen saturation. At the moment about 15% of the Dutch forest soils is N saturated. At the same time, the nitrogen input, together with SO_x deposition, is causing considerable soil acidification. The present contribution of nitrogen to actual soil acidification is about 35%, and that of sulphur about 65%. The combined action of increased N availability and soil acidification has led to a decline in cation availability (nutrient deficiency). This is resulting in a greater risk of damage to forests caused by pests and plagues, frost and drougth.

As regards the effects of acidifying deposition in the Netherlands, soil acidification is the greatest risk factor. The research carried out in the context of the second phase of the Dutch Priority Programme on Acidification has confirmed the hypothesis that Dutch forest soils are degrading (radical physico-chemical changes in the long run) by deposition of acidifying substances. This hypothesis was proposed about five years ago on the basis of various measurements and data from other countries. Confirmation of this hypothesis was obtained from input-output budgets and model analyses, and through nation-wide monitoring of soil solution chemistry. A major concern are virtually irreversible changes in the soil caused by depletion of the Al buffer, and the consequences of these changes for soil pH. It is not possible to specify precisely the long-term consequences of a decline in pH (to between 2.8 and 2.9) associated with aluminium depletion, which (in the event of unchanged deposition) is the expectation for Dutch forest soils. In any case, large changes in the soil and thus in the conditions of forest stand locations, are to be expected. This could lead to changes in vegetation and soil fauna.

If the current deposition trend continues, groundwater quality will further deteriorate. Finally there is the danger that the increased amount of stored organic N may be released more quickly as nitrate, if there is increasing mineralization and nitrification (in the event of a temperature rise) or a temporary reduction in N uptake (if forests disappear as a result of disease or large-scale felling).

The results and conclusions of research programmes in other countries mainly point in the same direction as those arrived at for the Dutch situation:

- visible symptoms (defoliation, yellowing) are not specific and not necessarily associated with air pollution;
- in general, forest decline is due to a complex set of factors (one being air pollution / acidification);
- the direct effects of gaseous pollutants on leaves and needles are probably less important than first assumed; this analysis is not relevant under the current conditions prevailing in some areas in the eastern European countries, such as the Erz mountains, where the direct effects are very likely to be more important;
- in contrast, the indirect effects on forest ecosystems (not only trees) may be more important; nutritional imbalances could become a majore mid-term problem.

5. EFFECTS ON HEATHLAND

Heathland can be roughly divided into two types: wet heathland (which is influenced by groundwater for at least some part of the year and in which the dominant species is Erica tetralix), and dry heathland (which is not influenced by groundwater and in which the dominant species is Calluna vulgaris). On a quantitative basis the dry heathland type is by far the most important in the Netherlands. The area of heathland has rapidly decreased over the years, due mainly to cultivation and afforestation. In the remaining heathland the dominant heathland species have been increasingly replaced by grasses for some decades now. In dry heathland Calluna vulgaris (heather) is being replaced by Deschampsia flexuosa (wavy-hair grass), while in wet heathland Erica tetralix (cross-leaved heath) is being replaced by Molinia caerulea (purple moor-grass). At the same time rare heathland types such as Arnica montana, Antennaria dioica (mountain everlasting), and Viola canina (heath dog violet) have almost disappeared altogether.

An inventory by means of satellite imagery has shown that about one third of the heathland in the Netherlands is still vital (> 70% covered by heathland species), about one third contains large amounts of grass and will probably change into grassland within the next 3-5 years, and about one third has already changed into grassland. The Dutch heathland is rapidly changing into grassland.

Although many possible causes of this degradation are reported (such as ineffective management, lowering of groundwater levels and stress from excessive recreation), it is obvious that air pollution and the resulting soil acidification and N eutrophication are key factors in this process.

The research carried out as part of the Dutch Priority Programme on Acidification was focused on direct and indirect effects of SO_x, NO_y, and NH_x on 'dominant' heathland species (Calluna and grasses) and on 'rare' species (Arnica montana, Viola canina). No research was done on the effects of ozone on heathland vegetation.

The most important findings of the heathland research are:

- At the ecosystem level, nitrogen input ultimately leads to the elimination of slow-growing species by fast-growing species, but Calluna will not be crowded out by grasses at nitrogen deposition levels up to 150 kg N $ha^{-1}yr^{-1}$ if its canopy remains closed. Opening of a Calluna canopy can be caused by stress factors such as frost, drought, heather beetle plagues, or by natural ageing. Under normal conditions in the Netherlands, the canopy will hardly ever be opened by natural ageing. The critical nitrogen load for the crowding out of an open Calluna canopy by grasses is about 10-15 kg N $ha^{-1}yr^{-1}$ (700 - 1100 mol_c $ha^{-1}yr^{-1}$). At this critical deposition level, vital heathland can be maintained with a sod-cutting frequency of once every 50 years. Both experimental research on Calluna and modelling work on Calluna (in competition with Deschampsia) and Erica (in competition with Molinia) indicate the same critical load of 10 -15 kg N $ha^{-1}yr^{-1}$. With grazing and very frequent sod-cutting (once every 10 years) a vegetation of Calluna or Erica, though without rare species, can be maintained at nitrogen deposition levels up to about 30 kg N $ha^{-1}yr^{-1}$. (Present N deposition on Dutch heathland is approximately 35 - 40 kg N $ha^{-1}yr^{-1}$.)

- At the individual plant level, nitrogen input (as NH_3 or $(NH_4)_2SO_4$) causes growth stimulation even at low dosages. In rare heathland species however, changes may occur that make them more sensitive to frost, drought, and plagues. A critical level of NH_3 cannot be exactly defined, but is probably in the range of 5-10 $g.m^{-3}$ (long term).

- The decline of rare heathland species is probably due to direct effects of gaseous SO_2 and soil acidification. Adverse effects of SO_2 on more than 5% of the heathland species can be expected at long-term average concentrations above a critical level of 8 $\mu g.m^{-3}$. Effects on dominant species (Calluna and grasses) will probably not occur at the current SO_2 levels in the Netherlands. Crowding out of Violion caninae by grasses can also take place, but is probably only important at nitrogen deposition levels above the current ambient level in the Netherlands. However, this level may affect the reproduction or establishment of Violin caninae.

- Extensive field research showed that many threatened plant species of wet and dry heathland and poor soils, such as Dactylorhiza maculata (heath spotted orchid), Thymus serpyllum (wild thyme), Pedicularis sylvatica (lousewort), and Arnica montana (Arnica), do not occur at pH levels (H_2O) under 4.2. The same research showed that there was no correlation between the occurrence of these species and the concentration of aluminium in the soil solution. Pot experiments and ecophysiological experiments showed that the disappearance of these species can be attributed to an indirect pH effect. Owing to the combination of soil acidification and N deposition, the NH_4/NO_3 ratio rises. This inhibits the uptake of base cations such as K and Mg. As a result of this the above-mentioned species die after germination, and already established plants show poor growth and flowering.

6. SCENARIO ANALYSES USING THE DUTCH ACIDIFICATION SYSTEMS MODEL (DAS)

6.1 Scenarios

The long-term developments with regard to acidification were explored with the Dutch Acidification Systems Model (DAS). With this model, concentrations and depositions of acidifying pollutants in the Netherlands can be calculated on the basis of different emission levels in the Netherlands and the rest of Europe. The deposition loads and concentration levels of acidifying pollutants are subsequently used to calculate the effects of acidification on a number of receptors.

Several scenario analyses were made with the DAS model. These analyses are explorations of possible future situations rather than forecasts of effects. The scenario calculations with the model have been divided into three categories:

- calculations based on historical emission data
- calculations based on emissions in the near future (the period until the year 2000) as estimated in the Netherlands Acidification Abatement Plan and the Netherlands Environmental Policy Plan + (NEPP+) together with the resulting (estimated) depositions in 2000
- calculations based on deposition targets (for the period 2000 - 2050).

The three variants made for this last category are presented in Table 2. The emissions which cause these depositions cannot be estimated in reverse with the current model. The three categories of calculations give three scenarios, which are, however, the same for the period until the year 2000.

Table 2. Deposition values for the period 2000 - 2050; the deposition targets of the scenarios are printed in **bold** type

	deposition in mol H^+ $ha^{-1}yr^{-1}$				
year	2000	2010	2010	2050	2050
receptor[1]	NL	NL	forest	NL	forest
variant 1	2200[2]	**2200**	2550	**2200**	2550
variant 2	2200[2]	**1400**	1600	1230	**1400**
variant 3	2200[2]	1230	**1400**	**700**	800

1. Deposition to forests is higher thant the average deposition in the Netherlands. The figures refer to the average deposition in the Netherlands and the average deposition to Dutch forests.
2. The average deposition level in the Netherlands in the year 2000 in the above-mentioned emission scenario for the Netherlands and Europe. The calculated value, 2240 mol H^+ $ha^{-1}yr^{-1}$, has been rounded down to 2200, because of the uncertainties.

For the period after the year 2000 the regional distribution and the relative share of each component up to the year 2000 as resulting from the emission based calculations are assumed to be constant. In order to calculate emissions, depositions and effects with the DAS model, the Netherlands was divided into 20 acidification areas. These areas are both emission and receptor areas. In order to calculate the influence of emissions from other countries in Europe, a subdivision of Europe was made for emission aggregation.

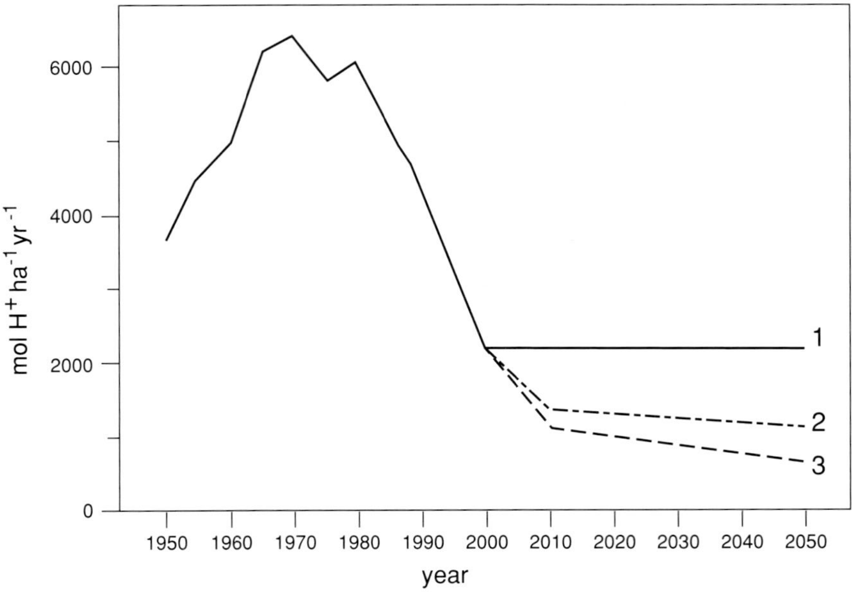

Fig. 1 Total acid deposition as a function of time for the three scenarios

6.2 Potential acid deposition until the year 2050

Based on historical emissions and the emission scenario up to the year 2000, the average deposition for the Netherlands in the year 2000 will be 2240 mol (rounded down to 2200) potential acid per hectare per year. This means a reduction of more than 50% compared to the potential acid deposition in 1989 (which was about 4800 mol H+). Apart from the average uncertainties in emission figures, the average uncertainty in the calculated deposition levels is at least 20%. Using the atmospheric transport model, potential acid deposition in the Netherlands in the year 2000 has been estimated at 1800 - 2600 mol per hectare. The target level of 2400 mol per ha for that year is within this range. Therefore, with the extra emission reductions included in the NEPP+ and with the expected emission reductions in other countries (Netherlands Acidification Abatement Plan) this (interim) target will be achieved. The potential acid deposition for the years 2000, 2010 and 2050 for the three variants is presented in Table 2. The complete scenarios until the year 2050 for total potential acid deposition are shown in Figure 1.

Figure 1 clearly shows that a very strong reduction of acid deposition is expected during the period 1990 - 2000. This is especially caused by a strong decrease in NH_x deposition, but considerable reductions are also expected for NO_y and SO_x. These reductions are, of course, the result of considerable emission reductions: NH_x emissions are expected to be reduced particularly in the Netherlands.
The total acid deposition levels in 1980 and 1989, as depicted in Figure 1, differ slightly from the values indicated in Section 3 of this summary. The main reason for this is the fact that the values given in Section 3 are based on deposition measurements and are determined for that specific year, taking into account the meteorological conditions in that year. The deposition levels in 1980 and 1989 in Figure 1 are calculated with emissionfigures which are associated with (large) uncertainties. Moreover meteorological conditions were averaged over 10 years.

6.3 Effects on forest soils

In order to gain an insight into the effects of the three emission-deposition scenarios on Dutch forest soils during a period of 60 years (1990 - 2050), the RESAM model was used. Presentation of the model results was confined to the results for pH, Al concentrations, Al/Ca ratio, the NH4/K ratio in the topsoil (20 to 30 cm), and pH, Al and NO_3 concentrations in the subsoil (the depth of the root zone). The parameters in the topsoil are important indicators of the effects on forests, while the parameters in the subsoil are important indicators of potential groundwater pollution. Critical values have been derived for most of these parameters (see Section 7).
The forests are represented by seven important tree species, Pinus sylvestris (Scots pine), Pinus nigra(black pine), Pseudotsuga menziesii (Douglas fir), Picea abies (Norway spruce), Larix lepto lepis / Larix kaempferi (Japanese larch), Quercus robur (oak) and Fagus silvatica (beech). The investigation of effects on forest soils was limited to acid sandy soils, on which 80 to 90% of the Dutch forests are to be found. These soils are susceptible to acidification. The forest/soil combinations investigated comprise almost 65% of the total Dutch forest area.

The most important conclusions are:

- Deposition reductions generally lead to a fast improvement in the soil solution chemistry, i.e. an increase in the pH value and a reduction in Al and NO_3 concentrations and the Al/Ca and NH_4/K ratios. The NO_3 concentration and the NH_4/K ratio, however, show a clear delay between reduced deposition and subsequent concentration reduction in the soil. This delay is mainly caused by N mobilization from the litter layer.

- A deposition reduction to 2200 mol H^+ $ha^{-1}yr^{-1}$ as an average for the Netherlands, which is predicted for all the scenarios for the year 2000, will reduce the exceedance of the critical Al concentration from about 75% of the forest soil area (the current situation) to 40%, and will reduce the exceedance of the critical Al/Ca ratio from about 65% to 40%.

- Reductions to an average of 1400 mol H^+ $ha^{-1}yr^{-1}$ for Dutch forests (Scenario 2) cause a further decrease in the exceedance of the critical Al concentration (0.2 $mol_c m^{-3}$) and the critical Al/Ca ratio (1.0) in the topsoil, and the critical NO_3 concentration of 0.8 $mol_c m^{-3}$ in the subsoil in 2010 (less than 20% of the forest-soil combinations investigated). In 2050 the exceedance of these critical values is insignificant. In Scenario 1, the critical Al concentration will be exceeded in a considerable area of forest (over 25%) in the year 2050; for the critical Al/Ca ratio this exceedance is about 10%. In Scenario 3, which gives a deposition of 1400 mol per ha on forests in the year 2010, and an average deposition of 700 mol per ha on the Netherlands as a whole in the year 2050, the exceedance of the critical Al concentration and nitrate concentration is reduced to an insignificant level markedly earlier than in the case of Scenario 2.

- Furthermore, deposition reductions according to Scenario 2 are sufficient to prevent depletion of the Al hydroxides buffer, which is not the case with Scenario 1.

6.4 Effects on growth of Douglas fir

The scenario analysis for forests was carried out with the SOILVEG model and was limited to Douglas fir, because a lot of research has been conducted on this tree species. Douglas fir has been chosen as "model tree" to illustrate the effects of the different scenarios.
SOILVEG is a dynamic simulation model which calculates the growth of Douglas fir as a function of the availability of carbon (via photosynthesis) and of the availability of N, K, Ca and Mg, (via atmospheric deposition and soil processes). The soil module is based on the RESAM model (see Section 6.3). Using SOILVEG it is possible to determine the direct and indirect effects of acidifying deposition on growth. The effects of water stress, diseases, pests and frost are not included. The results are presented in terms of biomass (stem and needles), N content of needles, organic matter, and NO_3 leaching.
The effects were calculated for sixty-year-old Douglas fir stands, which is about the average age of the current stands. A specific stand age was chosen, because the response of Douglas fir to air pollution and soil acidification is dependent upon the age of the forest. In addition an exposure period of 20 years was decided upon. Calculations were made for the main soil types of the sandy deposits in the Netherlands.

The most important conclusions are:

- The clearest positive effects of the reduction of N deposition are lower N content in the needles (there is an immediate and strong response of needle N content) and reduced nitrate leaching. N contents above 2% are calculated in all the forest areas in the south and south-east between 1970 and 1990, which makes the forests there particularly susceptible to frost damage and plagues. In 2050, the N contents in needles are lower than or equal to the optimal values in all scenarios.

- Douglas fir on Albic Arenosol (z27) is most sensitive to increasing soil acidification. A considerable reduction in needle mass and stem mass on this soil type occurs between 1970 and 2000, when deposition of N and potential acid are highest. The soil type Leptic podzol is least sensitive to the effects of soil acidification and increased N availability.

- In the south-eastern part of the Netherlands the increase of deposition of N and potential acid between 1970 and 1990 leads to a net decrease of needle mass (forest health) and wood production. In relatively clean areas, e.g. coastal regions, there is a positive effect of enhanced N-deposition on needle mass and wood production. When deposition and ambient concentration of N are lowered, a growth reduction may be expected in coastal regions, while a growth increase in the south-eastern half of the country is most likely.

- As far as the above-mentioned effects are concerned, deposition reduction before 2000 is much more important than reduction after 2000, except in coastal areas, where the availability of N will become the growth-limiting factor after 2030. In other cases, the changes (and uncertainties) in the reaction of the forest will be dominated by the strong reduction in deposition between 1990 and 2000.

6.5 Effects on heathland

To assess the effects of the different scenarios on the development of heathland vegetations, calculations have been made with the CALLUNA model for dry heathland and with the ERICA model for wet heathland, for 17 out of the 20 different Dutch acidification areas in which heathland is found. CALLUNA describes the competition between Calluna vulgaris (heather) and the grass Deschampsia flexuosa (wavy-hair grass). This competition is resulting in serious elimination of dry heathland vegetation in the Netherlands. The ERICA model describes the competition between Erica tetralix (cross-leaved heath) and the grass Molinia caerulea (purple moor-grass). With these two models the development of a heathland vegetation during the period 1950 to 2050 was simulated for each of the 17 areas, in the three different scenarios. On account of the stochastic nature of the occurrence of

heather beetle plagues, the results of 100 simulations were averaged for the CALLUNA model. This average was used to interpret the effects of the three scenarios for each area. The results of the calculations lead to the conclusion that Scenario 1 does not offer any prospects for the continued existence of the typical dry heathland vegetations that were still found at the beginning of this century. Scenario 3 offers good prospects for such vegetations. The results for the period 2000 - 2025 for the areas with the highest load (over 3000 mol N $ha^{-1}yr^{-1}$) vary, however. In the period after that (from 2025 to 2050) Scenario 3 meets the target for all areas (Calluna remains the dominant species). For the period 2000 - 2050 Scenario 2 gives results which are between those of the Scenarios 1 and 3.
With regard to wet heathland vegetations, calculation results of Scenarios 2 and 3 show that, at a sod-cutting frequency of once per 25 years, these vegetations can compete in all areas during the period 2000 - 2050. Calculations based on Scenario 1 seem to offer good prospects for the longer term (after the year 2025).

6.6 Synthesis of the scenario analyses

Scenario 2, with an average deposition of 1400 mol H^+ per ha for the Netherlands as a whole in the year 2010 and an average deposition of 1400 mol H^+ per ha for forests in the year 2050, seems sufficient to minimize the impact of acidification, as one of the factors of combined stress important for the health of forests and heathland. The critical values for Al concentration, Al/Ca ratio and NO_3 leaching in forest soils are exceeded in the year 2010 in less than 20% of the forest soils investigated. Scenario 2 results in hardly any exceedances of these critical values in the year 2050. Scenario 3, with a deposition of 1400 mol ha to forests in the year 2010, and a deposition of 700 mol per ha for the Netherlands as a whole in the year 2050, achieves a negligible exceeding of all critical values 10 to 15 years earlier than Scenario 2. Scenario 2 is sufficient to prevent depletion of Al hydroxides. For dry heathland Scenario 2 does not reach the target in areas which (now) have a high nitrogen load until the year 2025. This is also true for Scenario 3, however. Until the year 2000 the cover percentage of Deschampsia remains much higher than that of Calluna at the end of each period following sod-cutting. For wet heathland Scenario 2 reaches the target after the year 2000 (biomass Erica greater than biomass Molinia).

7. CRITICAL LOADS AND CRITICAL LEVELS

7.1 Introduction

Within the context of this report, the following (ECE) definitions of <u>critical load</u> and <u>critical level</u> are used.

Critical load: a quantitative estimate of an exposure (deposition) to one or more pollutants below which significant harmful effects on specified sensitive elements of the environment do not occur according to present knowledge.
Critical levels: concentrations of pollutants in the atmosphere above which direct adverse effects on receptors, such as plants, ecosystems or materials, may occur according to present knowledge.

For a correct interpretation of "critical loads" for forests it should be emphasized that exceeding these critical loads does not necessarily cause any visible effects on forests. If "critical loads" are exceeded temporarily, this will certainly not mean the death of (part of) the Dutch forests. Exceeding the critical load does, however, involve a certain risk to the vitality of forests, and this risk will increase as critical loads are exceeded further and for a longer period. When the critical loads are exceeded very considerably, as they presently are in certain parts of the Netherlands, the vitality of forests is endangered.
Table 3. gives a survey of the criteria that have been used to derive critical loads for terrestrial ecosystems.

Table 3. Critical values for Al concentration, Al/Ca ratio, NH_4/K ratio, NO_3 concentration and N content in forest soils (soil solution), groundwater, and needles

Criteria	Unit	Forest soils (soil solution)	Groundwater	Coniferous tree needles
Al 1)	$mol_c m^{-3}$	0.2	0.02	-
Al/Ca	mol mol^{-1}	1	-	-
NH_4/K	mol mol^{-1}	5	-	-
NO_3	$mol_c m^{-3}$	-	0.4 2), 0.8 3)	-
N	% of dry weight	-	-	1.8

1) refers to (inorganic) aluminium
2) refers to the target value for drinking water (25 $mg.l^{-1}$)
3) refers to the standard value for drinking water (50 $mg.l^{-1}$)

7.2 Critical loads

A survey of the average critical nitrogen load and total acid loads for the Netherlands is given in Tables 4. and 5.

Table 4. Average critical nitrogen loads ($mol_c\ ha^{-1}yr^{-1}$) for terrestrial ecosystems on well drained sandy soils in the Netherlands

	coniferous forests	deciduous forests	heathlands
vegetation changes	400 - 1400	600 - 1400	-
	(400 - 800)	(400 - 800)	(350 - 700)
elimination by grasses		-	700 - 1100
	-	-	(1400 - 2100)
frost damage/fungal diseases	1500 - 3000	-	-
	(1500)	(1500)	-
nutrient imbalances	800 - 1250[1)]		-
	(1000)	(1500)	
nitrate leaching to	900 - 1500	1700 - 2900	2000 - 3600
groundwater	(1000 - 1600)	(1600 - 2800)	(2000 - 3600)

1) worst case (total inhibition of nitrification)
( ) value, presented during the first phase of the Programme, and used as a basis for present abatement policy

Table 5. Average critical total acid deposition loads ($mol_c\ ha^{-1}yr^{-1}$) for terrestrial ecosystems on well drained sandy soils and surface waters in the Netherlands

Effects	Coniferous forests	Deciduous forests	Surface waters
root damage	1100 - 1400[1)]	1400 - 1700[2)]	
	(1400)	(1800)[3)]	
aluminium depletion	1200	1500	
aluminium leaching to groundwater	500[4)]	300[4)]	
	(200)	(200)	
fish dieback			-
			(400)

1) the first value is related to a critical Al concentration of 0.2 $mol_c\ m^{-3}$ and the second value to a critical Al/Ca ratio of 1.0
2) the first value is related to a critical Al/Ca mol ratio of 1 and the second value to a critical Al concentration of 0.2 $mol_c\ m^{-3}$
3) for deciduous forests on rich soils = 2400
4) these values are related to a depth of 2 m. (the average phreatic level)
( ) values presented in the first phase

Compared to the results of the first phase of the Priority Programme on Acidification (1985 - 1988) the range for heathland (for the criterion of elimination by grasses) has been lowered (see Table 4.). For frost damage/diseases and nutrient imbalances a range rather than one single value is now given. For leaching of nitrate to groundwater, the values have remained nearly unchanged. Insufficient scientific support has been provided for the critical value for deciduous forest with regard to the criterion frost damage (1500 mol_c $ha^{-1}yr^{-1}$) presented in the first phase of the programme. This value has therefore not been included in this report. Unlike during the first phase, a critical acid load related to the Al concentration has now been introduced. The increase in critical loads for groundwater as compared to the first phase is mainly due to the fact that the nitrogen uptake has been included in the calculations, and to more accurate data on base cation deposition and uptake. The critical loads for surface water which were reported during the first phase have not changed.
Evaluation of critical loads shows that most effects of nitrogen deposition will not occur below loads of 400 - 600 mol_c $ha^{-1}yr^{-1}$ (total nitrogen). With regard to critical acid loads, a lowest value of 300 - 500 $H^+ha^{-1}yr^{-1}$ can be derived (for the receptor groundwater, the criterion being the aluminium standard for drinking water). The latter values are about 10% of the present loads.

In comparison with the formulated critical loads, scenario calculations of the effects of total acid on forest floors seem to produce different results in certain respects. The critical load related to the Al/Ca ratio in the soil solution is 1400 mol_c $ha^{-1}yr^{-1}$ for deciduous and coniferous forests (Table 5.). Scenario calculations show that at a deposition level of 1400 mol_c $ha^{-1}yr^{-1}$ on forests hardly any harmful effects will occur in the long run. This result is rather consistent with the formulated critical loads. However, the model calculations also indicate that in many of the forests the Al/Ca ratio in the soil solution also meets the criterion at considerably higher deposition levels.
This can be explained by the fact that the model used to calculate the critical deposition does not distinguish between the different layers in the vertical soil profile. As a result the critical load values derived are relatively low. Calculations with a more sophisticated multi-layer model, like the one used for scenario calculations, indicate that at an average deposition of 2200 mol_c $ha^{-1}yr^{-1}$ as an average for the Netherlands, the Al/Ca ratio in the soil solution in the year 2000 will meet the criterion in about 50% of the forests on soils susceptible to acidification. This means that already at this deposition level these forests will be quite well protected, although some reserve is in order here owing to the problem of discrepancies between throughfall and deposition. The forest ecosystem is still not completely protected at this situation, however. At the abovementioned acid deposition level other harmful effects - such as eutrophication as a result of exceedingly high nitrogen deposition - will still occur.

The danger of aluminium depletion also continues to exist for all sandy soils that are susceptible to acidification.

7.3 Critical levels

For the present there is no reason to change the critical level values for forests presented during the first phase. These values are shown in Table 6.

With regard to heathland the following remarks can be made:

- The rare heathland species are sensitive to gaseous SO_2. Ecotoxicological risk analysis, carried out as part of the Dutch Priority Programme on Acidification, gave an (annual average) critical level of 8 $\mu g.m^{-3}$. This value is lower than the current annual average value for the Netherlands (about 10 $\mu g.m^{-3}$). Dominant species (Calluna and grasses) are hardly affected by SO_2 at current levels.
- Although O_3 is one of the most important air pollutants, there are hardly any data on the sensitivity of heathland vegetations to O_3. The Dutch Priority Programme on Acidification did not include any research on this.
- The effects of NH_3 on heathland vegetations were extensively studied as part of the Dutch Priority Programme on Acidification, but a critical level could not be indicated.

Table 6. Critical levels of O_3 and SO_2 for forests

	coniferous forest (poor sandy soils)	deciduous forest	explanation
O_3 concentrations* (daily average)	50	50	these critical levels are related to visible damage and to inhibition of translocation of assimilates to roots
SO_2 (yearly average)	25	25	these values are related to visible damage

* during growing season

1. INTRODUCTION

G.J.Heij (RIVM)

1.1 Research questions

In 1985 the Dutch Priority Programme on Acidification (Dutch abbreviation: APV) was established as a result of the awakening interest of policy makers in the field of the effects of air pollution on forests and other vegetations. The decline of forests in Germany and some parts of the Netherlands was probably one of the main stimuli in this development. Since 1985, almost all the research in the Netherlands on the effects of acidification has been co-ordinated within the Dutch Priority Programme on Acidification.

The first phase of the research programme (1985 - 1988) aimed to answer the following questions:

- Which substances are responsible for the damage and to what extent?
- How is the damage inflicted?
- What will be the effectiveness of abatement measures?

To a certain extent these questions have been answered for the Dutch situation in the evaluation report of the first phase (Schneider and Bresser, 1988).

Based on an interim evaluation performed in 1987, the Steering Committee for the programme decided to start a second phase of the research programme, which was carried out during the period 1988 - 1991.

In order to be able to advise on policy measures that lead to effective and efficient ways for abatement, the second phase programme had to answer the following questions:

- What is the quantitative contribution of O_3, SO_2, NO_x and NH_3 to the direct effects on vegetation?
- What is the quantitative contribution of SO_2, NO_x and NH_3 to the indirect effects (via the soil) on vegetation?
- What is the effectiveness of abatement measures in terms of environmental effects?

1.2 Research programme

1.2.1 Outline

In the first phase of the programme three main research areas have been identified:

- exposure-effect relationships for forests, cultivated crops and semi-natural vegetations (both direct and indirect effects), the main research effort;
- ammonia emissions;
- effectiveness of control measures.

These have been the main research items in the second phase too. The topics within the areas however differ from the first phase.

The study on effects on vegetation has been restricted mainly to Douglas fir on well drained, nutrient-poor sandy soils and to heathland. Douglas fir has been chosen as the "model tree", because of its resistance to pests and plagues and because, unlike most other coniferous species in the Netherlands, Douglas fir has needles with ages of several years, which is important in determining exposure-effect-relationships.
Heathland research included the heather itself, heathland species and competition with grasses.

The research area of ammonia emissions was selected because the uncertainty in the emission estimates of ammonia was too large to calculate transport and transformation in the atmosphere and the deposition. At the same time, measurements of the concentrations and deposition of ammonia and ammonium were almost absent, leaving a large area of uncertainty about the actual input of these substances.

The research on effectiveness of control measures was aimed at the development of an overall model to evaluate emission-deposition scenarios in terms of effects. The systems research on acidification in the first phase of the programme has led to a first version of the Dutch Acidification Systems Model (DAS). This model is intended to describe the quantitative relationships between emissions (in a large part of Europe), abatement techniques and associated costs, transport and transformation, concentrations and deposition, soil acidification and related effects on crops, trees, natural vegetations, small surface waters and materials (in the Netherlands). Under consideration are the acidifying substances SO_2, NO_x and NH_3, and also O_3. The first version of the DAS model has already been used in the evaluation of the first phase (for some effects).
Furthermore, attention has been paid to the determination of critical loads and levels, which is becoming more and more important; international cooperation (within the framework of the UN Economic Commission for Europe) is going on, to produce maps of critical loads and levels for Europe, in order to create a sound basis for policy measures. Especially the soil modelling work in the Dutch Priority Programme on Acidification has been closely related to the work on mapping of critical loads and levels.

Several topics in the field of acidification have not been covered within the second phase of this programme:

- effects of ozone and acid aerosols on public health;
- effects on agricultural crops, materials and monuments (No additional research has been funded within the programme; the evaluation of existing research, however, has been part of the integrated assessment study with the DAS model);

- effects on surface waters (Besides a limited study on heathland lakes (not yet finished) within the framework of the integrated assessment, these effects have not been studied.);
- corrective measures (This item has not been studied separately, within the programme; in some of the projects however (100, 118, 119, 124, 125, see annex), corrective measures have been taken into account in order to enlarge the range of situations and allow for more general conclusions; the measures themselves have not been subject of research.);
- effects on groundwater quality have not been incorporated in the programme; nevertheless some results of a separate research project are reported here; research on efflux from rootzone to groundwater has been part of a number of projects;
- only emissions of ammonia have been incorporated in the programme to a certain extent; emissions of SO_2 and NO_x have only been included in the integrated assessment study; information on SO_2 and NO_x emissions is highly compatible with UN-ECE data.

The studied cause-effect chain is depicted in Figure 1.1.

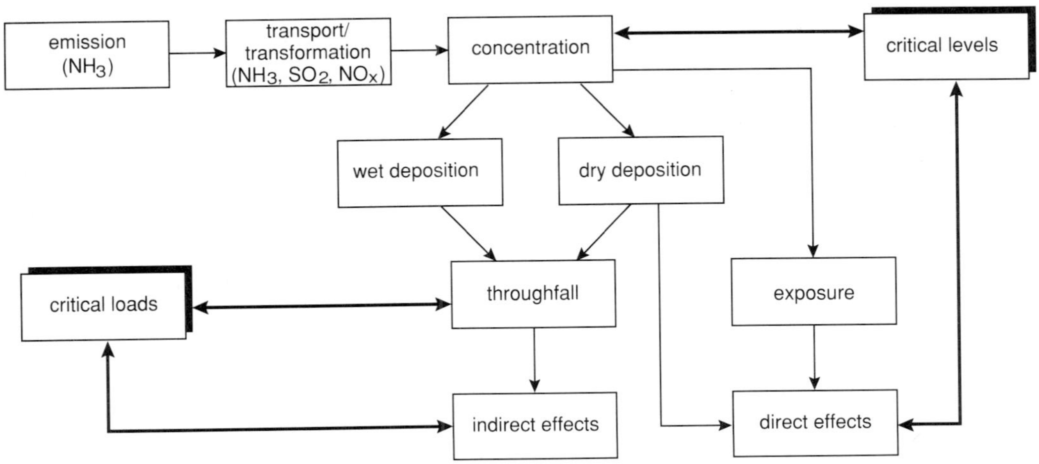

Fig. 1.1 The studied cause-effect chain

1.2.2 The projects

An overview of the research sites is given in Figure 1.2.

Effects

Within the programme much emphasis has been put on the integrated effects research, at the two main (forest) research sites in Speuld and Kootwijk (the ACIFORN - Acidification Research in Forests in the Netherlands - project) and at the heathland site Assel, studying the functioning of these types of ecosystems and the role of the stress factors air pollution, soil acidification and nutrient imbalances.

The projects related to the integrated forests sites study are: 100, 101.1, 101.2, 102.1, 102.2, 103, 104.1, 104.2, 105, 107, 108.1, 111.1, 111.2, 112, 116, 117, 118, 120, 128, 129, 190.1.

The numbers refer to the list in Annex 3. Some of these projects are fully incorporated within the ACIFORN programme, others are only partly within this integrated project.

An overview of the research items within the ACIFORN project is given in Figure 1.3.

During the summer of 1990 a project called CORRELACI (Correlative Evaluation of ACIFORN data) (203) was started, to try to get the maximum result out of the many data, resulting from the field monitoring project ACIFORN, concerning exposure-effect relationships. This project was finished by the end of 1990 and the main results have been incorporated in this report.

Based on the evaluation in 1988 it has been concluded that some subjects needed further elaboration:

- hydrology of the atmosphere-tree-soil-system with the following aspects:
  * the water conditions in the canopy (projects 104.1, 104.2, 117)
  * the unsaturated soil (projects 100, 102.2)
  * the trees themselves (projects 105, 111.1, 116);
- fluxes of nitrogen from the atmosphere to the trees (direct and indirect), the projects specifically dealing with nitrogen and trees being: 83, 100, 101.2, 102.1, 105, 106.1, 106.2, 106.3, 108.1, 108.2, 110, 115, 128, 129, 150, 190.1;
- tree physiology, projects under this item being: 103, 105, 107, 108.1, 108.2, 109, 110, 115, 150;
- morphology of needles and wood, projects under this topic being: 105, 111.1, 111.2;
- heathland research, projects dealing with heather and other species of nutrient-poor low vegetations being: 102.1, 102.2, 106.1, 106.2, 114.3, 118, 119, 122, 123, 124, 125;
- research on deposition, the projects being: 101.2, 102.1, 102.2, 104.1, 104.2, 117, 119, 120, 128, 129, 150, 190.1; during the execution of the second phase, it appeared

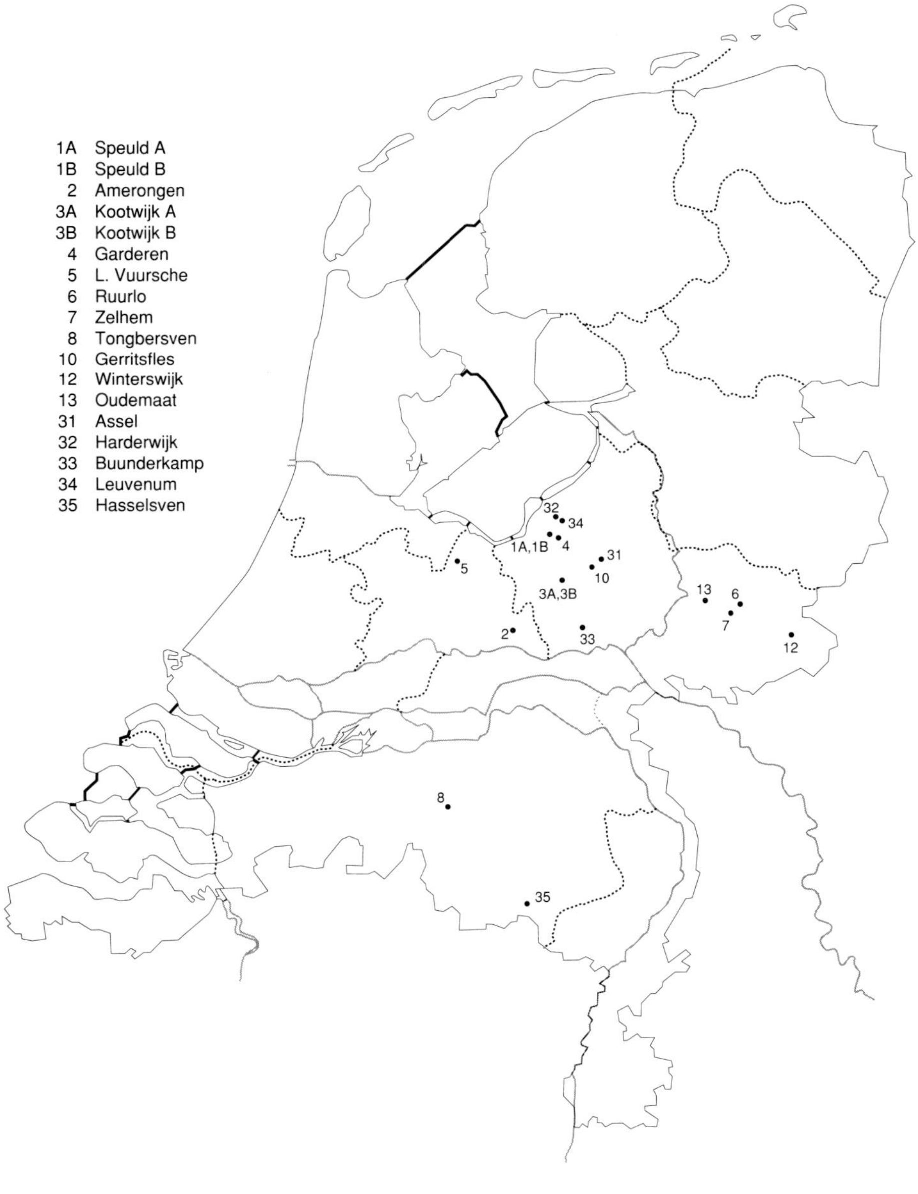

Fig. 1.2 An overview of the research sites

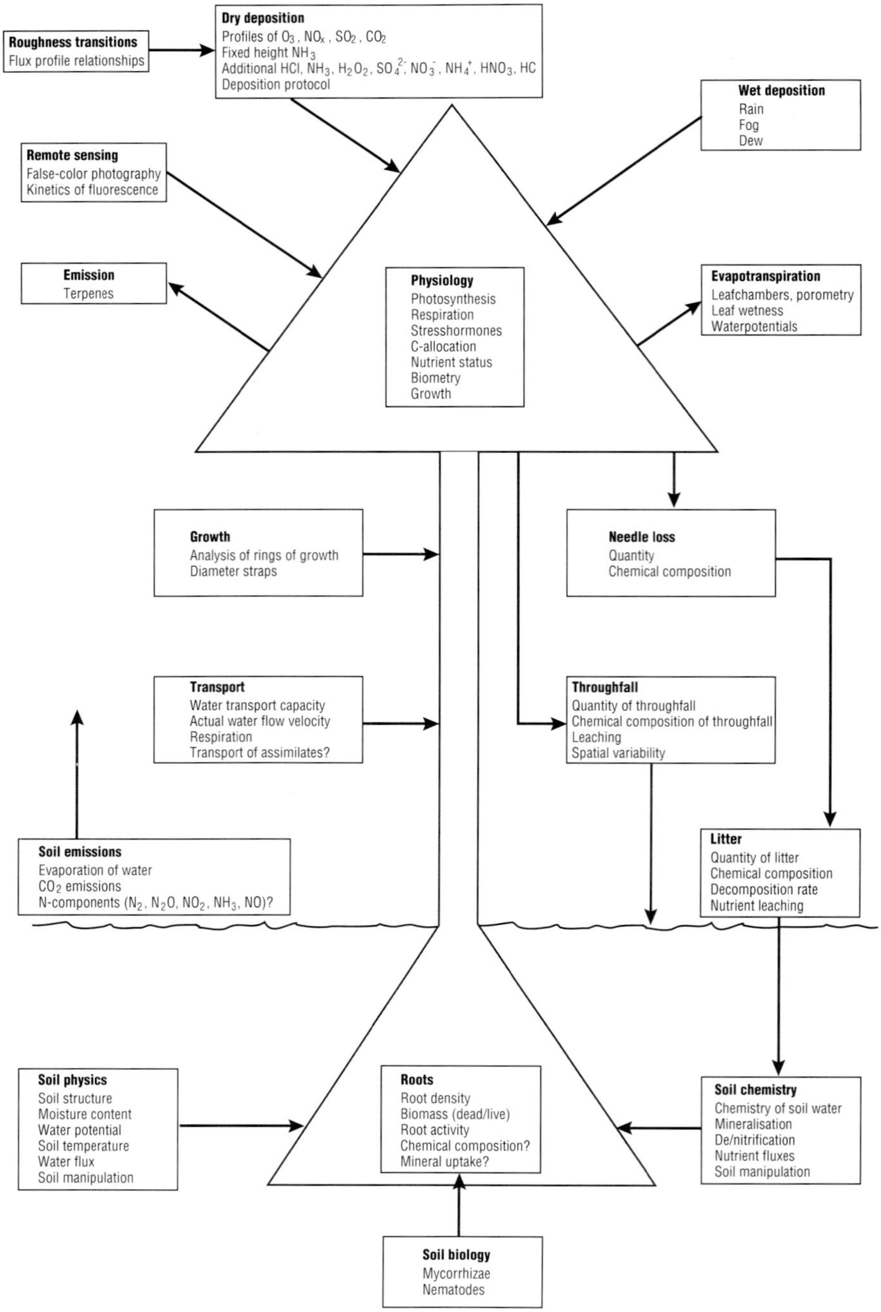

Fig. 1.3 An overview of the research items within the ACIFORN project

that a stronger emphasis was necessary on atmospheric input fluxes to different receptors. Therefore, in 1989 a project (191) was started (in fact a small programme in itself) aimed at the determination of deposition fluxes onto forest stands and heathland;

- exposition-effect research as part of a larger effort (assimilation chambers, open top chambers - OTC's -, field fumigation, field experiments, monitoring, modelling). A limited amount of research in OTC's has been judged valuable especially when combinations of stresses were investigated (air pollution, frost, water stress, pests). Relevant projects are: 107, 109, 110, 115, 125.

Ammonia emissions

Because of the environmental problems caused by the present agricultural practice in the Netherlands research on control measures has increased considerably. The Research Programme on Animal Manure and Ammonia deals with this research. Within the second phase Priority Programme on Acidification only a limited amount of money has been allocated to research on emissions from manured grasslands and arable land. Different techniques for the spreading of manure have been investigated. Projects in this research area are: 130, 131, 132, 133, 190.2, 200.1.

Effectiveness of control measures

The main emphasis in the second phase has been on the incorporation of the effect modules into the DAS model, and on the improvement of the input-output aspects (flexibility in use and presentation of results), on sensitivity, uncertainty analysis and validation, and on the use of the model in the evaluation of research in terms of policy recommendations. Projects related to this subject are: 112, 113, 114.1 (the main project, providing boundary conditions for the other modelling activities), 114.2 and 114.3. Within the soil modelling activities (project 113), attention has been paid to critical loads and levels too.

Other aspects

Beside the programme management itself (project 200) some other aspects needed attention:

- quality assurance. At some points within the programme the need for additional quality assurance became apparent in the first phase. Especially the measurements of ammonia concentrations in ambient conditions were a matter of concern. Also the value of throughfall measurements and some soil water parameters needed attention. The projects 126, 127 and 129 deal with these items. All projects incorporating measurements in the field participated in round robins.

- international exchange (project 202). Some funds have been allocated to allow for exchange between programmes from different countries on a researchers level and on the programme level.
- reviews (project 201). One of the disadvantages of an overall national research programme is the difficulty in obtaining independent reviews on the scientific results; funds have been allocated to provide for reviews by scientists from abroad.

1.2.3 Thematic reports and international review

In the course of 1990, the research results have been clustered into seven main themes:

- emissions of NH_3;
- atmospheric input fluxes;
- soil acidification / nitrogen cycling;
- biological and physiological effects;
- integrated effects (forests);
- integrated effects (low vegetation);
- integrated modelling;
- critical loads and levels.

For each of the themes, a separate report has been written, summarizing the results of all projects within each theme. These thematic reports can be found in Annex 1.

The thematic report on integrated modelling contains only model descriptions. The model results are presented in Chapter 6 of this final report.

In June 1990, an independent Review Team of eight researchers from abroad with an excellent scientific reputation has been convened, to provide a critical assessment of the second phase programme and also to indicate whether there is justification for future work in this area (see Annex 2).

1.3 The management of the programme

As in the first phase, the overall management of the Dutch Priority Programme, second phase, has been carried out by the National Institute of Public Health and Environmental Protection (RIVM). Programme Director was T.Schneider; the secretary was A.H.M. Bresser (until 1989-06-01) and is now G.J.Heij (since 1989-06-01). The Steering Committee has been responsible for the programme and for the funds available for subsidies. The Steering Committee acted under the chairmanship of G.J.R.Wolters, Director of the Air Directorate of the Ministry of Housing, Physical Planning and Environment (VROM). Responsibility for the scientific guidance of the programme was in the hands of the Programming Committee, under the chairmanship of K.Verhoeff, Director of Agricultural Research at the Ministry of Agriculture, Nature Management and Fisheries (LNV).

With the Programme Director, a Project Group of up to 13 researchers has been responsible for the co-ordination and evaluation of the research projects. Members of this group each covered a specific research subject.

The field research concentrated at the two main forest research sites (see Section 1.2.2) has been co-ordinated by a task force of research scientists called ACIFORN, under the leadership of one of the project leaders (P.Hofschreuder, Agricultural University). For research on ammonia emissions a task force has been assembled under the leadership of J.H.Voorburg (IMAG).

Steering Committee Acidification Research

Mr.G.J.R.Wolters (chairman)
Ministry of Housing, Physical Planning and Environment
Mr.V.G.Keizer (secretary)(Drs.H.Marseille)
Ministry of Housing, Physical Planning and Environment
Ir.N.D.van Egmond
National Institute of Public Health and Environmental Protection
Ir.G.A.L.van Hoek (Ir.B.A.Kleinbloesem)
Dutch Electricity Generating Board
Drs.A.W.M.Kleinmeulman (Drs.J.J.C.Karres)
Ministry of Agriculture, Nature Management and Fisheries
Drs.W.Koemans (Drs.C.L.A.Heijster)
Ministry of Transport and Public Works
Dr.Ir.N.van Lookeren Campagne (Ir.A.P.Gaasbeek)
Shell Nederland B.V.
Mr.D.F.W.T.Pietermaat
Ministry of Economic Affairs
Dr.Ir.T.Schneider
National Institute of Public Health and Environmental Protection
Drs.P.A.Scholten (Drs.A.Zweering)
Ministry of Economic Affairs
Prof.Dr.K.Verhoeff
Ministry of Agriculture, Nature Management and Fisheries

Programming Committee Acidification Research

Prof.Dr.K.Verhoeff (chairman)
Ministry of Agriculture, Nature Management and Fisheries
Ir.F.C.Zuidema (secretary)
National Council for Agricultural Research
Dr.Ir.J.H.Blom
KEMA, Research and Development, Testing and Certification, and Consultancy Services for the Electric Power Industry
Dr.C.A.M.van der Klein
Netherlands Energy Research Foundation
Ir.A.P.Gaasbeek
Shell Nederland B.V.
Ir.N.D.van Egmond
National Institute of Public Health and Environmental Protection

Mr.V.G.Keizer
Ministry of Housing, Physical Planning and Environment
Drs.R.A.Braakenburg van Backum
Ministry of Transport and Public Works
Drs.C.L.A.Heyster
Ministry of Transport and Public Works
Dr.Ir.T.Schneider
National Institute of Public Health and Environmental Protection

Project Group Acidification Research
Dr.Ir.T.Schneider (chairman)
National Institute of Public Health and Environmental Protection
Ir.G.J. Heij (secretary, since 1989-06-01) (Ir.A.H.M.Bresser until 1989-06-01)
National Institute of Public Health and Environmental Protection
Dr.R.M.van Aalst
National Institute of Public Health and Environmental Protection
Ir.J.C.A.M.Bervaes
Research Institute for Forestry and Landscape Planning "De Dorschkamp"
Prof.Dr.N.van Breemen
Agricultural University, Department of Soil Science and Geology
Ir.H.S.M.A.Diederen
MT-TNO
Drs.H.G.van Dobben
Research Institute for Nature Management
Ir.A.J.Elshout
N.V. KEMA
Ir.P.Hofschreuder
Agricultural University, Department of Air Pollution
Prof.Dr.A.C.Posthumus
Research Institute for Plant Protection
Dr.J.Slanina
Energy Research Foundation
Dr.J.A.van Veen
Research Institute ITAL
Prof.Dr.J.M.Verstraten
University of Amsterdam, Laboratory for Physical Geography and Soil Science
Ir.J.H.Voorburg
Institute of Agricultural Engineering
Ir.F.C.Zuidema
National Council for Agricultural Research

About 30 research institutes have been involved in the second phase, carrying out some 50 projects. The number of research scientists and professional assistants was about 300 (150 in each category). All project leaders had to report twice a year. Once every year the progress was also reported at a symposium for all the researchers involved. Proceedings of these symposia are available (some mainly in Dutch). Reports of individual projects have been published either by the institutes themselves or in a special series. Reports of interest to a wider audience abroad are in English or have an extensive English summary. Copies of reports can be obtained from or through the secretary. A list of available reports and publications has been given in Annex 3.

1.4 The programme budget

In Table 1.1 the sources of the subsidies available in the second phase of the research programme are presented.

Table 1.1 Available budget in the second phase programme (subsidies; in Dfl millions)

Source	1988	1989	1990	1991	total
Ministry of VROM (environment)	0.9	0.9	1.1	0.2	3.1
Ministry of LNV (agriculture, nature preserv.)	0.9	1.2	1.0		3.1
Ministry of EZ (economy)	1.0	1.0	1.1		3.1
Sep (electricity companies)	0.4	0.4	0.4		1.2
SHELL (refineries)	0.4	0.4	0.4		1.2
Ministry of V&W (public works/traffic)	0.2	0.3	0.4		0.9
total	3.8	4.2	4.4	0.2	12.6

The annual budget for subsidies was more or less the same as in the first phase of the programme. The total amount was less because the programme had a duration of 2.5 years (first phase 4 years).

As in the first phase programme, the contributions from the research institutes themselves were substantial. Table 1.2 presents an overview of the total budget for the programme.

Table 1.2 Total programme budget second phase (in Dfl millions)

	total
subsidies	
- manpower	9.2
- materials and subcontracting	3.4
contributions of institutes and other programmes[1]:	
- manpower	12.6
- materials and subcontracting	3.4
total	28.6

1) without RIVM contribution

A large part of the necessary equipment for research had already been financed in the first phase of the programme. Total budget (first plus second phase): Dfl 85 million.

1.5 Outline of the final report

The thematic reports contain all the scientific information resulting from the second phase of the Dutch Priority Programme on Acidification. They are compilations of the results of the individual projects. The project reports and publications are listed in Annex 3.

The final report is the translation of the scientific results into policy-oriented results and conclusions. The report deals with the following main aspects:

- ammonia emissions and abatement (Chapter 2);
- concentration and deposition of acidifying compounds (Chapter 3);
- effects of air pollution and acid deposition on forests and forest soils (Chapter 4);
- effects on heathland (Chapter 5);
- scenario analyses (effectiveness of policy measures) (Chapter 6);
- critical loads and critical levels for the environmental effects of air pollutants (Chapter 7).

Literature

Schneider T. and Bresser A.H.M., 1988
Summary report; Acidification research 1984 - 1988, Report nr. 00-06

2. AMMONIA EMISSIONS AND ABATEMENT

G.J.Heij, J.W.Erisman (RIVM) and J.H.Voorburg (IMAG)

2.1 Introduction

2.1.1 The importance of ammonia emissions in the Netherlands in relation to acidification

Initially, when it became evident that the environment is affected by the deposition of acidic compounds, the acidification was related to sulphur compounds. Later on it was established that also compounds originating from nitrogen oxides contribute to acid deposition.

During the eighties it became clear that deposition of atmospheric ammonia and ammonium causes acidification, through nitrification in the soil. The major emission source of ammonia would be animal manure. In the Netherlands, the contribution of ammonia to acidification was expected to be considerable, because the number of domestic animals per hectare is relatively high.

It became increasingly apparent that atmospheric ammonia is very important in relation to acid deposition in the Netherlands (see Chapter 3) and its related effects (Chapters 4 and 5). Its contribution to the nationwide average total potential acid deposition amounted to about 45% in 1989 (Chapter 3, Figure 3.13). Locally the contribution can be even higher.

NH_3 in the atmosphere is emitted by many low level sources with varying source strength. It has a short atmospheric residence time and is converted rapidly into ammonium, which can be transported over much longer distances.

In order to be able to decrease acid deposition in an effective way, it is important to ascertain how much ammonia is emitted and how the sources are distributed within the Netherlands.

Within the framework of the Dutch Priority Programme on Acidification, attention has been paid to ammonia emissions from livestock production, in order to get information on the emission of NH_3 per sector. Also the effectiveness of control techniques (abatement measures) has been taken into account.

The underlying aim was to obtain quantitative information on the whole cycle: emission, atmospheric transport, concentration and deposition of NH_3, to establish cause-effect relationships and critical deposition loads and concentration levels.

2.1.2 Sources of ammonia

The most important sources of atmospheric NH_3 are animal manure, fertilizers and some industrial activities. It is estimated that 80 to 90% of total anthropogenic NH_3 emissions comes from animal manure. Sources of minor importance include: traffic, natural soils, coal combustion, cats and dogs, human respiration, sewage sludge and wild animals. Three types of emissions from livestock animal manure can be distinguished (see Figure 2.1):

housing and storage emissions, referring to the NH_3 emissions during the period that the animals are in the stables, for example for cattle in winter and for pigs and chickens throughout the year; emission due to the spreading of manure, on pastures or arable land; emissions from grazing, referring to the emissions from cattle during the pasture period.
The emissions of NH_3 from livestock manure depend on several factors, including fodder composition, nitrogen and water content of the manure, meteorology, soil conditions, pH, etc. The emission is therefore dependent on type and age of animals, type of housing system and storage system, and the manner of application of manure. Thus the ammonia emission from livestock production is a very heterogeneous and diffuse process, spread over an area of about two million hectares. This explains why it is difficult to make an accurate estimate of the total emission and the contribution of the different sources.

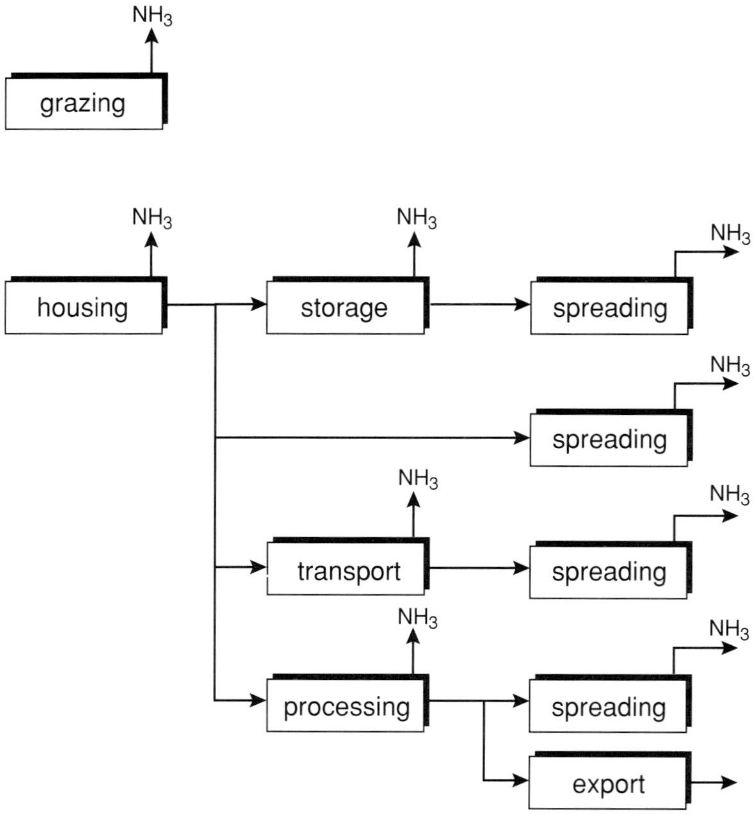

Fig. 2.1 Factors affecting agricultural emissions of ammonia from manure

2.1.3 Estimate of ammonia emission

In order to be able to quantify the total NH_3 emission and the spatial distribution of the emissions, the emission sources are divided over several categories and for each category emission factors are estimated. The categories are analogous to the yearly inventory of the number of animals for each municipality by the Central Bureau of Statistics (CBS, 1988). The emission factors (De Winkel, 1988, Van der Hoek, 1989), being the emission rate per source or activity, have been calculated up to now from the nitrogen mass balance, following the nitrogen flow from the "mouth to the surface". A distinction in factors is made between housing and storage emissions, emission due to manure spreading and emissions from grazing. Emission from sources other than livestock is taken directly from inventories made by Buijsman et al. (1984) and Erisman (1989).

Because few measurements were available concerning emission of ammonia from the different systems of livestock production and manure handling, within the Dutch Priority Programme on Acidification, attention was focused mainly on the <u>measurement</u> of the ammonia emission.

In the research projects concerning ammonia emissions the following aspects were studied:

- Emissions from animal housing systems and manure storage
- Emissions after manure spreading
- Effect of manure spreading techniques
- Modelling of emissions after spreading on arable land
- Emissions from grazing
- Emissions from agricultural regions

The study of these aspects also included the possibilities of emission reduction.

- Abatement of ammonia emissions with biofilters and air scrubbers

Since 1988 the research has been incorporated into the Research Programme on Animal Manure and Ammonia.

The thematic report (see Annex 1) summarises the results of the research on ammonia emissions carried out within the framework of the Dutch Priority Programme on Acidification. However, as far as possible, an attempt has been made to give a complete "state of the art" report, by including published results of the Research Programme on Animal Manure and Ammonia.

2.2 Ammonia emission factors

2.2.1 Measured emission factors

The most important result concerning measurements of ammonia emissions from livestock is a better insight into the factors contributing to ammonia volatilization from housing

systems, manure stores, grazing and spreading of manure. Table 2.1 lists emission factors resulting from the measurements. The measurements related to dairy cows were carried out in one stable over a period of only 4 months, with cows receiving rations geared to maximum utilization of proteins. The measurements related to pig housing cover a full year period. However, the housing system was somewhat modified and therefore not fully representative of usual pig housing. The measured emission factors for poultry (housing) are probably underestimated. Storage of solid manure from layers and broilers can cause considerable ammonia emission, which is not included in the measurements.
Emissions can not be measured easily for all housing systems, particularly for those which are naturally ventilated.
The research on emissions after landspreading shows a wide variation in ammonia losses during and after surface application of manure. Factors influencing the evaporation are weather conditions, soil conditions and physical properties of the manure. The research findings are:

* The ammonia emission during spreading is, for all types of spreading machines, less than 1% of the applied nitrogen.
* After surface application of manure, roughly 50% of the ammonia content of the manure evaporates, depending on weather, soil conditions and composition of the manure.
* About 50% of these losses takes place after spreading within the first six hours of application of the manure.
* A model has been developed describing the ammonia emission after application on arable land. This model gives an insight into the relative importance of the factors influencing ammonia losses.

No emission factors for application of manure were obtained from these measurements and modelcalculations.

Emissions from grazing proved to be smaller than previously estimated. The main part of the excreted nitrogen is in the urine, which easily penetrates into the soil. This explains why the emission is only in the order of 10% of the excreted nitrogen. Finally, the measurements show a positive correlation between emission and the amount of fertilizer nitrogen applied for grass production.

2.2.2 Estimated emission factors

The first emission factors from which the total NH_3 emission in the Netherlands was calculated were published by Buijsman et al. (1984). These factors were based on literature data, measurements and assumptions on evaporation rates of NH_3 from manure. Recently, new emission factors have been published by the "Werkgroep NH_3-emissiefactoren" (De

Winkel, 1988), in the following referred to as the Werkgroep. This working group was formed with representatives of the Ministry of Housing, Physical Planning and Environment and the Ministry of Agriculture, Nature and Fisheries to derive new, more detailed and up-to-date emission factors. So far this working group has established emission factors for the most important animal categories and housing systems only. Industrial emissions and emissions due to the application of fertilizers have not been taken into account.

The emission factors of the Werkgroep are listed also in Table 2.1 for those categories for which measurements were made in the framework of the acidification research.

Table 2.1 Measured ammonia emission in kg NH_3 per animal present (emission factors) and estimated emission factors for various animals and housing systems

	measurements (Acidification Programme)	estimates (Werkgroep NH_3-emissie-factoren)
Dairy cows[1)]	6	8.8
Young cows (< 2yr)	-	3.9
Veal calves	-	1.6
Breeding sows	-	8.1
Fattening pigs		
Fully slatted floor	-	2.5
Partly slatted floor	3[2)]	1.3-2.0
Layers		
Slurry storage under batteries	0.083	0.33
Manure belt, removal as slurry	0.034[3)]	0.29
Manure belt with drying	0.031[3)]	0.14
On litter	-	0.18
Broilers[3)]		
Floor not insulated	0.046	0.21
Floor insulated	0.040	0.14

1) housing period 180 days
2) measurements from one stable system
3) manure storage not included

2.2.3 Comparison of measured and estimated emission factors

In Table 2.1 both the measured as well as estimated emission factors for housing are listed. The measured factors are generally lower than those calculated by the Werkgroep, except for pig stables (partly slatted floor). The calculated emission factors are based on

assumptions regarding the evaporation rates of NH_3 from the manure under different conditions. The comparison with the measurements from the experiments clearly demonstrate that the uncertainty in the calculated emission factors is very large. Furthermore, the number of categories for which factors are measured is limited.
In general, the emission of NH_3 depends on several factors such as temperature, atmospheric turbulence, ambient NH_3 concentration, nitrogen and water content of the manure, humidity, soil pH, land type and manure composition. Large errors may be introduced when the complex emission process is described by a limited set of emission factors, assumed to be valid for the whole of the country on an annual basis.

Within the Dutch Priority Programme on Acidification it has been decided to estimate the total ammonia emission by using the consistent set of emission factors of the Werkgroep. The overall uncertainty in these factors has not been quantified so far.

2.3 The total national ammonia emission

The first inventory by Buijsman et al. (1984) resulted in a total livestock emission of about 130 kton in the Netherlands in 1982. Other sources (industry, fertilizers, remaining sources) contributed another 25 kton on an annual basis. Erisman (1989) derived a total livestock emission of 215 kton NH_3 in 1988 based on the emission factors by the Werkgroep (Table 2.1). This is the same order of magnitude as reported in the first phase of the Dutch Priority Programme on Acidification (Schneider and Bresser, 1988). This figure is somewhat different from that given by Oudendag and Wijnands (1989) in the thematic report on ammonia emissions, because they slightly changed the emission factors on statistical grounds. In the study by Erisman the total NH_3 emission due to other anthropogenic sources was calculated to be 33 kton. The differences between the estimated emissions for 1982 by Buijsman et al. (1984) and for 1988 by Erisman (1989) can not be attributed to the change in number of animals during those years, but mainly to the use of new emission factors. The total NH_3 emission from the different categories is listed in Table 2.2. From this table it is obvious that most ammonia is emitted by cattle. Furthermore, application of the manure is at present the largest source of the emissions. The calculated emission is based on the assumption that 50% of ammonia in the manure evaporates during and after spreading. As there is, amongst other things, no quantitative information about the conditions prevailing when farmers are spreading manure, it is not possible to make a more accurate calculation of the total ammonia emission due to manure spreading.
It is not possible to get more detailed and more accurate information on the total ammonia emission from livestock production from the research in the Dutch Priority Programme on

Acidification with the measurement-based emission factors and to produce a more sound estimate than published by Erisman. The measurements made so far are insufficient to give more reliable estimates. Within the framework of the Research Programme on Animal Manure, the inventory of the ammonia emission will be completed.

Tabel 2.2a Total emissions from livestock breeding in the Netherlands in kton $NH_3.yr^{-1}$ for 1980-1988. Distinction is made between stable, spreading and pasture emissions, taking into account the difference in grazing system for cattle

category	1980	1981	1982	1983	1984	1985	1986	1987	1988
cattle									
stable and storage	40.0	40.7	41.4	42.8	43.4	40.7	39.5	36.9	35.5
manure spreading	66.3	67.4	68.6	70.9	71.9	67.5	65.5	61.3	58.5
pastures	29.1	29.2	29.6	30.6	30.8	29.9	27.6	25.7	24.5
pig									
stable and storage	20.1	20.3	20.2	20.8	21.6	23.9	25.9	27.3	26.0
manure spreading	31.4	31.8	31.7	32.6	33.9	37.5	40.8	43.0	41.1
poultry									
stable and storage	18.6	19.5	20.1	17.7	16.6	16.5	17.0	17.9	17.1
manure spreading	9.3	9.8	10.2	10.1	10.6	11.9	12.2	12.8	12.3
subtotal	214.8	218.7	221.8	225.5	228.8	227.9	228.5	224.9	215.0
other sources (Table 2.2b)					32.8				
total	247.6	251.5	254.6	258.3	261.6	260.7	261.3	257.7	247.8

Table 2.2b Other anthropogenic NH_3 emissions in 1988, in kton $NH_3.yr^{-1}$

category		emission
sources incorporated in the emission map:		
sheep		2.17
horse		0.81
goat		0.12
duck		0.08
turkey		0.82
fertilizer		10.03
human		4.30
cat		1.24
dog		3.75
housekeeping		1.00
industry		7.56
subtotal	31.88	
sources not incorporated in the emission map:		
rabbit		0.35
mink		0.22
fox		0.02
sewage sludge		0.05
traffic		0.25
subtotal	0.89	
total	32.77	

The emission from **N fertilizers** has not been investigated within the framework of the Dutch Priority Programme on Acidificaton. The emission is mainly determined by the composition of the fertilizer (the chemical form of N and the presence of other chemicals). Other influencing factors are the circumstances during application of the fertilizer (soil conditions, meteorological conditions and presence of vegetation). For the Netherlands it has been estimated that only approx. 2% of the fertilizer-N is emitted as ammonia to the air (Thomas and Erisman, 1990). This rather low loss is due to the type of fertilizer applied. The yearly amount of ammonia emission from N-fertilizers in the Netherlands is estimated thus to be about 10 kton (Erisman, 1989).

The ammonia emission from **industrial sources** has not been investigated within the framework of the Dutch Priority Programme on Acidification. The emission is mainly due to the production of N-fertilizers, ammonia, nitric acid and urea. The emission depends on the production process, age of installation, abatement techniques, etc. The total industrial emission of NH_3 in the Netherlands in 1986 was approx. 6 kt (Thomas et.al., 1988).

Data on the number of animals within each category and their spatial distribution have been derived from publications by the CBS on a yearly basis. This also applies to the use of fertilizers.

Information on the emissions due to industrial processes is collected within the framework of the National Emission Inventory.

Figure 2.2 gives the emission developments during the period 1965 - 1988, based on calculations of Thomas and Erisman (1990). As can be shown from this figure, the ammonia emission has stabilized in the eighties due to the manure legislation and milk quotation, after a strong increase from 1950 to 1980.

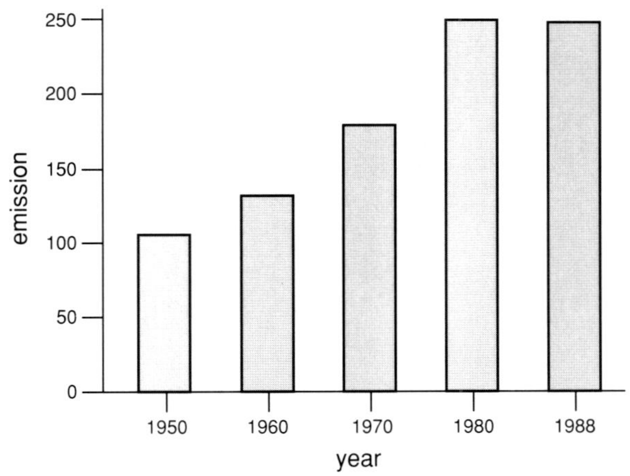

Fig. 2.2 Emission of ammonia in the Netherlands (kton NH_3)

2.4 Spatial distribution of the ammonia emission

In Figure 2.3 the spatial distribution of the animal NH_3 emission in the Netherlands for the year 1988 is presented (Erisman, 1989). Highest emissions are found in the southern and central parts of the Netherlands, as a result of intensive livestock breeding. Locally, NH_3 emissions can vary much more than is shown on the map.

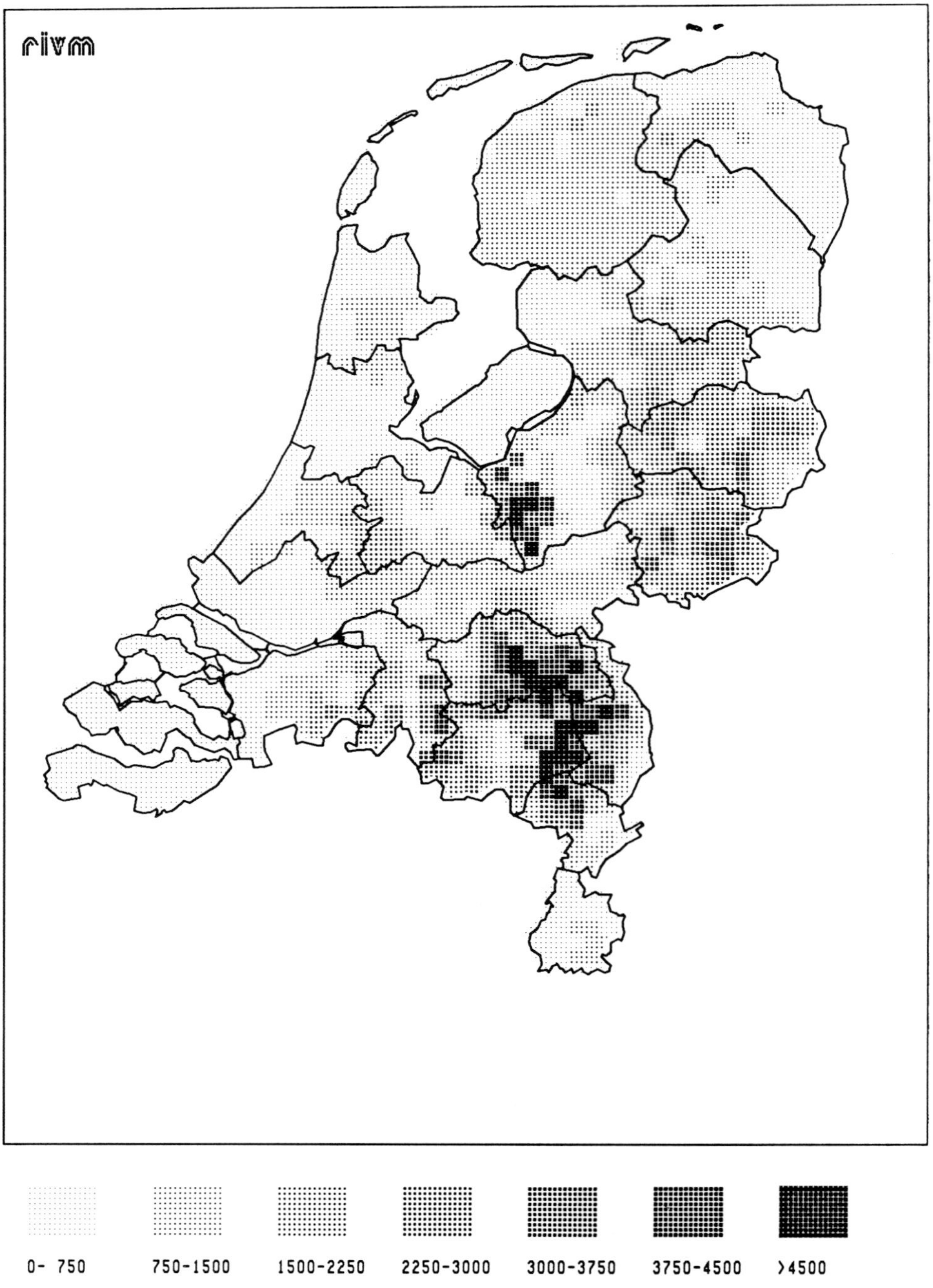

Fig. 2.3 Total livestock NH_3 emission in the Netherlands, in 1988 ($kg.yr^{-1}(5*5km)^{-2}$)

The calculations are based on the emission factors from the Werkgroep (see, for some factors, Table 2.1), together with those for the remaining animal categories (Van der Hoek, 1989). For the sources without new information on emission data, such as fertilizers, industry, etc., the data from Buijsman et al. (1984) have been used. The spatial distribution of the emissions is improved compared to the earlier published data by Buijsman et al. (1984), because of the exclusion of emissions in non-emitting areas like nature areas, water surfaces and cities. Through the use of different emission factors, the spatial distribution has also changed compared to the one presented by Buijsman et al.
The calculation method contains several errors. The total uncertainty in the NH_3 emission calculations for the 5x5 km^2 grid elements in the Netherlands was estimated to be at least 40% (Erisman, 1989). The largest uncertainty is formed by the emission factors and the location of the sources. Another uncertainty is introduced by the assumption that all manure produced in a municipality is also applied within that same municipality. The transport of manure outside the municipality via so called manure banks is still performed on a small scale. It is expected, however, that transports of this kind will increase in the near future due to governmental restrictions on the amount of manure spreading.

2.5 Emission reduction

Reduction of ammonia emissions from livestock production can be achieved by technical measures with respect to housing, storage and application of manure, by processing of manure and by optimizing the composition of fodder.
Within the Dutch Priority Programme on Acidification no attention has been paid to the influence of animal feeding on the ammonia emission. It is known, however, that a more balanced feeding reduces the amount of ammonia in the manure and thus, indirectly, the ammonia emission. This implies that the amount and type of N compounds in the fodder could be better adjusted to the need of the animals. Another promising option for increasing the utilization of proteins is the addition of certain synthetic amino acids to the fodder. This aspect is one of the priorities of the Research Programme on Animal Manure and Ammonia.

- Housing and storage systems

In general, reducing ammonia emissions from farm buildings is difficult, because of the diversity of animals and housing systems. Most housing systems have a manure storage under a slatted floor. It is evident that this storage cannot be covered and therefore it will continue to be a source of ammonia emission. Recent measurements , however, have shown that emissions from a stable floor polluted with freshly excreted urine are at least of the same importance.

It appears from experiments that in theory it is possible to reduce the emission from stables by more than 50%. However, systems should be developed for application under farming conditions.

In cattle houses, the ammonia emission from the manure under the slatted floor can be reduced by acidification of the manure. In an experiment with nitric acid the NH_3 emission from the stable was reduced by 40%. The main problem was the occurrence of denitrification. Experiments have been carried out to reduce the emission from the stable floor with scrapers and flushing systems. Flushing gives good results. Of course, the effect of acidification upon application of the manure as a fertilizer has to be taken into account.

Where it is not possible to realize sufficient reduction in NH_3 emission from a housing system by appropriate handling of manure, the treatment of exhausted air in a biofilter or air scrubber offers a technically sound solution. These systems can only be applied in houses with forced ventilation. The ammonia concentration in the exhaust air can be reduced by 80%.

However, measurements of air scrubbers and biofilters under conditions of livestock farms have shown that the ammonia reduction is often inefficient. This is mainly because of the inhibition of the nitrification process by free ammonia, nitrite or the pH value. This means that the efficiency strongly depends upon the removal of absorbed nitrogen, which makes the operation of air scrubbers and biofilters more complicated. Moreover, the disposal of effluent is an extra cost factor. A positive aspect is the odour emission reduction by properly working biofilters. Acid based air scrubbers generally do not reduce odour emission.

Although results are positive at certain conditions, biofilters are hardly used in livestock farming because of the high cost of the system, but they can be applied in a more cost-effective way in the manure-processing industry.

The covering of manure storage systems results in an important reduction of emissions of ammonia and odours. Depending upon the type of cover, a reduction of the emission during storage of approximately 80% is possible.

So far a relatively small part of the manure has been stored outside the stable. The largest part of the NH_3 emissions from manure storage consists of emissions from stables. Owing to the increasing storage capacity, in the future storage outside the stable will be of more importance. If these storages are covered, the manure will have a higher NH_3 content due to mineralisation during storage.

- Spreading of manure

The main source of ammonia volatilization is spreading of manure. Most of the ammonia is emitted within the first hours after spreading. Weather conditions, in particular, have a big influence on the emission from the land surface. The emission can be reduced considerably by minimizing the time the manure is at the soil surface. A number of techniques aiming at a fast incorporation into the soil have been developed. Machines with a wide variation in depth of injection and suitable for different soils and crop production systems have become available. Moreover, techniques of infiltration (dilution with water, before or immediately after application) can have a positive effect. With these techniques, the emission can be reduced by approx. 80%. Acidification of the manure also reduces the evaporation after manure spreading.
A disadvantage of these techniques is the danger of shifting the environmental problem from the air to soil and groundwater. Therefore, the amount of manure applied in this way should not exceed plant uptake during the growing season. In this context, adjustment of the applied amount of fertilizers is a possibility too.

The development of manure handling systems with little ammonia loss as a result of the factors listed above is arousing increasing interest and is being stimulated by the Government and the farmers organizations. Most attention is being paid to the development of injection techniques, acidification of manure, flushing systems and fast drying of poultry manure.

2.6 Conclusions

The most important result of the research which has been carried out is better information on the factors contributing to ammonia losses under farming conditions. This knowledge is a basis for the development of systems with reduced emissions. The main source of ammonia is spreading of manure. The emission can be reduced considerably by minimizing the time the manure is at the soil surface.
The number of measurements is too limited to form a basis for a better estimate of the total ammonia emission from livestock in the Netherlands. This estimate, together with the spatial resolution of the NH_3 emissions in the Netherlands, was therefore obtained from earlier published emission factors calculated by the N mass balance method. The measured factors are lower than the calculated ones, except for pigstables. The knowledge on emission factors can be improved by continuation of the measurements for different types of housing systems. The uncertainty of the estimate of the total NH_3 emission and its spatial distribution can mainly be reduced by reducing the uncertainty in the emission factors. Furthermore, detailed information on the type and exact location of housing

systems and the number of animals inside should become available. Finally, there is a great need for information on the emission pattern over the day and throughout the year, for deposition modelling purposes.

Literature

Buijsman E., Maas H. and Asman W., 1984
Een gedetailleerde ammoniak-emissiekaart van Nederland
Instituut voor Meteorologie en Oceanografie, Rijksuniversiteit Utrecht, rapport V-84-20, november 1984

CBS, 1988
Statistical Handbook (in Dutch)
Central Office for Statistics, Staatsuitgeverij, Den Haag

Erisman J.W., 1989
Ammonia emissions in the Netherlands in 1987 and 1988
July 1989, RIVM report nr. 228471006

Hoek K.W.van der, 1989
Evaluatie ecologische richtlijn 1989, Advies met betrekking tot emissiefactoren en toelichting
Ministry of Agriculture, Nature Mangement and Fisheries, Consultancy in General service for Soil, Water and Manure business for the Livestock breeding, Wageningen, the Netherlands
to be published

Oudendag D.A. and Wijnands J.A.M., 1989
Beperking van de ammoniakemissie uit dierlijke mest
LEI, onderzoeksverslag 56

Schneider T. and Bresser A.H.M., 1988
Summary report; Acidification research 1984 - 1988
Report nr. 00-06

Thomas R., Arkel W.G.van, Baars H.P., Ierland E.C.van, Boer K.F.de, Buijsman E., Hutten T.J.H.M. and Swart R.J., 1988
Emission of SO_2, NO_x, VOC and NH_3 in the Netherlands and Europe in the period 1950 - 2030
RIVM report 758472002

Thomas R. and Erisman J.W., 1990
Ammonia Emissions in the Netherlands: Emission Inventory and Policy Plans
Paper presented at the EMEP workshop on International Emission Inventory, Regensburg, FRG, 3-6 July 1990

Winkel K.de, 1988
Ammoniak-emissiefactoren voor de veehouderij
Publikatiereeks Lucht, nr. 76, Ministerie van VROM, 1988

3. CONCENTRATION AND DEPOSITION OF ACIDIFYING COMPOUNDS

J.W.Erisman and G.J.Heij (RIVM)

3.1 Introduction

3.1.1 Behaviour of SO_2, NO_x and NH_3 in the atmosphere

The most important acidifying compounds are sulphur dioxide (SO_2), nitrogen oxides (NO and NO_2, collectively known as NO_x), and ammonia (NH_3), together with their reaction products (see Figure 3.1). These reaction products are acids (HNO_2, HNO_3, H_2SO_4) and airborne particles or aerosols (NH_4^+, NO_3^-, SO_4^{2-}). Emission of SO_2 results mainly from combustion of sulphur-containing fuels, such as oil and coal, primarily in the processing industry (refineries) and in power stations, which are relatively high-stack sources. NO_x is also produced in combustion, by oxidation of the nitrogen in the air to nitrogen oxide and nitrogen dioxide. Important sources of nitrogen oxides are traffic, power stations and heating. The primary sources of NH_3 are the production and the spreading of animal manure (see Chapter 2).

After being emitted, the pollutants are transported and transformed in the atmosphere (see Figure 3.1). The most important reactions are oxidation and acid-base reactions. The concentrations and the deposition (wet and dry fluxes) of these acidifying pollutants and their reaction products cause an impact on the environment.

SO_2, NO_x and NH_3 each show a different behaviour in the atmosphere. The residence time of NH_3 in the atmosphere is short, due to its being emitted from ground-level sources, and to its high rate of dry deposition and rapid conversion to NH_4^+; it is thus rarely transported more than 250 km. The horizontal and vertical concentration gradients are large, so that concentrations and depositions can show considerable variations over relatively small distances. The above shows that a large part of the deposition of NH_3 in the Netherlands is caused by emissions in the Netherlands. Once being transformed into NH_4^+ (which has a much lower deposition rate) the distances of transportation are greater (more than 1000 km).

NO_x is emitted partly by low sources (traffic). However, the deposition rate is low and it takes relatively long before the gases are converted into rapidly depositing compounds (HNO_3, HNO_2). Therefore, NO_x compounds will be transported over long distances before disappearing from the atmosphere.

SO_2 is emitted mainly by high sources and can be transported over long distances. After

deposition, SO_2 is rapidly converted into sulphuric acid (H_2SO_4). NO_x and NH_3, together with their reaction products, are converted into nitric acid (HNO_3) in the air (for NO_x) or in the soil (for NH_3), and thus contribute to the acidification of the environment.
Ammonia is a basic gas which can, however, have an acidifying impact through nitrification after it has been deposited on the ground as a gas or as NH_4^+.

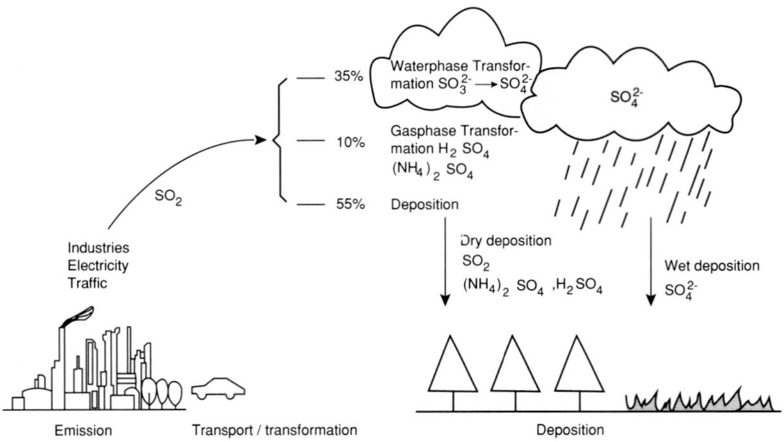

Fig. 3.1a Anthropogenic emission of SO_2 in the Netherlands (1989) and its various transport, transformation and deposition mechanisms

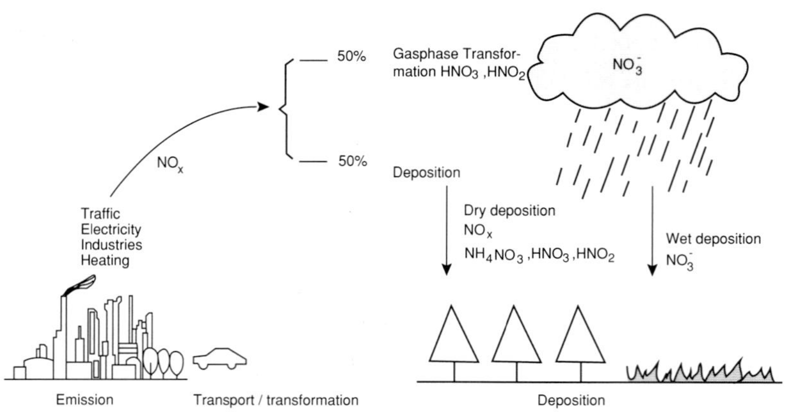

Fig. 3.1b Anthropogenic emission of NO_x in the Netherlands (1989) and its various transport, transformation and deposition mechanisms

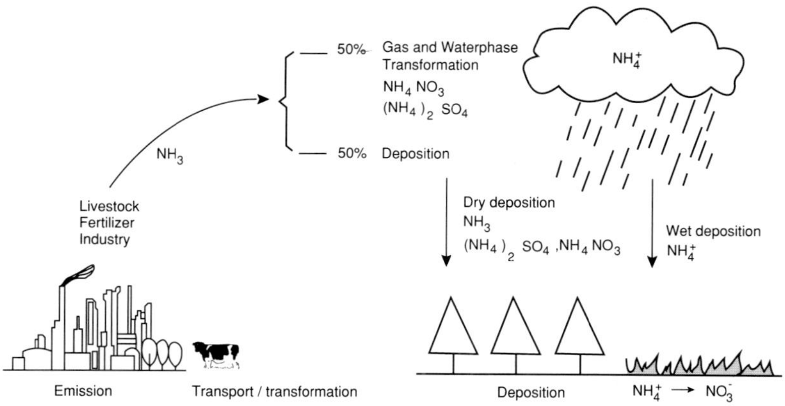

Fig. 3.1c Anthropogenic emission of NH_3 in the Netherlands (1989) and its various transport, transformation and deposition mechanisms

3.1.2 Deposition processes

Acid deposition to surfaces is determined by source characteristics, distance from the emission source, physical and chemical processes in the atmosphere and type of receptor. There are several types of deposition (see Figure 3.2). Deposition or absorption of gases and/or aerosols directly from the atmosphere is called dry deposition. Deposition of pollutants which have been incorporated into precipitation, is called wet deposition. Deposition of pollutants in fog and dew is called impaction or occult deposition (although deposition by dew in fact is also dry deposition). Total deposition is the sum of dry, wet and occult deposition fluxes.

Dry deposition

Dry deposition includes the absorption of gases by the soil surface as well as surfaces such as leaves/needles, stems or twigs, or directly by the stomata, as well as the deposition of particles (under the influence of factors such as diffusion).

How quickly gases and aerosols are removed from the atmosphere in these processes is called the dry deposition velocity. It depends on factors such as vegetation type / structure (i.e. roughness, which is determined by the height, density and uniformity of vegetation), soil characteristics, meteorological conditions and characteristics of the various compounds. Since e.g. SO_2 is absorbed via the stomata, absorption of this compound depends on the vegetation type. Furthermore, it shows a diurnal variation connected with the opening and closing of the stomata and the variability of the weather. The deposition velocity together with the concentration determines the deposition flux.

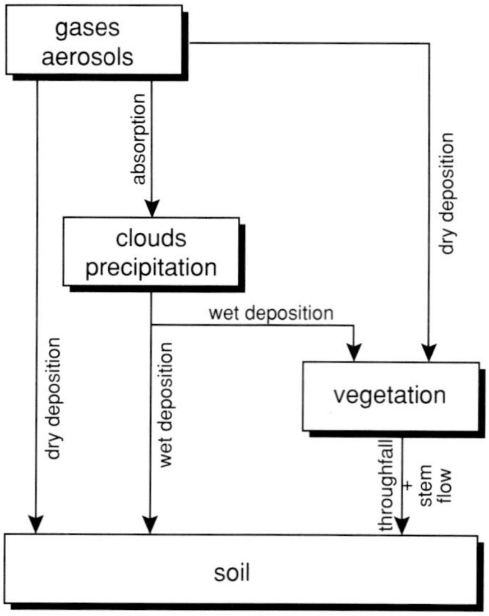

Fig. 3.2 Schematic representation of deposition

Wet deposition

Wet deposition is defined as deposition via rain, hail and snow. Gases or aerosols can be incorporated into drops of water. Particles can serve as condensation nuclei of waterdrops. When such waterdrops form precipitation, this is called rain-out. Waterdrops can also collect acidifying compounds during precipitation. This is called wash-out. Wet deposition is considered to be independent of the receptor surface.

Occult deposition

This type of deposition is less important under conditions such as in the Netherlands, but can be very important in mountainous areas such as in Germany and the eastern part of the USA (see Unsworth and Fowler, 1988 and NAPAP programme). Although in the Netherlands the quantity of occult deposition may not be very large, the frequency of occurrence of dew in particular is quite high.

Deposition on forests and heathland can be to the surfaces of leaves/needles and to the soil. Interaction between vegetation and the atmosphere takes place mainly via the surfaces of leaves/needles, continuously in the case of coniferous trees and seasonally in the case of deciduous trees. The receptor surface of forests is extremely large. The leaf area indexes (leaf area, projected to the soil surface) of the forest research sites of the Acidification Programme, for instance, are about 11 (Speuld) and 8 (Kootwijk). Rainwater which falls through the canopy onto the soil is called throughfall and water which flows along the stem is called stem flow. Throughfall contains both dry and wet deposition. Throughfall is an is called stem flow. Throughfall contains both dry and wet deposition. Throughfall is an indicator of the acid load on the forest/ heathland soil, whereas deposition (atmospheric input) is an indicator of the load on the entire system (soil + vegetation). After correction for ion exchange in the canopy in particular, the net throughfall (throughfall flux minus the flux of elements in rain on open land) is taken as a measure of dry deposition.

3.1.3 Potential and actual acid deposition

Since acid deposition consists of various substances, it is necessary to find a term which covers all these substances. In the Netherlands the term potential acid deposition, expressed in terms of a quantity of acid (mol H^+ $ha^{-1}yr^{-1}$), is used. In determining the potential acid deposition, the various substances are considered to contribute to acidification as follows:

- the deposition of oxidized sulphur compounds (SO_x) includes dry deposition of SO_2, and sulphate aerosol (SO_4), as well as wet deposition of sulphate (SO_4). 1 mol SO_x can lead to the production of 2 mol H^+;
- the deposition of oxidized nitrogen compounds (NO_y) includes dry deposition of NO, NO_2, HNO_2, HNO_3, and nitrate aerosol (NO_3), as well as wet deposition in the form of NO_3. 1 mol NO_y can lead to the production of 1 mol H^+;
- the deposition of reduced nitrogen compounds (NH_x) includes the dry deposition of NH_3 and NH_4 aerosol, as well as the wet deposition of NH_4. 1 mol NH_x can lead to the production of 1 mol H^+.

It should be mentioned that these figures indicate the maximum (potential) contribution. The actual acid load (H^+) depends on what happens to the compounds after deposition. NH_3 and NH_4, for instance, contribute to acidification only when they are converted into nitric acid through nitrification, as in the following example:

$$NH_4^+ + 2O_2 \rightarrow 2H^+ + NO_3^- + H_2O$$

A further discussion of this in relation to acidification of forest soils can be found in Chapter 4.

HNO_3 and H_2SO_4 can be neutralized completely or partly by NH_3, in which process NO_3 and SO_4 aerosols are formed. As a result, NH_3 contributes to acidification by 1 mol H^+.

3.1.4 Determination of deposition

One of the targets of the Dutch Priority Programme on Acidification was to determine the effectiveness of measures. The policy is aimed primarily at reducing emissions. This makes it necessary to choose an indicator of acid load which can be clearly associated with emissions. Policy targets are formulated in terms of deposition. The deposition to ecosystems (forests and heathland) can be estimated from throughfall measurements or from the atmospheric input. There is still a large gap between the results of these two methods (see Section 3.3.2). The relation between emissions and deposition can be determined only from the atmospheric input, not from throughfall measurements. From the atmospheric input estimates, the exact amounts of deposition and the contributions of the various sources can be estimated. Apart from that, it is important to have data on the load to the forest floor (throughfall) and to know whether it differs from the deposition, and if so, in what way. Critical deposition loads related to effects on various receptors are presented in Chapter 7. There critical concentration levels are also given. These loads and levels are used by policy-makers as a basis for deriving targets. In order to determine how far these critical levels and loads are exceeded, it is very important to determine the concentrations and the acid deposition as accurately as possible.

The value for the potential acid load (in mol $H^+ ha^{-1} yr^{-1}$) presented in the evaluation report on the first phase of the Priority Programme on Acidification (Schneider and Bresser, 1988) was based on:

2* dep. SO_x + dep. NO_y + dep. NH_x. (see also Figure 3.3)

With regard to the <u>effects on vegetation</u> (see Chapters 4 and 5) it is not only the direct above-ground exposure to certain concentrations and the above-ground uptake (dry deposition) which are important, but also the indirect impact via the soil.

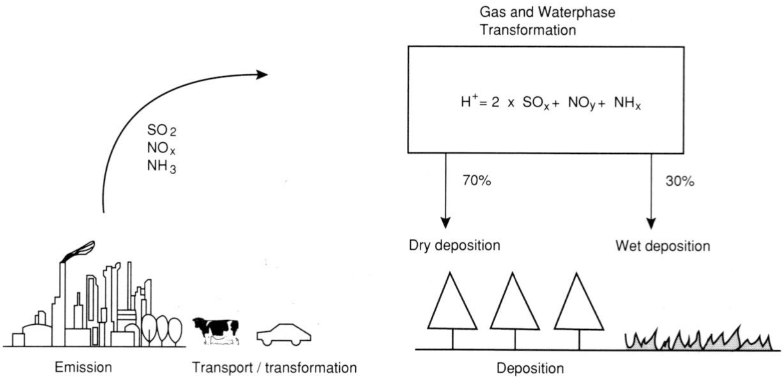

Fig. 3.3 The potential hydrogen flux by dry and wet deposition in the Netherlands (1989)

The potential wet acid deposition has so far been calculated in two ways, namely: H^+ + $2NH_4^+$ and NH_4^+ + NO_3 + $2SO_4^{2-}$. Wet acid deposition is calculated from measurement data supplied by the National Precipitation Monitoring Network. Adjustments are made for the contribution of calcium salts and, in the case of open collectors, for dry deposition in the collectors. When these adjustments are made, there is a negligible difference (of less than 1%) between wet deposition determined on the basis of H^+ + $2NH_4^+$ (correct in principle) and wet deposition determined on the basis of NH_4^+ + NO_3^- + $2SO_4^{2-}$ -$2Ca^{2+}$. The latter method is used in the Dutch Priority Programme on Acidification in order to distinguish between sources.

Dry acid deposition can be estimated on the basis of measurements/calculations of the deposition rate and measured/estimated concentrations. Especially over forests it is difficult to assess the rate of dry deposition experimentally. Fluxes of acidifying compounds were determined on various research sites of the Priority Programme on Acidification (Speuld, Assel, and Elspeet). Dry deposition processes were then parameterized on the basis of these measurements or from the literature. Model regionalization of deposition fluxes (on a 5 x 5 km^2 scale) was carried out using measured concentrations in the air and in rainfall, as well as rainfall amounts, and meteorological variables. Data were obtained from the National Precipitation Monitoring Network and the National Air Quality Monitoring Network. Because of insufficient measurement data, the spatial distribution patterns of NH_3 annual average concentrations were obtained by means of model calculations, the input being a detailed map of NH_3 emissions, supported by measurements at several locations.

Which countries the emissions originated in, and the relative contribution of the various source categories to the Netherlands' share (Table 4) were calculated by means of source-receptor matrices, derived from the model TREND (Van Jaarsveld and Onderdelinden, 1990; Asman and Van Jaarsveld, 1990). TREND is a statistical atmospheric transport model which can describe transport, conversion, and deposition of air pollutants from local sources as well as from more distant sources.

3.2 Concentration levels

Table 3.1 gives an overview of current annual average concentrations of acidifying compounds in the Netherlands. For details on locations and methods of measurements, as well as uncertainties in annual averages, see the Appendix to the relevant thematic report. Figure 3.4 shows the spatial distribution of SO_2 concentrations over the Netherlands for 1989. Since 1980 the average concentrations of SO_2 have shown a downward trend throughout the country (Annual Air Quality Report, 1989). Figures 3.5 and 3.6 show the spatial distribution of average concentrations of NO_2 and NO_x over the Netherlands for

1989. The average level of nitrogen oxides in the Netherlands has remained more or less the same since 1980. Finally, Figure 3.7 shows the spatial distribution of NH_3 concentrations over the Netherlands for 1988 (Asman and Van Jaarsveld, 1990). The distribution of concentrations has not been calculated for other years. Since 1980, however, the level of total emissions of NH_3 has been fairly constant, since the number of livestock has hardly changed (see Chapter 2). NH_3 concentrations in the years from 1980 onward can therefore be expected to be similar to the concentration shown in Figure 3.7.
Information on O_3 concentrations is also given here, effects on forests of O_3 are described in Chapter 4. Figure 3.8 shows the annual averages of the 98 and 50 percentile values of O_3 concentrations in the Netherlands. The 98 percentiles indicate the occurrence of periods with increased levels (episodes). The 1989 peak is caused by episodes which occurred in this year. The average 50 percentile in 1989 was not higher than that in 1988. The series of 10 years does not show a clear trend.

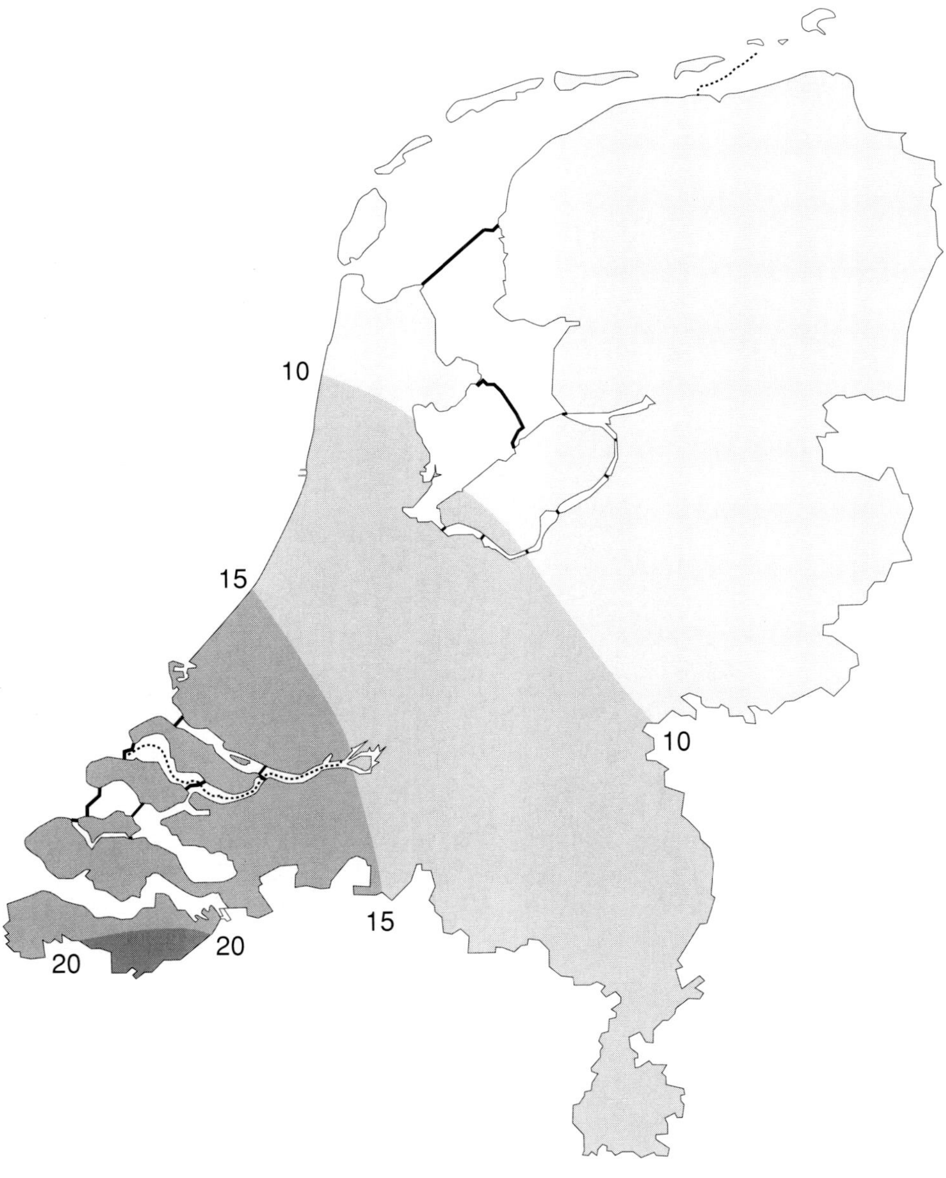

<u>Fig. 3.4</u> Average SO_2 concentration in the Netherlands in 1989 ($\mu g.m^{-3}$)

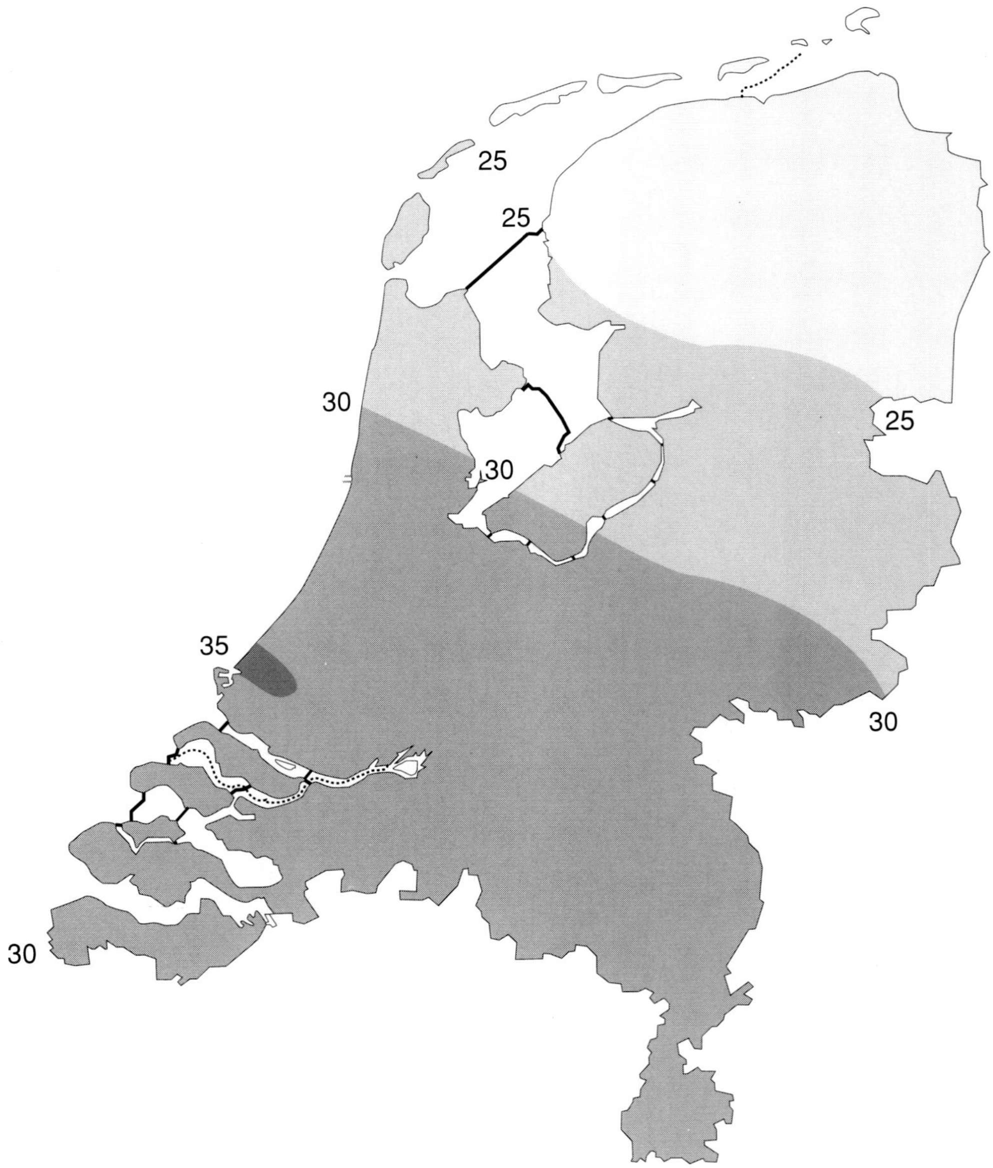

Fig. 3.5 Average NO_2 concentration in the Netherlands in 1989 ($\mu g.m^{-3}$)

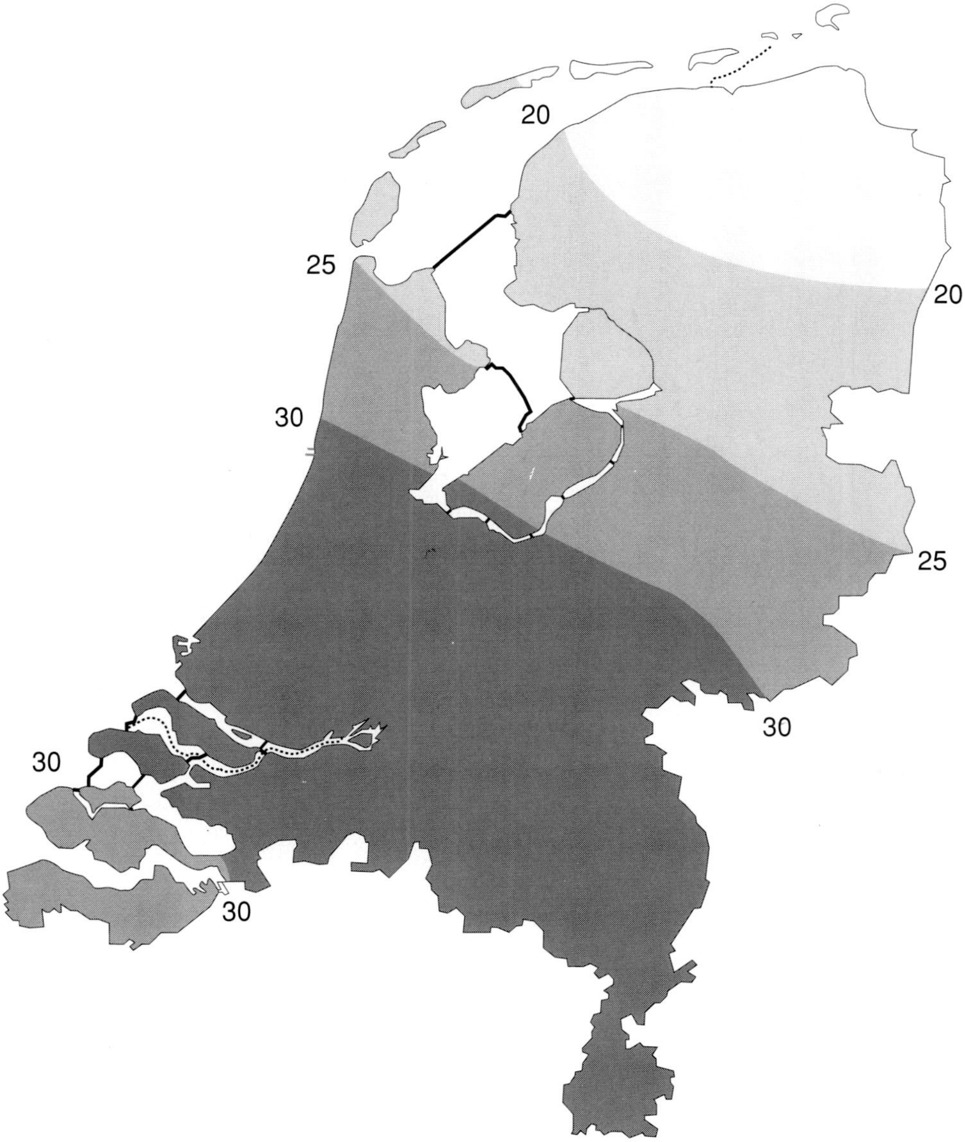

Fig. 3.6 Average NO_x concentration in the Netherlands in 1989 (ppb)

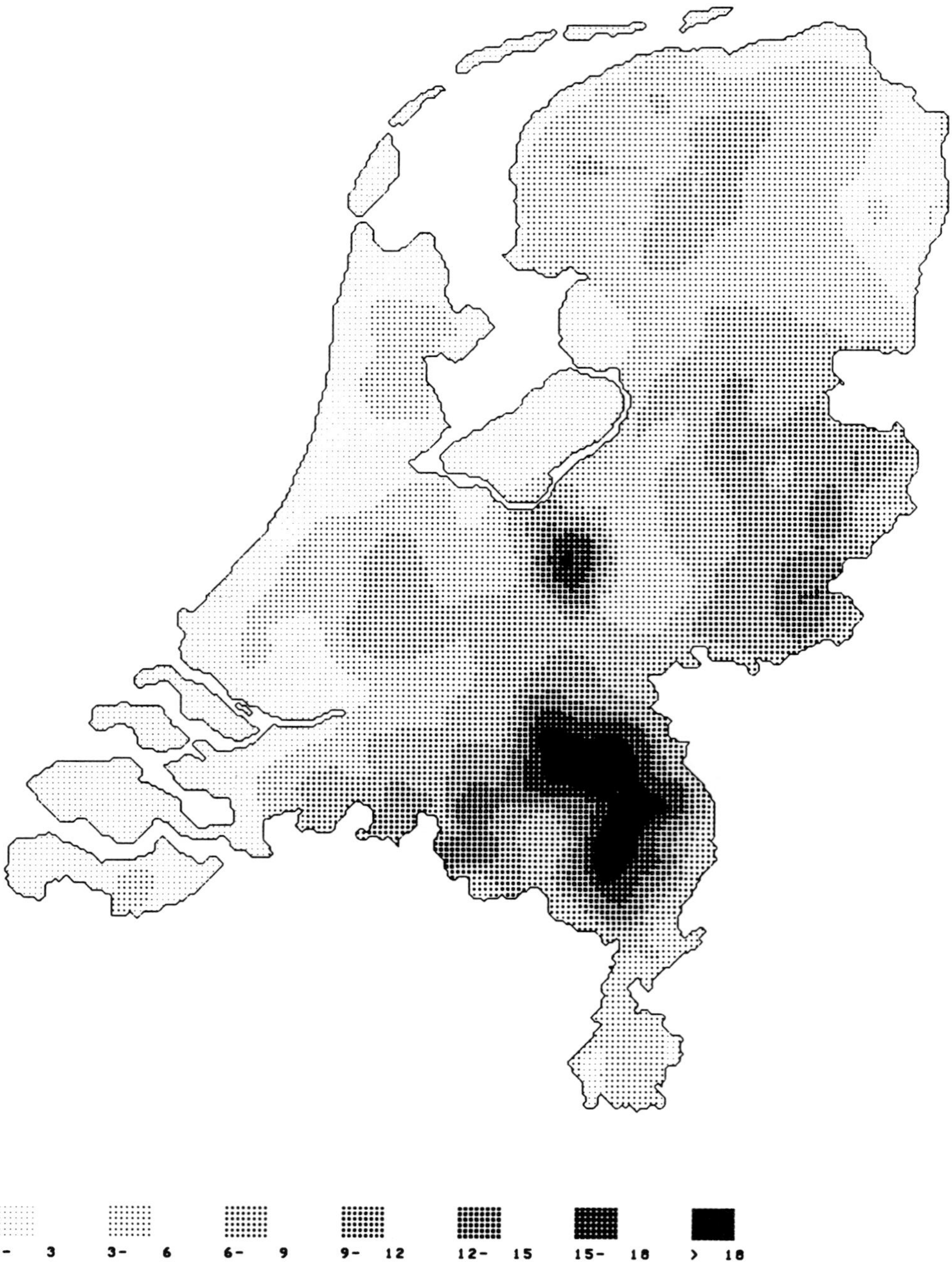

Fig. 3.7 NH_3 concentration in the air in the Netherlands in 1988 ($\mu g.m^{-3}$), calculated with the TREND model

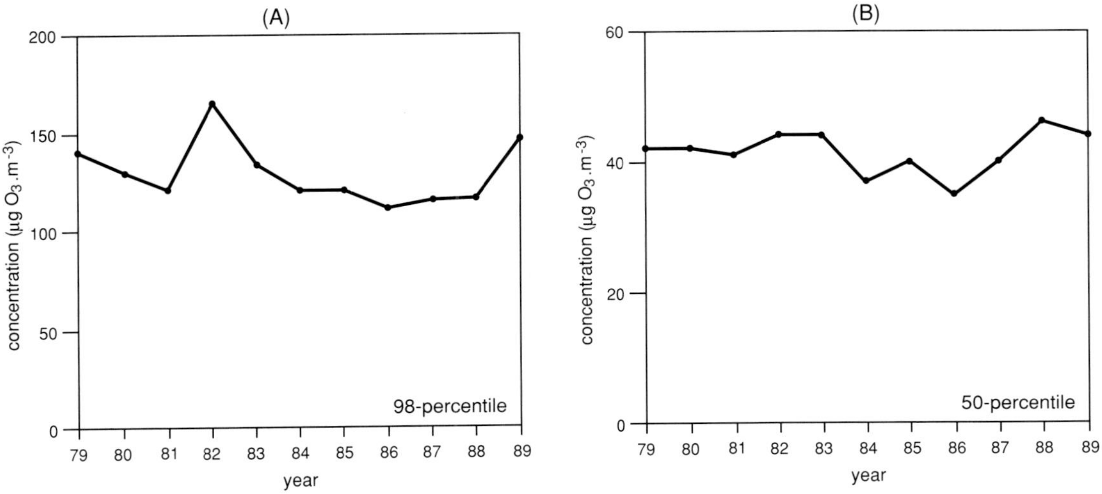

Fig. 3.8 Course of the average 98 percentile (A) and 50 percentile (B) values of O_3 concentration during the period 1979 - 1989 (based on percentiles of hourly data)

Table 3.1 Estimates of the annual average concentrations of acidifying compounds (in $\mu g.m^{-3}$), based on measurements at Elspeet, Speuld (two monitoring heights), Kootwijk and for the Netherlands as a whole in 1989

compound	Elspeet	Speuld 18 m	Speuld 30 m	Kootwijk	the Netherlands
SO_2	7.5	6.7	10.5	10.3	10
NH_3	4.0	5.1	6.3	6.7	5.5
NO			6.5	9.7	9
NO_2			26.5	27.8	25
HNO_2			0.6		1
HNO_3			0.6	0.9	1
HCl			0.6	1.1	1
NH_4				4.5	4.7
NO_3				4.3	5.6
SO_4				3.6	4.6
Cl				2.1	4.0

3.3 Deposition values

3.3.1 Local and regional values

Local values

Table 3.2 shows deposition fluxes measured at several sites (Speuld, Elspeet, and Assel) as part of the Priority Programme on Acidification. This table gives the fluxes obtained from micro-meteorological measurements as well as from throughfall measurements. (Considerable differences still exist between the results obtained from the two methods.) The three research sites are not located in heavily loaded areas. The fluxes are about equal to the national average.

The deposition flux of cloudwater and fog cannot be estimated. For most ions the concentrations in fog are much higher than in rain, but fog does not often occur in the Netherlands. Table 3.3 shows the concentrations of some important ions in fog, and its pH, measured at Speuld, Petten, and Arnhem. The figures refer to episodes.

Table 3.4 provides information on the average deposition of dew, as well as on the pH and the chemical composition of dew sampled at Speuld. The fluxes seem to be low in comparison with wet deposition or throughfall fluxes. On the whole, the contribution of fog and dew to total acid deposition is likely to be relatively low in the Netherlands.

Regionalization; values for the Netherlands

For the Netherlands as a whole deposition fluxes have been calculated empirically for the current situation (as described in Section 3.1.4, see Thematic Report, and Erisman, 1991). Wet deposition is assumed to be independent of the nature of the receptor surface. The influence of variations in roughness of the surface has been taken into account. This can lead to an increase in the dry deposition to forests, heathland an heathland pools in comparison with the deposition to an average Dutch landscape. For some components, surface resistance parametrizations have been taken into account. Vegetation-specific characteristics have not been taken into account, since too little is known on this subject.

Table 3.2 Annual average throughfall and deposition fluxes (mol $ha^{-1}yr^{-1}$) determined at Speuld, Elspeet and Assel for 1989

Speuld

	SO_x	NH_x	NO_y
throughfall	930	2100	790
wet deposition	310	630	330
dry deposition	600	2250	610
total deposition	910	2880	940

Elspeet

	SO_x	NH_x	NO_y
throughfall	1030	2150	940
bulk deposition	360	850	390
wet deposition	210	570	300
dry deposition	380	1160	-
total deposition	590	1730	-

Assel

	SO_x	NH_x	NO_y
throughfall	830	1050	380
bulk deposition	330	970	440
wet deposition	350	920	380
dry deposition	640	560	1080
total deposition	990	1480	1460

Table 3.3 Mean concentration ($\mu mol.l^{-1}$) of various ionic species and pH in fog. Values are rounded to two significant digits

compound	Speuld		Petten		Arnhem	
	1988	1989	1988	1989	1988	1989
pH	5.66	5.07	4.06	5.01	4.2	4.4
SO_4	1500	1900	1300	1200	1300	2000
NH_4	3700	5600	3100	2300	5100	7400
NO_3	1500	2000	1600	840	2600	1700

Table 3.4 Mean concentration (μmol.l^{-1}) and estimated annual flux of ions (mol ha^{-1}) in dew at Speuld, sampled on a foil

compound	concentration	flux
dew (82 mm)		
SO_4	59	48
SO_3	16	13
NH_4	189	155
NO_3	39	32
NO_2	36	30
Cl	64	53
H^+	0.8	0.7

Table 3.5 Comparison between deposition figures calculated during the first phase of the Dutch Priority Programme on Acidification and current estimates (mol ha^{-1}yr^{-1})

	1st phase		current figures	
compound	1980	1989	1980	1989
SO_x	1420	930	1570#	670
NO_y	1650	1480	1220	1160
NH_x	1350	1420	2330	2190
HCl dry	-	-	100	100
total acid ($2\ SO_x + NO_y + NH_x$)	5800	4800	6800	4800

\# Owing to the lack of specific meteorological observations in 1980, the dry deposition of SO_x is derived from 1981 figures

Table 3.5 shows the average deposition fluxes in the Netherlands for the years 1980 and 1989. This table also shows the values which were derived during the first phase of the Priority Programme on Acidification. The differences are discussed elsewhere in this section. Tables 3.6 and 3.7 show the deposition levels per compound per acidification area for 1980 and 1989. On the average wet deposition contributes for about 30% to the total deposition. Figure 3.9 shows the spatial distribution of the total depositions of SO_x, NO_y and NH_x in 1989 and Figure 3.10 shows the corresponding situation for the total Nitrogen (N) deposition. The spatial distribution of the total potential acid deposition in 1980 and

1989 is shown in Figure 3.11. In some areas deposition levels are higher than in the rest of the country; the relationship between these levels and the pattern of NH_3 emissions (Figure 2.3, Chapter 2.) is obvious. In the industrialized Rijnmond area and in the densely populated Randstad area, the high deposition levels of SO_x and NO_y respectively are also striking. Higher deposition levels are, moreover, found in town areas and forested areas on account of local roughness characteristics.

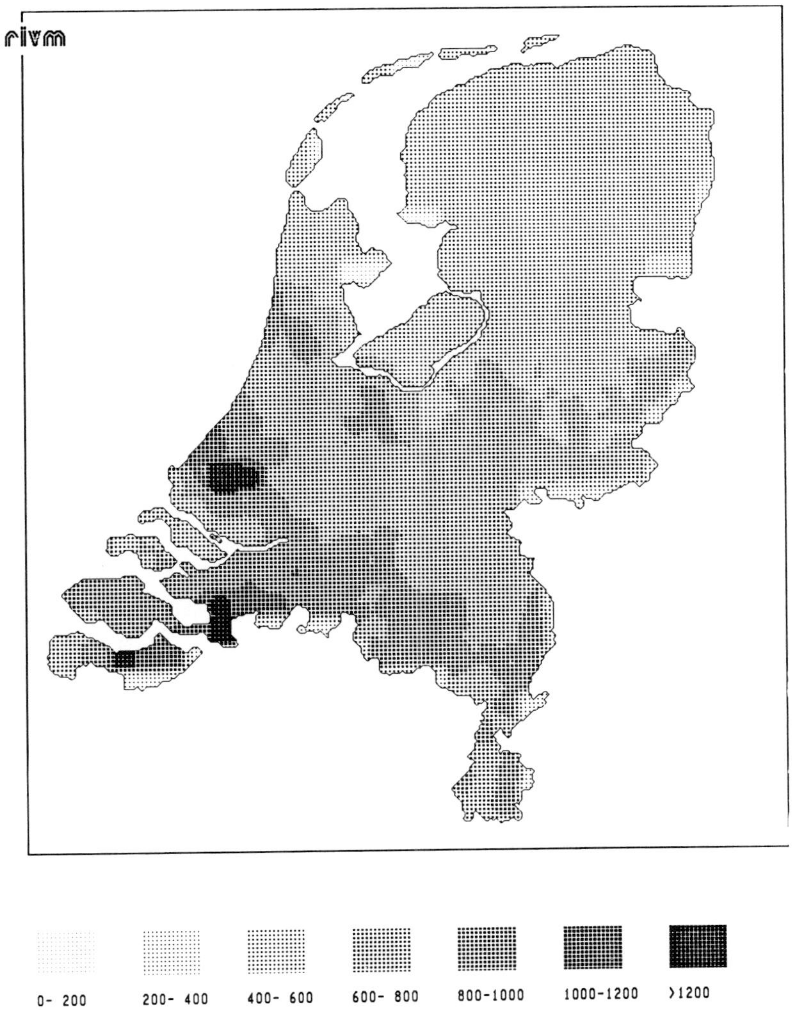

Fig. 3.9A Total deposition of SO_x in 1989 (mol $ha^{-1}yr^{-1}$)

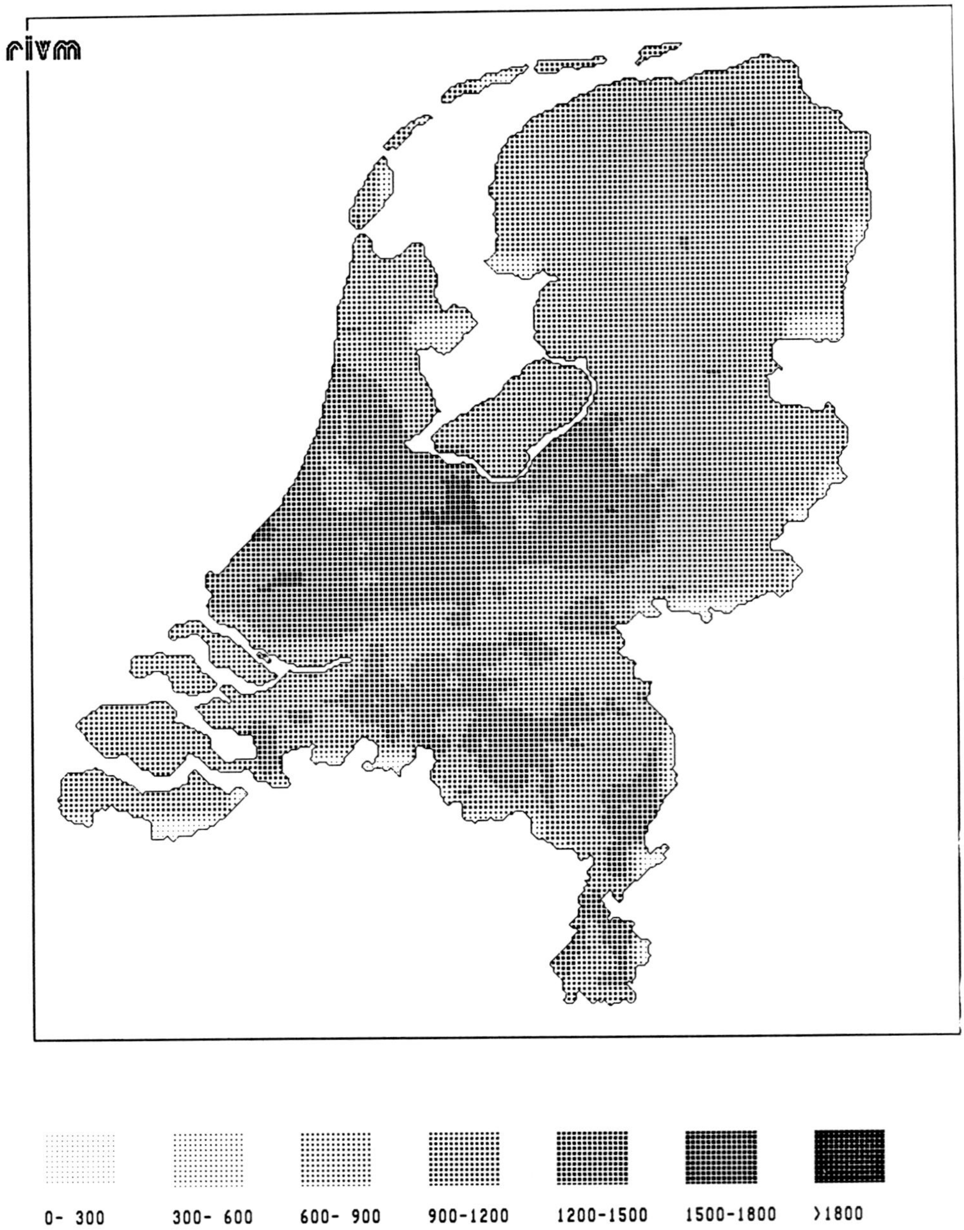

Fig. 3.9B Total deposition of NO_y in 1989 (mol $ha^{-1}yr^{-1}$)

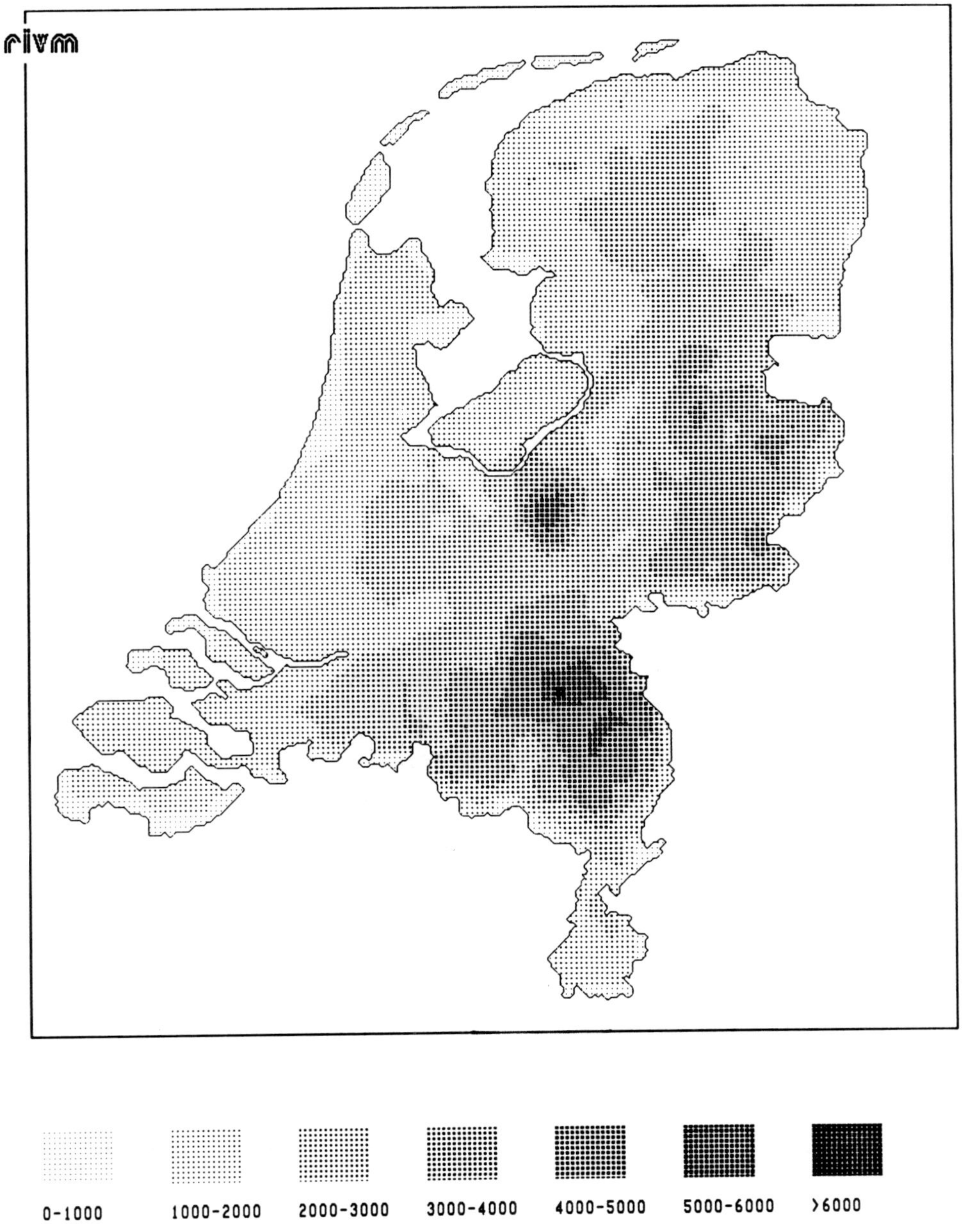

Fig. 3.9C Total deposition of NH_x in 1989 (mol $ha^{-1}yr^{-1}$)

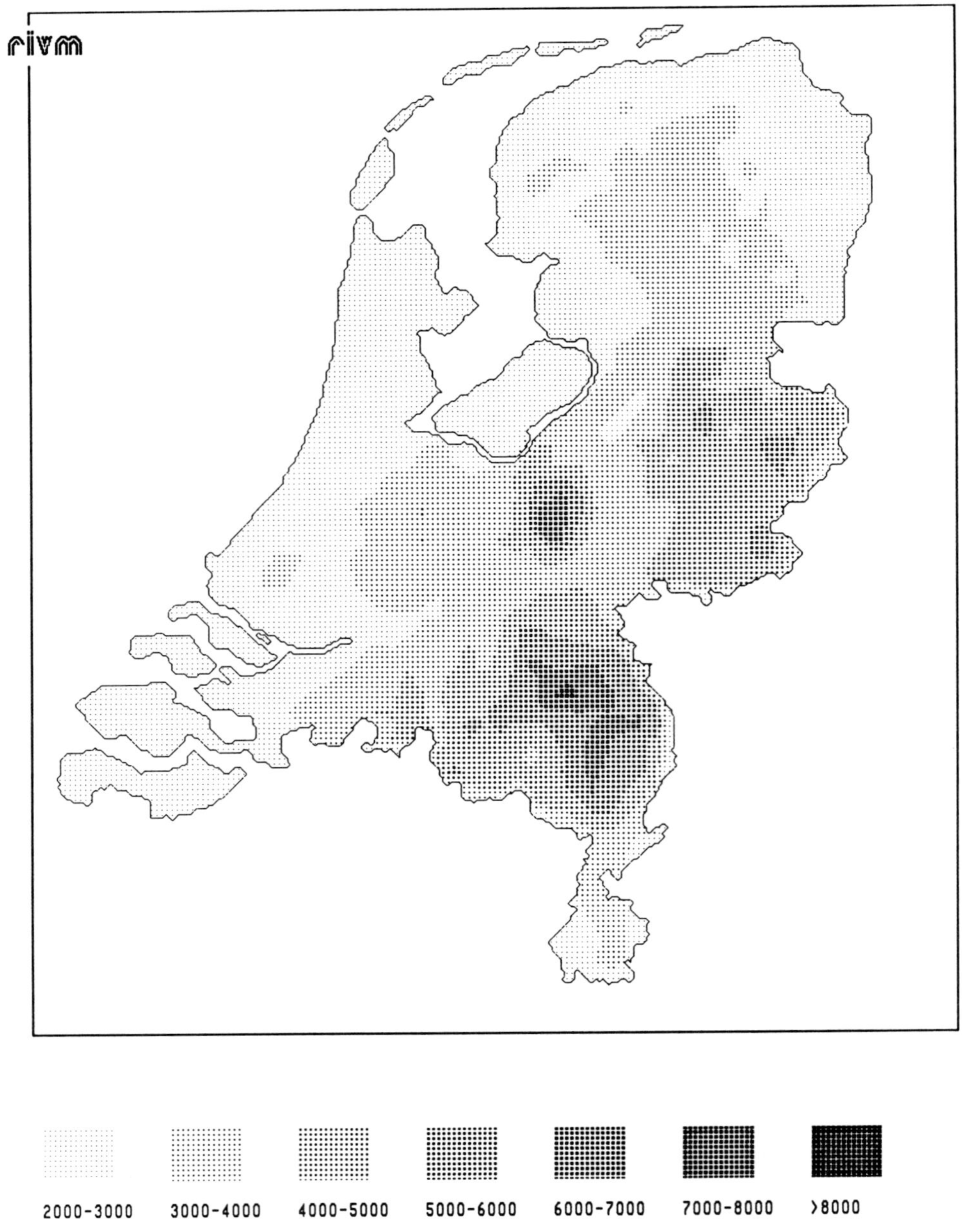

Fig. 3.10 Total N deposition in 1989 (mol $ha^{-1}yr^{-1}$)

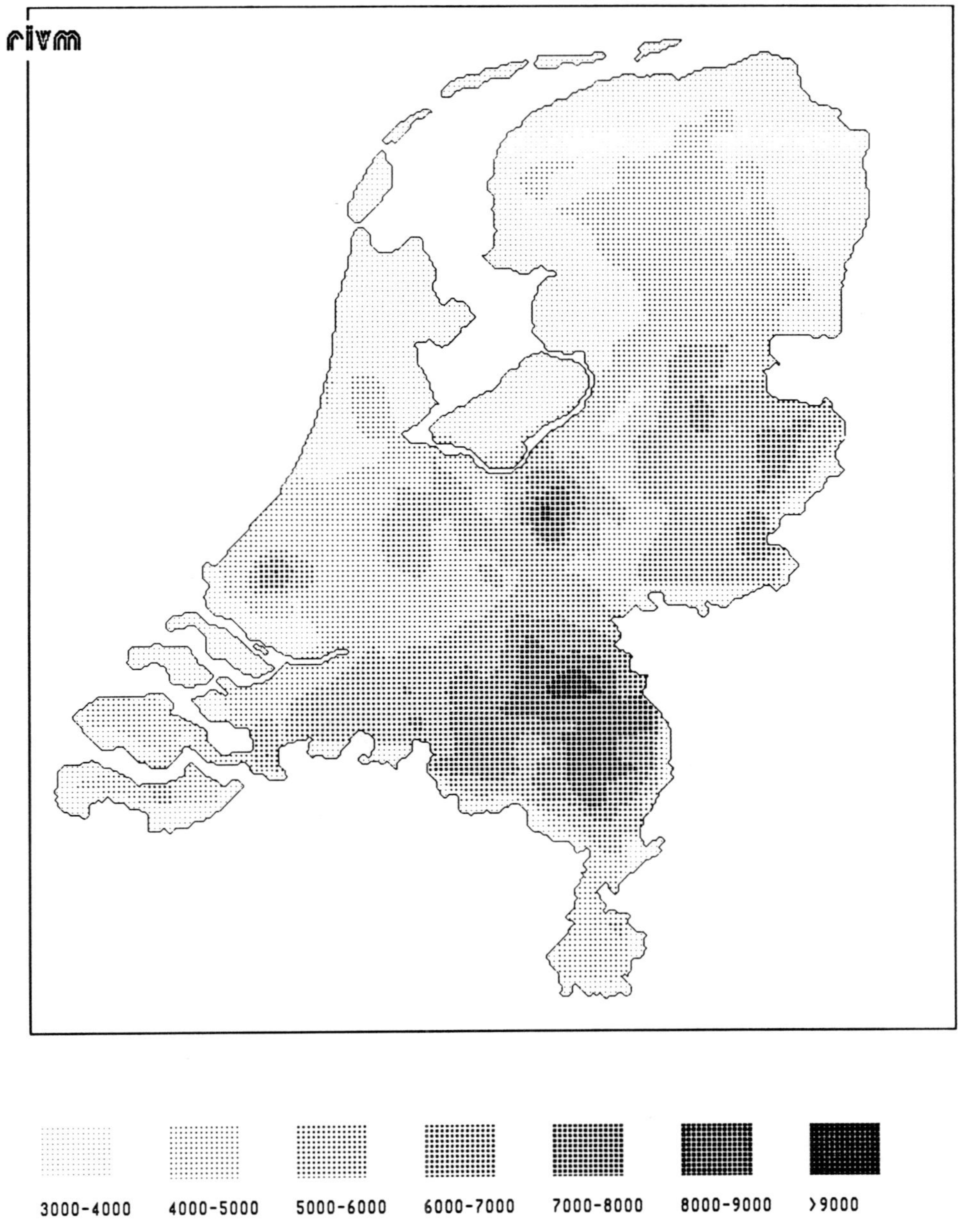

Fig. 3.11A Total potential acid deposition in 1989 (mol H^+ $ha^{-1}yr^{-1}$)

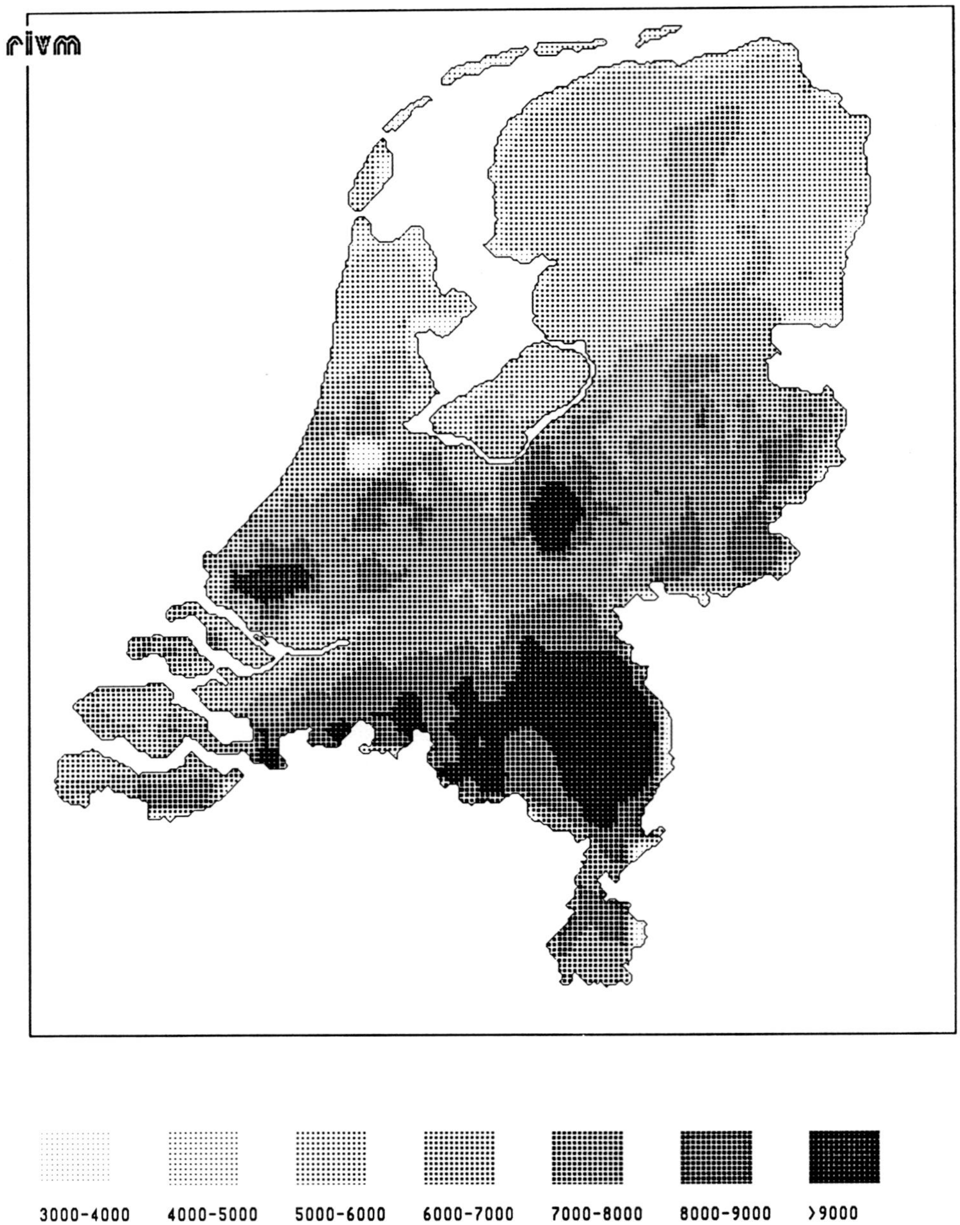

Fig. 3.11B Total potential acid deposition in 1980 (mol H^+ $ha^{-1}yr^{-1}$)

The contribution of wet and dry deposition of the various components to the total potential acid deposition in the Netherlands for the years 1980 and 1989 is shown in Figure 3. 12. The composition of the total acid deposition in 1989 is shown again in a different way in Figure 3.13. The dry deposition of NH_3 contributed about 25% to the total acid deposition in 1989. This contribution is large because NH_3 is emitted from low sources and is rapidly deposited (see also section 3.1.1). The contribution of NH_x to the total acid deposition in the Netherlands was about 45% in 1989.

The trend in total deposition has been determined on the basis of calculations for the years 1980 up to and including 1989 and is presented in Figure 3.14 for SO_x, NO_y, NH_x and total potential acid deposition. The total potential acid deposition as an average for the Netherlands decreased from about 6800 mol H^+ $ha^{-1}yr^{-1}$ in 1980 to about 4800 mol H^+ $ha^{-1}yr^{-1}$ in 1989. The decrease during this period can be attributed to the decrease in SO_2 emissions in Western Europe. Metereorological variability also plays a part. In the years 1988 and 1989 meteorological conditions caused relatively low deposition fluxes.

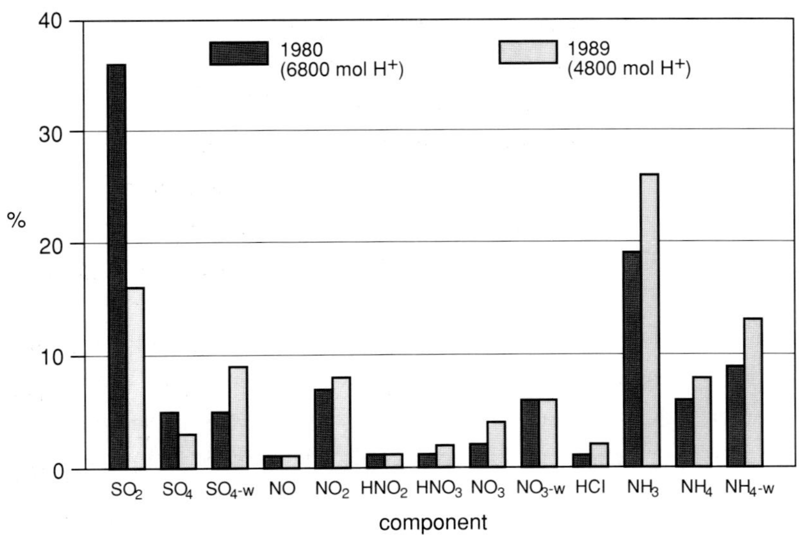

Fig. 3.12 Relative contribution of the various components to total potential acid in the Netherlands in 1980 and 1989

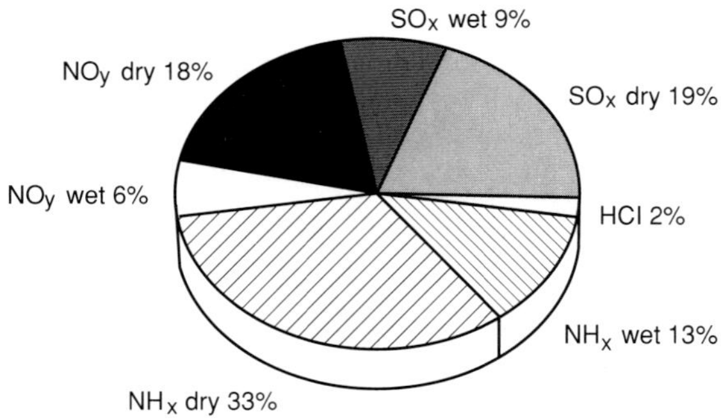

Fig. 3.13 Relative contribution of the various components to total potential acid in the Netherlands in 1989

The total deposition to forests, heathland and heathland pools cannot be assessed. An estimate was made on the basis of 5 x 5 km^2 grid-average depositions of SO_x, NO_y and NH_x. The dry deposition of these compounds for each grid element containing forests, heathland or heathland pools was compared with average levels for the various acidification areas. The resulting dry deposition ratios, together with the source-receptor matrices, were used for computing scenarios with the DAS model (see Chapter 6). The ratios which are averaged over several meteorological years, are shown in Table 3.8, as an average for the Netherlands. With these ratios and the deposition levels as shown in Tables 3.6 and 3.7 it is possible to determine the deposition fluxes to extensive forests and heathlands per acidification area. Table 3.9 shows the deposition fluxes to forest for every acidification area. As a result of increased roughness, dry deposition to forests and heathland in the Netherlands will be about 20% and 10% respectively higher than the deposition to an average landscape. Apart from roughness, other factors such as vegetation type, soil characteristics, surface wetness, etc., also influence the deposition rate. The surface resistance serves as an indicator. Since quantative data on the spatial variability of component-specific surface resistances are still lacking, this value is assumed to be constant for the Netherlands as a whole in the assessment of the deposition factors. Another limitation is the fact that the increased turbulence over transitions in roughness was not accounted for. On a local scale, for instance in the case of forest edges, such transitions can cause a further increase in deposition. Because of the great number of small forest plots in the Netherlands, a relatively large share of the forest can be regarded as forest edge. Finally, also local sources (farms or roads) near to the forests or heathlands have not been taken into account. Figure 3.15 illustrates the dispersion of Dutch forests; the forested surface area per square kilometer element served as a classification criterion (Meijers, 1990).

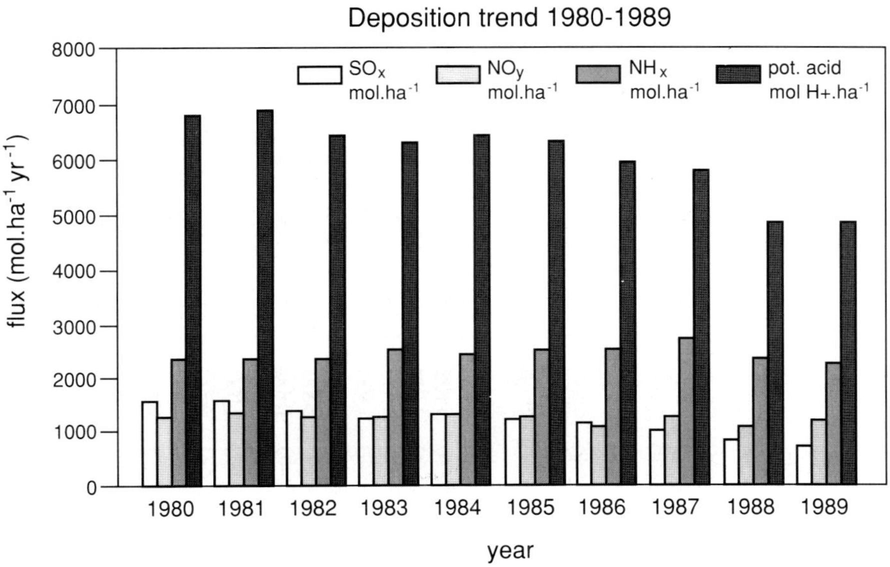

Fig. 3.14 Deposition trend 1980 - 1989

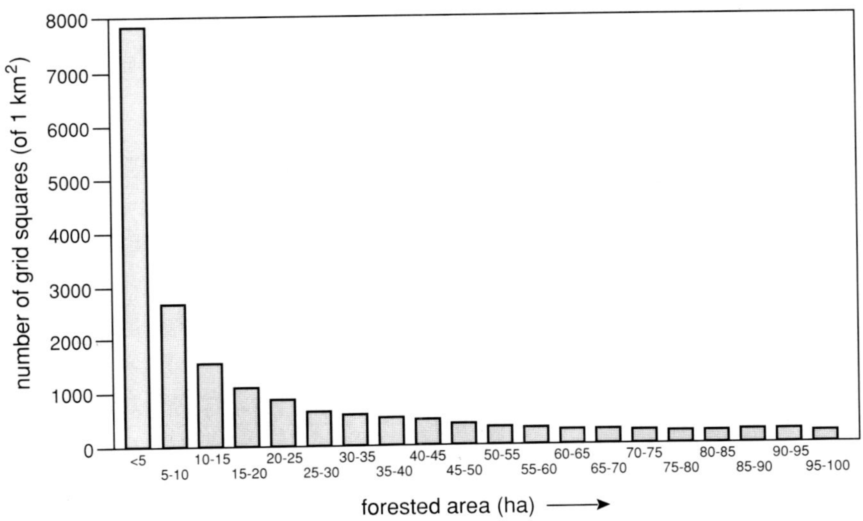

Fig. 3.15 Number of squares (of 1 km^2) with a certain forested area

Table 3.6 Deposition figures for each acidifying compound and each acidification area ($mol.ha^{-1} jr^{-1}$), in 1980 (HCl not included)

acidification region	SO_x			NO_y			NH_x			total nitrogen			potential acid ($2SO_x+NO_y+NH_x$)		
	dry#	wet	total	dry	wet	total	dry	wet	total	dry	wet	total	dry	wet	total
Nederland	1250	320	1570	830	390	1220	1690	640	2330	2500	1000	3600	5000	1700	6700
Groningen	1110	280	1390	700	360	1060	1090	600	1690	1800	1000	2800	4000	1600	5600
Friesland	1000	240	1240	700	330	1030	1330	530	1860	2000	900	2900	4000	1400	5400
Drenthe	1070	290	1360	700	370	1070	1390	620	2010	2100	1000	3100	4200	1600	5800
W/NO-Overijssel	1000	290	1290	730	400	1130	1610	650	2260	2300	1100	3400	4300	1700	6000
ZO-Overijssel	980	320	1300	790	380	1170	2530	680	3210	3300	1100	4400	5300	1700	7000
NW-Gelderland,Flevo.	1230	360	1590	880	430	1310	2310	790	3100	3200	1200	4400	5700	1900	7600
NO-Gelderl.	1230	350	1580	790	400	1190	2610	760	3370	3400	1200	4600	5900	1900	7800
Z-Gelderland	1240	380	1620	860	410	1270	1870	780	2650	2700	1200	3900	5200	2000	7200
Utrecht	1260	350	1610	980	410	1390	1680	670	2350	2700	1100	3800	5200	1800	7000
N-Nrd-Hol.	1240	250	1490	790	400	1190	710	430	1140	1500	800	2300	4000	1300	5300
Z-Nrd-Hol.	1230	290	1520	860	430	1290	790	630	1420	1700	1100	2800	4200	1700	5900
N-Zd.-Hol.	1500	330	1830	1010	430	1440	1020	560	1580	2000	1000	3000	5000	1700	6700
Z-Zd.-Hol.	1550	340	1890	950	420	1370	970	570	1540	1900	1000	2900	5000	1700	6700
Zeeland	1570	350	1920	860	470	1330	720	510	1230	1600	1000	2600	4700	1700	6400
W-Nrd.Brab.	1700	400	2100	860	370	1230	1280	700	1980	2100	1100	3200	5500	1900	7400
Mid-Nrd.Br.	1410	410	1820	860	350	1210	2150	730	2880	3000	1100	4100	5800	1900	7700
No-Nrd.Br.	1210	360	1570	890	360	1250	3540	680	4220	4400	1000	5400	6800	1700	8500
Zo-Nrd.Br.	1390	350	1740	900	350	1250	3290	660	3950	4200	1000	5200	7000	1700	8700
N-Limburg	1310	360	1670	1010	370	1380	3540	690	4230	4600	1100	5700	7200	1800	9000
Z/Mid.Limb.	1410	400	1810	950	460	1410	1980	780	2760	2900	1200	4100	5700	2000	7700

Specific meteorological information lacked in this year, dry deposition is taken equal to that in 1981

Table 3.7 Deposition figures for each acidifying compound and each acidification area (mol.ha^{-1} jr^{-1}), in 1989 (HCl not included)

acidification region	SO_x			NO_y			NH_x			total nitrogen			potential acid ($2SO_x+NO_y+NH_x$)		
	dry	wet	total	dry	wet	total	dry	wet	total	dry	wet	total	dry	wet	total
Nederland	450	220	670	850	310	1160	1550	640	2190	2400	1000	3400	3300	1400	4700
Groningen	370	130	500	790	290	1080	1040	520	1560	1800	800	2600	2500	1100	3600
Friesland	340	140	480	760	300	1060	1290	510	1800	2100	800	2900	2800	1100	3900
Drenthe	340	200	540	750	330	1080	1320	640	1960	2100	1000	3100	2800	1400	4200
W/NO-Overijssel	330	200	530	750	310	1060	1540	590	2130	2300	900	3200	3000	1300	4300
ZO-Over-ijssel	350	240	590	800	280	1080	2430	860	3290	3200	1100	4300	3900	1600	5500
NW-Gelder-land,Flevo.	370	230	600	930	320	1250	2130	650	2780	3100	1000	4100	3800	1500	5300
NO-Gelderl.	370	240	610	810	280	1090	2430	880	3310	3200	1200	4400	3900	1700	5600
Z-Gelderland	420	280	700	880	330	1210	1700	760	2460	2600	1100	3700	3400	1700	5100
Utrecht	490	260	750	1000	350	1350	1530	650	2180	2500	1000	3500	3500	1500	5000
N-Nrd-Hol.	370	160	530	780	340	1120	660	510	1170	1400	900	2300	2100	1200	3300
Z-Nrd-Hol.	430	210	640	870	300	1170	730	490	1220	1600	800	2400	2500	1200	3700
N-Zd.-Hol.	610	230	840	1000	310	1310	920	490	1410	1900	800	2700	3100	1300	4400
Z-Zd.-Hol.	650	230	880	950	310	1260	870	520	1390	1800	800	2600	3100	1300	4400
Zeeland	670	240	910	820	260	1080	650	580	1230	1500	800	2300	2800	1300	4100
W-Nrd.Br.	700	270	970	890	290	1180	1150	590	1740	2000	900	2900	3400	1400	4800
Mid.Nrd.Br.	560	270	830	930	280	1210	1910	630	2540	2800	900	3700	3900	1400	5300
No.Nrd.Br.	450	300	750	880	320	1200	3120	890	4010	4000	1200	5200	4900	1800	6700
Zo.Nrd.Br.	520	310	830	890	310	1200	2870	890	3760	3800	1200	5000	4800	1800	6600
N-Limburg	470	320	790	860	330	1190	3110	1000	4110	4000	1300	5300	4900	1900	6800
Z/Mid.Limb.	540	270	810	910	300	1210	1760	680	2440	2700	1000	3700	3800	1500	5300

Table 3.8 Coefficients for dry deposition flux on forest, heathland and heathland pools (averages for the Netherlands as a whole)

forest			heathland			heathland pools		
SO_x	NO_y	NH_x	SO_x	NO_y	NH_x	SO_x	NO_y	NH_x
1.2	1.2	1.2	1.1	1.1	1.1	2.2	0.4	1.9

The deposition levels for NH_x mentioned in this report are different from those which were reported earlier (during the first phase of the Priority Programme on Acidification). During this first phase, deposition figures were computed with a transport model (Asman and Maas, 1986) using a detailed emission map from Buijsman et al. (1984). The figures in the original map of livestock emissions included in Buijsman et al. were increased by 40%, in view of the imbalance between emission and deposition values for NH_x on the European scale (Buijsman, pers.comm.).

Current estimates are based on the TREND model. This model has been extensively checked against measurements of distribution and deposition of SO_x and NO_y on the national and on the European scale (Van Jaarsveld and Onderdelinden, 1990). The parameters for NH_3 in this model are based upon the latest insights (Asman and Van Jaarsveld, 1990). In mapping NH_x deposition, detailed emission maps for the Netherlands (Erisman, 1989) and for Europe (Asman, 1990), were used, to which the emission factors from the Working Group on NH_3 Emission Factors (De Winkel, 1988) were applied. The absolute values of these emissions are still under discussion (see Chapter 2). The values are about 60 % higher than the values which were orginally assessed by Buijsman et al. and also strong regional differences occur. As a result, the TREND model gives higher values for deposition in the Netherlands than the Asman and Maas model, also because the overall dry deposition velocity of NH_3 is assumed to be higher (1.2 instead of 0.8 cm s^{-1}, see Thematic Report). Because of the regional differences in NH_3 emissions, the regional distribution of NH_x deposition has changed too. The results of simulations with the TREND model show a fairly good agreement with measured concentrations of NH_3 and NH_4 aerosols in rain and in the atmosphere, at measuring sites in the Netherlands as well as in other European countries (Asman and Van Jaarsveld, 1990). For the Netherlands, the values obtained using the Asman and Maas model are systematically lower than those from measurements, in spite of corrected emission figures.

<u>Table 3.9</u> Deposition fluxes onto forests (mol $ha^{-1}yr^{-1}$) for each acidification region, in 1989 (HCl not included)

acidification region	SO_x			NO_y			NH_x			total nitrogen			potential acid ($2SO_x+NO_y+NH_x$)		
	dry	wet	total	dry	wet	total	dry	wet	total	dry	wet	total	dry	wet	total
Nederland	530	220	750	990	310	1300	1890	640	2530	2900	900	3800	4000	1400	5300
Groningen	380	130	510	800	290	1090	1100	520	1620	1900	800	2700	2700	1100	3700
Friesland	430	140	570	940	300	1240	2030	510	2540	3000	800	3800	3800	1100	4900
Drenthe	420	200	620	930	330	1260	1690	640	2330	2600	1000	3600	3500	1400	4800
W/NO-Overijssel	420	200	620	950	310	1260	2540	590	3130	3500	900	4400	4300	1300	5600
ZO-Over-ijssel	410	240	650	960	280	1240	3210	860	4070	4200	1100	5300	5000	1600	6600
NW-Gelder-land,Flevo.	430	230	660	1070	320	1390	2370	650	3020	3400	1000	4400	4300	1400	5700
NO-Gelderl.	450	240	690	1010	280	1290	2890	880	3770	3900	1200	5100	4800	1600	6400
Z-Gelderland	510	280	790	1080	330	1410	1950	760	2710	3000	1100	4100	4100	1600	5700
Utrecht	590	260	850	1190	350	1540	1750	650	2400	2900	1000	3900	4100	1500	5600
N-Nrd-Hol.	450	160	610	920	340	1260	520	510	1030	1400	800	2200	2400	1200	3500
Z-Nrd-Hol.	470	210	680	960	300	1260	940	490	1430	1900	800	2700	2800	1200	4000
N-Zd.-Hol.	760	230	990	1200	310	1510	580	490	1070	1800	800	2600	3300	1200	4600
Z-Zd.-Hol.	620	230	850	990	310	1300	950	520	1470	1900	800	2700	3200	1300	4500
Zeeland	720	240	960	850	260	1110	730	580	1310	1600	800	2400	3000	1300	4300
W-Nrd.Br.	1030	270	1300	1090	290	1380	1560	590	2150	2600	900	3500	4700	1400	6100
Mid.Nrd.Br.	680	270	950	1100	280	1380	2760	630	3390	3900	900	4800	5200	1500	6700
No.Nrd.Br.	540	300	840	1030	320	1350	4020	890	4910	5100	1200	6300	6100	1800	7900
Zo.Nrd.Br.	610	310	920	1020	310	1330	3300	890	4190	4300	1200	5500	5500	1800	7300
N-Limburg	560	320	880	1000	330	1330	3860	1000	4860	4900	1300	6200	6000	2000	7900
Z/Mid.Limb.	660	270	930	1060	300	1360	2570	680	3250	3600	1000	4600	5000	1500	6500

Other significant amendments to the methods of calculation used during the first phase of the programme include the introduction of a roughness map for the Netherlands, the diurnal variation in the moisture-dependent surface resistance of SO_2, a decreased HNO_3 concentration (derived from measurements at Petten and Speuld) and a higher surface resistance for NO and NO_2. On the whole this has resulted in lower NO_y deposition values and an increase in the SO_x and NH_x depositions. The roughness map leads to an increase in dry deposition, since the concentrations are related to the spatial extent of the roughness. The wet deposition of SO_x has decreased on account of the adjustment for Ca introduced by Buijsman (1990). Table 3.5 shows the changes in deposition for 1980 and 1989.

- Estimates of origin and contributions of source categories

The most accurate calculation of deposition in the Netherlands with the TREND model can be made for the year 1980. For this year accurate estimates of emissions and location of sources are available for the Netherlands and for Europe. The values for 1980 derived with the TREND model show a fairly good agreement with those derived from measurements (Erisman, 1991).

Relative contributions of foreign emissions per component and for the total deposition in 1980 and 1989 are shown in Figure 3.16. Figure 3.17 shows the origins of potential acid separately. Values are based on the distribution of sources for 1980, and changes in the location of sources between 1980 and 1989 have not been accounted for in the calculation of the 1989 values. In 1989, the Netherlands contributed almost 55% to the deposition in the Netherlands. The contributions to the deposition in the Netherlands of the surrounding countries (Great Britain, France, Belgium, and Germany) were more or less comparable.

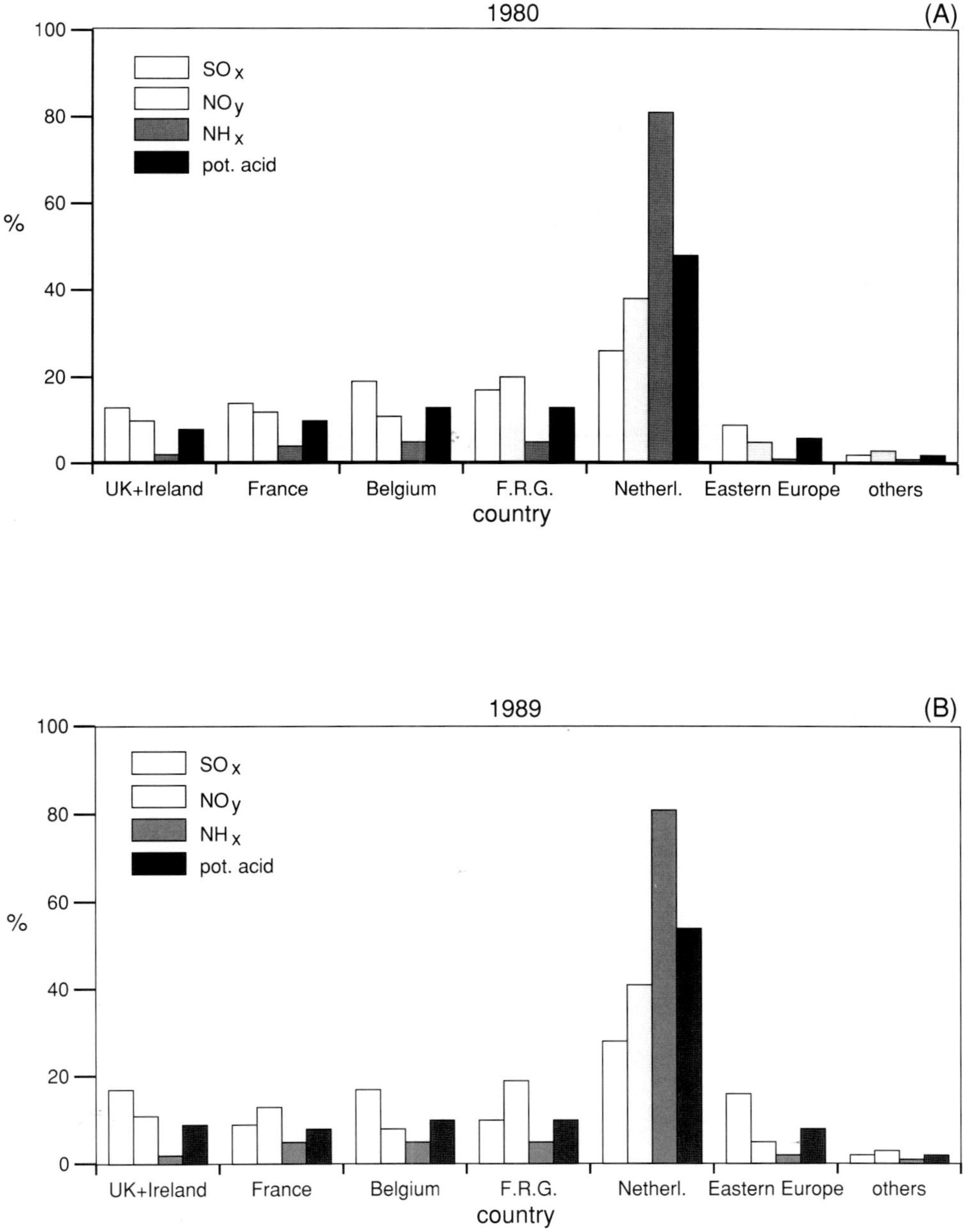

Fig. 3.16 Origins of each component and of the total acid deposition in 1980 (A) and 1989 (B)

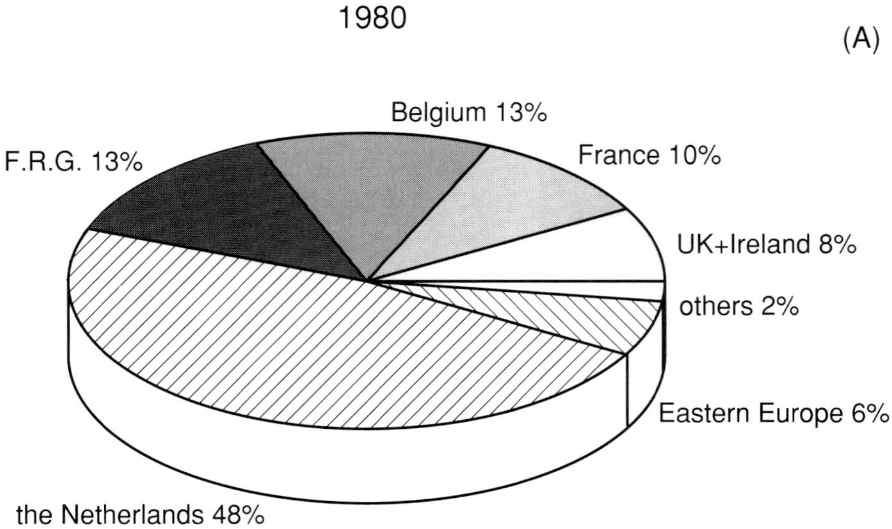

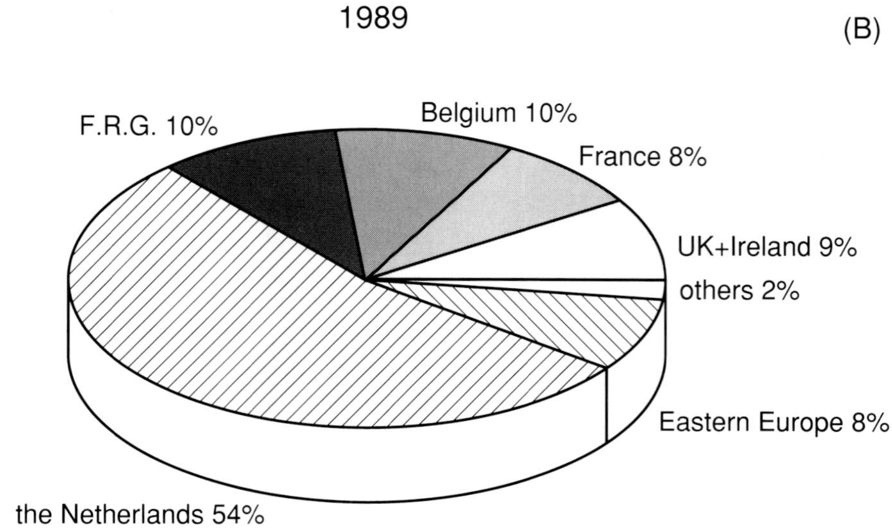

Fig. 3.17 Origins of total potential acid for 1980 (A) and 1989 (B)

Figure 3.18 shows the contributions of the different Dutch source categories to the total deposition and to the deposition of the different compounds for 1980 and 1989. Figure 3.19 shows the contribution to the total potential acid deposition separately. Calculations for both years were again based on the distribution matrix for sources for 1980. The largest contribution among Dutch source categories to the total deposition in 1989 is from agriculture.

3.3.2 The relationship between deposition and throughfall

In the Dutch Priority Programme on Acidification two different methods were used to estimate the deposition of acidifying pollutants on forests. Throughfall measurements were used to assess the load to the forest floor, whereas deposition measurements, estimates based on concentration measurements and estimates of the dry deposition rate were used to assess the atmospheric deposition to various receptors. The results of these two methods of assessment can show significant regional differences (Schneider and Bresser, 1988). The results of the two methods could never be compared on a local scale, since accurate small-scale estimates could not be made. As was described in section 3.3.1, however, estimates of deposition fluxes can now be made on a scale of 5 x 5 km^2, although with some measure of uncertainty (see section 3.3.3). The comparison between throughfall and measured deposition for three research sites has already been given in section 3.3.1.

The results of recent throughfall measurements, which were carried out in several forest stands distributed over the Netherlands, have been compared with estimates for the grids in which these stands are situated, for the period 1987 - 1989 approximately (Erisman, 1990). Comparisons were made of the deposition fluxes of potentially acidifying compounds (the acids H_2SO_4, HNO_2, HNO_3 and HCl, and the precursors SO_2, NO, NO_2, NH_3 and NH_4) with throughfall fluxes of NO_3, SO_4 and NH_4. Neutral aerosols and sea salt are not regarded as potentially acidifying. Adjustments for these compounds were made analogous to the adjustments made in assessing the wet potential acid deposition flux from wet deposition measurements (see section 3.1.4; Buijsman, 1990; Erisman, 1990). In the same way, adjustments were made for dry deposition of gases and aerosols on the funnels of open rain collectors, which are also used for throughfall measurements.

Comparison between the adjusted throughfall measurements and recent deposition measurements/estimates from inference shows a better agreement of the values for SO_4, NH_4 and potential acid fluxes calculated with the two methods. The adjustments have little influence on the values for NO_3 fluxes. The adjusted thoughfall measurements and deposition estimates show a significant spatial correlation. The same is true of net throughfall fluxes (total throughfall minus open field measurements) and dry deposition estimates. Throughfall measurements (which refer to local values), however, show, on average, significantly higher values (45%) for SO_4, 33% lower values for NO_3, and 13% higher values for NH_4. In general for the ratio NH_4/SO_4 a value of about 1 is observed in throughfall. The values for total potential acid are, on average, 14% higher when derived from throughfall measurements, which is not a significant discrepancy, however. For the 24 stands in which throughfall was measured, the average flux of total potential acid is 7800 mol H+ $ha^{-1}yr^{-1}$ based on throughfall measurements and 6800 mol H+ $ha^{-1}yr^{-1}$ based on measured concentrations.

The significant discrepancies between flux values derived from throughfall measurements and those derived from inference may be attributable to the deposition of fog and coarse particles, which are not taken into account by the model, and to exchange processes in vegetation which influence the composition of throughfall. These exchange processes must be studied in more detail before the causal link between throughfall and atmospheric deposition can be elucidated.

3.3.3 Uncertainties

The uncertainties in the annual average fluxes for the Netherlands as a whole and on the 5 x 5 km^2 scale are shown in Table 3.10. The figures refer to uncertainties in the total deposition of SO_x, NO_y, NH_x and potential acid with both average and worst-case figures being given. The worst-case approach reflects the uncertainty when a complete correlation is assumed between concentrations and deposition rates and between dry and wet deposition fluxes. The average values are based on partial correlations.

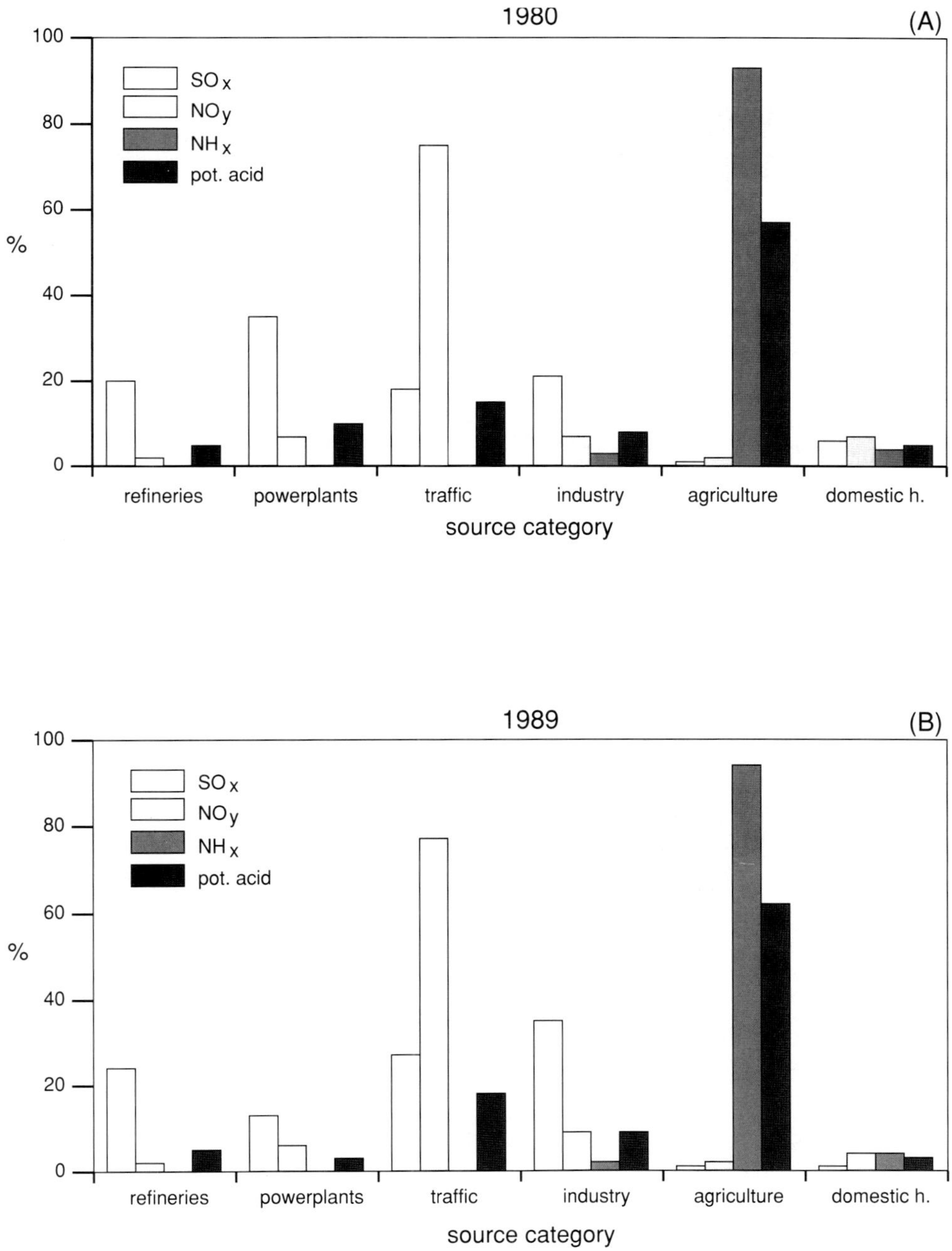

Fig. 3.18 The contribution of the various source categories to the Dutch part of the deposition in the Netherlands for 1980 (A) and 1989 (B)

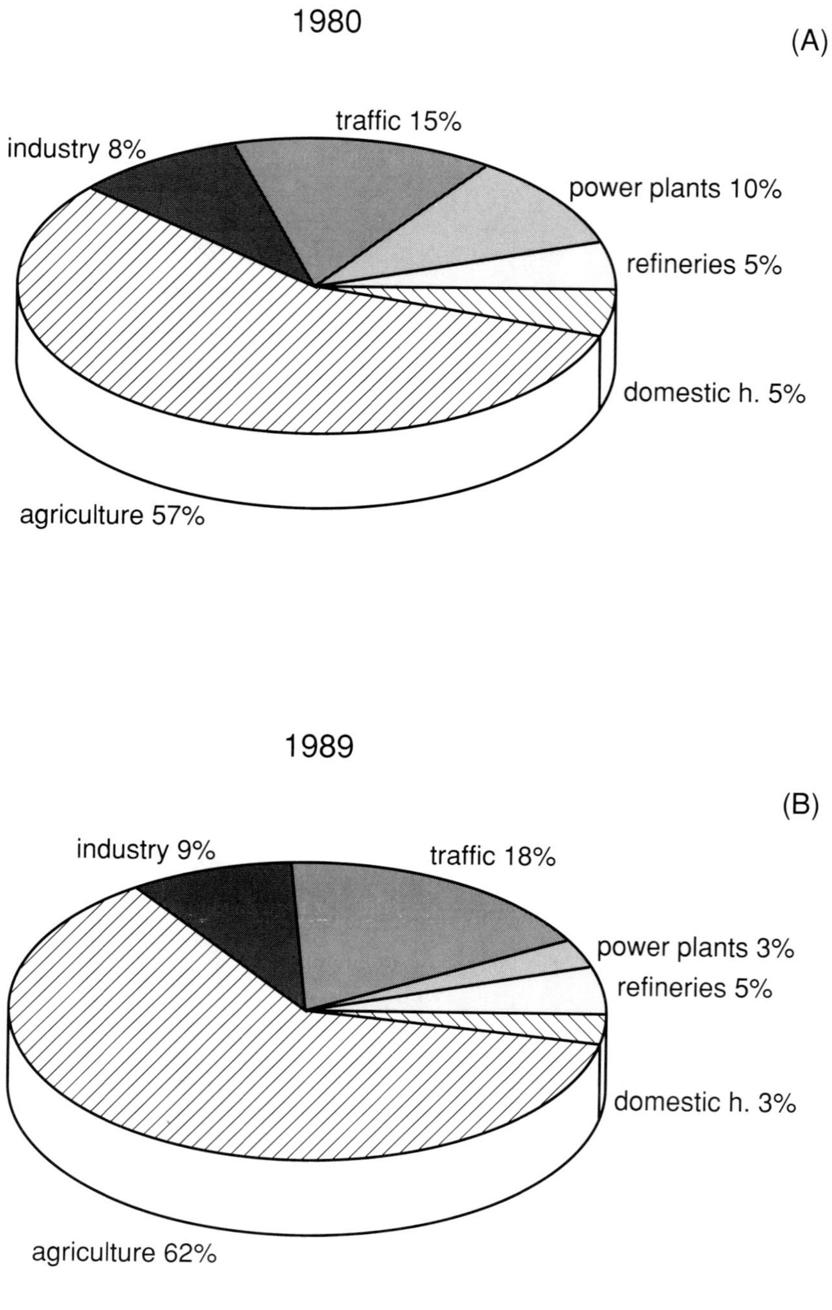

Fig. 3.19 The contribution of the various source categories to the Dutch part of the total potential acid deposition in the Netherlands for 1980 (A) and 1989 (B)

Table 3.10 Total uncertainty (%) in yearly average total deposition flux on different spatial scales (1988) (Erisman, 1991)

	$5x5\ km^2$		the Netherlands	
component	average	worst case	average	worst case
SO_x	30	40	15	25
NO_y	60	95	25	65
NH_x	85	100	30	100
pot. acid	45	80	15	50

Wet deposition can be assessed more accurately than dry deposition. For deposition on extensive nature areas without any transitions in roughness, the same uncertainty as for the 5 x 5 km grids can be assumed. Within these grid cells variations may be very large, however. The uncertainty in deposition on a forest within a grid cell, for instance, will probably be larger than the average uncertainty in deposition in the entire grid cell. Factors such as local sources, roughness transition zones, and vegetation-specific characteristics are of great influence. For the year 1988 the uncertainty in the total potential acid deposition varies from 45% to 80% on the 5 x 5 km scale and from 15% to 50% on the national scale. The largest contribution to the total uncertainty in the potential acid deposition comes from the uncertainty in NH_x and NO_y on the 5 x 5 km scale and in NH_x for the Netherlands as a whole. The large uncertainties for NH_x deposition fluxes are mainly caused by the following gaps in our knowledge:

- The emission factors for NH_3 for the various categories are insufficiently known, so that the national emission figures and the spatial distribution of emissions over the Netherlands are very inaccurate model inputs.
- There are insufficient data on concentrations and deposition parameters to test model results.
- The variation in the deposition velocity for the various receptor types (emitting or non-emitting) is unknown.

In fact the validation of the model of the whole system of emissions-atmospheric transport-concentration of NH_3 should be improved. For this purpose a so-called NH_3 crash programme has been set up which aims at reducing existing uncertainties in emission and concentration levels in natural areas in the Netherlands by means of measurements. Results however, cannot be expected until the end of 1991. Research on dry deposition of ammonia is also being continued.

The accuracy of annual average throughfall data was estimated for Douglas fir forest at Speuld at 27% for SO_4^{2-}, 17% for NO_3^- and 23% for NH_4^+ (95% confidence limit). Uncertainties are assumed to be somewhat lower for other tree species.

3.3.4 Deposition from natural sources and comparison of the situation in the Netherlands with other countries

Table 3.11 gives an estimate of the deposition from natural sources. In the current situation (1989), the total potential acid deposition exceeds the estimates of background deposition by a factor of about 16. The figures of deposition from natural sources are processed in the deposition estimates (which are based on measurements).
At the international level the deposition programmes of the various countries show great differences. In Europe and in the USA, deposition is mapped by means of transport models which have detailed emission data as input. In the USA the very advanced RADM model is used. This model is validated and supported by a lot of experimental (process-oriented) research and relatively little monitoring. Attention is focused mainly on the deposition of SO_x and, to a lesser extent, on NO_y components. NH_3 is not regarded as an important component of acid deposition. During the NAPAP congress in February 1990, the Dutch delegation advised that more attention should be paid to the deposition of NH_3 in the USA (Schneider and Heij, 1990). It is also being recognized in other countries that NH_3 is an important acidifying compound.

Table 3.11 Wet-, dry- and total deposition in the Netherlands due to natural sources (mol H^+ ha^{-1} yr^{-1})

component	wet deposition	dry deposition	total deposition
SOx	84	24	108
NOy	36	13	49
NHx	27	48	75
halogen	3	3	6
RCOOH	30	30	60
total potential acid	180	118	300

Model computations are also used in Europe, but the emphasis is more on monitoring (e.g.

EMEP), although there is some experimental (process-oriented) research. In Europe throughfall is measured at a relatively large number of locations.

In the Netherlands a great deal of monitoring is done (National Air Quality Monitoring Network, National Precipitation Monitoring Network) and this appears to be a good basis for the assessment of deposition. Process-oriented research, however, is only done on a very modest scale.

Table 3.12 shows the total deposition of SO_x, NO_y and NH_x for the USA and for Europe based on computations with the RADM model (for the USA) and the EMEP model (for Europe). Finally, Figures 3.20 to 3.23 show the depositions of SO_x, NO_y, NH_x and potential acid deposition in 1988, based on computations with the TREND model (Van Jaarsveld and Onderdelinden, 1990)

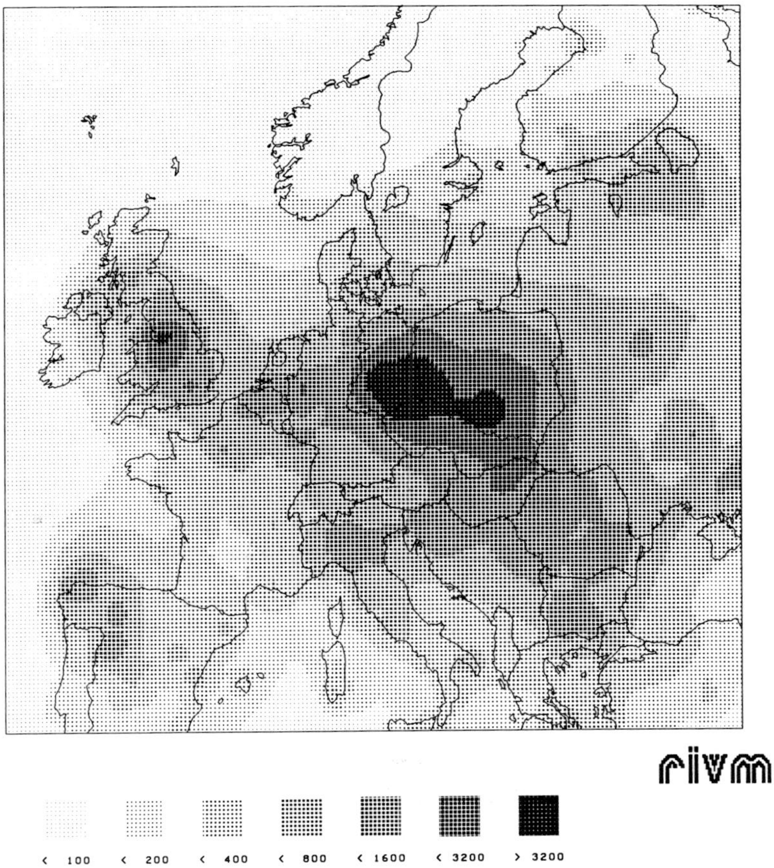

Fig. 3.20 Deposition of SO_x in Europe, in 1988 (mol $ha^{-1}yr^{-1}$)

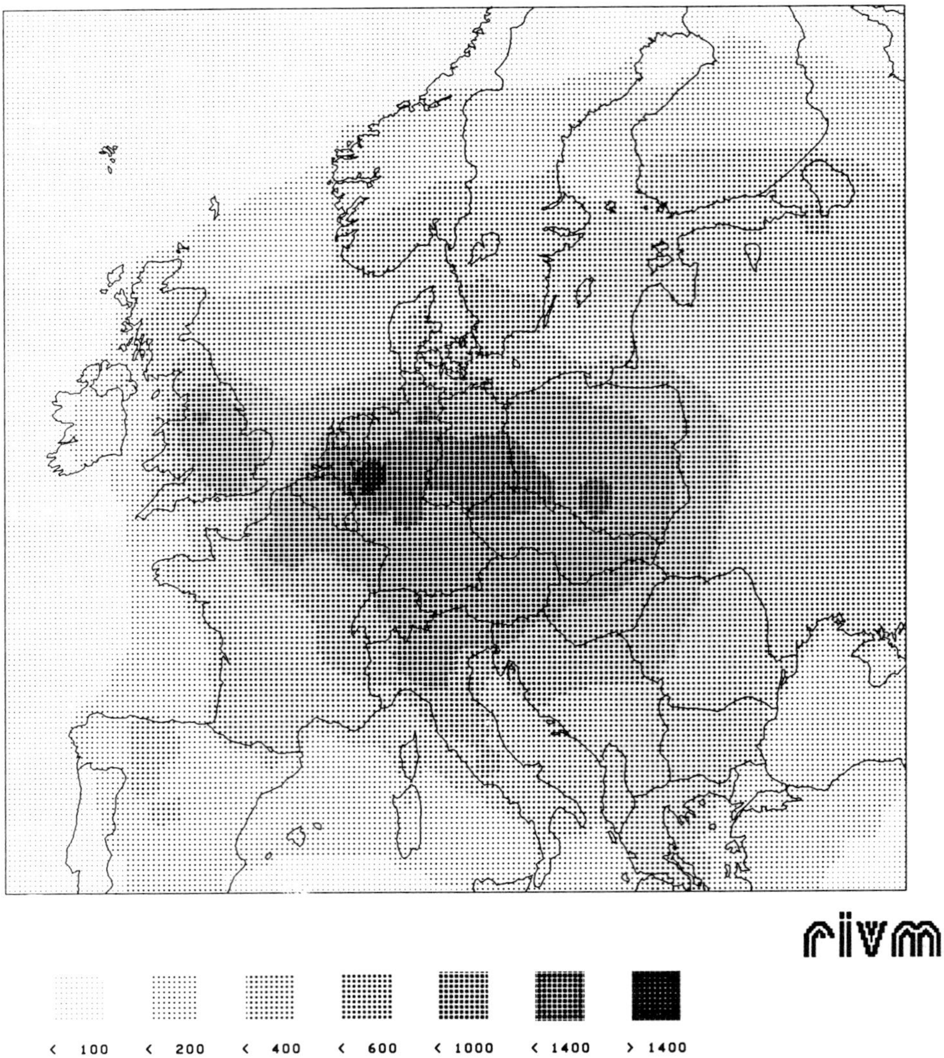

Fig. 3.21 Deposition of NO_y in Europe, in 1988 (mol $ha^{-1}yr^{-1}$)

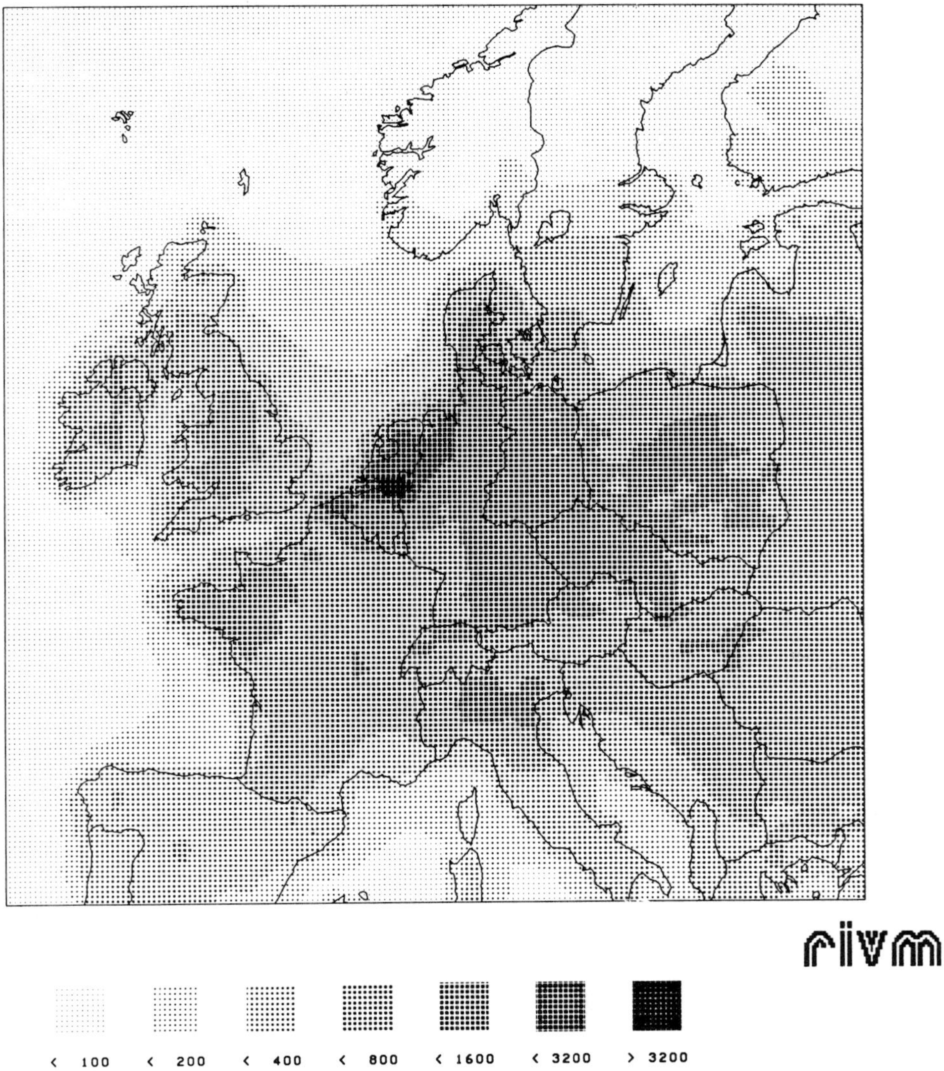

Fig. 3.22 Deposition of NH_x in Europe, in 1988 (mol $ha^{-1}yr^{-1}$)

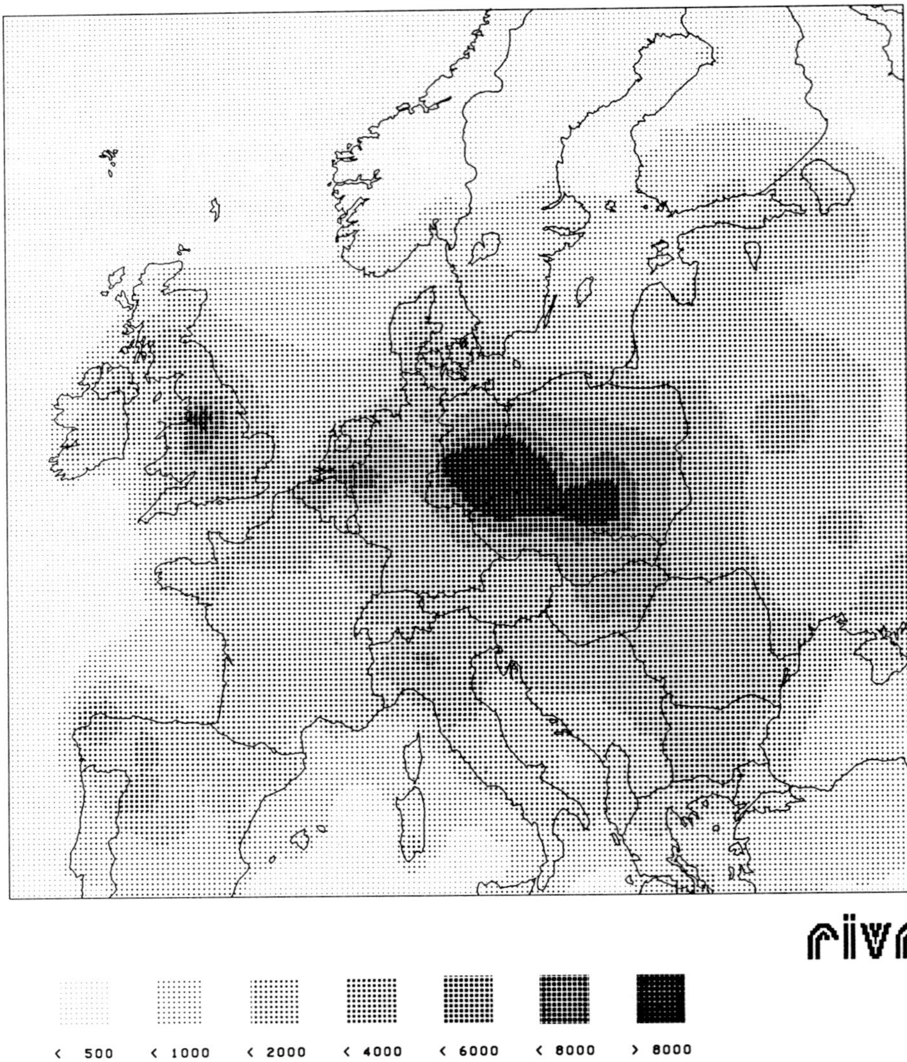

Fig. 3.23 Potential acid deposition in Europe, in 1988 (mol $ha^{-1}yr^{-1}$)

Table 3.12 Deposition figures (mol $ha^{-1}yr^{-1}$) of SO_x, NO_y and NH_x for the US and Europe in 1985

	US 1)	Europe 2)4)
SO_x		
dry	140	320
wet	130	160
total	270	480
NO_y		
dry	160	185
wet	165	95
total	325	280
NH_x 3)		
dry	-	265
wet	-	135
total	-	400

1) data derived from NAPAP
2) calculated with the EMEP model (NO_y estimate probably too low)
3) not estimated in the US
4) dry/wet ratio: 2:3 (derived from the literature)
Note: fluxes differ depending on the surface area assumed for US and Europe. These figures are probably relatively low in comparison with the Netherlands because water is included in the surface area

3.3.5 Conclusions

The situation in connection with acid load in the Netherlands can be described as follows:

- Two thirds of the acid deposition is formed by dry deposition.
- Since 1980 the total acid deposition (as an average for the Netherlands) has shown a downward trend. The decrease from about 6800 mol H^+ $ha^{-1}yr^{-1}$ in 1980 to about 4800 mol H^+ $ha^{-1}yr^{-1}$ in 1989 can be attributed to the decrease in SO_2 emissions in Western Europe. Meteorological variability also plays a part.
- The Netherlands contributed almost 55% of the total deposition in the Netherlands, in 1989.
- The largest Dutch contribution to the total deposition in the Netherlands in 1989 was made by agriculture.

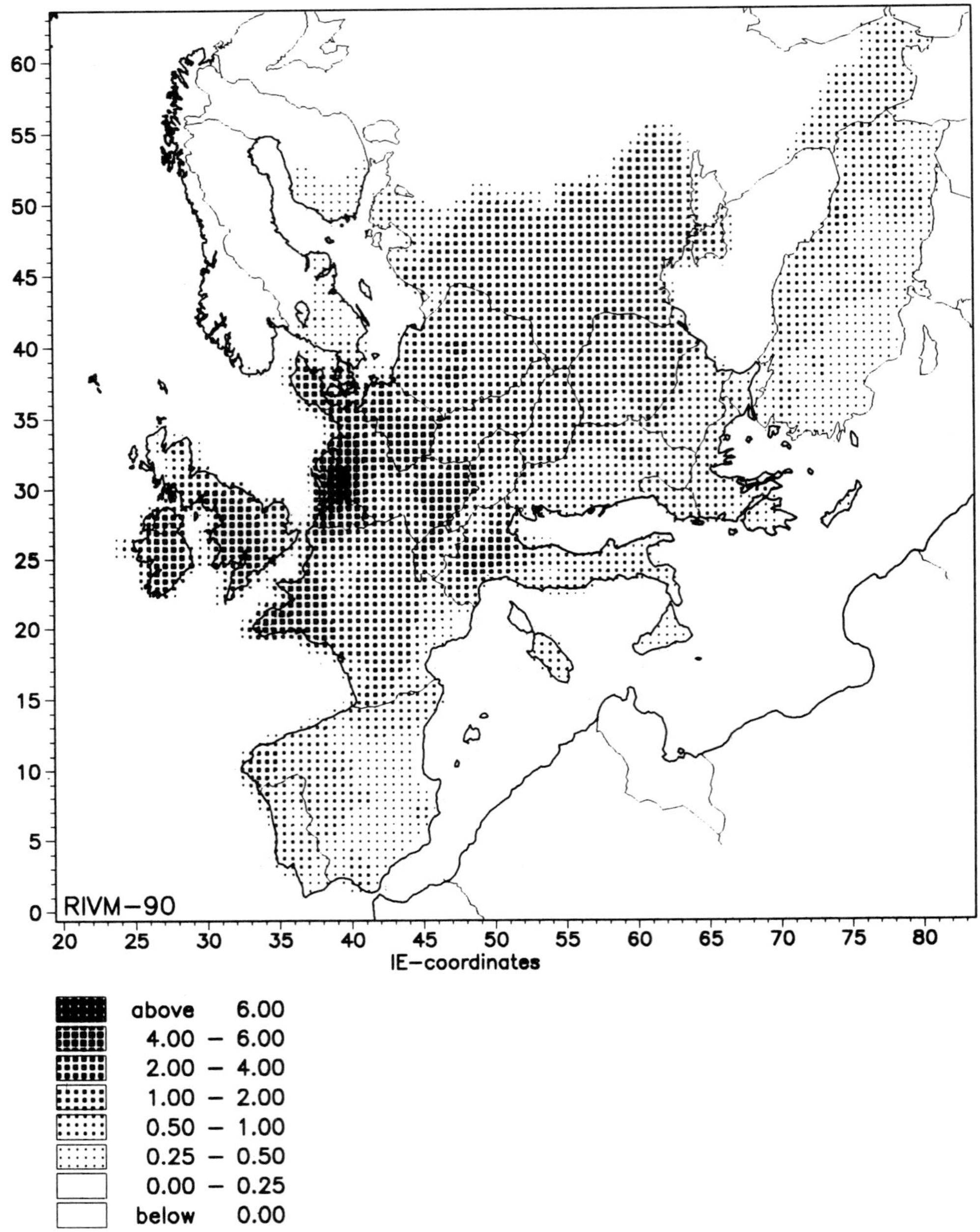

Fig. 3.24 NH_3 emission density in Europe (ton NH_3 $km^{-2}yr^{-1}$)

- The dry deposition of NH_3 contributed about 25% to the total potential acid deposition in 1989.
- The share of NH_x in the total acid deposition was on average about 45% in the Netherlands in 1989; about 80% of this originated from the Netherlands and 20% from outside the country.
- At roughness transition zones depositions may be higher. This is especially the case for the great number of small forest plots in the Netherlands.

In the Dutch Priority Programme on Acidification much attention has been paid to problems related to NH_3. This has been inspired by the fact that NH_3 is the most important component of acid deposition in the Netherlands. Figure 3.24 shows the NH_3 emissions for Europe and illustrates that the NH_3 problem in the Netherlands is greater than in other countries (Asman and Van Jaarsveld, 1990).

Literature

Annual Air Quality Report, 1989 (in Dutch)
Rapport nr. 222101006, Laboratory for Air Research
National Institute of Public Health and Environmental Protection, 1990

Asman W.A.H., 1990
Ammonia emission in Europe: updated emission and emission variations
Report 228471008, National Institute of Public Health and Environmental Protection

Asman W.A.H., Jaarsveld J.A.van, 1990
A variable-resolution statistical transport model applied for ammonia and ammonium
Report no. 228471007, National Institute of Public Health and Environmental Protection

Asman W.A.H., Maas J.F.M., 1986
Schatting van de depositie van ammoniak en ammonium in Nederland t.b.v. het beleid in het kader van de hinderwet (Estimation of the deposition of ammonia and ammonium in the Netherlands, in Dutch)
Report R-86-8, Institute for Meteorology and Oceanography, State University Utrecht, the Netherlands

Buijsman E., 1990
De berekening van de natte, zure depositie: een vergelijking van een aantal berekeningswijzen
Report nr. 228703011, National Institute of Public Health and Environmental Protection

Buijsman E., Maas H., Asman W., 1984
Een gedetailleerde ammoniakemissiekaart van Nederland
Instituut voor Meteorologie en Oceanografie, Rijksuniversiteit Utrecht, rapport V-84-20, november 1984

Buijsman E., Maas J.F.M., Asman W.A.H., 1987
Anthropogenic NH_3 emissions in Europe
Atmospheric Environment 21, 1009-1022

Erisman J.W., 1989
Ammonia emissions in the Netherlands in 1987 and 1988, 1989
Report 228471006,National Institute of Public Health and Environmental Protection

Erisman J.W., 1990
Atmospheric deposition of acidifying compounds onto forests in the Netherlands: throughfall measurements compared to deposition estimates from inference
Report 723001001, National Institute of Public Health and Environmental Protection

Erisman J.W., 1991
Acid deposition in the Netherlands
Report 723001002, National Institute of Public Health and Environmental Protection

Jaarsveld J.A.van, Onderdelinden D., 1990
Trend; an analytical long-term deposition model for multi-scale purposes
Report no. 228603009, National Institute of Public Health and Environmental Protection

Meijers R., 1990
Parametrisatie van de structuur van Nederlandse natuurgebieden
Instituut voor Ruimtelijk Onderzoek, Geografisch Instituut, Vakgroep Fysische Geografie, Rijksuniversiteit Utrecht

Schneider T., Bresser A.H.M., 1988
Summary report; Acidification research 1984 - 1988, Report nr. 00-06

Schneider T., Heij G.J. (eds.), 1990
Verslag NAPAP International Conference on Acidic Deposition; State of Science and Technology, 11 - 16 February 1990
Report nr. 200-06, Dutch Priority Programme on Acidification

Unsworth M.H., Fowler D. (eds.), 1988
Acid deposition at high elevation sites
Proceedings of the NATO advanced research workshop on Acid Deposition Processes at High Elevation Sites; Edinburgh, Scotland, 8 - 13 September 1986
Kluwer Academic Publishers

Winkel K.de, 1988
Ammoniak-emissiefactoren voor de veehouderij
Publicatiereeks Lucht nr. 76, Ministerie van Volkshuisvesting, Ruimtelijke Ordening en Milieubeheer, 1988

4. EFFECTS OF AIR POLLUTION AND ACID DEPOSITION ON FORESTS AND FOREST SOILS

G.J.Heij (RIVM), W.de Vries (Winand Staring Centre), A.C.Posthumus (IPO) and G.M.J.Mohren (De Dorschkamp)

4. 1 Introduction

Effects of air pollution and acid deposition can be direct, through above-ground uptake by vegetation, or indirect, through uptake via the soil. Apart from immediate damage, acid loads may cause **changes**, such as a disturbance of the mineral balance, which could lead to damage in the longer term.

Controlled laboratory and field experiments on needles, leaves, branches, and entire plants and trees have already shown that air pollution and acid deposition can affect the growth and development of vegetations in different ways. The results of such experiments, however, cannot easily be applied to field conditions, under which an ecosystem is exposed to all kinds of other stress factors. These stress factors include diseases and plagues, constantly changing weather conditions, in which the temperature and water supply are often far from ideal, as well as genetic factors. The problem is, therefore, a complex one: a combination of deterministic and stochastic factors (combined stress) affects the health/functioning of a tree or a forest stand. The different factors can reinforce each other. The sensitivity to frost and the sensitivity to plagues and diseases, for instance, can be influenced by the level of exposure to air pollution/acid deposition through a disturbance of the nutrient balance in soil solution and needles.

It is difficult to find a direct correlation between deposition levels and the vitality of a forest, since effects are usually caused by a combination of stress factors, as has been argued above. If a direct correlation was found it would not give any indication of the way in which SO_2, NO_x, and NH_3, both seperately and in combination, affect trees. It might also lead to an overestimation of critical loads, because even before there is any visible damage, the health of forest stands is affected in many different ways through changes in the nutrient status of the tree and of the soil. Therefore, the relationship between the moisture and nutrient status of the soil and the tree, and health affecting aspects (such as root development, root-shoot ratio, occurrence of mycorrhizal fungi, and visual effects) should first be studied through process-orientated or correlative research. After this relation has been established, critical deposition levels (see Figure 4.1) can be based on the relation between moisture and nutrient status, and deposition levels. Stress can cause visual effects such as leaf/needle loss, leaf/needle discolouration, and reduced growth. It can also cause morphological damage (such as damage of the wax layer of needles). The visual effects result from changes in various processes in the tree and in the soil, such as photosynthesis and the supply, uptake and allocation of nutrients and are also related to the availability of water.

The health of forests in the Netherlands has been monitored since 1983. To describe the state of health of a forest the term "vitality" is used. Assessments are based on the parameters loss of leaves/needles and discolouration of leaves/needles. The vitality as defined in this monitoring is determined by "traditional" factors such as fertility of the soil, origin of the trees, frost and drought, and pests and plagues, as well as by "new" anthropogenic factors such as air pollution and acid deposition. The annual vitality survey serves mainly as an indicator. It gives no more than an indication as to the possible occurrence of combined stress. The inventory does not give much insight into the causes of changes in vitality and usually does not give any indication of the relative share of the various causes. Figure 4.2 shows vitality data for the years 1983 through 1990.

A distinction between direct and indirect effects of air pollution is useful for the analysis of effects. For a synthesis it is less useful, however, because effects are usually the result of both direct and indirect influences or of the mutual interference of these. Nevertheless, the distinction between direct and indirect effects will be used, to give a systematic survey. This distinction is especially useful with regard to abiotic changes (such as changes in the composition of the soil solution).

This chapter gives a synthesis of the results of both controlled experiments and experiments under field conditions, as well as of model calculations. It is a compilation of results of research on the following subjects (separately described in the Thematic Reports): soil acidification and nitrogen cycling, biological and physiological effects, and integrated effects (of atmospheric input) on forests. Findings of research which was not carried out as part of the Dutch Priority Programme on Acidification were also included, in order to give a complete survey of the current situation.

First the direct effects are described. After that, the indirect effects are discussed. Attention will be paid to the current situation and extrapolations to the longer term will be made on the basis of the present loads and use of the soil. Finally, attention will be paid to the integrated effects of air pollution. Chapter 6 discusses model forecasts on a regional and national scale (scenario analysis).

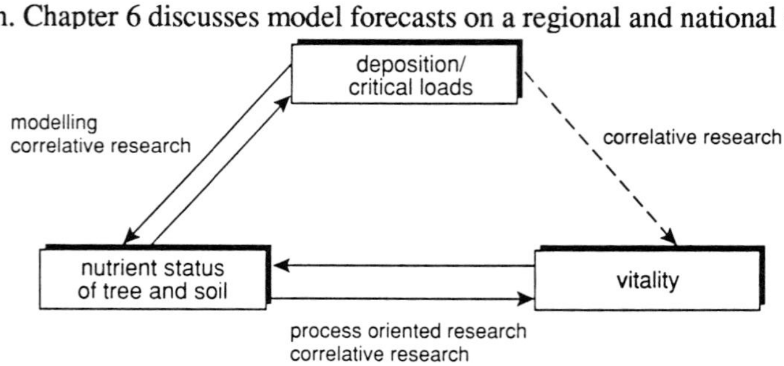

Fig. 4.1 Research methods for establishing critical deposition loads for nitrogen and sulphur

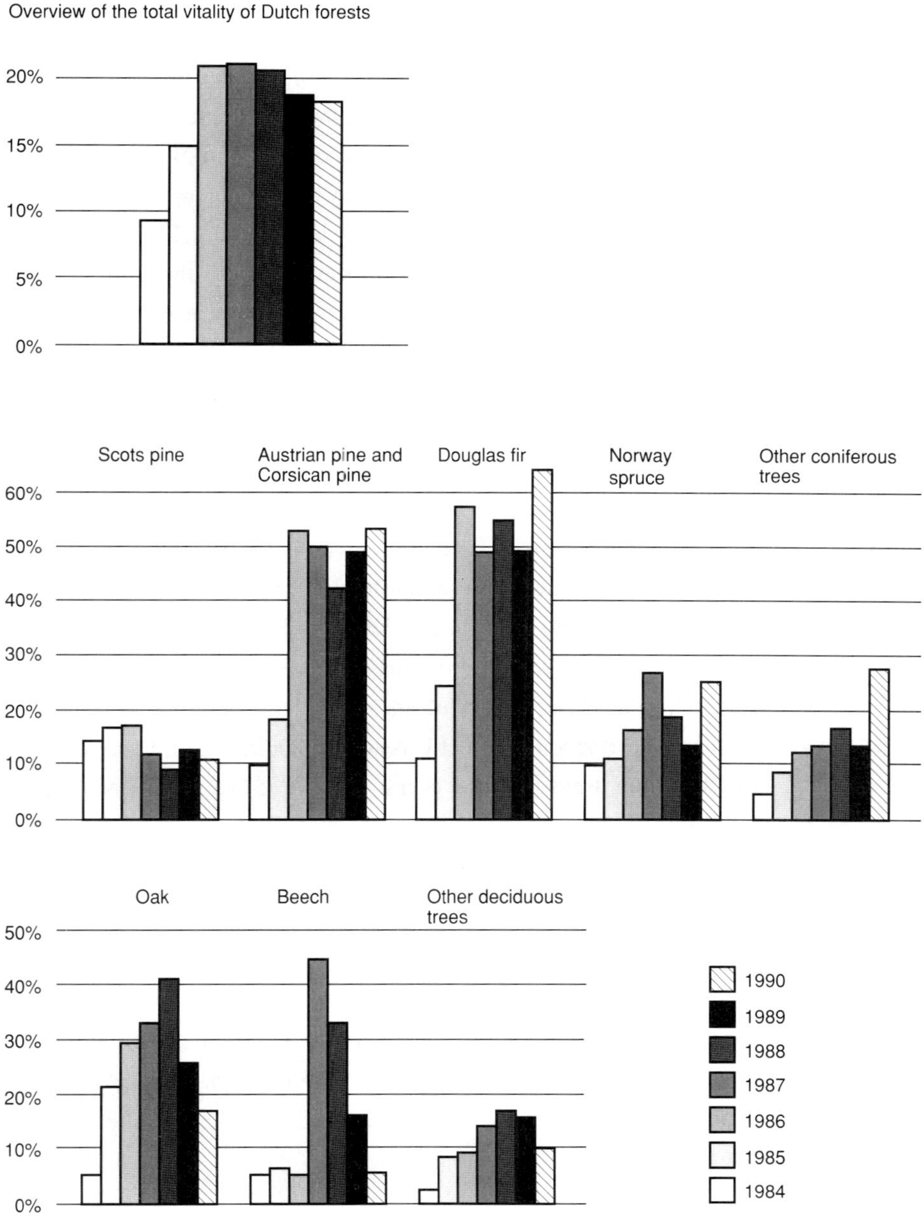

Fig. 4.2 Percentage of non-vital or almost non-vital forest between 1984 and 1990 (Source: Ministry of Agriculture, Nature Management and Fisheries)

4.2 Direct effects

4.2.1 Introduction

Direct effects are related to the uptake of air pollutants by the tree canopy. This uptake is controlled by variables which directly influence tree growth, such as radiation balance, relative humidity, turbulent exchange, and pollutant concentrations, but also by variables which have an indirect influence. Soil water tension indirectly influences tree water status, which in its turn determines the stomatal opening and thus the direct uptake of pollutants. The effects become even more complicated, if considered on a physiological level. The nutrient status of a plant compartment is determined by the direct uptake from the atmosphere, the uptake of nutrients by the roots, and also by leaf leaching. Direct uptake of pollutants is known to influence photosynthesis, and reallocation and synthesis of the type of assimilates, which eventually will affect growth of underground parts of the tree and related mycorrhizas.

The mutual interference of direct and indirect effects will be more pronounced for long-term effects than for short-term effects. Up to now, critical levels have been estimated mainly on the basis of short-term effects. Revisions may be needed when more subtle, physiological effects are studied and long-term effects become clear.

Another point of caution in interpreting the information given in the paragraphs on direct effects, is the variability in species. A clear distinction should be made between deciduous and coniferous species. This is not only because of leaf loss in winter and a different resistance to frost, but also because of the different maximum photosynthesis rates, with large differences in resistance to uptake of air pollutants.

In the next paragraphs, short-term and long-term effects and the additional influence of sub-optimal growing conditions on trees will be given for O_3, SO_2, NH_3, NO_2 and combinations of these pollutants. Also, attention will be paid to effects on epicuticular needle wax morphology of Douglas fir.

4.2.2 Effects of O_3

Concerning the effects of O_3 on trees the Dutch Priority Programme on Acidification has been carried out on Douglas fir (Pseudotsuga menziesii) only. In a more general context, research has been done on the effects of O_3 on poplar trees (Mooi, 1985), and especially on agricultural crops and herbaceous wild plants, which were used as indicators of effects of O_3 (Posthumus and Tonneijck, 1982).

- Effects of short-term exposure (under optimal conditions)

Unlike herbaceous plants, Douglas fir is relatively insensitive to short-term exposure to O_3 (days/weeks). Experiments under field conditions with fully grown Douglas fir showed that daily ozone levels of 85 - 170 μg.m^{-3} reduced the photosynthesis by at least 8 - 25%. Over the entire research period of three weeks (in May 1989), a minimum decrease of 2 - 5% was found (Steingröver et al., 1991). In principle, such a change in the photosynthesis has not any effects on the annual growth.

Exposure of young Douglas fir to O_3 concentrations over 200 μg.m^{-3} (for 10 to 17 hours) can result in a considerable reduction in photosynthesis. This reduction could amount to, for instance, 30% at a concentration level of 400 μg.m^{-3}. The respiration will then increase by 50%. The effect will stabilize within days. After exposure has been stopped, there will be a recovery. O_3 concentration levels over 100 μg.m^{-3} for a period of 9 days (8 hours a day) were found to inhibit the phloem loading of the young Douglas fir (less than 3 years old), which resulted in an inhibition of the translocation of assimilates to the roots (Gorissen and Van Veen, 1988; Gorissen et al., 1991a). This was inferred from measurements of the amount of 14C labelled CO_2 which was produced in the soil/root compartment of a closed chamber in which the tree grew in air containing $14CO_2$. At higher O_3 concentrations (up to 500 μg.m^{-3}) the inhibition increased. After exposure to the highest concentrations, no recovery was observed after the exposure was stopped. Recovery was observed, however, after exposure to concentrations of 200 μg.m^{-3}.

Similar effects were witnessed after exposure of 25 year-old Douglas fir to different concentrations of O_3 (in branch chambers) under field conditions. During this experiment first-year needles were exposed to different O_3 concentrations (of 0, 200, and 400 μg.m^{-3}) for 9 days, after which period the photosynthesis and translocation were studied by means of $14CO_2$. Photosynthesis decreased when trees were exposed to O_3 concentrations of 200 or 400 μg.m^{-3}. It was also found that 4 hours after uptake of $14CO_2$, the proportion of 14C assimilates (transported from the needles to the branches) had decreased.

At O_3 concentration levels over 200 μg.m^{-3}, visible damage in Douglas fir can occur after a few days. Not much research has been done on the sensitivity of other species to O_3 in the Netherlands. However, deciduous species are probably more affected during O_3 episodes than coniferous species (Reich, 1987). Poplars show leave loss during episodes of O_3 pollution (Mooi, 1985).

- Long-term effects

The effects on Douglas fir of exposure to increased O_3 levels for more than one year are unknown. The threshold values for long-term exposure to O_3 are therefore based on assumptions.

The most recent findings of several years of research on other coniferous species indicate that the threshold values for O_3 will probably come out lower, if long-term effects or effects on needles from different years are taken into account. At present quantitive data are still

lacking.
Research on coniferous species demonstrates that low O_3 concentrations can result in considerable damage of needles from earlier years. Such damage is often seen as an effect of O_3 on needle ageing. The findings of Wallin et al. (1990) prove that exposure of Picea abies to seasonal average concentrations of 57 and 95 $\mu g.m^{-3}$ (7-hour values) for a period of two seasons results in a reduction of photosynthesis of 35 and 59% respectively. Similar results are reported by Küppers and Klumpp (1988) for Picea abies and by Richardson et al. (1990) for Pinus taeda. Some years ago Reich and Amundson (1985) already described the effects of low concentrations on leaf ageing.

- Effect of suboptimal growing conditions

Water stress counters the effect of O_3 on Douglas fir. In periods of drought the stomata close, which diminishes the uptake and also the effects of O_3. The translocation of carbon in the plant, however, is inhibited when there is a lower water content in the soil. (Gorissen et al., 1991b, in press).
The regulating mechanism of the stomata largely determines the efficiency of the plant in using the water supplied. In Douglas fir this mechanism is not disturbed by O_3 concentrations of about 140 $\mu g.m^{-3}$. An increased shoot-root ratio, caused by a disturbed translocation of assimilates, makes the plants more sensitive to drought. A slower development of the cuticula owing to exposure to O_3 (140 $\mu g.m^{-3}$) increases the transpiration and can also make plants more sensitive to drought.
A marginal nutrients supply greatly increases the sensitivity of Douglas fir to O_3. Similar effects were described by Sellinger et al.(1986).

4.2.3 Effects of SO_2

- Short-term effects (under optimal conditions)

At the SO_2 concentration levels currently found in the Netherlands, there are hardly any short-term effects, except in the case of local high concentrations combined with low temperatures and stable weather (especially in winter) near emission sources (Kropff et al., 1990). Inhibition of the photosynthesis in Douglas fir only occurs at very high concentrations (over 1000 $\mu g.m^{-3}$), but other species are more sensitive. Current SO_2 concentration levels in the Netherlands do not cause any visible damage to Douglas fir.

- Long-term effects

Like NH_3 and NO_2, SO_2 can cause internal acidification, leading to ageing and shedding of leaves as a long-term effect. These effects are connected with the supply of nutrients (low pH and high ammonium concentrations of the soil accelerate the effects). Current SO_2

concentration levels do not cause any visible damage in Douglas fir (Kropff 1990, in press).

- Influence of suboptimal growing conditions

In periods of drought the uptake and also the effects of SO_2 decrease. At low SO_2 concentrations the stomata open further and the evaporation increases. Under the influence of SO_2 the shoot-root ratio also increases, which has a similar effect (Darrall, 1989).
There is probably a considerable interaction with the supply of nutrients via the internal pH regulation (Kropff, 1990).
At low temperatures the sensitivity to SO_2 increases owing to an increased concentration of sulphite (Kropff et al., 1990).

4.2.4 Effects of NH_3 (under optimal conditions)

Nitrogen is taken up partly above the ground, through uptake of NH_3 via the stomata. Especially at high ammonia concentrations, in slow-growing plants with a low nitrogen demand, this can be a relatively large part of the total amount of nitrogen uptake. Under those circumstances the uptake of nitrogen via the roots is not sufficiently regulated.
The most important effect of NH_3 is growth stimulation. In quite a few species only the growth of shoots is stimulated. This leads to a higher shoot-root ratio, which makes the plant more sensitive to drought and nutrient deficiencies. Growth stimulation shows a linear relationship with NH_3 concentrations at background levels.
A higher level of atmospheric NH_3 stimulates the maximum assimilation of CO_2 (through an increase in the chlorophyl content). It enhances the stomatal transmission (the stomata remain more open), and thus also the uptake of NH_3 (Van Hove, 1989).
Direct toxic effects of NH_3 are not very common. However, damage to the epicuticular wax layer of leaves and needles (Van der Eerden et al., 1989) and an increased occurrence of pests and plagues are serious additional effects of NH_3 (Nihlgard, 1985).

- Effects of NH_3 (suboptimal growing conditions)

An increased shoot-root ratio, diminished stomatal resistance, and damage to the epicuticular wax layer under the influence of NH_3 result in a greater sensitivity to drought.
High NH_3 loads lead to premature nutrient deficiency (Kaupenjohann et al., 1989).
NH_3 increases the sensitivity to frost (Dueck et al., 1990, in press).

4.2.5 Effects of NO_2

NO_2 stimulates plant growth at low concentrations (nitrogen fertilization), but causes visible damage at concentrations above approximately 1000 $\mu g.m^{-3}$. Plants are more sensitive to

NO_2 during the night than during the day. As far as the combination of SO_2 + NO_2 is concerned, NO_2 was found to increase the negative effect of SO_2 on the maximum CO_2 assimilation rate, whereas NO_2 in itself had no effect.

4.2.6 Effects on epicuticular wax layers of Douglas fir

The structure of the wax layer and the amount of wax on the needles of Douglas fir has been studied in the field and in NH_3 fumigation experiments with young trees in open-top chambers.

The amount and structure of amorphous wax on the stomata of dried needles were assessed with Scanning Electron Microscopy (SEM). No differences were found between needles from the two locations and from the two vitality categories which had been sampled. In fumigation experiments with different NH_3 concentrations, serious corrosion of crystalline wax was found in one-year-old needles which had been exposed constantly. Fumigation for a period of 5 weeks had no effect on the wax morphology of needles of the current year, however.

It was concluded that the morphology of the wax layer on needles cannot be regarded as an indicator of damage from air pollution.

4.3 Indirect effects

4.3.1 Introduction

During the past few years an increasing disturbance of the nutrient balance of both leaves/needles and the soil solution has been observed. There has also been strong acidification of the soil, which has particularly led to the mobilization and leaching of large amounts of aluminium together with nitrate and sulphate.

The following elements and element ratios in the tree and in the soil solution were found to be important indicators in the sequence of air pollution/acid deposition - nutrient balance - health/functioning of the tree:

- in the soil solution: pH, Al , Al/Ca, NH_4/Mg, NH_4/K;
- in the needles: N, N/P, N/Mg, N/K.

Figure 4.3 shows the fluxes of these elements through the soil and the tree as well as the most important influential processes. The indicators and the effect-related threshold limit values are discussed in Chapter 4, section 3.2. The current situation with regard to the indicators and developments which can be expected at unchanged atmospheric input are discussed in Chapter 4, section 3.3 and 3.4. The (indirect) effects are discussed in Chapter 4, section 3.5.

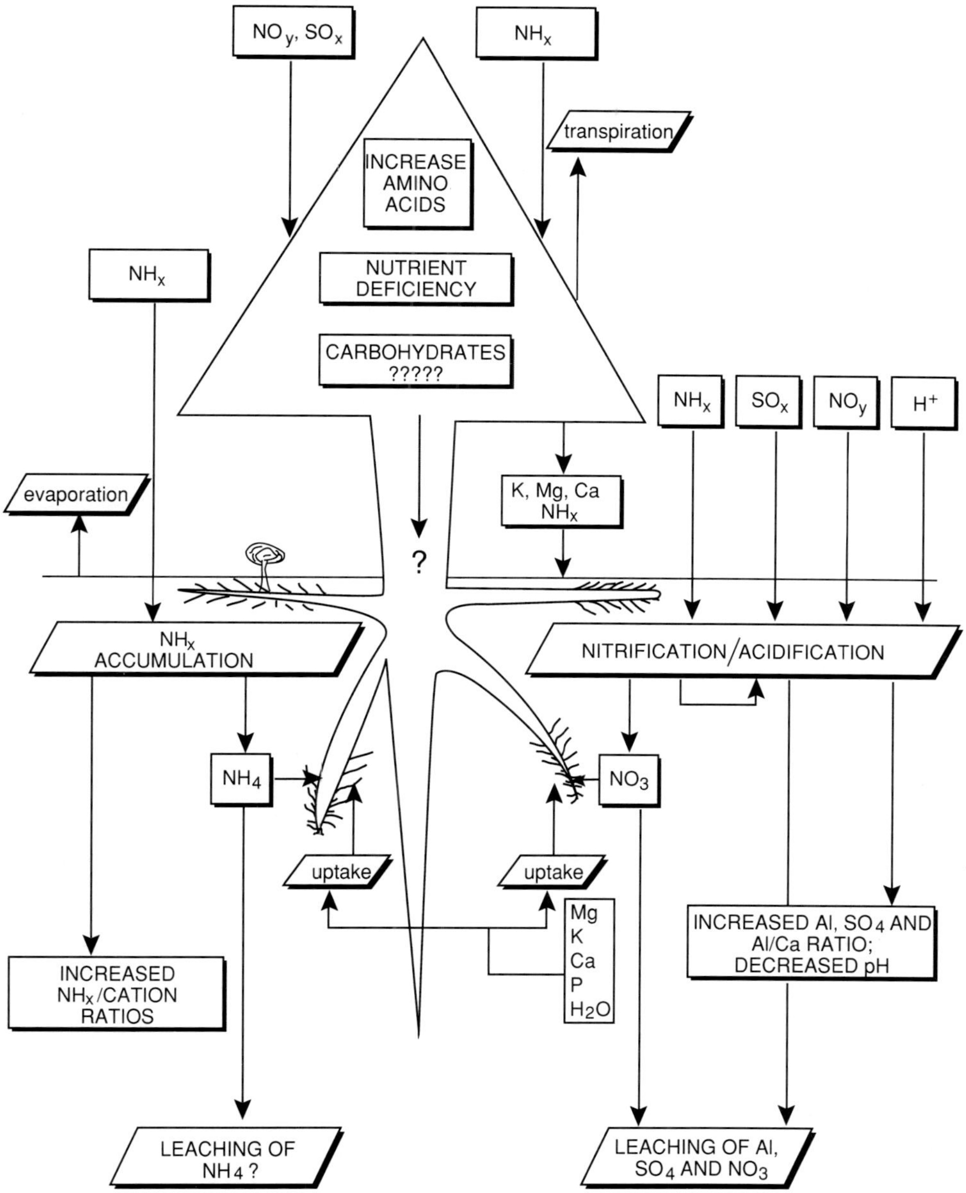

Fig. 4.3 Schematic overview of a number of important processes and parameters influencing the composition of the soil solution of Dutch forest soils

4.3.2 Key parameters in soil solution and needles, and threshold limit values in relation to the effects

pH level

The pH level of the soil is important for biological processes such as the growth of plants, bacteria, fungi, mycorrhizas and soil fauna (such as nematodes). The pH tolerances of tree species in the Netherlands are given in Van den Burg, 1981. The current pH is extremely low, also for most of the acid-tolerant species on/in the Dutch forest floors (see the Thematic Report by Van Breemen and Verstraten). However, the pH value is not a suitable indicator of acidification of the soil, since the pH of forest soils in the Netherlands is often naturally very low.

Apart from the direct adverse effect of a low pH (which probably occurs at a solution pH < 3.5 and almost certainly at a pH < 3), it is almost certain that it will cause an extreme P deficiency owing to the formation of insoluble iron and aluminium phosphates.

Aluminium content

High aluminium concentrations cause a decrease of fine roots and an increase of coarse roots. This has been observed in hydrocultures as well as in the field. A wide range of Al toxicity thresholds for various tree species has been reported in the literature varying between < 1.5 $mg.l^{-1}$ to > 30 $mg.l^{-1}$ (De Vries, 1991). Results are all based on experiments with seedlings growing either in solution cultures or in greenhouse pots. The tolerance to Al toxicity has been related generally to root and/or shoot growth, and sometimes to the degree of mortality.

In principle a wide range of sensitivities of trees to Al can be expected as sensitivity varies with solution pH, Al speciation, calcium concentration, overall ionic strength , the form of inorganic nitrogen (NH_4, or NO_3), mycorrhizal interactions, soil moisture, etc. Research of the ALBIOS (Aluminium in the Biosphere) project carried out in eastern North America and northern Europe indicates that red spruce is the most sensitive tree species, with statistically significant biomass reductions starting to occur near 5 mg l^{-1} of total aluminium or 2 mg l^{-1} of labile inorganic aluminium, which is toxic to roots. Other moderately sensitive species are sugar maple, Douglas fir, larch and European beech, whereas Scots pine and oak are relatively insensitive to Al. Many symptoms of Al stress could also indicate P deficiencies though (Keltjens, 1990). The critical value for Al in the soil solution is considered to be 2 $mg.l^{-1}$.

Al/Ca ratio

The experiments with Norway spruce, carried out by Rost-Siebert (1985), showed that it is not the Al in itself, but the Al/Ca ratio which has an important influence on the root development. The same conclusion can be drawn from experiments with Corsican pine by Boxman and Van Dijk (1988). They found that Al/Ca ratios higher than 1 can give an increase in short, relatively thick roots and a decrease in fine roots (Figure 4.4). This effect can influence the water and nutrient supply, since it reduces the total absorptive capacity (Arp and Stricker, 1989). The importance of the Al/Ca ratio has not only been shown in several controlled experiments but also under field conditions. A linear relationship was observed, for instance, between the number of root tips in the field and the Ca/Al molar ratio in the soil solution of plots with a varying degree of needle discolouration (Figure 4.5, Meyer et al., 1987). Laboratory research has shown that, at an Al/Ca ratio of 1 in the soil solution, the uptake of bivalent cations is seriously inhibited (Boxman and Van Dijk, 1988) and the roots are affected (Meyer et al., 1986). Even at a ratio below 1 effects were observed.

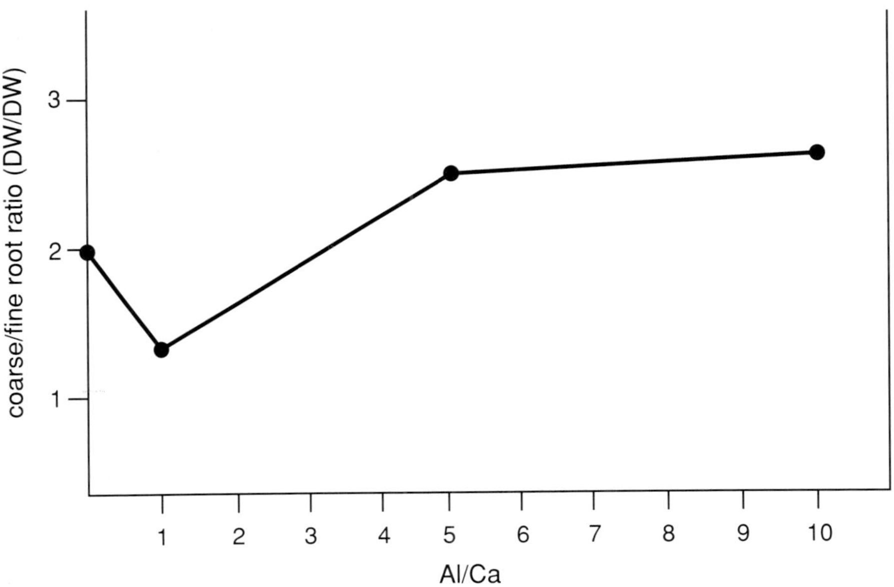

Fig. 4.4 Effects of aluminium on the ratio of coarse to fine roots in the case of Corsican pine (dry weight basis)

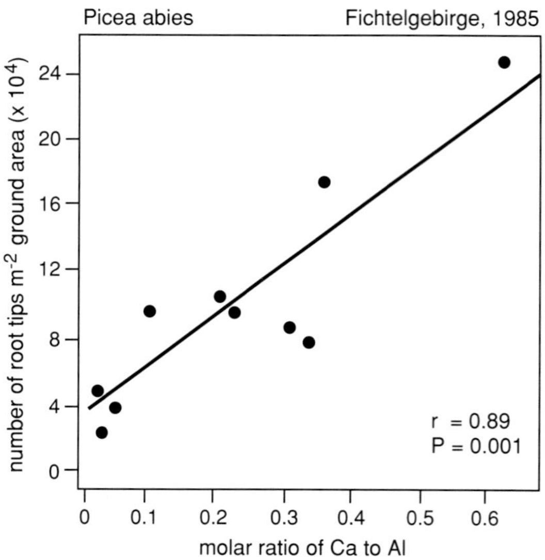

Fig. 4.5 Number of root tips of Picea abies as related to the molar ratio of Ca/Al

Under field conditions, however, there are several complications, such as the interaction with mycorrhizal fungi and the effect of nitrogen deposition on the development of roots. For correlative field research, Roelofs et al. (1985) used a soil analysis based on samples of soil extracted with water. This method only gives an indirect picture of the composition of the soil solution. The values of the Al/Ca ratio using this water extraction method, are an order of magnitude higher than those measured in the soil solution itself (Verhagen and Diederen, 1991), and should, therefore, not be compared to the critical value of 1 for soil solution.

Table 4.1 shows the values estimated by Roelofs et al. (1985). Although there is a relationship between the low vitality of some of the forest stands and the Al/Ca ratio of the extracts, there is no complete correspondence. Under field conditions, dissolved organic C (DOC), the pH level and Fe content of the soil solution might be important. The soil solution in vital forests with very high Al/Ca ratios, for instance, was found to contain large amounts of DOC. Complexation with organic C may reduce the Al toxicity. This is being studied at present.

Until the results of this research are known, a provisional critical value of 1 has been set for the Al/Ca ratio in the soil solution, as in the Evaluation report (Schneider and Bresser, 1988).

NH_4/K and NH_4/Mg ratios

Laboratory and greenhouse experiments with two-year old Corsican pine, Scots pine, and Douglas fir, showed that the uptake of K, Ca and Mg is strongly inhibited at high NH_4/cation ratios. The degree of mycorrhiza infestation and the proportion of fine roots to thick roots were also found to decrease (Boxman et al., 1991, Van Dijk et al., 1990). The relative share of fine roots decreased. Since nutrients are mainly taken up via the fine roots, such a decrease may lead to nutrient deficiencies. Hydroculture experiments with young Corsican pines showed that, at net NH_4/Ca and NH_4/Mg ratios of 50 and 100 respectively, even Ca and Mg are released. (Boxman and Roelofs, 1986). The resulting nutrient deficiency is further increased by deposition of NH_4 ions on the needles which stimulates the leaching of nutrients from the needles. Experiments have shown that coniferous trees take up NH_4 through the needles and compensate for this by releasing K and Mg.(Roelofs et al., 1985). All this increases the sensitivity to pests and plagues (for instance Sphaeropsis).
Experiments such as those with Corsican pine (Boxman et al., 1991) have shown that NH_4 increases the shoot/root ratio by affecting the nutrient balance. This has a positive effect on the above-ground growth and leads to an increased sensitivity to drought.

Field research has shown a demonstrable correlation between the values of NH_4/K and NH_4/Mg ratios in water extracts from soil samples and the condition of several tree species such as Douglas fir (Table 4.1; Roelofs et al., 1985). Table 4.1 shows that there is a higher correlation between these ratios and effects than between the Al/Ca ratio and effects. It also shows that the NH_4/K ratio has a lower correlation with effects than the NH_4/Mg ratio. Mg deficiency is found relatively frequently. A high amount of NH_4 in the soil can result in a strong acidification of the rhizosphere, which may lead to increased activity of Al especially around the roots. In most plants the occurrence of Al in the soil leads to a lower uptake of Mg.

Tabel 4.1 The ratios of several nutrients in water extracts from soil sampled in A) vital B) less vital, and C) non-vital Pinus nigra and Pseudotsuga menziesii stands ($mol.mol^{-1}$)

		NH_4^+/K^+			NH_4^+/Mg^{2+}			Al^{3+}/Ca^{2+}		
	n*	aver.	min.	max.	aver.	min.	max.	aver.	min.	max.
Pinus nigra										
A)	21	4.7	0.5	14.0	6.4	1.1	24.3	2.0	0.4	5.6
B)	17	9.2	0.8	36.8	10.0	1.8	26.3	1.3	0.2	2.8
C)	21	11.3	1.9	51.8	22.1	1.6	57.2	5.5	1.7	16.7
Pseudotsuga menziesii										
A)	10	3.8	0.5	11.8	4.5	0.6	10.0	6.6	0.8	40.9
B)	10	8.7	1.5	31.8	19.3	2.0	51.1	8.9	0.7	46.7
C)	11	18.2	3.8	64.5	47.6	7.9	118.0	15.6	1.4	54.0

* number of locations

Laboratory research gives a range of 1 to 10 ($mol.mol^{-1}$) as the critical ratio for NH_4/Mg in the soil solution (Roelofs et al., 1987). At present the data are lacking to determine the critical threshold limit value more accurately. The NH_4/Mg ratio will therefore not be used to determine a critical load. The NH_4/K ratio will be used for this purpose. Another reason why the NH_4/Mg ratio is not used, is that the values for the NH_4/Mg ratio based on the water extraction method are higher than those measured in the soil solution. With regard to the NH_4/K ratio, there are no such differences and both methods give more or less the same results. On the basis of laboratory experiments and field research, a critical value of 5 was set for the NH_4/K ratio ($mol.mol^{-1}$).

Nitrogen content of needles

With regard to the nutrient status, a nitrogen content of the needles between 1.6 and 2.0 % was considered to be optimal until recently (e.g. Van Goor, 1967). Van den Burg (1988), however, states that the optimal value is 1.8 - 2.5%. This adjustment of the criteria to determine the mineral nutrient balance of coniferous species was considered to be necessary owing to the changed conditions of the coniferous forests in the Netherlands (an important factor being the high nitrogen input). Values exceeding the threshold limit value of 2.5% do not cause immediate damage, but can lead to a growth reduction. Such a high nitrogen content of the needles, combined with a K and/or Mg deficiency resulting from a disturbance of the nutrient balance of the soil, can also lead to increased needle loss and increased sensitivity to other stress factors such as drought, frost, pests and plagues.

Already at values of 1.7 - 2.0% the sensitivity to extreme weather conditions and to pests and plagues increases strongly. On the basis of a fertilization experiment in Sweden, Aronsson (1980) for instance concluded that frost damage to needles of Scots pine increased strongly at a nitrogen content above 1.8%. Around this value the sensitivity to fungi such as Sphaeropsis sapinea and Brunchorstia also appears to increase strongly. This can be derived from correlative field research on the relationship between the nitrogen content of needles of Corsican pine and damage from Sphaeropsis (Roelofs et al., 1985; Boxman and Van Dijk, 1988; Van den Burg et al., 1988). In this connection, Van den Burg et al. (1988) suggest a critical nitrogen content of 1.6%. For Scots pine the critical value was found to be slightly higher (about 2.2%). On the basis of the above-mentioned research the critical value for nitrogen content was set at 1.8%.

4.3.3 The current composition of the soil solution and the nitrogen content of needles

Input-output budgets

Intensive monitoring of chemical and hydrological characteristics of deposition (throughfall) and soil solution, for a period of at least one year, was carried out at seventeen sites. By means of these data the input-output budgets were estimated (Table 4.2).

Table 4.2 Input-output budgets for SO_4, N, and Al (input negligible) for seventeen Dutch forest sites. For references see Table 2.1, Thematic Report Van Breemen and Verstraten. Periods are hydrological years, generally from April to April

		SO_4		NH_4		NO_3		Al	$H^{+\#}$	N_{out}/N_{in}	S_{out}/S_{in}
		kmol$_c$ ha^{-1}yr^{-1}								kmol.kmol^{-1}	
location	period	in	out	in	out	in	out	out			
1a Speuld A	'87/88	2.4	1.6	2.9	0.11	0.87	2.0	3.0	4.2	0.6	0.7
1b Speuld B	'88/90	2.3	3.1	2.8	0.01	0.79	2.2	4.6	4.2	0.6	1.3
2 Amerongen	'87/88	3.4	2.9	3.7	0.45	0.74	4.5	4.3	7.0	1.1	0.9
3a KootwijkA	'87/88	2.4	3.0	3.0	0.30	0.83	1.7	4.0	3.6	0.5	1.2
3b KootwijkB	'88/90	2.1	1.7	2.6	0.00	0.75	1.7	3.1	3.6	0.5	1.5
4 Garderen	'87/88	2.5	2.8	3.1	0.18	0.83	2.0	1.9	3.6	0.5	1.1
5 L.vuursche	'87/88	4.8	6.0	4.5	0.80	1.6	5.9	10.0	8.0	1.1	0.8
6 Ruurlo	'87/88	3.2	3.3	3.9	0.80	0.69	4.6	7.4	7.0	1.2	1.2
7 Zelhem	'87/88	3.8	5.5	5.1	1.7	0.99	6.4	9.4	8.8	1.3	1.4
8 Tongbersv.	'83/84	3.8	2.8	4.2	0.2	0.65	0.85	3.3	4.2	0.2	0.7
10 Gerritsfles	'83/84	3.9	4.0	4.1	0.31	1.3	1.5	4.6	4.0	0.3	1.0
12 Wintersw.	'79/85	3.4	3.6	3.1	0.05	0.57	0.63	0.05	3.1	0.2	1.1
13 OudemaatA	'81/87	2.9	2.7	2.8	0.00	1.06	4.9	0.1	6.6	1.3	0.9
13 OudemaatB	'81/87	2.7	2.5	2.9	0.00	1.20	2.7	4.3	4.4	0.7	0.9
13 OudemaatC	'81/87	2.2	2.2	2.2	0.00	0.95	1.8	2.4	3.1	0.6	1.0
33 Buunderk.	'88/89	1.9	1.1	1.4	0.14	0.49	0.28	1.5	1.1	0.2	0.6
34 Leuvenum	'89/90	2.5	3.2	3.7	0.50	0.87	1.2	3.9	3.5	0.4	1.3
Average										0.7	1.0

\# H^+ produced in N transformations (net NH_4 uptake plus net NO_3 drainage)

These input-output budgets show that the forests examined are saturated with SO_4 and that SO_4 almost behaves as a tracer (input and output are about the same). They also show that an average 70% of the leached NO_3 and SO_4 is accompanied by Al and that the leaching of Al is approximately equal to the H+ produced in N cycling. The fixation of N shows great variation. On average 1.5 kmol (($\pm$ 20 kg) N ha^{-1}yr^{-1} is fixed, via processes such as uptake by the forest stand and immobilization in organic matter, and/or disappears from the system through denitrification. The remaining part of the N input is leached out, mainly in the form of NO_3. However, in the case of extremely high N input (> 4 kmol ha^{-1}yr^{-1} (56 kg)) the net fixation can become negligible and a net N leaching may even occur as a result of a decreasing net immobilization. This situation is called nitrogen saturation. It seems to occur at 5 out of the 17 sites which were studied.

Concentrations in the soil solution

Table 4.3 shows the pH values, the Al, SO_4, and NO_3 concentrations as well as the NH_4/NO_3 molar ratio for the soil solution at depths of 0 - 10 cm and 50 - 100 cm for almost the same sites. The NH_4/NO_3 ratio gives information about the rate of nitrification. Nitrification is an important process for regulating the proportion of NO_3 to NH_4 in the soil at a given NH_x deposition. Acid-tolerant, chemolithotrophic bacteria are probably the most important producers of NO_3 in acid forest soils. The nitrifying capacity of the soil depends on factors such as the C/N ratio of the litter layer, the moisture content and the pH of the soil. Especially at pH levels under 3 acid-tolerant nitrification is strongly inhibited. At pH levels between 3 and 5 there is a hardly demonstrable net effect of the pH on the degree of nitrification. The nitrification rate varies considerably with soil type but also with depth. With regard to the NH_4/NO_3 ratio the depth at which measurements are carried out is, therefore, extremely important. The depth is chosen in relation to the aspect studied: the litter layer, the root zone (interaction with the vegetation) or the leaching from the root zone to the groundwater. Eventually almost all NH_4 input appears to be nitrified in the root zone, but the development of the NH_4/NO_3 ratio with depth and in time depends on the rate of nitrification, together with the input of atmospheric N and the amount of N from mineralization (Figure 4.6). However, in the case of extremely high input of NH_4, the concentration of NH_4 can be quite high. Concentrations of 10 - 20 mg l^{-1} were found, for instance, under the root zone of Scots pine and Douglas fir (Kleijn et al., 1989).

On the basis of the results of intensive monitoring (Table 4.3) Van Breemen and Verstraten (Thematic Report on N cycling and Soil acidification) concluded that, at 14 out of the 16 sites that were studied, the NH_4/NO_3 ratio is lower than 1 in the topsoil (at a depth of 0 - 10 cm). At all sites that were studied, this ratio is lower than 1 in the subsoil (at a depth of 50 - 100 cm).

Table 4.3 Mean field pH in the soil, and mean concentrations of Al, SO_4, NO_3, and NH_4/NO_3 in the soil solution ($mmol_c.1^{-1}$) at sixteen sites

location	pH		Al		SO_4		NO_3		NH_4/NO_3	
	0-10	50-100	0-10	50-100	0-10	50-100	0-10	50-100	0-10	50-100
	depth (in cm)									
1A Speuld A	3.2	3.7	0.6	1.4	0.9	1.3	1.4	1.2	0.32	0.04
1B Speuld B	3.2	3.9	0.3	2.0	0.7	1.0	0.5	1.2	0.72	0.02
2 Amerongen	3.4	3.7	1.1	3.2	0.7	2.0	0.5	2.4	0.72	0.09
3A Kootwijk A	3.5	3.6	1.1	1.9	1.4	1.8	2.0	0.9	0.68	0.22
3B Kootwijk B	3.4	4.0	0.3	1.6	0.6	1.0	0.3	0.7	1.30	0.11
4 Garderen	3.5	3.7	0.7	1.0	0.9	1.1	1.0	0.7	0.30	0.08
5 L.vuursche	3.3	3.3	0.8	9.8	2.8	6.9	2.1	5.3	0.92	0.15
6 Ruurlo	3.2	3.5	0.7	5.5	3.2	3.4	3.0	2.0	0.68	0.22
7 Zelhem	3.3	3.3	4.1	12.6	2.6	6.9	4.2	6.4	0.47	0.09
8 Tongbersven	3.4	4.1	0.1	1.8	1.1	1.5	0.4	0.5	2.10	0.18
10 Gerritsfles	4.0	4.2	0.8	1.7	0.9	1.2	1.0	0.8	0.90	0.02
12 Winterswijk	3.5	4.4	0.2	0.2	0.5	1.6	0.7	0.5	0.10	0.05
13 Oudemaat ABC	3.6	4.1&	0.6	1.5&	0.7	1.4&	1.1	1.5&	0.18	0.09&
32 Harderwijk	4.3	4.9	0.6	0.5	0.5	0.5	0.1	0.2	0.90	0.01
33 Buunderkamp	3.9	4.3	0.3	0.6	0.5	0.5	0.2	0.2	0.85	0.12
34 Leuvenum	3.3	3.9	0.2	0.9	0.7	0.8	0.5	0.5	0.83	0.23

& mean for three acid soil plots at the Oudemaat site

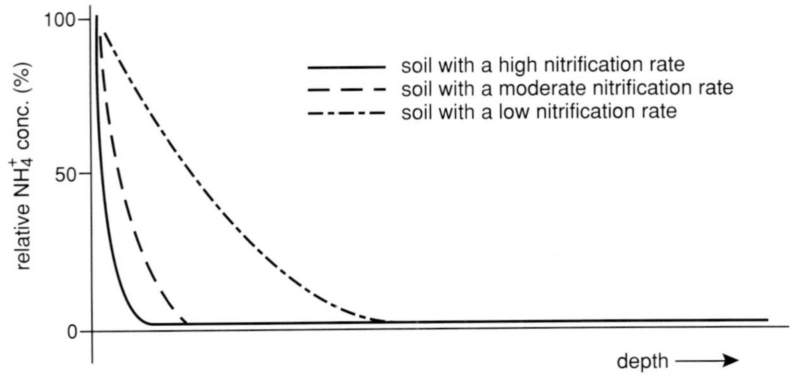

Fig. 4.6 Schematic relation between soil water NH_4^+ concentration and nitrification rate in relation to depth

With regard to atmospheric deposition, the 16 sites that were studied are not representative of the Dutch forest. In the spring of 1990, however, the soil solution was sampled by the Winand Staring Centre at 150 forest and nature sites once-only. The results of this sampling can be considered to be representative. They provide an almost nation-wide picture (as for non-calcareous soils).

The sampling of the soil solution was carried out in the spring of 1990 (during the period February until May). The average composition of the soil solution was assessed on the basis of mixed samples. Samples were taken from the topsoil, at a depth of 0 - 30 cm (the layer in which most of the roots are usually found), and from the subsoil, at a depth of 60 - 100 cm. The samples were only taken at sites on non-calcareous sandy soils throughout the country. The tree species studied were Scots pine (45), Corsican pine (15), Douglas fir (15), Norway spruce (15), larch (15), oak (30), and beech (15). The samples were taken in spring, because in this period the composition of the soil solution corresponds with the annual average of leaching soil solution. The depths of sampling are connected with the expected relationship between the quality of the soil solution and the rhizosphere and subsoil.

Tables 4.4 and 4.5 show the median values of the main elements and element ratios in the layer at 0 - 30 cm and in the subsoil (60 - 100 cm) depending on the tree species.

Table 4.4 Median values for the main elements and element ratios in the topsoil (0 - 30 cm) of 150 forest stands (data obtained from the Winand Staring Centre)

Tree species	pH	Al	Ca	SO_4	NO_3	NH_4/ NO_3	NH_4/ K	Al/Ca	NO_3/ SO_4
	$mol_c.m^{-3}$					$mol.mol^{-1}$			$mol_c.mol_c^{-1}$
Scots pine	3.60	0.81	0.39	1.09	0.56	0.35	1.00	1.22	0.62
Corsican pine	3.67	0.62	0.36	0.80	0.27	0.75	1.19	1.27	0.46
Douglas fir	3.41	1.43	0.90	2.56	0.92	0.70	2.56	1.42	0.44
Norway spruce	3.43	1.16	0.58	2.04	0.50	1.18	2.36	1.38	0.34
Larch	3.55	0.81	0.47	1.02	0.71	0.46	1.67	1.15	0.74
Oak	3.70	0.40	0.51	0.83	0.54	0.25	0.54	0.63	0.78
Beech	3.67	0.52	0.31	0.65	0.33	0.60	0.60	1.09	0.49
All species	3.59	0.69	0.45	1.03	0.55	0.45	1.12	1.13	0.59

Table 4.5 Median values for the main elements and element ratios in the subsoil (60 - 100 cm) of 150 forest stands

Tree species	pH	Al	Ca	SO_4	NO_3	NH_4/ NO_3	NH_4/ K	Al/Ca	NO_3/ SO_4
	$mol_c.m^{-3}$					$mol.mol^{-1}$			$mol_c.mol_c^{-1}$
Scots pine	3.87	0.68	0.29	1.11	0.55	0.17	0.86	1.48	0.64
Corsican pine	3.98	0.50	0.24	0.71	0.24	0.43	1.05	1.63	0.47
Douglas fir	3.77	1.70	0.44	1.86	1.25	0.08	1.24	2.82	0.70
Norway spruce	3.80	1.44	0.47	1.80	0.61	0.24	1.04	2.05	0.44
Larch	3.89	0.85	0.38	0.88	0.48	0.22	1.04	1.55	0.67
Oak	4.03	0.28	0.98	1.56	0.26	0.29	0.75	0.21	0.42
Beech	3.98	0.27	0.19	0.94	0.17	0.57	1.17	1.02	0.26
All species	3.89	0.58	0.40	1.11	0.48	0.22	0.94	1.42	0.53

The pH values in the topsoil are, without exception, low, but this is not necessarily an effect of acid deposition, because most of the sandy soils in the Netherlands have a naturally low pH level.

The lowest pH values and the highest element concentrations and ratios were found in the topsoil under Douglas fir and Norway spruce. The reverse situation was found for oak and beech. Scots pine, Corsican pine and larch occupy an intermediate position. In all probability, these results are related to the differences in dry deposition and transpiration between these species. But the results indicate that the composition of the soil solution under Douglas fir, on which research under the Dutch Priority Programme on Acidification has been particularly focused, is certainly not representative of the composition of the soil solution of Dutch forests as a whole. Median values for all the forest stands is most consistent with the values for Scots pine.

The average NH_4/NO_3 molar ratio in the topsoil is clearly less than 1.0. Analysis of all the results indicates that only 20% of the NH_4/NO_3 ratios (about 30 of the 150 stands) are higher than 1.0. This is consistent with the results of intensive monitoring at the 16 forest sites (see Table 4.3) and indicates that the nitrification rate in forest soils is considerable.

The average NO_3/SO_4 ratio in the subsoil is about 0.5 on an equivalent basis. Assuming an equivalent N/S ratio of at least 1.0 in the input (see section 3.3.2.), this means that an average of about 50% of the nitrogen input is fixed or disappears as a result of uptake, immobilization and denitrification. Analysis of all the results indicates that in about 15% of cases the NO_3/SO_4 ratio is higher than or equal to 1.0, which is an indication of the percentage of N-saturated forest ecosystems.

The average NH_4/K ratio remains below the value of 5.0 (with this value being exceeded in only 6 cases), while the average Al/Ca ratio is just above the value of 1.0. The low NH_4/K ratio is consistent with the relatively rapid nitrification of NH_4 to NO_3, and with the fact that the average quality of the top 30 cm of the soil has been taken into consideration. The NH_4/K ratio in the top 10 cm of the soil (where most of the fine roots occur) often has a higher value than that reported here (Thematic Report Van Breemen and Verstraten, Kleijn et al., 1989). In low vital and non-vital forests (in terms of needle loss and discolouration), the NH_4/K ratio in the top 10 cm of the soil profile is often above the critical value of 5. In forests within the vitality range of reasonably vital to vital, the NH4/K ratio is below the critical value (Roelofs et al., 1987). The relatively low Al/Ca ratio is mainly a result of the high Ca concentration. This is true not only for the topsoil but also for the subsoil, and is probably the result of either the high Ca input from the atmosphere or past liming/fertilization. The increased input can be explained by strong filtering of base cations by the forest canopy, whereby local input is much higher (probably about 2.5 to 3.5 times higher) than wet deposition. The high Na (and Cl) concentrations (not included in Tables 4.4 and 4.5) are also an indication of this (see also Chapter 6, section 3). The median values of the various elements and element ratios in the subsoil are, for all the 150 sites, very little different from the comparable values in the topsoil, except for the pH and the NH_4/NO_3 ratios, which in the subsoil are higher and lower respectively.

N content in needles

Analysis of the data from Speuld and Kootwijk reveals that, in the period of the research, the N content in six-month-old needles was at about the critical level for Douglas, at both sites (Speuld had an average of 1.80% in 1988 and 1.85% in 1989, and Kootwijk an average of 1.75% in 1988 and 1.80% in 1989), and that N accumulation is still occurring at both sites. Van den Burg and Kiewiet (1989) found that, in the period 1956 - 1988, the N content of the needles of Scots pine and Douglas fir in the Peel area increased to 1.8 - 2.5% (which is an optimum level from a nutritional point of view). A nation wide research programme revealed that the N content of the needles of Corsican pine, Scots pine and Douglas fir (sampled in December 1987) was not extremely high, given the N deposition (Houdijk, 1990). However, extreme situations were not examined.

Conclusions

The final conclusion is that in Dutch forest soils (which mainly have a naturally low pH level) there is always a predominance of NO_3 (NO_3 leaching) beneath the root zone, as a result of preferred uptake of NH_4 and/or nitrification. There is hardly any NH_4 leaching, because it is either taken up or nitrified. There is a possibility of NH_4 predominance in the litter layer, but in the rest of the root zone there is a strong predominance of NO_3.

The above conclusion also means that there is no reason to assume - in the calculation of the potential contribution of nitrogen to production of acid in the soil - that nitrification is incomplete. Given the large-scale occurrence of considerable nitrification, it is not right to make an adjustment for this in the calculation of the potential acid load for the total soil profile. The difference between potential acid load and actual acidification of the soil profile (= reduction of the buffer capacity) is mainly the result of the disappearance of N (and possibly S) through uptake, reduction processes (such as denitrification) and immobilization. As far as immobilization is concerned, however, one must realize that hereby an extra potential source of protons and nitrogen is created (which, as a result of the felling or dying of trees, for example, can be partly released again).

In about 15% of cases the NO_3/SO_4 ratio in the water that leaches below the root zone is greater than or equal to 1.0. This is an indication of an N-saturated system (insignificant immobilization). The average NO_3/SO_4 ratio is about 0.5 (see Table 4.5). This means that the average contribution of N to actual acidification is about 35%. The rest (65%) is SO_4. The uptake and retention of sulphate is generally insignificant, so that sulphate leaching is virtually the same as the sulphate load. This means that the potential acid load from sulphur is generally 100% achieved, in contrast to the potential acid load from nitrogen.

4.3.4 Expected long-term development of the composition of the soil solution in the event of unchanged deposition

In the event of unchanged deposition, N saturation at the ACIFORN sites will probably occur after a period of some decades, depending upon the organic matter content of the soil and the growth rate of the trees (see the Thematic Report by De Vries and Kros, and Chapter 6). Trends in the soil solution of a number of the earlier mentioned 16 stands indicate that the NO_3 content is increasing in stands which are still able to fix part of the atmospheric N input. Unless there are drastic emission reductions, it is to be expected that N leaching in forest ecosystems on dry sandy soils will increase, and that N-saturation situations will occur more often.

Mobilization of Al is currently the main buffer in the sandy forest soils in the Netherlands. But the Al buffer in the soil is limited in size. How long the Al buffer (in the root zone) will last depends upon the quantity of Al extractable from pyrophosphate or oxalate in the soil. The quantity of these is generally about the same. At current levels of deposition, the Al buffer in the top 10 cm of most sandy forest soils will last between 10 and 100 years (from 1990 onwards) (Thematic Report De Vries and Kros). When there is no longer an Al buffer, the pH of the soil solution in this zone may decline, which will cause the Al concentration to decline and the Fe concentration probably to rise. Furthermore, increasing P deficiency may occur as a result of the formation of iron phosphates along with phosphorus-aluminium compounds. As a result of the decline in the pH below the 3.0 level, the nitrification will decrease and the NH_4 concentration will increase. The pH of the litter will not decline, because it is buffered by organic acids. It will therefore be possible for nitrification to continue in the litter, and as a result of the acid hereby produced the pH of the mineral soil may decline to 2.8 to 2.9.

Model results indicate that, at current levels of deposition, the pH value of non-calcareous dune soils will decline considerably (by some whole units) in the coming decades, because these soils have hardly any substantial cation-exchange and Al-buffer capacity. As a result, a large percentage of the valuable plant species are being threatened (Thematic Report by De Vries and Kros; De Vries et al., 1991).

The above-mentioned changes in the soil chemistry will be difficult to reverse, because Al hydroxides dissolve in the top layer of the soil profile and (to a small extent) are deposited lower in the soil profile. Thereby, there is a permanent change in the root zone. In theory, the buffer capacity and the base availability can be increased by liming and fertilization. However, this involves a danger of increased mineralization and N leaching (Denier van der Gon, 1990).

These developments are irreversible, unless the deposition of acidifying substances is considerably reduced in the coming years. In this connection, scenarios have been developed (partly extrapolations of the current situation with regard to acid load, and partly already implemented policy), the results of which, in relation to the soil chemistry and so forth, are presented in Chapter 6.

What in any case appears clear from field experiments (Van Dijk et al., 1991) and calculations with soil models (Van Grinsven et al., 1989, Thematic Report De Vries and Kros, Chapter 6, section 3) is that reductions in deposition lead, almost immediately, to a reduction in the SO_4 and Al concentrations in the soil and thus to an improvement in the quality of the soil solution. The results of model calculations show a relatively delayed reaction of NO_3 to a reduction in the deposition compared with that of SO_4 and Al, which can be attributed to N mobilization from the litter layer. Interim results from current experiments suggest that reduction of the deposition results in a reduced N content of the needles within one year (Van Dijk et al., 1991).

4.3.5 Observed effects on Dutch forests and future expectations (in the event of unchanged deposition)

4.3.5.1 Influence on growth and vegetation change

In terrestrial ecosystems in which N is the growth-limiting factor, the input of extra N will lead primarily to increased production. Further increase in the N load will lead to a change in the composition of these ecosystems (Ellenberg, 1985). Species that have adapted to a nitrogen-poor environment will be replaced by nitrogen-loving species.

An increase in the availability of N strongly stimulates the growth of nitrogen-loving grasses and herbaceous plants, particularly in forests where the canopy is open as a result of needle loss. Grass species such as Molinia caerulea and Deschampsia flexuosa are not only crowding out heathland vegetation in the Netherlands (see Chapter 5), but are also increasingly to be found in forests (Van Breemen and Van Dijk, 1988, Hommel et al., 1990). It should be remarked here, however, that dominance of grasses can be a natural phase in the succession of young forest ecosystems on sandy soils. On the basis of long-running experiments with high nitrogen doses, it is to be expected that this increased growth of nitrogen-loving plants will continue (in the event of unchanged N deposition) (Van Dobben and Dirkse, 1989).

4.3.5.2 Sensitivity to drought, nutrient deficiency, frost, diseases and pests

Besides changes in the diversity of species in vegetations, it is to be expected that, as a secondary effect of extra input of nitrogen, there will be a greater possibility of damage and dieback from drought, frost, diseases and pests.

No extreme nutrient deficiencies have so far been observed at either of the ACIFORN

monitoring sites, Speuld and Kootwijk, and the growth rates are high. However, root density is not high at either stand. A further decline in this respect, because of a shift in the root/shoot ratio as a result of increased N availability or on account of soil acidification and Al effects (see section 4.3.2), will reinforce the effect of drought and/or nutrient deficiency, because the total absorption capacity is reduced (Arp and Stricker, 1989). Most of the Dutch forests are on poorer and often drought-sensitive soils and are generally subject to higher levels of deposition (via throughfall) than Speuld and Kootwijk (Thematic Report Van Aalst and Erisman). As a result they are more sensitive to water stress and nutrient deficiency (in comparison with the ACIFORN sites).
It is not possible to precisely specify the consequences for the vegetation of a long-term decline in pH (to between 2.8 and 2.9) associated with Al depletion, which, in the event of unchanged deposition, is the expectation for Dutch forest soils (see Chapter 4, section 3.4). There may be a P deficiency, as a result of the formation of iron phosphates and phosphorus-aluminium compounds (Kottke and Oberwinkler, 1990).

4.3.5.3 Effects on development and functioning of mycorrhizas
In forests, ectomycorrhizas are very important for the uptake of nutrients (Dighton, 1990), for the nutrient cycle (Fogel & Hunt, 1983) and probably also for trees' resistance to disease. It is possible that mycorrhiza infestations protect the roots and the tree against aluminium. This has acquired plausibility through the work of Kamminga - Van Wijk (see Thematic Report by Posthumus and Jansen) and the work of Kottke and Oberwinkler (1990).
In the last decades there has been a strong decline in the population of, and in the number of species of ectomycorrhizal fungi in Dutch forests. This is probably a result of the increase in acidifying deposition, according to Arnolds (1988). Other authors, for example in Germany and Czechoslovakia, also make this connection (see Jansen and Dighton, 1990). Apart from the toadstools pertaining to these fungi, also the fungi in the soil itself have declined. In pot experiments it has been observed that pH reduction leads to a shift in species (Jansen and Dighton, 1990). Species with a narrow pH range will be particularly sensitive to acidification.
Various researchers (for example, Boxman et al., 1986) have found that high NH_4 concentrations lead to a reduction in the radial growth of mycorrhiza-forming fungi. Al and heavy metals have the same effect (Boxman and Van Dijk, 1988). Pot experiments (with Douglas fir) reveal an optimum development of mycorrhizas when there is an optimum supply of N. Mycorrhiza development was greater after fertilization with 50 kgN $ha^{-1}yr^{-1}$ than after fertilization with 5 kgN $ha^{-1}yr^{-1}$. Application of 200 kgN $ha^{-1}yr^{-1}$ resulted in a strong decline in development in relation to the 50 kg application (Gorissen,

Jansen & Olsthoorn, in preparation; Meyer, 1986). Pot experiments with Corsican pine (Van Dijk et al., 1990), however, revealed the highest mycorrhiza percentages and numbers of fruit bodies in the case of no application of N or a very low application. But pot experiments cannot be automatically translated to the field situation. Theoretically, there will be an optimum supply of N for optimum mycorrhiza growth in the field situation also. However, the supply need not be the same for all mycorrhizal fungi. Field observations indicate that this optimum for mycorrhizal fungi in nutrient-poor sandy soils in the Netherlands is lower than 50 kgN $ha^{-1}yr^{-1}$ (see Thematic Report by Posthumus and Jansen). According to Jansen, mycorrhizas in pioneer forests (young plantations) usually have rapidly-growing fungi which are highly resistant to extreme environmental circumstances and which are adapted to litter and a humus-poor substrate. During the development of the forest to a young medium and final forest stage, in undisturbed situations ectomycorrhizal fungi which are adapted to a more stable forest, with a thicker litter and humus layer, increase in proportion. In the Netherlands, however, it is precisely the mycorrhizal fungi of more mature forests that appear to have declined considerably (Thematic Report by Posthumus and Jansen, Termorshuizen, 1990, Jansen & Dighton, 1990).

It can generally be concluded that in the field situation a negative correlation has been signalled between air pollution and all the mycorrhiza parameters (such as number of species, number of fruit bodies and degree of infestation). The mechanism involved, however, is still not completely clear. Reduction of pH in combination with strong nitrogen enrichment would seem to be the most important causes (Jansen & Dighton, 1990). In general, quantified statements concerning the effects of pH, aluminium and nitrogen on the occurrence and functioning of mycorrhizas in forests are not (yet) possible. The effects of an impoverished mycorrhiza population, with a different composition of species, on the nutrient supply of trees can not yet be quantified, either.

4.3.5.4 Effects on soil organisms

In the Dutch Priority Programme on Acidification no research has been carried out on soil fauna. In contrast to the process research in the context of the acidification programme, in which research was carried out on Douglas fir forests, research on soil fauna in the Netherlands has been mainly carried out in Scots pine forests (Pinus sylvestris). Results from this research have shed some light on the effects of acidification and nitrogen deposition.

In all probability, the microfauna plays a key role in decomposition and mineralization processes (Schouten and Van der Brugge, 1989). It appears that in soils which are sensitive to acidification there is a shift in species, and that, in general, the diversity and density of population of soil organisms declines (Denneman et al., 1986; Van Straalen et al., 1988). An established secondary effect of this is the decrease in the decomposition of organic material, as a result of the disappearance of the species needed for this, such as bacteria and the earthworm (Manuel et al., 1984).

Free-living nematodes (eelworms) are found in large populations and with a rich variety of species in virtually every soil type. Their precise function in ecosystems is still very unclear. With regard to the direct effects of pH and Al on free-living nematodes, it appears from laboratory experiments that Al has a flat concentration-effect curve and that the effects of H^+ are acute and occur within a small concentration range. It seems that many species of nematode have already reached the limit of their tolerance capacity (Schouten and Van der Brugge, 1989). The results of these experiments indicate that, if soil acidification continues unabated, it will have serious consequences for the survival chances of nematodes in acid forest soils.

4.3.5.5 Effects on the groundwater under forests

Some results of the sampling of shallow groundwater, as carried out by the RIVM at 150 sites, are given in Table 4.6 and Figures 4.7 and 4.8. This research showed that the nitrate content was higher than the drinking water standard in almost 30% of the coniferous forest sites investigated. In the case of deciduous forests this was true in 13% of the sites (Boumans et al., 1990). The NO_3 standard in drinking water is 50 mg l^{-1} (0.8 mmol.l^{-1}). It is true that in the subsoil (on the way to pumping wells) denitrification can occur, but where (under what conditions) and to what extent that occurs is still not known.

In addition, there are signs that Al will become a problem for the drinking water supply (Mulder et al., 1990), particularly in the case of shallow, private wells. The Al content of the shallow groundwater at the above-mentioned 150 sites is often above the drinking water standard (7 μmol.l^{-1}), as is illustrated by Figure 4.8. This is true for almost 90% of the coniferous forest sites investigated and for 70% of the deciduous forest sites. But it must be mentioned here that these data provide an indication of the water quality trend of the shallow groundwater under forest areas. The water that is withdrawn at a pumping site comes from a particular area around that site, called the recharge area, where, in general, various forms of soil use are to be found, as a result of which there will be a mixing of various types of water. In addition, there is an important retention of Al in the deep subsoil (Mulder et al., 1990). The composition of the water pumped up reflects the diversity in land use and, also, the characteristics of that part of the subsoil that the water has passed on its way to the wells.

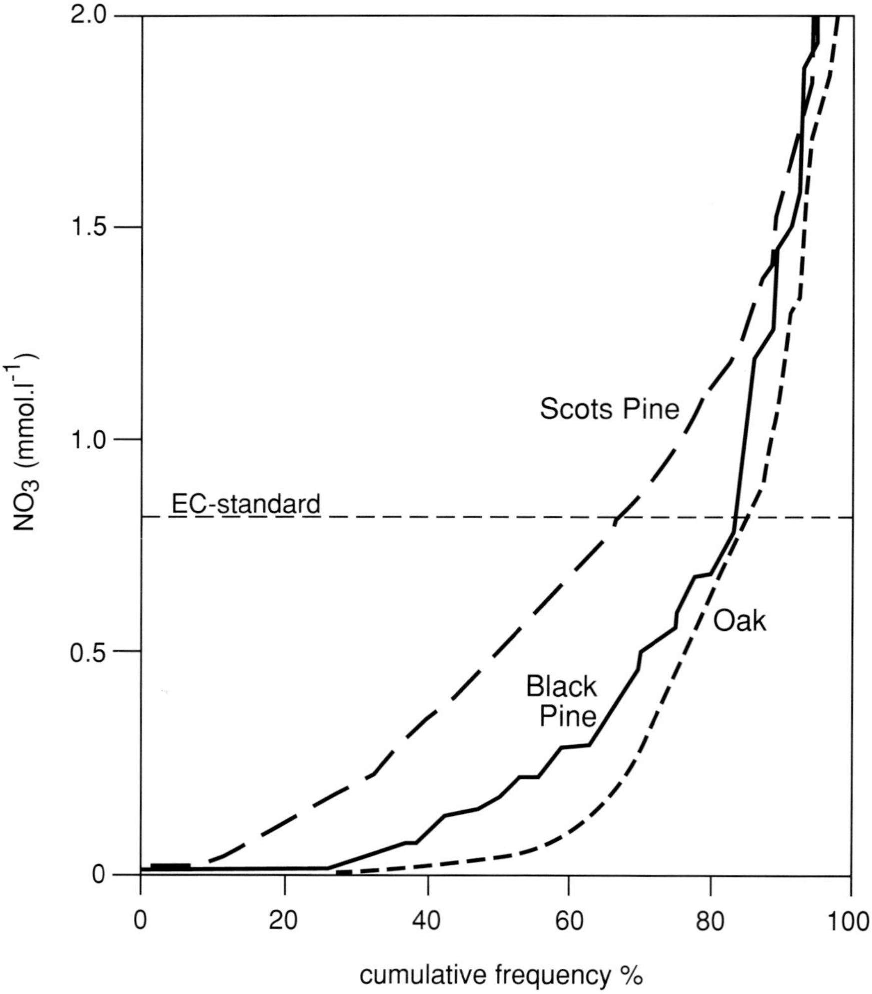

Fig. 4.7 NO_3 content of shallow groundwater for various tree species at 150 sites

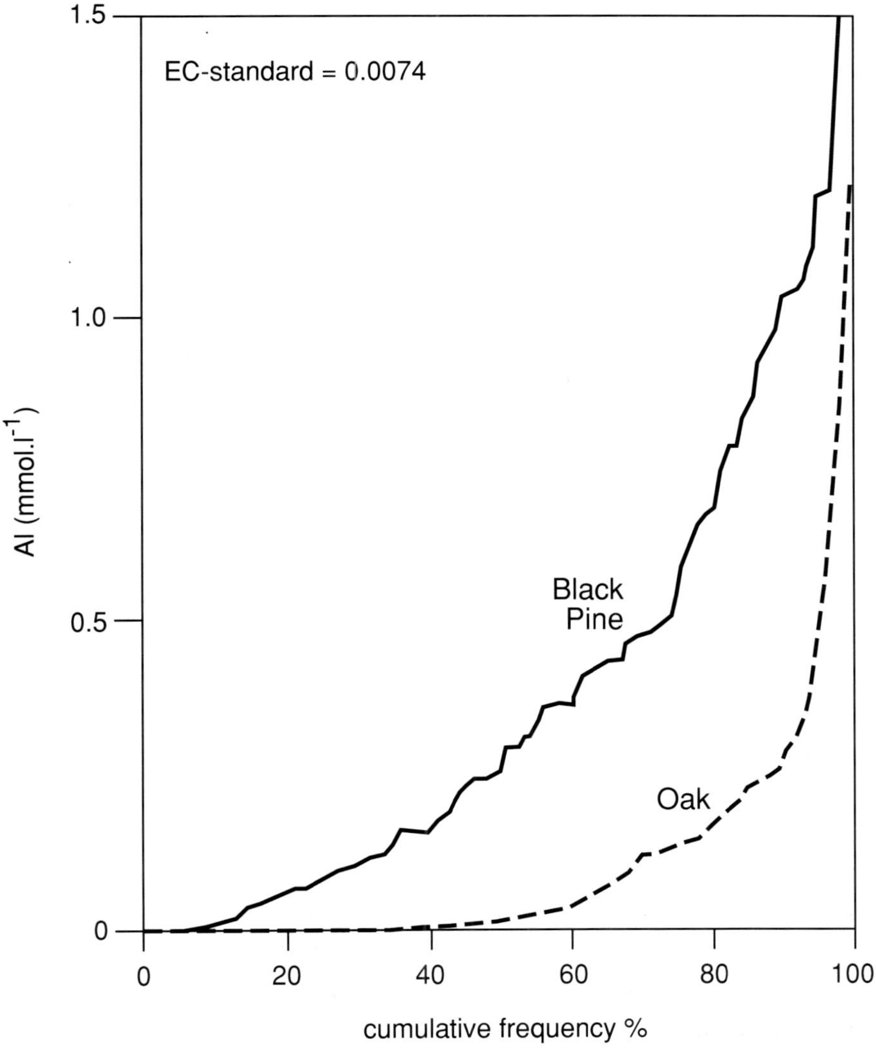

Fig. 4.8 Al content of shallow groundwater for various tree species at 150 sites

Table 4.6 Median values for the pH and the concentrations ($mol_c.m^{-3}$) of Al, SO_4 and NO_3 of shallow groundwater, for various tree species

Tree species	pH	Al	SO_4	NO_3
	$mol_c.m^{-3}$			
Scots pine	4.3	0.28	0.48	0.50
Corsican pine	4.2	0.24	0.46	0.16
Douglas fir	4.2	0.27	0.45	0.48
Norway spruce	4.3	0.17	0.51	0.12
Larch	4.3	0.16	0.41	0.20
Oak	4.8	0.02	0.55	0.02

Finally, it can also be stated that, in the event of the large-scale disappearance of above ground parts of forest ecosystems (for example, as a result of large-scale felling, diseases, etc.), an increased input of nitrogen to the groundwater is to be expected, on account of a large change in the nitrogen uptake. The same effect may manifest itself in the event of a temperature rise and a change in the pattern of rainfall, as a result of a climatic change. In that case the equilibrium will be disturbed, as a result of the fact that the mineralization flux may be greater than the N uptake. The pH level may play a role here.

4.4 Integrated effects of air pollution

In this section, the combined effects of direct and indirect influences on the level of the physiological processes of sap velocity and growth are discussed. These effects have been analysed for individual trees through experimental research in the laboratory and in field situations. The effects of air pollution at the level of physiological and biochemical basic processes, have been translated into the consequences for forest stand growth with the aid of the forest-stand model FORGRO.

4.4.1 Effects on the growth and sap flow velocity of individual trees

Because no direct relationship can be established between acid deposition and features of vitality, as employed in the vitality survey, a search was made within the Dutch Priority Programme on Acidification for possible effects of air pollution on other health-related aspects: (secondary)growth, analysis of annual rings and sap flow velocity. The results are discussed in this section.

As a result of historical development in land use, Dutch forests are often situated on poor soils. Nutrient deficiency (K, Mg, P) and water stress in dry periods are quite common on the poorest sites. Therefore an influence of air pollution on basal area growth is very

difficult to detect and not easy to separate from other factors.
Analysis of 2700 tree cores from 11 forest stands only resulted in indications of pollution stress for one stand with Douglas fir and Norway spruce. Excessive spreading of manure in the direct vicinity could have contributed to the phenomena observed at this stand. Results for other stands were more or less diffuse. All Scots pine stands and one oak stand showed an increased sensitivity (with respect to wood production per year) to total rainfall during the growing season. This phenomenon has arisen since the sixties. It rather points to drying out of top soils than to effects of air pollution. Generally, the results for Scots pine an oak do not show clear effects of air pollution. However, this does not imply that effects of air pollutants, be it above ground or below ground, are completely absent.
Studies on the year ring increment, wood production and water transport capacity of the Douglas fir trees at the ACIFORN sites Kootwijk and Speuld, revealed no significant differences between the trees of the two sites. No recent decline in radial, axial or volume growth was shown and a normal amount of sapwood was present, with only slightly sub-maximal moisture content.
Research on xylem sap velocity of 12 Douglas fir trees at three different plots at the ACIFORN site Speuld has given only relative values: velocities have not been converted to quantities as this is not required for correlation analysis. Between the plots, much variation occurred and no significant difference between the plots could be detected. In each plot, most variation corresponded to variation in vapour pressure deficit, which determines to a large extent the transpiration rate of a forest stand. No correlation with pollution variables such as O_3 and SO_2 could be found.
In conclusion it must be said that effects of air pollution on growth and sap flow velocity, in so far as they occur, appear mainly concealed by combination stress. Therefore, they are difficult to detect.

4.4.2 Modelling of forest stand growth

Overall reduction in forest growth, resulting from air pollution and acid deposition, can be due to disturbances such as reduced primary production, an increase in the cost of repairs, or changes in assimilate allocation, and combinations thereof. It may result from a reduction in CO_2 assimilation or increased ageing and turnover of the foliage and root biomass. These in turn can be caused by either direct damage, by nutrient and water stress or by combinations of these. Nutrient stress caused by soil acidification should be considered as an indirect effect of air pollution.
Simulation models are essential aids in the quantification of the consequences of air pollution for the growth and development of forests. Controlled experiments on the complete "forest" system are only possible on a limited scale and for a relatively short

period of time, and not all the relevant factors can be completely controlled. It therefore seems more obvious to study effects of air pollution at the level of physiological and biochemical basic processes, and subsequently to integrate these descriptions into a more comprehensive model that will allow a translation to be made to the level of the whole tree or the whole forest stand. In this way, the model functions as a bridge between the experiments and the system. It is along these lines that, within the context of the Dutch Priority Programme on Acidification, the forest-stand model FORGRO has been further developed and applied to both the ACIFORN monitoring sites, in order to quantify the influence of air pollution and soil acidification on forest growth, and in order to relate this influence to the influence of the traditional growth-site factors such as climate and soil. In this way, a model is used as a research instrument, whereby insights can be tested and information is obtained concerning the essential links in the system.
To achieve a quantitative estimate of the magnitude of the reduction of forest growth due to traditional factors versus the reduction due to air pollution and soil acidification, the carbon-balance model FORGRO was extended to include effects of air pollution and soil acidification. The simulation results indicate that direct short-term effects of air pollutants, notably SO_2 and O_3, are virtually absent at the ACIFORN sites. From the calculations, it was concluded that total stomatal uptake of pollutants is rather small; as a result the short-term direct effects are rather limited in scope. However, the possibility of acute damage under episodic high concentrations remains, especially when buffering capacity of the trees decreases, for example owing to ammonium nutrition. Gaseous uptake of NO_2 may have the same effect (Lange et al., 1989). When only short-term direct effects of air pollutants on photosynthesis and respiration are studied, the resulting growth reduction appears to be negligible for pollution conditions such as determined in Speuld and Kootwijk in 1988 and 1989; this is confirmed in the field studies.
Long-term effects have not yet been fully incorporated in the model, as no data were available to calibrate a dose-response relationship. Preliminary analysis of possible long-term direct effects on stand growth, however, indicated that long-term effects of ozone may very well be of importance under conditions in the Netherlands, and should be investigated further. When expressed in terms of percentage reduction of dry matter increment, model calculations indicated a possible loss of some 3-5% in the case of medium to high LAI under ambient ozone. At lower LAI, these tentative results indicate a somewhat larger effect, possibly around 10-20% loss. These values are in consistent with estimates reporter earlier (Van der Eerden et al., 1989; Tonneijck, 1989).
The nutrient supply as estimated for Speuld and Kootwijk, using the soil acidification model RESAM in combination with a carbon balance model to simulate nutrient uptake, was sufficient to maintain the nutrient concentrations in the biomass at the present level.

For this purpose, nutrient uptake was estimated from nutrient demand by the plant and nutrient supply expressed in terms of concentrations in the bulk soil solution, according to De Willigen & Van Noordwijk (1987). Calcium and magnesium uptake depends on the number of actively growing root tips, and is sensitive to high concentrations of aluminium and NH_4^+.

At both the ACIFORN sites, the major growth-influencing factors that determine the actual growth levels appear to be the availability of water and mineral nutrients. No extreme nutrient deficiencies have occurred so far on either site, and growth rates are high. Nutrient availability interferes with water availability to the extent that sufficient water must be available to facilitate nutrient uptake from the soil. In dry years, limitations in water supply overrule nutrients as the main growth-limiting factor. Root density in both stands is low, and a further decrease, either through a shift in root/shoot ratio due to increased nitrogen availability, or through soil acidification and aluminium toxicity, will increase the effect of low water and nutrient availability. On the ACIFORN monitoring sites, this has so far not resulted in mineral deficiencies. The majority of the forested area is located on sites of less fertility and mostly with water deficiency with in general a higher atmospheric input (throughfall) than at Speuld and Kootwijk, which renders them more susceptible to the disturbances described above.

4.5 Conclusions

With the possible exception of ozone, direct short-term effects owing to uptake of gaseous air pollutants appear to be of minor importance, except for situations near local sources and during episodes with high concentrations, where effects on photosynthesis and even visible damage may occur (NH_3, SO_2 and O_3). Yet, long-term (more than one year) chronic effects of uptake of air pollutants occurring in low concentrations are expected to be more important, although the effects are not known and the mechanisms not understood at all. It is suspected and indicated by preliminary data that long-term effects of chronic exposure to low concentrations will lead to increased ageing of the foliage, and to increased foliage loss. Present concentration levels of ozone have reached values that may well result in damage to the interior of the foliage, necessitating additional photosynthesis products for repair. Average levels of air pollution measured at the ACIFORN sites correspond well with results from the national monitoring network; the exposure to ozone and the expected damage applies to most of the forested area.

As a result of the widespread high nitrogen deposition, most forest ecosystems are in transition from a nitrogen-deficient state to a nitrogen-saturated state. Concurrently, the nitrogen input has caused considerable soil acidification, together with the deposition of SO_2. The interplay between increased nitrogen availability and soil acidification has led to increased aluminium concentrations in the rooted soil profile. At present, the soil solution in the main part of the rooted soil profile (the topsoil, with most of the fine roots that take up water and nutrients) of the majority of forests in the Netherlands is characterized by Al/Ca ratios around a critical value of 1 below the litter layer and further downwards. The NH_4/K ratio is high in the litter layer, approaching critical values of 5, but decreases with increasing depth owing to high nitrification in the top soil. As a result of nitrification, the NH_4/NO_3 ratio is on average below 1 immediately below the first centimeters of mineral soil, implying that NO_3^- is dominant over NH_4^+. Consequently, the majority of nitrogen uptake by the trees is still in the form of NO_3^- rather than NH_4^+. Rhizosphere acidification increases when nitrogen is taken up as NH_4^+ rather than NO_3^-, enhancing the toxic effects of aluminium on base cation uptake.

Major concerns are the limited uptake of nutrients due to adverse effects of Al on roots and the depletion of the aluminium buffer. As a consequence, pH can be expected to decrease, with decreasing buffering capacity within the aluminium range. Without deposition reductions, it is expected that major changes in soil conditions for plant growth will occur in the forseeable future. Also, at ongoing deposition and soil acidification at the present rate, groundwater quality will further deteriorate. Nutrient imbalances resulting from high nitrogen deposition have increased the vulnerability to frost, to insects and pathogen damage, as well as the susceptibility to gaseous air pollutants. Without deposition reductions, a permanently increased risk of damage from pests and diseases, frost and drought will occur, owing to an increased contribution of soil acidification to the combined effect of both traditional factors, gaseous air pollution, and mineral deficiencies.

In addition, the accumulated amount of nitrogen within the ecosystem, as a result from ongoing high deposition, poses a future threat to the groundwater system as the organically bound nitrogen may be released in the event of increasing mineralization (such as through gradual temperature rise) or by a decrease in uptake in case of forest harvesting, or in case of forest damage and declining growth rates.

Considering the possible interplay between nutrient status and the metabolic detoxification capacity of the foliage which it determines, the combination of direct effects (SO_2, NH_3 and NO_2) and indirect effects (soil acidification and NH_4^+ versus NO_3^- uptake) may lead to internal acidification and disruption of the internal pH regulation of the plant, leading to foliar damage and increased litter loss. As a result, the vulnerability to long-term effects may increase with increasing nutrient imbalance, and this aspect requires an integrated assessment to achieve a realistic long-term evaluation of damage risks.

4.6 Conclusions from acidification research programmes in other countries

Findings from acidification research programmes in other countries are more or less identical, as appeared at the fifth Meeting of Acid Rain (Research) Coordinators (MARC V), held in France, September 1990. From this meeting, with representatives from 10 countries (Federal Republic of Germany, Switzerland, France, Norway, Netherlands, Sweden, Finland, United Kingdom, United States, Austria), the following conclusions concerning the contribution of air pollution to forest damage have been reported:

- Obviously, there are far fewer discrepancies in the perception of forest decline in the different countries than there were 5 years ago. Ground monitoring of visual damage has shown a general, uniform evolution over the past few years. The idea of steadily increasing damage could not be scientifically evidenced. Many of the earlier statements such as "50% of the forest is damaged" or "forests will die within 5 years" proved to be unfounded. Visible symptoms (defoliation, yellowing) are not specific (this is widely acknowledged) and not necessarily associated with air pollution.
- The idea that forest decline is due to a complex set of factors is accepted by every country. The remaining difference seems to be between countries considering that "air pollution is only one of the causes but is a necessary ingredient to a greater or lesser extent" and those considering that "some features of decline or the damage observed in some areas have probably nothing to do with air pollution".
- Altogether, the direct effects of gaseous pollutants on leaves and needles are probably less important than first thought. The role of ozone is not clear for major species like spruce ("no effect" or "some effect"). In contrast, the indirect effects on the forest ecosystem (not only trees) may be more important. A consensus has nearly been reached on the role of acidic deposition in cation depletion and the experts of many countries now believe that nutritional imbalances, if not already present, could become a major mid-term problem. The importance of soil as a basic requirement for internal resources is widely recognized.
- The analysis is not relevant under the current conditions in some areas in the eastern European countries, such as the Erz mountains, where the direct effects are very likely to be more important.

Literature

Anonymus, 1985 - 1990
De vitaliteit van het Nederlandse bos
Ministerie van Landbouw, Natuurbeheer en Visserij

Arnolds E., 1988
The changing macromycete flora in the Netherlands
Trans.Br.mycol.Soc. 90: 391 - 406

Aronsson A., 1980
Frost hardiness in Scots pine
II Hardiness during winter and spring in young trees of different mineral nutrient status
Studia Forestalia Suecica 155: 1 - 27

Arp P.A. and Stricker I., 1989
Water uptake by black spruce seedlings from rooting media (solution, sand and peat) treated with inorganic and oxalated aluminium
Water, Air and Soil Pollution, 44; 57 - 70

Boumans L.J.M. and Beltman W., 1990
Kwaliteit van het bovenste freatische grondwater in zandgebieden van Nederland, onder bos- en heidevelden
RIVM rapport (in prep.)

Boxman A.W., Sinke R.J. and Roelofs J.G.M., 1986
Effects of ammonium on the growth and K^+ (^{86}Rb) uptake of various ectomycorrhizal fungi in pure culture
Water, Air, Soil Pollut., 32, 517 - 522

Boxman W.A. and Dijk H.F.G.van, 1988
Het effect van landbouwammonium-deposities op bos- en heidevegetaties, pp. 1-96
Vakgroep Aquatische Oecologie en Biogeologie, KUN

Boxman A.W., Krabbendam H., Bellemakers M.J.S. and Roelofs J.G.M., 1991
Effects of ammonium and aluminium on development and nutrition of Pinus nigra in hydroculture
Environmental Pollution
(in press)

Breemen N.van and Dijk H.F.G.van, 1988
Ecosystem effects of atmospheric deposition of nitrogen in The Netherlands
Environmental Pollution 54 (1988) 249 - 274

Burg J.van den, 1981
pH en boomgroei; een literatuuronderzoek
Rapport nr. 282, De Dorschkamp

Burg J.van den, Evers P.W., Martakis G.F.P., Relou J.P.M. and Werf D.C.van der, 1988
De conditie en de minerale voedingstoestand van opstanden van grove den (Pinus silvestris) en Corsicaanse den (Pinus nigra var.Maritima) in de Peel en op de zuidoostelijke Veluwe, najaar 1986
Instituut voor Onderzoek in de Bos- en Landschapsbouw, "De Dorschkamp", Rapport nr. 519

Burg J.van den, 1988
Voorlopige criteria voor de beoordeling van de minerale voedingstoestand van naaldbomen op basis van de naaldsamenstelling in het najaar
Rijksinstituut voor Onderzoek in de Bos- en Landschapsbouw "De Dorschkamp", report 522, Wageningen, 20 pp.

Burg J.van den and Kiewit H.P., 1989
Veebezetting en de naaldsamenstelling van grove den, Douglas en Corsicaanse den in het Peelgebied in de periode 1956 tot en met 1988
De Dorschkamp, rapport nr. 559

Darral N.M., 1989
The effect of air pollutants on physiological processes in plants
Plant, Cell and Environment 12, 1 - 30

Denier van der Gon H.A.C., 1990
De samenstelling van bodemvocht en grondwater onder de bosbekalking- en bosbemestingsproefvelden te Harderwijk en St.Anthonis
Rapport nr. 724903003, RIVM

Denneman C.A.J., Kraak M.H.S. and Straalen N.M.van, 1986
Bodemdieren als indicator van de vitaliteit van dennebossen op de Veluwe
Nederlands Bosbouwtijdschrift 58: 204 - 209

Dighton J., 1990
Acquisition of nutrients from organic resources by mycorrhizal autotrophic plants
Experientia
(in press)

Dijk H.F.G.van, Louw M.H.J.de, Roelofs J.G.M. and Verburgh J.J., 1990
Impact of artificial ammonium-enriched rainwater on soils and young coniferous trees in a greenhouse, part II: Effects on the trees
Environmental Pollution 63 (1990), 41 - 59

Dijk H.F.G.van, Boxman A.W., Roelofs J.G.M., 1991
Effects of a decrease in atmospheric deposition of S and N on the mineral balance and vitality of a Scots pine stand in the Netherlands
For.Ecol.Manage, 1991 (in press)

Dobben H.van and Dirkse G.M., 1989
Effects of experimental fertilization on forest undergrowth in northern Sweden
Acta Botanica Neerlandica 38, 389 - 390

Dueck Th.A., Dorel F.G., Horst R.ter and Eerden L.J.van der, 1990
Effect of ammonia, ammonium sulphate and sulphur dioxide on the frost sensitivity of Scots pine (Pinus sylvestris, L.)
Water, Air and Soil Pollution (in press)

Eerden L.J.van der, Lekkerkerk L., Smeulders S.M. and Jansen A., 1989
Effects of ozone and ammonia on Douglas fir (Pseudotsuga menziesii)
In: Bucker H. and Bucher Wallin, I. (Eds.), Proc.Congr.Air Pollution and Forest Decline, IUFRON P 2.05, Interlaken, Oct. 1988, 538 - 540

Ellenberg H., 1985
Veränderungen der Flora Mitteleuropas unter dem Einfluss von Dungung und Immissionen
Schweiz.Z.Forstwes. 135: 19 - 39

Fogel R.and Hunt G., 1983
Contribution of mycorrhizae and soil fungi to nutrient cycling in a Douglas fir ecosystem
Can.J.For.Res. 13: 219 - 232

Goor C.P.van, 1967
Bemestingsvoorschrift voor naaldhoutculturen
2e, herz.druk, Korte Mededeling Bosbouwproefstation "De Dorschkamp", Wageningen nr. 56

Gorissen A. and Veen J.A.van, 1988
Temporary disturbance of translocation of assimilates in Douglas firs caused by low levels of ozone and sulphur dioxide
Plant Physiol., 88, 559 - 563

Gorissen A., Schelling G. and Veen J.A.van, 1991a
Concentration dependent effects of ozone on translocation of assimilates in Douglas fir
(in press)

Gorissen A., Joosten N.N., Smeulders S.M. and Veen J.A.van, 1991b
Effects of ozone on carbon economy of Douglas fir, Part I: Juvenile Douglas firs at two soil moisture levels
(in press)

Gorissen T., Jansen A.E. and Olsthoorn A.F.M.
Influence of ammonium sulphate on growth of juvenile Douglas fir, mycorrhizal frequency and bacteria in the rhizosphere
(in prep.)

Grinsven J.J.M.van, Kros J., Breemen N.van, Riemsdijk W.H.van and Eek E.van, 1989
Simulated response of an acid forest soil to acid deposition and mitigation measures
Neth.Journal of Agricultural Science 37 (1989) 279 - 299

Hommel P.W.F.M., Leeters E.E.J.M., Mekking P. and Vrielink J.G., 1990
Vegetation changes in the Speulderbos (the Netherlands) during the period 1958 - 1988
Winand Staring Centre for Integrated Land, Soil and Water Research, report nr. 23

Houdijk A.L.F.M., 1990
Effecten van zwavel- en stikstofdepositie op bos- en heidevegetaties
Rapport Vakgroep voor Aquatische Oecologie en Biogeologie, Katholieke Universiteit Nijmegen

Hove L.van, 1989
The mechanism of NH_3 and SO_2 uptake by leaves and its physiological effects
Diss.University Wageningen, 153 pp.

Jansen A.E. and Dighton J., 1990
Effects of air pollutants on ectomycorrhizas; A review
Air Pollution Research Report 25, CEC Brussels

Kaupenjohann M., Dohler H. and Bauer M., 1989
Effects of N immissions on nutrient status and vitality of Pinus sylvestris near a hen-house
Plant and Soil 113, 279 - 282

Keltjens W.G., 1990
Effects of aluminium on growth and nutrient status of Douglas-fir seedlings grown in culture solution
Tree Physiology 6: 165 - 175

Kleijn G.E., Zuidema G. and Vries W.de, 1989
De indirecte effecten van atmosferische depositie op de vitaliteit van Nederlandse bossen
2. De bodemvochtsamenstelling van 8 Douglasopstanden
Stichting voor Bodemkartering, Wageningen, rapport no. 2050

Kropff M.J., Smeets W., Meijer E. and Zalm A.J.A.van der, 1990
Effects of sulphur dioxide on photosynthesis. The role of temperature and humidity
Physiol.Plant (in press)

Kottke I. and Oberwinkler F., 1990
Pathways of elements in ectomycorrhizae in respect to Hartig net development and endodermis differentiation
In: A.Reisinger and A.Bresinsky, Abstracts of the Fourth International Mycological Congress, 28 Aug. - 3 Sept. 1990, Regensburg: 81

Kropff M.J., 1990
Long-term effects of sulphur dioxide on plants, SO_2 metabolism and regulation of intracellular pH
Plant and Soil (in press)

Küppers K. and Klumpp G., 1988
Effects of ozone, sulfur dioxide, and nitrogen dioxide on gas exchange and starch economy in Norway Spruce (Picea abies (L.) Karsten)
Geo.Journal 17.2, 271 - 275

Lange O.L., Heber U., Schulze E.D. and Ziegler H., 1989
Atmospheric pollutants and plant metabolism
In: E.D.Schulze, O.L.Lange and R.Oren (eds.); Forest decline and air pollution. Ecological Studies 77, Springer Verlag, Berlin, pp. 238 - 276

Manuel A.R., Aalst R.M.van, Bastiaans H., Bresser A.H.M., Don A. and Zoetelief J., 1984
Verzuring door atmosferische depositie
Evaluatierapport, Publicatiereeks Milieubeheer, rapport nr. VROM 836551/1-84

Meyer F.H., 1986
Root and mycorrhizal development in declining forests
In: Indirect effects of air pollution on forest trees: Root-Rhizosphere Interactions
Proc.Int. EEC Workshop, Jülich, FRG, pp. 139 - 151

Meyer J., Oren R., Werk K.S. and Schulze E.D., 1986
The effect of acid rain on forest tree roots
In: Proceedings of the Jülich COST workshop, Commission of the European Communities 16 - 30

Meyer J., Schneider B.U., Werk K., Oren R. and Schulze E.D., 1987
Performance of Picea abies (L.) Karst. at different stages of decline. V.Root tips and ectomycorrhiza development and their relation to above-ground and soil nutrients
Oecologia (in press)

Mooi J., 1985
Wirkungen von SO_2, NO_2, O_3 und ihrer Mischungen auf Pappeln und einige andere Pflanzenarten
Die Holzzucht 39, 8 - 12

Mulder J., Beek C.G.M.van and Dierx H.A.L., 1990
Acidification of groundwater in forested sandy deposits in The Netherlands due to acid atmospheric deposition
Report SWE 90-015, KIWA

Nihlgard B., 1985
The ammonium hypothesis: An additional explanation to the forest dieback in Europe
Ambio 14, 2 - 8

Posthumus A.C. and Tonneijck A.E.G., 1982
Monitoring of effects of photo-oxidants on plants
In: Steubing L. and Jäger H.J. (eds.), Monitoring of air pollutants by plants, Dr.W.Junk Publishers, The Hague, 115 - 119

Reich P.B. and Amundson R.G., 1985
Ambient levels of ozone reduce net photosynthesis in tree and crop species
Science 230, 566 - 570

Reich, P.B., 1987
Quantifying plant response to ozone: a unifying theory
Tree Physiology 3, 63 - 91

Report of the 5th Meeting of Acidification Research Coordinators
Champenoux/Nancy, September 25 - 27, 1990

Richardson C., Sasek T., Fendick E., Bevington S. and Kress L., 1990
Carry-over effects of acid rain and ozone on the physiology of multiple flushes of Loblolly pine (Pinus taeda L.) seedlings
Conference Abstracts, International Conference on Acidic Deposition, Its Nature and Impacts, Glasgow 16 - 21 September 1990, p. 179

Roelofs J.G.M., Kempers A.J., Houdijk A.L.F.M. and Jansen J., 1985
The effect of airborne ammonium sulphate on Pinus nigra var.maritima in the Netherlands
Plant and Soil 84: 45 - 56

Roelofs J.G.M. and Boxman A.W., 1986
The effects of airborne ammonium sulphate on pine forests
In: Wissenschaftliches Symposium "Neue Ursachenhypothesen", 16. und 17. Dezember 1985, Berlin, Texte, Umweltbundes Amt: 280 - 283

Roelofs J.G.M., Boxman A.W. and Dijk H.F.G.van, 1987
Effects of airborne ammonium on natural vegetation and forest
Ammonia and acidification. Eurasap symposium, Bilthoven 1987, 266 - 276

Rost-Siebert K., 1985
Untersuchungen zur H- und Al-Ionentoxidität an Keimpflanzen von Fichte (Picea abies, Karst.) und Buche (Fagus sylvetica, L.) in Lösungskultur
Ber.d.Forschungszentrums Waldökosystemen/Waldsterben, Vol. 12, 1 - 219

Schneider T. and Bresser A.H.M., 1988
Summary report; Acidification research 1984 - 1988
Report nr. 00-06

Schouten A.J. and Brugge I.R.van der, 1989
Acute toxiciteit van aluminium en H^+-ionenconcentratie voor bodemnematoden uit een zuur en kalkrijk dennenbos
I.: Ontwikkeling en toepassing van een toets in waterig milieu
Rapport nr. 718603001, National Institute of Public Health and Environmental Protection

Sellinger H.von, Knoppink D. and Ziegler-Jöns A., 1986
Einfluss von Mineralstoffernährung, Ozon und saurem Nebel auf Photosynthese-Parameter und stomatäre Leitfähigkeit von Picea abies (L.) Karst.
Forstw.Cbl. 105, 239 - 242

Steingröver E., Swart W., Evers P., Vermetten A. and Meulen E.van der, 1991
Effects of ozone, PAR, temperature and VPD on photosynthesis
In: CORRELACI. Identification of traditional and air pollution related stress factors in a Douglas fir ecosystem: the ACIFORN stands. A correlative evaluation of monitoring data on the carbon, nutrient and water cycles
Rapport nr. 623, De Dorschkamp

Straalen N.M.van, Kraak M.H.S. and Denneman C.A.J., 1988
Soil micro-arthropods as indicators of soil acidification and forest decline in the Veluwe area, the Netherlands
Pedobiologia 32, 47 - 55

Termorshuizen A.J., 1990
Decline of carpophores of mycorrhizal fungi in stands of Pinus sylvestris
Thesis Agricultural University, Wageningen

Tonneijck A.E.G., 1983
Foliar injury responses of 24 bean cultivars (Phaseolus vulgaris) to various concentrations of ozone
Neth.J.Pl.Path. 89, 99 - 104

Tonneijck A.E.G., 1989
Evaluation of ozone effects on vegetation in the Netherlands
In: T.Schneider, S.D.Lee, G.J.R.Wolters and L.D.Grant (eds.): Atmospheric ozone research and its policy implications; Amsterdam, Elsevier, Studies in Environmental Science 35, p 251 - 260

Verhagen H.L.M. and Diederen H.S.M.A., 1991
Vergelijkingsmetingen van de analyse- en monsternemingsmethode van de vaste en de vloeibare fase van bodemmonsters
TNO-report
in prep.

Vries W.de, 1991
Assessment and policy implications of average critical loads for nitrogen and sulphur in the Netherlands
(in prep.)

Walling G., Schärby L. and Selldén G., 1990
Long-term exposure of Norway Spruce Picea abies (L.) Karst., to ozone in open-top chambers
I. Effects on the capacity of net photosynthesis, dark respiration and leaf conductance of shoots of different ages
New Phytol. 115, 335 - 344

Willigen P.de and Noordwijk M.van, 1987
Roots, plant production and nutrient use efficiency
Ph.D.thesis, Wageningen Agricultural University, the Netherlands

5. EFFECTS ON HEATHLAND

H.van Dobben (RIN)

5.1 Introduction

Heathland can be roughly divided into two types: wet heathland (which is influenced by groundwater for at least some part of the year and in which the dominant species is Erica tetralix), and dry heathland (which is not influenced by groundwater and in which the dominant species is Calluna vulgaris). On a quantitative basis the dry heathland type is by far the most important in the Netherlands. The area of heathland has rapidly decreased over the years, mainly owing to cultivation and afforestation (Table 5.1). In the remaining heathland the dominant heathland species have been increasingly replaced by grasses for some decades now. In dry heathland Calluna vulgaris (heather) is being replaced by Deschampsia flexuosa (wavy-hair grass), while in wet heathland Erica tetralix (cross-leaved heath) is being replaced by Molinia caerulea (purple moor-grass). At the same time rare heathland types such as Arnica montana, Antennaria dioica (mountain everlasting), and Viola canina (heath dog violet) have almost disappeared altogether.

An inventory by means of satellite imagery has shown that about one third of the heathland in the Netherlands is still vital (> 70% covered by heathland species), about one third contains large amounts of grass and will probably change into grassland within the next 3 - 5 years, and about one third has already changed into grassland (Table 5.2). It can be concluded that the Dutch heathland is rapidly changing into grassland and that the rate of change will probably increase with time.

Although many possible causes of this degradation are reported (such as ineffective management, lowering of groundwater levels and stress from excessive recreation), it is obvious that air pollution and the resulting soil acidification and N eutrophication are key factors in this process.

5.2 Effects of atmospheric input

Dominant species are found to be hardly affected by gaseous SO_2 or soil acidification. This fact was already known before the second phase of the programme started. Atmospheric deposition of nitrogen, however, eventually leads to replacement of slow-growing heathland species which are adapted to nutrient-poor conditions by fast growing grasses which use the available nitrogen more efficiently. Crowding out does not take place immediately, however. A closed canopy intercepts so much light that strong growth of grasses can only take place if gaps are formed in the canopy. The opening of the Calluna canopy can be caused by frost, drought, or plagues of the heather beetle. A high nitrogen deposition was found to increase the sensitivity of Calluna to frost and drought, and the risk of heather beetle plagues.

Table 5.1 Approximate total area of Dutch heathland in the course of time

1800	800 000 ha
1833	600 000 ha
1907	450 000 ha
1940	100 000 ha
1970	61 000 ha
1983	42 000 ha

Table 5.2 Percentage distribution of Dutch heathland over the degradation classes

Main classification:		
no degradation	(0- 30% grass)	42%
intermediate degradation	(31- 50% grass)	24%
strong degradation	(51- 70% grass)	20%
full degradation	(71-100% grass)	14%
		100%
Additional classes:		
vital heathland	(>70% heather)	32%
no grass	(0% grass)	11%

With regard to the impact of nitrogen deposition, three different processes can be distinguished, each having a different time scale:

(1) growth stimulation by nitrogen: for all species (time scale: weeks);
(2) morphological and physiological changes in species adapted to nutrient-poor conditions, resulting in increased sensitivity to secondary stress factors (time scale: approximately 1 year);
(3) crowding out of slow-growing species by fast-growing species (time scale: years).

Competition can also contribute to the decline of rare species, but soil acidification is probably the most important factor in this decline. Rare species, moreover, prove to be sensitive to direct effects of gaseous SO_2.

5.3 Research carried out and methods of research

The research carried out as part of the Dutch Priority Programme on Acidification was focused on direct and indirect effects of SO_x, NO_y, and NH_x on 'dominant' heathland species (heather and grasses) and on 'rare' species (Arnica montana, Viola canina). No research was done on the effects of ozone on heathland vegetation.

Several scenario analyses were made (Chapter 6) to study the competition between heathland vegetation and grasses at different levels of atmospheric N. Scenario analysis requires the use of a calculation method of to predict effects based on quantified cause-effect relationships. Three different methods were used:

1. Nutrient balances
   Elimination of heathland species by more efficient nitrogen-users can be expected if there is an accumulation of nitrogen in the soil. This accumulation can be estimated as the difference between input by atmospheric deposition and output by uptake in biomass, leaching to the groundwater, and removal by grazing or sod-cutting. Because estimates were available for all output terms, a maximum value for the input could be derived.
2. Dynamic simulation of growth and competition
   This method implies a detailed description in small time intervals of nutrient uptake and growth of individual species, and competition between species. For the Erica-Molinia system a competition model was developed by Berendse & Aerts (1984). Their model will be adapted for the Calluna-Molinia system by Conijn (1991), while a model for the Calluna - Deschampsia system was developed by Heil (1991).
3. Risk assessment based on a toxicological model
   Harmful effects determined in laboratory experiments can be extrapolated to the field using safety factors. This can also be done for a group of species, even if not all species have been individually tested. In this approach results of fumigation experiments were used to determine the exposure levels which protect a given percentage of a community's species against, for example, visible injury, a decrease in biomass production, or competitive strength.

Scenarios can only be evaluated against the background of intended management policies. In the Netherlands, heathland can only be preserved by human intervention. With traditional heathland management (grazing and sod-cutting) there is a constant removal of nitrogen so that the vegetation can compete. Without such management there will be an accumulation of nitrogen even at low deposition levels, resulting in a transition to grassland and finally to forest. High atmospheric deposition levels can, however, be partly compensated by more intensive management. In practice this means a higher frequency of sod-cutting, since removal of nitrogen by grazing is not very effective. A higher frequency of sod-cutting, however, would hamper the development of rare species (of plants, but also of animals such as amphibians and reptiles).

5.4 Research results

The most important findings of the research can be summarized as follows:

- At the ecosystem level, nitrogen input ultimately leads to the elimination of slow-growing species by fast-growing species, but Calluna will not be crowded out by grasses at nitrogen deposition levels up to 150 kg N $ha^{-1}yr^{-1}$ if its canopy remaining closed. Opening of a Calluna canopy can be caused by stress factors such as frost, drought, plagues of the heather beetle or by natural ageing. Under normal conditions in the Netherlands, the canopy will hardly ever be opened by natural ageing. The critical nitrogen load for the crowding out of an open Calluna canopy by grasses is about 10-15 kg N $ha^{-1}yr^{-1}$ (700 - 1100 mol_c $ha^{-1}yr^{-1}$). (Present N deposition on Dutch heathland is approximately 35 - 40 kg N $ha^{-1}yr^{-1}$.) At this critical deposition level, vital heathland can be maintained with a sod-cutting frequency of once every 50 years. Both experimental research on Calluna and modelling work on Calluna (in competition with Deschampsia) and Erica (in competition with Molinia) indicate the same critical load of 10 -15 kg N $ha^{-1}yr^{-1}$. With grazing and very frequent sod-cutting (once every 10 years) a vegetation of Calluna or Erica, though without rare species, can be maintained at nitrogen deposition levels up to about 30 kg N $ha^{-1}yr^{-1}$.
- At the individual plant level, nitrogen input (as NH_3 or $(NH_4)_2SO_4$) causes growth stimulation even at low dosages. Above-ground nitrogen uptake by Calluna is very efficient, and linearly related to concentration in a range from zero up to 400 $\mu g.m^{-3}$ NH_3 or 200 $\mu mol.1^{-1}$ $(NH_4)_2SO_4$. However, due to uptake, metabolic or morphogenetic changes may occur in rare heathland species that make them more sensitive to frost, drought, and plagues. A critical level of NH_3 cannot be exactly defined, but is probably in the range 5-10 $\mu g.m^{-3}$ (long-term).
- The decline of the Violon caninae species is probably due to direct effects of gaseous SO_2 and soil acidification. Adverse effects of SO_2 on more than 5% of the heathland species can be expected at long-term average concentrations above a critical level of 8 $\mu g.m^{-3}$. (This is slightly under the current annual concentration for the Netherlands, see Chapter 3.) Effects on dominant species (Calluna and grasses) will probably not occur at the current SO_2 levels in the Netherlands. Crowding out of Violion caninae by grasses can also take place, but is probably only important at nitrogen deposition levels above the current ambient level in the Netherlands. However, this level may affect the

reproduction or establishment of Violin caninae.

- Extensive field research showed that many threatened plant species of wet and dry heathland and nutrient-poor soils, such as Dactylorhiza maculata (heath spotted orchid), Thymus serpyllum (wild thyme), Pedicularis sylvatica (lousewort), and Arnica montana (Arnica), do not occur at pH levels (H_2O) under 4.2. The same research showed that there was no correlation between the occurrence of these species and the concentration of aluminium in the soil solution. Pot experiments and ecophysiological experiments showed that the disappearance of these species can be attributed to an indirect pH effect. Owing to the combination of soil acidification and N deposition, the NH_4/NO_3 ratio increases. This inhibites the uptake of base cations such as K and Mg. As a result the above-mentioned species die after germination, and already established plants show poor growth and flowering.

Figure 5.1 shows several results of the model CALLUNA, which simulates the competition between Calluna and Deschampsia (Heil, 1991). Calluna then increases until there is a heather beetle plague, which reduces the vegetation to practically zero. It depends on the degree of nitrogen accumulation at that moment whether Calluna shows a recovery, or Deschampsia becomes dominant. The latter will occur at a deposition of 20 kg N $ha^{-1}yr^{-1}$ in about 20 years' time. The heather beetle plagues are generated at random by the model, the risk of a plague depending on the N load within the system (see Chapter 6 for further details).

5.5 Uncertainties

Many heathland sites have been studied quite extensively and there is no reason to suppose that the Assel site is not representative for the Dutch heathland. At present, field data do not indicate any effects from ozone (although laboratory experiments have shown such effects). Further research into this would seem to be indicated.
The various processes of impact/competition have been studied and described quite thoroughly. The uncertainties are minimal. The critical levels are sufficiently supported by quantified data. The largest uncertainties are related to the soil compartment (the mineral balance of the Violion caninae locations and N accumulation / mineralization in the litter layer), as well as to the function of ericoid mycorrhizae.

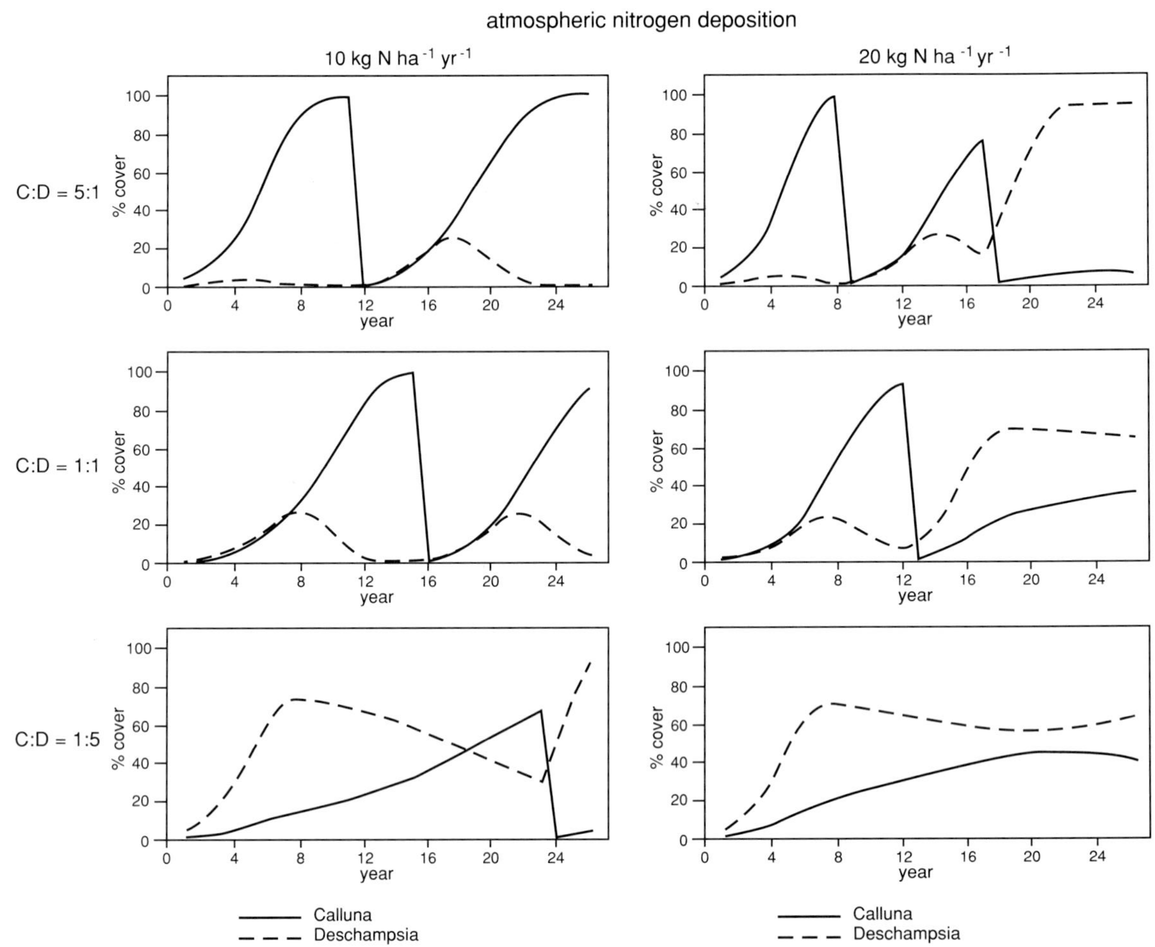

Fig. 5.1 Model results of interaction between Calluna and Deschampsia at two levels of atmospheric nitrogen deposition. C:D=ratio between Calluna and Deschampsia at the start of the simulation

Literature

Berendse F. and Aerts R., 1984
Competition between Erica tetralix L. and Molinia caerulea (L.) Moench as affected by the availability of nutrients
Acta Oecol./Oecol.Plant. 5, 3-14

Conijn J.G., 1991
De simulatie van de concurrentie tussen Calluna vulgaris en Molinia caerulea in droge heidevelden
RIN-report, in prep.

Heil G.W., 1991
Critical load of atmospheric nitrogen deposition in natural ecosystems: an example of heathlands
in prep.

6. SCENARIO ANALYSES USING THE DUTCH ACIDIFICATION SYSTEMS MODEL

6.1 Introduction

T.N.Olsthoorn (RIVM)

The previous chapters gave an overview of the present situation with an emphasis on the local situation at the different research sites. Some extrapolations were made on the basis of the present situation. This chapter pays attention to possible future situations (on the basis of emission - deposition scenarios for the period 1990 - 2050) and aims to give a picture of the national situation (on the basis of regionalized model calculations for the different acidification areas).

One of the goals of the Dutch Priority Programme on Acidification was to evaluate the effectiveness of abatement strategies. For this, a coherent model of the acidification in the Netherlands is a necessary tool.

The research, carried out simultaneously by many institutions in the Dutch Priority Programme on Acidification, was aimed at acquiring insights into phenomena, parameter values, relationships between phenomena, the gathering of information, the interpretation of historical data, the development of detailed models to explain observations and to forecast developments. The results of this research have been used largely, either directly or indirectly, in the development of the Dutch Acidification Systems Model (DAS), a regionalized high-level systems model, aimed at calculating the effect of abatement strategies on a number of receptor systems.

Model development was coordinated by RIVM, and carried out at the Winand Staring Centre, the Dutch Research Institute for Nature Management, the Research Institute for Forestry and Urban Ecology, the Centre for Agrobiological Research, Resource Analysis, and RIVM itself.

The basic structure of DAS was laid down in 1985. Since then, intermittently, the model has been developed further. In 1990 the model reached the stage in which it was used to calculate effects of abatement strategies for the Netherlands on a regional scale, for all submodules.

The spatial resolution of the model for the Netherlands is about 60 x 60 km^2, resulting in a subdivision of the country into 20 areas (Figure 6.1). The areas used within DAS have become widely known as "Dutch acidification areas". They are treated both as emission and as receptor areas.

DAS focuses on effects within the Netherlands only. Hence, outside the Netherlands, a subdivision of Europe was needed for emission aggregation only. This was done as coarsely as possible, keeping in mind the relative contribution of the individual areas to the total deposition and concentration in the Netherlands. Europe, excluding the Netherlands, was divided into 19 areas (see Figure 6.2).

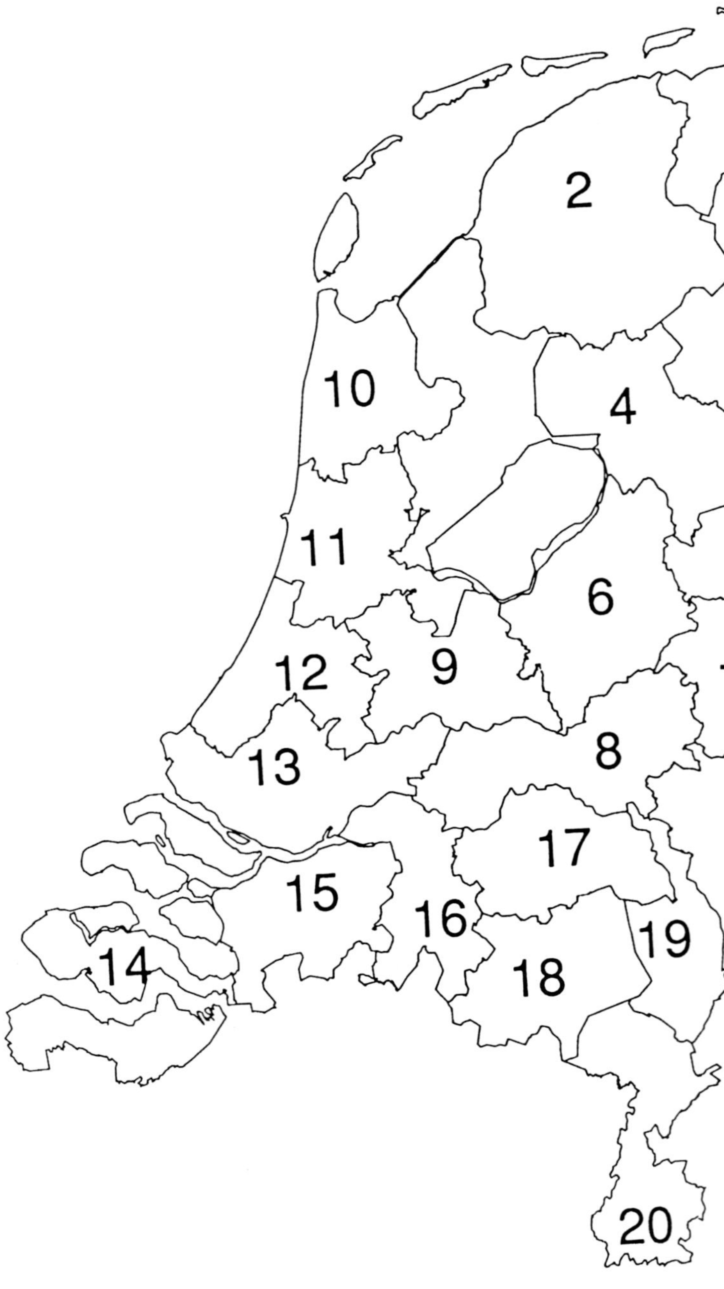

Fig. 6.1 The 20 Dutch acidification areas used by DAS

Fig. 6.2 The 19 European emission areas outside the Netherlands used by DAS

In this way, DAS was regionalized into 39 emission areas, 20 of which lay in the Netherlands.

The time-span for the model is a century, with time steps of one year. 1950 is taken as a starting point for simulations.

The air pollution compounds included are the acidification agents NH_3, NO_x and SO_2 and their derivatives. O_3 is included because it is used by some of the effect modules as a damaging agent.

DAS consists of modules that can run independently. Specified emission scenarios are translated into emissions per compound and per area by the EMISSIONS module. These emissions are used by the air and transport model SRM to produce year-averaged concentrations and depositions.

Three elaborate effect modules exist, which use complex dynamic models to calculate effects on soils (RESAM), on forests (SOILVEG and FORGRO) and on heathland pools (AQUACID). The effect modules to calculate the effects on heathlands (ERICA and CALLUNA) are relatively simple dynamic simulation models. The modules to calculate effects on agricultural production, construction materials and monuments made of natural stone (AGRIPROD, MATERIALS and MONUMENTS) only use dose-effect relationships. Only the modules FORGRO and AQUACID can not yet be used for regionalized scenario analysis.

The specifications of the Dutch Acidification Systems model have been described earlier (Bakema et al., 1990; Thematic Report DAS, Olsthoorn).

This chapter describes the effects of three different emission/deposition scenarios on forest soils, forest growth and natural heathlands, and their effects on agricultural production, construction materials and monuments as predicted with the model.

Literature

Bakema A.H., Boer K.F.de, Bultman G.W., Grinsven J.J.M.van, Heerden C.van, Kok R.M., Kros J., Minnen J.G.van, Mohren G.M.J., Olsthoorn T.N., Vries W.de and Wortelboer F.G., 1990
Dutch Acidification Systems Model-Specifications
Dutch Priority Programme on Acidification, report nr. 114.1-01

6.2 Emission and deposition scenarios for SO_2, NO_x, and NH_3

K.F. de Boer and R. Thomas (RIVM)

6.2.1 Introduction

The scenario calculations using the DAS model have been divided into three categories: calculations based on historical emission data, calculations based on expected emissions in the near future (the period until 2000) and calculations based on deposition targets (for the period 2001 - 2050). For this last category three variants were made. Together this yields three scenarios, of which the parts up to the year 2000 are identical.

For some effect modules data on growing-season and daily average ambient ozone concentrations are needed. So far ozone concentrations cannot be assessed with model computations based on NO_x and VOC emissions. Therefore, expert knowledge is used. This is discussed in Section 2.4.

Finally, for some modules data are needed on the deposition of several base cations and chloride. Further details are given in Section 2.5.

6.2.2 Emission - deposition scenarios

6.2.2.1 Historical emission figures

Historical emission figures, for the Netherlands as well as for other countries, were based on the literature and on estimates. Further details are given in Thomas et al. (1988). Since "new" emission figures for the past are published quite regularly, the historical emission figures for SO_2 and NO_x used for DAS calculations differ slightly from those recently published by EMEP. The differences are very small, however. For ammonia, the historical emission figures for other countries (Thomas et al., 1988) were corrected on the basis of an emission inventory for 1987 (Asman, 1989).

The uncertainty in historical emissions can be large, particularly when the distance in time and/or space is great.

Table 6.1 shows some recent emission figures for the Netherlands.

Table 6.1.a SO_2 emissions for the Netherlands after 1980 in ktons SO_2

Source	1985	1987	1988	1989
Traffic	31	35	35	32
Industry	72	74	64	58
Refineries	96	83	94	96
Power stations	64	64	65	44
Other sources	10	18	19	19
Total	273	274	277	249

Table 6.1.b NO_x emissions for the Netherlands after 1980 in ktons NO_2

Source	1985	1987	1988	1989
Traffic	327	344	344	349
Industry	66	68	55	55
Refineries	17	20	20	22
Power stations	80	83	87	76
Other sources	45	48	45	46
Total	535	563	551	548

Tabel 6.1c NH_3 emissions for the Netherlands after 1980 in ktons NH_3

Source	1985	1986	1987*	1988*
Agriculture	238	228	240	230
Industry	6	6	8	8
Housholds	9	9	10	10
Total	253	243	258	248

*) These figures were obtained from Erisman (1989); since the basic assumptions differ from those of Thomas et al. (1988), particularly for non-agricultural emissions, these figures were not used in the calculations, but estimates have been made by means of interpolation between the figures for 1986 and 1994.

6.2.2.2 Expected emissions for the period 1990 to 2000 (Scenario "NEPP+")

The scenario for emissions in the Netherlands during the period 1990 - 2000 was based on base-line measures from the CPB medium growth scenario and the corresponding coal variant for generation of extra electricity, described in Thomas et al. (1988), and on the measures from the Netherlands Acidification Abatement Plan+ and the National Environmental Policy Plan Plus (NEPP+) (which includes the SEP covenant). The assumption is made that the emission targets for the year 2000 set by the NEPP+ will be met. For target groups for which a range of values is given as a target, the lower end of the range is chosen. The resulting emissions are summarized in Table 6.2.[1)].

Table 6.2 Expected emissions in the Netherlands (for the period 1990 - 2000)

Economic sector	SO_2 (kton)		NO_x (kton)		NH_3 (kton)
	1994	2000	1994	2000	2000
Industry	45	9	66 *	26 *	3
Refineries	56	25			
Housholds and public sector	8	4	24	8	9
Road traffic	31	17	210	112	
Power stations	30	18	55	30	
Other sources	5	1	80	62	
Agriculture	-	-	-	-	70
Total	175	74	435	238	82

* including refineries

1) it has to be noted here that there are indications that the underlying emission reductions are rather optimistic (Second National Environmental Survey, in prep.)

The emission figures in other countries are based on Appendix 5 to the Netherlands Acidification Abatement Plan. The figures for 1994 are rather high but the targets for 2000 are fairly optimistic. Table 6.3, which was derived from the above-mentioned Appendix, shows the emission reductions per country in percentages of the 1980 level. In the emission scenario up to 2000, it is assumed that these reduction targets will be met.

Tabel 6.3 Expected emission reductions in other countries as compared to 1980 (for the period 1990 - 2000)

Country	SO_2 (%) 1994	SO_2 (%) 2000	NO_x (%) 1994	NO_x (%) 2000	NH_3 (%) 1994	NH_3 (%) 2000
FRG	70	80	30	50	10	25
Belgium	35	80	10	50	10	25
France	35	80	5	50	10	25
Great Britain	30	45	5	50	10	25
Northern Europe	60	80	15	50	10	25
Southern Europe	10	30	0	30	10	25
Central Europe	70	80	30	50	10	25
Eastern Europe	30	40	10	30	10	25

6.2.2.3 Some fictitious emission variants

As an exploratory exercise, several alternatives for international emissions for the year 2000 were estimated. The resulting deposition estimates are presented in this chapter, Section 2.3.1. The emission figures for 2000 were estimated on the basis of the following four assumptions:

a) The emission per unit area is the same for each country; data were obtained from WRI (1988). The assumption was made that there will be no changes in area between 1985 and 2000.
b) The emission per inhabitant is the same for each country; data were obtained from the CEC reports (1990a and 1990b) and the US Bureau of the Census (1990).
c) The emission per GNP unit is the same for each country. Starting point is the 1987 GNP data derived from the Statistical Abstracts (US Bureau of the Census, 1990). The GNP growth between 1987 and 2000 was based on the CEC report (1990 a).
d) The emission per energy unit (PJ) is the same for each country. Data on EC countries were taken from the CEC report (1990a), and data on the Eastern European countries were taken from the CEC report (1990b) and from the Environmental Data Compendium (OECD, 1989).

In all these assumptions "the same" means the same as in the Netherlands in the year 2000 according to the NEPP+ scenario. This gives four emission variants.

Estimates of area, number of inhabitants, energy use, GNP and expected emissions per country for the year 2000 are given in Appendix 3.

6.2.2.3.1 Emissions based on different units

The emissions for the four variants based on different units are given in Appendix 4. Figure 6.3 shows diagrams of these emissions.

In these diagrams Scandinavia includes Denmark, Norway, Sweden, and Finland. Southern Europe includes Italy, Switzerland, Austria, Spain, and Portugal. The East European countries are not included in the diagrams.

It should be noted that the variant of equal emissions of NH_3 per PJ is not very logical. For reasons of consistency, however, this variant was included.

6.2.2.4 Deposition trends for the period 2000 - 2050

For the period after 2000 no emission policy has been developed so far. Some (long-term) targets have been formulated, but only for deposition levels. No emission scenarios have been calculated on the basis of these deposition levels. It is assumed that these levels will be met. As to the way in which they are to be met, no pronouncements have been made. The calculations were made in order to evaluate the effect of such depositions on the different receptors. Assumed deposition trends are based on three variants (see Table 6.4).

Table 6.4 Deposition trends (for the period 2000 - 2050). The official target levels are printed in bold type

year	2000	2010	2010	2050	2050
	deposition in mol H^+ $ha^{-1}jr^{-1}$				
receptor:[1]	NL	NL	forest	NL	forest
Variant 1	2200 [2]	**2200**	2550	**2200**	2550
Variant 2	2200 [2]	**1400**	1600	1230	**1400**
Variant 3	2200 [2]	1230	**1400**	**700**	800

1) Deposition to forests is higher than the average deposition in the Netherlands. Figures represent the average deposition in the Netherlands and the average deposition to forests (in general) in the Netherlands. See also Section 2.3.

2) This figure represents the average value for the Netherlands for 2000, estimated with the above-mentioned emission scenario for the Netherlands and Europe. This value was calculated at 2240 mol $H^+ ha^{-1} yr^{-1}$. In view of the uncertainties this has been rounded down to 2200.

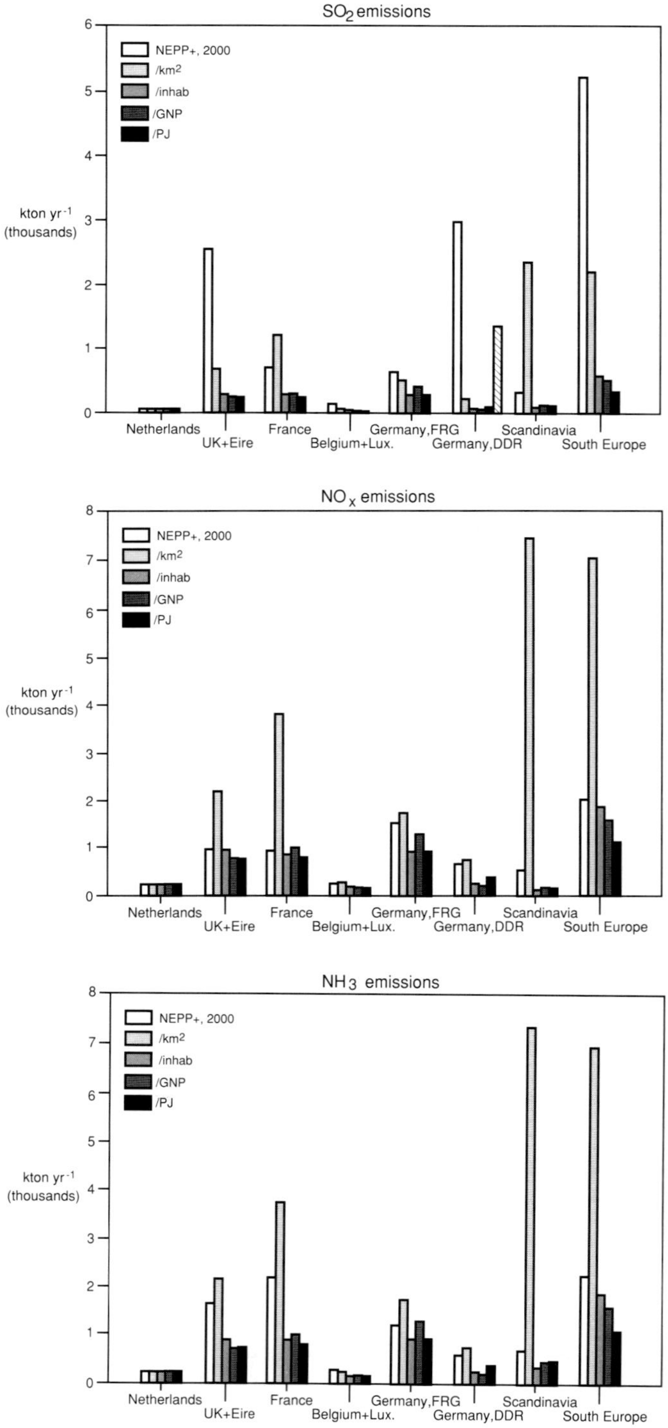

<u>Fig. 6.3</u> International emissions in 2000 for NEPP+ and four variants

The regional distribution of deposition and the relative share of each component as resulted from the emission-based calculations up to 2000 are assumed to be constant. Concentrations were assessed in the same way. No physical background exists for these assumptions.

6.2.3 Scenario results

Based on historical emissions and the emission scenario up to 2000, the average deposition to the Netherlands in 2000 will be **2200** mol potential acid per hectare per year. Apart from the uncertainties in emission figures, the average uncertainty in the calculated deposition levels is at least 20%. Using the atmospheric transport model, potential acid deposition in the Netherlands in the year 2000 has been estimated at 1800 - 2600 mol per hectare. The target level of 2400 mol per ha for the year 2000 is within this range. A slightly different value will be reached using the TREND model, which forms the basis of the DAS air transport model and uses a 5x5 km^2 grid instead of the 20 acidification areas. However, this value will be within the range of uncertainty mentioned above.

As a result of the unification of Germany, the emissions in East Germany could be significantly lower. The influence of this on acid deposition in the Netherlands is currently estimated at about 3%.

The above-mentioned three emission-deposition scenarios give the following results for depositions and concentrations. For easy reference Figure 6.4 only shows the trend in (weighed) averages for the Netherlands. Moreover, for concentrations and depositions of the different compounds only the values for Scenario 3 are presented. In the other scenarios the relative share of each compound is the same as in 2000.

Appendix 1 contains tables with the total deposition of potential acid and the total deposition of nitrogen in the 20 acidification areas.

The figures show that a strong reduction in acid deposition is estimated for the period 1990 - 2000. This is due to a strong reduction in NH_x deposition in particular, although considerable reductions are also expected in NO_y and SO_x.

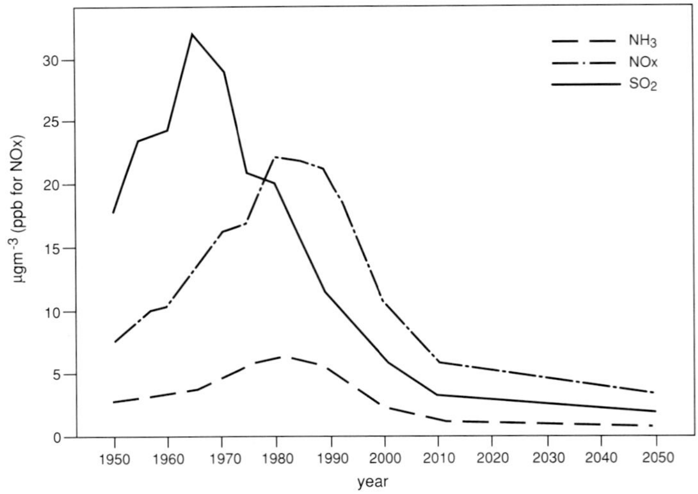

Fig. 6.4.A The trend of concentrations of different compounds as a function of time, in Scenario 3. Values for the period after 2000 were scaled (see Section 2.2.4)

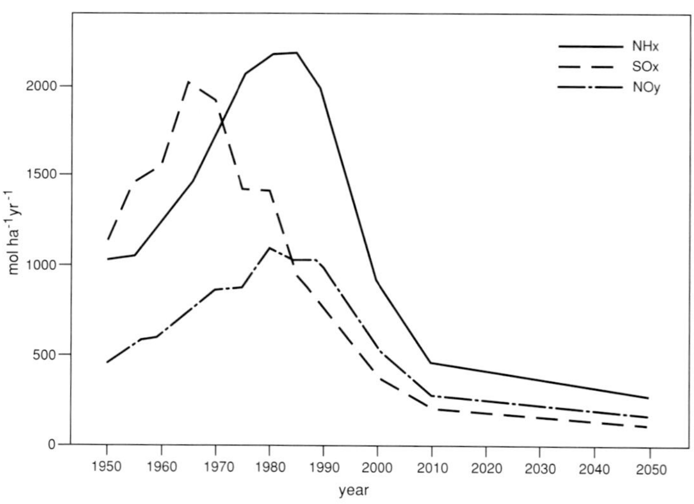

Fig. 6.4.B The deposition trend of different compounds as a function of time, in Scenario 3. Values for the period after 2000 were scaled (see Section 2.2.4)

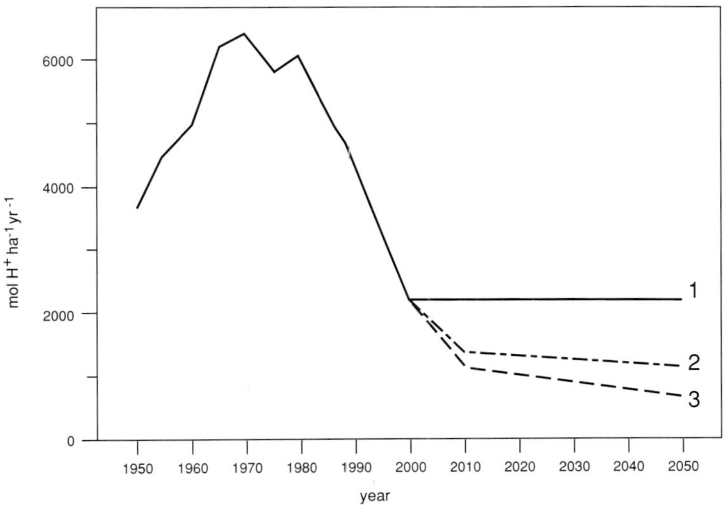

Fig. 6.4.C The trend of total acid deposition as a function of time for the three scenarios

The depositions of total potential acid for 1980 and 1989 slightly deviate from those indicated in Chapter 3. This is mainly caused by the following differences:

Chapter 3	Chapter 6
based on deposition measurements	based on emissions which are associated with large uncertainties
local roughness length (takes forests into account)	average roughness length
for one specific year, taking into account the meteorological conditions in that year	meteorological conditions averaged over 10 years

The tables in Appendix 1 give some insight into the regional distribution of the deposition levels. It is evident that there is a relatively high N deposition in the south and south-east of the country (with the exception of Limburg). This is caused by relatively high emissions of NH_3 in the south, combined with a short transport route of NH_3; it has been pointed out on other occasions that ammonia is mainly a local problem.
The deposition of total potential acid also turns out to be highest in the southern part of the country. Apart from the above-mentioned effect of ammonia, the industrial areas nearby (the Ruhr area and Antwerp) and the prevailing wind direction are important factors.

The transport models that were used calculate the average deposition to a particular area. However, the receptor systems do not receive the average deposition. The actual deposition is different because of roughness and the present position of sources and receptors. The dry position is therefore multiplied by correction factors for deposition to forest, heathland, and heathland lakes. These are different for every compound and region (Erisman, 1990).

6.2.3.1 Depositions resulting from emissions based on different units

If we input the emissions based on the assumptions in paragraph 6.2.2.3 into the air transport model, the resulting national average depositions of potential acid will be as indicated in Table 6.5. For purposes of comparison, the result of the emission scenario based on the NEPP+ target for 2000 is also shown.

A further comment on uncertainties is appropriate here. The uncertainties in the estimated emission figures are very large already. In addition to this there is the uncertainty in the air transport model. Altogether the uncertainty must be estimated at 30% at least. These margins are incorporated in the following table.

Table 6.5 Estimated depositions in 2000 based on emissions per unit

Area	scenario NEPP+	per km^2	per inhabitant	per mld $ GNP	per PJ
average	2240	2850	1490	1540	1490
rounded	**2200**	**2900**	**1500**	**1500**	**1500**
range of uncertainty	1800-2600	2000-3800	1000-2000	1000-2000	1000-2000

6.2.4 Ozone

During the period 1980 - 1989, growing-season and daily average ozone concentrations (Growing season: May - September; Daily average: between 10.00 and 17.00 hrs) of about 75 - 105 $\mu g.m^{-3}$ were observed (see Figure 6.5). During the exceptionally good summers of 1982 and 1989 the highest values were observed. If these summers are not taken into account, the average value for the Netherlands is 78 $\mu g.m^{-3}$.

For the period 1950 -1980, an increase in the ozone concentration of about 1% per year seems likely (Guicherit, 1989; Bojkov, 1988). With regard to the situation in the Netherlands, there is insufficient information for a regional differentiation of the "historical" ozone concentrations. For all 20 acidification areas the growing-season average ozone concentration (in $\mu g.m^{-3}$) can be estimated from:

$$O_3(t) = \frac{7800}{100 + (1985 - t)} \qquad 1950 < t < 1980$$

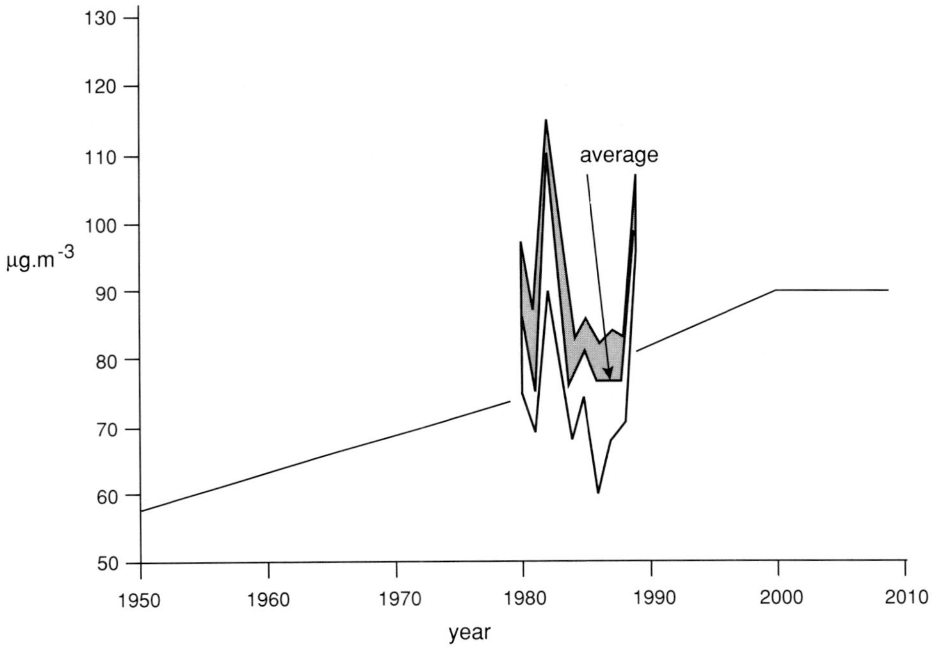

Fig. 6.5 The growing-season average ozone concentration. The regional variability is indicated by a minimum and a maximum

A comparison of the calculated concentration for 1979 based on this equation and the measured concentration in 1980 (see Figure 6.5) gives a discontinuity. However, this discontinuity is of the same magnitude as the deviations owing to meteorological variability in the observations between 1980 and 1989.

On the basis of various sources (Isaksen and Hov, 1987; Hough and Derwent, 1990; de Leeuw et al., 1990; Gery et al., 1989; de Leeuw and van Rheineck Leyssius, 1990) and data, it can be concluded that the ozone concentration in the Netherlands will show an increase in the coming decades. Extrapolation of the observed increase of 1% per year to the year 2000 is not inconsistent with forecasting studies. There are insufficient data to indicate the spatial distribution within the Netherlands. For the period 1990 - 2000 the growing-season average ozone concentration for all acidification areas can be estimated as:

$$O_3(t) = 78 * \left(1 + \frac{(t-1985)}{100}\right) \qquad 1990 < t < 2000$$

For the period after 2000 the uncertainties have become so large that there has been chosen for the very optimistic assumption of a constant level of ozone concentration.

6.2.5 Base cations and chloride

In addition to data on acidifying compounds and ozone, deposition figures for cations such as calcium, potassium, magnesium, sodium and chloride are needed for some effect modules. These deposition figures cannot yet be obtained through model calculations and must therefore be based on measurements of the National Precipitation Chemistry Network. The average values observed between 1978 and 1985 are used for all years. These values are given in Appendix 2.

Owing to the effect of sea-spray, the deposition of the nutrients potassium, calcium, and magnesium is higher in the coastal areas. This deposition decreases with increasing distance from the sea.

Literature

Alcamo J., Shaw R. and Hordijk L.(eds), 1990
The RAINS model of acidification, IIASA. Kluwer Academic Publishers, Dordrecht

Asman W.A.H., 1989
Ammonia Emissions in Europe. Updated emission and seasonal emission variation
National Environmental Research Institute (DMU), Roskilde, Denmark; DMU-LUFT-A132

Bojkov, R.D.,1988
Ozone changes at the surface and the free troposphere
In: Tropospheric Ozone (Isaksen,I.S.A., editor), Reidel, Dordrecht, pp. 83-96

Commision of the European Communities, DG for Energy, 1990a
Energy in Europe (main report), Luxembourg

Commision of the European Communities, DG for Energy, 1990b
Energy in Europe
(Working documents 3.1 en 13), Luxembourg

Erisman J.W., 1989
Ammonia emissions in the Netherlands in 1987 and 1988, RIVM report nr. 228471006

Erisman J.W., 1990
Acid deposition in the Netherlands. RIVM report nr. 723001002

Gery M.W., Edmond R.D. and G.Z. Whitten, 1989
Potential effects of stratospheric ozone depletion and global temperature rise on urban photochemistry. In: Atmospheric ozone research and its policy implications (Schneider,T. et al., editors) Elsevier, Amsterdam, pp. 365 - 375

Guicherit R.,1989
Concentrations and patterns of ozone in Western Europe
In: Atmospheric ozone research and its policy implications (Schneider T. et al., editors)
Elsevier, Amsterdam, pp. 167-176

Hough A.M. and R.G. Derwent, 1990
Changes in the global concentration of tropospheric ozone due to human activities. Nature, 344, 645-648

Isaksen I.S.A. and Hov O., 1987
Calculation of trends in the tropospheric concentrations of O , CO, CH and NO . Tellus 39B, 271-285

Leeuw F.A.A.M. de, and Rheineck Leyssius H.J.van, 1990
Sensitivity of oxidant concentrations on changes in UV-radiation and temperature
Atmospheric Environment, accepted for publication

Leeuw, F.A.A.M. de, Rheineck Leyssius H.J.van and Builtjes P.J.H., 1990
Calculation of long term averaged ground level ozon concentrations
Atmospheric Environment, 24A, 185-193

Ministry of Housing, Physical Planning and Environment, 1989; the Netherlands Acidification Abatement Plan. Second Chamber of the States General, Session 1988-1989, 18225, no.31

Ministry of Housing, Physical Planning and Environment, 1990; Nationaal Milieubeleidsplan-plus. Second Chamber of the States General, Session 1989-1990, 21137, nr.20

OECD, 1990: Environmental Data Compendium 1989. OECD, Paris

Olsthoorn, T.N. Thematic report on Integrated Modelling. In: Thematic Reports
Dutch Priority Programme on Acidification, nr. 200-07

Thomas R., Arkel W.G.van, Baars H.P.-, Ierland E.C.van, Boer K.F.de, Buysman E., Hutten T.J.H.M. and Swart R.J., 1988: Emission of SO_2, NO_x, VOC and NH_3 in the Netherlands and Europe in the period 1950 - 2030. RIVM rapport 758472002

US Bureau of the Census, 1990: Statistical Abstracts of the US 1990, 110th Edition, Washington DC

Volz A. and Kley D., 1988
Evaluation of the Montsouris series of ozone measurements in the nineteenth century. Nature, 332, 240-242

World Resources Institute, 1988: World Resources 1988-1989; New York

Appendix 1
Scenario results

Total acid Area	Emission scenario 1990	1994	2000	Scenario 1 2010	2050	Scenario 2 2010	2050	Scenario 3 2010	2050
1	3240	2660	1700	1700	1700	1060	930	930	530
2	3130	2550	1590	1590	1590	1000	880	880	500
3	3800	3100	1990	1990	1990	1240	1090	1090	620
4	3930	3210	2040	2040	2040	1280	1120	1120	640
5	5290	4290	2690	2690	2690	1680	1480	1480	840
6	4320	3500	2130	2130	2130	1330	1170	1170	670
7	5500	4430	2800	2800	2800	1750	1540	1540	880
8	5010	3950	2450	2450	2450	1540	1350	1350	770
9	5110	4210	2600	2600	2600	1630	1430	1430	810
10	2770	2350	1510	1510	1510	950	830	830	470
11	4620	4320	2600	2600	2600	1630	1430	1430	810
12	4860	4080	2520	2520	2520	1580	1380	1380	790
13	4950	3740	2290	2290	2290	1430	1260	1260	720
14	3500	2830	1910	1910	1910	1200	1050	1050	600
15	4390	3470	2230	2230	2230	1400	1230	1230	700
16	5300	4230	2630	2630	2630	1650	1440	1440	820
17	6160	4830	2790	2790	2790	1750	1530	1530	880
18	5850	4670	2810	2810	2810	1760	1540	1540	880
19	6140	4930	3030	3030	3030	1890	1660	1660	950
20	4780	3940	2590	2590	2590	1620	1420	1420	810
Average	4420	3590	2240	2240	2240	1400	1230	1230	700

Total nitrogen Area	Emission scenario 1990	1994	2000	Scenario 1 2010	2050	Scenario 2 2010	2050	Scenario 3 2010	2050
1	2200	1770	1100	1100	1100	690	600	600	340
2	2170	1720	1020	1020	1020	640	560	560	320
3	2670	2150	1340	1340	1340	840	730	730	420
4	2710	2180	1350	1350	1350	850	740	740	420
5	3930	3140	1910	1910	1910	1200	1050	1050	600
6	2940	2330	1370	1370	1370	860	750	750	430
7	4040	3230	1980	1980	1980	1240	1090	1090	620
8	3360	2650	1620	1620	1620	1010	890	890	510
9	3380	2750	1700	1700	1700	1070	930	930	530
10	1660	1360	860	860	860	540	470	470	270
11	2630	2200	1390	1390	1390	870	770	770	440
12	2820	2420	1560	1560	1560	980	860	860	490
13	2630	2130	1350	1350	1350	840	740	740	420
14	1780	1480	1040	1040	1040	650	570	570	330
15	2430	1990	1310	1310	1310	820	720	720	410
16	3440	2760	1700	1700	1700	1060	930	930	530
17	4540	3510	1940	1940	1940	1210	1060	1060	610
18	4140	3260	1900	1900	1900	1190	1040	1040	600
19	4460	3530	2100	2100	2100	1320	1160	1160	660
20	2860	2370	1580	1580	1580	990	870	870	500
Average	2910	2340	1440	1440	1440	900	790	790	450

Appendix 2
Deposition of cations (in mol $ha^{-1}jr^{-1}$)

Area	Ca	K	Mg	Na	Cl
1	110	43	120	970	1150
2	130	55	190	1610	1870
3	96	37	84	690	830
4	130	44	110	930	1080
5	97	30	62	480	580
6	120	37	90	740	860
7	100	32	69	550	660
8	110	33	78	630	740
9	120	36	96	790	920
10	150	65	260	2250	2600
11	140	49	140	1160	1360
12	170	56	130	1070	1310
13	180	60	160	1320	1580
14	280	110	470	3980	4610
15	170	58	190	1550	1820
16	120	34	78	580	700
17	110	29	64	480	580
18	110	28	59	440	540
19	130	30	59	420	520
20	280	47	64	360	450

Appendix 3
Estimates of area, number of inhabitants, energy use, GNP and expected emissions for the year 2000

Country	surface area 1000 ha	number of inhabitants	energy use	GNP mld $	emission in 2000 SO_2	NO_x	NH_3
Netherlands	3392	15.7	3162	299	75	238	82
United Kingdom	24155	57.9	10423	928	2553	974	573
Ireland	6889	3.67	510	39			
France	54563	57.9	10730	1231	710	933	769
Belgium	3282	9.62	2274	202	161	220	97
Luxembourg	259	0.37	154	9			
Germany, Fed. Rep.	24428	61.05	12303	1647	666	1536	431
German Dem.Rep.	10557	16.54	5102	243	3000	676	206
Denmark	4237	5.16	954	131	90	124	108
Finland	30547	5.05	1406	122	117	140	46
Norway	30786	4.29	1404	115	28	91	29
Sweden	41162	9.38	2863	222	100	159	56
SC-total	102495	18.72	5673	459	245	390	130
Hungary	9234	10.48	1643	108	979	189	134
Czechoslovakia	12540	16.2	4061	187	1860	843	164
Poland	30449	39.93	6834	320	2460	1050	421
EE-total	52223	66.61	12538	615	5299	2082	719
USSR	2227200	311.64	75701	2915	7680	1953	1157
Turkey	77076	68.56	2133	94	166	123	430
Yugoslavia	25540	25.04	2606	75	705	133	183
Greece	13080	10.27	1195	67	240	89	83
Bulgaria	11055	9.05	1685	80	620	105	92
Romania	23034	24.38	3715	180	960	273	290
Albania	2740	3.79	366	7	30	6	24
SE-total	152525	141.09	11700	503	2721	729	1103
Italy	29402	57.96	7093	1085	2660	1036	326
Switzerland	3977	6.7	1399	257	25	98	51
Austria	8273	7.66	1439	167	71	108	80
CE-total	41652	72.32	9931	1509	2756	1242	458
Spain	49941	40.5	4040	450	2275	665	274
Portugal	9164	10.57	748	61	186	116	57
SW.-total	59105	51.07	4788	511	2461	781	331

Appendix 4
Emissions based on emission levels in the Netherlands (in kton)

Country	emissions based on NL emission per km²			emissions based on NL emission per inhabitant		
	SO_2	NO_x	NH_3	SO_2	NO_x	NH_3
Netherlands	75	238	82	75	238	82
United Kingdom	534	1695	584	277	878	302
Ireland	152	483	167	18	56	19
France	1206	3828	1319	277	878	302
Belgium	73	230	79	46	146	50
Luxembourg	6	18	6	2	6	2
Germany, Fed. Rep.	540	1714	591	292	925	319
German Dem.Rep.	233	741	255	79	251	86
Denmark	94	297	102	25	78	27
Finland	675	2143	738	24	77	26
Norway	681	2160	744	20	65	22
Sweden	910	2888	995	45	142	49
SC-total	2266	7192	2478	89	284	98
Hungary	204	648	223	50	159	55
Czechoslovakia	277	880	303	77	246	85
Poland	673	2136	736	191	605	209
EE-total	1155	3664	1262	318	1010	348
USSR	49245	156272	53842	1489	4724	1628
Turkey	1704	5408	1863	328	1039	358
Yugoslavia	565	1792	617	120	380	131
Greece	289	918	316	49	156	54
Bulgaria	244	776	267	43	137	47
Romania	509	1616	557	116	370	127
Albania	61	192	66	18	57	20
SE-total	3372	10702	3687	674	2139	737
Italy	650	2063	711	277	879	303
Switzerland	88	279	96	32	102	35
Austria	183	580	200	37	116	40
CE-total	921	2923	1007	345	1096	378
Spain	1104	3504	1207	193	614	212
Portugal	203	643	222	50	160	55
SW-total	1307	4147	1429	244	774	267

Country	emissions based on NL emission per $ bn BNP			emissions based on NL emission per PJ		
	SO_2	NO_x	NH_3	SO_2	NO_x	NH_3
Netherlands	75	238	82	75	238	82
United Kingdom	233	739	255	247	785	270
Ireland	10	31	11	12	38	13
France	309	980	338	255	808	278
Belgium	51	161	55	54	171	59
Luxembourg	2	7	2	4	12	4
Germany, Fed. Rep	413	1311	452	292	926	319
German Dem. Rep.	61	193	67	121	384	132
Denmark	33	104	36	23	72	25
Finland	31	97	33	33	106	36
Norway	29	92	32	33	106	36
Sweden	56	177	61	68	215	74
SC-total	115	365	126	135	427	147
Hungary	27	86	30	39	124	43
Czechoslovakia	47	149	51	96	306	105
Poland	80	255	88	162	514	177
EE-total	154	490	169	297	944	325
USSR	731	2320	799	1796	5698	1963
Turkey	24	75	26	51	161	55
Yugoslavia	19	60	21	62	196	68
Greece	17	53	18	28	90	31
Bulgaria	20	64	22	40	127	44
Romania	45	143	49	88	280	96
Albania	2	6	2	9	28	9
SE-total	126	400	138	278	881	303
Italy	272	864	298	168	534	184
Switzerland	64	205	70	33	105	36
Austria	42	133	46	34	108	37
CE-total	379	1201	414	236	747	258
Spain	113	358	123	96	304	105
Portugal	15	49	17	18	56	19
SW-total	128	407	140	114	360	124

6.3 Effects on forest soils

W.de Vries, J.Kros, C.van der Salm and J.C.Voogd
The Winand Staring Centre for Integrated Land, Soil and Water Research

6.3.1 Introduction

6.3.1.1 The model RESAM

In order to gain insight into the long-term effect of acid deposition on Dutch forest soils, a process-oriented Regional Soil Acidification Model (RESAM) has been developed. RESAM describes changes in soil chemistry, both in the solid phase (minerals and adsorption complex) and liquid phase, due to natural and man-induced processes. It includes the major elements in forest soils, i.e. H, Al, Ca, Mg, K, Na, NH_4, NO_3, SO_4, Cl, HCO_3 and RCOO. The temporal resolution of model input and output is one year, as the model is intended to give insight into the long-term soil chemical response of forest ecosystems to acid deposition.

RESAM utilizes mechanistic descriptions for all processes in the vegetation canopy, litter layer and mineral soil horizons which significantly influence the concentration of major ions in the soil solution. An overview of the various process formulations, including the compounds involved in each process, is given in De Vries and Kros (1989) and in the thematic reports by Olsthoorn and De Vries and Kros.

6.3.1.2 Forest soils

In order to limit both data acquisition and computation time, the regional application of RESAM has been restricted to tree species and soil types (receptors) of major importance.

Forests are represented by seven important tree species, i.e. Pinus sylvestris (Scots pine), Pinus nigra (Black pine), Pseudotsuga menziesii (Douglas fir), Picea abies (Norway spruce), Larix lepto Lepis (Japanese larch), Quercus robur (oak) and Fagus silvatica (beech).

Forest soils are confined to acid sandy soils, as these cover 80 to 90% of the Dutch forest area. Furthermore, these soils are sensitive to acidification. Acid sandy soils of major aerial importance in the Netherlands are the leptic podzol (Entic Haplorthod), Gleyic podzol (Typic Haplorthod), Humic podzol (Typic Haplohumod), Humic Gleysol (Typic Humaquept), Plaggept and Albic Arenosol (Typic Udipsamment). The soils are characterized by 14 representative profiles based on a recent 1 : 250 000 soil map of the Netherlands. They cover a broad range in soil properties, such as the organic matter content, texture and groundwater level, which influence major hydrological, biochemical and geochemical processes in the soil. In the regional application, the soil profile is confined by the root zone. The vertical heterogeneity is taken into account by differentiating between soil layers (horizons). An overview of the designation and thickness of the horizons in the various soil profiles is given in De Visser and De Vries (1989).

The forest/soil combinations that have been included comprise nearly 65% of the total Dutch forest area, of which more than 50% is covered by Pinus sylvestris. The remaining 35% comprises tree species such as Populus (poplar) and Betula pendula (birch) (approximately 20%) and soil types such as calcareous sandy soils, clay soils, loess soils and peat soils (approximately 15%). Information on the distribution and area of forest-soil combinations in the 20 deposition areas was derived by overlaying a forest database, with tree species information in a 500 x 500 m^2 grid and a soil database, with soil type information in a 100 x 100 m^2 grid.
Average model parameters have been used for tree species and/or soil type, irrespective of the deposition area. Results of an uncertainty analysis have shown that a simulation run with average soil parameters gave nearly similar trends in soil solution chemistry as the mean of 200 simulation runs in which the variability of input data was included (Kros et al., 1990). The various data are based on (1) field research (Kleijn et al., 1989), laboratory experiments and model calibration (soil data), (2) literature surveys (for example De Vries et al., 1990; forestry data) and (3) calculations with a separate hydrological model (De Visser and De Vries, 1989; hydrological data). The element concentrations and the adsorbed quantity of cations in all soil layers of the considered forest soil combinations in all deposition areas in 1990 have been derived by running the model for a period of 25 years (1965-1990) using historic emmission-deposition data. Element quantities in primary minerals and in hydroxides are kept constant during this period.

6.3.1.3 Model output

Here, presentation of the model results is restricted to pH, Al concentration, Al/Ca ratio and NH_4/K ratio in the topsoil (from 20 to 30 cm) and pH, Al and NO_3 concentration in the subsoil (at the bottom of the root zone). The parameters in the topsoil are important indicators of forest stress, whereas the parameters in the subsoil are important indicators of potential groundwater pollution. For most of these parameters, critical concentration levels have been defined (Chapter 4). More information is given in the thematic report by De Vries and Kros and in De Vries et al. (in prep.).

6.3.2 Results

6.3.2.1 Comparison of model results and field data

In order to gain insight into the reliability of the model predictions, model results of the soil solution chemistry in 1990 were compared with soil solution measurements in 150 forest stands during the period March to May in the same year. The tree species included in the field survey are similar to those included in the simulations (see also Chapter 4, section 4..3.3). A comparison of median values of the parameters mentioned above is given in Table 6.6. The SO_4 concentration has been added to get insight into the relative contribution of S and N in soil acidification as predicted by RESAM.

Table 6.6 Median values of important soil solution parameters as measured in the field and predicted by RESAM, in 1990

Parameter		Topsoil measured	predicted	Subsoil measured	predicted
pH	(-)	3.6	3.7	3.9	3.8
Al	($mol_c\ m^{-3}$)	0.7	0.4	0.6	1.2
Al/Ca	(-)	1.3	1.7	-	-
NH_4/K	(-)	1.7	2.7	-	-
NO_3	($mol_c\ m^{-3}$)	-	-	0.5	0.7
SO_4	($mol_c\ m^{-3}$)	-	-	1.1	1.0

The agreement between model results and field data is good (difference < 10%) for pH and for SO_4, reasonable (difference between 10-30%) for the Al/Ca ratio and NO_3 concentration, and poor (difference > 30%) for the NH_4/K ratio and Al concentration.

It should be kept in mind that one measurement in early spring can give rise to both lower and higher concentrations compared to the flux weighted annual average value, depending upon the difference in actual 1990 and 30 year averaged meteorological data. Comparison between model results and field data for Na and Cl (which can be considered as tracers) in both topsoil and subsoil shows that the model results are always (slightly) lower, especially in the topsoil. In the subsoil, the correspondence is better. The difference in hydrological data is an explanation for the underprediction of Al concentrations in the topsoil. The overprediction of Al in the subsoil can partly be explained by an overestimation of the actual acid load, which is mainly determined by the NO_3 and SO_4 concentration. The remaining

remaining difference is most probably due to the long-term effect of liming (Ca) and fertilization (mainly K) and/or a higher base cation input from the atmosphere, which will cause an increase in base cation concentration and a decrease in Al concentration. This also explains the overestimation of the Al/Ca and NH_4/K ratio by RESAM.

6.3.2.2 Future trends in soil solution chemistry

6.3.2.2.1 Overall trends

Trends in median values of the soil solution parameters considered before, in response to three emission - deposition scenario's (Paragraph 6.2.2) are given in Figure 6.6. The Al concentration in the subsoil is not given because of the poor agreement with field data.

Between 1990 and 2000 there is no difference in trends for the three scenarios because the deposition values are similar. In this period the average deposition in the Netherlands drops from approximate 4400 to 2200 mol_c ha^{-1} yr^{-1} (Paragraph 6.2). In response to this deposition reduction, a large reduction in H concentration occurs in this period. However, the pH increase is most pronounced after 2000, since pH is a logarithmic value (Figure 6.6.a and 6.6.b).

Similarly, there is a considerable decrease in the Al concentration (Figure 6.6.c) and Al/Ca ratio (Figure 6.6.d) in the topsoil. Both medians drop below a critical value of 0.2 mol_c m^{-3} and 1.0 mol mol^{-1} respectively before the year 2000. Between 2000 and 2050 there is still a small decrease in Al concentration and Al/Ca ratio for Scenario 1, although the deposition level remains the same. For the Scenarios 2 and 3 there is still a substantial decrease in median values, especially up to 2020.

The behaviour of the "N related" parameters, i.e. the NH_4/K ratio in the topsoil and NO_3 concentration in the subsoil, is markedly different from that of the "Al related" parameters. This is especially the case with the NH_4/K ratio, which shows an initial increase up to 2000 - 2010, before it drops between 2000 - 2010 and 2050. The initial increase is due to a relatively small decrease in NH_4 concentration and a relatively large decrease in K concentration. The small decrease in NH_4 is due to N mobilization from litter, because of a decrease in the N content in litterfall (needles), which is simulated as a function of N deposition. The relatively large decrease in K is due to the fact that the K concentration in the topsoil is mainly determined by foliar exudation and this process is modelled as a function of NH_4 uptake, which in turn is determined by the NH_4 deposition. Consequently,

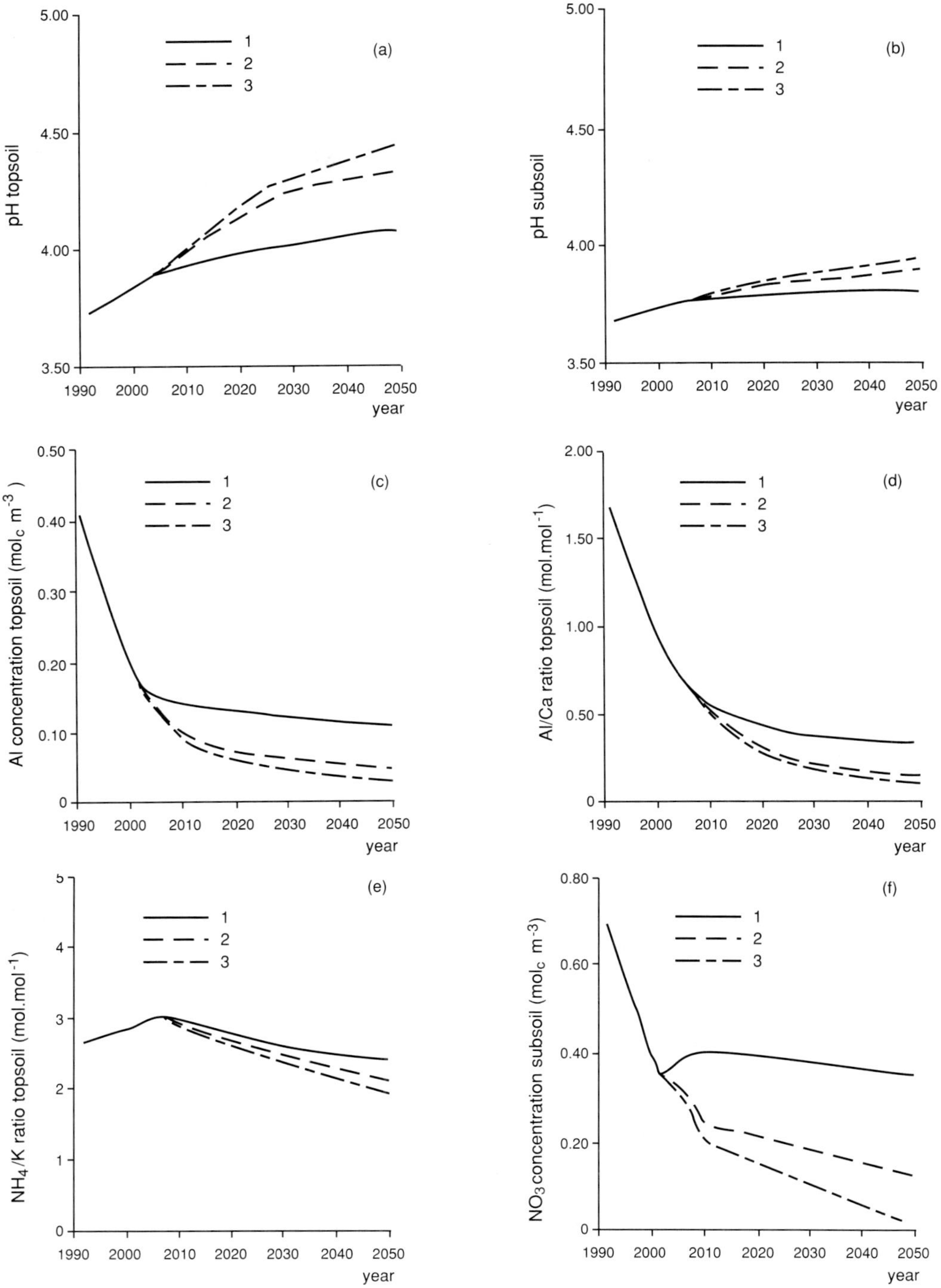

Fig. 6.6 Future trends in median values of important soil solution parameters in response to three emission - deposition scenarios. (a) pH of the topsoil; (b) pH of the subsoil; (c) Al concentration in the topsoil; (d) Al/Ca ratio in the topsoil; (e) NH_4/K ratio in the topsoil and (f) NO_3 concentration in the subsoil

the impact of the deposition scenario on the NH_4/K ratio is relatively small (Figure 6.6.e). Contrary to the NH_4/K ratio, the NO_3 concentration in the subsoil decreases between 1990 and 2000 but not to the same considerable extent as the Al concentration and Al/Ca ratio (Figure 6.6.f). This is due to the above mentioned N mobilization from litter, which causes a time lag between the decrease in N deposition and NO_3 concentration.

The median NH_4/K ratio in the topsoil is already below a critical level of 5 in 1990. The median NO_3 concentration in the subsoil is also below a critical value of 0.8 $mol_c\ m^{-3}$ (standard value) in 1990, and it even drops below 0.4 mol_c m-3 (target value) between 2000 and 2010 for Scenario 2 and 3. For Scenario 1 it remains at or just below the target value up to 2050.

For all parameters, the difference between Scenario 2 and 3 is much smaller than that between Scenario 1 and 2. This is consistent with the difference in average deposition values in 2050, i.e. 2550, 1400 and 880 $mol_c\ ha^{-1}yr^{-1}$ on forests respectively (see Section 6.2).

Trends in the percentage of the area where critical loads are exceeded are given in Figure 6.7 for Al and Al/Ca in the topsoil and NO_3 in the subsoil.

Trends for the NH_4/K ratio are not given since the actual exceedance of this ratio in the field is almost negligible in 1990. This is mainly due to the applied model schematization (topsoil 20 to 30 cm, see Section 6.3.1.3). In the field situation, NH_4/K ratios generally are higher in the first 10 cm of the soil (the layer with the most fine roots).

The figure shows that the area exceeding a critical Al concentration of 0.2 $mol_c\ m^{-3}$ and a Al/Ca ratio of 1.0 mol mol^{-1} is approximately 75% and 65% in 1990 respectively. In the year 2000, the percentage of forest soils exceeding both values is still approximately 40%. In the year 2050 this area is negligible for Scenarios 2 and 3. For Scenario 1 it remains approximately 25% and 5% for the Al concentration and Al/Ca ratio respectively.

It is important to stress that a deposition reduction according to Scenario 2 is sufficient to avoid exceedances in critical Al concentration or critical Al/Ca ratio in forest topsoils. The Al hydroxide depletion also stops for this scenario between 2020 and 2030 (not presented

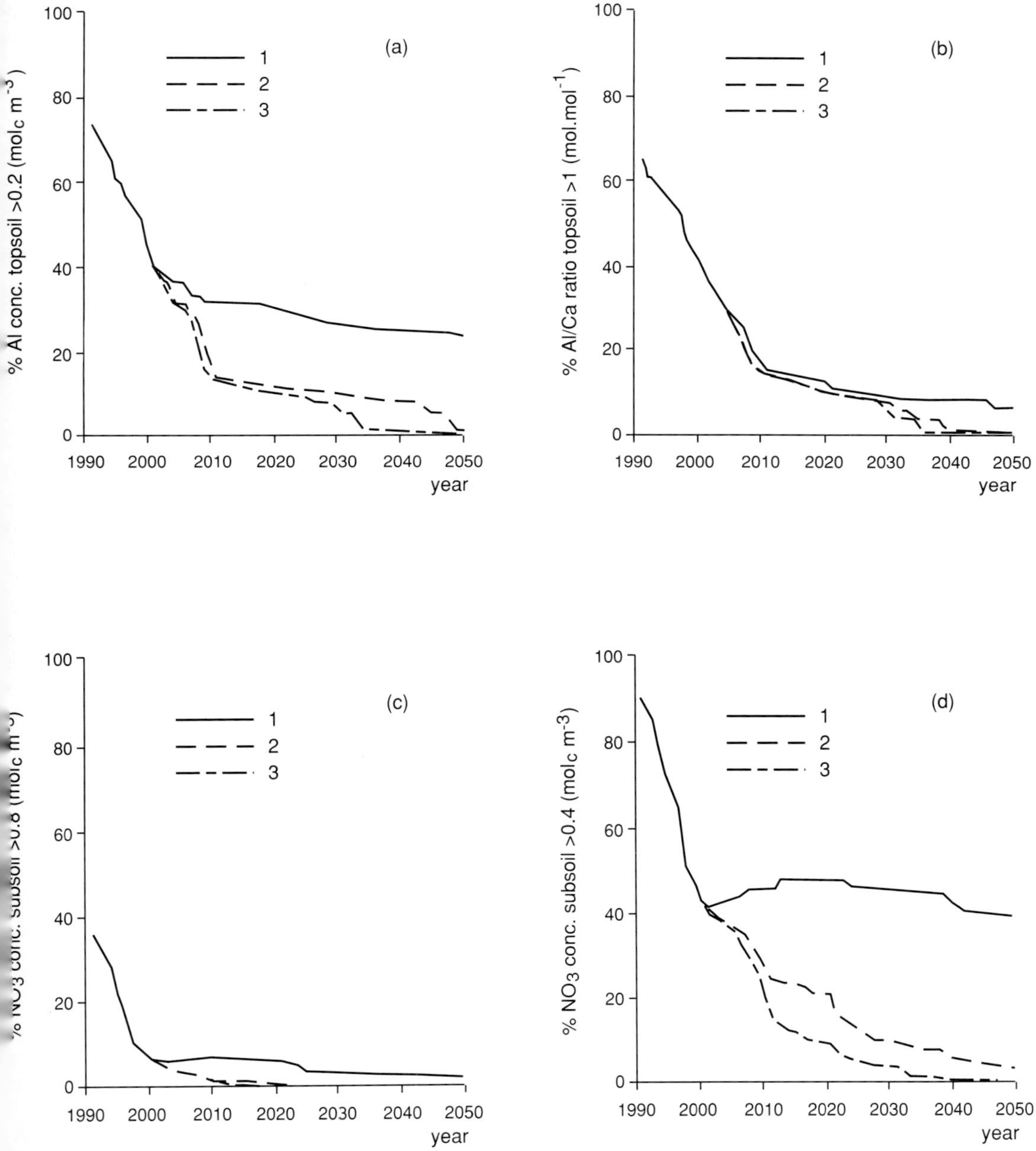

Fig. 6.7 The percentage of considered area of non-calcareous sandy forest soils exceeding critical Al values in the topsoil and critical NO_3 values in the subsoil in response to three emission - deposition scenarios. (a) Al concentration of 0.2 $mol_c\ m^{-3}$; (b) molar Al/Ca ratio of 1.0; (c) NO_3 concentration of 0.8 $mol_c\ m^{-3}$ (standard) and (d) NO_3 concentration of 0.4 $mol_c\ m^{-3}$ (target value)

here). The average deposition to forests at this time is close to 1400 mol_c $ha^{-1}yr^{-1}$, which is the average critical load derived for the effects of aluminium on forests (De Vries, 1991; Chapter 7).

The area exceeding the standard and the target value for NO_3 in 1990 is approximately 35% and 90% respectively, and it drops to (nearly) zero for both values before 2050 for Scenarios 2 and 3. For Scenario 1, the area exceeding the standard and target value in 2050 still remains 3% and 40% respectively (Figure 6.7.c and 6.7.d).

6.3.2.2.2 Trends in different deposition areas

The soil solution chemistry predictions will vary for the different deposition areas in response to different acid loads. This is illustrated in Table 6.7 for deposition area 3 (Drenthe, see Figure 6.1) and 18 (Eindhoven). Both areas have a relatively high proportion of forests, but the average potential acid load on forests in area 18 is much higher (6900 mol_c $ha^{-1}yr^{-1}$, in 1989, to forests, see Chapter 3) than in area 3 (4700 mol_c $ha^{-1}yr^{-1}$).

Table 6.7 Median values of important soil solution parameters in 1990, 2010 and 2050 for deposition area 3 and 18 in response to Scenario 3

Parameter	Compartment	1990		2010		2050	
		3	18	3	18	3	18
pH (-)	topsoil	3.8	3.5	4.1	3.7	4.5	4.4
	subsoil	3.8	3.7	3.9	3.8	4.2	4.1
Al (mol_c m^{-3})	topsoil	0.4	0.7	0.1	0.1	0.0	0.0
	subsoil	1.1	1.6	0.4	0.5	0.1	0.1
Al/Ca (-)	topsoil	3.1	4.3	1.1	1.6	0.3	0.3
NH_4/K (-)	topsoil	2.1	2.5	2.2	2.8	1.6	1.9
NO_3 (mol_c m^{-3})	subsoil	0.6	0.9	0.2	0.4	0.0	0.1

In 1990 the median pH in area 18 is lower than in area 3 (especially in the topsoil), while all concentrations and ratios are approximately 20-50% higher. In 2010, and especially in 2050, the differences between both areas is very low, which is consistent with the decrease in deposition differences for both areas.

6.3.2.2.3 Trends for different tree species

The influence of the receptor on the soil solution chemistry is illustrated in Table 6.8 for three different tree species i.e. Douglas fir, Scots pine and oak.

Table 6.8 Median values of important soil solution parameters in 1990, 2010 and 2050 for Douglas fir (DF), Scots pine (SP) and oak (OA) in response to Scenario 3

Para- meter	Com- partm.	1990 DF	SP	OA	2010 DF	SP	OA	2050 DF	SP	OA
pH	topsoil	3.6	3.6	3.8	3.9	3.9	4.2	4.4	4.4	4.5
	subsoil	3.7	3.8	3.8	3.9	4.0	4.0	4.1	4.2	4.4
Al	topsoil	0.8	0.6	0.3	0.2	0.1	0.1	0.0	0.0	0.0
	subsoil	1.7	1.3	0.8	0.6	0.4	0.3	0.1	0.1	0.1
Al/Ca	topsoil	3.7	3.1	1.5	1.2	1.0	0.4	0.2	0.2	0.1
NH_4/K	topsoil	2.6	2.8	2.4	2.9	3.1	2.6	1.9	2.1	1.7
NO_3	subsoil	1.1	0.7	0.5	0.4	0.2	0.2	0.1	0.1	0.0

The pH increases and all concentrations and ratios decrease according to Douglas fir > Scots pine > oak. This can mainly be explained by a decrease in evapotranspiration for these tree species. The differences between the tree species are in agreement with the field data except for the NH_4/K ratio, for which the differences appear to be larger in the field (see Chapter 4, Section 3.3). This might be due to differences in the forest filtering of nitrogen compounds which is not included in the model.

The tree species appear to have a larger influence on the soil solution chemistry than the differences in deposition between the areas. Like to the deposition areas, the influence of tree species decreases with time, due to the decrease in deposition differences.
It is important to note that the solution chemistry below Douglas fir, which has been the tree receiving the major research focus within the Dutch Priority Programme on Acidification, is certainly not representative for Dutch forests. Like to the field data, the solution chemistry below Scots pine appears to be most comparable to the overall results (see also Chapter 4, Section 4.3.3).

6.3.3 Conclusions

Deposition reductions generally lead to a fast improvement of the soil solution chemistry, i.e. an increase in pH and a decrease in the Al and NO_3 concentration and the Al/Ca and NH_4/K ratio. The NO_3 concentration and the NH_4/K ratio, however, show a clear delay between reduced deposition and subsequent concentration reduction, which is mainly due to N mobilization from the litter layer.

A deposition reduction to 2200 $mol_c\ ha^{-1}yr^{-1}$ as an average for the Netherlands, which is predicted for all scenarios in the year 2000, will reduce the exceedance of the critical Al concentration from about 75% of the forest soil area (current situation) to 40%.

Reductions up to 1400 $mol_c\ ha^{-1}yr^{-1}$ as an average for Dutch forest areas (Scenario 2) cause a further decrease in the exceedance (less than 20% of the considered forest-soil combinations) of a critical Al concentration of 0.2 $mol_c\ m^{-3}$ and a critical Al/Ca molar ratio of 1.0 in the topsoil and a critical NO_3 concentration of 0.8 $mol_c\ m^{-3}$ in the subsoil in 2010. In 2050 the exceedance of these critical values is negligible. For Scenario 1 there is still a considerable forest area (> 25%) exceeding the critical Al concentration in 2050; for the critical Al/Ca ratio this exceedance is about 10%. In Scenario 3, which gives a deposition of 1400 mol per ha on forests in the year 2010, and an average deposition of 700 mol per ha on the Netherlands as a whole in the year 2050, the exceedance of the critical Al concentration and nitrate concentration is reduced to an insignificant level markedly earlier than in the case of Scenario 2. Furthermore, deposition reductions according to Scenario 2 are also sufficient to avoid depletion of Al hydroxides buffer, contrary to Scenario 1.

The influence of tree species on the soil solution chemistry appears to be more important than the influence of the differences in deposition between areas. The dependance of soil solution chemistry on tree species and deposition areas decreases in future predictions due to deposition reductions.

Literature

Kleijn C.E., Zuidema G. and Vries W.de, 1989
De indirecte effecten van atmosferische depositie op de vitaliteit van Nederlandse bossen
2. Depositie, bodemeigenschappen en bodemvochtsamenstelling van acht Douglasopstanden
Wageningen, STIBOKA rapport 2050

Kros J., Janssen P., Vries W.de and Bak C., 1990
Het gebruik van onzekerheidsanalyse bij modelberekeningen: een toepassing op het regionale bodemverzuringsmodel RESAM
Wageningen, Staring Centrum, rapport 65

Olsthoorn T.N., 1991
DAS Overview of the Dutch Acidification Systems Model
Thematic Report

Visser P.H.B.de and Vries W.de, 1989
De gemiddeld jaarlijkse waterbalans van bos-, heide- en graslandvegetaties
Wageningen, STIBOKA rapport 2085

Vries W.de and Kros J., 1989
Lange termijn effecten van verschillende depositiescenario's op representatieve bosbodems in Nederland
Wageningen, Staring Centrum, Rapport 30.

Vries W.de, Hol A., Tjalma S. and Voogd J.C., 1990
Voorraden en verblijftijden van elementen in een bosecosysteem: een literatuurstudie
Wageningen, Staring Centrum, Rapport 94

Vries W.de and Kros J., 1991
Assessment of critical loads and the impact of deposition scenarios by steady state and dynamic soil acidification models. Thematic Report

Vries W.de, 1991
Assessment and policy implications of average critical loads for nitrogen and sulphur in The Netherlands
Submitted to Water, Air, Soil Poll.

Vries W.de, J. Kros, C. van der Salm and J.C. Voogd, 1991
The long-term impact of three emission deposition scenarios on Dutch forest soils
Water, Air, Soil Poll. (in preparation)

6.4 Effects on growth of Douglas fir

J.J.M.van Grinsven, J.G.van Minnen (RIVM), C.van Heerden (RIN)

6.4.1 Introduction

The SOILVEG Model

SOILVEG is a dynamic simulation model which calculates growth of Douglas fir as a function of the availability of carbon from photosynthesis and N, K, Ca, Mg from atmospheric deposition and soil processes. The soil module is based on the RESAM model (Section 6.3). SOILVEG accounts for direct and indirect effects of acid atmospheric deposition on growth. High ambient concentration of O_3 and SO_2 will reduce the photosynthetic carbon input to the tree. High levels of Al^{3+} and H^+ will reduce the uptake of nutrients by the roots and will also directly reduce the fine root mass (Figure 6.8 and Chapter 4, Section 4.3).

Important model simplifications of the atmosphere-plant-soil interactions are:

- no consideration of phosphorus
- no explicit calculation of photosynthesis
- no dynamic water balance, no consideration of water stress
- no consideration of plagues,
- no consideration of frost damage

Uncertainty analysis of SOILVEG revealed that uncertainties of model outputs are dominated by uncertainty about the decomposition kinetics of organic matter (Van Grinsven et al., 1991). The vegetation module of SOILVEG was calibrated using a compilation of literature data for tissue concentrations of N, K, Ca and Mg in Douglas fir, and standard growth curves for Douglas fir on prevailing Dutch sandy soils. The soil module was calibrated using solute concentrations and budgets of the sixteen intensively monitored Dutch forest sites (Table 4.3). SOILVEG was provisionally validated with forest productivity data for a specific site near Apeldoorn (Berdowski et al., 1991).

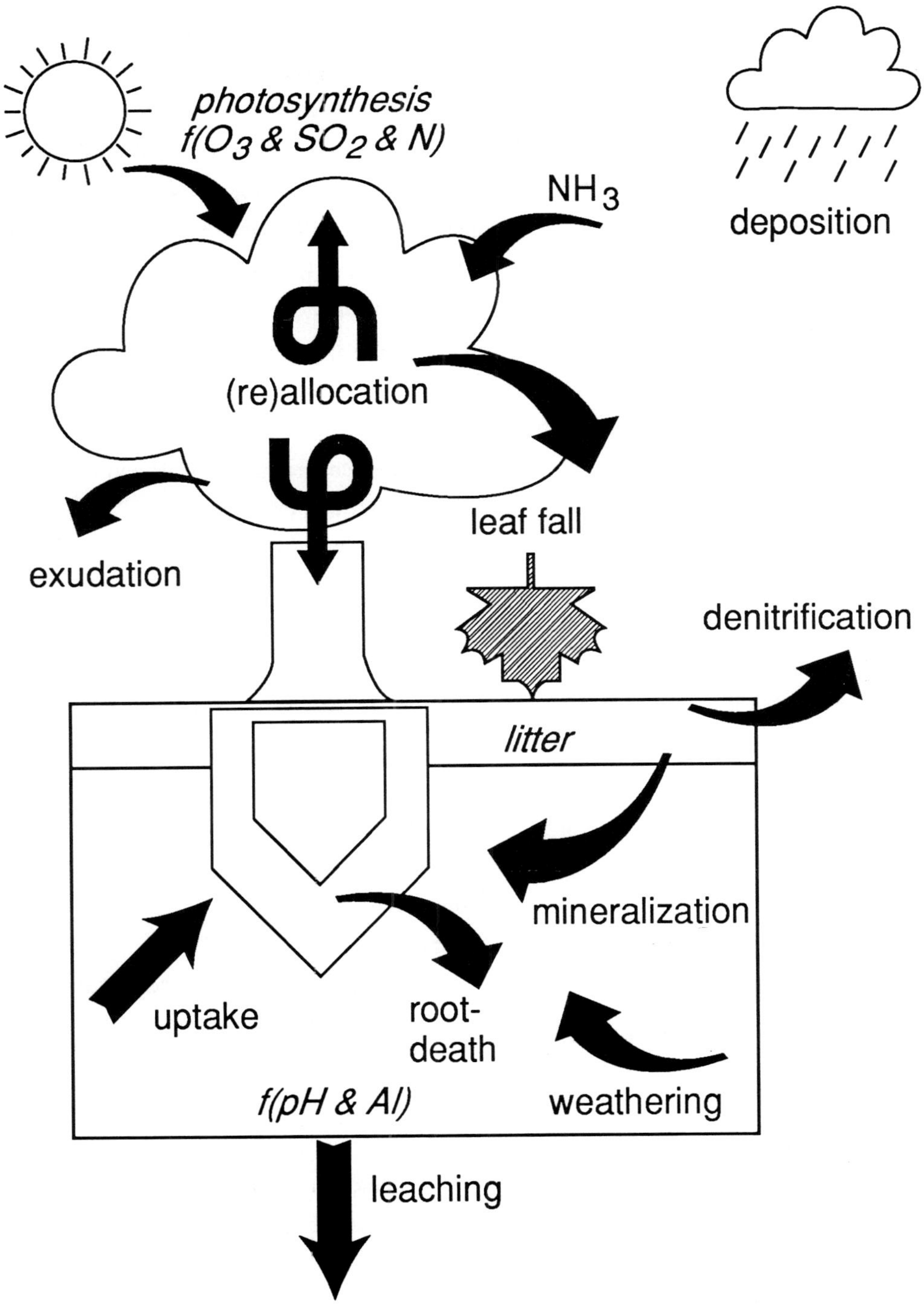

Fig. 6.8 Schematic overview of major processes in SOILVEG

The role of nitrogen in forests

Nitrogen is an essential nutrient for plant growth, and very often it limits forest growth in unpolluted areas. There are various direct and indirect effects of nitrogen on forest growth, which can be both positive or negative. The external input determines whether the net effect of increased N deposition and ambient NH_3 concentrations is positive or negative. If growth is not limited by photosynthesis, and concentrations of K, Ca or Mg in the plant are higher than their minimum values, increased availability of N will stimulate growth. However, growth stimulation of the leaf mass, will cause dilution of K and Mg, which will enhance leaf fall (Berdowski et al., 1991). So increased N availability can both increase and decrease leaf mass. A high N content in the tree will enhance respiration, which will decrease the carbon availability for growth.

High deposition of NH_x, after nitrification in acid soils will eventually increase the concentrations of H and Al in soil solution. Increased concentrations of H and Al in soil solution will have an adverse effect on the uptake of nutrients by roots.

In contrast to N deposition, the effect of increased S deposition in SOILVEG is always adverse or non-existent.

Output parameters

Six key output variables of SOILVEG were chosen to characterize the quality of the forest and the ecosystem:

SOILVEG PARAMETER	ENVIRONMENTAL SIGNIFICANCE
1. stem mass	forest productivity
2. needle mass	forest health
3. N content of needles	susceptibility to frost damage and plagues
4. Mg content of needles	efficiency of root uptake / forest health
5. Nitrate leaching	N saturation / nitrate contamination of ground-water
6. total N in litter	N saturation / risk of future nitrate contamination of groundwater (after increase of mineralization or decrease of uptake)

Effects are shown for 60-year-old Douglas fir stands, which is about the average age of present stands. A fixed forest age was chosen because the physiology of Douglas fir and its response to air pollution and soil acidification depend on the age of the forest. The effects of age would therefore complicate the interpretation of the effect of deposition scenarios. The trees are exposed to environmental conditions for a period of 20 years. Every five years 5-8% of the Douglas stand is felled and removed, which means that, unless there is regrowth wood mass will decrease.

Regional application

The actual area of forest land in the Netherlands covered by Douglas fir is small compared to, for example that of oak and Scots pine. However, a relatively large amount of research has been carried out on Douglas fir. Only 9% of the Netherlands is covered by forest. Most forests occur on sandy soils, which prevail in the south-eastern half of the Netherlands. SOILVEG calculations were carried out for the Leptic podzol (Entic Haplorthod, z06), the Humic podzol (Typic Haplohumod, z12) and the Gleyic podzol (Typic Humaquept, z08) and for Albic Arenosols (Typic Udipsamment, z27), which are the main soil types, that have developed in sandy sediments. Codes refer to the Dutch soil classification system. So far the formulation of the model, and the interpretation of model results have been focused on forest productivity.

The objective of the presented regional application of SOILVEG is to demonstrate the combined effects of atmospheric deposition and soil processes, assuming Douglas fir to be a representative tree. Only atmospheric deposition is regionalized. Maps of results (Figure 6.11 and 6.12), therefore, are by no means a realistic representation of the distribution of soil types and Douglas fir.

Relevant differences between soil types

The suitability of predominantly sandy soils for production forests is, is in part, judged on the basis of structural and hydrologic properties. These properties are not considered in SOILVEG. However, if field data for different soils provide evidence for differences in productivity for some soil, which is not a result of differences in soil acidity or availability of nutrients, it is accounted for by adjusting (calibrating) the soil specific root uptake efficiencies for various nutrients.

All considered soil types have a low natural fertility. Soil z06 (Leptic podzol) is a soil with a relatively good structure and water holding capacity. Soil z08 (Gleyic podzol) is a relatively wet soil, with a poor soil structure. Soil z12 (Humic podzol) is a relatively dry soil, in which the B-horizon may be poorly permeable to water and roots. Soil z27 (Albic Arenosol) is a soil with low organic matter content and a low water holding capacity. Some soil properties are given in Table 6.9.

Table 6.9 Soil-type differences which are considered in SOILVEG

Soil type	z06	z08	z12	z27
mean groundwater level (m):	2.5	1.1	1.5	2.5
primary minerals				
- dissolution rates	++	+	+++	++++
- pools	+	+++	+++	+++
amorphous Al hydroxide				
- dissolution rates	+	++	+++	+
- pools	++	+++	++++	+

('+,++,+++,+++' indicate the relative proportion)

6.4.2 Results

Effects of deposition in relation to soil properties and acidification area

A major objective of regional scenario analysis with SOILVEG is to compare the effectiveness of the three suggested deposition scenarios (Section 6.2.2) with respect to forest productivity and N saturation. Only the total effect of changes in S and N deposition will be discussed.

The effectiveness strongly depends on soil type and location in the country. For this purpose, results were compared for Groningen (area 1), as an example of a relatively clean area, and Venray (area 19), as an example of a relatively polluted area with respect to atmospheric deposition of N and S compounds.

Needle mass reponds stronger to changes in deposition than stem mass. The most important observation is that time trends of needle mass and stem mass in the relatively clean area 1 are opposite to trends in the relatively polluted area 19. The general increase of N deposition up to 1990, leads to an increase of both needle and wood mass (Figures 6.9 and 6.10) in area 1, but to a decrease of needle and wood mass in area 19. Apparently, in area 1 the positive fertilizing effect of increased N deposition dominates, while in area 19 adverse effects of increased N deposition, being enhanced soil acidification and respiration, dominate. The adverse effects of increased N deposition and deposition of acidity in area 19, are most evident for the most sensitive soil z27, in particular for wood mass (Figure 6.10).

In view of the different N-status and soil acidification of forests in coastal areas as compared to south-eastern areas, the strong decrease of N deposition and acid deposition after the year 1990, cause a decrease of needle and wood mass in area 1, while they cause an increase of needle and wood mass in area 19.

Irrespective of area and scenario, forest productivity is highest for the Leptic podzol (z06) and lowest for the Albic Arenosol (z27) (Figures 6.9 and 6.10).

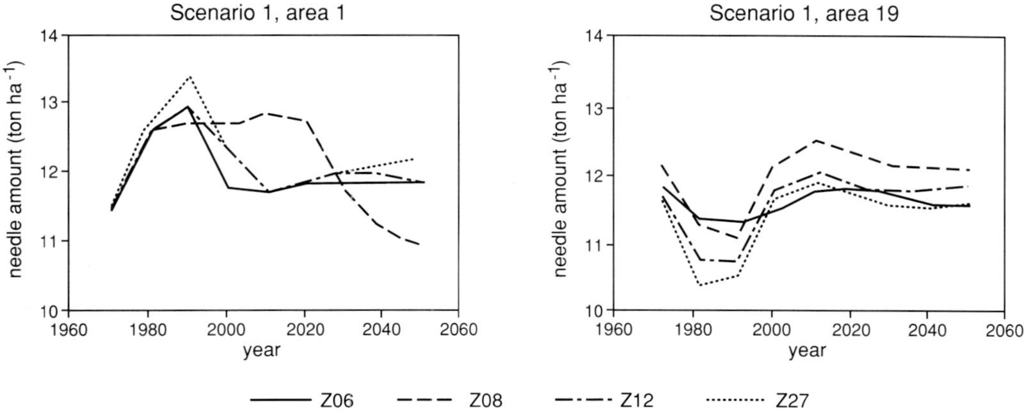

Fig. 6.9 Trends of needle mass in area 1 (Groningen) and area 19 (Venray), for Scenario 1 in the four soils

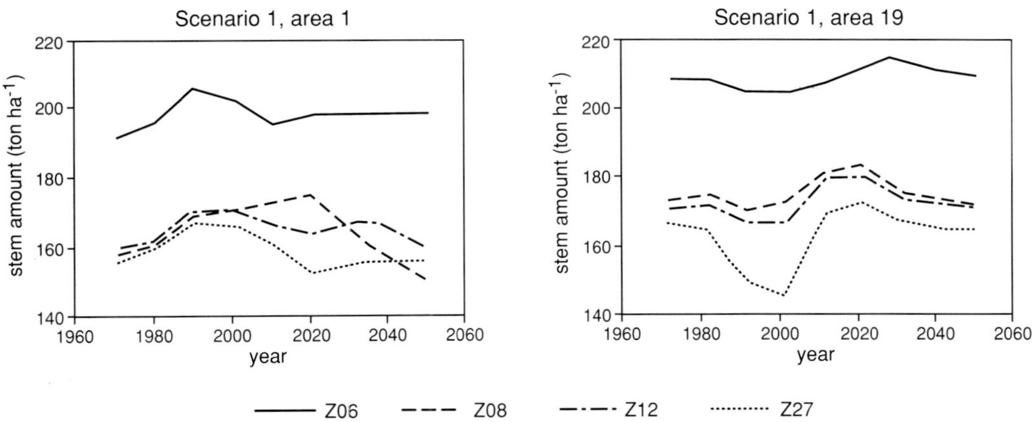

Fig. 6.10 Trends of wood mass in area 1 (Groningen) and area 19 (Venray), for Scenario 1 in the four soils

The most direct response of Douglas fir to decreased deposition and air pollution is a decrease of the N content of the needles (Figure 6.11). This direct response is, in part, a result of considerable needle uptake of gaseous NH_3. Between 1970 and 2000 simulated N contents are well above the critical level of 1.8% (Table 7.1), in the southern and south-eastern areas. If occurrence of plagues and frost damage had been included, for example in a comparable manner as was the case for Calluna (Chapter 5), the risk of a substantial decrease of needle and wood mass by forest die-back would have been high.

Nitrogen leaching is, in general, lowest under the soils with a shallow groundwater table, in particular the Gleyic podzol (z08) and to a lesser extent the Humic podzol (z12). In these soils there is a stronger accumulation of N in the organic matter and a larger loss of N to the atmosphere due to denitrification. Differences in leaching of N owing to differences in forest growth are minor.

Regional differences in deposition are, in general, reflected in forest and soil properties. The increase of base cation deposition from sea-spray when approaching the sea, is reflected in increasing Mg and K content of the needles. Increasing deposition of N when moving to the south-east is easily recognized in the regional patterns of needle N content (Figure 6.11), N storage in organic matter and nitrate leaching (Figure 6.12). As the nitrate leaching flux in fact represents the N surplus of the ecosystem, the regional pattern even amplifies the regional pattern of N deposition (see Figure 3.10).

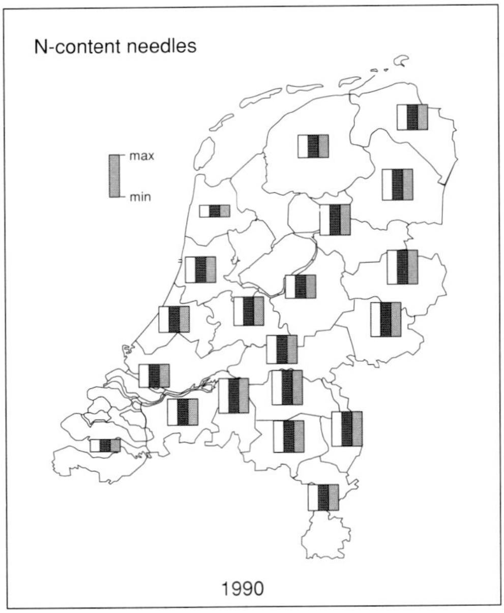

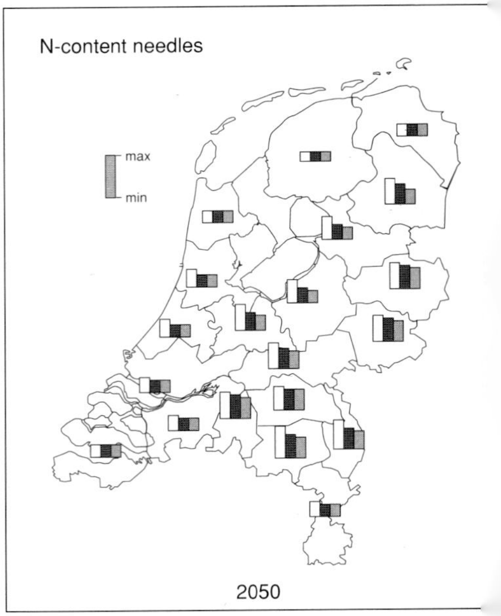

Fig. 6.11 Regional patterns for the N content in needles, comparing Scenario 1, 2 and 3 in 1990 (no difference) and 2050. Content is shown as bars (Scenario 1 is left bar)

N content: Minimum=0% Maximum=3.0%

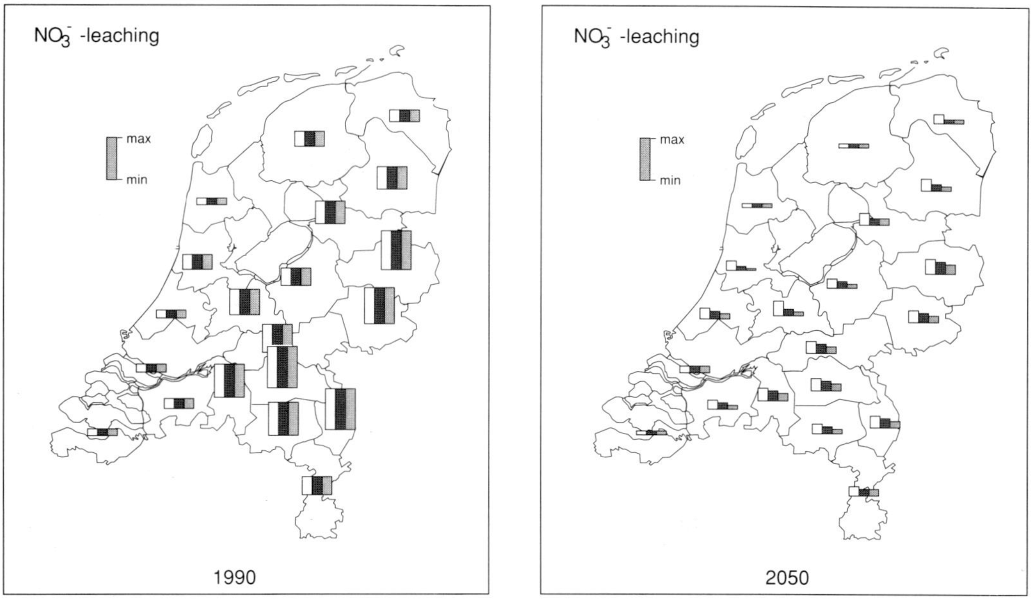

Fig. 6.12 Regional patterns of NO_3-N leaching, comparing Scenarios 1, 2 and 3 in 1990 (no difference) and 2050. Leaching is shown as bars (Scenario 1 is left bar)
NO_3 leaching: Minimum=0 Maximum=60 kg $ha^{-1}yr^{-1}$

The effectiveness of selected scenarios

When deposition of N is strongly decreased in area 1, SOILVEG predicts an increasing risk of loss of needle mass and wood production, as a result of absolute shortage of N for forest growth. For the least reductive Scenario 1, substantial growth reduction is only predicted (Figures 6.11 and 6.12) for z12 in area 1. With additional reduction of N deposition as in Scenarios 2 and 3, a strong decrease of needle and wood mass is predicted for all other soil types in area 1. Scenario 3 leads to a total collapse of the forest on z08, z12 and z27 in coastal areas (Figure 6.13). For area 19, SOILVEG does not predict a reduction of growth due to N-shortage.

The most marked beneficial effects of reduction of deposition of N and S compounds are a large decrease in needle N content and nitrate leaching. In the most polluted areas needle N content decreases from above critical levels (> 1.8% N) to optimum levels (± 1.5% N) (Figure 6.11). In the relatively clean areas needle N contents decrease from optimum levels to levels close to the physiological minimum (± 0.9% N) in accordance with the large growth reduction. Without additional reduction of N deposition after 2000 (Scenario 1), needle N content in the polluted areas becomes lower than, but remains close to the critical value. The decrease of needle N content and nitrate leaching after reduction of deposition is slow in polluted areas, because of continued N release from N-rich organic matter, and because in 1990 needle N content is close to its physiological maximum.

In terms of NO_3-N leaching, assuming a realistic average precipitation surplus of 150 mm.yr^{-1} for Douglas fir (Reurslag et al., 1990), the drinking water standard is 17 kg ha^{-1} yr^{-1} and the target value 8.5 kg ha^{-1} yr^{-1}. In the clean areas, NO_3 leaching is close to the standard, and irrespective of the scenario meets the target value after 2010 (Figure 6.12). In the polluted areas, Scenario 1 is insufficient to meet the drinking water standard. For Scenarios 2 and 3, several decades are needed before the deposition reductions bring NO_3-leaching below the target value.

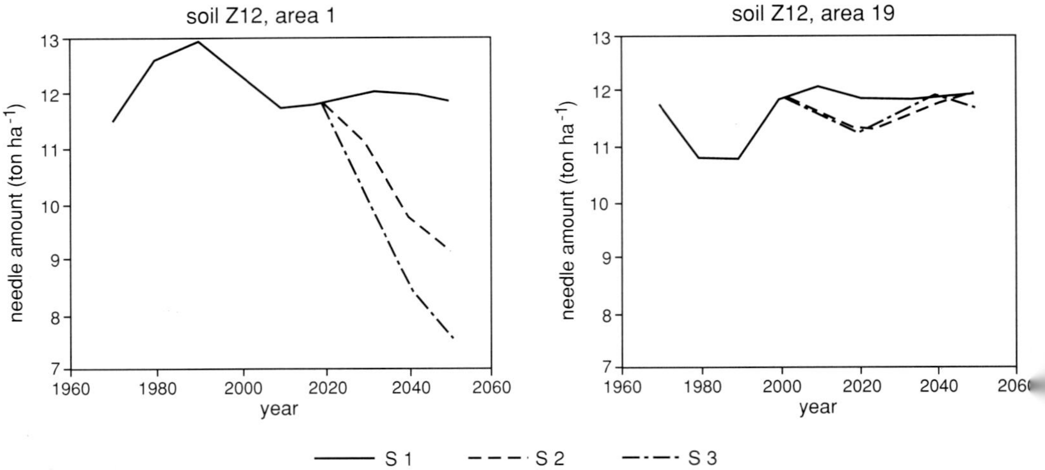

Fig. 6.13 Trends of needle mass in area 1 (Groningen) and area 19 (Venray), in soil type z12 for the three scenarios

6.4.3 Conclusions

Once more it is emphasized that all results and conclusions apply to a production forest of Douglas fir which has an optimum supply of water, is not subject to frost damage and plagues, and does not suffer from enhanced leaf exudation of nutrients. The major conclusions are:

I. In the south-eastern part of the Netherlands the increase of deposition of N and potential acidity between 1970 and 1990 leads to a net decrease of needle mass (forest health) and wood production. In relatively clean areas, e.g. coastal regions, there is a positive effect of enhanced N deposition on needle mass and wood production. When deposition and ambient concentration of NH_x and NO_x are lowered, a growth reduction may be expected in coastal regions, while a growth increase in the south-eastern half of the country is most likely.

II. There is an immediate and strong response of needle N content to N deposition and ambient concentration of NH_3. N contents above 2% are common in areas in the south-east between 1970 and 1990, which makes these forest areas particularly susceptible to frost damage and plagues.

III. Leaching of nitrate is highest under Leptic podzols (z06) and Albic Arenosols (z27). Low nitrate leaching under Humic (z12) and Gleyic (z08) podzols is caused mainly by stronger retention in organic matter, and to a lesser extent by larger denitrification losses.

IV. Douglas fir forest on Albic Arenosol (z27) is most sensitive to increased soil acidification. Drastic reduction of needle mass and stem mass of Douglas on this soil type occurs between 1970 and 2000, when deposition of N and potential acid are highest. The Leptic podzol is least sensitive to effects of soil acidification and increased N availability. Simulated needle mass and wood mass on Leptic podzol (z06) are similar both in the simulation with unchanged deposition after 1950 and with real deposition after 1950.

V. The clearest positive effects of reduction of deposition are decreased N content in needles and decreased nitrate leaching, but this effect is rather independent of the deposition scenario. In 2050 N content in needles is near or below optimum values in all scenarios. In 2050 nitrate concentratiosn in soil leachates will, in general, meet drinking water standards, except for Scenario 1 in the most polluted areas, under z06 and z27 soils.

VII. Comparison of different scenarios after 2000 is rather meaningless, except for coastal areas where N is growth-limiting after 2030. For the other cases, changes (and uncertainties) of forest response are dominated by the presumed large reduction of deposition between 1990 and 2000.

Finally, it is stated that the model predictions with SOILVEG would be improved by including submodels for (1) a dynamic soil water balance, (2) stochastic prediction of plagues, frost damage and drought stress and (3) stochastic prediction of phosphorus availability and uptake.

Literature

Bakema A.H., Boer K.F.de, Bultman G.W., Grinsven J.J.M., Heerden C.van, Kok R.M., Kros J., Mohren G.M.J., Olsthoorn T.N., Vries W.de and Wortelboer F.G., (1990) Dutch Acidification Systems Model - Specifications. Dutch Priority Programme on Acidification rep. nr. 114.1-01

Grinsven J.J.M.van, Janssen P.J.M., Minnen J.G.van, Heerden C.van, Berdowski J.J.M. and Sanders R., (1991). SOILVEG: A model to evaluate effects of acid atmospheric deposition on soil and forest. Volume 2. Uncertainty analysis. Dutch Priority Programma on Acidification rep. no. 114.1-03

Minnen J.G.van, Grinsven J.J.M. van and Heerden C.van, (1991)
SOILVEG: A model to evaluate effects of acid atmospheric deposition on soil and forest. Volume 3. Regional analysis of long-term effects of various deposition scenarios on Douglas fir. RIVM/RIN report. Dutch Priority Programma on Acidification rep. no. 114.1-04

Olsthoorn T.N., 1990
Integrated Modelling. In: Thematic reports. Report Dutch Priority Programme on Acidification nr. 200-07

Reurslag A., Zuidema G. and Vries W.de, 1990
De indirekte effekten van atmosferische depositie op de vitaliteit van Nederlandse bossen. 3. Simulatie van de waterbalans van acht Douglasopstanden. Wageningen, Staring Centrum, Rapport 76

SOILVEG: A model to evaluate effects of acid atmospheric deposition on soil and forest, (1991)
Volume 1. Model principles and application procedures. Dutch Priority Programme on Acidification rep. no. 114.1-02
Part A. Model structure, vegetation processes and preliminary results. Berdowski J.J.M., Heerden C.van and Minnen J.G.van
Part B. Soil processes. Van Grinsven J.J.M., Minnen J.G.van, Heerden C.van, Berdowski J.J.M. and Vries W.de
Part C. Users guide. Van Minnen J.G. and Heerden C.van

6.5 Effects on heathland

G.W.Heil (Resource Analysis), F.Berendse (CABO) and A.H.Bakema (RIVM)

6.5.1 Introduction

The Dutch Acidification Systems model contains two models, CALLUNA and ERICA, to describe the effects of acidification on heathland. CALLUNA describes the effects on dry heathland and ERICA describes the effects on wet heathland (see also Chapter 5). Both models have been adjusted so that regionalized deposition figures from the DAS model can be used.

CALLUNA (Heil, 1991) is based on data from the literature. This model has only been validated to a limited extent. No uncertainty analysis has been made so far. The model was not described in the report Dutch Acidification Systems Model - Specifications (Bakema et al., 1990). A short description is therefore given below.

ERICA is a new implementation of the NUCOM2 model (Berendse, 1988), and is described in more detail in Bakema et al., 1990.

The values of the parameters used in ERICA are mainly based on field experiments made during the period 1981 to 1985. The model was validated with data from the same experimental plots in which the model parameters were determined, as well as with data from other experimental plots (Berendse, 1988).

6.5.2 Description of the model and input parameters

6.5.2.1 Description of the model

The ERICA model has been described before, as was mentioned above. We shall therefore confine ourselves to a short description of the CALLUNA model.

CALLUNA describes the competition between Calluna vulgaris and the grass Deschampsia flexuosa. The dry heathland vegetations in the Netherlands are becoming increasingly overgrown with this grass. A schematic representation of the model is given in Figure 6.14.

A vegetation one year after sod-cutting was chosen as a basis for the simulations which were used for DAS scenarios (Section 6.2.2). The cover percentage of Calluna vulgaris was assumed to be 1 and that of Deschampsia flexuosa 0.1. These initial cover percentages are based on differences in seed supply and in germination and establishment of Calluna vulgaris and Deschampsia flexuosa after sod-cutting.

Atmospheric deposition forms the only nitrogen input into the model. Several nitrogen compartments are distinguished in the model: biomass compartments, consisting of the biomass of the two species and a soil compartment comprising several subcompartments.

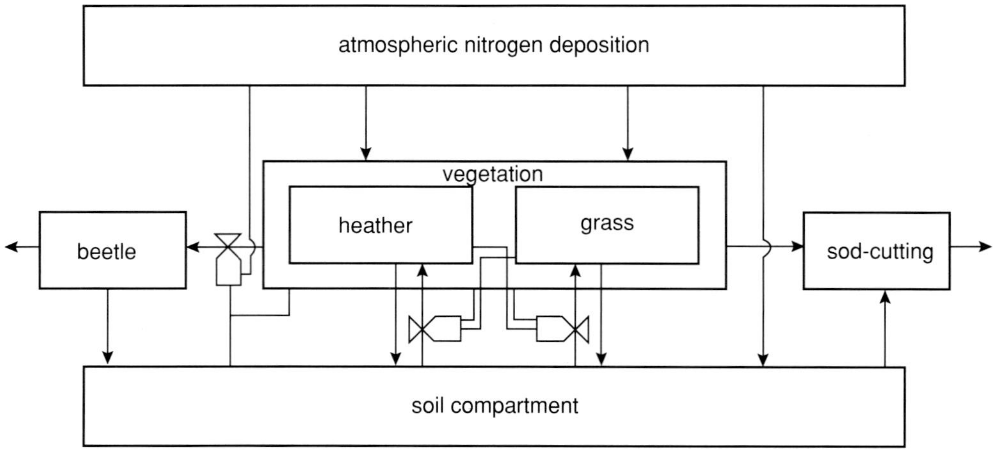

Fig. 6.14 Schematic representation of the CALLUNA model

The composition of the soil compartment depends on the vegetation. Via the various soil subcompartments the nitrogen is gradually released again so that it can be taken up by the vegetation. The nitrogen which is not taken up by the vegetation is immobilized. Nitrogen is removed by sod-cutting. After sod-cutting the nitrogen supply in the various compartments is assumed to be zero and the cover percentage of the various species is initialized again.
In the simulations heather beetle plagues can occur, if the cover percentage of Calluna is 50 or higher. The risk of a plague increases with the amount of available nitrogen in the system, according to:

$$P_{plague} = 0.006 + 0.006 * N_{available}$$

In which:

P_{plague} the risk of a heather beetle plague ($year^{-1}$)

$N_{available}$ the available nitrogen from deposition and mineralization (kg N ha^{-1} $year^{-1}$)

The occurrence of a plague reduces the cover of Calluna to 1% of the original cover. The Calluna which has died passes into the soil compartment.
Calculations with CALLUNA are based on cover percentages of the various species, rather than on biomass, on which calculations with ERICA, the model for effects on wet

heathland, are based.
Availability of nitrogen influences the growth and competitive strength of the various species. The cover percentage is used to assess the competition of the species. The model results refer to the Calluna cover percentage in relation to the Deschampsia cover percentage.

6.5.2.2 Sod-cutting

In the simulations a heathland management consisting of sod-cutting every 25 years is assumed. This frequency is based on the original use of heathland at the beginning of this century.
In order to assess the effects of atmospheric N deposition over a period of 100 years, the frequency of sod-cutting is assumed to be constant. After every period of 25 years an evaluation is made of the competitive strength of the heathland vegetation during this period. Although from the point of view of nature conservation other forms of management might be preferred, such forms have not been taken into account owing to lack of uniformity.

6.5.2.3 N deposition data

The total nitrogen deposition on heathland was calculated as the sum of dry and wet deposition of NH_x and NO_y, the dry deposition first being multiplied by the area-dependent deposition factor for heathland (see Section 6.2.2).
The data on the deposition of nitrogen in the different acidification areas show an overall change in the level of deposition. In 1980 the deposition reached a maximum in all areas. From that year onwards the deposition has shown a gradual decrease.
For a good interpretation of the results, the acidification areas were classed into three groups, depending on the deposition level for the heathland in the particular area in 1990. Group I contains the areas with a deposition to heathland below 3000 mol N per hectare per year. The areas 1, 2, 3, 10 and 15 fall under this group. The areas in group II have a deposition between 3000 and 4000 mol N per hectare per year. This group consists of the areas 4, 6, 8, 9, 11, 16 and 20. The remaining areas come into group III, having a deposition of over 4000 mol N per hectare per year, namely the areas 5, 7, 17, 18 and 19.

6.5.2.4 Method of calculation

To evaluate the effects of the different scenarios on the development of heathland vegetations, calculations have been made with CALLUNA and with ERICA for 17 out of the 20 different Dutch acidification areas. The areas 12, 13 and 14 (Leiden, Rijnmond, and Zeeland) were not taken into account, owing to the fact that there is hardly any heathland in these areas.

With these two models the development of a heathland vegetation during the period 1950 to 2050 was simulated for each of the 17 areas, in three different scenarios.

On account of the stochastic nature of the occurrence of heather beetle plagues, the results of 100 simulations with CALLUNA were averaged. This average was used to interpret the effects of the three scenarios for each area.

6.5.3 Results

6.5.3.1 Examples of model calculations

By way of example, the results of the model calculations for the areas 6 (Veluwe) and 18 (Eindhoven) are presented in Figures 6.15 to 6.18. Area 6 falls under group II and has a medium nitrogen deposition level, whereas area 18 has one of the highest nitrogen deposition levels.

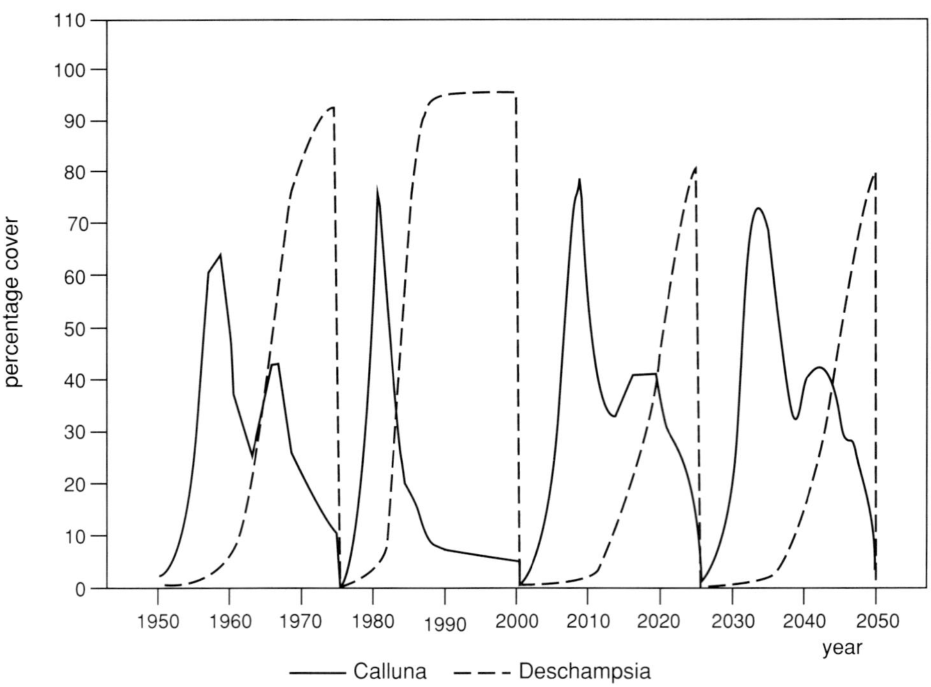

Fig. 6.15 Average result of 100 simulations with the CALLUNA model for area 6 and Scenario 1

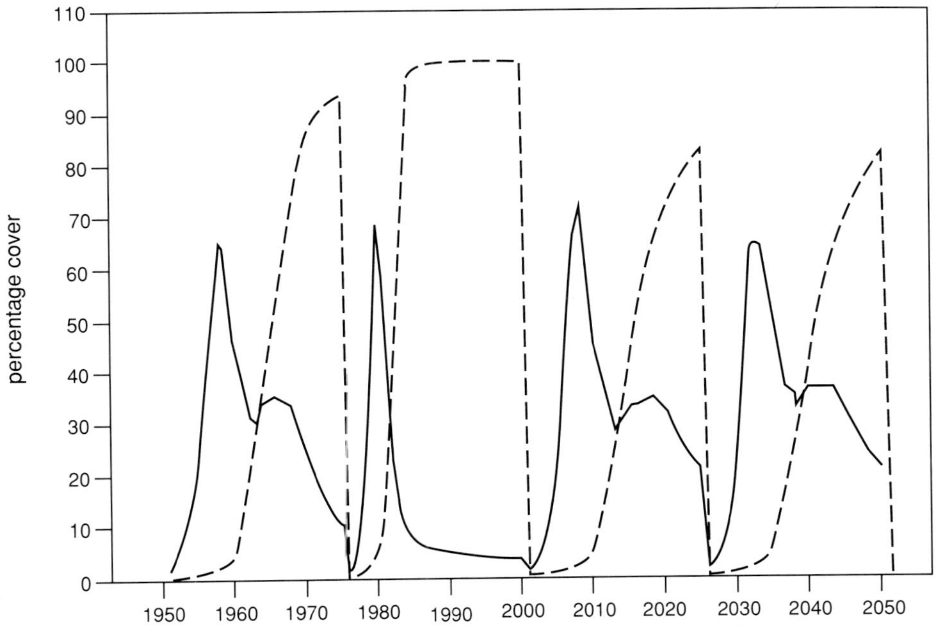

Fig. 6.16 Average result of 100 simulations with the CALLUNA model for area 18 and Scenario 1

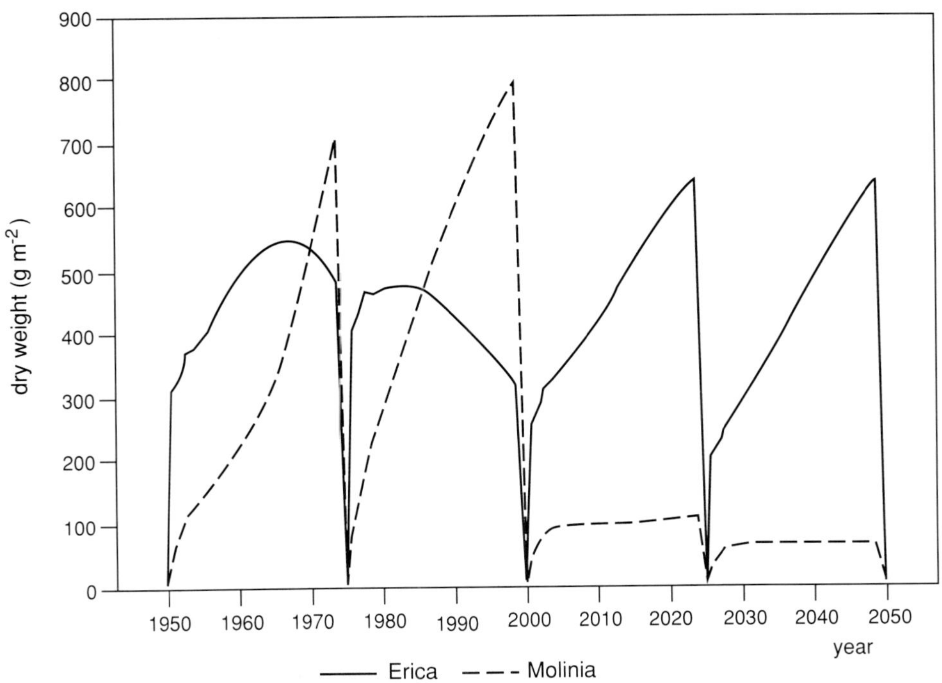

Fig. 6.17 Results of a simulation with the ERICA model, for area 6 and Scenario 1

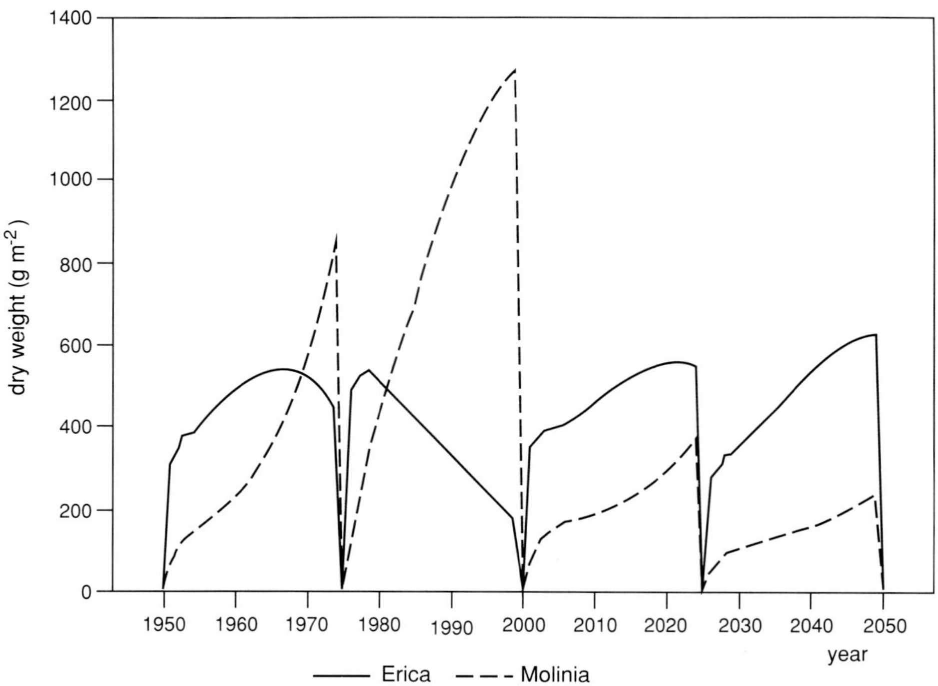

Fig. 6.18 Results of a simulation with the ERICA model, for area 18 and Scenario 1

The results of the model for effects on wet heathland, ERICA, are not entirely comparable with the results of the model for dry heathland, CALLUNA. This is caused by the influence of the occurrence of heather beetle plagues in dry heathland vegetations, the different units used in the calculations (ERICA calculates in biomass, while CALLUNA calculates in cover percentages) and a difference in the quantity of organic matter which is removed in sod-cutting (in CALLUNA all organic matter is removed, whereas in ERICA 25% remains after sod-cutting).

6.5.3.2 Scenario results

CALLUNA

The results of CALLUNA are evaluated on the basis of the cover percentage of Calluna vulgaris in relation to the cover percentage of the grass Deschampsia flexuosa at the end of each period following sod-cutting (Table 6.10). Although a higher cover percentage of grass does not mean that the heathland was overgrown by grass during the entire period, the classing of areas into one group in which the cover percentage of Calluna is lower than that of grass, and another group in which the cover percentage of Calluna is higher gives a good idea of the differences between the scenarios.

Table 6.10 Scenario results for Calluna

	Group I	Group II	Group III
1950 - 1975			
Scenarios 1, 2, 3	-	-	-
1975 - 2000			
Scenarios 1, 2, 3	-	-	-
2000 - 2025			
Scenario 1	-	-	-
Scenario 2	++	-	-
Scenario 3	++	+	-
2025 - 2050			
Scenario 1	-	-	-
Scenario 2	++	+	-
Scenario 3	++	++	++

Group I	area 1, 2, 3, 10 and 15
Group II	area 4, 6, 8, 9, 11, 16 and 20
Group III	area 5, 7, 17, 18 and 19
++	cover percentage of Calluna higher than cover percentage of Deschampsia in all areas
+	cover percentage of Calluna higher than cover percentage of Deschampsia in half the number of areas or more
-	cover percentage of Deschampsia higher than cover percentage of Calluna in more than half the number of areas

Table 6.10 shows that until the year 2000, at the end of each period following sod-cutting, the cover percentage of Deschampsia will always be (much) higher than the cover percentage of Calluna.

For the period 2000 to 2025 Scenario 1 does not meet the target for Group I that Calluna must be the dominant species after 25 years. Scenario 2 meets the target for Group I but not for Group II. Scenario 3 meets the target entirely for Group I and partly for Group II.

For the last period, 2025 - 2050, Scenario 1 gives an outcome which is similar to the outcome for the previous period (the deposition level remains the same in this scenario). For this last period Scenario 2 entirely meets the target for Group I and partly for Group II. Scenario 3 meets the target for all areas, for this period.

ERICA

The results of computations with ERICA were evaluated on the basis of the biomass of Erica compared to the biomass of Molinia, in different scenarios, for four different periods of 25 years following sod-cutting.

Table 6.11 Results of calculations with ERICA, based on three different scenarios

Group	I	II	III
1950-1975			
Scenario 1, 2, 3	-	-	-
1975-2000			
Scenario 1, 2, 3	-	-	-
2000-2025			
Scenario 1	++	++	+
Scenario 2, 3	++	++	++
2025-2050			
Scenario 1, 2, 3	++	++	++

Group I area 1, 2, 3, 10 and 15
Group II area 4, 6, 8, 9, 11, 16 and 20
Group III area 5, 7, 17, 18 and 19
++ biomass of Erica greater than biomass of Molinia in all areas
\+ biomass of Erica greater than biomass of Molinia in half the number of areas or more
\- biomass of Molinia greater than biomass of Erica in more than half the number of areas

In the period 1950 to 2000, at a sod-cutting frequency of once per 25 years, the biomass of Erica exceeds that of Molinia in only two areas, namely area 1 and 10. These are the areas with the lowest nitrogen deposition. In all other areas the biomass of Molinia is usually (much) greater at the end of the period following sod-cutting.

In the period 2000 to 2050 Erica usually has the greatest biomass in almost all areas. Only in calculations based on Scenario 1 for the period 2000 - 2025 is Molinia dominant in the areas 7, 17 and 19. However, the difference between the two species is rather small, and the heathland vegetation is certainly not entirely overgrown with grass.

Erica seems to be able to compete quite well in that period, at a sod-cutting frequency of once per 25 years. The calculations based on Scenario 2 and those based on Scenario 3 do not show any great differences.

6.5.4 Conclusions

CALLUNA

The results of calculations with CALLUNA lead to the conclusion that Scenario 1 does not offer any prospects for the continued existence of the typical dry heathland vegetations that were still found at the beginning of this century. Scenario 3 offers good prospects for such vegetations. The results for the period 2000 - 2025 for the areas in Group II and III vary, however. For the period 2025 - 2050 Scenario 3 meets the target for all groups. For the period 2000 - 2050 Scenario 2 gives results which are between those of the Scenarios 1 and 3.

ERICA

Model calculations based on Scenarios 2 and 3 show that at a sod-cutting frequency of once per 25 years, wet heathland vegetations can compete in all areas during the period 2000 - 2050. Calculations based on Scenario 1 seem to offer good prospects for the longer term (after 2025).

Literature

Bakema A.H., Boer K.F.de, Bultman G.W., Grinsven J.J.M.van, Heerden C.van, Kok R.M., Kros J., Minnen J.G.van, Mohren G.M.J., Olsthoorn T.N., Vries W.de and Wortelboer F.G., 1990
Dutch Acidification Systems Model-Specifications
Dutch Priority Programme on Acidification, report nr. 114.1-01

Berendse F., 1988
De nutriëntenbalans van droge zandgrondvegetaties in verband met de eutrofiëring via de lucht. Deel 1. Een simulatiemodel als hulpmiddel bij het beheer van vochtige heidevelden
Centrum voor Agrobiologisch Onderzoek, publikatie m743

Heil G.W., 1991
Critical load of atmospheric nitrogen deposition in natural ecosystems: an example of heathlands
in prep.

6.6 Effects on crops, materials and monuments

A.H.Bakema and F.G.Wortelboer (RIVM)

6.6.1 Introduction

DAS contains damage functions for three types of damage from acidification: the damage to agricultural and horticultural produce (yield reduction), damage to materials (higher costs of maintenance and earlier replacement), and damage to stone monuments (earlier loss of cultural assests owing to faster ageing).

The damage functions used for calculating damage to crops are described in Van der Eerden et al. (1986, 1988) and those used for calculating damage to materials and monuments in Gosseling et al. (1990).

A more detailed description of the models can be found in Bakema et al., 1990.

No uncertainty analysis of the models has been carried out so far. Nor have the damage relationships been validated with measurements yet. All dose-effect relationships were based on experiments. The quantities of exposed crops, materials and monuments and the distribution over the Netherlands were assumed to be constant during the whole period 1950 to 2000.

Gosseling et al. (1990) report that no reliable damage relationships could be derived for stone monuments. The relationship used was based on only one experiment and refers to only one type of stone. It therefore gives no more than an indication of the effects that are to be expected. The equations used in calculating the damage to materials and crops also show large uncertainties. (Gosseling et al., 1990, Van der Eerden et al., 1988).

6.6.2 Results

The effects of Scenarios 1, 2, and 3 on damage to crops, materials, and monuments were calculated using these modules. Differences in calculated effects with respect to the different scenarios only occur for damage to monuments. The results of the calculations are briefly discussed below.

6.6.2.1 Damage to crops

The damage is expressed as an average percentage yield reduction for all acidification areas in the Netherlands. The contributions of sulphur dioxide and of the combination of sulphur dioxide and ozone to this yield reduction are shown separately. The spatial distribution of crops over the Netherlands is based on Van der Eerden et al. (1985). The calculations made indicate the yield reduction which would occur if during the period 1950 to 2050 in the entire Netherlands the same crops were grown as in 1985. The results are shown in Figure 6.19.

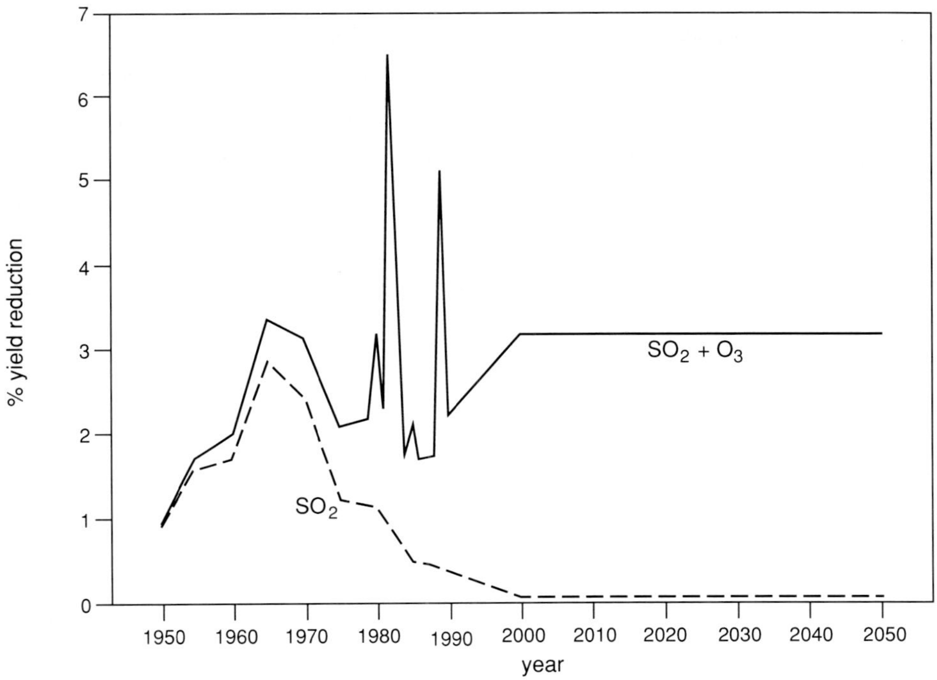

Fig. 6.19 The average damage to crops calculated on the basis of Scenario 1, 2 and 3, as a percentage of yield reduction for the areas with crops in 1985. The damage caused by sulphur dioxide and that caused by sulphur dioxide and ozone are indicated separately

It is clear that the effect of sulphur dioxide is of major importance during the period until 2000, whereas the effect of ozone increases from the year 1980 onwards. The significant fluctuations in the damage from ozone during the period from 1980 to 1990 are due to the fact that the calculations for this period were based on measured ozone concentrations.

The yield reduction averages a few percentage points and remains at a constant level of 3 % after 2000. After this year the effect of sulphur dioxide is negligible. The damage from sulphur dioxide decreases to zero after 2000, when the scenarios are going to differ. Since all three scenarios are based on the same ozone concentrations, the results are the same in each scenario (6.2.4).

6.6.2.2 Damage to materials

The damage function used to calculate the damage to materials is based on the damage from

sulphur dioxide concentrations. The damage is expressed as a percentage increase of the costs of maintenance and replacement owing to increased corrosion of galvanized steel, painted steel, duplex steel, zinc and galvanized objects.
The values refer to the average damage for the entire Netherlands. The spatial distribution of the materials over the Netherlands and the quantity of materials exposed are based on data for the years 1977 to 1986 inclusive (Gosseling et al., 1990). In the absence of any damage from air pollution, the total costs of maintenance and replacement for this period are estimated at about 750 million guilders a year. Expressed as a percentage increase of the costs, the damage is independent of the absolute quantity of the different materials. Only the interrelationship of the different materials and the spatial distribution influence the calculated values. Some results are shown in Figure 6.20.

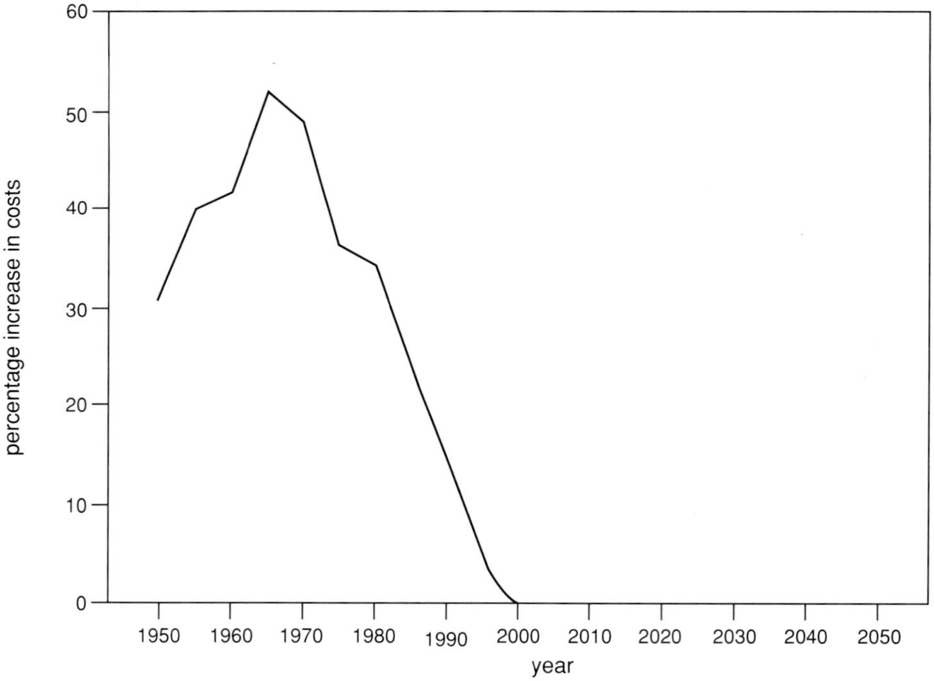

Fig. 6.20 The calculated damage to materials for Scenarios 1, 2 and 3 as a percentage increase in costs of maintenance and replacement

6.6.2.3 Damage to monuments

The damage function for calculating the damage to stone monuments is based on measurements of the dry deposition of sulphur dioxide and shows a large uncertainty. The damage is caused by a decrease in the amount of stone in the monuments. Figure 6.21

shows the estimated damage as a percentage increase of the rate of natural weathering. Since the damage is assumed to have a linear relationship with the dry deposition of sulphur dioxide, and since the contribution of each pollutant to the total acid deposition for the period from 2000 onwards is equal in the three scenarios, the ratios of the damage in these three scenarios are equal to the ratios of the total acid deposition in them.

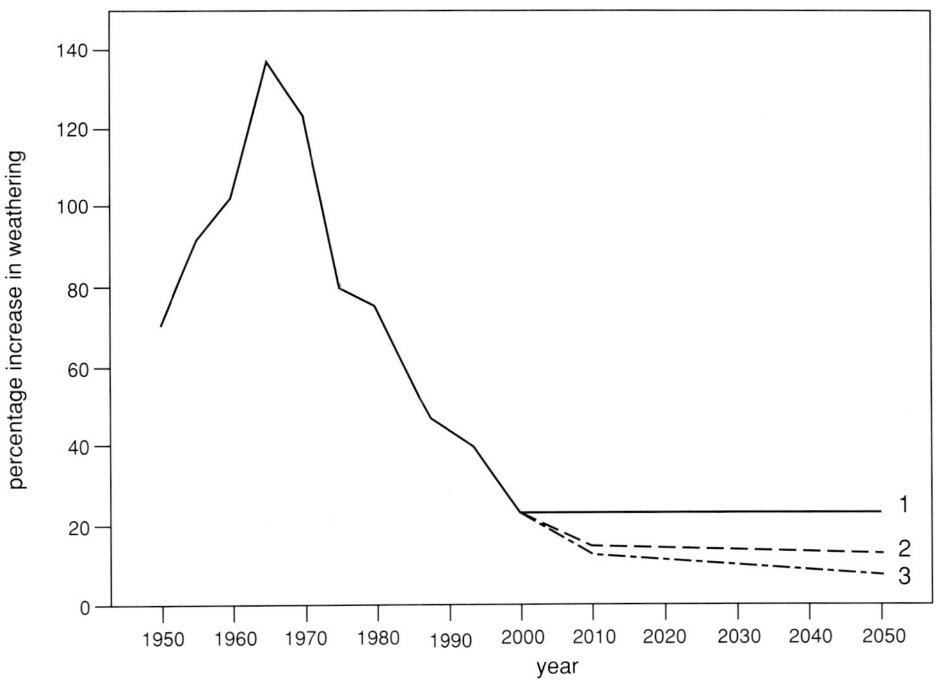

Fig. 6.21 The damage to monuments calculated as a percentage increase of the weathering rate, as compared to a situation with no deposition of sulphur dioxide, for the three scenarios.

6.6.3 Conclusions

In the scenarios which were used, the effects of sulphur dioxide on crops are estimated to be

practically zero after the year 2000. The effect of ozone remains considerable, however. Since in all three scenarios the same assumptions are made with regard to future ozone concentrations, the damage to crops is estimated to be the same in these scenarios.
On the basis of the damage functions used, the damage to materials shows a strong decrease up to the year 2000. For all scenarios the damage is reduced to zero after the year 2000. However, the damage functions show large uncertainties.
With regard to the damage to stone monuments the results only serve as an indication. Up to the year 2000 the damage shows a strong decrease. The differences in damage after the year 2000 fit in with the ratios of total acid deposition in the three scenarios.

Literature

Bakema A.H., Boer K.F.de, Bultman G.W., Grinsven J.J.M.van, Heerden C.van, Kok R.M., Kros J., Minnen J.G.van, Mohren G.M.J., Olsthoorn T.N., Vries W.de and Wortelboer F.G., 1990
Dutch Acidification Systems Model-Specifications
Dutch Priority Programme on Acidification, report nr. 114.1-01

Eerden L.J.van der, Tonneijck A.E.G.and Wijnands J.H.M., 1985
Schade door luchtverontreiniging aan de agrarische produktie in Nederland
Concept rapport IPO-LEI

Eerden, L.J.van der, Tonneijck A.E.G. and Wijnands J.H.M., 1986
Economische schade door luchtverontreiniging aan de gewasteelt in Nederland
IPO rapport R 324

Eerden L.J.van der, Tonneijck A.E.G. and Wijnands J.H.M., 1988
Crop loss due to air pollution in The Netherlands
Env. Pollution 53: 365-376

Gosseling H.J., Olsthoorn A.A. and Feenstra J.F., 1990
Economische schade aan materialen en monumenten door verzuring
IVM rapport W-90/002

7. CRITICAL LOADS AND CRITICAL LEVELS FOR THE ENVIRONMENTAL EFFECTS OF AIR POLLUTANTS

W.de Vries (Winand Staring Centre) and G.J.Heij (RIVM)

7.1 The concept of critical loads and critical levels

Political decisions regarding control of air pollution require scientific determinations of the loads and levels at which pollutant deposition and ambient concentrations cause adverse environmental effects. The critical load concept constitutes, together with a technology-based approach, the two track approach that is the basis for the development and adaption of the Netherlands' acidification abatement policy. The Dutch government presented this policy in its present form in the 1989 National Environmental Policy Plan, and elaborated upon it in the Netherlands' Acidification Abatement Plan in 1989.

Within the context of this report, the following definitions of critical load and critical level are used (Convention on Long Range Transboundary Air Pollution, 1990):

Critical Load: a quantitative estimate of an exposure (deposition) to one or more pollutants below which significant harmful effects on specified sensitive elements of the environment do not occur according to present knowledge.

Critical Levels: concentrations of pollutants in the atmosphere above which direct adverse effects on receptors, such as plants, ecosystems or materials, may occur according to present knowledge.

In deriving critical loads, a long-term perspective has to be considered, as most of the effects are results of accumulated depositions.

The term "critical load" should not be confused with "target load", which is related to political decisions. The critical load is an inherent property of the environment, whereas the target load is less restrictive and also implies other considerations. The value of the latter depends on the presence of sensitive receptors, the economical and/or intrinsic value put on these receptors, the available abatement technology and the costs of emission reductions.

7.2 Critical loads

7.2.1 Method and criteria to derive critical loads

In the Netherlands, critical loads have mostly been derived in an indirect way (see Chapter 4., section 4.2). In this approach, critical chemical values for ion concentrations or ion ratios in the ecosystem, based on dose-response relationships between these criteria and the ecosystem status, play an important role. Use has also been made of empirical data which directly relate loads to effects.

An overview of the various chemical criteria that have been used to derive critital loads for terrestrial ecosystems is given in Table 7.1. These criteria have been discussed in Chapter 4.

The critical Al concentration and Al/Ca ratio have been used to derive a critical acid load, whereas the NH_4/K ratio, the NO_3 concentration and the N content of the needles have been used to derive a critical nitrogen load, independent of the acidification aspect. However, there is a relationship between critical acid and critical nitrogen loads (see Thematic Report by De Vries and Kros, and De Vries, 1991)

Table 7.1 Critical values for Al concentration, Al/Ca ratio, NH_4/K ratio, NO_3 concentration and N content in forest soils (soil solution), groundwater, and needles

Criteria	Unit	Forest soils (soil solution)	Groundwater	Coniferous tree needles
Al 1)	$mol_c m^{-3}$	0.2	0.02	-
Al/Ca	mol mol^{-1}	1	-	-
NH_4/K	mol mol^{-1}	5	-	-
NO_3	$mol_c m^{-3}$	-	0.4 2), 0.8 3)	-
N	% of dry weight	-	-	1.8

1) refers to (inorganic) aluminium
2) refers to the target value for drinking water (25 $mg.l^{-1}$)
3) refers to the standard value for drinking water (50 $mg.l^{-1}$)

7.2.2 Assessment of critical loads with steady-state models

The critical loads presented in this chapter have been derived with a one-layer steady state mass balance model. The model is based on the implicit assumption that dynamic processes such as cation exchange, adsorption/desorption of sulphate and ammonium, and mineralization/immobilization of nitrogen, sulphur and base cations are unimportant for the assessment of long-term critical loads. The model only includes dry deposition of base cations (seasalt corrected), base cation weathering, net uptake of nitrogen and base cations, and a critical leaching rate of acidity (H + Al - HCO_3). It ignores the effect of nutrient cycling by litterfall, mineralization and root uptake on the soil solution composition in the rootzone. Further information is given in De Vries, 1991 and in the Thematic Report by De Vries and Kros.

7.2.3 Average critical loads for forests, heathlands and groundwater

An overview of the assessment of critical nitrogen and acid loads to forest ecosystems, heathlands, groundwater and surface water in the Netherlands is given in De Vries (1991) and in the Thematic Report by De Vries and Kros. The average critical loads for forests given in this paragraph, are mainly based on these reports. Values are related to well drained non-calcareous sandy soils. Most Dutch forests (ca. 80%) occur on these soils.

A survey of the average critical nitrogen loads on well-drained sandy soils in the Netherlands, is given in Table 7.2, together with data derived during the first phase of the Dutch Priority Programme on Acidification (Schneider and Bresser, 1988).

Table 7.2 Average critical nitrogen deposition levels ($mol_c\ ha^{-1}yr^{-1}$) for terrestrial ecosystems on well-drained sandy soils in the Netherlands

	coniferous forests	deciduous forests	heathlands
vegetation changes	400 - 1400	600 - 1400	-
	(400 - 800)	(400 - 800)	(350 - 700)
elimination by grasses		-	700 - 1100
	-	-	(1400 - 2100)
frost damage/fungal diseases	1500 - 3000	-	-
	(1500)	(1500)	-
nutrient imbalances	800 - 1250 1)		-
	(1000)	(1500)	
nitrate leaching to groundwater	900 - 1500	1700 - 2900	2000 - 3600
	(1000 - 1600)	(1600 - 2800)	(2000 - 3600)

1) worst-case (total inhibiting of nitrification)
( ) value, presented in the first phase and used as a basis for present abatement policy

Regarding vegetation changes in forests, the first values are model-based, using an average net nitrogen uptake of 300 - 500 $mol_c ha^{-1}yr^{-1}$, and a natural nitrate leaching rate of 100 mol_c $ha^{-1}yr^{-1}$. The value of 1400 is derived from experimental data on loads above which vegetational changes do occur (Tyler, 1987). The critical loads related to elimination by grasses refer to Caluna vulgaris; at these loads, many heathland vegetation types have disappeared already (Thematic Report by Van Dobben, 1990).

The critical loads on coniferous forests related to frost damage and fungal diseases are based on a dynamic soil-vegetation model (first value, De Vries, 1988), and on empirical data (second value, Van den Burg and Kiewit, 1989) assuming a critical nitrogen content in needles of 1.8% of the dry weight (see Chapter 4.). From the Dutch model SOILVEG (Chapter 6., section 4) a value of 2000 might be derived, for coniferous forests.

The values related to nutrient imbalances refer to NH_3-N only and are also based on both experimental data (first value, Boxman et al., 1988) and a simple steady state nutrient-ratio model (second value, De Vries, 1991), using a critical NH_4/K ratio of 5 in the topsoil and assuming no nitrification in the topsoil (worst-case option).

The two values for nitrate leaching underneath forest soils are based on a net nitrogen uptake of 300 - 500 mol_c $ha^{-1}yr^{-1}$ and a leaching rate related to a critical nitrate concentration of 25 mg l^{-1} and 50 mg l^{-1} respectively (see Table 7.1) and a precipitation surplus of 150 and 300 mm yr^{-1} for coniferous and deciduous forests respectively. For heathlands a precipitation surplus of 400 mm.yr^{-1} has been taken and a net nitrogen uptake of 400 $mol_c ha^{-1}yr^{-1}$ (De Vries, 1988; 1991).

Compared to the first phase of the Dutch Priority Programme on Acidification, the range for heathland (for the criterion of elimination by grasses) has been lowered, resulting from the heathland research (see Thematic Report by Van Dobben). For vegetation changes in heathland no values have been presented in Table 7.2 (resulting from the second phase) because of lack of sound research results.

For frost damage/fungal diseases and nutrient imbalances a range is given now, based on model caculations and on empirical data, instead of one value. For nitrate leaching to groundwater the values have remained nearly unchanged. The critical load for deciduous trees for the criterion frost damage (1500 mol_c $ha^{-1}yr^{-1}$) reported in the first phase (Schneider and Bresser, 1988) has no sufficient support from research and has been left out of this report.

It can be concluded that most effects of nitrogen deposition will be prevented at loads below 1000 mol_c $ha^{-1}yr^{-1}$. At this load vegetation changes might still occur in the long-term. However, effects will be slowed down considerably compared to the effects caused by the present nitrogen input in the Netherlands.

An overview of the derived critical acid loads on well-drained sandy soils is given in Table 7.3, together with data derived during the first phase of the Dutch Priority Programme on Acidification.

Table 7.3 Average critical total acid loads (mol_c $ha^{-1}yr^{-1}$) for terrestrial ecosystems on well-drained sandy soils and surface waters in the Netherlands

Effects	Coniferous forests	Deciduous forests	Surface waters
root damage	1100 - 1400[1)] (1400)	1400 - 1700[2)] (1800)[3)]	
aluminium depletion	1200	1500	
aluminium leaching to groundwater	500[4)] (200)	300[4)] (200)	
fish dieback			- (400)

1) the first value is related to a critical Al concentration of 0.2 mol_c m^{-3} and the second value to a critical Al/Ca ratio of 1.0
2) the first value is related to a critical Al/Ca mol ratio of 1 and the second value to a critical Al concentration of 0.2 mol_c m^{-3}
3) for deciduous forests on rich soils = 2400
4) these values are related to a depth of 2 m. (the average phreatic level)
( ) values presented in the first phase

All values have been derived by the simple mass balance model (Chapter 7., section 2.2). Although this steady state model has been developed for application on forest soils, it can also be used to derive critical loads for groundwater. The only difference in the application for the various receptors is a change in system boundaries, i.e. the root zone for forest soils and the unsaturated zone (depth to phreatic level) for groundwater. This influences the areal weathering rate (in mol_c $ha^{-1}yr^{-1}$) by a difference in the depth of the considered soil profile. Apart from weathering there is also a difference in critical acidity leaching for the various receptors, due to different criteria for the critical acidity value. In forest soils, the critical acidity is positive. It thus leads to an increase in the critical load value by allowing a certain rate of Al mobilization (leaching). Contrary to forest soils, the critical acidity value for groundwater is negative, since the critical pH for this system is above 5- causing HCO_3 production.

For all receptors the waterflux has been taken to be equal to the precipitation surplus or runoff, i.e. the water draining from the root zone. It should be noted that the precipitation surplus is not a very adequate value for the waterflux in forest soils. The critical chemical levels for Al and Al/Ca (Table 7.1) are related to the first 30 cm, where most of the fine roots occur. Consequently, it would be better to use the annual waterflux at this depth. However, this also affects the weathering rate and uptake of base cations and nitrogen (expressed in mol_c $ha^{-1}yr^{-1}$), thus requiring the use of multi-layer models. Since the one-layer mass balance model will be used to derive critical load maps for Europe, a multi-layer approach was not used for reasons of consistency (see also Chapter 7., section 2.4).

The values in Table 7.3 regarding forests are all based on:

- a seasalt corrected dry base cation deposition of 300 mol_c $ha^{-1}yr^{-1}$ for coniferous forests and 150 mol_c $ha^{-1}yr^{-1}$ for deciduous forests;
- a net uptake of base cations equal to 300 mol_c $ha^{-1}yr^{-1}$ for coniferous forests, and 350 mol_c $ha^{-1}yr^{-1}$ for deciduous forests ;
- a net uptake of nitrogen equal to 300 mol_c $ha^{-1}yr^{-1}$ for coniferous forests, and 500 mol_c $ha^{-1}yr^{-1}$ for deciduous forests (see before);
- a base cation weathering rate of 200 mol_c $ha^{-1}yr^{-1}$ for a root zone of 1m.;
- a critical H leaching rate of 300 mol_c $ha^{-1}yr^{-1}$ for coniferous forests and 600 mol_c ha^{-1} yr^{-1} for deciduous forests; this is calculated by multiplying a critical H concentration of 0.2 mol_c m^{-3} (pH = 3.7) at a depth of 30 cm. with the value of the precipitation surplus (150 mm yr^{-1} and 300 mm yr^{-1} for coniferous and deciduous forest respectively).

The critical Al leaching rate related to root damage is calculated from a critical Al/Ca molar ratio of 1.0 and a critical Al concentration of 0.2 mol_c m^{-3} (Table 7.1). The critical Al leaching rate related to depletion of Al hydroxides is based on the weathering rate of Al from primary minerals (see Thematic Report by De Vries and Kros; De Vries, 1991).

The values for Al leaching to groundwater have been derived by using the same values for base cation deposition, base cation uptake, nitrogen uptake and precipitation surplus. For the weathering rate a value of 400 mol_c $ha^{-1}yr^{-1}$ was used, related to a groundwater depth of 2 m. The acidity value used equals -0.14 mol_c m^{-3} (De Vries, 1991).
Compared to the first phase, critical loads related to root damage are very similar (a bit more stringent). Differences are due to the use of different models and data (cf. De Vries; 1991). Furthermore, a critical load related to Al depletion has been introduced in contrast with the first phase. Values are very similar to those derived for root damage. The increase in critical loads for groundwater compared to the first phase are mainly due to the present inclusion of N uptake in the calculation and more accurate data about base cation deposition and base cation uptake. The critical loads for surface water as reported in the first phase, i.e. 400 - 700 mol_c $ha^{-1}yr^{-1}$, have not changed.

7.2.4 Uncertainties

The uncertainty in the average critical load values derived before can be large and is mainly determined by the uncertainty in (1) critical chemical values set for the receptor, (2) calculation methods and (3) data (Thematic Report by De Vries and Kros; De Vries, 1991; Sverdrup et al., 1990).

(1) Uncertainties in critical chemical values can be very large, especially for critical acid loads to forests (soils), since the range in Al toxicity appears to be very large for different tree species (De Vries, 1988; De Vries, 1991). A change in critical Al concentration directly influences the critical load by the critical acidity leaching term. The importance can be illustrated as follows: taking an average precipitation surplus of 200 mm yr^{-1}, which is a reasonable value for the forests in the Netherlands, and assuming a critical Al concentration of 0.2 $mol_c\ m^{-3}$, which corresponds to an acidity value of about 0.4 $mol_c\ m^{-3}$, leads to a critical acidity leaching of 800 $mol_c\ ha^{-1}yr^{-1}$. However, when a critical value of 5 mg l^{-1} is assumed, the acidity value equals about 0.8 $mol_c\ m^{-3}$ and the leaching term becomes 1600 $mol_c\ ha^{-1}yr^{-1}$, thus increasing the critical load by 800 $mol_c\ ha^{-1}yr^{-1}$. For groundwater the uncertainty in critical acidity leaching is much less, since the range in critical chemical levels for alkalinity is much lower.

The uncertainty in critical nitrogen loads to forests related to frost damage and nutrient imbalances can also be large, due to uncertain criteria, whereas the load related to nitrate leaching to groundwater is much more reliable, since 50 mg l^{-1} of NO_3 is an accepted standard value.

(2) Uncertainties in the calculation methods relate to the model structure and the assumptions that have been made to simplify the "real world". Unlike the uncertainty in critical chemical values and data, it is rather difficult to quantify the uncertainty due to modelling assumptions. The underlying premise in an uncertainty analysis is that the model structure is correct, or at least represents current knowledge adequately (see also Hettelingh, 1990, pp. 49 - 55).

In this context, it is important to note that the use of a one-layer model will most likely cause an underprediction of critical acid loads. The annual average waterflux at 30 cm depth (see Chapter 7., section 2.3) is much higher than the precipitation surplus, thus affecting the critical acidity leaching. This difference is approx. 100 - 150 mm yr^{-1}, which causes an increase in acidity leaching of 400 - 600 $mol_c\ ha^{-1}yr^{-1}$. The increase in critical load will be in the same order of magnitude, considering that the overall effect on the sum of weathering and uptake of nitrogen and base cations is small (cf. De Vries,

1991). The increase in critical load, if Al concentrations and Al/Ca ratios in the topsoil are considered, is illustrated by the results of the RESAM application (Chapter 6., section 3). Results of this multi-layer model show that approximately 50% of the forests on non-calcereous sandy soils is below critical soil solution levels in 2000 at an average load of 2200 mol_c $ha^{-1}yr^{-1}$. This can be regarded as an indication of the median critical acid load considering forest topsoils.

A large source of uncertainty may occur when deriving critical loads in specific forest areas due to the occurrence of N fixation, denitrification or a complex hydrology including seepage or surface runoff. However, the assumptions that have been made regarding these processes are reasonable for well drained forests and heathlands.

Finally, the uncertainties in critical loads for groundwater are high when using a simple mass balance model. Because of uncertainties regarding weathering and denitrification in the saturated zone, the phreatic level (a groundwater depth of 2 m) has been taken as the target, for the assessment of both critical acid and nitrogen loads related to nitrate leaching. At the depth to groundwater extraction (generally more than 10 m) critical loads may be much higher due to the occurence of these processes.

(3) Uncertainties in data are due to measurement errors, and non-representativity. The values that have been used for weathering, uptake and precipitation surplus are (long-term) averages for the receptors given. The uncertainty for the critical load at a specific location may be in the order of 50% due to non-representativity in these data.

Evaluation of the critical loads leads to the conclusion that below a load of 400 - 600 mol_c $ha^{-1}yr^{-1}$ total N no significant harmful effects will occur (related to vegetation changes). Concerning the critical acid loads, a lowest value of 300 - 500 mol H^+ $ha^{-1}yr^{-1}$ for groundwater (the criterion being the aluminium standard for drinking water) can be derived. The last mentioned value is about 10% of the present inputs (Chapter 3).

Uncertainties in the critical loads for forest ecosystems are large, due to the complexity of these ecosystems. The critical loads mainly have a warning function. For a correct interpretation of these critical loads it should be emphasized that exceeding these critical loads does not necessarily cause any visible effects on forests. If critical loads are exceeded temporarily, this will certainly not mean the dieback of (part of) the Dutch forests. Exceeding the critical load does, however, involve a certain risk to the health of forests, and this risk will increase as critical loads are exceeded further and for a longer period. When the critical loads are exceeded very considerably, as they presently are in large parts of the Netherlands, the health of forests is endangered.

7.3 Critical levels

Critical levels are used in relation to direct effects of air pollution.

The exposure of the vegetation to pollution is expressed in terms of a concentration and the duration of the exposure (to this concentration). Critical levels are therefore given as related values for the concentration and for the time during which this concentration may occur.

In order to assess critical levels, information is needed on quantative relations between exposure and effect. In many cases this information is not available because no experiments have been made.

With regard to <u>forests</u> there is presently no reason to deviate from the values which were presented in the evaluation report on the first phase (Schneider and Bresser, 1988). These values are shown in Table 7.4.

<u>Tabel 7.4</u> Critical levels of O_3 and SO_2 for forests

	coniferous forest (poor sandy soils)	deciduous forest	explanation
O_3-concentrations* (daily average)	50	50	these critical values are related to visual damage and the inhibition of translocation of assimilates to the root
SO_2 (annual average)	25	25	these values are related to visual damage

* during the growing season

With regard to <u>heathland</u> the following remarks can be made:

- The rare heathland species are sensitive to gaseous SO_2. Ecotoxicological risk analysis, carried out as part of the Dutch Priority Programme on Acidification, gave an (annual average) critical level of 8 $\mu g.m^{-3}$ (see Thematic Report by Van Dobben). This value is lower than the current annual average value for the Netherlands (about 10 $\mu g.m^{-3}$). Dominant species (Calluna and grasses) are hardly affected by SO_2 at current levels.
- Although O_3 is one of the most important air pollutants, there are hardly any data on the sensitivity of heathland vegetations to O_3. The Dutch Priority Programme on Acidification did not include any research on this.
- The effects of NH_3 on heathland vegetations were extensively studied as part of the

Additional Programme on Acidification, but a critical level could not be indicated.

Literature

Boxman A.W., Dijk H.F.G.van, Houdijk A.L.F.M. and Roelofs J.G.M., 1988
Critical loads for nitrogen with special emphasis on ammonium
In: J.Nilsson and P.Grennfelt (eds.), Critical loads for sulphur and nitrogen; report from a workshop held at Skokloster, Sweden, 19 - 24 March, 1988
Miljø rapport 1988: 15, Nordic Council of Ministers , København: 295 - 322

Burg J.van den and Kiewiet H.P., 1989
Veebezetting en de naaldsamenstelling van grove den, Douglas en Corsicaanse den in het Peelgebied in de periode 1956 tot en met 1988
De Dorschkamp, Rapport nr. 559

Convention on Long-Range Transboundary Air Pollution, 1990
Manual on methodologies and criteria for mapping critical levels/loads and geographical areas where they are exceeded
Prepared by the Task Force on Mapping with the assistance of the Secretariat of the United Nations Economic Commission for Europe (UN-ECE)

Hettelingh J.P., 1990
Uncertainty in modeling regional environmental systems; the generalization of a watershed acidification model for predicting broad scale effects
Reprint of Ph.D.Thesis, IIASA, Laxenburg, Austria, RR-90-3

Schneider T., Bresser A.H.M., 1988
Summary report; Acidification research 1984 - 1988
Report nr. 00-06

Sverdrup H.U., Vries W.de and Henriksen A., 1990
Mapping Critical Loads, A guidance to the criteria, calculations, data collection and mapping of critical loads
Prepared for the Task Force on Mapping, with the assistance of the Secretariat of the United Nations Economic Commission for Europe (UN-ECE)
Miljørapport 1990:14
Nordic Council of Ministers

Tyler G., 1987
Probable effects of soil acidification and nitrogen deposition on the floristic composition of oak (Quercus robur L.) forest
Flora 179: 165 - 170

Vries W.de, 1988
Critical deposition levels for nitrogen and sulphur on Dutch forest ecosystems
Water Air Soil Poll. 42: 221 - 239

Vries W.de, 1991
Assessment and policy implications of average critical loads for nitrogen and sulphur in the Netherlands
(in prep.)

ANNEX 1 THEMATIC REPORTS

EMISSIONS OF NH_3

J.H. Voorburg[1)]
G.J. Monteny[2)]

1) Institute of Agricultural Engineering, Wageningen
2) Agricultural Research Department, Wageningen

INTRODUCTION

Within the framework of the Dutch Priority Programme on Acidification Research, attention has been paid to ammonia emissions from livestock. Because very little information was available about the contribution of ammonia from the different systems of livestock production and manure handling to the acidification problem, attention has been paid mainly to measuring the ammonia emission. It has appeared impossible to make a complete inventory of ammonia emissions.
Since 1988 the research has been incorporated into the Research Programme for Animal Manure, and within this framework the inventory will be completed. Moreover, research into the prevention and abatement of ammonia has been intensified.
This review summarizes the results of the research into ammonia emissions carried out within the framework of the Priority Programme on Acidification Research. As far as possible a complete "state of the art" report is given by including published results of the Research Programme on Animal Manure.

1. THE RESEARCH PLAN

One of the main problems of the research into ammonia appears to be the measurement of ammonia emissions. This is relatively simple for livestock houses with forced ventilation because ventilation rates can be measured and sampling to establish the ammonia concentration can be done at the exhaust openings. But even then, quite an effort is necessary to take into account fluctuations in the amounts of exhaust air and ammonia concentrations.
For field conditions two measurement techniques are available, namely the tunnel method and the micro-meteorological mass balance. Examples of the two methods are given by Klarenbeek and Bruins (1990). The tunnel method is suitable to compare systems. To obtain quantitative information it is important to create conditions in the tunnel which are comparable with the ambient situation.
Much resarch into ammonia has been started. In projects concerning ammonia emissions the following aspects have been studied:
- Emissions from animal housing systems (IMAG)
- Effect of land spreading techniques (IMAG)
- Modelling of emissions after land spreading on arable land (IB)
- Emissions from grazing (CABO and NMI)
- Emissions after land spreading (PAGV, NMI, MT-TNO, CABO and IMAG)
- Emissions from agricultural regions (MT-TNO)

- Abatement of ammonia emissions with biofilters and air scrubbers (IMAG).

The coordination between research projects is enhanced through contacts between project leaders in the working group on ammonia emissions as well as during the symposia on acidification research which are organised yearly.

The results of the research projects have been published in English, or in Dutch with an extended English summary. This paper reviews the main results. When possible they are compared with research reports from other countries.

2. EMISSIONS FROM ANIMAL HOUSING SYSTEMS

Referring to Buijsman et al. (1984) the main source of ammonia emissions in the Netherlands is cattle. Housing systems for cattle are mostly naturally ventilated, which makes it very complicated to measure the ammonia emissions. It was decided to start with less complicated research, namely into ammonia emissions from poultry housing systems, which could be performed at the Research Centre for Poultry Production at Beekbergen, where facilities are available for comparing the effects of different poultry housing systems on ammonia emissions.

2.1 Poultry housing systems

Kroodsma et al. (1990) compared three housing systems for layers and two for broilers. The layers were kept in batteries with different manure collection and handling systems in each compartment, namely:

a. Droppings collected and stored in pits below the cages and handled as a slurry.

b. Droppings collected on a belt, removed twice a week and stored and handled as a slurry.

c. Droppings collected and dried on a belt, removed once a week and stored and handled as a dry manure.

The broilers were kept on a concrete floor with litter. In one of the compartments, the floor was insulated and heated.

2.1.1 Layers

The results of the experiments with layer droppings are given in Table 1, which shows that system A has the highest ammonia emission. The emissions from systems B and C are nearly equal.

The composition of the manure gives little information about differences in losses. In fact it was not possible to calculate the ammonia losses from an input-output balance of manure-N for the various compartments. This is explained by the fact that the ammonia losses amount to 5-10% of the production of N in the manure. Moreover, different sampling methods had to

be used for the different systems. The samples from B and C were taken twice a week, or weekly when the belt was emptied. The samples from system A were taken from an increasing stock of slurry. Good mixing was only possible when the pit was emptied.
Nevertheless, Table 1 gives important information which can explain the smaller ammonia losses from systems B and C. Most nitrogen excreted by birds is in the form of uric acid. The uric acid break-down is supposed to be delayed by the high dry matter content of the manure, especially in system C. The percentage of N in the form of NH_4-N supports this theory. The amount of uric acid in the manure was not measured.

Table 1. Manure composition and ammonia losses from different housing systems for layers (Kroodsma et al., 1990)

		Slurry under batteries (A)	Removal belt slurry (B)	Removal belt dry (C)
Total solids	(%)	12.6	25.8	47.3
Ash	(% of TS)	29.0	24.7	23.6
NH_4-N	(mg/g)	6.5	5.3	5.8
N-kj	(mg/g)	10.3	16.1	29.2
P	(mg/g)	3.5	6.2	10.0
K	(mg/g)	4.2	4.8	9.0
NH_4-N	(% of N-kj)	63.1	32.9	19.9
NH_3 loss	(g/bird.year)	83.2	33.9	31.2

It should be realized that the ammonia emissions from the different housing systems are only part of the total emission from poultry production. Besides, the losses after spreading and during storage of manure should be taken into account. They are included in system A (storage of slurry below the cages), but not in systems B and C.
During the storage of dry manure from system C, composting of the manure is to be expected. The ammonia losses from this storage were not measured. By simulating this composting process, Kroodsma et al. calculated the ammonia evaporation during the storage of dry manure from layers at 112.4 g per bird per year (50-75% dry matter) and at 201.0 g per bird per year (35-65% dry matter). This means that the ammonia emissions during storage can be 3 to 7 times higher than the emissions from the housing system. In a simulation experiment with broiler manure (Kroodsma et al., 1990) an ammonia loss of 29 g per bird place per year, or 50% of the loss from the housing system, was found. Hardly any reliable data are available on the composition of dry poultry manure after a storage period. The scarce data available suggest losses up to 50% of the total nitrogen content of the manure during storage. This should be taken into account in studying ammonia losses from housing systems (Table 1) and after land spreading (Table 6). As this is a specific problem in storage

facilities for dry poultry manure, there is a direct solution, namely immediate forced drying of the manure.

2.1.2 Broilers

In an earlier experiment, the ammonia emission from a broiler house with an insulated floor appeared to be lower than the emission from a comparable house with an uninsulated floor. It was supposed that condensation of water on the uninsulated floor caused a lower dry matter content in the litter manure mixture and thus stimulated the mineralization of uric acid. In order to verify this theory, an experiment was set up during six fattening periods. The results are summarized in Table 2. It cannot be concluded from these experiments that floor insulation and heating give a reliable reduction in emission. Also the storage of broiler manure will give rise to ammonia evaporation, depending on the composting process. On the basis of experiments in which composting was simulated, Kroodsma et al. (1990) calculated an NH_3 emission from the stored manure of 19.9 g per bird per year. In these experiments losses tended to be lower when the dry matter content of the manure was higher.

Table 2. NH_3 emission in g per broiler per fattening period

Period	Uninsulated	Insulated
November/December	9.51	4.76
January/February	5.64	4.01
March/April	7.30	6.78
May/June	8.42	8.15
July/August	6.41	7.39
September/October	8.49	9.08
Total/year	45.8	40.2
Average per period	7.63	6.70

A large number of emission measurements with different animals and housing systems has been carried out by Oldenburg (1989). Taking into account that the measurement techniques and housing systems in his report are not completely comparable with those in the Netherlands, his emission data are generally the same as those obtained in Dutch experiments. However, this does not hold for broilers. Oldenburg does not calculate the emission per bird (place) per period. From the detailled information it can be concluded that from day 20 to day 35 the emission is more than 1 g NH_3 per broiler per day or at least 15 g per fattening period. In the Netherlands a fattening period is usually 7 days longer than on broiler farms where the German measurements were taken.

2.2 Pig housing systems

At the Research Station for Pig Production at Rosmalen, three pig housing systems with 80-100 fattening pigs each were compared as to ammonia emission. The systems were chosen because of an expected reduction in ammonia emission from pig houses and manure storage. The storage of manure under the slatted floor was supposed to be the main source of ammonia emission. Therefore, the following housing systems were compared:

A. Partly slatted floor: manure storage below the slatted floor.

B. Partly slatted floor: manure collected in channels under the floor and removed once a week.

C. Partly slatted floor: the channel below the floor slopes down, so that the urine runs off immediately. A dung scraper removes the faeces twice daily.

The main results of one year of emission measurements are given in Table 3.
The results show that there is no difference in emission between the three housing systems. Frequent removal of the manure from the pig house had no positive effect. So it is suggested that not the amount of manure in the house is the main factor related to ammonia emission but the surface that is polluted with manure. The house floor and the manure storage are the most important surfaces. Also pollution of the pigs or of the inside walls and storage should be taken into account.
These conclusions have led to further experiments with flushing systems. The aims of flushing are the frequent removal of manure and the covering of the polluted area with a flushing liquid containing no or little ammonia.

Table 3. NH_3-emission per fattening place per year

Housing type	kg NH_3/pig.year
A	3.0
B	3.0
C	3.1

Ammonia losses from pig houses were also measured by Burton and Beauchamp (1986). They measured nitrogen losses from three different pig houses by means of an input-output balance and by direct measurements of flow rates of ammonia in ventilated air. Except for one barn they found a good agreement between both methods.
The direct measurement resulted in an ammonia loss of 2.7 kg N per pig per year from a barn with daily manure scraping. The other barn had a slatted floor over a storage pit with a capacity for six months. This barn had an ammonia loss of 1.34 kg per pig per year.

Therefore, also in this experiment the system of manure handling was more important than the amount of manure stored in the house.

2.3 Dairy cows

As 70-80% of cattle are accomodated in cubicle houses, ammonia measurements were concentrated on this housing system. To overcome the problems related with measurements in naturally ventilated houses, forced ventilation was installed in the house. It is a house for 40 dairy cows with cubicles and slatted floors. The manure is stored underneath the slatted floor and underneath the cubicles. The results of these measurements are given in Table 4.
The emission in the period from January till April is approx. 1 kg per cow per month. The emission tends to increase in spring and summer related with high outside temperatures. In spite of the fact that the cows are grazing at daytime in May and June the emission is higher than in the actual housing period. During the day the emission falls by 50-60% because the slatted floor dries up and no fresh urine is added. As soon as the animals enter the house the emission increases.

Table 4. NH_3 emission from a cattle house in kg per month for 40 dairy cows

Month	NH_3 emission per month (kg)
January	38.5
February	38.6
March	43.8
April	43.8
May	60.2
June	46.8

In addition to cubicle houses with slatted floors, there are cubicle houses with concrete floors from where dung scrapers take the slurry to outdoor storage. Comparative measurements have been carried out with the tunnel method (Kroodsma et al., 1990). From these measurements it was estimated that emissions from houses with concrete floors are of the same level as those from houses with slatted floors. Frequent scraping does not reduce the emission.
As stated, these measurements were carried out with a tunnel covering about 1 m^2 of a polluted surface. It is not possible from these measurements to calculate the total emission from a housing system.
Efforts have been made to estimate the ammonia emissions from houses with natural ventilation. The main problems are the calculation of the ventilation rate and the sampling of exhaust air. Hofman (1990) calculated the ventilation rate on the basis of the CO_2

concentration in the house and the CO_2 production per animal. This volume was multiplied by the NH_3 concentration measured with a Dräger tube. From six years' measurements in 274 loose housing systems he estimated an average loss of 5.9 kg N per animal per 185 days. Hollander (1989) estimated the ammonia emission on the basis of the NH_3 concentration outside the barn on the lee side. He estimated the emission at 5 to 6 kg NH_3 per animal per 210 days.

Though the estimates are in fair agreement with the estimates based on measurements from a house with forced ventilation, there are doubts about the methods applied. Further research to verify these methods is in progress.

2.4 Manure storage

Most housing systems have a manure storage facility underneath the slatted floor. In this situation the emission from the stored manure cannot be separated from that from the housing. Because of a stricter legislation allowing a shorter period ofmanure spreading, most farmers have to extend their storage capacities. This is often realized by installing storage tanks outside the building.

De Bode (1990) reports about odour and ammonia emissions from storage tanks with cattle and pig slurries. These measurements were carried out with mini-silos completely covered by the tunnel. Measurements on a farm-scale tank with a floating tunnel were also made. Moreover, ammonia losses were calculated on the basis of N-input and output from the mini-silos. The results are given in Table 5.

These results give rise to many questions that cannot be answered. The differences between cow and pig slurries cannot be explained from a difference in NH_3 content or pH of the slurry. It is also obvious that ammonia emissions from storage on a farm scale are different from those from storage in experimental mini-silos.

The calculation of ammonia losses on the basis of N-input and N-output seems less accurate as only 5-15% of the total amount of N is lost.

The ammonia losses of 300 mg/m^2.h from cattle slurry during the summer are of the same level as the losses from in-house storage measured with a tunnel by Kroodsma et al. (1990), but a m^2 of concrete floor polluted with slurry gives only half this emission.

De Bode aimed to measure the reduction in ammonia emission by covering the storage tanks. He proved that an adequately closing cover could reduce the emission by 70-90%. Crust formation on cattle slurry could be simulated by adding chopped straw to it. In this way the emission could be reduced by 60-70%.

Table 5. Ammonia emission from uncovered slurry tanks in mg $NH_3/m^2.h$

	Pig slurry		Cattle slurry	
	summer	winter	summer	winter
Tunnel	600	200	300	130
Balance calculation *	1800		600	
Farm scale		650		350

* input minus output

2.5 Biofilters and air scrubbers

Where it is not possible to realize sufficient reduction in the NH_3 emission from a housing system by appropriate handling of manure, the treatment of exhaust air in a biofilter or air scrubber can be an alternative solution. These systems can only be applied in houses with forced ventilation. Because of the high costs of the system, biofilters are hardly used in livestock farming. Scholtens and Demmers (1990) estimate these costs at Dfl. 12-19 per pig delivered.

Measurements of air scrubbers and biofilters under conditions of livestock farms have shown that the ammonia reduction is often insufficient. This is caused mainly by the inhibition of the nitrification process by free ammonia, nitrite or the pH value. It means that the efficiency strongly depends on the removal of absorbed nitrogen, which makes the operation of air scrubbers and biofilters more complicated. Moreover, the disposal of effluent is an extra cost factor. A positive aspect is the reduction in odour emission achieved by properly working biofilters and air scrubbers.

3. AMMONIA EMISSION FROM MANURE SPREADING AND GRAZING

The main source of ammonia emission is the spreading of manure on the land. The odours emitted from spreading indicate that there are considerable losses of volatile compounds.

More or less based on the experiences with odour emissions from manure spreading the reseach programme distinguishes between three aspects:

- ammonia emission during spreading;
- ammonia emission from the layer of manure on the field after spreading;
- ammonia emission from manure and urine excreted by grazing animals.

3.1 Ammonia emission during manure spreading

Ammonia losses during the application of manure are caused by the "cloud" of slurry drops thrown out by the machine and by the manure already lying on the field.

This section pays attention to ammonia losses caused by the "cloud" of drops and tries to answer the question on how far a more sophisticated way of surface spreading can reduce losses. Klarenbeek and Bruins (1990) show that in several experiments less than 1% of the NH_4-N in manure is lost in this way. For those who have smelt manure spreading activities, these losses are surprisingly low. One should realize, however, that the smell is an instantaneous sensation caused by the manure cloud and the manure lying on the field. The drops of manure are in the air for a few seconds only. A loss of 0.05% during this time means a 1% loss per minute and 60% per hour. In other words, the losses per second are much higher than from manure lying on the field. This is in agreement with sensoric experiments with land spreading of manure.

Thus it can be concluded that slurry spreading machines should be designed for a good distribution of manure, minimizing the emission after spreading. The emission during manure spreading should not be neglected, but is of minor importance.

3.2 Ammonia emission after manure spreading

Many data on ammonia emission after the spreading of manure were obtained in the past few years. Since more than 50% of the total ammonia emission from agriculture takes place after spreading, the introduction of techniques for applying manure to grassland and arable land with low ammonia emission can be very effective in reducing the ammonia emission nation-wide. These techniques are relatively cheap compared with, for example, housing systems with low ammonia emissions.

For both grassland and arable land, several techniques with low ammonia emissions have been developed.

3.2.1 Grassland

Almost all cattle slurry is used on grassland. Pig slurry is used both on grassland and arable land.

Depending on the weather and soil conditions most ammonia in slurry and manure is lost in the first day after surface spreading. Th emission can go on for several days, or even weeks afterwards. Figure 1 (Klarenbeek and Bruins, 1990) shows the emission-time relation. Several experiments have been made to measure the ammonia emission after spreading of manure on the land. Klarenbeek and Bruins (1990) describe measurements both with the tunnel and the micro-meteorological method. Untreated slurries were used. Vertregt and Selis (1990) used both treated (separation, dillution) and untreated manures. Table 6 shows the main results.

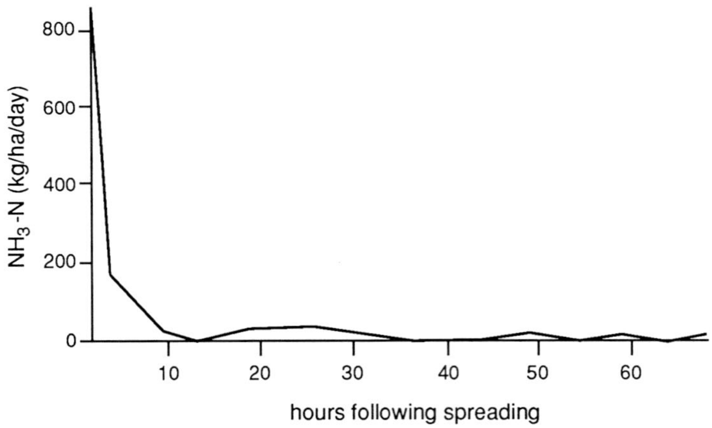

Fig. 1. Ammonia emissions following land spreading of pig slurry

Table 6. Ammonia losses from different types of slurry and dry manure

Slurry	Rate of application (m^3/ha)	Nitrogen applied (kg NH_4-N/ha)	Emission (in % of NH_4-N)	(in kg/m^3 slurry)
Klarenbeek and Bruins				
Cattle	62-69	155-180	42-45	1.1
Pig	39	134	33	1.1
	33-35	84-89	43-52	1.1-1.3
	26	54	82	1.7
	62	260	33	1.4
	30			
Broilers*	7	50	20	1.4
Laying hens*	6.5	56	17	1.5
Vertregt and Selis				
Cattle	30	76	40-80	
Pig	30	163	50-90	

* solid manure; application rate in tons/ha

The findings of Klarenbeek and Bruins show that pig and poultry slurries tend to produce a higher ammonia emission than cattle slurry. It is also obvious that the absolute emission depends on the application rate and the nitrogen content of the slurry. Weather conditions (wind speed, radiation and temperature), soil type and soil moisture content also play an important role, however.

In order to prevent ammonia emissions from spreading of animal manure, several application techniques with low ammonia emission have been developed, such as deep injection, sod injection, shallow injection and surface band spreading. The techniques can be used on normal grassland. Problems may come up on poorly drained or low lying grassland where the machines can only be used in dry periods of the year. In these situations the conventional vacuum tanker can be combined with irrigation, or diluted manure can be irrigated.

Table 7. Ammonia emission following application of slurry on grassland with different systems

Application system	Rate of application (m^3/ha)	Nitrogen applied (kg NH_4-N/ha)	Emission (% of NH_4-N)	Emission (kg/m^3 slurry)	Reduction* in % (control = 0%)
Deep injection					
Pig	40	233	0.2	0.01	98.7-100
Cattle	40	120	0.4	0.01	99-99.8
Sod injection					
Pig	20	112	4	0.22	91
Cattle	20	61	8	0.24	84-89
Irrigation (10 mm water)					
Pig	10	57	16	0.91	63-74
Cattle	10	30	13	0.39	55-89
Dilution (1:3)					
Pig	40	56	25	0.35	44-51
Cattle	40	38	19	0.18	67-73
Control (surface spreading)					
Pig	10	57	48	2.74	0
Cattle	10	30	65	1.95	0

* From: Bussink et al. (1990)

For the systems described, the above measurements on ammonia emission have been made (Klarenbeek and Bruins, 1990). In Table 7 the main results are shown. Table 7 shows that injection has the strongest effect on ammonia emission although irrigation and dilution can also achieve a moderate to high reduction in ammonia emission from land spreading.

In other European countries (UK, FRG) ammonia emission measurements on machines with low ammonia emission have also been performed. The results from these measurements are very much comparable with the Dutch results. The differences can mostly be explained by looking at the differences in soil type, weather conditions, rate of application, nitrogen content of slurry, dry matter content etc.

3.2.2 Arable land

Less (international) research has been performed on the reduction of ammonia emission on arable land than on grassland. The data of Table 8 were obtained from Anglo-Dutch experiments (Bruins and Huijsmans, 1989).

Table 8. Reduction of ammonia emission on arable land

Time between application and soil tillage (h)	Reduction with plough	rotary harrow	rigid tines
0	90	78	40
3	68	55	33
3	78	58	37
6	54	43	34

The results in Table 8 show that soil tillage immediately after spreading is the most effective measure. Mixing soil and manure intensively (plough and rotary harrow) gives the highest reduction.

3.2.3 Factors influencing ammonia losses after manure spreading

A simulation model for ammonia volatilization from the soil surface to the atmosphere was developed by van Faassen et al. (1990). In this study different situations are described, namely manure incorporated into the soil and manure lying on the soil surface. In the latter case a part of the liquid from the manure penetrates into the soil, depending on the composition of the manure, physical soil conditions and weather conditions (rainfall). For manure lying on the surface the meteorological conditions have the strongest influence on ammonia emission. Important factors are wind speed and temperature.

The volatilization of ammonia from a soil/manure mixture is less than from manure. Chemical processes between manure and soil prevent a high emission. The main conclusion is that the degree of interaction between soil and manure mostly determines the rate of emission. This is in agreement with the results obtained from experiments with spreading techniques. However, it is difficult to make a good link between the model and spreading techniques as it is not possible to quantify the aggregation between soil and manure.

Volatilization highly depends on weather conditions such as wind speed, soil surface temperature, rainfall and gas-filled pore fraction of the soil (humidity of the soil). For the practical farmer the weather situation is quite uncertain and difficult to take into account in his management decisions. For the measurement of ammonia emissions related to existing and new developments in manure spreading techniques, the relevant parameters should be described exactly for a better comparison of experiments. Though the model cannot predict what will happen after surface application of manure, it is an important contribution to gain insight into the conditions and technical measures influencing ammonia emission.

It explains how dry matter content and rainfall can influence emission and how separation techniques can reduce ammonia emission. Liquid manure containing few solids or fibres easily penetrates into the soil, as demonstrated by Vertregt and Rutgers (1988) for urine from cattle and by Bruins and Huijsmans (1989) for the liquid fraction of pig slurry. This seems to be in disagreement with findings of Vertregt and Selis (1990). They found ammonia emissions from the liquid fractions of cattle and pig slurries to be only slightly lower compared with the unfractionated slurries. This difference seems to be explained by the separation technique used. In the experiments by Bruins and Huijsmans the liquid fraction had a dry matter content of 2.4%, and Vertregt and Selis used liquid fractions with dry matter contents of 8.1 and 11.5%. The unfractionated slurries in those experiments had dry matter contents of 9.6 and 12.2%.

Moreover, it should be considered that Vertregt and Selis spread manure on grassland. The part of liquid manure sticking to the grass cannot penetrate into the soil.

An aspect of the spreading experiments which deserves more attention, is the duration of the experiments. Very often the ammonia emission is measured for four or five days after the spreading. It is found that there can be a slight emission for a few more days after the measuring period. It is tobe expected that this emission depends on the experimental conditions. In this way an extra error, however small, is introduced. The possibility of incorporating "rest emission" into the calculated data should be considered.

3.2.4 Grazed pastures

Over the past few years, several experiments have been performed into the ammonia emission from grazed pastures.

In some experiments (Vertregt, 1989) grazing was simulated with artifical urine patches. Measurements were taken in tunnels. Micro-meteorological measurements were also made during field experiments with grazing cattle.

The experiment with artifical urine was performed at an application rate of 600 kg N/ha and showed that between 6% and 19% (mean: 13%) of applied urine N was lost by ammonia emission. It also appeared that the ammonia emission is roughly 10% of the total N excretion in the urine and faeces. Furthermore, calculations have been made on the relationship between the level of N-supply and the ammonia emission on grazed pastures. The calculated emissions were 16 kg N/ha at an N-supply level of 100 kg/ha and 38 kg N/ha at a level of 500 kg N/ha.

In experiments under field conditions (Bussink et al., 1989) the relationship between the rate of N-supply with artificial fertilizer and the ammonia emission from grazed pastures has also been studied. Grassland management systems with 250, 400 and 550 kg N/ha.year were compared. The first results are shown in Table 9.

Table 9. Annual losses of NH_3-N from grazed swards

Management	Total loss (kg NH4-N/ ha.year	Losses per animal (g NH_3/grazing day	% of excreted N
.grass + 550 kg N/ha	39-42	40-45	7.6-8.4
.grass + 250 kg N/ha	8.1	14	3.0

The figures in Table 9 show that a reduction in N-input on grassland result in a substantially lower ammonia emission. They also seem to match reasonably well with the figures calculated by Vertregt (1988).

Also in the UK, measurements of ammonia emissions from grazed pastures have been made. Table 10 shows the main results.

Table 10. Annual losses of NH_3-N from grazed swards

Management	Total loss (kg N/ha.year)	Losses per animal (kg N/grazing day)	% of input
Rotational grazing			
.grass + 420 kg N/ha	25.1	18	6.0
.grass / 210 kg N/ha	9.5	9	4.5
.grass/white clover + 0 kg N/ha	6.7	5	4.2
Continuous grazing			
.grass + 420 kg N/ha	16.3	11	3.9

Comparison of these data with the results from the Dutch experiments show that they are very much the same. Figure 2 shows the relationship between N-input level and ammonia emission from grazed pastures from both Dutch and UK experiments (Jarvis and Bussink, 1990).

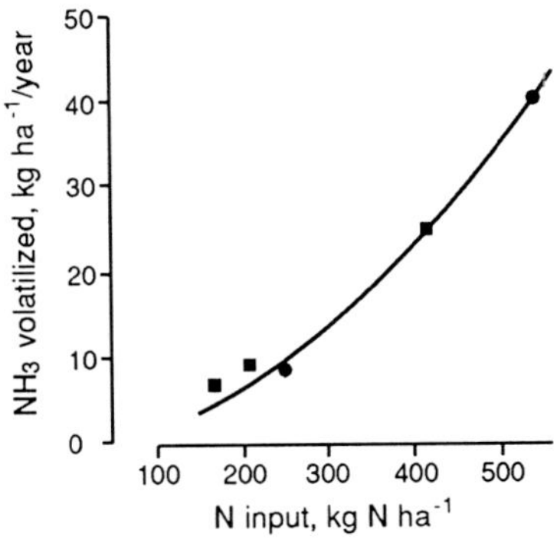

Fig. 2. Relationship between N input and NH_3-N loss at Hurly (■) and Lelystad (•)

The curve can be described by the equation:

$$y = 0.00347\ x^{1.854}\ (r = 0.81),$$

where y is the annual loss of ammonia and x is the N-input into the system.

4. CALCULATION OF THE NATIONAL AMMONIA EMISSION

Oudendag and Wijnands (1989) calculated the total ammonia emission in the Netherlands (Table 11). This emission is estimated on the basis of statistical data and assumed emission rates per source or activity.

Table 11. Calculated ammonia emission from most important livestock production in the Netherlands in 1986 in 1000 t NH_3

Animal	Housing and storage	Grazing	Manure spreading	Total
Cattle	42	25	73	140
Veal calves	1	0	3	4
Fattening pigs	14	0	31	45
Breedings sows	12	0	12	24
Layers	9	0	9	18
Broilers	8	0	2	10
Total	86	25	130	241

Table 11 does not include the ammonia emission from animals such as sheep, horses and ducks. The total manure production of these animals is estimated at about 2% of the national manure production. Therefore it can be expected that the influence of ammonia emission from these animals is much less than the measurement error in the data given in Table 11.

A more exact estimate of the ammonia emission in the Netherlands could be made if more exact data on ammonia losses per activity were available. Table 12 lists emission factors as used by Oudendag and Wijnands, and the results of the research reviewed in this report. There are considerable differences between most of the emission factors listed in the official table and the emission measurements of the experiments reviewed in this report. This could be a basis for a renewed calculation of Table 11. It should also be realized that the figures reviewed in this report have a limited value.

The measurements dealing with dairy cows were carried out in one house over a period of only 4 months with cows receiving rations aiming at maximum utilization of the proteins. The measurements for pig housing apply to a full year period. However, the housing system was somewhat modified and therefore not fully comparable to usual pig housing. Current research tends to result in emission factors which are more in agreement with the official table.

Table 12. Ammonia emission in kg NH_3 per animal per year for various animals and housing systems

		Offical table	Reviewed measurements
Dairy cows*		8.8	6
Young cows (< 2y)		3.9	-
Veal calves 1.6		-	
Breeding sows		8.1	-
Fattening pigs			
	Fully slatted floor	2.5	-
	Partly slatted floor	1.3-2.5	3
Layers			
	Slurry storage under batteries	0.318	0.083
	Manure belt, removal as slurry	0.187	0.034
	Manure belt with drying	0.030	0.031
	On litter	0.184	-
Broilers			
	Floor not insulated	0.208	0.046
	Floor insulated	0.144	0.040

* housing period 180 days.

For emission factors for poultry there are large differences between the offical table and the measurements referred to in this report. However, it should be realized that storage of solid manure from layers and broilers can cause a considerable ammonia emission (not measured).

The main part of ammonia losses is caused by spreading of manure on the land. The calculated emission is based on the assumption that 50% of ammonia in the manure is evaporated during and after spreading. The research reviewed in this report shows that there is a wide variation in ammonia losses during and after surface application of manure. Factors influencing the evaporation are weather conditions, soil properties and physical properties of the manure. As there is no quantitative information about the conditions prevailing when farmers are spreading manure, it is impossible to make a more accurate calculation of the total ammonia emission due to manure spreading.

5. EVALUATION OF THE RESEARCH

Though the total ammonia emission from livestock production amounts to roughly 200 000 t nitrogen, it is a very diffuse process spread over an area of about two million hectares. This explains why it is difficult to make a more accurate estimate of the total emission and the contribution of the various sources.

With reference to the aims of the research within the framework of the Priority Programm on Acidification Research to get more detailed and more accurate information on the total ammonia emission from livestock production, it can be stated that existing inventories such as published by Oudendag and Wijnands (1989) give a good estimate. There are indications that some sources are overestimated and others underestimated. The measurements made so far are not sufficient to serve as a solid basis for more reliable new estimations.

The most important result of the reviewed experiments is better information on the factors contributing to ammonia losses under farming conditions. This knowledge is a basis for developing systems with lower emissions. The most important achievements in research concerning livestock production with reduced ammonia losses are:

1. <u>Poultry housing</u>

   Fast drying of the manure to more than 70% dry matter reduces NH_3 formation.

2. <u>Pigs and cattle</u>

   Ammonia losses are not only caused by manure stored in the house but even more by the part of the house which is polluted with manure and urine. In the development of better housing systems, attention has to be paid to this aspect.

3. <u>Manure storage</u>

Covering of the manure storage results in an important reduction in emissions of ammonia and odour.

4. Land spreading
   - The ammonia emission during spreading is less than 1% of the nitrogen applied for all types of spreading machines.
   - After surface application of manure, roughly 50% of its ammonia content is evaporated depending on weather, soil conditions and manure composition.
   - The highest emission occurs immediately after land spreading.
   - About 50% of the emission takes place within the first six hours after application.
   - An important reduction in NH_3 evaporation can be obtained when the ammonia is transported into the soil. This can be realized by mechanically (injection, ploughing) or physically (dilution, irrigation).

The development of manure handling systems with low ammonia losses on the basis of the keys listed above has an increasing interest and is encouraged by the government and the farmers organizations.
Most attention is paid to the development of injection techniques, manure acidification and flushing systems. A group of researchers have acquired experience at measuring ammonia emissions and up-to-date measurement techniques have become available. These are the most important achievements, so far, of the Priority Programme on Acidification Research. They enable the further development of techniques and systems to reduce ammonia losses from livestock farming to be guided scientifically.
Stimulated by the EC COST 681 programme, it has been possible to start international co-operation in research into ammonia emission from livestock. This can be a good basis for further international co-operation in order to speed up the reduction in environmental pollution from animal husbandry.
Special attention has to be paid to the reliability of the measurement techniques used. More attention should be paid to the standardization of the total measurement procedure, the calculation of measurement errors and the comparability of different measurement techniques.

6. SUMMARY AND CONCLUSIONS

The research into ammonia emissions from livestock has resulted in a far better insight into the factors contributing to ammonia volatilization from buildings, manure stores, grazing and land spreading of manure.
The emission factors measured can contribute to better defining the emissions from the

different sectors and activities in livestock farming.
The number of measurements, however, is too limited to be a basis for a better estimate of the total ammonia emission from livestock. The most important result of the research summarized in this report is the better insight into the possibilities to reduce ammonia losses. The main source of ammonia volatilization is land spreading. Development has begun of a number of techniques aiming at a fast incorporation into the soil. A number of machines with a wide variation in depth of injection and suitable for different soils and crop production systems has become available. Moreover, techniques of infiltration and dilution can also have a positive effect.
The reduction in ammonia emission from farm buildings is a more difficult task, because of the diversity of animals and housing systems.
Most housing systems have manure storage under the slatted floor or cages. It is evident that this store cannot be covered and will be a source of ammonia emission. Recent measurements, however, have shown that a stable floor polluted with freshly excreted urine is at least of the same importance as a source.
A valuable result of the research within the framework of the Acidification Programme is the development of reliable measurement techniques. The researchers who have gained experience in ammonia measurements can give essential support to the development of systems and techniques with minimum ammonia losses.
Emissions cannot be easily measured from all housing systems. Most livestock houses are naturally ventilated so that a reliable measurement technique for these situations is needed.
A model describing the ammonia emission after the application to arable land has been developed. This model is a useful contribution to the knowledge of factors influencing ammonia losses.
An important factor is the weather. As long as it cannot be predicted accurately, it is impossible to predict the ammonia emission. More attention should be paid to measuring ammonia emissions from different systems or techniques. The model enables measurements and comparisons to become more reliable.

7. REFERENCES

Bode, M.J.C. de, 1990. Odour and ammonia emissions from manure storage. IMAG, Wageningen

Bruins, M.A. and J.F.M. Huijsmans, 1989. De reductie van de ammoniakemissie uit varkensmest na toediening op bouwland. IMAG, Wageningen. Rapport 225

Buijsman, E., H. Maas and W. Asman, 1984. Een gedetailleerde ammoniakemissiekaart van Nederland. Instituut voor Meteorologie en Oceanografie. Rijksuniversiteit Utrecht

Burton, D.L. and E.G. Beauchamp, 1986. Nitrogen losses from swine housings. Agricultural Wastes 0141-4607/86

Bussink et al., 1990. Ammoniakemissie bij verschillende toedieningsmethoden van dunne mest aan grasland. Rapport A 89.086, NMI, Den Haag

Faassen, H.G. van, W.J. Chardon, R. Vriesema and J. Bril, 1990. Ammonia volatilization from arable land after surface application or incorporation of cattle slurry. Final Report Project 112. IB, Haren

Hofman, J., 1990. De ammoniakemissie in Nederland. Gezondheidsdienst voor Dieren in Noord-Nederland

Hollander, J.C.Th., 1989. Emissie van ammoniak uit dierlijke mest. Emissie na toediening aan akkerbouwgrond en emissie uit een rundveestal. MT-TNO, rapportnr. R 89/215

Jarvis, S.C. and Bussink, D.W. (1990). Nitrogen losses from grazed swards by ammonia volatilization. Proceedings of European Grassland Federation Meeting, Czechoslovakia, 1990 (in press)

Klarenbeek, J.V. and M.A. Bruins, 1990. Ammonia emissions during and after land spreading of animal slurries. IMAG, Wageningen

Kroodsma, W., J. Huis in 't Veld and R. Scholtens, 1990. Ammonia emissions from dairy, pig and poultry housing systems. IMAG, Wageningen

Oldenburg, J., 1989. Geruchs- und Ammoniakemissionen aus der Tierhaltung. KTBL-Schrift 333

Oudendag, D.A. and J.A.M. Wijnands, 1989. Beperking van de ammoniakemissie uit dierlijke mest. LEI, Onderzoekverslag 56

Scholtens, R. and T.M.G. Demmers, 1990. Biofilters and air scrubbers in the Netherlands. IMAG, Wageningen

Vertregt, N. and B. Rutgers, 1988. Ammonia volatilization from grazed pastures. CABO, Wageningen. Report 84

Vertregt, N. and H.E. Selis, 1990. Ammonia volatilization from cattle and pig slurry applied to grassland. CABO, Wageningen. Report 130

ATMOSPHERIC INPUT FLUXES

R.M.van Aalst[1)]
J.W.Erisman[1)]

[1)] National Institute of Public Health and Environmental Protection, Bilthoven

PREFACE

In a very late stage of the Programme it became increasingly clear that more knowledge on the atmospheric input of nature areas should be gained. The results summarized in this report reflect a quick anticipation of many research groups to this problem. Therefore, following persons are acknowledged for their adequate contribution and enormous effort:

- MT-TNO,	H.S.M.A.Diederen, J.H.Duyzer, J.C.Th.Hollander, Th.R.Thijsse
- KEMA,	F.G.Römer, W.Ruijgrok, B.H.te Winkel, L.van den Beld, A.J.Elshout
- ECN,	J.Slanina, G.P.Wyers
- KNMI,	F.C.Bosveld
- LUW,	P.Hofschreuder, A.W.M.Vermetten, A.H.Versluis
- LUW,	A.W.J.van Pul
- LUW,	N.van Breemen, M.P.van der Maas
- RIN,	H.F.van Dobben, J.J.M.Berdowski
- RUU-BE,	R.Bobbink, G.W.Heil
- RUU-FG,	G.P.J.Draaijers, W.Bleuten, R.Meijers
- KUN,	J.Roelofs, A.L.F.M.Houdijk
- RIVM,	R.M.van Aalst, J.W.Erisman, F.A.A.M.de Leeuw, J.A.van Jaarsveld, G.M.F.Boermans, M.G.Mennen, B.G.van Elzakker, E.Buijsman
- UvA,	J.M.Verstraten

ABSTRACT

The thematic report on "Atmospheric input" is a compilation of results of research carried out within the framework of the Dutch Priority Program on Acidification. Research was focused on the estimation of the deposition of potentially acidifying components onto Dutch nature areas.

Measurements of concentration gradients and meteorological parameters have been carried out at two sites: Speuld, a Douglas fir forest, and Elspeet, a heathland. The knowledge gained from these experiments, together with some additional measurements at other locations in the Netherlands, has been used for generalization. Deposition patterns over the Netherlands for different compounds have been mapped.

Throughfall measurements have been made under forests and low vegetation at several locations in the Netherlands, including the two research locations Speuld and Elspeet. The throughfall measurements have been compared with deposition estimates for the research locations and different locations in the Netherlands. Uncertainties and gaps are identified.

1. INTRODUCTION

1.1 General introduction

In the early 1980's, the problem of acidification was acknowledged at governmental level in the Netherlands as a major environmental threat. The concern for damage to forests and ecosystems, and to materials and monuments, and for acidification of soils urged for policy development and stimulation of research [1]. Acid deposition was used as a characterization of the atmospheric input leading to these effects. As such, wet and dry deposition of acids, as well as dry deposition of acid precursors, such as sulphur dioxide and nitrogen oxides, were considered. At the same time, the important role of ammonia in the acidification of soils was acknowledged. Ammonia, in the atmosphere the most important acid neutralizing compound, was demonstrated to be nitrificated to an important extent in soil into nitric acid, and thus to contribute to acidification. This led to the concept of potential acid deposition, in which the deposition of actual acid (H^+) was added to that of potential acid (NH_3 + NH_4^+). Also, the role of atmospheric nitrogen input into ecosystems was recognized as a disturbing factor for natural ecosystems and forests on poor soils.

The Dutch Priority Programme on Acidification Research started in 1985 and focused on the following three main fields of research:

- the relationship between exposition and effects on plants in their immediate surroundings (air, soil), in particular the effects on forests, heather and crops;
- ammonia emissions and emission abatement techniques;
- the effectiveness of abatement measures in decreasing the effects.

The research efforts in this Programme to measure and to investigate atmospheric inputs, which are reported here, relate to these main fields in various ways. The atmospheric deposition is certainly not the only and not always the most relevant way to characterize the exposure of ecosystems to acidification. In order to understand the effects, detailed knowledge of concentration and deposition in various parts of the system (e.g. leafs, branches, stem, soil), its time evolution and the subsequent fate of the pollutants may be needed. However, these aspects are not dealt with in this report; the reader is referred to the thematic report on integrated effects. Here, we consider deposition as the total, long-term average input into the system as a whole from the atmosphere. In this report we consider deposition in the Netherlands as a whole and on the main target areas, forest and heather. Deposition on crops and agricultural areas and on materials and buildings is not considered explicitly.

As most measures to decrease the effects of acidification presently considered are emission abatement measures, it is essential that the relationship can be made between the emission of various pollutants and the deposition and exposure, and the related effects. This relationship is formally described in models. Models of the various compartments (air, soil, (soil) water, vegetation) are interconnected in the Dutch Acidification System model, described in the chapter on Integrated Modelling. In this system, there is, so far, no well-defined model relating the deposition and the input into the soil below vegetation. Experimentally, the input to forest floors and soils below vegetation has been characterized by measurement of throughfall, stemflow and litter flux. Due to interaction with vegetation, e.g. stomatal uptake and leaching, this flux may differ from the deposition flux. In this report these measurements are compared with deposition in order to gain understanding of the relation between these fluxes.

During the execution of the Programme, it became increasingly clear that the precision of total deposition estimates depends critically on dry deposition. Estimates of dry deposition from concentrations and literature values of deposition velocities indicated a large contribution of dry deposition (~ 70 %); uncertainty in these estimates was large, especially for forest and heather. In 1989, a coordinated effort was started to measure dry deposition, especially at the forest location Speuld and the heather location Elspeetse Veld. In this report, emphasis is placed on the - sometimes preliminary - results of this crash programme

1.2 Processes and definitions

Processes relevant to atmospheric input are listed below; a short characteristic/definition is given.

Atmospheric deposition: The total mass input per unit area and per unit time (flux) of air pollutants into the total system of soil, water and vegetation. Also: the process of this input.

Wet deposition: Atmospheric deposition carried by precipitation.

Occult deposition: Atmospheric deposition by interception of cloud or fog droplets.

Dry deposition: Atmospheric deposition by any other mechanism, e.g. direct uptake of gases by vegetation, soil or water (including incorporation into dew), sedimentation, impaction, interception or diffusion of aerosol.

Throughfall: The flux to the floor below a vegetation carried by water dripping down from the vegetation. This may include:

Stemflow: the flux carried by water flowing down the stems of vegetation.

The total input to the soil is the result of throughfall and stemflow, litter decomposition and dry deposition to the floor below the vegetation. This flux may be different from the atmospheric deposition due to canopy exchange.

Acidifying compounds: the pollutants considered are acids (H_2SO_4, HNO_3, HNO_2, HCl) and their precursor oxides (SO_2, NO, NO_2), as well as ammonia and compounds (NH_3,

NH_4^+). Organic acids are not considered.

Potential acid deposition: The summed deposition (expressed in moles acid) of H^+, acid precursor oxides (multiplied by the number of moles H^+ upon production of acid per mole) of NH_3, and of NH_4^+, the latter multiplied by two.*

* The assumption is that ammonia neutralizes acids (HX) either in aerosol or in atmospheric water:

$$NH_3 + HX \dashrightarrow NH_4^+ + X^-$$

Upon uptake in soil, (partial) nitrification takes place, schematically:

$$NH_4^+ \dashrightarrow 2H^+ + NO_3^-$$

In this process the neutralized acid is regenerated, and a second acid molecule is produced from ammonia. The actual production of acid from ammonia by nitrification is dependent on soil type and soil conditions. So, the potential acid deposition provides an upper limit value of the actual acid load to the soil.

2. MEASUREMENT OF AIR CONCENTRATION AND WET, DRY AND OCCULT DEPOSITION FLUXES

In this paragraph an overview will be given of the concentration and deposition measurements made at the different research locations. Furthermore, results from these measurements in terms of deposition fluxes and deposition parameters will be presented and discussed.

2.1 Methods

2.1.1 Concentrations

2.1.1.1 Speuld

At the research location Speuld, concentrations of SO_2, NO_x (NO and NO_2) and O_3 have been continuously monitored for 2 full years at five levels up to 30 m above the forest floor (2 levels above the canopy) by LUW. NH_3 was monitored at 30 m height. These measurements were supplemented by several measuring campaigns by LUW and TNO and semi-continuous measurements by ECN of NH_3, HNO_3, HCl, HNO_2 and H_2O_2. Aerosol measurements were also made during the campaigns. The coverage of these measurements in time is rather small. The yearly average concentrations at 30 m height for different components have been listed in Table 1. For a detailed description of the measuring methods and strategies, the reader is referred to the Appendix and [2,3].

2.1.1.2 Elspeet

At Elspeetsche Veld measurements of concentrations of SO_2 at 4 levels (RIVM) and of NH_3 at 2 levels (KEMA) were made over a period of one year. These were supplemented by measurements of NO_2 over a period of six weeks at 4 levels (TNO) and of NH_3 over two periods of 3 weeks by a photo acoustic spectrometer at 4 levels (KEMA). Details of the measuring methods and accuracy have been listed in the Appendix. Yearly average concentrations at a height of 4 m have been listed in Table 2.1.

2.1.1.3 The Netherlands

Within the framework of the Dutch Air Quality Monitoring Network (LML) hourly averaged concentrations of SO_2, NO and NO_2 are measured on a routine basis. From these measurements an accurate spatial distribution of concentrations over the Netherlands during the period 1980 - 1989 can be obtained.

Table 2.1 Yearly average concentrations of acidifying components measured at Elspeet, Speuld (at 18,5 m and 30 m, two seperate towers) and in the Netherlands in 1989 ($\mu g\ m^{-3}$)

component	Elspeet (4 m)	Speuld (18.5 m)	Speuld (30 m)	the Netherlands (4 m)
SO_2	7.5	6.7	10.5	10
NH_3	4.0	5.1	6.3	5.5
NO			6.5	9
NO_2			26.5	25
HNO_2		0.6		1.0
HNO_3		0.6	0.9	1.0
HCl		0.6	1.1	1.0
NH_4^+			4.5	4.7
NO_3^-			4.3	5.6
SO_4^{2-}			3.6	4.6
Cl^-			2.1	4.0
H_2O_2		0.1		

Daily averaged concentrations of SO_4^{2-}, NO_3^- and NH_4^+ aerosol are measured at a limited set of stations [29]. Only recently (by the end of 1988) have filter-pack measurements of total NO_3 (gaseous HNO_3 and particulate NO_3^-) and total NH_x (gaseous NH_3 and particulate NH_4^+) become available as daily averages for one out of eight days at 6 locations

in the Netherlands. Measurements by the Annual Denuder System (ADS) are made with the same frequency at 4-5 different locations by RIVM and LUW yielding concentrations of gaseous SO_2, HNO_2, HNO_3, HCl, NH_3 and particulate NH_4^+, NO_3^- and SO_4^{2-}. For a detailed description of the spatial distribution of HNO_2, HNO_3, HCl and NH_3 over the Netherlands, the measurements by filter-packs and ADS are insufficient.
In Table 2.1. an overview of the yearly average concentrations in the Netherlands have been listed. For details of measuring sites, measuring technique and uncertainties in yearly averages the reader is referred to the Appendix.

During 1988 and 1989 measurements of hydrocarbons were made by TNO at Kootwijk [4]. In the Dutch Acidification Programme no attention has been paid to the contribution of these components to acid deposition. The results of these measurements have therefore not been considered in this report.

2.1.1.4 Quality control
To investigate the calibration procedures of the continuous monitoring systems of the forest locations checks have been performed in the field under normal working conditions. In total, six campaigns were carried out at Speuld and at Kootwijk. These external controls had the additional advantage that corrections could be made for errors in the data processing. The checks were carried out using a mobile dynamic calibration system with primary standards (permeation devices). The conclusion of the calibration checks of the NO_2 monitors is that they are calibrated correctly. No significant deviation was found. For SO_2 a systematic deviation in the offset of about 3 ppb was found. The dataset has been corrected for this offset. After correction no significant deviation was found.
The response time of the NH_3 monitor at Speuld appeared to be very long (many hours). The scanning frequency at the different height levels for the other components was much faster. Because of this, the scanning procedure was cancelled and changed to measure the concentration of NH_3 at one height.
In the acid rain research programme six different NH_3 measuring techniques are used (see Appendix). During a two day campaign the different methods were compared. Simultaneous ambient measurements and measurements on a calibration system were carried out. The deviation of the results of the ambient measurements is large. At a level of about 3 $\mu g\ m^{-3}$ a deviation of a factor of two is found. The measurements of the calibration gas agreed much better. At a level of 7 $\mu g\ m^{-3}$ a relative standard deviation of 10% was found.
The tungsten oxide coated thermodenuders were tested before as well as after use at Elspeet (about one month). Typical results for the mean concentration values found in Elspeet are 10% relative standard deviation for a concentration of 3.5 $\mu g\ m^{-3}$ NH_3. For a test gas

mixture of about 22 $\mu g\ m^{-3}\ NH_3$ fed to both the thermodenuders as well as the photo acoustic spectrometer, maximal differences of c. 10% were found [19].

2.1.2 Wet deposition

2.1.2.1 Sampling and analysis

The concentrations of components in precipitation and the amount of precipitation are monitored routinely in the KNMI/RIVM Precipitation Monitoring Network on a monthly basis [5]. The potential acid deposition from precipitation can be calculated from the volume weighted average deposition of $H^+ + 2\ NH_4^+$. The origin is established from the amount of SO_4^{2-} (corrected for the sea salt contribution), NH_4^+ and NO_3^- in the samples. Fluxes of potential acid calculated by $H^+ + 2\ NH_4^+$ and $2\ SO_4^{2-} + NO_3^- + NH_4^+$ have been compared [6]. It was shown that the agreement is very good, provided a correction is made for the contribution of neutral Ca^{2+} salts to the sample:

$$\text{potential acid} = NH_4^+ + NO_3^- + 2\ SO_4^{2-} - 2\ Ca^{2+} \qquad (1)$$

The Cl^- compound in precipitation is almost completely associated with Na^+ and, therefore, of sea salt origin. The relation (1) is used as a basis for calculation of potential acid in this report.

Up to 1988 precipitation samples were collected by open samplers. In the beginning of 1988 these were replaced by wet-only samplers, except at 5 locations where parallel measurements have been made by both samplers. The results of these parallel measurements are not available yet, and so the 1980 - 1987 data from bulk samplers were corrected by 25% for Na^+, SO_4^{2-} and NH_4^+, by 15% for NO_3^- and by 55% for Ca^{2+} for the contribution of dry deposition of gaseous species and aerosols in the funnels of these samplers [7]. Wet-only samplers are provided with a device to exclude dry deposition onto the funnel during dry periods.

The origin of Ca^{2+} in rain water samplers was estimated from cluster analysis of the results of 8 wet-only samplers [8]. It has been estimated that for coastal sites 50% of Ca^{2+} is originates from SO_4^{2-} and 50% from NO_3^-, while all Ca^{2+} originates from SO_4^{2-} at the continental site [8]. These are all results obtained from measurements not influenced by dry deposition. Furthermore, SO_4^{2-} had been corrected for sea salt contribution in these measurements. Therefore, the procedure for correcting fluxes of SO_4^{2-} and NO_3^- by Ca^{2+} and sea salt (SO_4^{2-}) is by first correcting Ca^{2+}, Na^+, SO_4^{2-} and NO_3^- for dry deposition contribution (open samplers) and then applying the correction for Ca^{2+} and sea salt.

2.1.2.2 Quality control

In the research programme a large number of institutes are involved in measurements of wet deposition (see paragraph 4) alongside the Precipitation Monitoring Program. To investigate the comparibility of the results of the different laboratories two intercomparison experiments were held. In the first one artificial samples were used and in the second one ambient precipitation samples. The reproducibility (ISO 5725) of the analysis of eight laboratories is given in Table 2.2.

2.1.3 Dry deposition

Among the measuring methods for dry deposition [9] micrometeorological measurements are most suitable for dermining the dry deposition of gases and (fine) particles. Within these methods, the flux to the total system may be determined, and the relationship with air concentrations and meteorology is directly established. In flat homogeneous terrain the flux measured at a sampling point above the surface represents the average vertical flux over the upwind fetch, provided that the sampling point is in the constant flux layer. In the most direct of the micrometeorological methods, the eddy correlation method, the flux is derived from measurements of the vertical component of the wind velocity and the gas concentration. Another micrometeorological method is the flux-gradient technique, where the flux is derived from measurements of the vertical gradient of the air concentrations and meteorological variables.

Table 2.2 Reproducibility (%) (ISO 5725) of rain water samples

compound	artificial sample	ambient sample
NO_3^-	10	20
NH_4^+	10	30
K^+	10	60
Na^+	20	20
Mg^{2+}	20	20
Ca^{2+}	20	20
SO_4^{2-}	30	30
Cl^-	50	100

In the modified Bowen ratio technique, the flux is derived from the ratio of concentration and temperature/water vapour concentration differences at two heights and appropriate meteorological or heat/moisture flux measurements.

For more comprehensive theory and information on the use and limitations of these methods the reader is referred to [9,10]. Here, the actual implementation of these methods at the most important locations is described. Knowledge on deposition processes can be used to infer dry deposition fluxes from basic information on air concentration and meteorological data. The flux F is determined from the measured concentration C by:

$$F = V_d C \qquad (2)$$

where V_d, the deposition velocity, is determined as:

$$V_d = (R_a + R_b + R_c)^{-1} \qquad (3)$$

The aerodynamic resistance R_a and the quasi-laminar layer resistance R_b may be determined from meteorological measurements. The surface resistance R_c, which is dependent on the pollutant and the receptor is often estimated from literature data due to lack of representative and accurate measurements. The technique is also used to infer annual averaged deposition fluxes from incomplete measuring data.

2.1.3.1 Implementation over low vegetation

In the acidification programme several experimental designs were used in order to achieve optimal results for each trace gas. At the heathland location Elspeet three separate systems were used for NH_3 (KEMA), SO_2 (RIVM) and NO_2 (TNO) [18]. All systems used separate intake lines to draw air from 2 or 4 heights (between 0.5 and 4 m) to centrally located gas analysers. These analysers were a luminox NO_2 analyzer, a pulsed fluorescence Teco analyser for SO_2 and a tungsten oxide-coated automatic thermodenuder system for NH_3. Typical averaging time for the system was 20 min. for NO_2, 40 min. for SO_2 and 30 minutes for NH_3. NH_3 gradients were also measured at 4 heights with a photo- acoustical spectrometer [17].

The turbulence characteristics of the air for each measurement run of the SO_2 and NH_3 system were derived from windspeed and wind direction fluctuation at 4 m, according to a procedure outlined elsewhere [16,43]. For the NO_2 gradient system this information was obtained directly from eddy correlation measurements using a sonic anemometer at a height of 4 meter.

2.1.3.2 Implementation over forest

Intensive research on deposition fluxes was carried out at the location Speuld. A detailed description of the forest site is given in [2]. In this coniferous forest of ca. 16 - 18 m in height, a 30 m tower was erected for concentration and micrometeorological measurements. 4 minute average, half-hourly concentration measurements at the 20 m and 30 m level are

used for gradient measurements of SO_2, NO_x and O_3 by LUW. Heat flux and friction velocity were derived from wind speed and temperature measured at 20 and 30 m. On a second tower at some 100 m distance wind speed and temperature profiles between 18 and 36 m were measured over a period of 1 year by KNMI, together with momentum, heat and moisture fluxes at 30 m. From these measurement flux profile relations for this forest were established [13].
Incidently, during NH_3 gradient flux measurement campaigns by TNO, the LUW tower was extended to 36 m; NH_3 concentrations were measured by using 6 oxalic acid-coated denuders at each of 4 levels between 18 and 36 m. Through this procedure an error estimate for each gradient measurement is obtained. Friction velocity and sensible heat fluxes were determined by eddy correlation at 30 m. On a third tower hourly ammonia concentrations were continuously measured at 17 (19) and 26 (28) m between August 1989 and January 1990 by ECN, using automatic Vanadium pentoxide coated thermodenuders [3]. Fluxes were derived from these measurements using the LUW wind speed and temperature measurements at 20 and 30 m, and similar flux-profile relations.

2.1.3.3 Implementation over a maize crop
During the growing season (May to October) in the years 1987 - 1989, measurements of fluxes of O_3, NO_x, CO_2, momentum and sensible and latent heat were carried out over a field of forage maize [11]. 30-min average fluxes obtained from gradient measurements at heights of 2.5 m and 6.5 m, with meteorological measurements at 4.5 m and 6.5 m, were compared with modified Bowen ratio measurements at 2.5 and 6.5 m, and with eddy correlation fluxes of O_3, latent and sensible heat, and momentum, at a height of 7 m.

2.1.3.4 Quality control
Gradients at the 30 m forest tower are determined from measurements by monitors on the ground, with heated teflon inlet tubes up to 37 m long. Laboratory and field experiments were set up by LUW and TNO to investigate possible systematic effects of these systems on gradients and dry deposition fluxes.
It was concluded that systematic differences between the levels 20 and 30 m are unlikely to exceed a few percent for SO_2, and 1% for NO_2. For ozone, occasionally important inlet filter losses were found. These losses do not affect the concentration gradients by more than 1%. Similar experiments have been carried out by ECN for NH_3 [3]. It was concluded that measurements are contaminated and response times are lengthened when tubes longer than about 2 m are used.

2.1.4 Occult deposition, fog and dew

Fog was sampled at the forest location Speuld and at Petten near the seashore, using a fog cyclone, activated by an IR-fog detector [3]. Fog was also sampled at Arnhem during some events, using an automated CWP fog collector, in which fog droplets impact on a set of teflon strings. Occasionally, fog deposited on a Douglas fir branch was collected [12].
Dew was collected routinely over a period of 7 months in 1988 and 1989 in Speuld on a 14 m high tower near the top of the Douglas fir canopy. The dew was collected on an artificial foil surface, and on the branch of a Douglas fir. The dew from the branch was collected on a foil by shaking. The collection efficiency of this foil sampler was estimated at 35%, The dew amounts sampled are only a minor fraction of the amount determined by a total energy balance method [12]. Some experiments were also carried out inside and outside a maize canopy; dew was sampled on the glass plate of an automated sequential dew sampler [12]. Dew was sampled at Petten on a dew generator, and on a dew slide (cooled dew generator). Dew deposited on grass was sampled automatically by a "dew car". For comparison, dew was also collected from a perspex plate lying on the ground.

2.2 Results

2.2.1 Speuld

2.2.1.1 Flux-profile relations and spatial representativity

Flux-profile relations for momentum, heat and moisture were investigated by KNMI. Undisturbed flux-profile relations were obtained at the highest level (31 - 36 m), when a zero plane displacement of 12.5 m was assumed [13]. The complexity of the terrain was reflected in a clear wind direction dependence of the roughness length, ranging from 1.5 to 3 m. Despite this complexity, the flux profile relation for heat showed a well-defined behaviour as a function of stability, but progressively deviating from the Dyer and Hicks flux profile relation with decreasing height. Height dependent correction factors could be defined [13]. Flux-profile relations were also investigated for ozone. No definitive conclusions could be drawn from intercomparison of gradient and eddy correlation measurements.
From the monitoring tower in the forest a dataset comprising two years of continous measurements of NO, NO_2 and SO_2, and additional data is available for analysis. From the data analysis [15] it appeared that the parameterization used to derive u_* from wind speed and the roughness length gave results that compared well (typical scatter some 20 %) with the u_* measurements carried out on the second tower at 100 m distance. Heat fluxes derived from the temperature gradients also showed good average results although the scatter was of the order of 50 %.

2.2.1.2 Gradient analysis

Deposition velocity and flux estimates at this location were derived from gradients measured above the forests with flux-profile relations for heat established by KNMI. From the analysis of the individual half hour averaged gradients it was obvious that the random error in the flux and V_d is very high due to the scatter in the concentration gradients. This scatter can be attributed to instrumental noise, to differences in time average of measurements at two heights and to normal atmospheric fluctuations. As a result, the individual half hour values have little value for interpretation and a statistical approach had to be adopted.

The average flux and V_d for a whole year (approximately 10.000 values) are strongly influenced by some outliers, with physically unrealistic values. The median (50 percentile) flux and V_d of the different compounds appeared to be a more stable parameter. So far the exact relation between the median and (representative) average flux or V_d is not known. Model calculations using the resistance analogy with fixed R_c values showed distributions for V_d being reasonably representative for the average value. However, average fluxes were found to be significantly higher than median values. Other systematic errors in the flux and V_d might have been introduced by systematic errors in the displacement height, roughness lenght and flux profile relations.

2.2.1.2.1 NH_3

A number of campaigns were carried out by TNO over a two year period in summer and winter time to measure NH_3 fluxes to the forest [14]. In total, some 80 - 90 minute average runs were obtained, only three runs were carried out during the night. The average and median air concentrations were between 4 and 5 μg m^{-3}, with highest values of up to 25 μg m^{-3}. Preliminary results are presented here.

The median flux over the whole period was equivalent to roughly 2400 mol NH_3 $ha^{-1}a^{-1}$ (34 kg N $ha^{-1}a^{-1}$). In approximately 50 % of the runs the standard deviation in the observed gradients was low enough to derive a deposition velocity and surface (canopy) resitance for NH_3. Medians and average day time deposition velocities were quite high, around 5 cm s^{-1}. The median surface resistance of NH_3 is less than 10 s m^{-1}.

Gradient measurements were carried out by ECN continuously between August 1989 and January 1990 [3]. Preliminary results of the analysis are presented here. By combination with meteorological data from LUW, a set of roughly 1350 hourly valid measurement data was obtained. Measurements made during very stable or unstable meteorological conditions ($|z L^{-1}| > 0.5$) or low wind speed (< 1 m s^{-1}), were rejected. For the measuring period median deposition flux for NH_3 of 1860 mol $ha^{-1}a^{-1}$ (26 kg N $ha^{-1}a^{-1}$) was estimated. The deposition velocity was highly variable, with an average value of 9 cm s^{-1}, whereas the

median V_d was 3.2 cm s^{-1}.
A substantial fraction of the ECN measurements shows upward fluxes; TNO found very few significant upward fluxes, however, most of TNO measurements were made during daytime.

2.2.1.2.2 NO, NO_2, SO_2

For SO_2 the median deposition volocity for 1988 and 1989 was estimated to be 1.2 and 0.9 cm s^{-1}, respectively. The median flux was estimated 593 mol ha^{-1} y^{-1} in 1988 and 536 mol ha^{-1} y^{-1} in 1989. The uncertainty in these values is at least a factor two.
The data for NO and NO_2 showed large scatter with median deposition velocities of approx.0.1 cm s^{-1} and 0.2 cm s^{-1} for NO and NO_2 respectively. The error in these estimates is large. The 95% confidence intervals of the average V_d is roughly 1 cm s^{-1} for NO_2 and 0.3 cm s^{-1} for NO. The NO_x deposition flux was estimated 360 mol ha^{-1} y^{-1} and 270 mol ha^{-1} y^{-1} for 1988 and 1989 respectively.

2.2.2 Elspeet

The integrated study at the heathland location Elspeet (dominant species Calluna vulgaris (L.) Hull (heather) with a canopy height of c. 25 cm) was conducted from April 1989 to April 1990 [18]. Eddy correlation measurements of u_* showed very good agreement with u_* values derived from meteorological measurements of the SO_2/NH_3 gradient measuring system described in section 2.1.3. [16, 43]. R_c values for SO_2 of 70 ($\pm$ 90) s m^{-1} under dry conditions and of 20 ($\pm$ 21) s m^{-1} under wet conditions were calculated from the measurements of SO_2 gradients [16]. These values have been used for the estimation of the yearly dry deposition velocity and dry deposition flux, respectively 0.8 ($\pm$ 0.4) cm s^{-1} and 300 ($\pm$ 268) mol $ha^{-1}a^{-1}$.
The limited set of data from the measurements by the photo acoustic spectrometer for NH_3 showed net deposition fluxes. A R_c value of c. 100 s m^{-1} was derived from these measurements although the data vary considerably [17,18]. The measuring period was too short and circumstances varied too much to draw conclusions from these measurements. The results of the measurements of NH_3 by the Tungsten oxide-thermodenuders showed that this method was not accurate enough to measure the NH_3 concentration gradient [18,19]. Therefore, the yearly average dry deposition velocity and flux were estimated by the inference method using the 15 minute averaged concentration per half hour at 4 m height, together with R_a and R_b values obtained from the measurements of the dry deposition monitoring system. R_c values were taken from the measurements by TNO [20] at several heathland sites, stratified into wet circumstances (R_c = 9 s m^{-1}) and dry

circumstances (R_c = 28 s m^{-1}). The yearly average dry deposition velocity was estimated to be 1.1. cm s^{-1} and the flux 810 mol ha^{-1}a^{-1} (11 kg N ha^{-1}a^{-1}). The deposition velocity is lower than reported for other locations in the Netherlands (see paragraph 2.2.3.2.) because of the meteorological conditions during 1989, favouring low deposition.
The location Elspeet was less ideal for NO_2 flux measurements because of emissions from a road nearby. Therefore, it is hard to draw conclusions about representative annual averages from the data obtained during the measuring period of 6 weeks. The variation in results after application of rejection criteria (wind direction and wind speed) was very large. In the first half of the measuring period V_d ranged from 0.05 cm s^{-1} to 0.3 cm s^{-1}, with maximum values of 0.6 cm s^{-1} during the rest of the measuring period. A clear daily pattern was sometimes observed, with R_c ranging from 500 s m^{-1} during daytime and 2000 s m^{-1} at night. The average NO_2 flux for the whole period was 120 mol ha^{-1}a^{-1} (1.7 kg N ha^{-1}a^{-1}) [21].
The total deposition at location Elspeet can not directly be calculated from these measurements, because of lack of aerosol data and data on the input of nitrogen oxides. The deposition of these compounds was estimated by the inference method using results of aerosol, NO and NO_x measurements from the National Air Quality Monitoring Network. The results have been listed in Table 5.2 and 6.1.

2.2.3 Other sites

2.2.3.1 Sinderhoeve

The measurements over a maize crop in the growing season showed reasonable agreement for ozone fluxes determined by eddy correlation and fluxes by the modified Bowen ratio technique. Gradient measurements showed underestimation of the flux due to measuring the profile too close to the vegetation. The estimated accuracy in 30-minute averaged fluxes was 20 -40 %, in daily average uptake ~ 10 %. The daytime average deposition flux on 11 selected days ranged from 230 - 1140 mol ha^{-1}a^{-1}. Ozone uptake correlates well with water vapour exchange [11].

2.2.3.2 Dry deposition fluxes over heathland

Measurements have been conducted in the Netherlands by TNO from 1984 to 1986 in different seasons, mainly over heather/purple moor grass vegetation (Strabrechtse Heide, Fochteloerveen and Terletse heide) [20]. About 120 hourly measurements of the dry deposition flux were made. For NH_3, after rejection of invalid or unreliable data, half of these measurements resulted in an hourly value of the dry deposition velocity (error less than 50 %). A mean deposition velocity of 1.9 cm s^{-1} was observed during the measuring periods for all heathland locations. The average flux of NH_3 during these periods was equivalent to 850 mol $ha^{-1}a^{-1}$ (12 kg N $ha^{-1}a^{-1}$). The surface resistance for NH_3 is low. An average value of 23 s m^{-1} is derived from a fit of all the results. Wet vegetation obviously acts as a perfect sink for NH_3 (R_c = 9 s m^{-1}). No clear difference between the various locations is found. The nocturnal observations did not result in an accurate flux estimate. Therefore a surface- layer model was used to account for the influence of night-time meteorology. These measurements and calculations resulted in a 24 hour yearly averaged value of 1.6 cm s^{-1} for NH_3 over heathland.

For NH_4^+, only 10% of the measurements gave significant flux values, probably due to the low deposition velocity of NH_4^+ particles. The average flux measured was equivalent to 210 mol $ha^{-1}a^{-1}$ (3 kg N $ha^{-1}a^{-1}$). The best estimate for a 24 hour average deposition velocity was 0.17 cm s^{-1}.

During one year (August 1987 - July 1988) c. 100 4 hour averaged flux measurements of NH_3 were made at Asselse heide, complemented with a limited set of flux measurements of NO_2, NO, O_3 and SO_2 [22]. The results from this study agree with those obtained from the other NH_3 measurements made over heathland in terms of V_d and R_c. The yearly average flux for NH_3 at Asselse Heide was estimated 550 mol $ha^{-1}a^{-1}$ (7.7 kg N $ha^{-1}a^{-1}$).

During three intensive field campaigns, flux measurements were carried out over various locations (peat moorland and heathland). Fluxes of NO_x showed much scatter and were low, indicating deposition velocities not larger than 0.1 cm s^{-1}. NO fluxes were upward, about 230 mol $ha^{-1}a^{-1}$ (3 kg N $ha^{-1}a^{-1}$); NO_2 fluxes were often towards the surface with an average flux of 345 - 460 mol $ha^{-1}a^{-1}$ (4.5 - 6 kg N $ha^{-1}a^{-1}$). The deposition velocity was about 0.3 ± 0.2 cm s^{-1} [23].

2.2.4. Occult deposition

2.2.4.1 Fog

No estimate has so far been made of the flux by occult deposition of fog and cloud droplets. Concentration of most ions in fog is much higher than in rain, but the occurence of fog is less frequent. Concentrations of some major ions in fog and its pH in Speuld and Petten are listed in Table 2.3.

Concentrations of sulphate, ammonium and also S(IV) were higher in Speuld than in Petten, whereas in Petten Na^+ and Cl^- were higher, as is to be expected at a location close to the sea shore [3]. Fog measurements in Arnhem often showed much higher concentrations in fog collected from a branch than in fog collected by an automatic sampler. Higher concentrations of Cl^-, Na^+, K^+, Ca^{2+} and Mg^{2+} on the branch indicated leaching; higher concentration of NO_3^-, SO_4^{2-} and NH_4^+ indicate concentrations of dry deposition on the branch [12].

Table 2.3 Mean concentration (μmol/l) of various ionic species and pH in fog. Values are rounded to two significant digits [3]

compound	Speuld 1988	Speuld 1989	Petten 1988	Petten 1989	Arnhem 1988	Arnhem 1989
pH	5.66	5.07	4.06	5.01	4.2	4.4
SO_4^{2-}	1500	1900	1300	1200	1300	2000
NH_4^+	3700	5600	3100	2300	5100	7400
NO_3^-	1500	2000	1600	840	2600	1700

2.2.4.2 Dew

Table 2.4 summarizes data of average dew deposition, pH and chemical composition of dew sampled at Speuld. Similar concentrations were found in Arnhem; however, Cl concentrations were substantially lower. From Arnhem results it can be derived that the contribution of secondary aerosols (nitrate, sulphate, ammonium) to the concentration in dew is significant. Short studies indicated enhanced concentrations for dew on needles and tree branches. From the data, annual flux estimates were made for dew, based on a total annual dew flux as estimated from the energy balance measurement. The estimates are also shown in Table 2.4. Fluxes are low with respect to wet deposition or throughfall fluxes [12].

Table 2.4 Dew deposition (mm), mean concentration (μmol/l) and estimated annual flux of ions (mol ha^{-1}) in dew at Speuld, sampled on a foil [12]

compound	concentration	flux
dew (mm)		82
SO_4	59	48
SO_3	16	13
NH_4	189	155
NO_3	39	32
NO_2	36	30
Cl	64	53
H^+	0.8	0.7

In Petten concentrations of most ions in dew were found to be a factor of 10 higher than in rain. The pH of dew is close to 5 in most cases. High concentrations of NH_4^+ and S (IV) were related to continental air masses, whereas in marine air Cl^- was a dominant ion. H_2O_2 was hardly found in dew. Differences in dew composition sampled from grass and on perspex plate gave an indication of canopy interactions; sulphate, chloride and several cations were enhanced in dew from grass, and nitrogen compounds were found in this dew in very low concentrations only during spring and summer [3].

Measurements inside a maize canopy show that dew deposition fluxes of all acidifying components were decreased compared with dew fluxes outside the field. Decreased concentrations of NO_2^- and NH_4^+ inside the field were indicative of uptake by the vegetation [12].

2.2.4.3 Uncertainties and gaps

No standard methods are available for sampling and detecting dew, especially from vegetation. No method has been implemented to determine the amount of fog or cloud water collected by the vegetation.

Although the contribution of fog to the atmospheric input is not known, there is no evidence that it is large in the Netherlands. The relevance of this research is probably more directed to characterizing exposure of vegetation to the high concentrations of various ionic species in fog and dew.

3. GENERALIZATION OF MEASURING RESULTS

3.1 Methods

In order to obtain estimates of the deposition flux over the Netherlands, deposition on a 5 x 5 km^2 scale was inferred from measurements of concentrations in air, of concentrations in precipitation and amount of precipitation, and of meteorological variables [24].

Hourly concentrations of gaseous substances were obtained from interpolation of the measurements of the National Air Quality Monitoring Network (LML). Dayly average concentrations of aerosol components were obtained from the LML. For ammonia, an annual average concentration field was obtained from model calculations [51] in view of the lack of measured data. Meteorological measurements at LML stations and aerodynamic roughness estimates were used to calculate the aerodynamic resistance and the laminar layer resistance to dry deposition on an hourly basis. The roughness estimates on a 1 x 1 km^2 and 5 x 5 km^2 scale were derived from meteorological measurements and land use data and a forest characteristics inventory [25,26]. For each component specific choices were made for the surface resistance to dry deposition (Table 3.1). The surface resistance was assumed to be independent of the land use. This limitation was necessary in view of the lack of adequate data for the various surfaces.

Data on amounts and chemical composition of precipitation obtained from the National Precipitation Monitoring Network were interpolated to obtain wet deposition on the 5 x 5 km^2 grid. It has been assumed that wet deposition is independent of the nature of the surface. For further details and a discussion on the limitations the reader is referred to [24].

Table 3.1 Yearly average R_c values and dry deposition velocities in the Netherlands for gases [24]

component	R_c (s m^{-1}) day summer	day winter	night summer	night winter	V_d (cm s^{-1})	remarks
SO_2	54	68	67	75	1.0	parameterized for dry, wet conditions and on Q
NO	2000		2000		0.04	
NO_2	150	300	1000		0.2	
NH_3	30		30		1.2	TREND model [51]
HNO_3	0		0		2.5	
HNO_2	50		60		1.0	similarity with SO_2 assumed
HCl	0		0		2.5	
aerosols					0.4	parameterized on u_*

3.2 Results

Wet deposition fluxes for the years 1988 and 1989 are presented in Table 3.2. Local fluxes have been estimated assuming that the amount and composition of precipitation is independent of the structure of the earth surface. It could, however, be expected that in areas with large roughness elements (like forests or urban areas) the catchment of the small drops through increased turbulence and larger receptor surface may be more efficient than over smooth homogeneous surfaces. No data on this could be found in the literature.

Table 3.2 The yearly average wet deposition of SO_4^{2-}, NO_3^-, NH_4^+ and total potential acid in the Netherlands in 1988 and 1989 (mol $ha^{-1}a^{-1}$)

component	the Netherlands 1988	1989
SO_4^{2-}	220	230
NO_3^-	290	300
NH_4^+	640	660
potential acid	1370	1420 (eq H^+)

The spatial distribution of the total (dry and wet) deposition of SO_x, NO_y and NH_x in the Netherlands is presented in Figures 3.1, A, B and C, respectively. The total potential acid flux in 1981 and 1989 is presented in Figure 3.2 and 3.3, respectively.

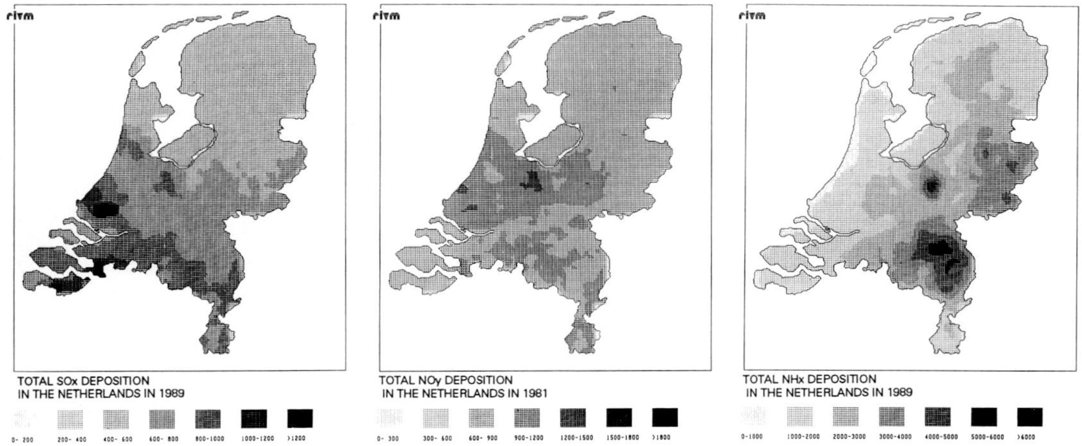

<u>Fig. 3.1</u> Total deposition of SO_x (A), NO_y (B) and NH_x (C)

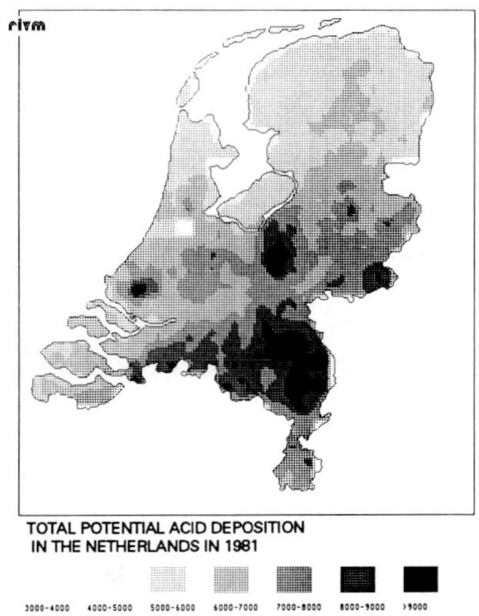

<u>Fig. 3.2</u> Total deposition flux of acid in the Netherlands in 1981 (mol H $ha^{-1}a^{-1}$)

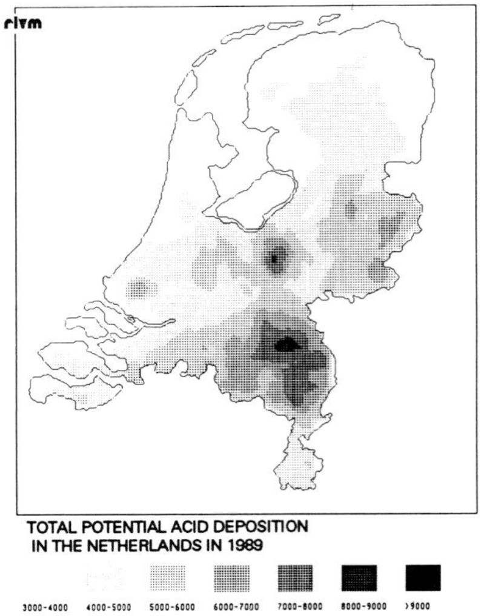

Fig. 3.3 Total deposition flux of acid in the Netherlands in 1989 (mol H $ha^{-1}a^{-1}$)

A clear gradient can be observed over the country with highest acid deposition in the South and lowest in the North. Local variations are mainly due to variations in dry deposition. These variations are not always fully represented in the Figures, because of a smoothing introduced by the interpolation of the concentrations from different stations of the monitoring network, and the lack of spatial variations of the concentrations of HNO_2, HNO_3 and HCl and of SO_4^{2-} and NO_3^- aerosols. However, some areas with increased deposition are noted: the Gelderse Vallei, the Peel area and the province of Overijssel, with enhanced NH_x deposition due to intensive livestock breeding in these regions; the industrialized Rijnmond area with enhanced SO_x deposition; and the densely populated Randstad area with high NO_y deposition. Furthermore, increased deposition is observed in urban and forested areas due to the roughness characteristics of these areas.

The contribution of wet and dry deposition of various compounds to the total potential acid deposition in the Netherlands in the years 1981 and 1989 is shown in Figure 3.4. Dry deposition of NH_3 accounts for about 27% of the total potential acid deposition in 1989. This contribution is large because NH_3 is emitted by low level sources and is rapidly deposited [27].

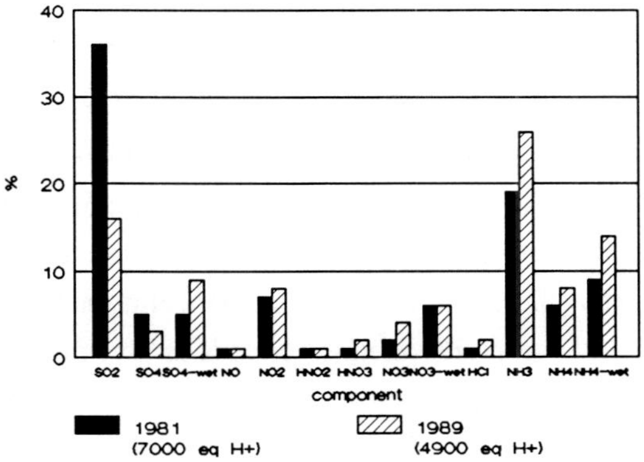

Fig. 3.4 Contribution of different components to total acid deposition for 1981 and 1989

The trend in the total deposition is established from the calculations made for the years 1981 to 1989. This trend has been plotted in Figure 3.5 for SO_x, NO_y, NH_x, as well as total potential acid deposition.

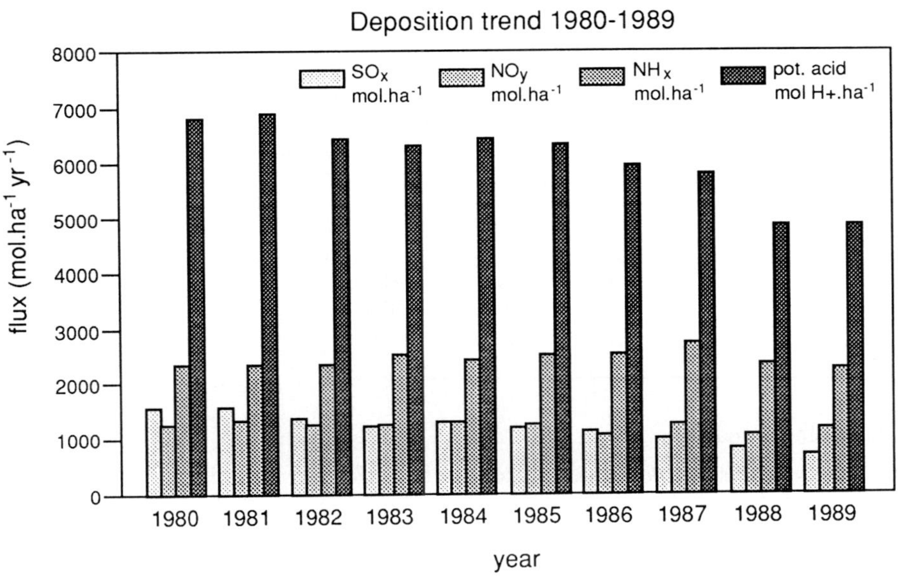

Fig. 3.5 Deposition trend in 1980 - 1989

Total acid deposition decreased from 7000 mol H^+ $ha^{-1}a^{-1}$ in 1981 to 5000 mol H^+ $ha^{-1}a^{-1}$ in 1988. The decrease of 40% is due to both meteorological variability as well as a decrease in emission of SO_2 in Europe. The meteorological variability is of great importance in 1988 and 1989. During both years meteorological conditions favoured low deposition fluxes.

The total deposition to forests, heathland and heathland pools can be estimated from the 5 x 5 km^2 grid average depositions of SO_x, NO_y and NH_x. The dry deposition of SO_x, NO_y and NH_x for each grid where forests, heathland and heathland pools are located has been related to the economic region averages. The resulting dry deposition ratios are used together with the source receptor matrices for scenario studies of acid deposition in the Netherlands by the DAS model (see the thematic report on Integrated Modelling). They represent averages over several meteorological years. For the Netherlands as a whole the ratios have been listed in Table 3.3. The ratios for heathland vegetation will be near to one, because roughness for these areas corresponds to that for the average 'Dutch landscape'. It is to be emphasized here that variations in R_c due to difference in vegetation, soil and other receptor characteristics were not taken into account. Furthermore, increase in turbulence due to roughness transition zones is also not accounted for. The depositions on heathland pools are indeed significantly higher than the economic region averages for SO_2 and NH_3. This is because of the high solubility of SO_2 and NH_3 in water with the pH range found in Dutch moorland pools (pH = 5 to 7), and because most heathland pools are situated in rough terrain (heathlands with isolated trees).

Table 3.3 Average ratio of dry deposition on heathland, forest and heathland pools to the dry deposition flux in the Netherlands. Also listed is the average dry deposition flux in 1989 (mol $ha^{-1}a^{-1}$) [24]

compound	forest	heathland	heathland pools	flux
SO_x	1.2	1.1	2.2	460
NO_4	1.2	1.1	0.4	850
NH_x	1.2	1.1	1.9	1650

3.3 Uncertainties and gaps

The uncertainty analysis of deposition fluxes can be performed by application of the error propagation theory to the calculation method. The uncertainty in the dry, wet and total deposition fluxes for each component can be split into random errors and systematic errors. Random errors are defined as errors which are not correlated and will not lead to systematic differences. Examples of random errors can be certain measuring artefacts, like analytical errors; or errors due to approximations or parametrizations of complex systems. All other errors, which are correlated and will lead to systematic differences, are defined as systematic errors. Examples of these errors are many. However, the extent of their systematic differences is mostly not known, otherwise corrections could be applied for such errors.

Examples of systematic errors might be: neglecting the difference in R_c values for different receptor surfaces, neglecting spatial variability in concentrations of HNO_2, HNO_3, HCl and aerosols, neglecting occult deposition, errors in the NH_3 emission factors (and resulting errors in NH_3 and NH_4^+ concentrations), neglecting roughness transition zones, neglecting difference in canopy structure, neglecting the influence of regional ground level emissions on vertical profiles for NO_x and correlations which have not be taken into account (e.g. the influence of NH_3 concentrations on the SO_2 dry deposition rate and vice versa).
The random errors in the final result will be a function of averaging time intervals and the spatial scale considered. Systematic errors might only be reduced to a very limited extent as a result of averaging in time and space.
Random errors and systematic errors in fluxes can be estimated from errors in basic quantities by the standard error propagation theory. For mathematical calculations like adding and subtraction, individual absolute random errors for each parameter are squared. The root of the sum of the squares is then the absolute random error in the result. For dividing and multiplying, the relative random error in the results are calculated likewise from the individual relative errors in the parameters. For systematic errors, this method can also be applied when a correlation term is added: two times the product of the individual relative errors times the correlation coefficient between the individual parameters.
The start of the propagation chain is formed by the estimation of random and systematic errors in the measurements (u, C, Q, T, mm, C, etc.) and parameters used in the calculation scheme (R_c, interpolation radius, z_o, d, etc.) from possible error sources. For some parameters the uncertainty in the basic parameters is much more difficult to determine than the uncertainty in the result. The latter can sometimes be obtained, for example from intercomparison studies. In that case the uncertainty in the result is used directly.
The correlation coefficient is very hard to determine and usually even unknown. A worst case approximation can be made by assuming 100% correlation. The uncertainty analysis has not been applied as thoroughly as proposed here. Such an analysis is published in [24], which will be available in the near future. In this report preliminary results of this study will be summarized.
The uncertainty in wet deposition due to random errors was established for the individual components based on intercomparison experiments (Table 2.2), comparison studies for wet-only and open samplers, studies with more than one sampler at a location and studies on the uncertainties due to network configurations (both by transport models as well as measurements). The results for the different components on a local scale (5 x 5 km^2) and for the Netherlands as a whole have been listed in Table 3.5. The uncertainty due to random errors in wet deposition flux is low.
Systematic errors in the wet deposition flux are introduced by dry deposition of gases and

aerosols onto the funnels of the measuring devices and by measuring artefacts (eg. the efficiency of collecting small drops and fog, analytical errors). An impression of the importance of systematic errors in fluxes on a local scale was obtained by comparison of interpolated wet deposition estimates to results of a local network [24]. The disturbing influence of nearby NH_3 sources on the measured flux was clearly demonstrated. Estimates for systematic errors have also been listed in Table 3.4.

Table 3.4 Yearly average R_c range (s m^{-1}) for forests, moorland and agricultural areas

component	forests		moorland		agricultural area	
	Rc min	Rc max	Rc min	Rc max	Rc min	Rc max
SO_2	40	200	40	200	20	100
NH_3	10	100	10	60	10	100
NO	1000	emission	1000	emission	1000	emission
NO_2	150	emission	150	emission	150	emission
HNO_2	40	200	40	200	20	100
HNO_3; HCl	-20*	20	-20*	20	-20*	20

* Formally, negative surface resistance has no meaning. The data here reflect a range of uncertainty in total deposition velocity

Throughout the calculation procedure for the dry deposition flux, several random errors may be introduced. Random errors can usually be neglected for most components when considered for the flux in the Netherlands as a whole. Random errors calculated from the inference method have been listed in Table 3.5 for 5 x 5 km^2 grids and as country averages. Systematic errors in the dry deposition fluxes are much harder to quantify. Systematic errors have been estimated for the fluxes on a 5 x 5 km^2 scale and for the Netherlands as a whole. The difference between systematic errors for the two scales considered is due to the contribution of processes only influencing the dry deposition flux on a local scale. This can be, for example many roughness transition zones and/or forest edges within one grid, or a grid mainly composed of arable land where several crops are grown during one year, local sources within a grid, etc. Apart from these local contributions to the systematic errors, errors will be independent of the scale considered. The magnitude of systematic errors was estimated by defining the sytematic errors in R_a, R_b, R_c and concentration C. The latter two were taken specifically for the components considered. Furthermore, the systematic error in R_c was defined depending on forest, heathland and agricultural landuse categories. These have been listed in Table 3.4.

From the yearly average absolute parameters and corresponding systematic errors, the systematic error in V_d and F was calculated by following the inference method. The obtained errors for the different categories were then weighted averaged for the whole country (based on the surface coverage in the country). For the systematic error in the 5 x 5 km^2 grid flux, the average systematic error was taken from the three categories. The estimated errors in C and the calculated errors in V_d and F have been listed for both scales in Table 3.5. In the systematic error estimates of total deposition it is assumed that there is no correlation. These estimates are therefore conservative. A worst case estimate has been made by assuming 100% correlation between dry and wet deposition and deposition of compounds among themselves. The fully correlated error estimates have also been listed in Table 3.5.

Table 3.5 Total uncertainty (%) in yearly average total deposition flux on different spatial scales, for all individual components (1988)

component		5 x 5 km^2					the Netherlands				
	F	random	systematic errors				random	systematic errors			
			$\Delta C/C$	$\Delta V_d/V_d$	$\Delta F/F$	$\Delta F/F$ worst		$\Delta C/C$	$\Delta V_d/V_d$	$\Delta F/F$	$\Delta F/F$ worst
	mol $ha^{-1}a^{-1}$	%	%	%	%	%	%	%	%	%	%
dry SO_2	490	15	20	40	45		0.4	15	20	25	
SO_4^2	80	47	20	50	54		1.2	15	30	34	
wet SO_4^{2-}	220	5			25		2			20	
total SO_x	790	11			30	40	0.6			17	25
dry NO	70	22	30	200	200		0.5	20	100	102	
NO_2	460	18	30	100	140		0.5	20	50	54	
HNO_2	50	26	50	60	78		0.7	40	40	57	
HNO_2	100	26	50	60	78		0.7	40	30	50	
NO_3^-	140	47	40	50	64		1.2	25	30	39	
wet NO_3^-	290	6			25		2			20	
totalNO_x	1110	10			61	96	6			25	46
dry NH_3	1470	148	50	30	58		17	40	20	45	
NH_4^+	400	53	50	50	71		1.3	40	30	50	
wet NH_4^+	640	13			40		6			30	
totalNH_x	2510	87			37	98	10			29	98
dry HCl	100	26	50	60	78	78	0.7	40	40	57	57

From Table 3.5 it is obvious that both for the Netherlands as a whole and for the 5 x 5 km^2 scale random errors are negligible compared with systematic errors, except for the random error in the NH_x on a 5 x 5 km^2 basis. The uncertainty in the total potential acid deposition can be estimated from the errors in the individual fluxes for SO_x, NOy and NH_x. The estimated uncertainty in the yearly average total potential acid deposition flux is 44% for the 5 x 5 km^2 grids (random errors taken for NH_3) and 16% for the Netherlands as a whole. The estimated uncertainty in the yearly average total potential acid deposition flux is 45% for large forest areas and 44% for large heathlands. These figures are not different from those for the 5 x 5 km^2 grids, because the error is completely determined by the random error in the NH_x flux . Uncertainty in the SO_x and NO_y fluxes for forests is about 10% higher than for 5 x 5 km^2 grids and uncertainty in fluxes for both components for heathland is about 5% higher than for the grids. The uncertainty in deposition estimates for individual nature areas is probably much larger than the values presented here.

If 100% correlation between systematic errors is assumed, errors are much larger, leading to 80% for total potential acid deposition for each 5 x 5 km^2 grid and 48% for the Netherlands as a whole. The best estimate for the errors in total fluxes will probably ly somewhere between the non correlated and fully correlated estimates.
For SO_x the dominant uncertainties are in the deposition velocity for SO_2 and sulphate. For grassland and heathland deposition uncertainty studies have provided much information for SO_2 deposition velocity, resulting in rather narrow error ranges [16,44]. Also, budget considerations are supporting present estimates of deposition fluxes [49]. In the case of SO_x, emissions, export, wet deposition and total air concentration are known with sufficient accuracy to estimate dry deposition as the only remaining unknown with reasonable accuracy. Unfortunately this is not possible for NO_y and NH_x. For NO_y, the uncertainty in concentrations of reaction products and their fate (e.g. organic nitrates, HNO_2) is large. For NH_x the emission is unknown by at least 40% [50]. Moreover, due to high spatial variability in the concentration it is hard to get an experimental estimate of the total content of NH_x in air. These two factors preclude a budget estimate of NH_x dry deposition.

The following points need attention in the coming years:
- more data on R_c values under Dutch environmental conditions;
- measurements on NH_3 concentrations and deposition processes;
- incorporation of edge effects or roughness transition zones;
- measurements on aerosol concentration and deposition velocities;
- measurements of spatial variation of HNO_2, HNO_3 and HCl concentrations;
- reference deposition measurements for model validation;
- the influence of forest structure on deposition velocities.

4. SOIL INPUT: MEASUREMENTS OF THROUGHFALL AND STEM-FLOW

4.1 Methods

Throughfall has been measured and monitored under deciduous and coniferous forest, and under low vegetation (heather, grass, maize). The measurements were usually combined with measurements of bulk precipitation in nearby open field locations. Sampling of bulk precipitation was usually carried out on a biweekly or monthly basis in polythene funnels (c. 200 - 400 cm surface area) placed at a height of 120 - 400 cm, equipped with nylon mesh filters and anti-bird crowns, and mounted on light-protected bottles. Throughfall under forest was sampled weekly or fortnightly in comparable systems, placed at a lower height (typically 50 cm above forest floor). Usually two samplers were used; in Speuld and Kootwijk. 7 - 12 samplers were used in random configuration. At some locations half-open drain pipes radially extending from the tree stem were used. Stemflow in forest was sampled at site 3, 12 and 13 by sampling water intercepted by various types of collars placed around the trunk (see Appendix E). In low vegetation, bulk precipitation was sampled with similar systems. Throughfall and stemflow (each 5 replicates) were gathered in dark bottles dug in the soil, with HgI_2 in the bottles to preserve the samples [40]. Throughfall under maize was sampled daily at 50 cm above the surface. Measurements of bulk precipitation were carried out using two wet-only rain samplers located nearby the maize field [48].

At the Kootwijk forest location, measurements with a wet-only throughfall sampler were carried out [3]. Measurements on an event basis have been carried out at several locations, viz. Speuld [3, 30] and Gelderse Vallei [31].

Comparison of fluxes in water running off a large polythene shelter/roof (240 m^2) placed below a forest canopy in Speuld indicated good spatial representativity of throughfall measurements [32].

Stability of the samples under various conditions was studied, especially for NH_4^+ and NO_3. In order to prevent photochemical conversion, light-tight bottles were used [7]; in order to prevent biological assimilation by micro organisms most groups added Hg-compounds to the samples. The influence of sample preservation (on line analysis, vs. freezing or storage at ambient temperatures) on the NH_4^+ content was found to be non-significant for weekly samples: the effect of preservation for one month was not investigated. This is the sampling time used for many measurements. Throughfall samples collected in open gauges in forest produced average NH_4^+ fluxes higher by 10 - 25% than samples from wet-only throughfall collectors. Care was taken by some groups to remove samples contaminated by bird excrements, as signalled by high P contents.

Chemical analysis was made by a variety of methods by the different groups. In order to investigate the comparability of the results of the different laboratories, two intercomparison experiments were held. In the first one artificial samples were used and in the second one ambient throughfall samples. The reproducibility (ISO 5725) of the analysis of eight laboratories is given in Table 4.1.

Table 4.1 Reproducibility (ISO 5725) of the analysis of throughfall samples (%)

compound	artificial sample	ambient sample
SO_4^{2-}	5	20
Cl^-	20	20
NO_3^-	10	30
NH_4^+	10	10
K^+	5	10
Na^+	10	10
Mg^{2+}	20	20
Ca^{2+}	20	10

Because statistical outliers were removed they did not affect the results given above. In reality outliers will affect in a negative way the comparability of the results. The reproducibility of the analysis of nitrate is unsatisfactory. The differences in the results originating from different laboratories may hinder generalisation.

4.2 Results

An overview of the results of throughfall measurements for NH_4^+, NO_3^- and SO_4^{2-} on the locations Speuld, Kootwijk, Elspeet and other locations in the Netherlands is presented in the Appendix. Also shown in the Appendix are data for throughfall of other ions in Speuld, Kootwijk and the heathland sites Elspeet and Assel.

The locations where throughfall fluxes have been measured, presented in the Appendix, have been plotted in Figure 4.1. For a discussion of these results, the reader is referred to the thematic report on soil acidification, and the various references indicated in the Tables in the Appendix. Interesting issues are:

- modelling and sequential sampling of throughfall fluxes and research into the mechanism of throughfall and wash-off of dry deposition [31];
- research into the influence of canopy structure, particulary LAI, on throughfall of several ions [47];
- strong correlation and near stoechiometry of NH_4^+ and SO_4^{2-} in throughfall, an indication of co-deposition of NH_3 and SO_2 [40].

Fig. 4.1 Map of the Netherlands with the locations where throughfall fluxes have been measured

Stemflow samples were obtained at few locations. The stemflow flux of nutrients represents typically only 5 - 10% of the total flux under trees, depending on type and age of the trees and ions considered. The numbers in the Appendix do not always include stemflow. For a discussion of the contribution of litter decomposition to soil input fluxes, the reader is referred to the soil input acidification thematic report. No attempts were made to estimate below-canopy dry deposition.

4.3 Uncertainty and gaps

The accuracy of annual average throughfall data was estimated for Douglas fir forest at Speuld at 27% for SO_4^{2-}, 17% for NO_3^- and 23% for NH_4^+ (95% confidence limit, [30]). Uncertainties are assumed to be somewhat lower for other tree species.

Research in the future should be focused on parallel sampling of throughfall (event basis, wet-only) and air concentrations. Furthermore, attention should be paid to process research of canopy exchange.

5. COMPARISON OF THROUGHFALL AND DEPOSITION

5.1 Introduction

In studying the relation between throughfall and total (dry + wet) deposition, two approaches have been taken. First, experimental data of throughfall and deposition have been directly compared. This was possible for the locations Speuld and Elspeet, where concurrent measurements of bulk precipitation, throughfall, stemflow and dry deposition have been carried out. The preliminary results, based on incomplete data, are presented in paragraph 5.2 and 5.3. Sometimes assumptions have been formulated by which canopy exchange terms can be ascertained. The resulting values of throughfall, corrected for canopy exchange, have also been compared with measured deposition fluxes. This approach is also presented in paragraph 5.2 and 5.3. For the majority of throughfall locations, measured deposition data are lacking. The measurements for these locations are compared in 5.4. with deposition estimates from inference, as given in 3.2.

5.2 Speuld

In Table 5.1 measured throughfall fluxes at Speuld are compared with deposition estimates derived from the measurements reported in paragraph 2. Fluxes by fog or cloud water are not known (see paragraph 2.2.4.1.). Fluxes carried by dew are incorporated in the measured dry deposition. The contribution of wet and dry atmospheric deposition and canopy leaching to the throughfall fluxes was estimated, using bulk- and wet-only open field precipitation and throughfall, and making various assumptions about the behaviour of different solutes [30]. These assumptions were: (1) no canopy leaching of Na^+; (2) the ratio of dry to wet deposition of all base cations would be similar; (3) the resulting calculated canopy leaching would be equivalent to the uptake of NH_4^+ and H^+, which are taken up in a ratio equal to the mean ratio of H^+ and NH_4^+ in throughfall and bulk open fields; (4) the resulting calculated dry deposition of H^+ and NH_4^+ would be matched by dry deposition of Cl^- (from the difference between bulk open field deposition plus calculated dry deposition and throughfall), and of NO_3^- and SO_4^{2-} in a ratio similar to that of the differences between throughfall and bulk open field input of NO_3^- and SO_4^{2-} after correction for seasalt SO_4^{2-} (SO_4^{2-}/Na^+ in seasalt is 0.12 on equivalent basis); (5) the rest term is the canopy uptake of SO_4^{2-} and NO_3^-; (6) remaining charge differences are attributable to weak acid anions. The results are also shown in Table 5.1.

Table 5.1 Throughfall and deposition fluxes (mol ha^{-1} a^{-1}) at Speuld in 1989

	SO_4^{2-}	NH_4^+	NO_3^-
throughfall	930	2100	790
deposition (see table 6.1)	800	2780	910
wet deposition, (wet only, [30])	307	631	332
dry deposition**	595	2254	606
dry deposition, calculated*	710	2000	490
total deposition, calculated*	1000	2600	820

* calculated from bulk precipitation and throughfall data [30]
** including aerosol deposition from inference

The comparison of experimental dry deposition estimates and throughfall measurements show resonable agreement. The agreement is even better if assumptions on canopy exchange are taken into account in the deposition estimates from throughfall. It must be emphasized that the uncertainty ranges in the data listed in table 5.1 are still very large. Canopy exchange seems to have a moderate effect on SO_4^{2-} and NO_3^- fluxes; the effect on the NH_4^+ flux is substantial in this approach. Experimental dry deposition estimates are not sufficiently accurate to validate these results.

5.3 Elspeet

Throughfall and deposition fluxes for Elspeet are summarized in Table 5.2. Dry deposition fluxes were estimated for Elspeet from throughfall and bulk precipitation data [40]. Based on a comparative study of throughfall below artificial and natural heather, it was shown that canopy exchange for SO_4^{2-} did not occur. Furthermore, stoechiometric co-deposition for SO_4^{2-} and NH_4^+ was found [40]. The result, also given in Table 5.2, indicate reasonable agreement for NH_4^+, but a large discrepancy for SO_4^{2-} with deposition measurements. No representative data for dry deposition are available for NO_3^-, due to lack of fetch in the same directions at the heathland location.

The large throughfall fluxes of Ca^{2+} found on this location (see Table E.2 in Appendix E) indicate interference of wind-blown soil dust and sea salt on the throughfall. It was argued that throughfall fluxes of potential acid deposition should be corrected for contributions of Ca sulphate and nitrate, and sea salt. Also, corrections have to be made for dry deposition on throughfall samplers, which is not related to dry deposition onto the canopy [42]. If these corrections are applied, the resulting dry depositions calculated by the method in [40] agree much better with experimental deposition fluxes.

Table 5.2 Elspeet annual average deposition and throughfall fluxes (1989) in mol $ha^{-1}a^{-1}$

	SO_4^{2-}	NH_4^+	NO_3^-
throughfall	1000	2100	940
deposition	575	1700	-*
wet deposition	210	570	300
dry deposition***	375	1160	-*
dry deposition, calculated**	680	1300	550
idem, corrected***	470	1150	?

* no representative value available
** derived from bulk precipitation and throughfall data [40]
*** derived from throughfall corrected for neutral Ca components and sea-salt, and dry deposition on samplers
**** including aerosol deposition from inference

5.4 Comparison between throughfall and deposition from inference

Throughfall fluxes on various locations (see paragraph 4) have been compared to deposition estimates from inference (see paragraph 3) [42]. Because of the local character of throughfall fluxes, this comparison had been made on the basis of 1 x 1 km^2 inference results. Throughfall fluxes were corrected for the contribution of neutral Ca^{2+} salts and for sea-salt, and for dry deposition on the throughfall samplers. Correction factors for dry deposition have been obtained from parallel measurements between open and wet-only samplers both below a canopy and in open field. Corrected throughfall values are about 82 %, 94 % and 85 % of uncorrected values for SO_4^{2-}, NO_3^- and NH_4^+, respectively. The dataset for which the data necessary for correction were available comprises 24 locations, and is indicated in Table E.1 in the Appendix.

Corrected throughfall fluxes have been plotted versus the total (wet + dry) deposition fluxes for SO_4^{2-}, NO_3^-, NH_4^+ and potential acid in Figure 5.1 (A) to (D).

The correlation between total deposition and throughfall fluxes for SO_4^{2-} is high (0.84). This high correlation reflects the correlation between the spatial distribution of throughfall and deposition of SO_4^{2-}. Throughfall fluxes are higher, however, by up to a factor 2. Correlation for the components NO_3^- and NH_4^+ is poor. NO_3^- fluxes are estimated at about a factor 2 higher by the inference model.

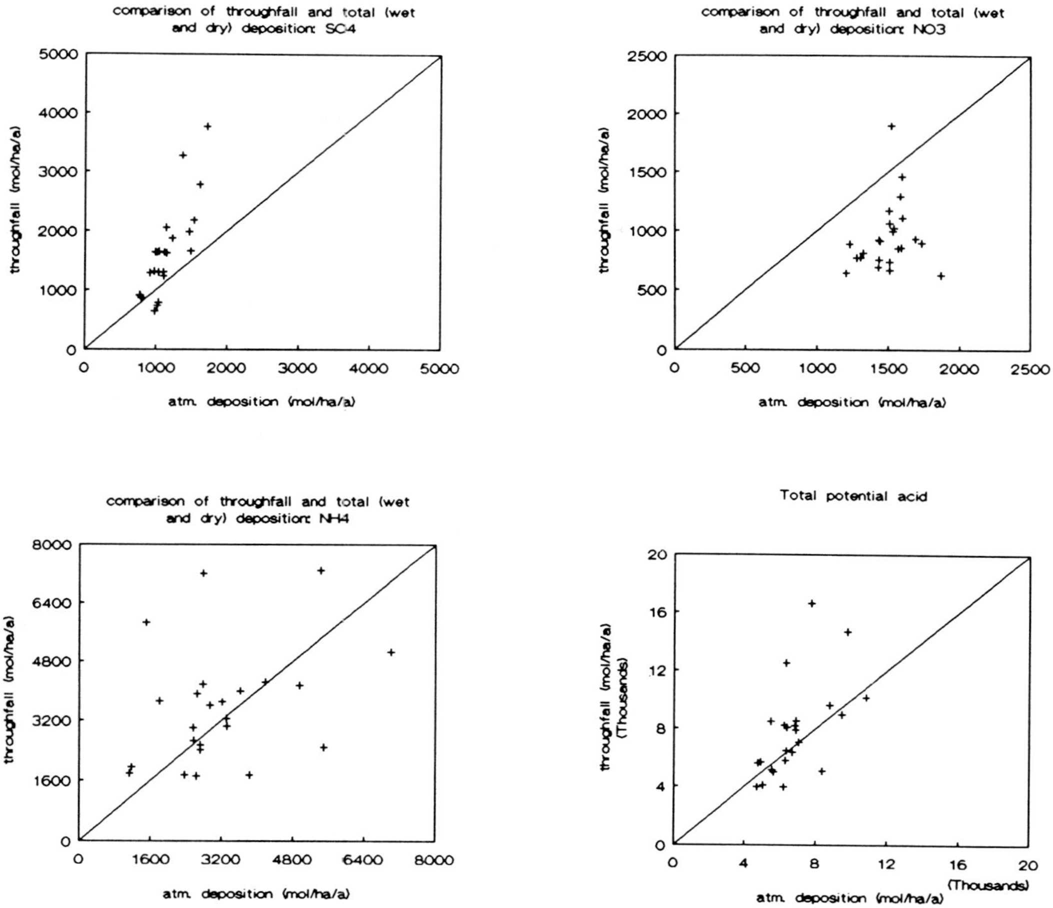

<u>Fig. 5.1</u> Comparison of total deposition for SO_4^{2-} (A), NO_3^- (B), NH_4^+ (C) and total potential acid, estimated by inference and throughfall

For NH_4^+ no systematic difference between throughfall and deposition is found. The agreement between estimates of total potential acid and throughfall is reasonable, although throughfall measurements show somewhat higher fluxes than inference estimates. The yearly average total potential acid deposition for the forests considered is 7800 mol H^+ $ha^{-1}a^{1}$ for throughfall measurements and 6800 mol H^+ $ha^{-1}a^{-1}$ for model estimates. The difference in fluxes obtained by the two methods lay within the uncertainty ranges (see section 3.3. and 4.2.).
Significance tests showed that corrected throughfall fluxes and deposition estimates differ significantly for SO_4^{2-}, NH_4^+ and NO_3^- on the 0.05 level of significance. Total corrected throughfall fluxes agreed significantly with deposition model estimates for the total potential acid deposition flux. All uncorrected fluxes differ significantely from model estimates.
Some factors possibly contributing to remaining differences are [42]:

- errors in deposition estimates for large particles;
- contribution of occult deposition to throughfall;
- canopy interaction, such as stomatal uptake and canopy leaching.

From the results obtained in this study it was concluded that there are still inexplicable differences between flux estimates by throughfall measurements and the deposition model. Research on the processes causing these differences is necessary. Without this research a causal relation between throughfall and deposition can not be established.

6. CONCLUSIONS AND RECOMMENDATIONS

6.1 Conclusions

In this report, attention has been focused on annual averaged potential acid deposition onto forest and heathland in the Netherlands and on the Netherlands as a whole.
Concentration and deposition estimates have been presented for a coniferous forest (location Speuld) and a heathland (location Elspeet) and some other locations, based on measurements of air concentrations, meteorological parameters, concentrations in precipitation and precipitation amount, and dry deposition fluxes, using micrometeorological measuring techniques. An overview of all flux estimates and measurements at several locations in the Netherlands is given in Table 6.1.

Table 6.1 Flux measurements and estimates by different methods and for different locations in the Netherlands (mol $ha^{-1}a^{-1}$). The estimates have been presented more accurate than they are in fact, because of the intercomparison

location	comp.	year	height (m)	F=Vd C inter-ference	F measured above can. ****	F calc. model	throughfall measured	through-fall*	through-fall**
Speuld									
dry	SO_2	1988	30	-	593	485		797	555
	SO_4^{2-}			76***	-	95			
wet	SO_4^{2-}			324	324	215		324	177
total	SO_x			993	-	795	964	1120	732
dry	NH_x			3144***	-	1465		2113	1090
dry	NH_4^+			385***	-	535			
wet	NH_4^+			629	629	590		629	630
total	NH_x			4158	-	2590	2067	2742	1720
dry	NO_x			-	360	410			
	HNO_3			160***	-	265		444	624
	HNO_2			47***	-	175			
	NO_3^-			48***	-	155			
wet	NO_3^-			340	340	340		274	274
total	NO_3			955	-	1345	671	718	898
Speuld									
dry	SO_2	1989	30	-	536	350		713	486
	SO_4^{2-}			59***	-	70			
wet	SO_4			200	200	200		306	207
total	SO_x			795	-	620	933	1019	693
dry	NH_3			-	1860	1335		2001	1278
	NH_4^+			394***	-	511			
wet	NH_4^+			530	530	530		631	605
total	NH_x			2784	-	2376	2110	2632	1883
dry	NO_x			-	270	480			
	HNO_3			180***	-	158		487	410
	HNO_2			47***	-	55			
	NO_3^-			109***	-	228			
wet	NO_3			300	300	300		332	333
total	NO_3			906	-	1221	790	819	743

location	comp.	year	height (m)	F=Vd C inter-ference	F measured above can. ****	F calc. model	throughfall measured	through-fall*	through-fall**
Elspeetsche Veld									
	NH_3	1989	4		810	1210		1300	1147
	NH_4^+			350		546			
wet	NH_4^+			570		570		846	635
total	NH_x			1730		2326	2146	2146	1882
	SO_2	1989	4		300	240		680	470
	SO_4^{2-}			75		75			
wet	SO_4^{2-}			210		210		355	310
total	SO_x			585		525	1035	1035	780
dry	NO_x			-	-	460			
	HNO_3			-	-	150		548	541
	HNO_2			-	-	55			
	NO_3			-	-	225			
wet	NO_3			-	-	310		387	329
total	NO_3			-	-	1200		935	870
Assel	NO_3	1987	1.5	-	552	1185			
Terl.heid	NH_3	+	1	-	647	1015			
Focht.vn	NH_3	+	1	-	765	1275			
Strab.hei	NH_3	+	1	-	2059	1800			
Koningsh.	NH_3	+	1	-	588	1015			
Cabauw	HNO_3	1986	200	-	215	175			
Zegveld	SO_2	1988	4	-	585	440			

* separated into wet- and dry deposition by the method described in [30]
** separated into wet- and dry deposition by the method described in [42]
*** fluxes calculated from inference with measured yearly averaged concentrations and estimated Vd [15]
**** data for dry deposition represent median values, these may differ from averages
\- not measured, or not available yet
\+ measurements made in small parts of the year, uncertainty reflects translated yearly averages

Experimental estimates have been obtained for NH_3, SO_2 and NO_x over forests and heathland. The uncertainty in the fluxes over forests are large. In the table median values have been presented. These can differ from the averages. Although pollutant concentrations in fog have been measured, no flux estimates for occult deposition are available. Estimates for fluxes in dew have been established seperately. However, these are included in dry deposition as measured. Data for aerosol dry deposition are lacking.

Deposition estimates for the Netherlands have been made on a 5 x 5 km^2 scale. These results were produced by inference methods on the basis of experimental concentrations and meteorological data, and landuse information. The uncertainty in the deposition was estimated at c. 40% on this scale. However, for individual nature areas, the uncertainty could be much larger.

Throughfall has been monitored on a number of locations under forest, heathland, grassland and maize. Although throughfall and total deposition do not show dramatic differences, there are important systematic differences for SO_4^{2-} and NO_3^-, and largely varying deviations for NH_4^+. Comparative studies of throughfall and deposition, as well as exploratory studies for dew and fog, indicate moderate to strong canopy interactions.
The dry deposition parameters used for the inference methods are reasonably consistent with the results in this table. However, NO_y fluxes are estimate somewhat higher by this method than by other methods.

6.2 Research needs

Research needs are most urgent for dry deposition measurements. There is insufficient accurate knowledge of dry deposition (processes) on forest and heathland. Application of micrometeorological techniques requires more accurate (< 1%) rapid (t < 0.1 s), and sensitive methods for concentration measurements, which are to be further developed and made operational. This is especially urgent for NH_3, SO_2, NO, NO_2 and aerosols components (sulphate, nitrate and ammonium, basic cations). No methodology for determining (coarse) particle deposition fluxes has been developed in the Netherlands. Throughfall measurements, theoretical considerations as well as recent measurements, indicate that particle fluxes can be sizable, especially for forests.
It is not yet clear which measuring strategy should be followed in determining annual average fluxes. Both campaign-type measurements and also continuous monitoring show low coverage in time, disregarding certain periods in which fluxes are hard to determine (low winds, extreme stability). Inference methods can help to construct annual averages from incomplete measuring sets. However, inference depends on detailed description of deposition parameters under different conditions. It is recommended that systematic measurements of dry deposition over forest and heathland are continued, with improved methods. Additional research is needed for accurate dry deposition parametrizations under various conditions. Attention has to be paid to research on the influence of combinations of trace gases on deposition processes. In particular the co-deposition phenomenon of SO_2 and NH_3 must be studied.
Monitoring of throughfall with all necessary analytical and sampling precautions (wet-only samplers, event basis, quick analysis, prevention of biological and photochemical conversion) has to be carried out, together with measurements of air concentrations and

basic meteorological parameters. In order to improve our understanding of throughfall, wash-off and canopy interactions, more research is needed on these processes. Therefore concurrent monitoring of throughfall and deposition is recommended. It is necessary to include both biological and atmospheric scientists for the comparison between methods and for the synthesis and interpretation of results. A special point of interest is presently the origin of various basic cations in throughfall and the relation with (acid) deposition.
Measurements of occult deposition fluxes are presently lacking. There is presently no evidence that these fluxes form a significant fraction of the total atmospheric input flux in the Netherlands.
Modelling of dry deposition is presently semi-empirical (resistance analogy), well matched to the experimental possibilities. In-canopy deposition models have not been applied in this study. It is recommended that explicit modelling of particle dry deposition and occult deposition in forest be carried out, and be tested against measurements, to fill the gap in our knowledge in this respect. Simple models for wet deposition are available and well matched to the available experimental information. Models for dew have been developed and tested, so far not very successfully.
Generalization of deposition results obtained at only some measuring locations requires:
- more information on surface resistance for various landscape/landuse types, as a function of time of the year and of the day, and of the conditions controlling this parameter;
- more information on spatial variations of air concentrations, especially for HNO_2, HNO_3, HCl, NH_3 and aerosol components.
- more information on vegetation structure, including edge effects and roughness transition zones, and its influence on turbulence and deposition parameters.

Up to now no reliable method for deposition estimates by inference on a small scale (100 x 100 m) seems to be available.
Finally it is recommended to proceed with research on the uncertainty of deposition and throughfall, both by inference methods and by direct experimental methods (e.g. measurements in duplo).

7. REFERENCES

[1] van Aalst R.M. (1984), Verzuring door atmosferische depositie - Atmosferische processen en depositie. TNO, Delft, the Netherlands

[2] Vermetten A.W.M., Hofschreuder P., Versluis A.H., Bij E.S.v.d., Tongeren J.v., Molenaar E., Houthuyzen J.D., Veld F.in 't (1990) Air pollution in forest canopies. Report no. R-424, Wageningen Agricultural University, Wageningen, The Netherlands

[3] Slanina J., Keuken M.P., Arends B., Veltkamp A.C., Wyers G.P. (1990) Report on the contribution of ECN to the second phase of the Dutch Priority Programme on Acidification. ECN, Petten, the Netherlands

[4] Thijsse Th.R., Bijlsma W.B., Hoogeveen A.W. (1990) Aanvullende koolwaterstofmetingen op de lokatie Kootwijk. Report no. R 90/056, TNO, Delft, The Netherlands

[5] KNMI/RIVM (1988) Chemische samenstelling van de neerslag over Nederland, jaarrapport 1987. KNMI report no. 156-10; RIVM report no. 228703005, De Bilt, the Netherlands

[6] Buijsman E. (1990) De berekening van de natte, zure depositie: een vergelijking van een aantal berekeningswijzen. Report no. 228703011. National Institute of Public Health and Environmental Protection, Bilthoven, the Netherlands

[7] Ridder T.B., Baard J.H., Buishand T.A. (1984) De invloed van monstermethoden en analysetechnieken op gemeten chemische concentraties in regenwater. Technical report T.R.-55. Royal Netherlands Meteorological Institute, De Bilt, the Netherlands

[8] Slanina J. Römer F.G., Asman W.A.H. (1982) Investigation of the source regions for acid deposition in the Netherlands. Proc. CEC Workshop on Physico-Chemical behaviour of atmospheric pollutants, 9 Sept., Berlin, 1982

[9] Hicks B.B., Draxler R.R., Albritton D.L., Fehsenfeld F.C., Hales J.M., Meyers T.P., Vong R.L., Dodge M., Schwartz S.E., Tanner R.L., Davidson C.I., Lindberg S.E., Wesely M.L. (1989) Atmospheric processes research and procecess model development. State of Science/Technology report no. 2. National Acid Precipitation Assessment Programme

[10] Fowler D. and Duyzer J.H. (1990) Micrometeorological techniques for the measurement of trace gas exchange. In: Exchange of trace gases between terrestrial ecosystems and the atmosphere (Eds. Andrae M.O. and Schimel D.S.), John Wiley and Sons

[11] Pul W.A.J.van (1990) Deposition fluxes above a crop during a growing season. Report project 72, Department of Meteorology, Agricultural University Wageningen, the Netherlands

[12] Römer F.G., Winkel B.H.te, Ruijgrok W., Steenkist R., Wakeren J.H.A.van (1990) The chemical composition of dew and the deposition flux of water vapour: field measurements and modelling. Final report. Report no. 50583-MOC 90-3411, KEMA, Arnhem, the Netherlands

[13] Bosveld F.C. (1990) Turbulent exchange coefficients over a Douglas fir forest. Report no. Royal Dutch Meteorological Institute, De Bilt, the Netherlands

[14] Duyzer J.H., Verhage H.L.M., Weststraten J.H., Bosveld F.C., Measurement of the dry deposition flux of NH_3 into coniferous forests

[15] Duyzer J.H., Diederen H.S.M.A., Weststraten J.H., Bosveld F.C., Vermetten A.W.M., Versluis A.H., Wyers G.P. (1990) Dry deposition of acidifying compounds into coniferous forest; results of a long-term monitoring programme, 1990, in press

[16] Erisman J.W., Elzakker B.G.van, Mennen M.G. (1990) Dry deposition of SO_2 over grassland and heather vegetation in the Netherlands. Report nr. 723001004, National Institute of Public Health and Environmental Protection, Bilthoven, the Netherlands

[17] Rooth R.A., Verhage A.J.L., Wouters L.W. (1990) Photoacoustic measurement of ammonia in the atmosphere, influenced by water vapour and carbon dioxide Submitted to Applied Opics

[18] Erisman J.W., Elzakker B.G., Mennen M.G., Duyzer J.H., Römer F.G., v.d.Beld L., Verhage A.J.L., Rooth R.A. (1991) The Elspeet experiment: Micrometeorological and throughfall measurements on the deposition of acidifying components to a heathland, in press

[19] Beld L.van den, and Römer F.G. (1990) Ammonia measurements at Elspeet in the period May 1989 - April 1990. Report 90389-MOC 90-3415, KEMA, Arnhem, The Netherlands

[20] Duyzer J.H. and Diederen H.S.M.A. (1989) Measurements of dry deposition velocities of NH_3 over heathland and forest. Report P 89/023. TNO, Delft, The Netherlands

[21] Duyzer J.H., Verhage H.L.M., Weststraten J.H. (1990) Monitoring depostition of nitrodioxide on the Elspeetse Veld heathland, MT-TNO report, in press

[22] Duyzer J.H., Verhage H.L.M., Erisman J.W. (1989) De depositie van verzurende stoffen op de Asselse heide. Report no. R 89/29, TNO/RIVM, Delft, The Netherlands

[23] Duyzer J.H., Fowler D., Meixner F., Dollard G., Johanson C., Gallagher M. (1990) The Halvergate trace gas experiment on surface exchange of oxides of nitrogen: preliminary results. Contribution to the COST 611 workshop: Field measurements and interpretation of species derived from NO_x, NH_3 and VOC emissions in Europe, 12-14 March 1990, Madrid, Spain

[24] Erisman J.W. (1990) Acid deposition in the Netherlands. Report no. 723001002, National Institute of Public Health and Environmental Protection, Bilthoven, The Netherlands

[25] Erisman J.W. (1990) Estimates of the roughness length at Dutch Air Quality Monitoring Network stations and on a grid basis over the Netherlands. Report no. 723001003, National Institute of Public Health and Environmental Protection, Bilthoven, the Netherlands

[26] Meijers R. (1990) Parametrisatie van de structuur van Nederlandse natuurgebieden. Department of Physical Geography, State University of Utrecht, Utrecht, The Netherlands

[27] Erisman J.W., Leeuw F.A.A.M.de., Aalst R.M.van (1989) Deposition of the most acidifying compounds in the Netherlands during 1980-1986. Atmospheric Environment, 23, 1051-1062

[28] RIVM (1989) Jaarrapport 1988. Report no. 228702015., National Institute of Public Health and Environmental Protection, Bilthoven, the Netherlands

[29] RIVM (1990) Jaarrapport 1989. Report no. 222101006., National Institute of Public Health and Environmental Protection, Bilthoven, the Netherlands

[30] Maas M.P.van der (1990) Hydrochemistry of two Douglas fir ecosystems in the Veluwe, the Netherlands. Final report no. 102.1

[31] Ivens W.P.M.F., Draaijers G.P.J., Bos M.M., Bleuten W. (1988) Dutch forests as air pollutant sinks in agricultural areas. Report no. AD 1988-01, Department of Physical Geography, State University of Utrecht, Utrecht, the Netherlands

[32] Dijk H.F.G.van, Boxman A.W., Roelofs J.G.M. (1990) Effects of a decrease in atmospheric deposition on the mineral balance and vitality of the Dutch forest. Progressreport, project RIVM 118, VROM 641330, Catholic University Nijmegen

[33] Kleijn C.E., Zuidema G. Vries W.de (1989) De indirecte effekten van atmosferische depositie op de vitaliteit van Nederlandse bossen. 2. Depositie, bodemeigenschappen en bodemvochtsamenstelling van 8 Douglas opstanden. Report no. 2050, STIBOKA, Wageningen, the Netherlands

[34] van Dobben H.F., Mulder J., van Dam H., Houweling H. (1990) The impact of acid atmospheric deposition on the biochemistry of moorland pools and surrounding terrestrial vegetation, Pudoc, Wageningen, in prep.

[35] Verstraten J.M., Tietema A., Bouten W., Dopheide J.C.R. (1990) Field monitoring for research on the role of acid atmospheric deposition in the biogeochemical balance of an oak-beach forest ecosystem near Winterswijk, the Netherlands. Project no. 01, Laboratory of Physical Geography and Soil Science, University of Amsterdam, The Netherlands

[36] Booltink H.W.G., Pape Th., Breemen N.van (1988). Report no. 02-01

[37] Draaijers G.P.J., Ivens W.P.M.F., Bos M.M., Bleuten W. (1989) The contribution of ammonia emissions from agriculture to the deposition of acidifying and eutrofying compounds onto forests. Environ. Pollut.,60,55-66

[38] Houdijk A.L.F.M. (1990) Effecten van zwavel- en stikstof depositie op bos- en heide vegetaties (Effects due to sulphur and nitrogen deposition to forest and heather vegetation, in Dutch). University of Nijmegen, the Netherlands

[39] Jansen P. (1989) Belang van de strooisellaag voor de totale depositie in een dennenbos. Landschap,2,147-161

[40] Bobbink R., Heil G.W., Raessen M.B.A.G. (1990) Atmospheric deposition and canopy exchange in heathland ecosystems. Department of plant Ecology and Evolutionary Biology, University of Utrecht, the Netherlands

[41] Visser P.H.B.de and Breemen N.van (1990) Manipulation of water and nutrient supply in two forest ecosystems in the Netherlands. Report no. 8, project 100, Dutch Priority Programme on Acidification, Wageningen, the Netherlands

[42] Erisman J.W. (1990) Atmospheric deposition of acidifying compounds onto forests in the Netherlands: throughfall measurements compared to deposition model estimates. Report no.723001001, National Institute of Public Health and Environmental Protection, Bilthoven, the Netherlands

[43] Erisman J.W. and Duyzer J.H. (1990) A micrometeorological investigation of surface exchange parameters. Report no.723001005, National Institute of Public Health and Environmental Protection, Bilthoven, the Netherlands

[44] Erisman J.W., Versluis A.H., Verplanke T.A.J.W., Haan D.de, Anink D., Elzakker B.G.van, Aalst R.M.van (1989) Monitoring the dry deposition of SO_2 in The Netherlands. Report no.228601002, National Institute of Public Health and Environmental Protection, Bilthoven, the Netherlands

[45] Hicks B.B., Baldocchi D.D., Meyers T.P., Hosker Jr R.P., Matt D.R. (1987) A preliminary multiple resistance routine for deriving dry deposition velocities from measured quantities. Water, Air and Soil Pollut.,36,311-330

[46] Breemen N.van, Visser W.F.J., Pape Th. (1988) Biochemistry of an oak-woodland ecosystem in the Netherlands affected by acid atmospheric deposition. Agr. Res. Report 930, Pudoc, Wageningen, the Netherlands

[47] Heil G.W., Bobbink R., Dam D.van, Heijne B. (1989) Lai of grasslands: A measure for ammonium deposition from polluted air. In: Man and his ecosystem. Proc. 8th World Clean Air Congress 1989, The Hague, the Netherlands

[48] Stolk A., van Pul W.A.J., Römer F.G. (1990) Throughfall water measurements in a crop canopy (In Dutch). Report 51514-MOC 90-3406

[49] Jaarsveld J.A.van (1989) Een operationeel atmosferisch transportmodel voor prioritaire stoffen; specificatie en aanwijzingen voor gebruik. Report no.228603008, National Institute of Public Health and Environmental Protection, Bilthoven, the Netherlands

[50] Erisman J.W. (1989) Ammonia emissions in the Netherlands in 1987 and 1988. Report no.228471006, National Institute of Public Health and Environmental Protection, Bilthoven, the Netherlands

[51] Asman W.A.H. and Jaarsveld H.A.van (1990) A variable-resolution statistical transport model applied for ammonia and ammonium. Report no. 228471007. National Institute of Public Health and Environmental Protection, Bilthoven, the Netherlands

Appendix

A) Monitoring, Speuld and Kootwijk. 1988 and 1989

analysers:

SO_2 Thermo Electron 43W, range: 0 - 750 ppb, detection limit: 1 ppb
Thermo Electron 43A, range: 0 - 100 ppb, detection limit: 1 ppb
Measuring principle: Pulsed fluorescence

NO_x Monitor Labs 8840, range: 0 - 200 ppb, detection limit: 1 ppb

NO Measuring principle: Chemoluminescence

NO_2 Is calculated as NO - NO (ppb)

O_3 Bendix 8002, range: 0 - 200 ppb, detection limit: 1 ppb
Measuring principle: Chemoluminescence

O_x Is calculated as NO + O (ppb)

NH_3 Monitoring Labs 8840, range 0 - 200 ppb, detection limit: 1 ppb
+ stainless steel converter at 800 C.
Measuring principle: conversion of NH to NO , evaluation the difference

Sampling strategy: one analyser for each component switches between isolated and lightly heated (3 - 5 K above ambient T) teflon tubing (diam: 12 mm) to five levels with teflon entrance filters, pore size 5 µm, at 5, 10, 15, 20 and 30 m (NH_3: only 30 m). Residence times are typically 3 - 5 seconds, air is pulled through contineously. Signal loss due to pressure drops in the tubing (20 -50 hPa) amounts to 0 -5 % and is corrected for afterwards. Wall and filter losses: on average < 2 %. This cycle is repeated twice an hour: at each level 1 Hz samples are averaged over 4 minutes. Dead time (for switching between levels) is 1.5 minutes. Time coverage for each level is + 13 %.

Accuracy of individual 4 minute averages: depends on linearity and the accuracy of pressure correction. For all analysers better than 5 % or 1 ppb.

Precision of individual 4 minute averages: depends on systematic errors such as losses, analyser zero, sensitivity drift, etc. For SO_2 precision is less, due to a correction for a calibration offset which could not be determined within 0.5 ppb. For NO, zero drift can rise to 2 ppb just before servicing (on average < 0.5 ppb). For O_3, filter losses may be important under certain circumstances (on average < 5 %). For NH_3 signal damping due to absorption and desorption processes plays an important part, so that no precision can be given for the minute averages.

The uncertainty in the calculated monthly and annual averages depends on time coverage (80 - 90% of the halfhourly data are available, except for Dec 1989: 40 - 50 %).

For SO_2 monthly average concentrations were quite low, so that the uncertainty will largely be determined by the uncertainty in the correction for the calibration procedure (0.5 ppb). Averaging over one year will reduce the uncertainty to + $\mu g/m^{-3}$ (+ 10 %).

For NO and NO_x the uncertainty in monthly averages will also be about 0.5 ppb (or 5 % at higher concentrations). Annual averages have an uncertainty of 5 % or 1 $\mu g/m^{-3}$.

For O_3 monthly average concentrations are generally higher, so that the accuracy (better than 5 %) is more important here. Annual averages have an uncertainty of + 5 %.

For NH_3 the uncertainty in monthly and annual averages is estimated at 20 - 25 % or 1 $\mu g/m^3$.

B) Campaigns, Speuld

Other components and measuring techniques:

NH_3

1. triple (Ferm -type + annular) denuders, coated with oxalic acid, sample time 2 - 6 hours. Uncertainty of individual samples: + 15 %.
   Number of samples: + 40 (1988,1989)
2. denuder-filter packs (ECN/LUW), denuder coated with phosphoric acid, sample time 2 - 6 hours. Flow: 10 l/min. Uncertainty of individual samples 15%.
   Number of samples: + 40 (1988,1989)
3. thermodenuder (ECN), aug - dec 1989, continuously
4. wet denuder (ECN), 1989 (Tuesdays only)
5. six-fold (Ferm-type) denuders (TNO-MT), gradient measurements above the canopy, in campaigns in 1988 and 1989

HNO_3, HCl

1. denuder-filterpacks, denuder coated with sodium fluoride.
   Uncertainty 15 % or 0.1 $\mu g/m^{-3}$. Sample time: 2 - 6 hours.
   Number of samples: + 40
2. wet denuder (ECN), Tuesdays 1989

H_2O_2, HNO_2 1. wet denuder (ECN), Tuesdays 1989

aerosols
1. denuder-filterpacks, triple filter (mitex + backup filters for NH_3 and HNO_3)
   Number of 2 - 6 hour samples: + 40 (1988,1989)
2. cyclones, aerosol, tunnel samplers (LUW, april 1990)
   Components: K, Ca, Mg, Na, Cu, Zn, SO_4, NO_3, NH_4.

C) Monitoring at Elspeet

SO_2
Sampled by 6 m long light protected teflon tubes by:
Thermo Environmental Instruments Inc. Model 43 W. Version 2 (modified).
2 monitors: one at 4,2,1 and 0.5 m one continous at 4 m accuracy 1 $\mu g/m^3$

NO_2
luminol-chemoluminescence
1 monitor: 4,2,1 and 0.5 m
accuracy: 0.1 $\mu g/m^{-3}$

NH_3
Sampled through 15 m long lightly heated (5 K above ambient T) teflon tubing by:
Tungsten oxide coated automatic thermodenuder connected to a monitor Labs analyser, Model Kema
2 instruments at 4 and 1 m
accuracy: 0.5 ppb
Photo acoustic spectrometer
1 monitor at 4,2,1 and 0.5 m
accuracy: 0.1 ppb

D) Monitoring in the Netherlands (Elskamp, 1990)

SO_2
Thermo Environmental Instruments Inc.
Model 43 W. Version 2
Detection limit 1 ppb
Precision (80 % full scale) 1 %
83 locations in 1989

NO/NO_x
Philips PW 9762/00
Detection limit < 1 ppb

Precision (80 % full scale) < 1 %
25 locations in 1989

NH_3 Annual denuder system, filterpacks, conventional denuder, thermodenuders, wet denuder (see Speuld)
2-5 locations over recent years

Aerosol Low volume sampler, Annual denuder system, filterpacks
6 locations in 1989

HNO_2, HNO_3, HCL
See Speuld
2 locations in 1989

Appendix E. Throughfall measurements in the Netherlands

Table E1 Throughfall and precipitation measurements (mol ha a) in the Netherlands at different locations. (N being type of vegetation: DF = deciduous forest; CF = coniferous forest; H = heatland; G = grassland and M = maize). Locations used for comparison to deposition model results have been marked by: *

						precipitation			troughfall			
location	X	Y	start	end	N	NH_4	NO_3	SO_4	NH_4	NO_3	SO_4	ref.
1. Speuld*	175	475	8705	8805	CF	1231	457	347	2918	874	1177	[33]
2. Amerongen*	160	447	8705	8805	CF	1375	382	463	3673	743	1678	
3. Kootwijk*	178	463	8705	8805	CF	1253	471	374	2995	833	1198	
4. Garderen*	176	474	8705	8805	CF	1547	463	381	3074	831	1225	
5. Lagevuurse*	146	466	8705	8805	CF	1732	472	406	4478	1572	2404	
6. Ruurlo*	226	454	8705	8805	CF	1207	370	392	3912	690	1618	
7. Zelhem*	223	450	8705	8805	CF	1391	451	345	4813	955	2097	
8. Tongbersven	142	397	8301	8401	CF	783	336	465	4772	827	2173	[34]
8. Tongbersven	142	397	8401	8501	CF	849	394	455	4149	403	1843	
9. Boxtel	144	398	8101	8201	CF	745	389	491	3719	906	1953	
9. Boxtel	144	398	8201	8301	CF	853	366	457	3937	912	1858	
9. Boxtel	144	398	8301	8401	CF	783	336	465	3740	922	1766	
9. Boxtel	144	398	8401	8501	CF	849	394	455	3741	505	2077	
10. Gerritsfles	186	461	8201	8301	CF	673	338	371	3849	1075	1837	
10. Gerritsfles	186	461	8301	8401	CF	881	451	484	4584	1403	2082	
10. Gerritsfles	186	461	8401	8501	CF	741	437	422	3755	868	1785	
11. Kiplo	226	537	8201	8301	CF	707	401	384	3528	1015	1550	
11.Kiplo	226	537	8301	8401	CF	749	411	412	3979	1269	1740	
11. Kiplo	226	537	8401	8501	CF	685	392	410	3806	713	1923	
12. Winterswijk	239	446	7904	8004	DF	980	410	600	1900	530	1590	[35]
12. Winterswijk	239	446	8004	8104	DF	1010	390	610	1740	500	1690	
12. Winterswijk	239	446	8104	8204	DF	1020	500	600	2450	600	1990	
12. Winterswijk	239	446	8204	8304	DF	1140	430	660	2950	530	2050	
12. Winterswijk	239	446	8304	8404	DF	1380	450	770	2820	610	1970	
12. Winterswijk	239	446	8404	8504	DF	1110	460	620	2520	680	1700	
12. Winterswijk	239	446	8504	8604	DF	1420	510	560	2550	560	1420	
12. Winterswijk	239	446	8604	8704	DF	1150	500	460	2550	620	1370	
13. Oude Maat	215	457	8104	8204	DF	1230	480	675	2918	900	1686	[36]
13. Oude Maat	215	457	8204	8304	DF	1010	480	565	2640	863	1396	
13. Oude Maat	215	457	8304	8404	DF	1330	660	715	2995	1115	1630	
13. Oude Maat	215	457	8404	8505	DF	1270	620	580	2893	1373	1436	
13. Oude Maat	215	457	8504	8604	DF	1340	610	615	2543	1040	1230	
13. Oude Maat	215	457	8604	8704	DF	1610	560	720	2520	1455	1266	
1. Speuld*	175	475	8701	8801	CF	879	501	412	2360	818	1120	[30]
1. Speuld	175	475	8801	8901	CF	751	323	325	2067	671	964	
1. Speuld	175	475	8901	9001	CF	748	374	342	2110	790	933	
3. Kootwijk*	178	463	8701	8801	CF	964	470	523	2151	671	996	
3. Kootwijk	178	463	8801	8901	CF	786	297	335	2091	676	919	
3. Kootwijk	178	463	8901	9001	CF	702	360	305	2068	789	861	
3. Kootwijk*	178	463	8601	8701	CF	1006	371	513	6490	1330	2330	[37]
14.Terschell.*	150	601	8611	8807	CF	518	349	421	2157	1391	2463	[38]
15. Schoorl*	107	522	8611	8807	CF	551	431	447	2356	1144	1892	
16. Bakkeveen*	211	562	8611	8807	CF	947	426	614	5032	814	2092	
17. Diever*	217	547	8611	8807	CF	827	430	607	3203	716	1625	

location	X	Y	start	end	precipitation N	NH4	NO3	SO4	troughfall NH4	NO3	SO4	ref.
18. Orvelte*	242	543	8611	8807	CF	970	486	501	4720	983	2098	
19.Hardenberg*	234	507	8611	8807	CF	1091	422	501	4455	995	2118	
20. Lunteren*	172	456	8611	8807	CF	1291	465	556	6100	1073	2687	
21. Loenen*	194	457	8611	8807	CF	1015	465	505	4341	1261	2483	
22. Heumen*	165	419	8611	8807	CF	945	420	706	4991	916	2467	
23. Hulst*	57	367	8611	8807	CF	956	385	591	7052	1194	3669	
24.Ossendr.*	84	397	8611	8807	CF	1128	463	542	3619	792	2238	
25. Breda*	110	398	8611	8807	CF	891	395	426	8667	2039	4762	
26. Maarheeze*	173	371	8611	8807	CF	900	408	402	5105	1098	2873	
27.Groote Peel*	184	399	8611	8807	CF	2238	364	809	8762	924	4003	
28. Valkenburg			8601	8701	G			455			1055	
28. Valkenburg			8701	8801	G			445			935	
29 Leusden			8509	8609	CF	1140	220	405	4630	520	2035	[39]
30. Elspeet	182	477	8904	9004	H	846	387	355	2146	935	1035	[40]
31. Assel	187	469	8810	8908	H	966	379	329	1048	437	829	
32.Harderwijk*	174	482	8801	8901	CF	842	322	405	2207	642	881	[41]
32. Harderwijk	174	482	8901	9001	CF	842	322	405	2101	1002	861	
33. Sinderhoeve			8806	8811	M	725	350	300	560	410	380	[48]

Table E.2 Throughfall data for several cations measured at Speuld, Kootwijk and Elspeet (mol $ha^{-1}a^{-1}$)

location	period	Na	K	Ca	Mg	Al	Fe	Mn	Cl	H
Speuld	3/87-3/90	994	484	212	168	6	15	30	1384	74
Kootwijk	3/87-3/90	789	409	163	139	3	8	17	1091	55
Elspeet	4/89-4/90	3409	511	462	172				3999	
Assel	10/88-8/89	979	385	271	180				1226	

SOIL ACIDIFICATION / N CYCLING

N.van Breemen[1)]
J.M.Verstraten[2)]

[1)] Agricultural University Wageningen, Department of Soil Science and Geology
[2)] University of Amsterdam, Laboratory of Physical Geography and Soil Science

1. INTRODUCTION

In the Dutch Priority Programme on Acidification, soil acidification and related processes have been studied at various sites in forests and heathlands, mostly in association with related research dealing with atmospheric deposition and with various affects of acid deposition on biota. In this thematic report we will summarize the most important findings related to soil acidification. While the emphasis will be on projects in the second phase of the Dutch Priority Programme on Acidification, results of projects from the first phase and from related research in the Netherlands will be presented too in order to sketch a comprehensive picture. The emphasis will be on a description of the soil acidification and nitrogen cycling of the sites where hydrochemical monitoring has taken place. The insights gained by the data from these activities have been used for modelling aimed at scenario analysis and estimation of critical loads. This aspect has been dealt with in a separate thematic report by De Vries and Kros (1990). However, soil acidification modelling aimed at interpreting results from the monitoring sites of luw and luw/rin (see Table 2.1), and at comparing the situation in Dutch and German forest soils is summarized in this report.

2. RESEARCH SITES AND METHODOLOGY

2.1 General outline of sites and methodology

Table 2.1 shows the forested sites where soil studies have been performed, and will be referred to in this report. The names and numbers are similar to those given in Table 2 of Van Aalst and Erisman,1990. From about 40 heathland sites, samples were taken for research on nitrification (De Boer and Tietema,1990). These sites have not been further specified in this summary report.

The major monitoring sites of the second phase of the "Dutch Priority Programme on Acidification", the sites of the ACIFORN projects, are locations 1B, 3B and 31. Detailed work on N cycling in forests was done at sites 1B, 3B, 12, 33 and 34. Materials and methods of the work at these sites will be given in some detail in paragraph 2.3. In this paragraph we give a more general outline relevant for all research sites.

The soils of the forested sites in Table 2.1 were characterized physically and chemically using various standard procedures as described in the original reports. All soils are light textured. Most are sand to loamy sand, and only at sites 1B and 13 (locally) and site 100 (throughout the stand) are sandy loam textures present in the surface 50 cm of the soil profile. Groundwater classes at the various sites are GT VII* (sites 1A, 1B and 2), GT VII (sites 3A, 3B, 4 and 7) and GT VI (sites 5 and 6) which indicates that only sites 5 and 6 will not have any waterstress during (long) dry periods.

Table 2.1 Soil monitoring research sites in the Netherlands, covered in this report

location	X	Y	ref.	soilclass	vegetation	yr	throughfall NH_4 kmol	NO_3 (+/-)	SO_4 /ha.yr
1A Speuld A	175	475	staring	holtpodzol	Douglas	1	2.9	0.87	2.4
1B Speuld B	175	475	luw/uva	vorstvaag	Douglas	3	2.2	0.76	2.1
2 Amerongen	160	447	staring	holtpodzol	Douglas	1	3.7	0.74	3.3
3A Kootwijk A	178	463	staring	haarpodzol	Douglas	1	3.0	0.83	2.4
3B Kootwijk B	178	463	luw/uva	holtpodzol	Douglas	3	2.1	0.71	1.9
4 Garderen	176	474	staring	haarpodzol	Douglas	1	3.1	0.83	2.5
5 L.vuursche	146	466	staring	veldpodzol	Douglas	1	4.5	1.57	4.8
6 Ruurlo	226	454	staring	veldpodzol	Douglas	1	3.9	0.69	3.2
7 Zelhem	223	450	staring	duinvaaggr	Douglas	1	4.8	0.96	4.2
8 Tongbersven	142	397	luw/rin	haarpodzol	Scots p.	3	4.5	0.62	4.0
10 Gerritsfles	186	461	luw/rin	duinvaaggr	Scots p.	3	4.1	1.12	3.8
12 Winterswijk	239	446	uva	vlakvaaggr	oak/beech	8	2.4	0.58	3.4
13 Oudemaat	215	457	luw	holtpodzol	Oak/birch	6	2.8	1.13	2.9
31 Assel	189	468	luw/uva	haarpodzol	Calluna	1	1.0	0.54	1.5
32 Harderwijk	174	482	luw	duinvaaggr	Scots p	2	2.2	0.82	1.7
33 Buunderk.	181	448	uva/ioo	z.enkeerdgr	Oak	1	1.4	0.49	1.9
34 Leuvenum	176	481	uva/ioo	duinvaaggr	Douglas/Sc.p.	1	3.2	0.87	2.5
35 Hasselsven	163	371	luw	haarpodzol	Calluna	3	1.4	0.3	1.7
100 Solling (FRG)			luw/ug	parabr.erde	N.spruce	12	1.1	1.1	5.2

References:
staring: Kleijn et al, 1989;
luw/uva: Tiktak et al, 1988; Van der Maas, 1990; Tiktak and Bouten, 1990
luw/rin: Van Dobben et al, draft 1988
uva (Winterswijk): Verstraten et al, 1988; Verstraten et.al 1990.
luw (Oudemaat); Van Breemen et al, 1988
luw (Harderwijk): De Visser and van Breemen, 1990
uva/ioo: De Boer and Tietema, 1990
luw/ug: Matzner,1989; Wesselink, 1990
Approximate international equivalents of national soil classification taxa: holtpodzol: Dystrochrept (USDA) and Leptic Podzol or Dystric Cambisol (FAO/Unesco)
vorstvaaggrond: Dystrochrept (USDA) or Ortic Podzol (FAO/Unesco)
haarpodzol: Haplorthod (USDA) or Podzol (FAO/Unesco)
veldpodzol: Aquod (USDA) or (poorly drained) Podzol (FAO/Unesco)
duinvaaggrond: Psamment (USDA) or Arenosol (FAO/Unesco)
vlakvaaggrond: Albaquult (USDA) or Dystric Planosol (FAO/Unesco)
zwarte enkeerdgrond: Plaggept USDA) or Anthrosol (FAO/Unesco)
parabraunerde: Hapludult (USDA) or Dystric Luvisol (FAO/Unesco)

pH values of the forest floors and of the surface mineral soil horizons are invariably very low and close to 3.5 (pH in H_2O) to 3 (pH in 0.01 M $CaCl_2$ or M KCl extracts). Base saturation (as % of the CEC in unbuffered extracts) are invariably below 20 % and generally below 10 % in the surface layers, and often still lower in the subsurface horizons, to increase slightly at greater depth. pH values (measured in H_2O) of most of the surface organic layers of the heathland soils (sites 31, 35 and all sites sampled for nitrification studies) are distinctly higher than those of the forest soils: 3.5 to 4.6.

Atmospheric inputs at the sites of Table 2.1 were estimated in different ways from throughfall and open field deposition. In general, throughfall fluxes were assumed to represent atmospheric inputs of SO_x, NO_x and NH_x, while open field bulk deposition was taken as representative of inputs of other solutes. Because NH_3 is presumably taken up in appreciable quantities by the canopy, these values can be regarded as lower limits of atmospheric input. For sites 1B, 3B, 12 and 100, more sophisticated estimates of atmospheric input were made on the basis of throughfall, and open field deposition, as described e.g. by Van der Maas,1990 and Duysings et al, 1990.

At all sites of Table 2.1, soil solution samples were taken regularly, with frequencies ranging from 4 times per year (staring sites) to monthly (luw and luw/rin sites) or fortnightly (uva and luw/uva sites), for a number of years indicated in Table 2.1 under yr. At the staring sites, soil solution was collected by centrifuging soil samples. At all other sites, soil solutions came from permanently installed ceramic cups or suction filter plates. Differences in solute concentrations observed by these different methods, and by aqueous exracts as used by the University of Nijmegen, are discussed in Appendix 1. The number of replicates of soil solution sampling at any one depth in the monitoring programmes generally varied from two to five, with occasionally more intensive campaigns. Spatial variability of soil solution concentrations is generally very large, with the lowest variability in deciduous stands, higher values in Scots pine, and highest values under Douglas fir. Under Douglas, standard deviations of solute concentrations in centrifuged solutions for a given site and soil depth were in the order of 30 to 80 % of the mean (Kleijn et al,1989). At sites 1B and 3B the 95 % confidence intervals of soil solution concentrations, based on 25 replicated cup samples collected in early spring 1990 (van der Maas,1990), were 1 to 2 % for pH, 15 to 40 % for the major solutes (Al, NO_3, SO_4), base cations (Na, K, Ca, Mg) and Cl, 40 to 80 % for NH_4, and 20 to 130 % for H_2PO_4.

For most sites solute fluxes were estimated. These were obtained by multiplying water fluxes with representative solute concentrations. Water fluxes were measured directly, by gauging surface runoff, at site 12, the only watershed involved. At all other sites, water fluxes were estimated, either by means of sophisticated soil hydrological models using more (sites 1B and 3B, 12, model "SWIF") or less (sites 8, 10, 13 and 35, model "SWATRE") intensive soil hydrological monitoring. For the "staring" sites (1A, 2, 3A, 4, 5, 6 and 7), monthly drainage flux estimates were modelled by Reurslag et al (1990), correcting drainage fluxes by equating Cl^- input to Cl^- output (pers. comm. W.de Vries). The Cl input/output ratio before correction varied from 0.6 to 1.9, with an average value of

1.3. For sites 33 and 34, input/output budget estimates were based completely on Cl-budgets from fortnightly soil solution samples. For the Aciforn forest sites, the derivation and results of the input output budgets will be discussed in more detail later in this report.

2.2 Materials and methods at the ACIFORN sites

- Site description and chemical soil properties

Both forest plots are located in the Veluwe area, a large undulating area with forests and heathlands, in the central part of the Netherlands. Soil chemical and textural data are in Table 2.2.1 and 2.2.2. The Speuld site is 2.5 ha. The monoculture Douglas fir (Pseudotsuga menziesii) stand, planted in 1962, is situated on top of an ice-pushed ridge. The surface is undulating very gently. The Kootwijk site is a 1.2 ha. Douglas fir stand that is nine years older than the Speuld stand. It is located in a coversand plain. The topography is almost flat.

At Speuld the groundwater is found at a depth of about 40 m. The soils are well drained. At Kootwijk the soils are excessively well drained. The groundwater level is about five meters below the soil surface.

At the Speuld site the soil is a Typic Dystochrept (Soil Survey Staff, 1975) on sandy loam and loamy sand textured Rhine sediments of Middle Pleistocene age. The soil can also be classified as an Orthic Podzol (FAO/UNESCO, 1974) and as a Holtpodzol (De Bakker en De Schelling, 1966)).

The soil at Speuld is rather heterogeneous. The texture of the soil shows a strong spatial variability, that is related to the elongated, parallel outcrops of layers of different textures typical of an ice-pushed ridge.

At the Kootwijk site the soil is a Typic Dystrochrept (Soil Survey Staff, 1975) developed on loamy fine sandy coversand deposits of Weichselian age. The soil can also be classified as a Leptic Podzol (FAO/UNESCO, 1974) and as a Holtpodzol (De Bakker en De Schelling, 1966). The parent material is well sorted fine quartz sand with a median grain size of 150 - 170 μm which contains some loam. Due to former charcoal production a high iron content is found locally.

At each of the two forest locations, three soil profiles were described and sampled in great detail for morphological, chemical and physical properties. These profile pits were chosen at random and used to install equipment for monitoring of the soil solution chemistry.

Table 2.2.1 Soil chemical and textural data for the Speuld (1B) and Kootwijk (3B) sites

Depth (cm)	pH H_2O	pH KCl	%C	% N	CEC* mmol/kg	Base S. (%)	Texture (%) <2µm	2-16µm	16-2000µm	>2000 µm
Speuld										
0- 5	3.63	2.83	7.3	0.30	59	-	1	3	70	1
5- 10	3.70	3.00	2.9	0.11	39	-	0	3	87	0
15- 20	3.87	3.70	2.0	0.07	42	-	0	4	94	1
30- 35	4.15	4.27	0.8	0.04	21	-	0	3	95	2
50- 55	4.22	4.38	0.3	0.02	13	-	0	2	97	0
90- 95	4.22	4.28	0.2	0.01	24	-	0	2	95	0
Kootwijk										
0- 5	3.45	2.90	3.9	0.21	51	8	0	0	90	0
15- 20	3.78	3.40	1.3	0.05	27	7	0	0	98	0
30- 40	4.25	3.90	0.9	0.04	20	5	0	0	98	0
50- 60	4.53	4.32	0.6	0.03	14	14	0	0	99	0
90-100	4.70	4.58	0.2	0.01	8	13	0	0	99	0

*CEC: Calculated as sum 0.5 M $BaCl_2$ extractable Ca, Mg, K, Na, plus 1M KCl extractable Al, H, and NH_4

The heather site near Assel is located in the Veluwe area, a large undulating area with forests and heathlands, in the central part of the Netherlands. The 3 ha open heathland site, situated on a very gentle convex slope (2%) of a gently undulating ice-pushed moraine, is surrounded by forests except to the west, where land is used for agriculture. The vegetation consists of Calluna vulgaris (54%), Erica tetralix (11%), Molinia caerulea (11%) and Deschampsia flexuosa (5%). The elevation is 47 m a.m.s.l.; the actual ground water is probably many meters deep, but drainage is somewhat poor due to the presence of a poorly permeable thin iron pan (placic horizon). The soil parent material is a fluvio-glacial sand in which a Haplorthod (USDA), Podzol (FAO/Unesco) or Haarpodzol (De Bakker and Schelling, 1966) has developed. Depths of the subsequent horizons (mean from about 100 observations) are: Oh -5-0 cm; Ah 0-7 cm; E 7-14 cm; B2h 14-21 cm; B2ir (with iron pan of 4 mm on top) 21-28 cm; B3 28-68 cm. Base saturation is relatively high indicating that the soil has been affected only little by acid atmospheric deposition, presumably because of the low interception capacity of the heather vegetation.

- Soil physical properties

The bulk density of the soil has been used to distinguish between three individual "soil physical horizons" (Table 2.2.3.). The forest soils have a litter and fermentation layer ("forest floor") and the heathland soil (Assel) has a litter, fermentation and humus layer (Table 2.2.3).

Table 2.2.2 Soil chemical and textural data of the Assel heathland site

Horizon (depth;cm)	pH-H_2O	pH-KCl	%C	%NO	CEC* mmol/kg	BS %
Oh/Ah	4.29	2.87	16.9	0.52	193	70
E	4.71	3.40	1.2	0.04	16	50
B2h	4.42	3.51	10.5	0.35	128	20
B3(30-40)	4.25	4.20	0.9	0.02	15	20
B3(70-80)	4.41	4.51	0.3	0.01	7	43
C(110-120)	4.53	4.63	0.3	0.01	5	60

	<2 µm	2-16 µm	16-2000 µm	>2000 µm
	%			
Oh/Ah	0	0	59	0
E	0	0	93	1
B2h	0	0	84	4
B3(30-40)	0	0	88	2
B3(70-80)	0	0	95	0
C(110-120)	0	0	96	1

* CEC is calculated as the sum of 1 M LiEDTA exchangeble Ca, Mg, K and Na and 1 M KCl exchangeble H, NH4 and Al.

Table 2.2.3 Horizon boundaries based on physical properties of the organic (O) and mineral soil horizons (A, B and C)

horizon	$^{b}\rho_s$ gram/cm^3	depth [cm] Speuld	Kootwijk	Assel
O	0.11	8/4 - 0	5/3 - 0	1 - 0
A	< 1.2	0 - 17	0 - 15	0 - 5
B	1.2 - 1.5	17 - 70	15 - 60	5 - 15
C	> 1.5	> 70	> 60	> 15

Laboratory water retention refers to desorption of almost saturated soil, a situation that is hardly ever reached in the field. In this study the water holding capacity of the soil is overestimated by a factor of up to 2 when using the laboratory retention. Therefore, retention characteristics obtained from water contents and pressure heads simultaneously measured in the field were used (Figure 2.2.1). Further details are given in Tiktak and Bouten (1990) for the forest soils and Bizot et al. (1990) for the heathland soils.

The soil at Speuld has a higher water holding capacity than at Kootwijk (90 and 78 mm, respectively), resulting mainly from significant differences in the water holding capacity of

the forest floor and the A horizon. For greater depths, differences are not significant at the 90% level.
The water holding capacity of the forest floor is higher at Speuld than at Kootwijk (6.55 and 4.07 mm, respectively), caused by differences in the thickness of the forest floor (5.01 and 3.04 cm, respectively).
The soil at the Kootwijk site is rather homogeneous. On the contrary, the Speuld soil is heterogeneous. The soil physical monitoring plot was found to be representative for the finer textured soils at that site. The water holding capacity of these soils is relatively high (Figure 2.2.2). The water holding capacity at the soil monitoring plot, situated in the Eastern part, is almost 75% higher than the water holding capacity in the Western part (105 mm respectively 50 to 65 mm). Because the water holding capacity sets clear limits to forest transpiration, it is expected that the transpiration for the whole stand is lower and the drainage to the subsoil is higher than the values predicted from these monitoring data only.

- Monitoring of solutes in inputs and soil solution
At Speuld and Kootwijk, bulk precipitation and throughfall were collected by 400 cm^2 black polythene funnels with a sharply tooled knife edge and a straight-sided drop of 10 cm into a 5 litre opaque polythene bottle at 150 cm (precipitation funnels) or 75 cm (throughfall funnels) above the ground. The funnels were provided with a filter in the drain opening to minimize contamination by needles and insects, and a crown to prevent perching by birds. Bulk precipitation was collected from two open areas, 200 to 1300 meters from the sites. Throughfall was collected with twelve replications, placed randomly at each site. Wet-only open field precipitation was sampled with a wet-only apparatus manifactured by ECN, Petten, the Netherlands. A cap over a funnel identical to the bulk open field funnels (but with rounded edges) is automaticaly removed when rainfall starts, and closed again after rainfall has stopped. The wet-only funnels were placed at 150 cm above the ground surface, 2 meter apart from one of the bulk open field samplers.
Sampling and analysis was done weekly. All bottles were taken to the laboratory and replaced. Throughfall from each site was pooled to one volume-weighted sample before analysis. Precipitation samples were analyzed separately. Throughfall samples from individual devices were analyzed periodically to estimate spatial variability. Amounts of water were determined by weighing the bottles (estimated accuracy ± 0.05 mm rainfall).

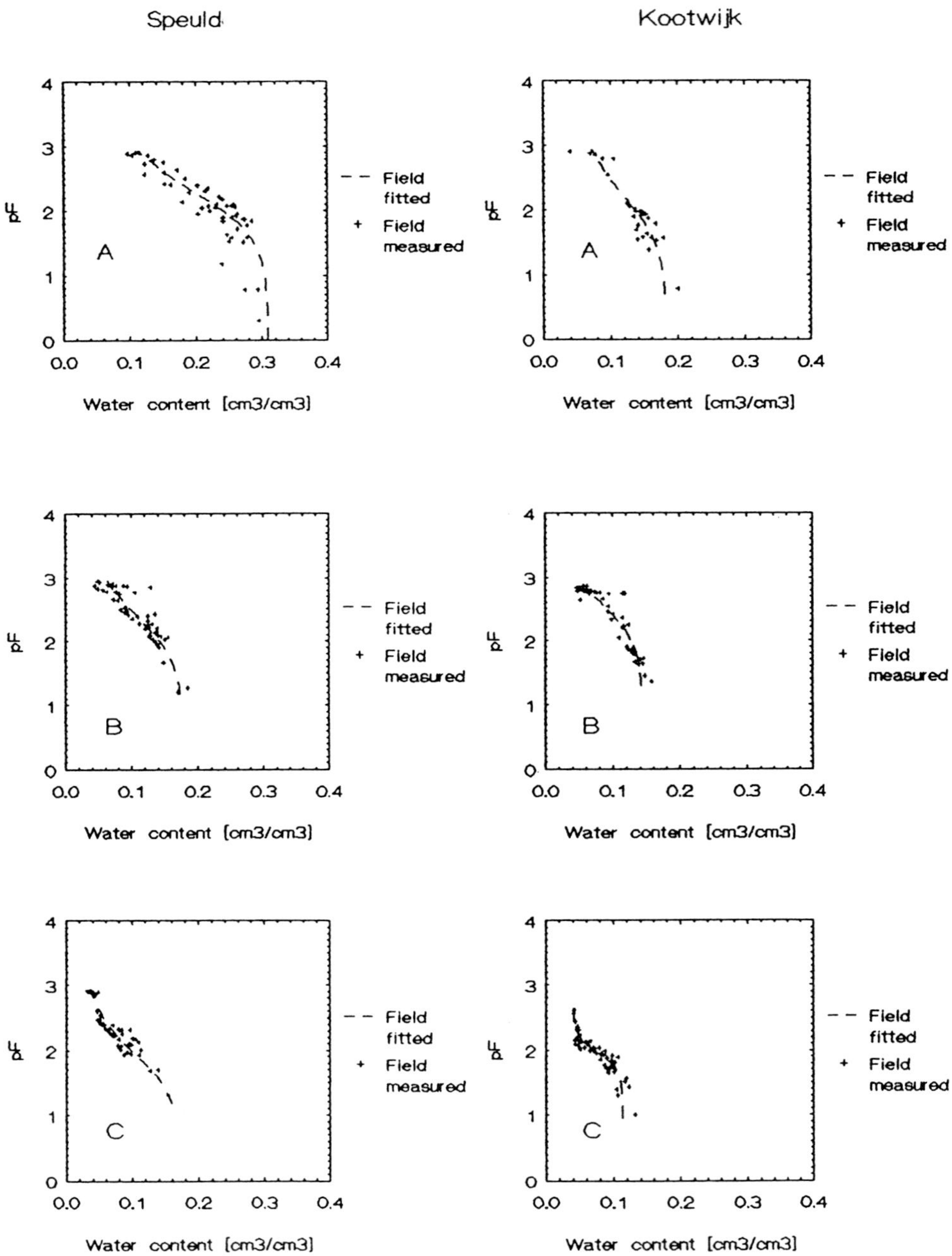

Fig. 2.2.1 Field water retention characteristics of the forest soils

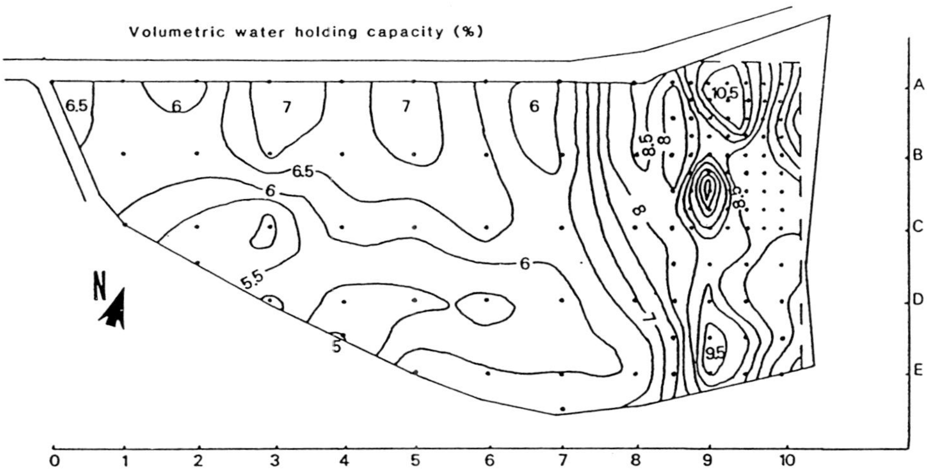

Fig. 2.2.2 Map showing the distribution of the volumetric water holding capacity (m^3/m^3) of the upper 50 cm of the soil at the Speuld monitoring site

Bulk open field and wet-only samples contaminated by bird excrements (indicated by H_2PO_4 concentrations over 2 mmol/m^{-3}) had high levels of K, NH_4 and C_t, and were removed from the data base. Throughfall samples apparently contaminated by bird droppings had increased P-contents but did not have significantly higher concentrations of other elements, and were retained in the data base.

The 95 % confidence interval of concentrations and the sum of cations or anions were determined during 13 (site 1B; total of 203 mm waterflux) or 5 (site 3B; total of 130 mm waterflux) weekly campaigns in which twelve individual throughfall samples were analysed. Confidence intervals for sums of anions or cations (as percentage of the mean) were 25 % for Kootwijk and 22 % for Speuld. The confidence intervals for the individual components were between 20 and 30 %, except for elements present at low concentrations (Al and P, 50 to 70 %). No significant differences in throughfall composition and flux were detected between Speuld en Kootwijk.

At the Assel heathland site, precipitation and throughfall were monitored as described by Bobbink et al.(1990). Throughfall was collected by means of fork-shaped troughs.

At the forest sites, litter fall was collected into 1 m^2 wooden trays with sharp 15 cm high edges and horizontal plastic netting (mesh size 1.5 mm) about 10 cm above the ground surface. Each plot had 6 collectors, evenly distributed, that were collected every four weeks. Samples were freeze-dried immediately and kept frozen, and are pooled over one year for chemical analysis. Soil solutions were collected, from porous ceramic cups (Soil Moisture Equipment Co., type 655X1-B1M3), at the depths of 10, 20, 40, 60, and 90 cm into evacuated 250 ml glass bottles.

In the forest sites, soil solution was also collected quantitatively by an automated continuous sampling procedure, using filter plates and a suction which equals the pressure head in the soil. The latter procedure is used to collect flux-weighted soil solution samples from underneath the litter layer (in duplicate) and at a depth of 10 and 120 cm. These samples were collected fortnightly. The filter plates are constructed from porous polythene plate, covered by a Versapor 200 filter (0.2 μm), according to Driscoll et al. (1985). At zero cm depth, ceramic material (Masse P80; Fa. Staatliches Porzelan Manufactur, Berlin) was used to prevent biological damage to the collector plates. The soil solution is sampled by a suction which is regulated electronically to equal the soil pressure head at the depth of the porous plate (measured with a tensiometer) and into a 500 ml polythene bottle. The sampling bottle was placed in a glass buffer tank.

At both forest sites this sampling scheme of five cups and four plates was replicated three times by installation from three individual soil pits. Installation was done in December 1986 at Speuld, June 1987 in Kootwijk, and in early 1988 at Assel. At the Assel heathland site, soil solution was monitored by fortnightly duplicate sampling from 3 depths, using porous cups as described for the forest locations.

Water samples were stored in 100 or 250 ml flasks and stored cool and in the dark. Ec and pH were determined within 24 hours of sampling. pH was measured with an Orion Research pH/mV meter (model 801A) and an Orion combination pH electrode (no. 91-55), using buffer solutions pf pH 4.00 and pH 7.00, made up from ampules (Merck, Titrisol no. 9884 and 9887). A few drops of saturated potassium chloride were added to weakly buffered samples (rain water) before pH measurement.

Inorganic carbon was determined with a Total Organic Carbon Analyzer (T.O.C. analyzer. Beckman type 915 B). Chloride, fluoride, sulphate and nitrate were determined by ion chromatography. Bicarbonate was calculated from pH and inorganic carbon. Phosphate, silica and all major cations (K, Na, NH_4, Ca, Mg, Al, Fe, Mn) were determined by spectrophotometry. Quality control was carried out according to Stuyfzand (1983) and by comparing the sums of cation and anion charge, and comparing calculated and measured electrical conductivity, whereby the concentration of organic anions is calculated with the formula of Oliver et al. (1983).

- The soil hydrological monitoring programme

Figure 2.2.3 gives a summary of the equipment used and the soil properties which are monitored. The closed arrows represent measured flows; open arrows represent flows which were calculated with model SWIF (Tiktak et al., 1990)). Data were collected for the three hydrological years 1987, 1988 and 1989. For a complete description see Tiktak and Bouten, 1990.

Precipitation was monitored continuously with a rain-gauge (surface area 400 cm^2; height 60 cm) at open areas close to the research sites. Throughfall amount was measured weekly in funnels connected to 5-litre polythene bottles. The collectors had a surface area of 480 cm^2 and the rim of the funnels was placed at a height of 30 cm. The number of collectors was 25 in Kootwijk and 36 in Speuld.

Soil water pressure heads were automatically measured twice daily by means of tensiometers (Burt, 1987) using an Arcom microcomputer.

The water content of the forest floor (L and F horizons) was determined weekly by gravimetry (sampling of 500 cm^2 using four replicates per site, measuring the thickness of the forest floor; oven-dried at 60^oC).

Soil water content to a depth of 260 cm was measured by the neutron scattering method (Gardner and Kirkham, 1952) using a CAMPBELL neutron probe. The probe was calibrated three times according to NEN 5782. The neutron probe readings are inaccurate for depths less than 50 cm.

Water content was also measured weekly using Time Domain Reflectometry (TDR; Topp et al., 1980). A number of probes were horizontally installed in a wall of an excavated soil pit to a depth of 70 cm. Thus, TDR measurements provide information about the water content at depths where the neutron scattering method is inaccurate. Furthermore, in Speuld 155 sensors were installed vertically for measuring average water contents of the upper 50 cm of the soil to study spatial variability of soil characteristics.

From 1988 water contents were measured every two hours by means of capacitive probes (Hilhorst, 1984) with eight sensors to a depth of 60 cm. From 1989 soil water contents were also automatically recorded to a depth of 70 cm by means of 36 TDR probes (Heimovaara and Bouten, 1990).

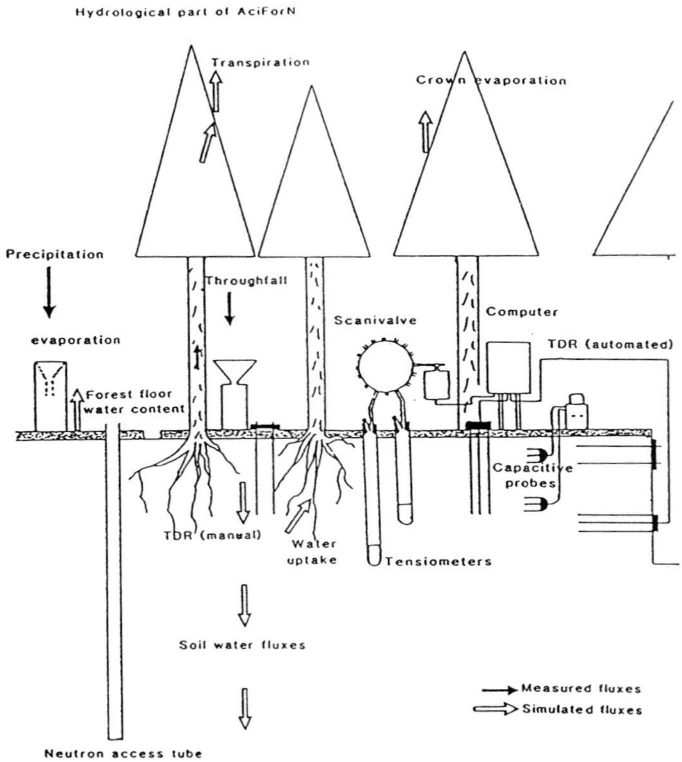

Fig. 2.2.3 Equipment for soil hydrological monitoring used at the Aciforn sites

2.3 Materials and methods of the N-cycling studies

Nitrogen cycling was studied in detail at locations 1B, 3B, 12, 33 and 34 by Tietema et al. (1990b).

Nitrification and mineralisation rates were estimated with PVC-tubes with a length of 25 cm and an internal diameter of 7 cm to sample intact soil cores, consisting of the (L+F+H)layer and the top 7 cm of the mineral soil. On each sampling date, two soil cores were taken within 10 cm distance of each other. One of the samples, the reference sample, was taken directly to the laboratory for analysis. The other, the incubation sample, was enclosed at the top and the bottom of the tube by PVC-lids and placed back in the soil for further incubation. Aeration of the soil samples during the incubation period was made possible through 2 holes (dia.=14mm) in the upper part of the tube . The incubation period was 4 weeks, at the end of which the soil sample was collected again and transported to the laboratory. At the same time, new samples (reference and incubation) were taken for the following period. In total, 6 reference samples and 6 related incubation samples were taken per incubation period.

In the laboratory, the soil cores were divided into the L+F+H and top 7 cm of the mineral soil (Ah). After determining the fresh-weight, the samples were deep frozen (-18°C), before being freeze-dried. The dried samples were milled (L+F+H) or ground (Ah). Roots were removed from the latter samples, milled and mixed again with the mineral sample. Stones (>2mm) were removed from the mineral samples. The water content of the samples was determined gravimetrically. After extraction (5g L+F+H, 10g Ah sample) in 100 ml KCl (1 mol L-1) and filtration (0.45 m), concentrations of NH_4-N, NO_3-N and NO_2-N were determined colorimetrically by means of an autoanalyser. The differences in NO_3-N and in total inorganic N concentration of the incubation samples and the related reference samples were considered to be caused by nitrification and mineralization during the incubation periods.

Denitrification was estimated by N_2O emission rates which were determined using the closed chamber method according to Hutchinson and Mosier (1981). Duplicate measurements were performed on a weekly basis from February 1987 to December 1987. Both chambers were inserted in the topsoil not more than 1 m apart. Air samples were taken using Venoject vacumized tubes in a sequence of 0, 2, 4, 8, 16, 32, 48, 64, 96, 128 and 192 minutes after intersection.

Parallel to the N_20 flux measurements, CO_2 measurements were performed as well, using the same closed chamber method. These results were compared with CO_2 flux measurements using the ventilated chamber method carried out simultaneously. This latter method gives more realistic results because the produced CO_2 is removed from the volume underneath the chamber by a constant air flow through a $Ba(OH)_2$ solution. From this comparison it was concluded that the CO_2 fluxes with the ventilated chamber method were 1.7 times higher compared to the results obtained with the closed chamber method (Bouten, unpublished results). Regarding the strong similarities between the N_2O and CO_2 molecules (mass diffusion constant), the results from the N_2O flux measurements were corrected with this factor (for details see Tietema and Verstraten, 1988; Tietema et al, 1990c).

All air samples were analysed on N_2O within one day after collection, using a CARLO ERBA gas chromatograph, equipped with a 2m*1/8" stainless steel Poropak Q column (80/100 mesh) and a 63 Ni electron capture detector (ECD). Helium was used as a carrier gas at a constant flow rate of 15 ml min-1. The temperature of column and detector were 50 and 200°C, respectively.

2.4 Modelling of soil acidification

The model WATERSTOF, developed by Wesselink (1990) aims at simulating the dynamics of soil solution chemistry on a time scale of months, seasons and years as an aid in interpreting results of forest ecosystem monitoring studies (van Breemen et al., 1989, Matzner, 1989).

Only one-dimensional flow of water and chemical constituents is considered. The soil profile is divided in discrete layers (soil horizons), that are considered as continuously stirred reactors from which only convective mass flow takes place. The model consists of a hydrological and a (bio)chemical module in which the dissolved H, K, Na, Ca, Mg, Al, NH_4, NO_3, Cl, SO_4, H_2PO_4 and organic anions (ORG) are considered. A simple empirical hydrological module that needs little computation time, was applied for effective calibration and scenario studies.

Locations 8, 10, 13 and 100 of Table 2.4.1 have been used for simulation runs of WATERSTOF.

- Model stucture

Processes in the hydrological module are root water uptake, percolation, capillary rise (from each soil layer) and evaporation (from the first layer only). These fluxes are calculated as a function of soil moisture content using parameters including water uptake distribution with depth, 2 parameters in the percolation function for each layer and moisture contents at field capacity (theta fc) wilting point (theta wp) saturation theta s) and reduced oxygen availability (theta ra). Daily values were available for boundary -input- fluxes of throughfall and potential evapotranspiration, so the time step in the waterbalance was set at one day. Simulations start at the beginning of a hydrological year (1 April where soil moisture content is assumed to be at field capacity).

Chemical processes included in the model are cation exchange, sulphate adsorption, biocycling (uptake and mineralization), complex formation of $Al(OH)^{2+}$ and $AlSO^{4+}$, nitrification and mineral weathering. The original time step in the chemistry module of one day was increased to 5 days with minor effects on calculated solution chemistry and saved considerable computation time. The input requirements of the model are summarized in Table 2.4.1.

Total annual plant uptake was assumed to be equal to the sum of increment in biomass (net uptake), litterfall and leaf leaching. Uptake and mineralization of nutrients were assumed to vary seasonally with a sinusoidal pattern. The nutrient uptake is independent of transpiration in the model for theta>theta wp. For theta< theta wp nutrient uptake is stopped. N-uptake was assumed to be in ammonium form until ammonium levels decreased to zero, when N-uptake was assumed to be in nitrate form.

Table 2.4.1 Input requirements WATERSTOF

throughfall chemistry:	
	measured monthly cumulative throughfall chemistry is scaled down to 5-day values proportional to the 5-day throughfall-water flux.
hydrology:	
	5-day values for water fluxes and soil moisture contents (output from waterbalance).
biocycle:	
	yearly uptake equals mineralization. The size of the biocycle is based on litterfall measurements. 5-day values are calculated from annual values based on a sinusoidal seasonal pattern.
initial conditions:	
	estimated from measurements: size of mineral pools, CEC, exchangeable fractions and soil solution chemistry at start of simulation. initial sulphate adsorption: from calibration.
parameters:	
	from calibration: nitrification rate constant, water and nutrient uptake distribution with depth, N-mineralization distribution with depth, exchange coefficients (partly) from measurements: exchange coefficients from simultaneous observations of solution chemistry and exchangable concentrations.

Phosphate adsorption was modelled with a linear adsorption isotherm, while for sulphate either a Freundlich or Langmuir isotherm was applied.

For the base cations K, Ca, Mg and Na weathering is described as a function of the mineral mass present and [H+]. The common undersaturation with gibbsite in surface soils of European forests has been attributed to (1)equilibrium with a solid organic-phase in organic-rich horizons (Mulder et al.,1989; Cronan et al.,1986), or (2) kinetically constrained weathering in organic poor horizons (Matzner,1989). Case (1) Al is modelled in WATERSTOF as Al-H exchange by means of the Gaines-Thomas equation. Case (2) by rate controlled gibbsite dissolution, depending on the degree of of undersaturation with gibbsite.

Cation exchange is described with a Gaines-Thomas type exchange equation.

The Gaines-Thomas equation has been developed for permanent charge clay-mineral surfaces. Cation exchange on organic material probably needs a different approach, accounting for (1) variable charge surfaces and (2) specific binding of cations to organic groups (Tipping and Hurley, 1988).

Nitrification was modelled with a Michaelis-Menten type equation with the concentration of ammonium as the variable.

3. RESULTS

3.1 Atmospheric inputs

- General

Table 3.1 shows the mean annual throughfall inputs of the various sites. These values can be regarded as approximate estimates of the atmospheric inputs of NH_x, NO_x and SO_x. In the individual forest sites considered, inputs are generally in the order of 2 to 5 kmol(+/)/ ha.r for NH_4 and SO_4, and 0.6 to 1.6 kmol(+/-)/ha.yr for NO_3. Appreciably higher values were observed by Houdijk (1990) at thirteen sites throughout the country, measured between November 1986 and November 1987: averaged over the four regions Coastal, North, Central and South, inputs were 4.7 kmol(+/-)/ha.yr for NH_4, 5.0 kmol(+/-)/ha.yr for SO_4, and 1.1 kmol/ha.yr for NO_3. Extremely high values were observed at a fourteenth site (De Groote Peel) where throughfall inputs were 8.7 kmol(+/-)/ha.yr for NH_4, 8.0 kmol(+/-)/ha.yr for SO_4, and 0.92 kmol(+/-)/ha.yr for NO_3. The generally high values observed by Houdijk may partly reflect true differences in inputs between her sites and the monitoring sites, but are probably higher than the long-term mean annual deposition at her sites would be, because the 1.5 year measuring period is much wetter than normal (ca. 35%).

- Aciforn sites

Inputs at the Aciforn forest monitoring sites 1B and 3B are among the lowest observed in forests in the Netherlands. The values are particularly low, because Douglas fir tends to have distinctly higher throughfall fluxes of ammonium than Scots pine, Corsican pine and oak (Houdijk,1990). These low atmospheric inputs at sites 1B and 3B are probably related to the fact that those sites are in the center of (for dutch standards) very extensive, closed forest land, one of the prerequisites for the atmospheric measurement activities within Aciforn.

Table 3.1 Three-year mean fluxes of dry atmospheric deposition (DAD), total atmospheric deposition (TAD), throughfall (T) and canopy leaching (CL) for Speuld and Kootwijk (mol(+/-).$ha^{-1}.y^{-1}$); 1 January 1987 - 31 December 1989)

	Speuld				Kootwijk			
	DAD	TAD	T	CL	DAD	TAD	T	CL
Na	459	929	929	0	306	750	750	0
K	30	48	481	433	20	38	407	369
Ca	181	305	415	110	134	249	326	78
Mg	128	256	332	67	90	211	275	64
Al	5	8	19	11	3	6	10	4
Fe	18	30	27	-3	8	17	15	-2
Mn	4	7	57	50	3	6	32	26
H	-155	170	73	-97	-167	127	58	-69
NH_4	2087	2749	2179	-570	1856	2573	2103	-470
Cl	673	1263	1263	0	450	1008	1008	0
H_2PO	5	5	6	1	4	4	13	9
NO_3	425	787	760	-27	395	751	712	-39
SO_4	1523	2212	2123	-91	1356	2081	1940	-141
WA*	133	231	351	120	49	130	304	174

*WA: weak acid anions

Van der Maas (1990) estimated the contribution of wet and dry atmospheric deposition and canopy leaching to the throughfall fluxes, using bulk- and wet-only open field precipitation and throughfall, and making various assumptions about the behaviour of different solutes. His assumptions involved: (1) no canopy leaching of Na, (2) ratio of dry to wet deposition of all base cations similar;(3) resulting calculated canopy leaching equivalent to uptake of NH_4 and H, which are taken up in a ratio equal to the mean ratio of H and NH_4 in throughfall and bulk open field; (4) resulting calculated dry deposition of H and NH_4 matched dry deposition of Cl (from difference between bulk open field deposition and throughfall), and of NO_3 and SO_4 in a ratio similar to that of the differences between throughfall and bulk open field input of NO_3 and SO_4 after correction for seasalt-SO_4 (SO_4/Na in seasalt is 0.12); (5) the rest term is the canopy uptake of SO_4 and NO_3 (6) remaining charge differences are attributed to weak acid anions.

The results of these calculations (Table 3.1) illustrate the great importance of dry deposition, in particular of NH_4 and of SO_4, and suggest that appreciable amounts of ammonium and much smaller amounts of SO_4, H and NO_3 are taken up. The high canopy uptake of NH4 and low canopy uptake of NO_3 are in agreement with recent experimental work using ^{15}N in the FRG.

3.2 Soil hydrology at the Aciforn sites

Table 3.2.1 summarizes the main annual water balance terms for both forest sites. The Table gives information about the monitoring years (1987-1989) and about the very dry hydrological year 1986.

The hydrological budgets presented in this chapter are primarily based on simulation results. The model was calibrated with data which were determined in 1989.

Table 3.2.1 Main annual water balance terms [mm] for both forest sites. First and last dates of each hydrological year are March 1 and February 28, respectively. P represents open field precipitation, T throughfall, I interception loss, E^*_r reference evapotranspiration, E^*_s potential forest floor evaporation, E_s simulated actual forest floor evaporation, E^*_{pl} potential transpiration, E_{pl} simulated actual transpiration, D simulated drainage at 200 cm depth, and DS net storage change

Speuld year	P	T	I	E^*_r	E^*_s	E_s	E^*_{pl}	E_{pl}	D	ΔS
1986	790	442	368	546	89	44	390	317	96	-15
1987	1104	622	482	501	79	54	316	315	261	-8
1988	815	459	356	508	80	61	332	316	98	-16
1989	936	527	409	601	94	47	476	380	102	-2
Kootwijk year	**P**	**T**	**I**	**E^*_r**	**E^*_s**	**E_s**	**E^*_{pl}**	**E_{pl}**	**D**	**ΔS**
1986	790	498	292	546	85	41	368	282	169	+6
1987	1104	708	396	501	75	51	298	296	366	-5
1988	850	539	311	508	77	58	313	285	180	+16
1989	865	549	316	601	89	45	450	340	161	+3

The four years were quite different from each other.The summers of 1986 and 1989 were dry, while 1987 and 1988 had wet summers. From the table this cannot be read directly. Annual throughfall was approximately 63% of open field precipitation at Kootwijk and 56% at Speuld. Although the standard deviation of the annual throughfall collected in the separate funnels exceeded 25% of the mean, differences between Speuld and Kootwijk were significant at the 99% level.

Annual actual transpiration ranged from 315 to 380 mm for Speuld and 282 to 340 mm for Kootwijk (Table 3.2.1). For Speuld the actual transpiration was 2% lower than the potential transpiration in 1988 and 20% lower than the potential transpiration in 1989. In Kootwijk transpiration reduction was 9% and 24%, respectively. The study shows that drought stress occurred from June until October in dry years. The higher transpiration reduction for Kootwijk is primarily caused by the lower water holding capacity of the forest floor and the A horizon. The lower water holding capacity for Kootwijk also causes a higher maximal transpiration reduction. In Kootwijk the highest transpiration reduction found was almost 80%, whereas the highest reduction found for Speuld was only 40%. Water stress is thus an important factor of combination stress.

Table 3.2.2 Reduction of transpiration by evaporation of intercepted water [mm]. The second column gives the transpiration reduction as a percentage of the annual Makkink reference evaporation

Year	Speuld		Kootwijk	
1986	12.4	2.3%	11.1	2.0%
1987	55.4	11.1%	52.8	10.5%
1988	45.2	8.9%	41.8	8.2%
1989	19.2	3.2%	19.2	2.9%

In Speuld 50% of the annual transpiration is taken up from the upper 20 cm of the soil. In Kootwijk, only 35% is taken up from this layer (Table 3.2.3). This is primarily caused by the high water holding capacity of the forest floor and the A horizon in Speuld, resulting in a concentration of roots in the upper parts of the soil profile. At both monitoring sites, up to 18% of the total water uptake is from depths between 80 cm and 200 cm. Water uptake from greater depth becomes more important in periods when the top soil is dry. Immediately after larger storms, however, transpiration reduction decreases strongly and water uptake from the upper parts of the soil profile is stimulated.

Total drainage from below the rootzone over the years 1987-1989 is 53% higher for Kootwijk than for Speuld, resulting mainly from the lower interception loss and from the higher transpiration reduction, caused by the lower water holding capacity for Kootwijk.

In the topsoil, directly below the fermentation layer, downward water flow only occurs during rainfall, thus generating a very dynamic temporal behaviour. At 50 and 100 cm depth water fluxes are lower and show a less dynamic behaviour. Net downward water fluxes below 50 cm depth are almost zero between May and October 1989. The water flow pattern is quite similar for the years with a marked dry period. In the wettest year (1987) water flow at 50 cm depth occurred during almost the entire summer period.

Table 3.2.3 Annual water uptake for 7 layers from March 1, 1986 to February 28, 1989. The left part of the table shows the annual water uptake in [mm], the right part of the table gives the uptake of the layer as a percentage of the actual transpiration. Minimum and maximum depths are for layer: I: -3/0 cm, II: 0/10 cm, III: 10/20 cm, IV: 20/40 cm, V: 40/80 cm, VI: 80/160 cm and VII: 160-240. In Speuld the boundaries of the first layer are -5/0 cm.

Speuld Year	I	II	III	IV	V	VI	VII	I	II	III	IV	V	VI	VII
1986	20	89	48	56	48	25	31	6	28	15	17	15	8	9
1987	21	92	50	61	53	20	18	7	29	16	19	17	6	6
1988	21	95	50	57	48	21	24	6	30	15	18	15	6	7
1989	25	110	58	65	56	29	37	6	28	15	17	14	7	9
Kootwijk Year	**I**	**II**	**III**	**IV**	**V**	**VI**	**VII**	**I**	**II**	**III**	**IV**	**V**	**VI**	**VII**
1986	20	47	40	64	58	35	18	7	17	14	23	21	12	6
1987	21	57	44	72	70	33	11	7	15	15	24	23	11	4
1988	21	48	44	69	60	32	14	7	17	15	24	21	11	5
1989	25	62	55	80	60	38	21	7	18	16	23	16	11	6

Daily fluxes appear to be quite different for both monitoring sites. In Speuld upward fluxes are larger than in Kootwijk. In the subsoil of Speuld, downward fluxes are spread more evenly throughout time and, as a result, daily values are lower. In Speuld drainage increases gradually from autumn to winter, whereas in Kootwijk drainage starts more suddenly, indicating a sharper moisture front. The more dynamic temporal behaviour of fluxes in Kootwijk mainly results from the lower water holding capacity of the Kootwijk soil and from the extremely low conductivity at low pressure heads (< 10^{-5} cm/day).

3.3 Nitrogen cycling and nitrogen transformations

Nitrogen is the most abundant nutrient in all ecosystems. In many systems nitrogen is a limiting factor for production of biomass. However, during the last 30-40 years forest and heathland ecosystems in NW Europe, especially in the Netherlands, have been subjected to an increasing input of nitrogen by atmospheric deposition leading to changes in these systems and in some cases to disturbances (Nilsson and Grennfelt, 1988).

Input, output and transformations of nitrogen are critical factors for the production and sustenance of ecosystems and play an important role in the proton budget of terrestrial ecosystems and soil acidification (Van Breemen et.al., 1988; Tietema and Verstraten, 1988). The main processes in the nitrogen cycle in forested and heathland ecosystems are given in Figure 3.3.1.

In this chapter the most important N-transformations, the net mineralization/ net immobilization, nitrification and denitrification in relation to input and output fluxes by leaching will be discussed. N2-fixation by vegetation is neglected, which is a reasonable assumption for most trees (except Alnus) in Dutch forests.

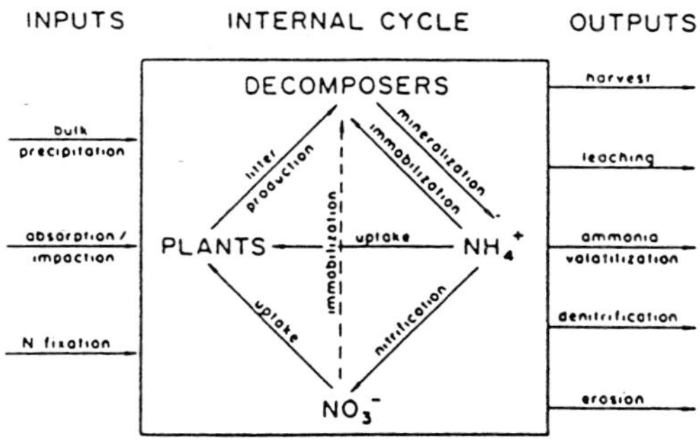

Fig. 3.3.1 A simplified model of nitrogen cycling in forest and heathland ecosystems (Vitousek, 1981)

3.3.1. Nitrification

Heathland soils

In heathland soils the process of nitrification is mainly restricted to the organic (humus) horizons and the organic rich part of the mineral soil. First microbiological aspects will be discussed based on the papers of De Boer et al.(1989 a and b), De Boer and Tietema (1990), and Stams et al. (1988). In table 3.3.1.1. some data on nitrate production and N mineralization in heathland humus samples are presented.

Table 3.3.1.1 Some data on N mineralization, nitrate production and properties heathland humus samples (*incubation of sieved, field-moist humus at 20°C)

Locations	pH	moisture (%) b	total N (%) b	C/N	Accumulation of mineral -N and nitrate -N (mg N/kg humus/wk) mineral N	NO_3 -N in humus *	NO_3-N in suspenstions pH4	pH6	ratio pH6/ pH4	pattern
Callunalocations										
Assel	3.8	67	1.45	21.9	20.3	7.4	70	76	1.1	IV
Dwingeloo	4.4	65	1.17	21.0	15.5	0.7	19	483	25.8	III
Ede	3.9	60	1.14	23.9	19.3	8.2	60	100	1.7	III
Ginkel	3.8	66	1.40	21.6	42.5	13.7	155	120	0.8	IV
Hoorneboeg	3.9	64	1.28	22.6	18.1	0.3	0	120	∞	II
Kampina 1	3.8	67	0.98	26.3	0.3	0.0	0	0	-	I
Kampina 2	3.8	53	1.15	21.7	27.5	8.3	107	145	1.3	IV
Loon/Drunen	4.3	50	0.35	23.7	3.2	0.0	0	0	-	I
Molenveld	3.8	65	1.23	23.5	0.8	0.0	0	0	-	I
Reemsterveld	4.4	66	1.12	22.4	14.6	0.6	19	194	10.4	III
Terlet	4.1	52	0.78	18.7	24.0	8.0	26	24	0.9	IV
Wolfheze	4.1	44	0.48	n.d	12.1	8.3	37	47	1.3	IV
Ericalocations										
Assel	3.8	73 c	1.26	25.2	2.1	-0.8	0	0	-	I
Balloo	4.0	65	1.20	20.8	17.9	0.0	0	0	-	I
Deelen	3.7	84 c	1.41	26.0	4.8	0.0	0	0	-	I
Dwingeloo	4.3	70	0.98	25.1	8.4	2.7	39	45	1.1	IV
Hoenderloo	3.8	69	1.45	23.5	26.1	1.1	10	20	2.0	III
Kampina	3.8	65	1.13	24.6	18.1	9.1	72	149	2.1	III
Kampshei	3.9	80 c	1.80	19.3	13.2	0.0	0	0	-	I
Kootwijk	3.8	64	1.14	24.6	16.6	-0.1	0	0	-	I
Molenveld	3.8	69	1.25	23.4	5.0	-0.2	0	0	-	I
Uddel	3.7	80 c	1.71	21.3	10.0	0.1	0	0	-	I
Deschampsialocations										
Ede	4.0	59	1.20	17.4	32.3	14.9	107	163	1.5	III
Ginkel	4.1	56	1.04	20.7	36.1	13.7	100	128	1.3	IV
Hoorneboeg	3.9	63	1.02	24.7	24.3	0.2	0	0	-	I
Kootwijk	4.4	66	1.63	16.1	46.8	7.7	42	152	3.6	III
Molenveld	3.8	58	1.20	18.8	27.3	1.5	0	35	∞	II
Reemsterveld	4.0	54	0.85	22.9	19.0	7.6	4	77	1.7	III
Terlet	4.5	57	1.02	15.9	77.3	15.9	6	121	2.0	III
Wolfheze 1	4.2	63	1.56	17.6	34.1	7.2	40	45	1.1	IV
Wolfheze 2	4.2	52	1.12	16.9	38.0	11.9	36	36	1.0	IV
Molinialocations										
Assel	3.5	64	1.30	22.1	14.4	12.	225	224	1.0	IV
Balloo	3.9	68	1.16	81.4	21.9	11.1	82	153	1.9	III
Dwingeloo	4.1	64	1.16	22.1	24.2	5.3	55	64	1.2	IV
Ede	3.6	50	1.00	20.7	22.3	10.8	197	235	1.2	IV
Kampina	3.7	58	0.91	21.1	20.0	11.5	223	251	1.1	IV
Kampshei	3.9	79	1.44	21.8	9.90	0.0	0	0	-	I
Kootwijk	3.6	59	1.22	20.8	20.9	10.6	165	178	1.1	IV
Reemsterveld	3.6	61	1.34	20.1	21.2	11.3	100	137	1.4	IV
Terlet	4.6	53	0.91	21.8	72.4	7.8	27	103	3.8	III
Uddel	3.8	85 c	1.98	17.3	4.9	0.1	0	12	∞	II

- = not determined.
a = (wet weight-dry weight)/wet weight x 100.
b = on basis of dry humus.
c = water saturated samples.

For the 41 humus samples of Table 3.3.1.1 nitrate production showed a significant ($p < 0.005$) positively correlated with N mineralization (r=0.59) and a significant negative correlation with the (initial) moisture content (r=-0.54) and the C/N ratio (r=-0.51). Nitrate production was not significantly correlated with pH and initial mineral N content.

From a comparision of the net nitrate production in suspensions of the humus samples at pH 4 and pH 6 a differentiation into four patterns could be made for Dutch heathland soils:

I. **No nitrate production** at both pH values studied (n=12)

II. **Acid-sensitive nitrate production:** nitrate production at pH 6 but not at pH 4 (n=3)

III. **Acid-tolerant, pH-dependent nitrate production:** nitrate production at both pH 6 and pH 4 with the production at pH 6 being at least 1.5 times faster than at pH 4 (n=10)

IV. **Acid-tolerant, pH independent nitrate production:** nitrate production at both pH 6 and pH 4 with the production at both pH values being almost equal (n=16)

Most samples, that did not show accumulation of nitrate in suspensions at either pH value, orginated from Erica-dominated sites. Acid-tolerant nitrification, that is nitrate production at pH 4 (patterns III and IV) is wide-spread among Dutch heathland soils and was not restricted to humus of a certain vegetation type. Acid-sensitive nitrification (pattern II) is uncommon.

All humus samples that accumulated nitrate during the incubation of field-moist material also did in suspensions, with the exeption of Dwingeloo-Calluna.

The effect of urea on nitrification in suspension at pH 5 was considered to be stimulating when the accumulation of nitrate in 3 weeks was at least 1.5 times higher in suspensions with urea addition than in those with ammonium addition. Humus samples showing increased nitrate production by urea also produced more nitrate at pH 6 than at pH 4 (patterns II and III). However, some humus samples with pH-dependent nitrate production (pattern III) did not show urea-stimulated nitrate production.

In all samples, acetylene (0.06%) completely blocked the nitrate production indicating that the nitrification is caused by chemolithotrophic bacteria (De Boer et al., 1989a). Therefore, nitrate production by fungi does not seem to be of quantitative importance. Using the same acetylene blocking technique combined with antibiotic treatments, indicated that nitrate production in acidic forest soils in the USA is mainly caused by fungi (Adams, 1986;

Killham, 1986). At the moment, it is unclear which factors determine whether nitrification in acid soils is mainly autotrophic or heterotrophic.

Acid-tolerant chemolithotrophic nitrifying bacteria are widespread in Dutch heathland soils (De Boer et al., 1990). Acid-tolerant nitrite oxidizing bacteria have been isolated (Hankinson and Schmidt, 1988; De Boer and Laanbroek, 1989). However, thus far no acid-tolerant ammonium-oxidizing bacteria have been isolated and all information about their physiology has to be deduced from suspension experiments.

The difference between group III and group IV may be due:

1) to the presence of predominantly acid-tolerant ammonium-oxidizing strains with growth rates that are more (pattern III) or less (pattern IV) affected by an increase in pH from 4 to 6 or
2) to the presence of both acid tolerant and acid-sensitive strains in humus samples of group III.

In the second case, nitrate production at pH 6 is probably the result of a combined activity of acid-tolerant and acid-sensitive strains whereas at pH 4 only the acid-tolerant strains are active. The presence of both acid-sensitive and acid-tolerant strains in one humus sample may be indicated by those humus samples in which nitrate production in suspensions was both pH-dependent and stimulated by urea. Previously, it was shown that stimulation of the nitrate production by urea may indicate the presence of acid-sensitive, ureolytic ammonium-oxidizing bacterias (De Boer et al., 1989b). Urea stimulated nitrate production was not detected in any of the humus samples with pH-independent nitrate production (pattern IV) indicating that the acid-tolerant ammonium-oxidizing bacteria may not be ureolytic.

Some humus samples deviated from those discussed above and showed pH-dependent, acid-tolerant nitrate production (pattern III) that was not stimulated by urea (e.g. Kootwijk-Deschampsia). This type of nitrification may be the result of combined activities of acid-tolerant - and non-ureolytic, acid-sensitive ammonium-oxidizing strains.

Nitrite was never detected in the suspension experiments indicating that nitrite-oxidation did not limit nitrate production. Therefore, the suspension experiments do not give information about the effects of a rise in pH on the process of nitrite-oxidation. It is obvious that acid-tolerant nitrite-oxidizing bacteria must be present in humus samples showing nitrate production at pH 4 but it is unknown wheter acid-sensitive, nitrite-oxidizing bacteria do contribute to nitrate production at pH 6. Hankinson and Schmidt (1988) showed that acid-tolerant - and acid-sensitive nitrite-oxidizing bacteria do coexist in an acid forest soil.

Acid-sensitive nitrate production, as opposed to acid-tolerant nitrate production, apparently requires special conditions (suitable pH or the presence of urea). Therefore, acid sensitive nitrate producers will be less important than the acid tolerant producers in heathland humus.

Indeed, it was observed that humus samples with little or no nitrate formation in suspensions at pH 4 with much nitrate formation at pH 6 produced only a little nitrate during incubation of field-moist material.
Obviously pH is not a factor that determines whether or not chemolithotrophic nitrifying bacteria can exist in heathland humus. Acid-tolerant nitrifying bacteria grew exponentially even at pH values as low as 3.5 (De Boer et al. 1989a). In addition, nitrate production in suspensions at pH 4 with acid-tolerant nitrification (pattern III and IV, n=26) is negatively correlated (r=-.726) with humus-pH indicating that at the time of sampling the highest numbers of acid-tolerant, ammonium-oxidizing bacteria were present in the most acid humus. This may imply that the activities of acid-tolerant ammonium-oxidizing bacteria have resulted in a drop of the humus-pH, although a low in situ pH might also be explained by other processes.

Finally some additional results on the nitrification rate in heathland soils based on in situ incubations of intact soil cores (Tietema, not published data) will be presented. The method is described in Tietema et al. (1990a). For the monitoring site Assel under mono-culture of Calluna vulgaris a nitrification rate of the organic (LFH) and the Ah (0-5 cm depth) horizons was estimated of 0.5 respectively 0.2 kg N $ha^{-1}yr^{-1}$ (Table 3.3.1.2). These figures represent very low nitrification rates which are often reported in the literature.

Forest soils

Soil samples of the 0-5 cm layer (LFH included) of all forest soils studied so far (n=20) showed an 80 to 100% inhibition of nitrate production upon incubation in the presence of 0.06% acetylene. Furthermore, soil samples from most of these soils contained acid-tolerant nitrate producing micro-organisms (production of nitrate at pH 4 in suspensions). Therefore, it seems that acid-tolerant, chemolithotrophic nitrification is widespread in Dutch acid forest soils too. The lower pH-limit of acid-tolerant nitrification in suspensions of a Douglas fir soil (organic layer) was found to be approximately pH 3. This may imply that acid-tolerant nitrate production can be a major source of acidification in many strongly acid forest soils.
Stams et al. (1988) found heterotrophic nitrification in the forest floor of an oak-birch ecosystem (site 13).
From the in situ incubations of intact soil cores it can be concluded that nitrification rates ranged from 7 to 65 kg N ha^{-1} yr^{-1} in the organic layer and the top 5 cm of the mineral soil (Table 3.3.1.2.). Results from the Winterswijk site revealed that nitrification below this organic rich part of the mineral soil was neglectable. On an areal basis both organic and organic rich parts of the mineral soil contribute about equally to the total nitrification rate.

An exeption was found in the Leuvenum site, where the upper 5 cm of the mineral soil consisted of only 1-2 cm Ah material (Tietema and Verstraten, 1988; Tietema et al, 1990a, b, c and d; Tietema and Verstraten, 1990).

Table 3.3.1.2 Nitrification and net mineralization rates in kg N ha^{-1} yr^{-1} (II) and nitrogen solute fluxes in thoughfall and leaching water (beneath the rooting zone) in kg N ha^{-1} yr^{-1}

	Oudemaat				Winters-wijk	Buun-kamp	Leuve-num	Koot-wijk	Speuld	Assel
location	13A	13B	13C	13D	12	33	34	3B	1B	31
nitrification rate										
LFH					30	4	8	13	13	0.5
Ah					25	3	4	10	10	0.2
total					55	7	12	23	23	0.7
net nitrogen mineralization rate										
LFH					62	85	100	31	41	49
Ah					29	9	30	12	32	7
total					91	94	130	43	73	56
through-fall N	55	56	45	63	42	27	52	40	42	
leaching N	78	28	22	88	10*	6	24	23	31	

* denitrification occurs (> 20 kg N ha^{-1} yr^{-1})

No clear correlation between the nitrification rate and the net mineralization rate can be noticed. This might be caused by the fact that in these nitrogen (near) saturated forests, the nitrification is not substrate limited. This is illustrated by the fact that annual NH_4^+-N concentrations in the litter layer of the oak-beech forest ecosystem of Winterswijk fluctuate between 60 and 200 mg N kg^{-1} (Tietema et al., 1990a) and NH_4^+ concentrations in the litter leachate from the same forest range from 40 to 600 mol l^{-1}.

Tietema et al (1990c) found a strong correlation between nitrogen transformations (nitrification and net nitrogen mineralization) and soil temperature at location 12 (Winterswijk; Table 3.3.1.3). This correlation was strongest at a gravimetric water content of more than 200% and 50% in the organic layer and in the top of the mineral soil, respectively. Because in this relatively wet site these conditions occur frequently, it can be concluded that in this forest soil temperature largely determines the nitrogen transformation rates. A preliminary analysis of the available results at Speuld (site 1B) revealed a similar relation of nitrification and watercontent and soil temperature in the organic layer (Table 3.3.1.3; Tietema et al 1990b). A linear relation was found between nitrification at a water content less than 200%, while soil temperatures determine the nitrification rate when the watercontent exceeds 200%. From these results it can be generally concluded that in locations with high nitrification rates, water content and soil teperature can be considered as regulating factors if the nutrient status of the soil is not too low.

Table 3.3.1.3 Relation between nitrogen transformations (NI and NM in mg N kg-1 4 weeks-1) and soil temperature (T in °C) and gravimetric water content (W in weight %), based on in situ incubation measurements with intact cores

layer	process	equation	R2
12) Winterswijk			
LH (W>200%)	nitrification	NI= 7.34*e 0.19T	0.78
	net mineralization	NM= 21.71*e 0.15T	0.82
Ah (W>50%	nitrification	NI= 0.52*e 0.22T	0.72
	net mineralization	NM= 1.47*e 0.13T	0.49
1B) Speuld			
LH (W<200%)	nitrification	NI= 0.24*V-18.6	0.79
LH (W>200%)	nitrification	NI= 2.88*T+ 0.86	0.58

As the tree composition in the different forests are not completely comparable, no conclusions can be drawn as to the relation between nitrification and vegetation.

The nitrogen deposition, measured as throughfall, in the different forest sites of Table 3.3.1.2. is rather low compared to most others (see Table 2.1). The low N deposition at Buunderkamp (site 33) can be explained by the favourable exposition compared to the surrounding arable lands, the relatively low amount of precipitation in the year of measurement and the structure of the canopy. The relatively low deposition rates on the other sites might be explained by the location of the sites in the middle of a relatively forested area. The data on nitrogen deposition reveal no correlation with nitrification rates: high nitrification rates coincide with high nitrogen depositions at Winterswijk (site 12), while at Leuvenum (site 34) a combination of low nitrification rates and relatively high deposition rates was observed.

3.3.2 Denitrification

Denitrification can be an important process that contributes to the decrease of the nitrate load and to the neutralization of the acid load of soils and groundwater. There seems to be a relationship between the denitrification rate and the hydrological regime of the soil system and the groundwater-level fluctions (Figure 3.3.2.1.). Unfortunately these figures have been obtained from agricultural soils in the Netherlands.

With respect to forested and heathland soils, only field and laboratory data of the Winterswijk site (groundwaterclass V; fluctuations between 50-200 cm) are available

(Tietema and Verstraten, 1988; Tietema et al, 1990a; Tietema et al, 1990b). The main results are given here. Laboratory measurements with undisturbed soil cores from this oak-beech ecosystem indicated that despite the high acidity (pH 3.0-3.5) in the top soil, further reduction of N_20 to N_2 was quantitatively important. This implies that the field measurement N_20-fluxes represent minimum values of nitrogen losses by denitrification.
The field measured N_20 fluxes (Figure 3.3.2.2.A) revealed high temporal variability and ranged from non-detectable values (< 0.1 mg N m^{-2} day^{-1}) to the highest value of 68 mg N $m^{-2}day^{-1}$. The 2 peak values (Figure 3.3.2.2.A) of N_20 compared 63% of the extrapolated annual N_20 loss, which amounted to 20 kg N/N_20 $ha^{-1}yr^{-1}$. This high temporal variability made the extrapolation of weekly measured rates to annual nitrogen losses very difficult.
From the nitrogen budget of this oak-beech ecosystem (Tietema and Verstraten, 1988) it can be concluded that the nitrogen loss by denitrification has to be 65 kg N $ha^{-1}yr^{-1}$.
In the Netherlands most forest and heathland soils are dry soils with (very) low groundwater-tables. Consequently, the contribution of denitrification to the nitrogen budget will be limited. However, temporal conditions of wetness in the litter layer are common (Tiktak and Bouten, 1990; Bizot, Tiktak and Bouten, 1990); and might present favourable conditions for denitrification.

3.3.3 Mineralization/immobilization of nitrogen

All the forest and heathland ecosystems that were studied show a net mineralization rate measured in the field with in situ incubations of intact soil cores (Tietema et al, 1990b). These measured net mineralization rates range from 43 (Speuld) to 130 (Leuvenum) kg N $ha^{-1}yr^{-1}$ for forest soils. Net mineralization is mainly found in the organic layer (Table 3.3.1.2) and no clear correlation could be observed between the net mineralization and nitrification rates.
For the heathland site (Assel) a net N mineralization rate of 56 kg N $ha^{-1}yr^{-1}$ was established (Table 3.3.1.2).

3.3.4 Nitrogen leaching

As is shown by Table 3.3.1.2, and further elaborated in paragraph 3.4.2, many forested areas in the Netherlands show high rates of nitrate leaching to the shallow groundwater. From this table it can be concluded that appreciable nitrate leaching occurs in these forested systems at both high and (relatively) low nitrogen deposition levels. Obviously, the critical nitrogen load for these systems has been exceeded leading to an inefficiency in the nitrogen budget and an output to the groundwater of 20 % or more (>100%) of the input values. These results are confirmed by the data from the Peel area (Roelofs et. al 1988). Although insufficient data with respect to the nitrogen status of the (forested) soils are available in the

Netherlands, the increasing amounts of nitrogen leaching to the shallow groundwater give a very strong indication of the (near) saturation of these soils for nitrogen (see also paragraph 3.4.2 and table 3.4.2.1).

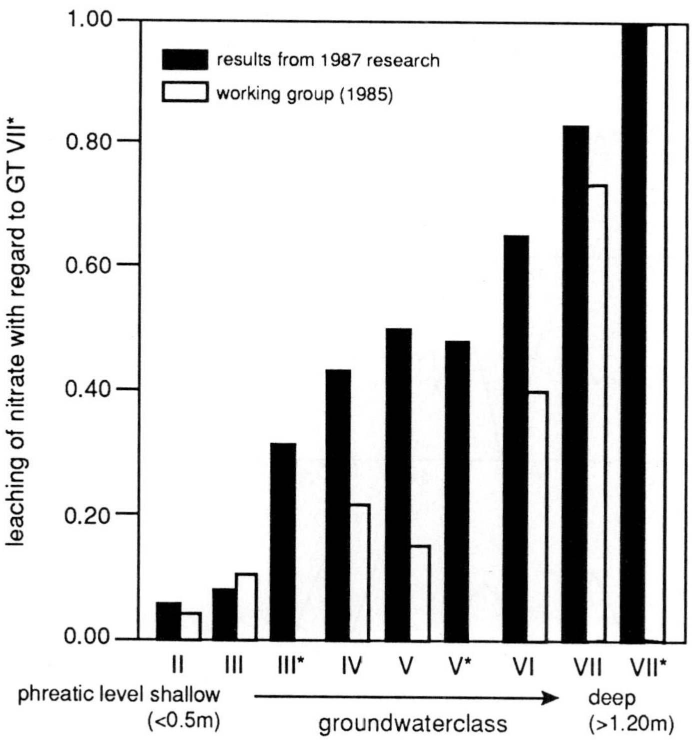

Fig. 3.3.2.1 Relationship between groundwaterclass and NO_{3^-} leaching

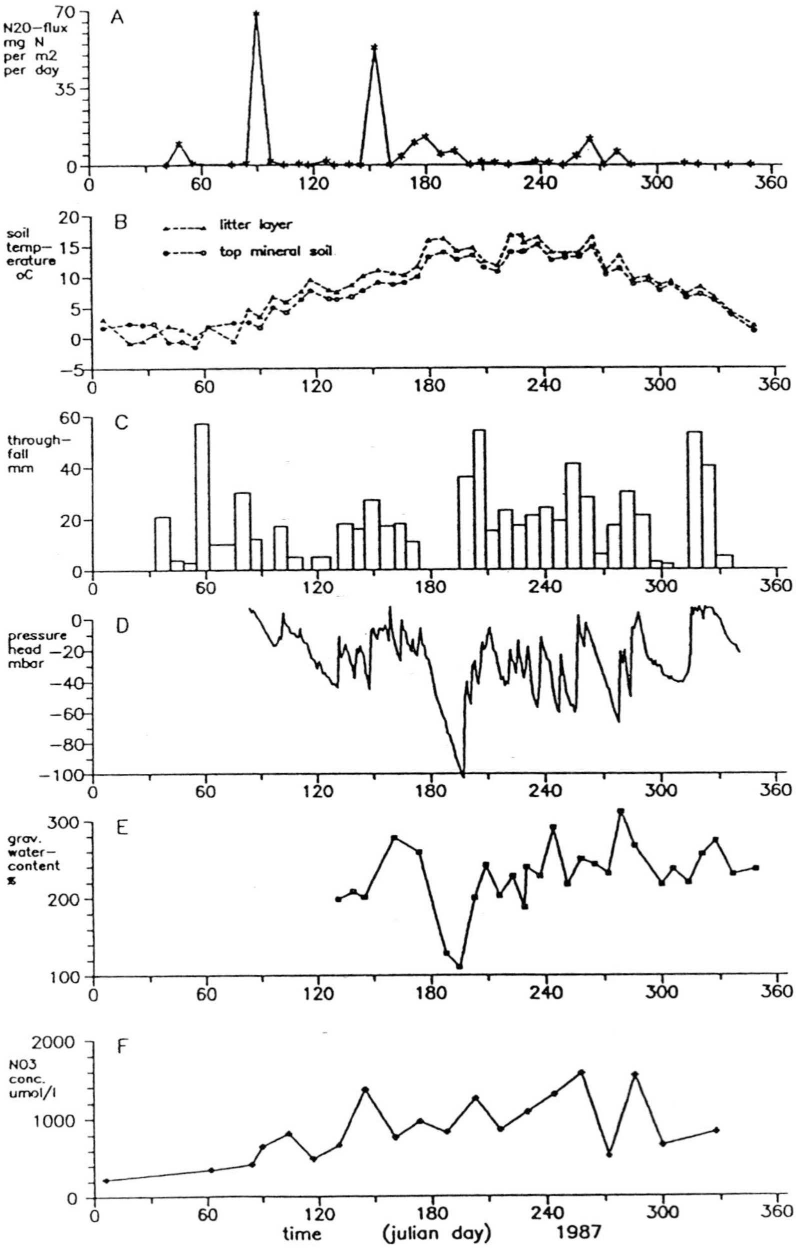

Fig. 3.3.2.2 N_2O fluxes (A) at site 12 (Winterswijk) in relation to soil temperature (B), throughfall (C), pressure heads in the top of the mineral soil (D), gravimetric water content of the litter layer (E), and mean NO_3^- concentrations in the soil water directly beneath the litter layer (F)

3.4 Soil acidification

3.4.1 Soil acid base status

Soil acid base status, general

Table 3.4.1.1 summarizes some data from essentially all forest and heathland sites in the Netherlands (plus one site in the FRG), where the soil solution composition has been monitored for at least a year within the past decade. At all forest sites (1) the soil is highly acid, with (mean annual soil solution) pH values in the surface mineral soil generally below 4, and base saturations between 3 and 20 %, (2) the major dissolved cation is Al^{3+}, balanced by sulphate and nitrate in roughly equal proportions, (3) Al concentrations increase with depth to values generally above 1 mmol(+/-)/l, and sometimes up to 10 mmol (+/-)/l, and (4) ammonium to nitrate ratios in the surface soils are mostly well below 1, and invariably below the ratios in throughfall, indicating that nitrification is widespread. The predominance of dissolved Al, NO_3 and SO_4, observed earlier by Van Breemen et al (1982 and 1988) and Verstraten et al (1990) as signs of excessive soil acidification caused by high atmospheric inputs of sulphur and nitrogen at a few sites, thus appear to be typical for Dutch forests.

Using the aqueous extraction method, Roelofs and coworkers (e.g. Houdijk, 1990), almost invariably observed NH_4/NO_3 ratios between 1 and 5 exceeding in the upper mineral soil of a large number of coniferous and deciduous forest stands in the country (see e.g. Table 3 in Van Breemen and Van Dijk, 1988). They concluded that ammonium accumulation is more widespread in Dutch forest soils than is nitrification. The results of the sampling comparison experiment (Appendix 1), however, suggested that aqueous extraction strongly increases dissolved NH4 relative to dissolved NO3 compared to solute concentrations in soil solutions obtained from porous cups or centrifugation. This finding, in our opinion, resolves the nitrification/no nitrification debate, and strongly suggests that acidification by nitrification is predominant in Dutch forests. Forest sites with low rates of nitri-fication and with high levels of dissolved ammonium do exist, however, (the Ysselsteyn site used for the comparison experiment is an example), and may be underrepresented in the sites selected for monitoring. As can be concluded from section 3.3, and also from table 3.4.1 the situation is the reverse for heathlands, where low rates of nitrification predominate. Also, heathland soils are generally less acidic and have lower solute concentrations than forest soils, which can be attributed to a combination of lower atmospheric input due to less interception of pollutants by the canopy, and lower rates of evapotransiration.

Table 3.4.1.1 Mean field pH and base saturation (%) in the soil, and mean concentrations of Al, SO_4, NO_3, and NH_4/NO_3 in the soil solution (mmol(+/-)/l) at the sites specified in Table 2.1

location		pH	b.sat.	Al	SO_4	NO_3	NH_4/NO_3
		0-10 cm depth					
1A	Speuld A	3.2	13	0.6	0.9	1.4	0.32
1B	Speuld B	3.2	--	0.3	0.7	0.5	0.72
2	Amerongen	3.4	9	1.1	0.7	0.5	0.72
3A	Kootwijk A	3.5	6	1.1	1.4	2.0	0.68
3B	Kootwijk B	3.4	8	0.3	0.6	0.3	1.3
4	Garderen	3.5	13	0.7	0.9	1.0	0.3
5	L.vuursche	3.3	11	0.8	2.8	2.1	0.92
6	Ruurlo	3.2	10	0.7	3.2	3.0	0.68
7	Zelhem	3.3	9	4.1	2.6	4.2	0.47
8	Tongbersven	3.4	3	0.1	1.1	0.4	2.1
10	Gerritsfles	4.0	12	0.8	0.9	1.0	0.9
12	Winterswijk	3.5	15	0.2	0.5	0.7	0.10
13	Oudemaat abc	3.6	8	0.6	0.7	1.1	0.18
31	Assel	4.7	50	0.04	0.4	0.0	16
32	Harderwijk	4.3	20	0.6	0.5	0.1	0.9
33	Buunderkamp	3.9	n.d.	0.3	0.5	0.2	0.85
34	Leuvenum	3.3	n.d.	0.2	0.7	0.5	0.83
35	Hasselsven	3.5	5	0.4	1.1	0.0	2.
100	Solling	3.6	10	0.7	0.7	0.5	0.29

location		pH	b.sat.	Al	SO_4	NO_3	NH_4/NO_3
		50-100 cm depth					
1A	Speuld A	3.7	11	1.4	1.3	1.2	0.04
1B	Speuld B	3.9	--	2.0	1.0	1.2	0.02
2	Amerongen	3.7	9	3.2	2.0	2.4	0.09
3A	Kootwijk A	3.6	10	1.9	1.8	0.9	0.22
3B	Kootwijk B	4.0	14	1.6	1.0	0.7	0.11
4	Garderen	3.7	8	1.0	1.1	0.7	0.08
5	L.vuursche	3.3	7	9.8	6.9	5.3	0.15
6	Ruurlo	3.5	6	5.5	3.4	2.0	0.22
7	Zelhem	3.3	10	12.6	6.9	6.4	0.09
8	Tongbersven	4.1	--*	1.8	1.5	0.5	0.18
10	Gerritsfles	4.2	--*	1.7	1.2	0.8	0.02
12	Winterswijk (50 cm)	4.4	22	0.2	1.6	0.5	0.05
13	Oudemaat&	4.1	10	1.5	1.4	1.5	0.09
31	Assel(30 cm)	4.4	40	0.2#	0.3#	0.0#	0.41#
32	Harderwijk	4.9	9	0.5	0.5	0.2	0.01
33	Buunderkamp	4.3	n.d.	0.6	0.5	0.2	0.12
34	Leuvenum	3.9	n.d.	0.9	0.8	0.5	0.23
35	Hasselsven	4.3	20	0.4	0.6	0.0	2
100	Solling	4.2	5	1.5	1.5	0.2	0.19

* CEC too low to permit reasonable estimate of base saturation
& mean for three acid soil plots at the Oudemaat site
\# at 30 cm depth; n.d. = not determined.

Soil acid base status, Aciforn monitoring sites

Table 3.4.1.1 shows that the Aciforn sites (locations 3B and 3B) have the lowest levels of dissolved Al, SO_4 and NO_3 in the surface soil of all Douglas fir sites considered (locations 1 to 7). This parallels the comparatively low atmospheric inputs at the Aciforn sites on account of their sheltered position. Nevertheless, the soils are strongly acidic, with the typical signs of enhanced acidification by atmospheric deposition.

The three replicates in soil water sampling for the monitoring programme resulted in very high standard deviations for the concentrations of components. To better characterize the soil solution composition, a sampling campaign was done in early 1990 using 25 cups at 10, 20 and 90 cm depth at both sites in a 12.5 * 12.5 m (Speuld) or 10 * 10 m grid (Kootwijk). The same ceramic cup lysimeters were used as in the standard monitoring programme. The cups were washed according to Meiwes et al (1984), installed in November and December 1989, and flushed twice by applying suction in January. Results for the samples obtained in early March 1990 are given in Table 3.4.1.2.

Table 3.4.1.2 Mean soil solute concentrations (mmol(+/-).m-3) at 10, 20 and 90 cm depth for Speuld (SP) and Kootwijk (KW) in the beginning of March 1990 (n=25).* indicates a significant difference between Sp and KW

	------SP-----			--------KW------			s.l.=0.05		
Depth	10	20	90	10	20	90	10	20	90
pH	3.22	3.35	3.91	3.38	3.48	3.96	*	*	
C_t	3011	2603	1105	2483	2396	1321			
Si	197	267	319	100	173	268	*	*	*
K	82	48	29	56	30	25		*	
Na	356	357	662	414	395	560			
Ca	274	257	183	181	168	120	*	*	*
Mg	127	150	206	116	122	144			
Al	278	795	1992	292	623	1590			
Fe	18	29	4	14	13	3	*	*	*
Mn	45	44	114	21	25	43	*	*	*
NH_4	396	239	21	364	238	73			
Cl	840	857	1035	924	807	957			
NO_3	548	555	1225	278	308	674	*	*	*
SO_4	709	840	1022	637	747	1005			
H_2PO_4	5	1	0	10	5	0	*	*	*
Org.anions	80	74	39	71	72	49			
	-------------molar ratio's------------								
Ca/Al	1.85	0.65	0.14	1.54	0.52	0.14			
NH_4/K	4.09	4.30	0.84	5.26	5.89	1.41			
NH_4/Mg	6.10	3.09	0.22	4.72	3.21	0.64			

pH-values in the surface soil are significantly lower at Speuld than at Kootwijk, which is probably a result of the higher NO_3-production as is suggested by the higher NO_3-concentrations at Speuld. The higher NO_3-concentrations are accompanied by higher concentrations of Si, K, Ca, Fe and Mn while the difference in Al-concentration at 90 cm depth is significant at 10 %.

Molar ratio's of Ca/Al<1, NH_4/K>5 and NH_4/Mg>10 (Ulrich, 1983; Roelofs et al 1988) are indicative of mineral imbalances. The sites Speuld and Kootwijk can be characterized as NO_3-Al systems since the Ca/Al ratio is the most critical ratio if the whole profile is considered (see table 3.1.1). The NH_4/K ratio might be critical in the first 10 to 20 cm particularly at Kootwijk. Availability of K is more likely at Speuld than at Kootwijk as is shown by the significant difference at 20 cm. The Ca/Al ratios indicate that a very critical situation exists with regard to acid toxicity and Ca and Mg uptake at 20 cm and lower depths. Further Ca availability throughout the profile is higher at Speuld. There is no significant difference in Al/Ca ratio or other ratios between Speuld and Kootwijk.
The P concentration in the soil solution is higher at Kootwijk parallel to the higher P content in the soil (Tiktak et al., 1988). Differences in K concentration between Speuld and Kootwijk were less than was expected from the standard monitoring with three replicates.

The heathland soil at Assel is only moderately acid, and has a relatively high base saturation throughout the soil (20 to 60 %), with base cations dominating over Fe and Al in the soil solution. The dominance of sulphate over other anions (including organic anions) shows the influence of acid deposition, however. Levels of dissolved inorganic N are very low, indicating that the heath ecosystem is able to assimilate practically all incoming nitrogen.

For Assel, mean soil solution concentrations at 5 (Ah-horizon), 20 (B2h-horizon) and 30 cm depth (B3-horizon) from July 1988 to March 1990 will be discussed (Table 3.4.1.3.).

Comparison of tables 3.4.1.2 and 3.4.1.3 illustrates the low solute concentrations in heathland compared to those under forest, in particular for inorganic N and Al, which are one (NH_4, Al) to two (NO_3) orders of magnitude lower under Calluna.

Table 3.4.1.3 Mean soil solute concentrations (simple means, mmol(+/).m^{-3}) at 5 (Ah-horizon), 20 (B2h-horizon) and 30 cm depth (B3-horizon) over the period 21 June 1988 to 20 June 1989 at Assel

Depth	pH	Si	K^+	Na^+	Ca^{2+}	Mg^{2+}	Al^{3+}	Fe^{2+}	Mn^{2+}	NH_4^+
5 cm	3.66	101	14	141	108	54	43	39	0	33
20 cm	3.81	100	10	135	128	44	88	18	1	11
30 cm	4.15	135	7	131	53	40	243	5	1	3

Depth	Cl^-	NO_3^-	SO_4^{2-}	$H_2PO_4^-$	WA[1)]
5 cm	200	2	385	0.36	59
20 cm	190	5	328	0.14	61
30 cm	192	8	311	0.09	27

[1)] WA: Weak Acid Anions (calculated as difference sum(+) minus sum(-)

The monitoring series at Assel was interrupted by a drought (second half of June and July) which coincided with a severe beetle infestation that practically destroyed the Calluna cover. These events dramatically influenced the solution chemistry of the soil (Figures 3.4.1.1 to 3.4.1.5). After the rains returned, concentrations of practically all solutes peaked, first in the uppermost horizons and later at greater depth, reaching values that were 2 to 20 times the maxima in the previous year. At the same time, pH in the Ah horizon dropped from about 3.7 to 3.2. Concentrations of SO_4, Ca, Mg and Al fell back to their previous levels over the winter of 1989/90. However, levels of NH_4 and, to a lesser extent of K and NO_3, stayed high until the end of the monitoring period in March 1990, when the soil solution pH in the Ah horizon reached an all-time high of 3.9.

These patterns can be explained by (1) accumulation of dry deposited SO_4 during the dry period, followed by entry into the soil, soil acidification and, next, dilution in the subsequent wet period, and (2) for NH4 and K by continued (or even enhanced) mineralisation in the absence of plant uptake after destruction of Calluna. The increase in pH towards the end of the monitoring period can be explained partly by protonation of mineralized ammonia at low or zero nitrification.

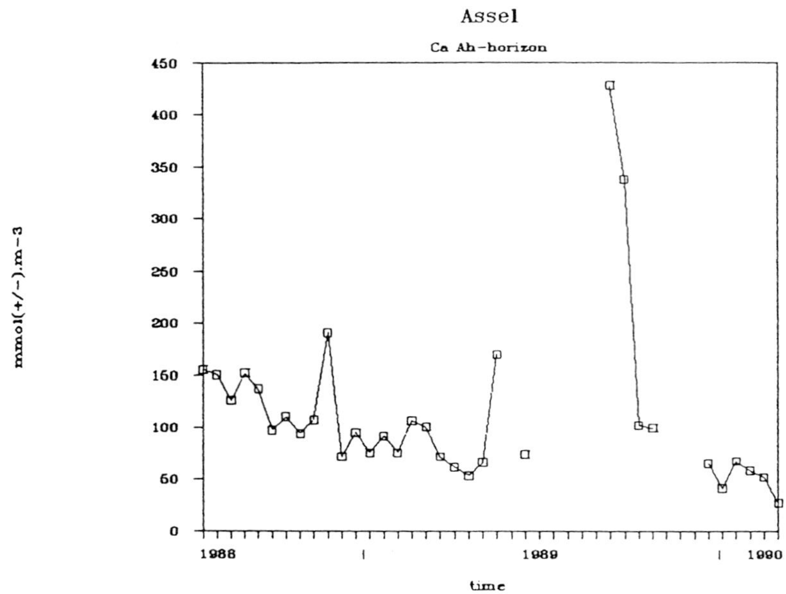

Fig. 3.4.1.1 Concentration of Ca^{2+} in the A_h horizon at the Assel heathland site between June, 1988 and March 1990

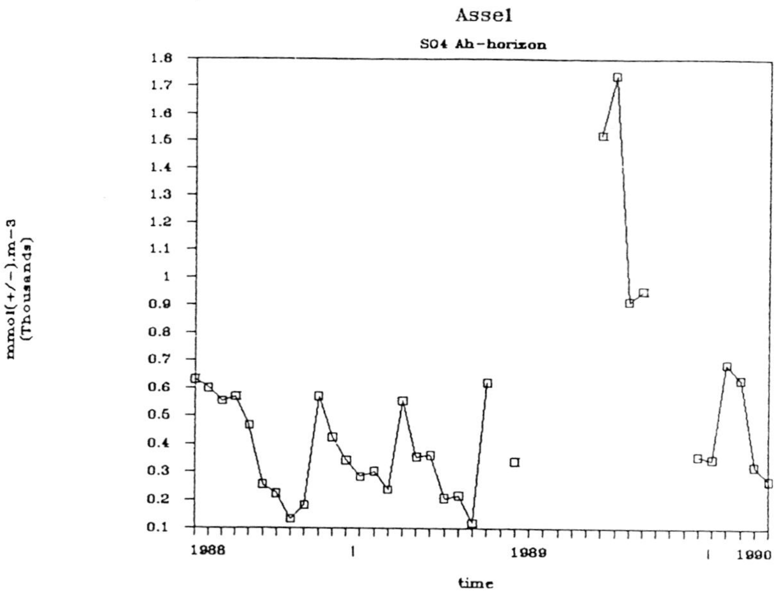

Fig. 3.4.1.2 Concentration of K in the Ah horizon at the Assel heathland site between June 1988 and March 1990

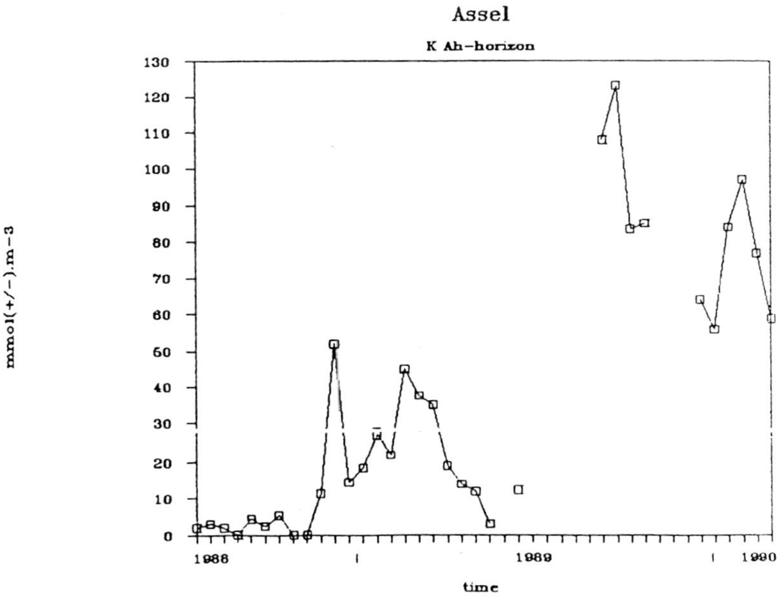

Fig. 3.4.1.3 Concentration of K in the Ah horizon at the Assel heathland site between June 1988 and March 1990

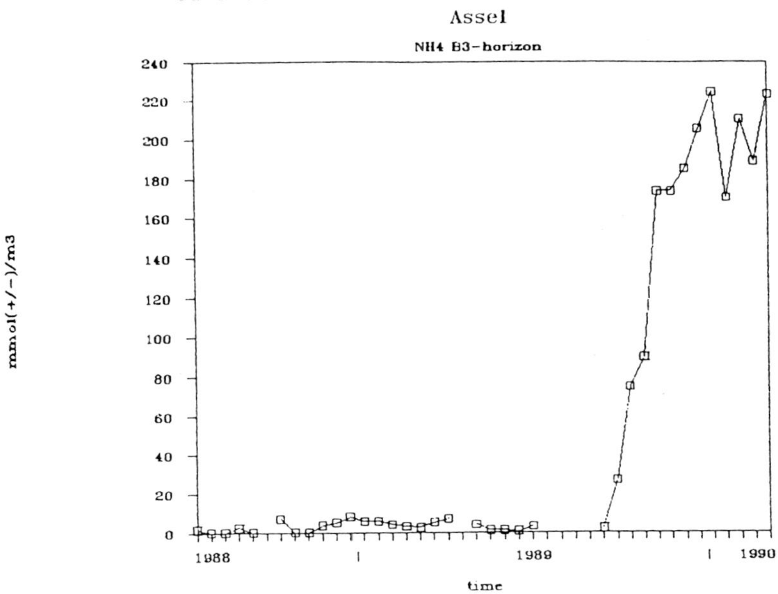

Fig. 3.4.1.4 Concentration of NH_4 in the B3 horizon at the Assel heathland site between June 1988 and March 1990

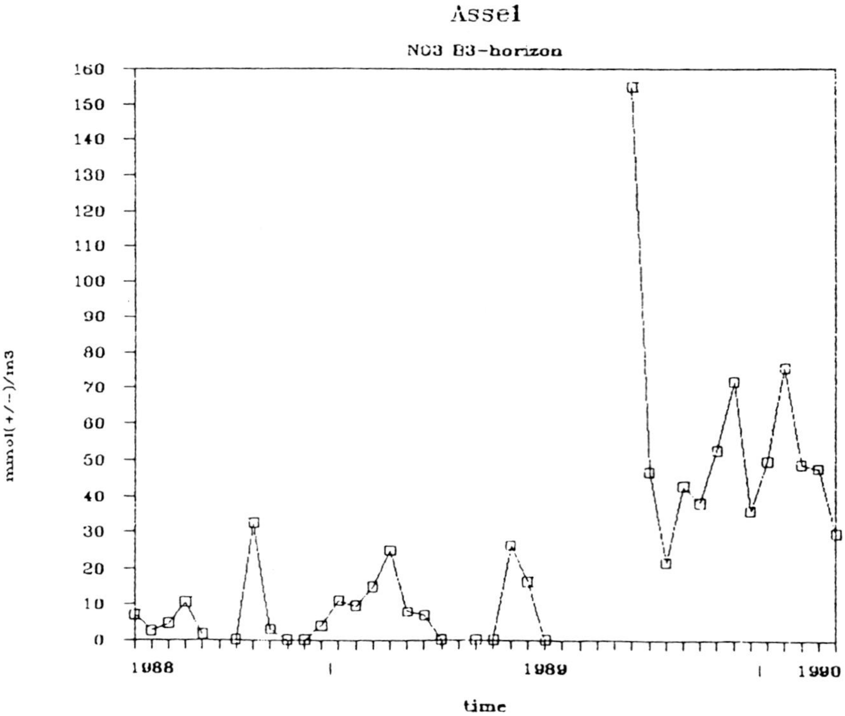

Fig. 3.4.1.5 Concentration of NO_3 in the B3 horizon at the Assel heathland site between June 1988 and March 1990

3.4.2 Input-output budgets

Input-output budgets, general

For fifteen of the Dutch sites presented in Table 2.1, input output budgets have been estimated for periods ranging from 1 to 6 years. Results that are most pertinent to soil acidification and N cycling have been summarized in Table 3.4.2.1.

The results show that (1) sulphate inputs are invariably similar to sulphate outputs, (2) appreciable amounts (up to 30% of the input) of ammonium are leached to the groundwater at very high ammonium inputs, (3) at all Dutch forest sites, nitrate output exceeds nitrate input, (4) in all Dutch sites except 13A and 12 (which have a calcareous subsoil), Al output is roughly equal to the acidity produced in the N cycle, and (5) at the forest sites from 20% to more than 100% of the N input is recovered in the drainage output. So, essentially all soils are sulphate-saturated, probably reflecting the effect of high levels of accumulated sulphur deposition. Five of the sites appear to be also N-saturated (N-out > N-in), indicating that the biota are unable to assimilate incoming N and that the pool of soil organic N has begun to contribute to net drainage of nitrate and the associated soil acidification. Net nitrification of ammonium plus soil organic N account for practically all soil acidification taking place, which involves replacement of H^+ inputs ($NH_4^+ \rightarrow NO_3^- + 2H^+$) by Al^{3+}

Table 3.4.2.1 Input-output budgets for SO4, N, and Al (input negligible) for fifteen Dutch and one West German site. All sites are forested, except Hasselsven which is a heathland. For references see Table 2.1. Units are kmol(+/-)/ha.yr. Periods are hydrological years, generally from April to April

location	period	SO_4 in	SO_4 out	NH_4 in	NH_4 out	NO_3 in	NO_3 out	Al out	$H^{+\#}$	N_{out}/N_{in}
1A Speuld A	'87/88	2.4	1.6	2.9	0.11	0.87	2.0	3.0	4.2	0.6
1B Speuld B	'88/90	2.3	3.1	2.8	0.01	0.79	2.2	4.6	4.2	0.6
2 Amerongen	'87/88	3.4	2.9	3.7	0.45	0.74	4.5	4.3	7.0	1.1
3A KootwijkA	'87/88	2.4	3.0	3.0	0.30	0.83	1.7	4.0	3.6	0.5
3B KootwijkB	'88/90	2.1	1.7	2.6	0.00	0.75	1.7	3.1	3.6	0.5
4 Garderen	'87/88	2.5	2.8	3.1	0.18	0.83	2.0	1.9	3.6	0.5
5 L.vuursche	'87/88	4.8	6.0	4.5	0.80	1.6	5.9	10.0	8.0	1.1
6 Ruurlo	'87/88	3.2	3.3	3.9	0.80	0.69	4.6	7.4	7.0	1.2
7 Zelhem	'87/88	3.8	5.5	5.1	1.7	0.99	6.4	9.4	8.8	1.3
8 Tongbersv.	'83/84	3.8	2.8	4.2	0.2	0.65	0.85	3.3	4.2	0.2
10 Gerritsfles	'83/84	3.9	4.0	4.1	0.31	1.3	1.5	4.6	4.0	0.3
12 Wintersw.	'79/85	3.4	3.6	3.1	0.05	0.57	0.63	0.05	3.1	0.2
13 OudemaatA	'81/87	2.9	2.7	2.8	0.00	1.06	4.9	0.1	6.6	1.3
13 OudemaatB	'81/87	2.7	2.5	2.9	0.00	1.20	2.7	4.3	4.4	0.7
13 OudemaatC	'81/87	2.2	2.2	2.2	0.00	0.95	1.8	2.4	3.1	0.6
33 Buunderk.	'88/89	1.9	1.1	1.4	0.14	0.49	0.28	1.5	1.1	0.2
34 Leuvenum	'89/90	2.5	3.2	2.9	0.50	0.86	1.2	3.9	3.5	0.5
35 Hasselsven	'83/84	1.8	1.8	1.6	0.00	0.70	0.00	0.95	0.9	0.0
100 Solling	'69/85	5.2	6.0	1.1	0.01	1.1	0.89	5.9	0.9	0.4

#H^+ produced in N transformations (net NH_4 uptake plus net NO_3 drainage)

from soil Al compounds. Most of the N-saturated sites also show appreciable drainage of NH_4^+, showing that ground water under forest is polluted by a combination of nitrate and ammonium.

Input-output budgets, Aciforn forest sites

The soil solute fluxes at Speuld and Kootwijk were calculated from fortnightly soil water fluxes (Aaldrik and Bouten, 1990; exact periods 4 March 1987 - 2 March 1988, 2 March 1988 - 1 March 1989 and 1 March 1989 - 28 February 1990) and soil solution concentrations averaged over all (three) samples for a given depth. The concentrations of the soil water samples are assumed to be representative for the week preceding and the week following the first day of sampling with evacuated bottles. Soil solute fluxes were calculated per fortnight and summed per year. Monitoring in Kootwijk started in July 1987. Concentrations between March and July 1987 were estimated by extrapolation based on the dynamics of 1988. In dry periods without water samples concentrations were estimated by lineair interpolation of mean concentrations. To account for the strong spatial variability the yearly soil solute fluxes of all components at different depths were corrected with factors produced by an annual Cl-flux at each soil depth that was equal to the throughfall flux. The results of this correction yielded (1) generally constant Na-and SO_4-fluxes with depth, (2) differences in throughfall- and 10 cm-fluxes for Ca and Mg in the same order of magnitude as the Ca and Mg input by litterfall (instead of a factor 2 to 3 higher before correction), and (3) the observed differences in Ca-fluxes by litterfall at Speuld and Kootwijk were reflected. The resulting flux data are given in Table 3.4.2.2 (three-year means). NH_4-fluxes strongly decreased in the litter layer and the first 10 cm of the profile. In general atmospheric input of Ca, Mg and K are equal to the leaching at 90 cm. Only the leaching values of K at Speuld tend to be higher than the atmospheric input. Production of Ca, Mg and K through weathering is low in comparison to atmospheric input for poor soils as in Speuld and Kootwijk. As a result accumulation of these cations for plant uptake is very limited. High concentrations of Al and H (caused by high acidic atmopheric input) result in high saturation of the exchange complex with these components (Tiktak et al, 1988). This reduces the possibility for adsorption of Ca, Mg and K, and facilitates their leaching. Phosphorus is not leached from the ecosystems but its availability for plants may be reduced because of the formation of Al-P precipitates.

At Kootwijk fluxes of NO_3 appear to have been much higher in 1987 than in 1988 and 1989 (see Van der Maas, 1990) suggesting that immobilisation is high in relatively dry years (1988 and 1989) and mineralisation and nitrificaton is higher in wet years (1987). Another explanation for the high solute fluxes in 1987 at Speuld and Kootwijk could be the reduction in the needle N-pool from November 1986 to November 1987 as was observed by Evers et al (1990). These data suggest high litter fall amounts which could have been followed by high N-mineralisation in this period resulting in high solute fluxes.

Table 3.4.2.2 Three-year mean atmospheric deposition (AD), throughfall (T) and drainage fluxes of water (mm) and solutes at 10, 20, 40, 60 and 90 cm depth for Speuld and Kootwijk after Cl-calibration (kmol(+/-).ha^{-1}.y^{-1}; periods: 3 March 1987 - 1 March 1988, 1 March 1988 - 28 February 1989, and 28 February 1989 - 27 February 1990)

Speuld	H^+	K^+	Na^+	Ca^{2+}	Mg^{2+}	Al^{3+}	Fe^{2+}	Mn^{2+}	NH_4^+
AD	0.17	0.05	0.99	0.29	0.27	0.01	0.02	0.01	2.80
T	0.07	0.48	0.99	0.42	0.34	0.02	0.03	0.06	2.20
10	1.40	0.35	0.98	0.84	0.60	1.92	0.09	0.14	0.99
20	1.03	0.25	0.90	0.70	0.51	3.18	0.07	0.16	0.23
40	0.28	0.11	0.97	0.34	0.47	4.18	0.01	0.15	0.04
60	0.18	0.06	1.05	0.19	0.38	4.44	0.00	0.18	0.00
90	0.22	0.13	1.03	0.34	0.34	4.57	0.00	0.19	0.01
	Cl^-	NO_3^-	SO_4^{2-}	$H_2PO_4^-$	WA 1)	Si			
AD	1.38	0.79	2.26	0.01	0.16	0.01			
T	1.38	0.76	2.14	0.01	0.35	0.02			
10	1.38	3.25	2.51	0.01	0.21	0.92			
20	1.38	3.22	2.22	0.00	0.22	0.85			
40	1.38	2.69	2.22	0.00	0.18	0.85			
60	1.38	2.65	2.36	0.00	0.12	0.59			
90	1.38	2.22	3.11	0.00	0.10	0.76			
Kootwijk	H^+	K^+	Na^+	Ca^{2+}	Mg^{2+}	Al^{3+}	Fe^{2+}	Mn^{2+}	NH_4^+
AD	0.12	0.04	0.79	0.24	0.22	0.01	0.01	0.01	2.60
T	0.06	0.41	0.79	0.33	0.28	0.01	0.02	0.03	2.12
10	0.78	0.03	0.83	0.48	0.47	2.43	0.04	0.06	0.29
20	0.52	0.02	0.83	0.34	0.37	2.66	0.03	0.05	0.23
40	0.16	0.01	0.70	0.10	0.23	2.44	0.01	0.05	0.00
60	0.14	0.01	0.73	0.22	0.24	3.10	0.00	0.07	0.00
90	0.15	0.02	0.79	0.19	0.28	3.05	0.01	0.07	0.00
	Cl^-	NO_3^-	SO_4^{2-}	$H_2PO_4^-$	WA 1)	Si			
AD	1.09	0.75	2.10	0.00	0.08	0.01			
T	1.09	0.71	1.93	0.01	0.30	0.02			
10	1.09	2.52	1.65	0.01	0.14	1.01			
20	1.09	2.16	1.66	0.00	0.13	0.78			
40	1.09	1.21	1.32	0.00	0.08	0.50			
60	1.09	1.60	1.72	0.00	0.08	0.48			
90	1.09	1.67	1.69	0.00	0.09	0.50			

1): WA: Weak Acids (HCO_3^- and organic acids)

The ecosystem proton budgets for Speuld and Kootwijk, as calculated by van der Maas (1990), using the information from Table 3.4.2.2, clearly indicate that external, atmospheric N is the main source of soil acidification at both sites, with internal production and biotic uptake contributing only slightly. With the information from Table 3.4.2.2 it can also be calculated that of the total external proton input at Speuld, 18 % originates from NO_x, 49 % from SO_2 (after SO_4-aerosol correction: 0.12*Na), and 33 % from NH_3. The percentages for Kootwijk are 21% NO_x, 55% SO_2 and 24% NH_3.

At Speuld and Kootwijk respectively 83 and 79 % of the proton production is buffered by dissolution of Al. Soil acidification amounts 4.40 for Speuld and 3.74 $kmol.ha^{-1}.yr^{-1}$ for Kootwijk.

Table 3.4.2.3 Proton budgets for the ecosystems Speuld (SP) and Kootwijk (KW).Results are calculated with mean values from the period 1 march 1987 to 1 March 1990, in kmol(+/). $ha^{-1}.yr^{-1}$

	H sources					
	external		internal			
	H	N	N	Bio[1]	WA[2]	Ad[3]
SP	0.17	4.22	0.00	0.27		0.86
KW	0.12	3.51	0.00	0.21	0.01	

	H sinks internal				
	Cd[4]	Ald[5]	Ap[6]	WA	H
SP	0.70	4.56		0.07	0.21
KW	0.32	3.04	0.38		0.15

1) Bio Biological processes for all nutrients exclusive of N (uptake, leaching, mineralisation)
2) WA Weak Acids
3) Ad Anion dissolution
4) Cd Cation dissolution (exclusive of Al)
5) Ald Al-dissolution
6) Ap Anion precipitation

3.4.3 Modelling of soil acidification

Sensitivity analysis

To identify model input that should be regarded with extra atten-tion and as a first step towards model calibration a sensitivity analysis has been done, using a Monte Carlo approach in combination with multiple linear regression. In the sensitivity analysis 51 model inputs were varied simultaneously around their nominal values. A total of 100 model runs were carried out, each time with a different input vector of 51 model inputs. The 100 input-vectors were obtained by sampling from the normal probability distribution of the parameters between extremes of +/- 5% of the nominal value using a Latin-Hypercube sampling technique (Gardner et al, 1983). Multiple linear regression is used to quantify the relation between model input and output from the 100 model runs.

The sensitivity analysis of WATERSTOF for the Oudemaat site (location 13) showed that the solution concentrations of NO_3, Ca and Al are most sensitive to (1) the nitrification rate constant, (2) the biocycle (size and distribution in depth), (3) N-mineralization (soil N-pool), and (4) initial size of the adsorbed pool.

Although the N biocycle is most probably not in a steady state, such a steady state has been assumed for lack of a better alternative. However, [NO3] and [Al]-concentrations are highly sensitive to perturbations from this steady state, and turning or "tuning" this model-feature can probably produce any [NO_3] or [Al]-concentration level wanted. It was decided to stay with the steady state concept. After calibration and simulation on the 4 forest-ecosystems the null-hypothesis: "was a steady state biocycle justified or not", probably can be answered.

Model calibration

The hydrology module in WATERSTOF was calibrated separately by hand. For the Gerritsfles and Tongbersven site it was calibrated on monthly water fluxes calculated with a Darcy model (SWATRE). The water balances for Solling and Oudemaat were calibrated on measured Cl-time series at various depths by running the model with chloride only. For the calibration of the chemistry the Monte Carlo technique was used. A number of 9 (Gerritsfles) , 8 (Oudemaat) respectively 3 (Solling) parameters were calibrated simultaneously in a large number of model runs (Monte Carlo). Each calibration run had a unique input vector of calibration parameters. Parameters were sampled from uniform distributions. Minimum and maximum values of the parameter distributions were obtained from:

- hand-calibration in the case of the nitrification rate
- observed root patterns (from profile descriptions or root sampling) in the case of the root uptake distribution in depth
- profiles of organic matter in the case of the mineralization distribution in depth
- measured minimum and maximum cation concentrations in the case of the exchange coefficients

Data from selected months over the whole monitoring period were used in the calibration procedure. Those data were not used for later validation. A first criterion for the calibration was a minimum of the sum of squared differences of average measured and calculated monthly soil solution concentrations. A second criterion used is based on the so called regional sensitivity analysis (RSA) developed by Hornberger and Spear (1980) and Cosby et al (1985).

Results

The results of the validation of WATERSTOF on the four different forest ecosystems, all four characterized by high acid atmospheric inputs and acid soils but differing in soil type, hydrology, mineralogy and forest stand, can be summarized as follows.

The simple empirical water balance in WATERSTOF allowed proper simulation of hydrology in Gerritfles, Oudemaat and Solling soils. This was shown by a good agreement of simulated and measured Cl-time series (Figure 3.4.3.1) justifying the use of Cl- as an 'internal standard' for element budget studies (cf. paragraph 2.1). Simulation of [Cl] in the Tongbersven soil did not match observations in summer periods. Water samples from dry summer periods are probably not representative for the calculated "average" hydrological status of the soil.
Identifiability of model parameters from observation data is an important criterion for model validations. Identifiability was tested by calibrating a number of parameters from the (bio)chemical module using a Monte Carlo technique. Uncertainty around parameter ranges of the Geritsfles site could be reduced substantially in the calibration procedure. Also simulations were in good agreement with observations. Thus, it can be concluded that the WATERSTOF model is valid for the Gerritfles location.
Efforts to calibrate parameters for the Oudemaat site were less successful. A global optimum for the most sensitive parameter Knitrif could be calibrated, but calibrated optima for uptake and mineralization distribution with depth were different for different element-layer observation sets. Gaines-Thomas exchange coefficients were poorly identifiable. Intensive calibrations on the Oudemaat site showed that present (simple) model description of N-dynamics allow simulation of average [NO_3] patterns, but is too crude to simulated incidental seasonal fluctuations of [NO_3].
Simulations for the Solling site are still in a preliminary phase, calibration of optimum parameter ranges was not carried out yet. Simulations of litter layer chemistry are in excellent agreement with observations. Monte Carlo carried out with 3 parameters (nitrification rate constant, uptake and mineralization distribution in depth) show that simulated concentrations in the litter layer are mainly throughfall determined. Annual trends and seasonal variations of simulated concentration patterns of [Al], [NO_3] and [SO_4] in the suboil (0-40 cm) agree well with observation (Figure 3.4.3.2, -4). Considering that simulations start in 1975 and validation on solution concentrations at 10, 20 and 40 cm depths starts in 1981. Calibration with more parameters and identification of optimum parameters ranges still have to be carried out in order to gain confidence in model predictions. Model predictions in the 40-100 cm suboil are rather poor for [SO4], [NO3] and [Al].
As with the to the Oudemaat simulations it is concluded that description of N-dynamics in the present model is too crude to simulate NO3 dynamics in the 40-100 cm layer realistically.

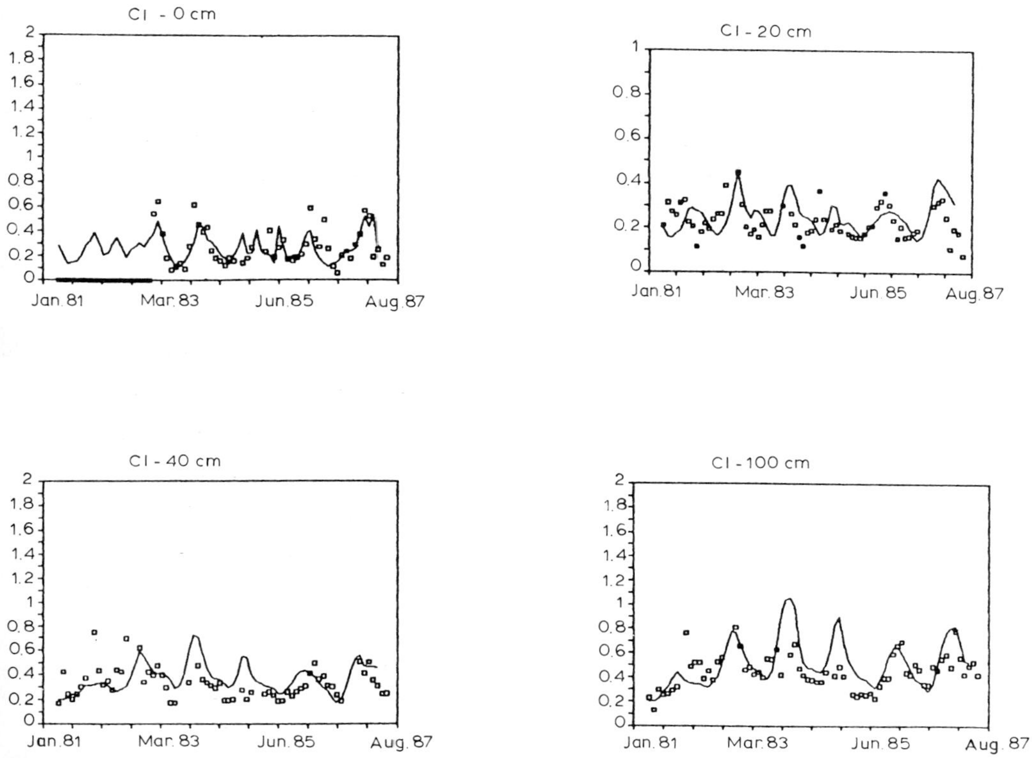

Fig. 3.4.3.1 Average measured (square) and simulated (line) Cl concentrations for Oudemaat, April 1981 to 1987 (location 13)

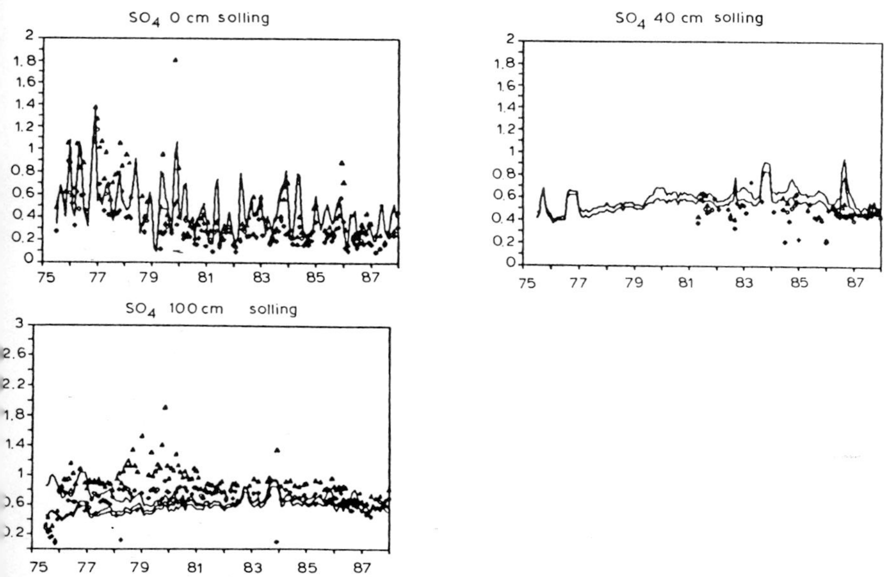

Fig. 3.4.3.2 Measured minimum and maximum total SO_4 concentration (resp. diamonds and triangles) and "uncertainty" ranges calculated with WATERSTOF, for Solling between 1973 and 1988 (location 100)

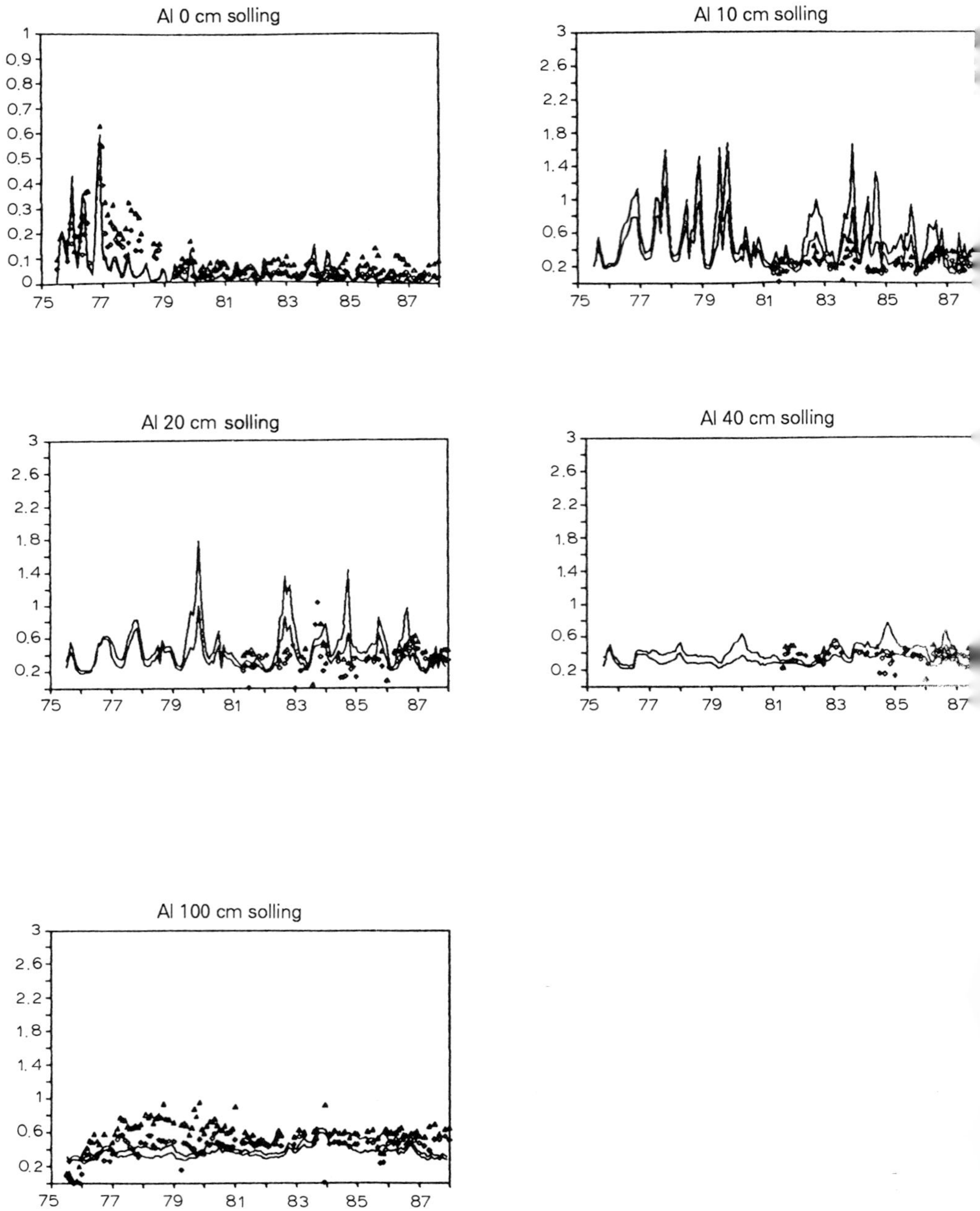

Fig. 3.4.3.3 Measured minimum and maximum total [Al] ([Al^{3+}] + [$AlSO_4^{+}$] + [$AlOH^{2+}$]) concentrations (respectively diamonds and triangles) and "uncertainty" ranges calculated with WATERSTOF, for Solling between 1973 and 1988 (location 100)

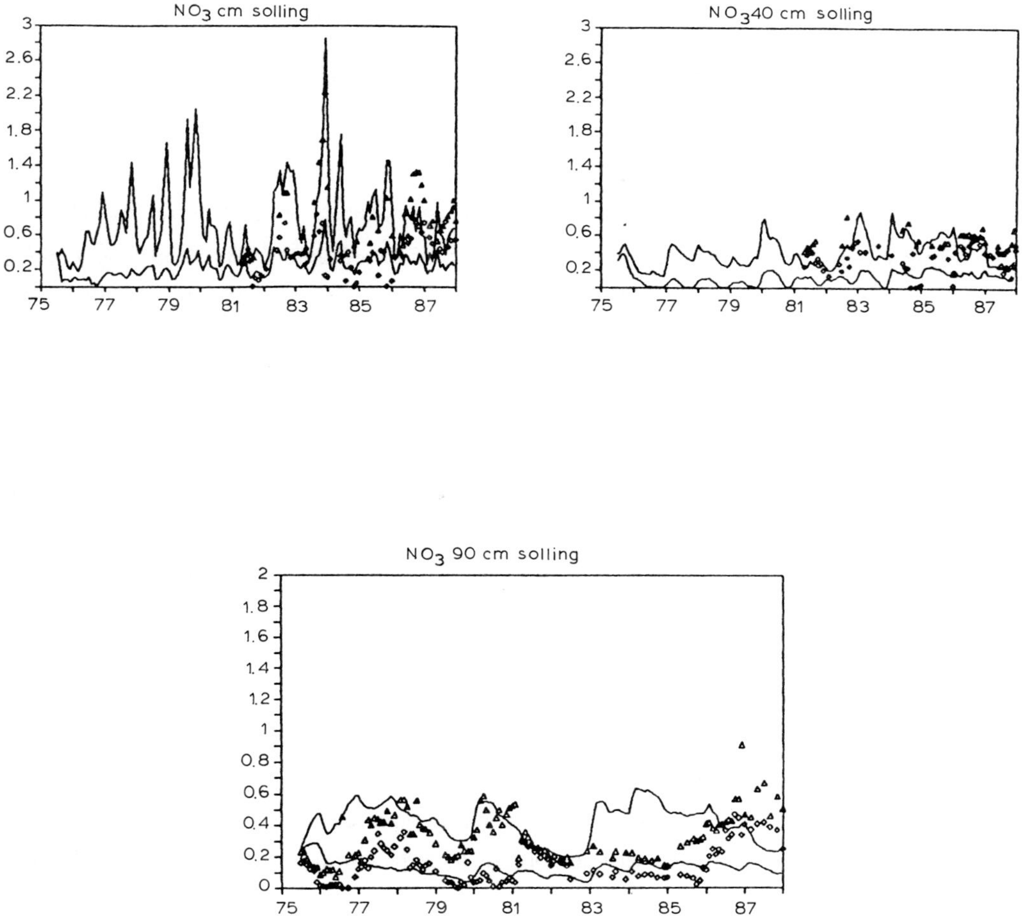

Fig. 3.4.3.4 Measured minimum and maximum NO_3 concentration (resp. diamonds and triangles) and "uncertainty" ranges calculated with WATERSTOF, for Solling between 1973 and 1988 (location 100)

3.4.4 Aluminium

From the foregoing it is clear that mobilization of aluminium is one of the major acid neutralization processes in sandy forest soils in the Netherlands impacted by acid deposition. In view of that, a few aspects of the chemistry of Al deserve special attention, viz. (1) the nature of dissolved Al, (2) the nature of the soil-Al fraction serving as a H^+ sink and as a source for dissolved Al, and (3) the future capacity of soils to buffer acidity by dissolution of Al. The data presented here are derived mainly from research on sites 8, 10, 13, and 35 by Mulder (1988), Van Grinsven (1988), and Mulder et al, 1989, 1990. In the mineral soil horizons of all acidic forest soils in the Netherlands, dissolved Al in mainly (> 95 %) monomeric, inorganic form. Polymeric dissolved Al is quantitatively insignificant, while organically complexed Al becomes dominant only in surface organic layers, and just below such layers in sites with relatively low atmospheric deposition,, i.e. in heathlands.

While most soil acidification models describe moblization of Al from mineral soil as dissolution of a gibbsite-like phase, a number of observations suggested control of dissolved Al by interaction of H^+ with Al-organics, rather than by solid $Al(OH)_3$: (1) equilibration between Al^{3+} and H^+ upon addition of strong acid is very rapid (e.g. within a few hours), (2) repeated additions of strong acid cause removal of pyrophosphate extractable (= presumably organically bound) Al, and (3) the saturation index for gibbsite (calculated from the $Al(OH)_3$ activity product in the soil solutions from surface horizons sampled in the field) gradually decreased over a six year period.

Because the (pyrophosphate extractable) fraction of soil Al that is effective in buffering H^+ constitutes only a small part of soil Al, the buffering capacity in the Al buffer range is considerably smaller than indicated by total soil Al. Indeed, there is strong evidence that the Al pool responsible for H^+ buffering is being depleted rapidly in the surface layers, and that lower pH values as well as lower levels of dissolved Al can be expected in forest surface soils in the coming years to decades.

3.4.5 Temporal variations

In this section we will briefly summarize the most important findings relating to seasonal and long-term temporal variations in soil chemical properties.

Solute concentrations usually vary seasonally. In surface soils, most solutes tend to become more concentrated in summer as a result of evaporative concentration and decreased leaching, while the soil solution is generally more dilute in winter and early spring as a result of dilution with throughfall water. The resulting patterns, both measured and simulated, can be be seen e.g. in Figs 3.4.3.1-.4. For nutrients, lowest concentrations may occur in summer, as is shown for nitrate in Oudemaat, site 13C (Figure 3.4.5.1.). Apparently, biotic uptake in summer exceeds the effect of evaporative concentration at this site. By contrast, at the nearby site 13A, where the trees grew very poorly, evaporative concentration exceeded uptake. For a proper interpretation of the causes behind these seasonal patterns, either detailed budgetting on a time scale of months, or modelling on an appropriate time scale is needed. An example of such detailed budgetting, taking into account evaporative concentration as well as seasonality of nutrient uptake, carried out for site 13A, is shown in Figure 3.4.5.2. The shaded areas in Figure 3.4.5.2. show the time-depth regions where solutes appear into the solution from the solid soil phases (for all solutes except NO3, where shaded areas refer to the a positive net value of mineralization + nitrification, and uptake).

The results show that dissolution (and, of course leaching) of base cations and aluminium takes place mainly in winter, and not in summer, when concentrations are generally higher. The reason for seasonal leaching obviously lies in the hydrological regime, involving increased drainage by decreased evapotranspiration in the fall. The coincidence of dissolution and increased drainage is not so obvious, however. Nitrate, the main anion accompanying Al and base cations, cannot, like HCO_3 or SO_4, be mobilized simply by dilution. Only increased nitrification, decreased nitrate uptake, or any combination thereof, could explain the increased net production of nitrate during the fall. Although conditions for nitrification itself probably do not improve when soil temperatures fall from 12-14°C around July to 5-8 °C in October (van Breemen et al,1988), increased supply of ammonia from decaying organic matter may increase over that period as a result of root turnover, and may enhance nitrification. Still more important may be strongly decreased consumption of NO_3^- by plants and soil microbes during the fall.

The monitoring periods available (maximally 8 years) are too short for definite trend analyses. Nevertheless, some preliminary conclusions can be drawn from the work by Stein and van Breemen (1990). To account for any long-term trend underlying the month-to-month variation associated with seasonal patterns in temperature, soil moisture, and related variations in biotic activity, at locations 8, 10, 13 and 35 (four or six years of soil solution monitoring) two approaches were used. First, soil solution pH and concentrations in Al, NO_3 and Cl were described as a functon of time, temperature in the preceding months and/or soil moisture. In a second analysis, values for similar months were compared directly using a rank correlation test. Generally speaking, the nitrate concentrations at most depths showed a significant long-term increase, except at the N-saturated site (13A). At most sites, pH decreased significantly in the surface soils, while Al concentrations tended to increase. Cl concentrations rarely showed a significant long-term trend. The results suggest that the already highly acidic soils at these sites undergo further strong soil acidification, and tend to approach nitrogen saturation. However, these time series may be too short for broad generalisations. For instances, the long time series at Solling shows that nitrate concentrations in the subsoil may show fluctuations with a period of several years. Furthermore, Wesselink (1990) could model the increase in nitrate concentrations over a four year period at location 10 assuming a steady state N cycling, and concluded that the observed trend can be explained by high inputs in the period 1984-87 compared to 1983-84, and lower annual precipitation in the first half of the monitoring period.

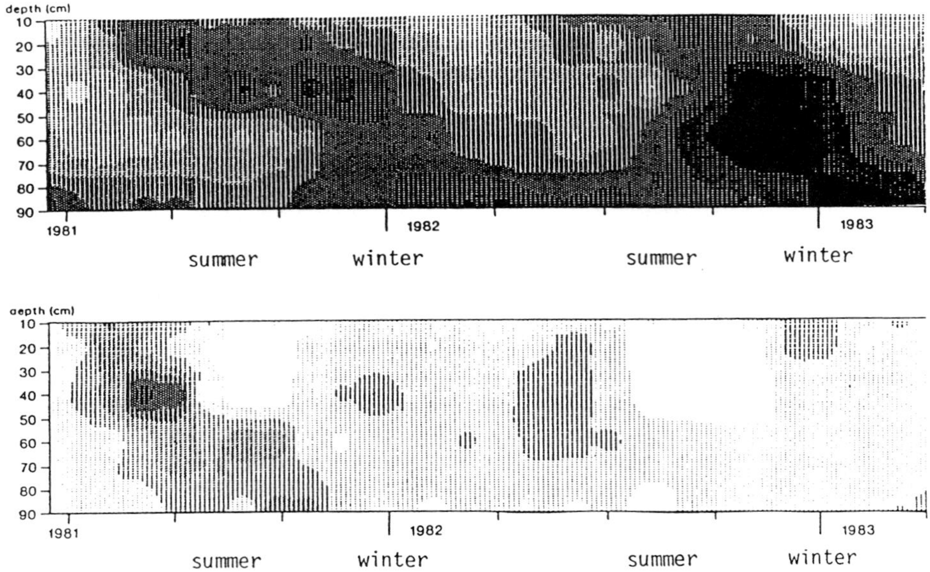

Fig. 3.4.5.1 NO_3^- concentrations vs time and depth at Oudemaat; poorly growing oak (site 13A); well growing oak (site 13C)

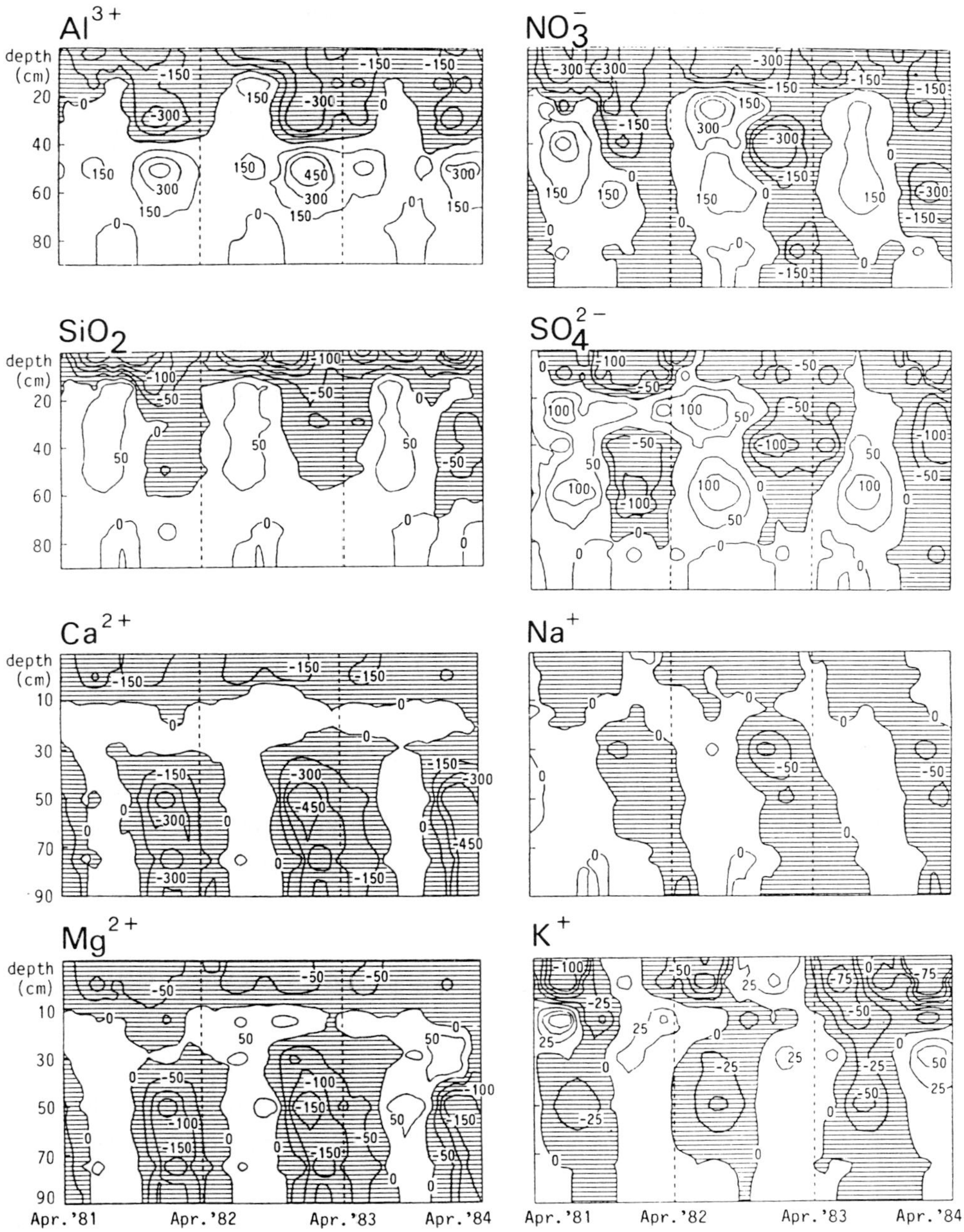

Fig. 3.4.5.2 Solute concentrations vs time and depth (site 13A); for explanation see text

4. SUMMARY AND CONCLUSIONS, SUGGESTIONS FOR FURTHER RESEARCH

More or less detailed soil hydrochemical monitoring at some twenty forest and heathland sites, sampling for microbiogical research at about forty heathlands, and simulation modelling of soil hydrochemistry at a selected number of sites has provided a comprehensive picture of soil acidification and nitrogen cycling under the influence of atmospheric deposition in the Netherlands.

Practically all forest soils are now strongly acidic, with soil solutions dominated by Al, NO_3 and SO_4. Rates of natural soil acidification are almost negligible compared to rates of acidification by oxidation of ammonium from atmospheric deposition. Chemolithotrophic acidophilic nitrate producing bacteria are widespread in forest and heathland soils, and appear to be mainly responsible for the rapid transformation of ammonium to nitric acid. As a result, nitrate drainage outputs are generally high (i.e. > 20 % of the N input) in forest soils, and nitrate pollution of groundwater under forest is the rule, rather than the exception. Undisturbed heathlands are still able to assimilate most of the N input, but may start leaking nitrogen after disturbance. Assuming steady state nitrogen cycling (i.e. assuming constant pools of mineralizable soil N) most features of the seasonal soil chemical changes can be simulated reasonably well by a relatively simple soil acidification model. Yet there are strong indications that in 5 out of 18 dutch sites studied here, NO_3-N drainage outputs are equal to or even exceed the atmospheric inputs, implying that the pool of soil N has begun to contribute to further soil acidification. Trends in soil solution chemistry over a number of years for other stands indicate that nitrate levels are on the increase in stands that are now still able to retain part of the incoming atmospheric N. In a number of locations, particularly in areas with very high N inputs, dissolved ammonium dominates over nitrate. In such locations drainage of ammonium to the ground water is a relevant problem, which may have been underrepresented in this programme.

Organically bound soil Al in Dutch forest and heathland soils is the main buffer against atmospheric acidity at present. However, the sizes of that reactive Al pool are generally small and they are rapidly being depleted. As a result, the soil chemistry will change to still lower pH levels (as wel as lower levels of dissolved Al and presumably higher levels of dissolved Fe) in the near future unless present inputs are cut back dramatically in the coming years.

Several aspects of soil chemistry are poorly understood and limit our ability to forecast the future changes in soil chemistry as a function of future deposition/management scenarios. A big black box is decomposition of soil organic matter and the related N transformations as a function of environmental variables, including N deposition.

To test the results of scenario modelling and to evaluate the effects of future changes in emissions, it is essential to establish a limited number of long-term hydrochemical soil monitoring plots in the country. In future acidification research, however, monitoring should be deemphasized, and replaced by manipulation studies aimed at validating simulation modelling of relevant deposition/management scenarios. The weakest part of the soil acidification research is probably the link to the vegetation. Compared with our understanding of the soil chemical processes, our understanding of the effects of acidification on biota lags far behind.

5. REFERENCES

Adams, J.A., 1986. Identification of heterotrophic nitrification in strongly acid larch humus. Soil Biol. Biochem. 18, 339-341

Bizot, N., Tiktak, A. and W. Bouten, 1990. Soil water balance research of a heathland ecosystem. A soil physical measuring programme in lysimeters on the Asselse heide, the Netherlands. Dutch Priority Programme on Acidification 102.2.02, RIVM, Bilthoven, the Netherlands

Bobbink, R., Heil, G.W., Gaessen, M.B., van Dam, D. and B. Heijne, 1990. Troughfall, canopy exchange and nutrient stress in heathland ecosystems. Report of project 119 of the Dutch Priority programme on acidification, Dept. Plant Ecol. Evol, Biology, University of Utrecht, 8 p.

Burt, T.P., 1978. An automatic fluid-scanning switch tensiometer system. Techn. Bull. (21). Brit. Geom. Res. Group, pp. 29

Cosby, B.J., Hornberger, G.M. and J.N. Galloway, 1985. Modelling the effects of acid deposition: Assesment of a lumped parameter model of soil water and stream water chemistry. Water Resources Research 21, 51-63

Cronan, C.S., Walter, W.J. and P.R. Bloom, 1986. Predicting aqueous aluminium concentrations in natural waters. Nature 324, 140-143

De Bakker, H. and J.Schelling, 1966. Systeem van de bodemclassificatie voor Nederland. Soil Survey Inst., Wageningen, the Netherlands, 217 pp.

De Boer, W., Duits, H. and H.J. Laanbroek, 1989b, Urea stimulated autotrophic nitrification in suspension of fertilized, acid heath soil. Soil Biol. Biochem. 21, 349-354

De Boer, W. and Laanbroek, H.J., 1989, Ureolytic at low pH by Nitrosospira spec. Arch. Microbiol. 152, 178-181

De Boer, W., Klein Gunnewiek, P.J.A., Troelstra, S.R. and Laanbroek, H.J., 1989a, Two types of chemolithotrophic nitrification in acid heathland soils. Plant and Soil. 119, 229-235

De Boer, W., and A.Tietema, 1990. Nitrification in Dutch acid forest- and heathland soils: mechanisms and effects of biotic and abiotic factors. Report of project 106.1., Dutch Priority Programme on Acidification,17 pp.

De Visser, P.H.B., and N. van Breemen, 1990. Manipulation of the water and nutrient supply in two forest ecosystems in the Netherlands. Report of project 100, Dutch Priority Programme on Acidification, Dept. of Soil Sci. and Geology, Agric University, Wageningen, May 1990, 28 pp. + Apps.

De Vries, W, and J. Kros, 1990. Assessment of critical loads and the impact of deposition scenarios by steady state and dynamic soil acidification models. Winand Staring Centre for Integrated Land, Soil and Water Research, Wageningen, the Netherlands, 45 pp.

Diederen, H.S.M.A. en H.L.M. Verhagen, 1990. Vergelijkingsmetingen van analyses en bemonsteringsmethoden van de vaste en vloeibare fase van de bodem. MT TNO Rapport (in prep)

Driscoll, C.T., Van Breemen, N. and J. Mulder, 1985, Aluminium chemistry in a forested spodosol. Soil Sci, Soc. Am. J. 49, 437-444

Duysings, J.J.H.M., Verstraten J.M., Bouten W, Tietema A, and M. Oostrom, 1990. Hydrochemical budget of a deciduous forest ecosystem in the Netherlands. Journal of Ecology (submitted)

Evers P. (ed.), 1990. Impact of air pollution on ecophysiological relations in 2 Douglas fur stands in the Netherlands. Dutch Priority Programme on Acidifications, projects 15, 105a and 105b, 250 pp.

Gardner, R.H., Röjde, C. and U. Bergström, 1983. PRISM: A systematic method for determing the effect of parameter uncertainties on model prediciton. Studsvik. Energieteknik ab, report NW - 83/555 Nyköping

Gardner, W.R. and D. Kirkham, 1952. Determination of soil moisture by neutron scattering. Soil Science 78, 340 - 391

Hankinson, T.R., and Schmidt, E.L., 1988, An acidophilic and neutrophilic Nitrobacter strain isolated from numerically predominant nitrite-oxidizing population of an acid forest soil. Appl. Environm. Microbiol. 54, 1536-1540

Heimovaara, T.J. and W. Bouten, 1990. A computer controlled 36 channel TDR system for monitoring soil water contents. Water Resour. Res. (Accepted)

Hilhorst, M.A., 1984. A sensor for the determination of the complex permittivity of materials as a measure for the moisture content. In: P. Bergveld (Ed.). Sensors & actuators. pp. 79-84, Kluwer Technical Books, Deventer, 1984

Hornberger and Spear, 1981. An approach to preliminary analysis of environmental systems. J. Environmental Management 12: 7-18

Houdijk, A.L.F.M., 1990. Effecten van zwavel- en stikstofdepositie op bos- en heidevegetaties. Rapport in opdracht van Directie Lucht, Ministerie van VROM, Vakgr. Aquatische Oecologie en Biogeologie, Katholieke Universiteit Nijmegen, 124 pp.

Hutchinsson, G.L. and A.R. Mosies, 1981, Improved soil cover method for field measurement of nitrious oxide fluxes. Soil Sc. Soc. Am. J. 45, 311-316

Hynes, R.K., and Knowles, R. 1982, Effect of acetylene on autrotrophic and heterotrophic nitrification. Can.J.Microbiol. 28, 334-340

Killham, K., 1987, A new perfusion system for the measurement and characterization of potential rates of soil nitrification. Plant and Soil, 97, 267-272

Kleijn,C.E., G.Zuidema and W.de Vries, 1989. De indirecte effecten van atmosferische depositie op de vitaliteit van Nederlandse bossen. 2. Depositie, bodemeigenschappen en bodemvochtsamenstelling van acht Douglasopstanden. Rept. 2050, Soil Survey Inst, Wageningen, the Netherlands, 96 pp.

Lache, D.W.,1976, Umweltbedingungen von Binnendünen und Heidegesellschaften im Nordwest Mitteleuropas. Scripta Geobotanica, 11, 96 pages

Maas, van der, M.P., 1990. Hydrochemistry of two Douglas fir ecosystems in the Veluwe, the Netherlands. Dutch priority Programme on Acidification 102.1.01, R.I.V.M., Bilthoven

Matzner, E., 1989. Acid precipitation: Case study in Solling, p39-83 in D.C. Adriano and M. Havas (Eds.). Acidic Precipitation, Vol.1. Adv. Envir. Sci., Springer Verlag, N.Y.

Meiwes K.J., Köning, N., Khamna, P.K., Prenzel J. and B. Ulrich, 1984. Chemische Untersuchingsverfahren für Mineralböden, auflage Humus und Würzeln zur Charakteriserung und Bewertung der Versaurung in Waldböden. Berichte des Forschungszentrum Waldökosysteme- Waldsterben, Universität Göttingen, Band 7, 109-121

Mulder, J. 1988. Impact of acid atmospheric deposition on soils: Field monitoring and aluminium chemistry. PhD thesis, Agric. Univ. Wageningen, the Netherlands, 163 pp.

Mulder, J., Van Breemen, N. and H.C.Eijck, 1989. Depletion of soil aluminium by acid deposition and implications for acid neutralization. Nature, 337: 247-249

Mulder, J.,Pijpers, M. and N. van Breemen, 1990. Aluminium solubility control by soil organics in mineral soils. Abstract, Int. Acid Precipitation Symposium, Glasgow, Sept. 1990

Nilsson, J. and P. Grennfelt (eds.), 1988. Critical loads for sulphur and nitrogen. Report workshop Skokloster, Sweden, 19-24 maart, 1998, 418 pp.

Oliver, B.G., Thurman, E.M. and R.L. Malcolm, 1983. The contribution of humic substances to the acidity of colored natural waters. Geochim. Cosmochim. Acta, 47: 2031-2035

Olsthoorn, A.F.M., 1990. Root research on Douglas-fir in the Aciforn project: Influence of soil acidification on fine root growth. Dutch Priority Programme on Acidification 103.01., R.I.V.M., Bilthoven, the Netherlands

Reurslag, A., Zuidema, G. and W. de Vries, 1990. De indirecte effecten van atmosferische depositie op de vitaliteit van Nederlandse bossen: 3. Simulatie van de waterbalans van acht Douglasopstanden, Staringcentrum Rapport 76, Wageningen

Roelofs, J.G.M., Boxman, A.W., Van Dijk, H.F.G., and A.L.F.M. Houdijk, 1988. Nutrient fluxes in canopies and roots of coniferous trees as affected by nitrogen-enriched air-pollution. In: Bervaes, J., P. Mathy and P. Evers (Eds.). Relationships between above and below ground influences of air pollutants on forest trees, C.E.C. Air Pollution Research Report 16, p 205-221

Stams, A.J.M., Booltink, H.W.G., Luthe-Schipholt, I.J., Beemsterboer, B., Woittiez, J.R.W. and N. van Breemen, 1988, Field study on the fate of atmospheric ammonium in acid forest soils with 15N. In: Rol van microbiële processen in de stikstofhuishouding van bosgronden. Project van het verzuringsonderzoek, RIVM, Bilthoven, 72 pp.

Stein, A., and N. van Breemen, 1990. Long-term (4-6 yrs) trends in dissolved nitrate in Dutch forest soils. Abstract Int. Acid Precipitation Symposium, Glasgow, Sept. 1990

Tietema A. and J.M. Verstraten, 1988. The Nitrogen budget of an oak-beech forest ecosystem in the Netherlands in relation to atmospheric deposition. Dutch Priority Programme on Acidification, report no. 04-01, 55 pp.

Tietema, A., Duysings, J.J.H.M, Verstraten, J.M. and J.W. Westerveld, 1990a. Estimation of actual nitrification rates in an acid forest soil. In Harrison, A.F., Imeson, P. and O.W. Heal (Eds), Nutrient cycling in terrestrial ecosystems: Field methods, applications and interpretation, 190-197. Elsevier Applied Science, London

Tietema, A., Riemer L. and J.M. Verstraten, 1990b. Nitrification in Dutch acid forest and heathland soils: II. Effects of biotic and abiotic factors. In De Boer W. and A. Tietema. Report of project 106.1, Dutch Prority Programme on Acidification, 11-16

Tietema, A., Bouten W. and P.E. Wartenbergh, 1990c. Nitrous oxide dynamic in an acid forest soil in the Netherlands. Forest Ecology and Management (Submitted)

Tietema, A., Verstraten J.M. and A.J. van Wijk, 1990d. The nitrogen cycle of an oak-beech forest ecosystem in the Netherlands at increased nitrogen deposition: 1. Biochemical nitrogen transformations and solute fluxes. Biogeochemistry (Submitted)

Tietema, A. and J.M. Verstraten, 1990. The nitrogen cycle of an oak-beech *forest ecosystem in the Netherland at increased nitrogen deposition: 2. The nitrogen and proton budget. Biogeochemistry (Submitted)

Tiktak, A. and W. Bouten, 1990. Soil hydrological system characterization of the two Aciforn stands using monitoring data and the soil hydrological model "SWIF". Dutch Priority Programme on Acidification 102.2.01. R.I.V.M., Bilthoven, the Netherlands, 62 pp.

Tiktak, A., W. Bouten and M. Schaap, 1990. SWIF: A simulation model of Soil Water In Forested ecosystems. Report nr. 000, Laboratory of Physical Geography and Soil Science, University of Amsterdam, 36 pp.

Tiktak, A., C.J.M. Konsten, M.P. van der Maas and W. Bouten,1988. Soil chemistry and physics of two Douglas-fir stands affected by acid atmospheric deposition on the Veluwe, the Netherlands. Dutch Priority Programme on Acidification, rep. nr 03-01, National Institute of Public Health and Environmental Hygiene (R.I.V.M.), Bilthoven, 93 pp.

Tiktak, A., Konsten, C.J.M., Bouten, W. and M.P. van der Maas, 1988. Soil chemistry and physics of two Douglas-fir stands affected by acid atmospheric deposition on the Veluwe, the Netherlands. Dutch Priority Programme on Acidification 03.01. R.I.V.M., Bilthoven, the Netherlands, 42 pp.

Tipping, E. and M.A. Hurley, 1988. Model of solid-solution interactions in acid organic soils based on ten complexation properties, J. Soil science 39, 509-519

Topp, G.C., J.L. Davis and A.P. Annan, 1980. Electromagnetic determination of soil water content: Measurement of coaxial transmission lines. Water Resour. Res. (16):574-582

Ulrich, B., 1983. Soil acidity and its relations to acid deposition. In: Ulrich, B. and J Pankrath (eds), Effects of accumulation of air pollutants in forest ecosystems, D. Reidel Publishing Company, Dordrecht, Boston, and London, p. 127-146

Van Aalst, R.M., and J.W. Erisman, 1990. Thematic report on Atmospheric input, Dutch Priority Programme on Acidification, RIVM, Bilthoven, the Netherlands, May 1990, 35 pp.

Van Breemen, N., Boderie, P.A.M. and H.W.G. Booltink, 1989. Influence of airborne ammonium sulphate on soils of an oak woodland ecosystem in the Netherlands, Seasonal dynamics of solute fluxes. p 209-236 in: D.C. Adriano and M. Havas (Eds.) Acidic Precipitation, Vol.1. Adv. Envir. Sci., Springer Verlag, N.Y.

Van Breemen, N., Visser, W.F.J. and Th. Pape (Eds.), 1988. Biogeochemistry of an oak-woodland ecosystem in the Netherlands affected by acid atmospheric deposition. Pudoc, Wageningen, the Netherlands, 197 p.

Van Breemen, N. and H.F.G. Van Dijk, 1988. Ecosystems effects of atmospheric deposition of nitrogen in the Netherlands. Environmental Pollution, 54, 249-274

Van der Maas, M.P., 1990. Hydrochemistry of two Douglas Fir ecosystems and a heather ecosystem in the Veluwe, the Netherlands. Preliminary Report 102.1-01, Dutch Priority Programme on Acidification, Wageningen, May 1990, 28 pp + Apps.

Van Dobben, H., J. Mulder, H. van Dam and H. Houweeling, 1988. The impact of acid atmospheric deposition on the biogeochemistry of moorland pools and surrounding terrestrial environment. draft, 149 p.

Van Grinsven, J.J.M.,1988. Impact of acid atmospheric deposition on soils: Quantification of chemical and hydrological processes. PhD thesis, Agric. Univ. Wageningen, the Netherlands, 213 pp.

Verstraten, J.M., Dopheide, J.C.R., Duysings, J.J.H.M., Tietema, A. and W. Bouten, 1990. The proton cycle of a deciduous forest ecosystem in the Netherlands and its implications for soil acidification. Plant and Soil (Accepted)

Vitousek, R.M., 1981. Clear-cutting and the nitrogen cycle. In clerk F.E. and T. Rosswall (Eds.) Terrestrial nitrogen cycles. Ecol. Bull. 33, 631-642

Wesselink, L.G., 1990. Modelling of soil acidification in Dutch and West-German forest ecosystems. Report of project 202.1 of the Dutch Priority programme on acidification, Forschungs Waldökosysteme, Göttingen, West Germany and Dept. of Soil Sci. and Geology, Agric. University, Wageningen, the Netherlands

Appendix 1.

QUALITY CONTROL OF CHEMICAL MEASUREMENTS IN SOILS AND SOIL SOLUTIONS

This appendix is a summary, partly based on Diederen (1990), of the results of three intercomparison experiments dealing with analysis of soil samples, and sampling and analysis of soil solutions, that were held in 1989 and early 1990. Results of comparisons dealing with artificially made up aqueous solutions (at concentrations typical for rain and throughall water) and with actual rain and throughfall water, are described by Van Aalst and Erisman (1990).

In the first experiment the analyses of soil samples and in the second one of samples of soil moisture are compared. The third experiment is set up to investigate the comparability of three different methods for sampling of soil moisture. The reproducibility (ISO 5725) of the analyses of the different components of soil samples of eight laboratories was as follows (exchangeable cations were determined with various extractants; $CaCl_2$, KCl, $BaCl_2$ and Ag Thioureum):

pH-KCl	5%
pH-H_2O pH-$CaCl_2$	10%
Org. C, total P	40%
Al exch	80%
total N, CEC, H exch, NH_4 exch	100%

The reproducibility (ISO 5725) of the analyses of the different components in cup-sampled soil solution by eight laboratories is given in the following table:

pH, NO_3, Mn	10%
Ca, Mg, Al, SO_4	20%
Cl, NH_4, Fe	40%
Na	50%
K	60%
H_2PO_4	80%

Because statistical outliers were removed they did not affect the results given above. In reality outliers can affect the comparability of the results in a negative way.

The reproducibility of the analyses of soil samples of Al, total N, CEC, H and NH_4 is unsatisfacory. With respect to the analyses of soil solution the reproducibility of the analyses of Cl, ortho P, Na, NH_4 and K is unsatisfactory. The differences in the results originating from different laboratories can obstruct generalisation.

In a third experiment three different sampling methods of soil solution were compared: the suction cup method used in most monitoring programmes, the centrifuge method employed by the Staring center (e.g. Kleijn et al, 1989), and the 1 hour 1:3 aqueous extraction used by the University of Nijmegen (e.g. Houdijk, 1990) for correlative field studies. The three methods were compared in the soils of two forest stands: at Speuld (location 3A in Table 2.1), and at Ysselstein. The three methods were employed at horizontal distance of 10 cm, at fifteen spots that were situated at least 2 m apart. Each set of samples was analysed only at the institute involved in the particular sampling technique. At the same time, however, three (natural) soil solution samples were analysed at all three laboratories. The analytical results of these three samples were very similar, with standard deviations (from the mean) generally below 10 % for the major solutes (Table A.1.1). So, any appreciable differences between sample types can be attributed mainly to the sampling technique. The results for centrifuged and suction cup samples are reported as analysed. Those for the water extract were recalculated as molar concentrations in the soil moisture present before extraction. Normally the latter procedure is not done, and water extracts analysis are expressed as a moles per mass of dry soil, and used to calculate ratios of various solutes. In view of the importance of ratios such as NH_4/NO_3 (to estimate the degree of nitrification of NH_4), NH_4/K (as a measure to the K availability to plants in high N environment), and Al/Ca (as a measure for Al toxicity to plants), special emphasis will be given to these parameters.

Tables A.1.2 to A.1.4 show the results for resp. NH_4 and NO_3, NH_4 and K, and Al and Ca. In general, concentrations of all solutes tend to be similar or higher (up to a factor of two to three) in centrifuged solutions in comparison with the cup solutions. This difference can be explained by heterogeneity of the soil solution at a scale of mm to cm. The soil solution will be more concentrated in the finer pores, where contact times between the water and the solid phases are longer due to slower water movement and, hence, longer residence time of the water compared to water in coarser pores. Centrifugation will yield a large fraction of such "slow" water. By suction, on the other hand, the relatively mobile and presumably less concentrated water in the largest waterfilled pores will be sampled preferentially.

An exception is Al at 60 cm depth in Speuld, which has a much higher concentration in the cup water than in the centrifuged solution.

Aqueous extraction (1:3) would mobilize solutes from the finest pores, and therefore may give still higher (recalculated) soil solution concentrations than centriguged samples. In addition, extraction may tend to increase or decrease concentrations relative to those in "true" soil solution, depending on the kind of interaction between different ions and the solid phases. Relatively inert ions such as NO_3 and Cl should have similar concentrations in centrifuged water and in extracts. This holds true for NO_3 (Table A.1.2), but, surprisingly, not for Cl, which is 1.5 (Speuld) to about 6 (!) (Ysselsteyn) times higher in aqueous extracts than in centrifuged solutions (results not shown here). Cations which undergo ion exchange but no dissolution/precipitation reactions are expected to react differently depending on their valence: due to increased preference of the exchange complex for higher valence cations over monovalent cations in more dilute solutions, the numbers of monovalent cations in solution should increase and those of e.g. divalent cations should decrease upon extraction by water. Indeed (recalculated) soil solution concentrations based on aqueous extracts are higher for NH_4 and K (Table A.1.3) and tend to be lower for Ca (Table A.1.4) than centrifuged solutions. Al is much higher in aqueous extracts than in the other sample types, which can be ascribed mainly to the relatively quick dissolution of solid Al fractions (Mulder et al, 1989), particularly in subsoils, where levels of mobilizable Al are generally higher than in the strongly acidified surface soils (Table A.1.4).

The net effect of the processes described above on the various solute molar ratios often used in the acidification programme can be summarized as follows: NH_4/K ratios of all the methods are similar and do not differ significantly at P = 95%; extraction and centrifugation produced the closest correspondence (Table A.1.3). However, cup and centrifuged solutions give similar NH_4/NO_3 and NH_4/Mg ratios, but Al/Ca ratios may differ by factors between 0.5 to 5. The aqueous extracts invariably give higher values of NH_4/Mg, NH_4/NO_3 (x 2 to x 4) and of (molar equivalent) Al/Ca ratios (generally x 5 to x 30) than the cup and centrifugation methods. In addition it has to be mentioned that agueous extracts also give much higher values of NH_4/Mg/molar equivalent).

Table A.1.1 Mean and standard deviation (as % of the mean) of solute concentrations in three soil solution samples from different depths analysed by three laboratories (mmol (+/-)/m^3)

	pH	K	Na	Ca	Mg	Al	Fe	Mn	NH_4	Cl	NO_3	SO_4
0 cm mean	3.26	66	510	377	132	168	6	46	122	920	350	602
st.d.	1	4	6	6	5	21	16	4	6	2	1	5
20 cm mean	3.42	104	673	266	198	1170	21	63	163	1173	829	840
st.d.	2	3	5	3	7	2	10	3	3	2	2	5
90 cm mean	4.09	38	561	95	112	1606	0	59	16	828	515	1118
st.d.	2	15	8	20	4	4	141	5	94	2	1	6

Table A.1.2 Comparison of three methods to obtain soil solution, (concentrations in mmol/l), based on 16 paired samples from two sites. Y = Ysselsteyn, Scots pine; Sp = Speuld, Douglas fir (site 3B)

	concentr. mmol/l		concentration by cup or by extract, relative to concentration by centrifuge, paired adjacent samples				areal mean NH_4/NO_3 ratio per method		
	centrifuge		cup/centr		extr/centr		centr	cup	extr
	NH_4	NO_3	NH_4	NO_3	NH_4	NO_3			
Y 10 cm									
average	1.11	0.92	1.00	0.93	2.25	0.96	1.25	1.33	2.74
st.dev.	0.66	0.22	0.47	0.43	0.34	00.6	0.76	0.52	1.43
Y 40 cm									
average	0.74	0.84	0.84	0.80	3.27	1.02	0.89	1.00	2.53
st.dev.	0.45	0.23	0.39	0.27	0.74	0.09	0.51	0.72	1.12
Sp 20 cm									
average	0.13	0.63	0.77	0.83	4.85	1.15	0.26	0.22	1.00
st.dev.	0.09	0.36	0.74	0.23	1.25	0.16	0.16	0.26	0.48
Sp 60 cm									
average	0.11	0.76	0.12	0.84	5.11	1.32	0.22	0.02	0.84
st.dev.	0.04	0.45	0.21	0.31	1.67	0.33	0.20	0.03	0.79

Table A.1.3 Comparison of three methods to obtain soil solution concentrations (in mmol/m^3), based on 16 paired samples from two sites. Y = Ysselsteyn, Scots pine; Sp = Speuld, Douglas fir (site 3B)

	areal mean concentr.		concentration by cup or by extract, relative to concentration by centrifuge (means for paired samples)				areal mean NH_4/K ratio per method		
	centrifuge		cup/centr		extr/centr		centr	cup	extr
	NH_4	K	NH_4	K	NH_4	K			
Y 10 cm									
average	1108	90	1.00	0.88	2.25	2.01	5.47	6.36	5.91
st.dev.	655	159	0.47	0.76	0.34	0.53	2.68	2.49	1.97
Y 40 cm									
average	742	147	0.84	0.74	3.27	2.99	5.42	7.36	5.27
st.dev.	447	47	0.39	0.45	0.74	0.52	4.19	6.72	2.93
Sp 20 cm									
average	134	89	0.77	0.35	4.85	4.12	1.53	2.83	1.79
st.dev.	87	48	0.74	0.30	1.25	1.44	0.57	2.82	0.55
Sp 60 cm									
average	109	87	0.12	0.28	5.11	5.63	1.31	0.36	1.22
st.dev.	43	28	0.21	0.21	1.67	2.17	0.46	0.41	0.41

Table A.1.4 Comparison of three methods to obtain soil solution concentrations (in mmol/m^3), based on 16 paired samples from two sites. Y = Ysselsteyn, Scots pine; Sp = Speuld, Douglas fir (site 3B)

	areal mean concentr.		concentration by cup or by extract, relative to concentration by centrifuge (means for paired adjacent samples)				areal mean Al/Ca ratio per method		
	centrifuge		cup/centr		extr/centr		centr	cup	extr
	Al	Ca	Al	Ca	Al	Ca			
Y 10 cm	360	515	0.46	0.91	6.72	0.84	0.71	0.38	11.02
Y 40 cm	579	463	0.70	0.84	13.2	1.57	1.28	1.02	23.18
Sp 20 cm	997	844	0.61	0.30	8.19	0.80	1.19	2.72	12.90
Sp 60 cm	633	712	2.76	0.36	19.1	1.85	1.26	8.06	9.66

BIOLOGICAL AND PHYSIOLOGICAL EFFECTS

A.C.Posthumus[1)]
A.E.Jansen[2)]

1) Research Institute for Plant Protection, Wageningen
2) Agricultural University Wageningen, Department of Phytopathology

1. INTRODUCTION

In order to study the possible effects of air pollution in the Netherlands on the vitality of Douglas fir (Pseudotsuga menziesii (Mirb.) Franco), it is necessary to know what the separate and combined influences of several air pollutants on the biology and physiology of Douglas fir trees are. This relates to both the above ground and the below ground parts of the trees, showing direct and indirect effects of atmospheric deposition respectively, and so effects in the rhizosphere have been included. Both the effects above and below ground have been studied in various research projects within the Dutch Priority Programme on Acidification (DPPA), the results of which are presented in this thematic evaluation report. Only technical evaluation of the research projects will be carried out, paying special attention to the methods and results, the underlying assumptions and the range of uncertainty of the results. Aspects to be dealt with especially are: quantitative relations between air pollution and effects, the role of diseases and pests, rhizosphere and mycorrhizas, and drought aspects.

Direct effects on the above ground parts of trees relate mainly to the effects on leaf or needle physiology and morphology. In particular the uptake of air pollutants by the stomata of leaves or needles and the inhibition of the photosynthesis under the influence of air pollutants are the processes studied most frequently. Reduced photosynthesis is probably the most important cause of reduction in growth of trees under the influence of air pollution, although disturbances of the hormonal balances of trees may not be excluded (Posthumus, 1991).

For practical reasons, experiments have been carried out with juvenile trees or only branches of these trees. To transfer the results of this research to full grown trees and total forest ecosystems is still a great problem, but the basic knowledge about effects of air pollutants on processes like photosynthesis may be used for the construction of dynamic simulation models of the growth and production of trees in a forest stand under realistic, ambient conditions. Furthermore, the normal processes of water transport and tree growth, and the structure and amount of needle wax have been investigated in trees of the field study plots of the DPPA.

Above ground effects of NH_3, SO_2, NO_2, O_3 and some combinations of pollutants (NH_3 + NO_2 and SO_2 + NO_2) have been studied on the physiology of needles of Douglas fir (projects 109, 110, 115), while only the uptake of NH_3 by needles has been determined in more detail (project 110). Apart from the effects of a certain pollution burden, the water transport velocity, year ring increment and water transport capacity of Douglas fir trees have been investigated (projects 116, 111-1). Also, needle surface wax structures have been studied, both of trees at the field study plots of the DPPA and of juvenile trees fumigated with different concentrations of NH_3 in open-top chambers at Wageningen (project 111-2).

In 1904 Hiltner introduced the term "rhizosphere" for the soil zone around the root where the concentration of bacteria was considerably higher than in the rest of the soil. Later, the term was redefined as the volume of soil around a root that physically, chemically and biologically differs from the bulk soil due to the effects of this root. From the rhizosphere of fine roots of Douglas fir trees, especially the pH was studied. Rhizosphere pH not only strongly depends on the functioning of the fine roots, but also influences the root physiology and thereby plant growth. Therefore, it is closely linked to nitrogen uptake by fine roots (project 83).

Uptake studies were not only done in pot experiments but also in hydroponic systems. Of course one cannot speak of a rhizosphere in such systems, but the observed processes of acidification or alkalization of the growing media appeared not to differ from those in soil. In the hydroponic experiments plant growth and nutrient uptake of mycorrhizal and non-mycorrhizal plants were compared. Field studies had shown that the occurrence of mycorrhizal fungi was consistently and inversely related to the load of atmospheric pollutants (Jansen, in press). The consequences of a decreased mycorrhizal infection for the nutrition of the tree were studied in these hydroponic systems.

Mycorrhizal fungi are, for their carbon source, at least partially dependent on the tree host. There are other microorganisms in the rhizosphere that also depend largely on carbon supply from the roots of the tree host, such as saprotrophic rhizosphere fungi (micromycetes) and rhizosphere bacteria. Plants strongly regulate the rhizosphere microbial populations by means of exudation or secretion of specific organic substances. One may hypothesize that both the mycorrhizal fungi and the rhizosphere organisms could be influenced by an altered carbon partitioning within the plant, which may result from exposure to air pollutants. A reduced transport of carbohydrates to the roots may result in qualitative changes (changes in species composition) and quantitative changes (reductions in development of mycorrhizas, mycorrhizal fungi and rhizosphere organisms, and also in production of fruit-bodies of mycorrhizal fungi).

On the other hand, one may also hypothesize that these organisms can be strongly damaged by the disturbed soil environment that results from atmospheric deposition, such as lower pH, higher concentration of aluminium ions, changed aluminium:calcium and aluminium:magnesium ratios, and higher concentration of ammonium or nitrate. These two possible pathways were put forward by projects 107, 108 and 201. Project 107 was devoted mainly to carbon partitioning, and project 108 to the effects of changed soil chemistry. Project 201 summarized both the carbon demands of a population of mycorrhizal fungi and the effects on such a population when the transport of carbohydrates is reduced, and also the effects of lowered pH, of pH induced changes in soil chemistry and of nitrogen on mycorrhizas.

It should be stated beforehand that it is important to study the integrated effects of several components of air pollution and other abiotic and biotic stress factors, under the influence of the relevant environmental conditions. But it is clear that many possibilities for investigating the integrated effects of air pollution on Douglas fir have not been used until now. It is obvious that only some of the problems in respect of the so-called novel type of forest decline in the Netherlands have been studied and that not all possible techniques have been applied. For example, realistic simulation experiments with full grown trees under the influence of the combination stress of several air pollutants and other abiotic and biotic factors have not been carried out. Still, it is remarkable that so much data have been collected and that quite a lot of new information and insights have been gained.

2. THE EFFECTS OF SEVERAL AIR POLLUTANTS SEPARATELY AND IN SOME COMBINATIONS ON THE NEEDLE PHYSIOLOGY AND TREE GROWTH OF DOUGLAS FIR

In project 115 ("Quantitative analysis of the uptake of SO_2 and O_3 by plants and effects on physiological processes and plant growth", W.L.M. Smeets et al.) the objective was to develop a dynamic (mechanistic) process simulation model for the uptake of the air pollutants SO_2 and O_3 and the effects thereof on the physiology of Douglas fir needles, in relation to water stress. It was intended to develop suitable exposure-response submodels to be built in the mechanistic models for the growth of crops and forest stands.

The uptake of the pollutants and the effects thereof on photosynthesis and the functioning of the stomata have been analysed in a system for gas exchange analysis. This system, which was computer controlled, was developed during the first phase of the DPPA (Kropff, 1989). Effects of SO_2 on plants may be subdivided into short-term and long-term effects, on the basis of the underlying effect mechanisms. The short-term effects of SO_2 consist of a rapid decline in photosynthesis within one hour after the start of the fumigation, followed by stabilisation and a rapid total recovery after stopping the fumigation. This short-term effect is produced by accumulation of sulphite (SO_3^{2-}) and bisulphite (HSO_3^-) in the leaves, inhibiting competitively the binding of CO_2 to RuBP carboxylase. Based on these insights, an uptake/effect model for SO_2 in Vicia faba L. (broad bean) was developed in a previous study (project 22 of the DPPA). This submodel has been validated and incorporated into the deterministic model for the growth and production of broad bean crops and Douglas fir stands, to estimate the uptake of air pollutants by the needles of a Douglas fir stand and to assess the influence of short-term SO_2 effects (see project 112 of the DPPA). Furthermore, the influence of air humidity and temperature on the sensitivity of the photosynthesis for SO_2 has been studied in broad beans. At higher humidity the stomata opened more and the SO_2 uptake and reduction of the photosynthesis increased. Decrease of the temperature from 20°C to 7°C increased the effect of SO_2 2.7 times, at the same level of SO_2 uptake in the leaves. This could be explained by an increased accumulation of SO_3^{2-} and HSO_3^- in the leaves, caused by a slower oxidation rate. Leaves fumigated at 7°C accumulated 4 times more SO_3^{2-} and HSO_3^- than leaves fumigated at 20°C. Model results were confirmed by experimental analyses of the concentrations of SO_3^{2-} and HSO_3^- in leaves after the fumigation, showing that more toxic SO_2 metabolites accumulated in leaves exposed at a lower temperature.

In field experiments long-term fumigations with SO_2 in low concentrations caused leaf/needle injury and leaf/needle abscission of broad beans and Douglas firs. This may be explained by the influence of SO_2 on the regulation of the intracellular pH. On the basis of this knowledge a conceptual model has been developed for the chronic effects of SO_2. The major part of the observed reduction in total dry matter production of broad beans could be explained by leaf injury, possibly caused by effects of SO_2 on intracellular pH regulation.

Short-term fumigations with O_3 on Douglas firs have been carried out with several O_3 concentrations. Slow reactions in photosynthesis, transpiration and respiration were the results, and the no-effect level was relatively high (ca 200 $\mu g.m^{-3}$). The insensitivity of Douglas fir needles for these fumigations could, among other things, be explained by the low stomatal conductance for O_3 (3-5 times lower than in the case of sensitive plants). Photosynthesis declined slowly after O_3 fumigation to a constant level after three days, with a simultaneous and proportional decrease of stomatal conductance and a simultaneous rise in dark respiration, and recovered also slowly after termination of the fumigation. Results made it possible to establish the effects of O_3 on the physiology of Douglas fir needles, and the observed exposure-response relation was incorporated into the stand growth model (project 112 of the DPPA).

Long-term, realistic fumigations with O_3 (140 $\mu g.m^{-3}$from 10.00 a.m. till 6.00 p.m. and 60 $\mu g.m^{-3}$ from 6.00 p.m. till 10.00 a.m.) on Douglas firs have been performed over a period of 6 months in fumigation greenhouses, to investigate the effects on photosynthesis, needle conductance and drought sensitivity. Photosynthesis at light saturation and dark respiration were not influenced, but the light use efficiency and cuticular transpiration increased as a result of the O_3-fumigation. The observed increase of cuticular transpiration in optimal conditions of soil moisture as a result of O_3 presented the only indication for a possible influence of O_3 episodes on the sensitivity of Douglas fir to water stress. Stomatal function was not influenced by realistic O_3 concentrations.

It has not to be excluded that in the field fumigation experiments with SO_2 possible combination effects of SO_2 + O_3 may have played a role, because of the normal summer smog episodes occurring in the relevant period of the year.

In project 110 ("The physiological effects of low concentrations of NH_3, SO_2 and NO_2 on Douglas fir (Pseudotsuga menziesii)", L.W.A. van Hove & M.G.J. Mensink) the objective was to study the mechanisms of uptake and the effects of gaseous air pollutants in plants, in order to build dynamic simulation models. In a previously developed leaf chamber (Van Hove et al., 1988), the uptake of NH_3 by needles of Douglas fir was studied, and the effects of long-term exposure to NH_3, SO_2, NO_2 and some combinations of these gases on photosynthesis and stomatal conductance of two year old seedlings of Douglas fir were determined. It was concluded that the NH_3 transfer into the needles could be calculated from the boundary layer resistance, the stomatal conductance for water, and the NH_3 concentration close to the needle surface. At low vapour pressure deficit a large quantity of NH_3 was found to be adsorbed to twig and external needle surfaces.
In long-term experiments Douglas fir seedlings were exposed over a period of 5 months to filtered air, 66 $\mu g.m^{-3}$ NH_3, 96 $\mu g.m^{-3}$ SO_2, 95 $\mu g.m^{-3}$ NO_2, 52 $\mu g.m^{-3}$ NH_3 + 82 $\mu g.m^{-3}$ NO_2 and 128 $\mu g.m^{-3}$ SO_2 + 129 $\mu g.m^{-3}$ NO_2 in well conditioned fumigation cabinets (Van Hove et al., 1989). After 3 months of exposure 50-70% inhibition of the maximal electron transport (Jmax) of the control was found for all exposure treatments, using methods of chlorophyll fluorescence measurement as described in project 109 of the DPPA. After 5 months of exposure further decline of Jmax was shown, except for the exposures to NH_3 and NH_3+ NO_2 (Figure 1). Only the exposures to SO_2 and SO_2+ NO_2 showed a reduction in the maximal net CO_2 assimilation (Pmax), similar to that of Jmax. Exposure to NO_2 caused no reduction of Pmax and a higher total chlorophyll content. But, on the other hand, exposure to NH_3 and NH_3 + NO_2 resulted in a higher Pmax and higher total chlorophyl content than in the case of the control. So, it is concluded that in some exposures, causing clear inhibition of the electron transport rate, the reduction in Pmax could be compensated or even over-compensated by the production of extra chlorophyll (Figure 2).

Furthermore, exposure to SO_2 and SO_2 + NO_2 reduced the stomatal conductance, while exposures to NH_3, NO_2 and NH_3 + NO_2 resulted in higher stomatal conductance. Exposure to NH_3 caused a higher transpiration rate in the dark, indicating incomplete stomatal closure, which makes the tree more sensitive to water stress.
Unfortunately, insufficient variation in the concentrations of the air pollutants was used to determine the total exposure-effect relation or some no-adverse effect levels at least.

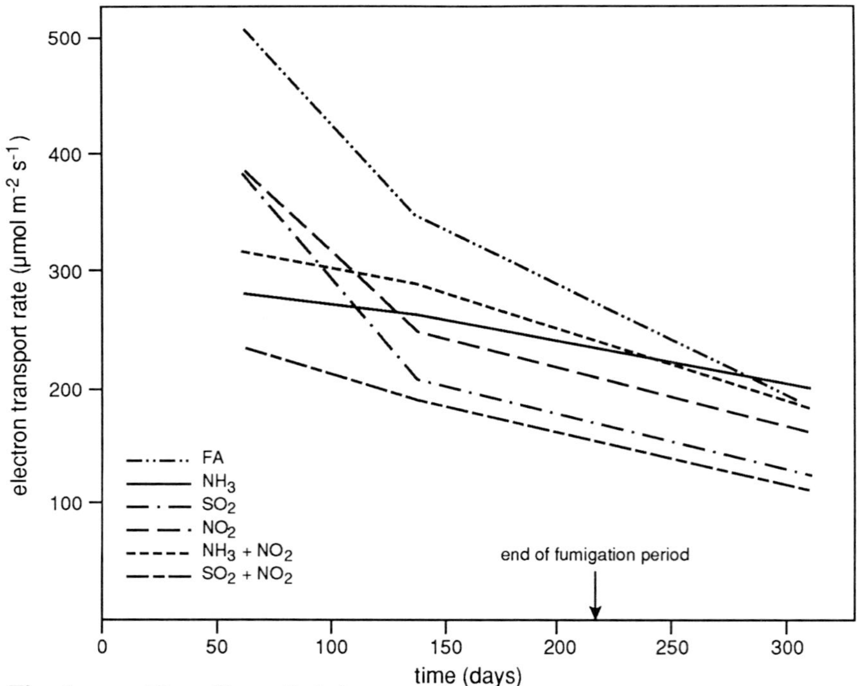

Fig. 1. The effect of air-borne pollutants on the maximal electron transport rates (J_{max}) in needles of Douglas firs. Plants were fumigated for 7 months (mean shoot age 220 days) and then grown in filtered air. Measurements were performed at 70, 140 and 310 days (mean shoot age). Mean ± SE is shown (n=10). (From Van Kooten & M.E.Bossen, project 109)

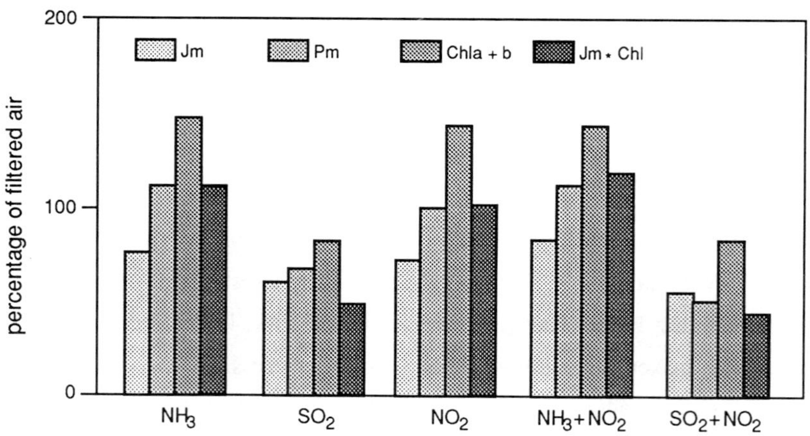

Fig. 2. The maximum electron transport rate (J_{max}), maximum net CO_2 assimilation rate (P_{max}), total chlorophyl content (Chl) and J_{max} x Chl of shoots exposed to NH_3, SO_2, NO_2, NH_3+NO_2 and SO_2+NO_2 as percentage of control shoots (filtered air). (From L.W.A.van Hove & M.G.J.Mensink, project 110)

Project 109 ("Effect of air-borne pollutants on the photosynthesis of Douglas fir (Pseudotsuga mensiesii)", O. van Kooten & M.E. Bossen) was carried out in close co-operation with project 110 of the DPPA, and was of vital importance for it. The specially adapted chlorophyll fluorescence measuring technique (Schreiber et al., 1986; Van Kooten & Van Hove, 1988) was of great use for the assessment of long-term effects of the air pollutants SO_2, NO_2, NH_3 and some combinations on the photosynthesis of Douglas fir needles.

Principal objectives were (1) the elucidation of the mechanisms by which SO_2, NO_2 and NH_3 interfere with the process of photosynthesis in needles of Douglas fir by the use of chlorophyll fluorescence, and (2) the development of a portable chlorophyll fluorescence measurement system (which has since become available and is being tested for use in the field).

Although the first objective was not reached really, a lot of information was gained on the influence of air pollutants on the photosynthetic apparatus of two-year-old seedlings of Douglas fir. The electron transport rate in photosystem II could be determined on the basis of chlorophyll fluorescence, according to Genty et al. (1989). The maximal electron transport rate (Jmax), the convexity of the light response curve (M) and the quantum yield at limiting light intensity (qy) were calculated from light response curves of photosynthetic electron transport rates. As has been reported by the authors of project 110 the fumigations with different pollutants and some combinations caused a lower Jmax in the shoots after 2 and 5 months, as compared with the control treatment of filtered air (Figure 1). The combination of $SO_2 + NO_2$ was most detrimental, showing an inhibition of 70%. Recovery from the inhibition of Jmax after the termination of the seven-month period of fumigation was not found, because at the time of measurement (10 months after the start of the fumigation) the Jmax was low and about equal for all treatments including the control. However, the cause of the low Jmax was not known.

Suggestions from the authors of project 115 that SO_2 is influencing photosynthesis by competitive inhibition of the binding of CO_2 to RuBP carboxylase could not be proven in these experiments.

Air pollution affects stomata, which alters the transpiration rate, thereby affecting xylem sap velocity. Xylem sap velocity is a potential measure of pollution induced transpiration stress.

The first step was carried out in project 116 ("Xylem sap velocity in a Douglas fir stand, spatial and time dependent variability.", F. Noppert). In this project the relative xylem sap velocity of twenty-seven-year-old Douglas fir trees was measured by means of the adapted Heat Pulse Velocity (HPV) technique (Visser et al., 1989). Hourly measurements were carried out from 27 May till 6 November 1989, at three plots with four trees each and two sensors per tree at the field research site Garderen (ACIFORN location Speuld). Spatial and time dependent variation were estimated.

HPV values were found in the range of 0-18 cm/h and the mean value was 2.6 cm/h. The largest contribution to spatial variation was found on plot level (80%), second largest contribution was found on sensor level (11%), and no significant ($p < 0.05$, nested Anova) added variance could be detected on tree level. On further inspection one plot showed HPV values twice as high as the other plots, 100 m away. The largest time dependent variance (68%) was found within a day (Figure 3), and then, in descending order,a week (14%), a month (11%) and above one month. The HPV values decreased throughout the season.

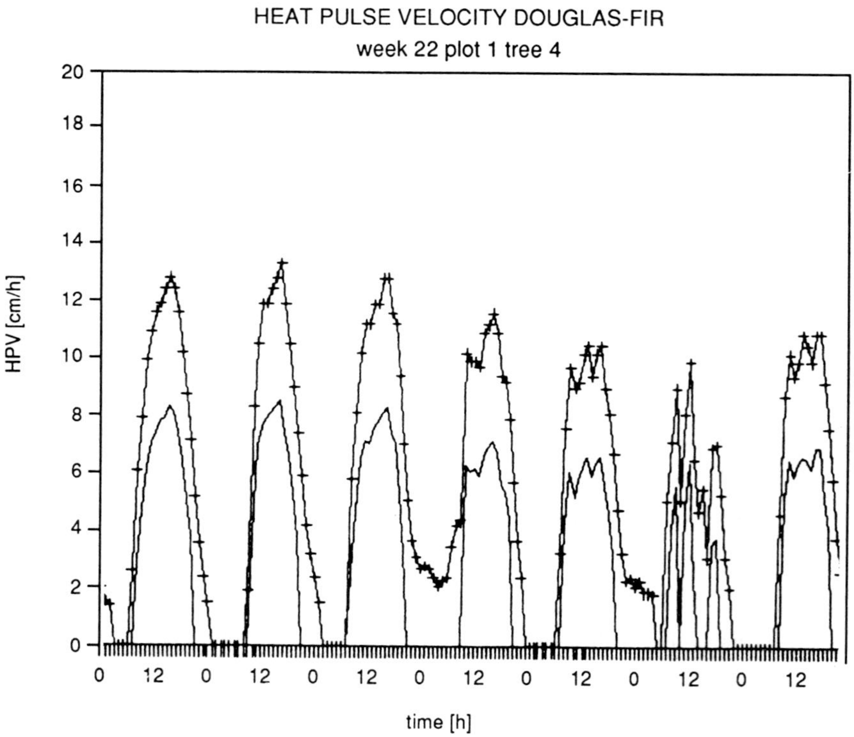

Fig. 3. In this figure the HPV values of two sensor-units within one tree are presented (lower curve, southern sensor-unit and upper curve, northern unit). Within this one week period the daily cycle of the HPV signal can be clearly seen. While the absolute values of HPV differ between the two units, the patterns are similar. Further it can be seen that nighttime HPV is not always nill, suggesting transpiration in the dark. Daytime dips resemble dips found with earlier KEMA research, caused by rainfall. (From F.Noppert, project 116)

The results suggest that, even at a distance of 100 m, variation in xylem sap velocity can be quite high. As a consequence, transpiration models estimating forest transpiration of large areas can give misleading results when input data of only one small plot are used. It can be concluded that possible effects of air pollution on transpiration will be masked easily by the high daily variability. Reducing this variability by statistical means enhances the chance of finding air pollution effects.

By matching HPV measurements and wood anatomy, it was hoped that absolute sap fluxes could be estimated. Unfortunately emboly around sensors and other problems produce an unknown, but absolute bias in HPV value. The bias will differ among sensors. Therefore, absolute values of individual sensors are hard to compare, but HPV dynamics proved to be reliable. At present HPV dynamics are used to interpolate the absolute values of other methods in order to generate the most probable transpiration of the Garderen site over the entire year.

In project 111-1 ("Wood production, stem growth and water transport capacity of Douglas fir in the ACIFORN stands in Kootwijk en Garderen", I. de Kort) the objective was to discover if there were any differences between the Douglas fir stands in Garderen (= Speuld) and Kootwijk with respect to the parameters indicated in the title of the project. At each location 5 selected trees were felled and disks were collected at 6 heights, and from 20 others increment cores were taken at breast height. On the basis of this material growth analysis of the trees, measurement of anatomical features of the wood and moisture content in the sapwood were performed. The selected trees were of different classes of vitality (1, 2 and 3 according to Anonymous, 1989), but the average vitality of the selected trees was equal in both stands.

Radial and volume growth of the trees was determined by assessment of several measures in the disks and increment cores. Also the amount of sapwood and moisture content were assessed by adequate methods. Results of axial growth analysis did not show much difference between the trees. Warm, dry summers (for example that of 1976) caused shorter internodes. There was not much difference between the average curves of both stands (Figure 4). Also ring width curves at all heights and, ultimately, total profiles of the trees were generated (Figure 5). On the whole there was not much difference between the trees and between the two stands. Also the stemwood volume was not significantly different for both stands. Since 1980 there has hardly been any difference between the average radial growth in the two stand, and on the whole there is not much difference between the stands in average annual basal area increment, in average wood volume and in average crown biomass. The absolute and relative amounts of basal sapwood area do not differ significantly in both stands. Average moisture contents have been determined and related to several other parameters. A small amount of sapwood appears to be compensated for by a higher average moisture content, due to a higher amount of early wood.

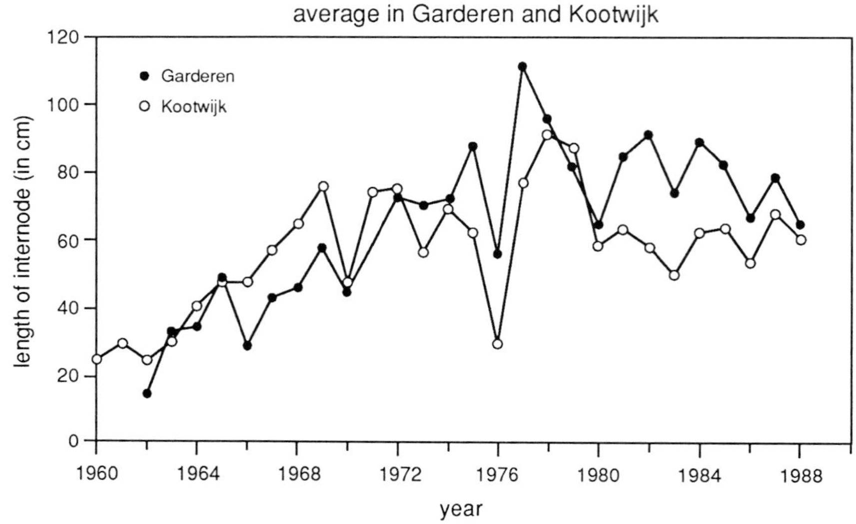

Fig. 4. Axial growth of trees in Garderen and Kootwijk. (From I.de Kort, project 111-1)

Results of this study showed no recent decline in radial, axial or volume growth and a normal amount of sapwood, related to the total amount of wood and crown biomass. Moisture content of the sapwood was only slightly below maximal. The investigated parameters were not different for the two stands, and did not indicate exposure of the trees to other than natural stresses. Air pollution level at the sites was obviously not high enough to cause any clearly visible effect on the wood production, stem growth and water transport capacity of Douglas fir trees. However, unfortunately no comparison of these trees with a really good reference sample from unpolluted air has been made, nor with a sample from the severely ammonia polluted air in the South-East of the Netherlands.

In a recent study (project 73 of the first phase of the DPPA) by Visser and Maessen (1990), on the basis of tree ring analysis of three coniferous species and inland oak, it was found that only in the case of Douglas fir and Norway spruce were there indications for the influence of pollution stress (manure activities). However, a changing response of all coniferous trees and one inland oak stand to the total amount of precipitation during the current year and the two preceding years was found.

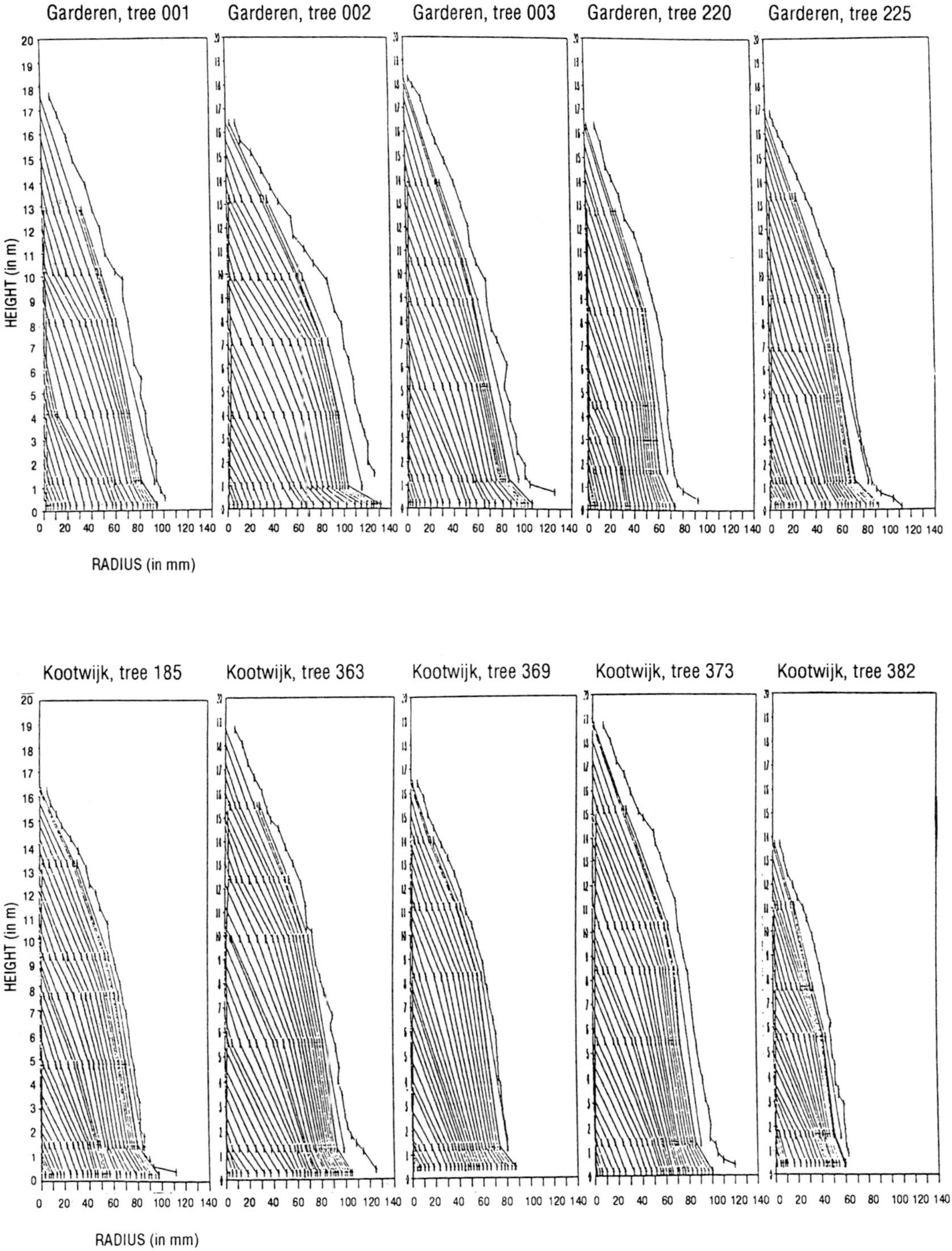

Fig. 5. Profiles of the stems (height in m; radius in mm).
a. Garderen
b. Kootwijk
(From I.de Kort, project 111-1)

In project 111-2 ("Natural and NH_3-induced variation in epicuticular needle wax morphology of Pseudotsuga menziesii (Mirb.) Franco", G. Thijsse & P. Baas) epicuticular wax morphology of Douglas fir needles was studied with Scanning Electron Microscopy (SEM) in current year and older needles of 10 representative trees (vitality classes 2 and 3) from each of the two research sites of the DPPA (Kootwijk and Garderen), to look for possible differences in place and time throughout the growing season of 1988. For comparison, material available from NH_3 fumigation experiments, including controls with filtered and non-filtered ambient air, was studied. From each tree, sampling date, and exposure level five different needles were chosen for estimating the amount of amorphous wax. Five different stomata of each needle were examined. The fine structure of the wax overlying the stomatal pores was very similar for all stomata of one needle. So, only one instead of five stomata per needle were examined for fine structure.

Methods for SEM were applied to the dry stored (at 4°C) needles. Air-dried samples were attached to aluminium specimen holders and coated in vacuum with a thin layer of gold, before examination in a scanning electron microscope. Micrographs were made at magnification x1000 (to quantify amorphous wax on the stomata) and at magnification x2600 (to study the fine structure of the wax). Degradation of the regular, three-dimensional porous structure of the wax layer was found already several weeks after bud break and reached a high level at the end of the first growing season. The increase in amorphous wax showed a similar, but slightly slower development. In one- and two-year-old needles these processes had progressed only slightly further. It was very remarkable that the rates of change in crystalline wax morphology did not differ significantly between sun-exposed and shaded needles, between sites and between the two vitality classes of the trees included in the samples. As an extra observation, the development of fungi on the needles from June on and the occurrence of cracks in the wax layers were found to be dependent on the age of the needles.

In juvenile trees (4-year-old) of a fumigation experiment within project 124/125 of the DPPA at the Research Insitute for Plant Protection in Wageningen, the effects of NH_3 on needle wax morphology were studied. Fumigations in open-top chambers with 25, 45 and 100 $\mu g.m^{-3}$ NH_3 during 5 weeks in 1989 did not affect the wax morphology in current year needles. One-year-old needles that had been exposed to different concentrations from bud break onwards showed a severe degradation of the crystalline wax. In two-year-old needles the effect of NH_3 could not be traced and seemed to be overshadowed by the natural ageing process. Ambient levels of O_3, SO_2 and NO_x in Wageningen did not affect the epicuticular wax morphology.

It is suggested that the variation in epicuticular needle wax morphology recorded for the two forest stands does not show effects of local pollution levels, although real proof can only be provided by comparison of trees in filtered and non-filtered ambient air. Other environmental factors (light intensity, humidity, temperature, precipitation and wind velocity may also influence development and degradation of the crystalline or structural wax. So, needle wax morphology cannot be considered as an indicator for the degree of damage caused by air pollutants, if no special measures have been taken to discriminate the possible causes.

3. INDIRECT EFFECTS ON FINE ROOTS, MYCORRHIZAS, AND INTERACTIONS IN THE RHIZOSPHERE

In project 83 ("The effect of high ammonium input to the soil on ionic uptake balance and rhizosphere pH of Douglas fir", A.J. Gijsman) the effects were studied of changes in nitrogen availability in the soil (due to increased deposition of NH_x) on the ionic uptake balance and the rhizosphere pH. Since nitrogen is the main factor in the ionic uptake balance, and nitrogen input by atmospheric deposition in the Netherlands is large and seen as a major detrimental factor, the amount and source of absorbed nitrogen received a lot of emphasis. The main questions were:

1. Do Douglas fir trees preferentially absorb NH_4^+ or NO_3^-?
2. Do Douglas fir trees get physiological problems when more ammonium than nitrate is absorbed?
3. How is the rhizosphere pH along the root axis affected by the ammonium or nitrate uptake?
4. What are the consequences of the different patterns in rhizosphere pH for the functioning of the root system?

These questions were studied in a series of pot experiments with juvenile Douglas fir trees, provenance Arlington 202, potted in soil (upper 20 cm after the litter layer was removed) originating from ACIFORN site Kootwijk. Plants got a basal fertilization plus $Ca(NO_3)_2$, $(NH_4)_2SO_4$ or NH_4NO_3 at various levels (e.g. 10, 50 and 100 mg/kg oven dry soil) and a nitrification inhibitor was added. When both ammonium and nitrate were available, Douglas fir was found to absorbe nitrogen preferentially in the ammonium form. But in relatively dry soil, nitrate was taken up at a higher rate than ammonium, as ammonium was less available in dry conditions due to higher adsorption onto the soil complex. Although ammonium is absorbed with preference, growth (dry matter production) decreased with increased ammonium supply in the absence of nitrate. From this very poor growth, and also high mortality, the author concluded that the plants suffered from physiological disorders. The net carboxylate production decreased (Figure 6), resulting in internal acidification of the plant and probably also high internal NH_4^+ concentration. Nitrogen uptake of ammonium fed plants was even lower than of not fertilized controls, and therefore the whole uptake physiology of the roots was apparently disturbed.

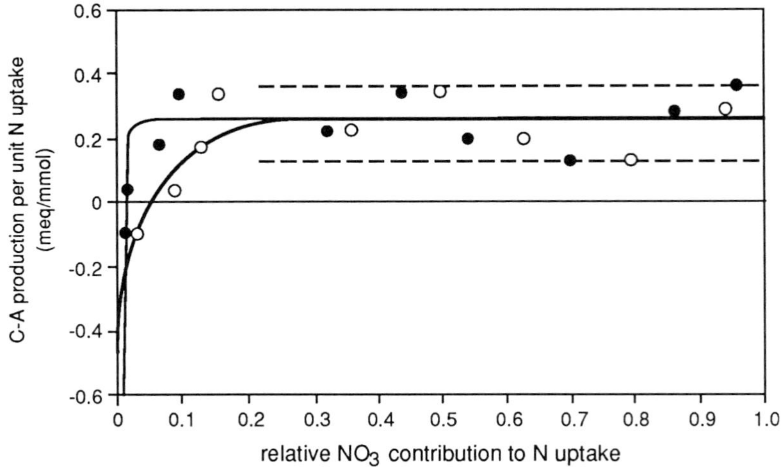

Fig. 6. Net carboxylate production in relation to the uptake of NO_3^- as proportion of total N uptake. (From A.J.Gijsman, project 83)

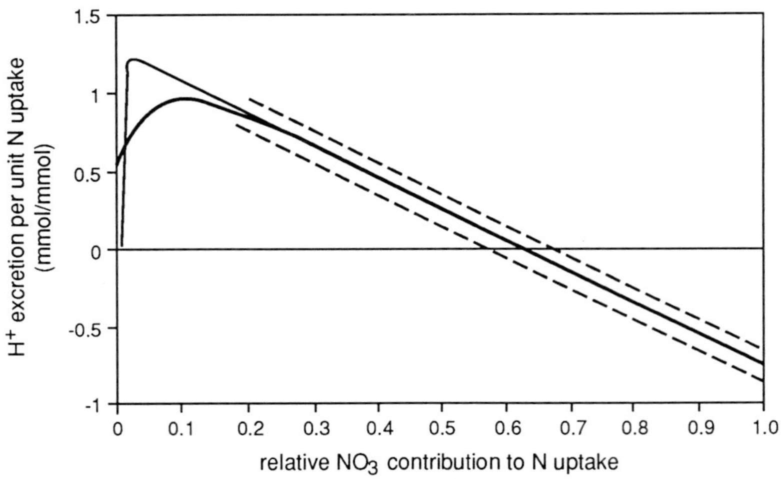

Fig. 7. Excretion of H^+ ions in the rhizosphere in relation to the uptake of NO_3^- as proportion of total N uptake. (From A.J.Gijsman, project 83)

When most of the nitrogen (>65%) is absorbed as nitrate, the rhizosphere pH is alkalized (Figure 7). With a decreasing proportion of nitrate uptake the roots excrete increasing amounts of protons, and the rhizosphere is acidified. When less than about 20% of the nitrogen is absorbed in nitrate form the rhizosphere is acidified.

In nitrate feeding, the alkalized zone at the root tip is 20-25 mm of the root, including the growth zone. With an increasing contribution of ammonium to N uptake, the alkalization at the root tip was smaller, the alkalized zone shorter and the acidification along older root zones stronger. This means that nitrate-fed plants protect their root growth zones from the severe acidity in strongly acid soils, while ammonium-fed plants are unable to do so. Ammonium-fed plants showed a reduced root growth, and shorter and thicker root tips were observed. This could, however, be an effect of raised aluminium levels in the rhizosphere due to the low pH. Water uptake is highest in the root apical zone, so a reduced root length can lead to a reduced water uptake capacity, also because the exploited soil volume is smaller. A reduced root length may also have consequences for the uptake of less mobile nutrients such as phosphate, especially when also the ectomycorrhizal fungi are disturbed (see project 108).

All these experiments were done on juvenile Douglas fir trees. The author argues that older trees may show a less extreme response to ammonium feeding, as they will have a much higher carboxylate content than young trees, and thereby a higher "buffering capacity" for unfavourable conditions.

The primary aim of project 107 ("Effects of air pollutants on the rhizosphere of Douglas firs", A. Gorissen) was to study the effects of atmospheric pollutants on the microbial community in the rhizosphere in combination with soil related factors such as moisture and nitrogen concentrations. The emphasis was laid on changes in carbon partitioning in juvenile Douglas fir trees when subject to various (realistic) concentrations of O_3 or SO_2. Effects on roots and rhizosphere organisms of ammonium treatment, and ammonium in combination with ozone, were studied and compared with observations in stands subject to different deposition.

Altered carbon partitioning was studied by treating Douglas fir plants with O_3 or SO_2 for different periods, followed by labelling with $^{14}CO_2$ for a period of 2-9 days, in controlled phytotron conditions with juvenile plants and in 'branch chambers' with 25-year-old trees. Fumigation with ozone (200 μg/m3, 3 days) affected the proportion of ^{14}C-assimilates in needles (increased) and the respiration in the root/ soil compartment of the trees (decreased). The total assimilation of CO_2 was not affected by the fumigation. So, the increase of ^{14}C-assimilates in the needles and the decrease of root/soil respiration was due to a disturbed translocation of assimilates. When subject to various levels of O_3 for a longer period (23, 81 and 169 μg/m3, 28 days, 8 h/day), the reduction in root/soil respiration was

found to be linear with the O_3 level, but an increased concentration of ^{14}C-assimilates in needles or other plant parts was not found.
Fumigation with SO_2 (53 μg/m3, 28 days) decreased the root/soil respiration, but no effect on the internal partitioning of ^{14}C-assimilates was found. After fumigation with O_3 or SO_2 was stopped, recovery of the tree, or adaptation, occurred within a few days in the case of O_3 and within ca. 2 weeks in the case of SO_2. Only when unrealistic concentrations of ozone (400 or 500 μg/m3, 3 days, 8 h/day) were used, did the root/soil respirations continue to be lower than in the control treatment. This indicates an interference of ozone with the carbon economy of the trees, resulting in a reduced total net uptake of $^{14}CO_2$ or a blocked transport of photosynthates from the needles.
Assimilated ^{14}C in needles was largely found in the free sugar fraction and 'rest'; the proportion of ^{14}C in the starch fraction was very small.
The influence of a disturbed soil environment was studied in a pot experiment by treating juvenile Douglas fir trees with ammonium sulphate in different concentrations (5, 50 and 200 kg N/ha/y) and for different periods (6, 18 and 23 months). The decrease of the size of the bacterial populations on the roots was higly significant at the highest nitrogen gift. The addition of ammonium sulphate also influenced the uptake of several nutrients: the concentration in the needles of P in particular was strongly reduced, to 23% of the control value in the 200 kg N treatment after 23 months, but also the concentrations of Mg, Ca, Fe and Zn decreased. The concentration of N in the plants strongly increased with the treatment.
In a combined O_3 (200 μg/m3, 28 days, 8h/day) and ammonium sulphate (5, 50 and 200 kg N/ha/y) treatment no effects on the size of the bacterial populations in the rhizosphere were found. Both exposure to ozone and ammonium in the soil clearly decreased the starch content of the needles as was formerly observed after short-term treatment with 400 μg O_3/m^3. This was explained by an enhanced hydrolysis of starch for repairing activities.

All mycorrhiza studies were bundled in one project 108 ("Acid rain and ectomycorrhiza of Pseudotsuga menziesii, an integration of laboratory and field studies") split up into the parts "field observations and greenhouse experiments" (project 108-1, part A in the report, A.E. Jansen & F.W. de Vries) and "laboratory experiments" (project 108-2, part B in the report, "Physiology of mycorrhizal and non-mycorrhizal trees", C. Kamminga-Van Wijk, and part C, "Physiology of mycorrhiza forming fungi", R.H. Jongbloed). Also in this project the main question was whether the mycorrhizas are decreasing because the trees are diseased or because the soil is becoming an unsuitable habitat. But in contrast to saprophytic rhizosphere microorganisms, a reduction in mycorrhizal flora might have consequences for the tree hosts, as the mycorrhizal association is generally thought to be mutualistic. All studied effects on mycorrhizal fungi, mycorrhizas and trees were presented in a scheme (Figure 8).

Project 108 B, "Physiology of mycorrhizal and non-mycorrhizal trees", examined partly the same questions as project 83. What is the growth of mycorrhizal and non-mycorrhizal trees like with nitrate or ammonium feeding, or mixed feeding, at limiting rates of nutrient addition? What are the nitrogen uptake characteristics at several concentrations? What is the effect of different pH values and how does the plant affect medium pH? Does addition of aluminium affect the growth of mycorrhizal and non-mycorrhizal Douglas fir trees? In general: growth of the mycorrhizal and non-mycorrhizal plants was studied under several 'acid rain conditions'.

The experiments were all done on Douglas fir seedlings grown in hydroponic systems where the nutrient availability was exactly controlled. The seedlings were mycorrhized, or not, with an isolate of the mycorrhizal fungus Laccaria bicolor (Maire) P.D. Orton (same isolate as used in 108 C). Uptake of NO_3^- or NH_4^+ was measured as disappearance from the uptake solution.

The plants had a higher affinity and higher uptake capacity for NH_4^+ compared to NO_3^-. Uptake of nitrate was completely suppressed in solutions where both nitrogen forms were given. Nevertheless, plants grew better on nitrate than on ammonium or mixed nitrogen source. But in contrast to this, the mycorrhizal development was best on a mixed nitrogen source, only slightly less on ammonium and very poor on nitrate. The mycorrhizal plants had generally the largest biomass, except in extreme low nutrient conditions. This is probably due to the 'costs' for the host plants to maintain the mycorrhizal fungus. Growth of the mycorrhizal fungus and mycorrhizal development was best at low nutrient concentrations.

Plants grown on ammonium contained more, and those from nitrate source less nitrogen compared to plants grown on a mixed nitrogen source. Ammonium-fed plants contained less calcium; the magnesium content did not differ and the potassium content only slightly. Much nitrogen and phosphate was accumulated in the mycelium.

Plants influenced the medium pH very strongly (Figure 9). On ammonium only, pH lowered from ca. 4.5 to ca. 3.2; on nitrate only pH increased to 7.8, while on mixed feeding pH remained ca. 4.5. Fresh weight production and nutrient uptake in buffered solutions were highest at pH 4, while at pH 5 the formation of side roots was highest. Growth at pH 3 was poor, and nutrient content of the plants lowest. However, a low pH in combinations with low nutrient concentrations did not cause problems in plants or mycorrhizas, even when ammonium was the only nitrogen source. A low pH together with high nutrient concentrations was very bad for both the Douglas fir plants and the mycorrhizal fungus, and mycorrhizal development was low.

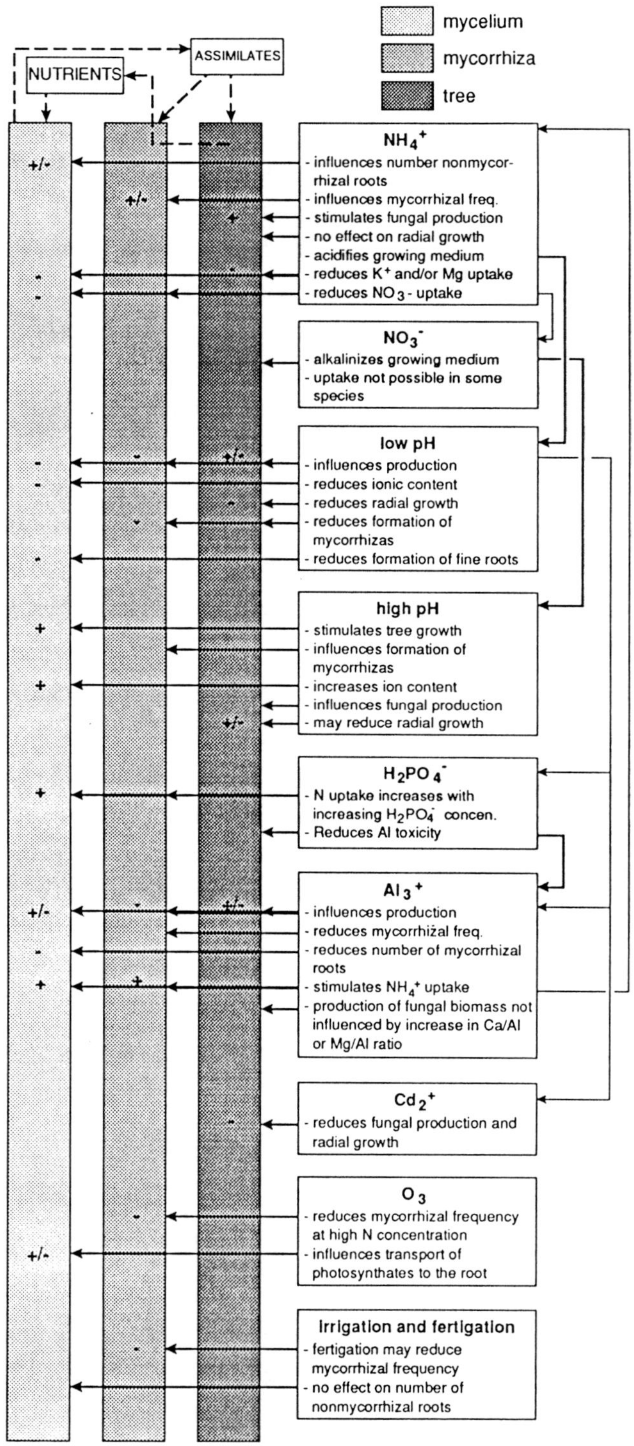

Fig. 8. Scheme indicating the positive or negative effects of several agents on the mycelium of mycorrhizal fungi, the mycorrhizas and the Douglas fir trees. (From A.E.Jansen et al., project 108)

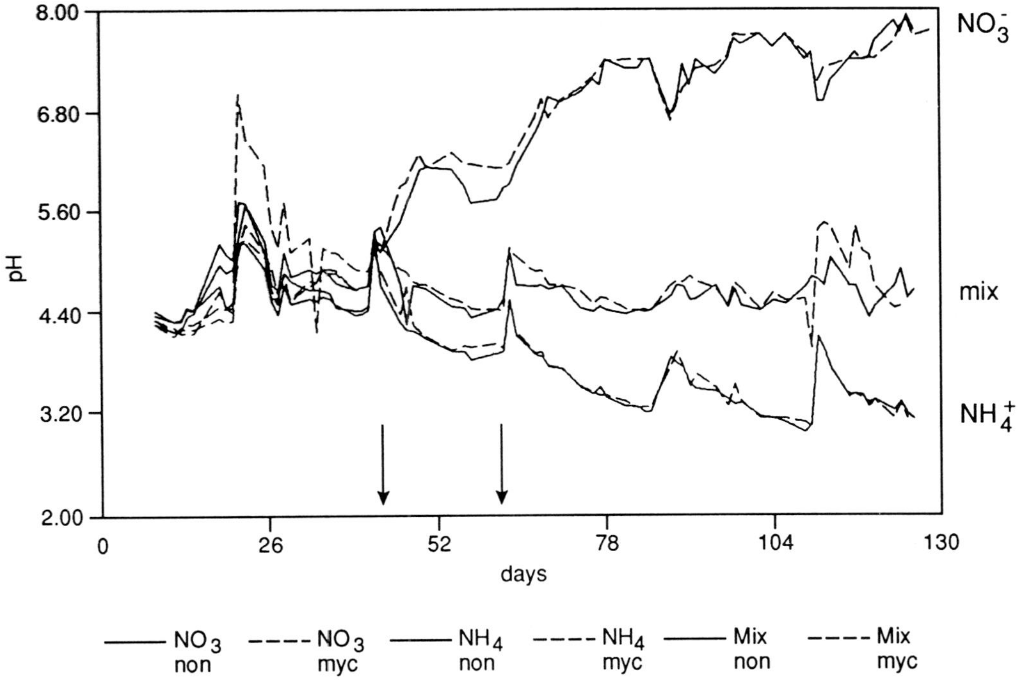

Fig. 9. Time course of the pH of the culture solution when mycorrhizal and non-mycorrhizal Douglas fir seedlings were grown on nitrate, ammonium or mixed feeding. All seedlings were kept on a mixed feeding (NO_3^- : NH_4^+ = 53:47) from day 1till 43; from day 43 till 63 plants got either NO_3^- only, mix 53:47 or mix containing NO_3^- : NH_4^+ = 25:75; from day 63 on plants got either NO_3^- only, mix 53:47, or NH_4^+ only. (From C.Kamminga-Van Wijk, project 108 B)

Low aluminium concentrations (0.1 mM, 6 weeks) did not damage the plants and even stimulated growth and ion uptake compared to no aluminium at all. Growth and ion uptake at 0.5 mM aluminium were nearly the same as in the case of the control. Negative effects on plant growth, ion uptake and especially mycelium growth and mycorrhiza formation were found at higher concentrations (1 and 2 mM). So, the mycorrhizal fungus was much more sensitive than the plant to high aluminium concentrations, but the author could not exclude an interaction with the relatively high, not limiting, nutrient concentrations used in this experiment. The mycelium accumulated aluminium in concentrations up to 10 times higher than in the plant roots. This could probably protect the tree root from detrimental effects of the aluminium. Calcium and magnesium were reduced, and uptake of nitrogen especially ammonium was stimulated by the aluminium, but the mechanism for this is unknown.

Project 108 C studied biomass growth, radial growth and nutrient uptake of 3 isolates of mycorrhizal fungi under several 'acid rain conditions'. What was studied were the influence of several pH values and ammonium concentrations on radial growth, biomass production and the uptake of ammonium, phosphate and potassium, and the aluminium and cadmium sensitivity of the fungi.
Selected for this study were three fungal species; one that had been decreased in the Netherlands during the past 20 years, one that had been increased, and one that had been more or less constant in occurrence: Lactarius rufus (Scop.ex) Fr., Lactarius hepaticus Plowr. ap. Boud. and Laccaria bicolor (Maire) P.D. Orton respectively. These 3 isolates were cultured from fruitbodies found in Douglas fir stands. The pure cultures were grown on solid MMN medium but with a reduced phosphate concentration, 4 mM instead of 6 mM, by replacing $(NH_4)_2HPO_4$ with an equivalent amount of NH_4Cl. Aluminium sensitivity was tested at pH 3.5 in solid or liquid MMN media where salts were added at one-fourth of the strength. Uptake studies were also done in liquid medium at one-fourth strength.
Radial growth in both Lactarius species was optimal at pH 4.0, and inceased in Laccaria bicolor with increasing pH up to 6.6. Biomass production had different optimum pH values, 3.0, 4.0 and 4.8 for Lact. rufus, Lact. hepaticus and Lacc. bicolor respectively.
All 3 isolates showed increased biomass production and decreased radial growth with increasing ammonium concentrations from 1 to 10 mM. During growth, the fungi acidified the medium; this acidification increased considerably with increasing concentrations of ammonium. The relative increase of biomass production with increasing ammonium concentrations is higher in buffered media than unbuffered media, as in unbuffered media the pH can drop easily below optimum values.
Toxic effects of aluminium (0.1 and 0.3 mM) were most severe in liquid medium with low phosphate concentrations (20 μM). Both Lactarius species were more sensitive than Laccaria bicolor. At high phosphate concentration (120 μM) the sensitivity to aluminium of the Lactarius species was reduced. The mechanism of the phosphate-aluminium interaction is not clear. Sensitivity to aluminium was not altered when calcium or magnesium was added, that is when the ratios of aluminium to calcium or magnesium were lowered. The effect of aluminium is greatly reduced when the medium is solidified with agar.
Sensitivity to cadmium is much higher than to aluminium. At 1 μM Cd radial growth and biomass production were reduced by 100% in the case of Lact. hepaticus, by 91 and 95%

respectively in the case of Lacc. bicolor and by 14 and 8% respectively in the case of Lact. rufus. In Lacc. bicolor the lag time was increased, but not so in the other fungi. The influence of cadmium was studied in one experiment only, as an example of the influence of heavy metals. It did not get more attention because the increase of the heavy metal concentrations in the soil due to 'acid rain' is of minor importance, compared with other changes in soil chemistry.
Both age of the culture and medium pH influences the uptake rate of potassium. Uptake rate is highest in mycelium from the exponential growth phase. The 3 fungi showed different potassium uptake at different pH, but for all a pH between 4 and 5 appeared to be optimal. In Lacc. bicolor the presence of ammonium did not affect the potassium uptake rate in short term experiments, however, inhibition occurred in long-term experiments. Phosphate uptake (at concentration of 5 μM) was reduced at aluminium concentrations above 100 μM.
If reduced radial growth occurs in natural forest stands too, it could imply a reduction in the capacity of the fungi to reach young, growing roots, and thus to form mycorrhizas. It also could imply a reduction in absorbing surface and nutrient uptake capacity.

Project 108 A comprised the field observations and greenhouse experiments of the mycorrhiza studies. It was building on the results of field studies from the first phase of the DPPA (1985-1987). From that study on 25 permanent plots under different load of air pollutants it was concluded that there were indeed less fruitbodies and less species of mycorrhizal fungi per unit area of Douglas fir stands and lower mycorrhizal frequencies in regions with a relative high load of atmospheric pollutants compared to relatively clean regions. The number and frequency of mycorrhizas and the numbers of species and fruitbodies were consistently and inversely related to the loads of atmospheric pollutants, except NH_3, but only some of the associations were statistically significant.
There were interactions between the effect of region, which is related to the load of atmospheric pollutants, and the effect of stand age. The mycorrhizal flora in young stands before canopy closure appeared to be not very much affected by atmospheric pollution. The impoverishment of the mycorrhizal flora was especially found in old forest stands in regions with a relative high load of atmospheric deposition.
Tree hosts in the more polluted regions were still producing fine root tips, and therefore potential sites for mycorrhizal fungi to form mycorrhizas, which, however, seemed to be unable to colonize these root tips. This suggests an effect of changed conditions in the soil, rather than a lowered transport of photosynthates to the fine roots.
The aim of the fieldwork in project 108 was to estimate the yearly carbon use by the population of mycorrhizal fungi. Ingrowth cores were installed in the top 5 cm of the soil (the layer with highest density of fine roots and mycorrhizas) and harvested after 10

months. Mycorrhizal development in these plots was also assessed regularly. Estimations were made of annual production of mycorrhizal fungi. The influence of irrigation and fertigation on fruitbody production and mycorrhizal development was studied in ACIFORN stand Kootwijk; the effects of ammonium sulphate, also in combination with O_3, on mycorrhizal development were studied in pot experiments; both in co-operation with other projects.

It seemed likely that all fine roots present in an ordinary root sample were younger than one year. In stands suffering from a higher load of atmospheric deposition, a very much higher proportion of newly formed root tips were found to die within one year than in less polluted stands.

Annual production of mycorrhizal fungi varies between 0.541 and 480 kg/ha/y. In stands in more polluted regions it was distinctly lower than in the less polluted regions. The energy comsumption needed for this annual production of the mycorrhizal fungi was estimated to vary between 1.1 and 1000 kg/ha/y, which is not very high when compared with values found in Douglas fir stands in their natural habitats (see also project 201).

The irrigated and fertigated plots did not show an increase in mycorrhizal status and fruitbody production. It was not clear whether the differences in root and mycorrhiza characteristics found were caused by the treatment or were reflecting the differences that existed before. An effect of irrigation, and especially fertigation, on mycorrhizal status and fruitbody production could not be demonstrated.

In the pot experiment with ammonium sulphate, mycorrhizas were significantly reduced by the treatments. Adding ammonium sulphate also caused a lowered pH of the bulk soil and an increased concentration of Al^{3+}. Using statistical techniques, it appeared likely that mycorrhizas were more affected by a low soil pH and less by high concentrations of NH_4^+ or Al^{3+}. But in contrast to mycorrhizas, non-mycorrhizal roots seemed to be reduced by high Al concentration. Mycorrhizal and non-mycorrhizal roots did not seem to be influenced by calcium:aluminium or magnesium:aluminium ratios. A calcium fertilization, often used as a remedial treatment for low pH or high Al concentration, is therefore likely not to be a remedy for an impoverished mycorrhizal flora.

Ozone-treated plants appeared to have a better mycorrhizal development than the controls. In the control plants the mycorrhizal frequency increased with increasing nitrogen addition. Ozone in combination with a high load of nitrogen was detrimental for the mycorrhizal development.

The review project 201 ("Effects of air pollutants on ectomycorrhizas, a review", A.E. Jansen & J. Dighton) aimed to discuss the effects of atmospheric pollutants on ectomycorrhizal fungi in the temperate zone of Europe. The main questions were: What are the effects of the different pollutants on fungi and thereby on trees? What are the effects for ecosystems? What is known about the processes? This project was a literature study and did not include practical research.

The morphology and occurrence of mycorrhizas in temperate climates are relatively well known. Although usually considered a mutualistic symbiosis, much of the physiology of mycorrhizas is still unknown, especially the contribution to tree growth in mature trees. Two theories on the regulation of mycorrhiza formation exist, the carbohydrate theory (Bjorkman, 1949) and the hormone theory (Slankis, 1951). How air pollutants affect mycorrhizas can not be directly derived from these theories, nor the effect on tree nutrition. Direct effects of gaseous pollutants on mycorrhizas are unlikely, as these gases will not penetrate the soil. Effects on mycorrhizas will be indirect effects, either soil-mediated or tree-mediated.

There are several soil-mediated effects. Atmospheric pollutants induce a lowered pH in the soil. As also mycorrhizal fungi have pH optima for their growth, a shift in soil pH may result in a changed species composition. This was indeed observed by several authors. A lowered pH induces changes in soil chemistry, which might also be detrimental for mycorrhizal fungi. An increased aluminium concentration was found to reduce hyphal growth, but not all mycorrhizal fungi are equally susceptible. The same is true for other metal ions released by lowered pH, but several data support the idea that mycorrhizal fungi are able to protect the roots against accumulation of toxic levels of heavy metals.

Atmospheric nitrogen deposition may acidify the soil, but input of nitrogen mainly gives a fertilization effect, especially on nutrient poor soils. The effects of nitrogen input are often studied in fertilization experiments. Sometimes an increase in fruitbody production was found, but mostly fruitbody production, and mycorrhizal development, decreased. The differences between the results of these fertilizer experiments may well be explained by the balance of available nutrients in the soil solution, and so depend also on the type of nitrogen used, the amounts and the method of application.

Decreased transport of photosynthates (whether photosynthesis is reduced or only the transport is not very relevant for the below ground compartment) may very well reduce the mycorrhizal development. The carbon use of a population of mycorrhizal fungi varies, from ca. 1 kg/ha/y in a diseased stand (see 108A) to 30,000 kg/ha/y in a natural Douglas fir stand (Fogel & Hunt, 1979). In western Europe, the carbon use by the mycorrhizal population may amount to 3 to 4 t/ha/y. Experiments have shown that reduced transport of photosynthates and reduced root respiration are consequences of fumigation with SO_2 and O_3. The interrelations between carbon allocation and the mycorrhizal development are, however, not clearly understood.

The influence of atmospheric pollutants on mycorrhizal fungi is not demonstrable in young stands before the canopy closure. Mycorrhizal fungi in mature stands are, however, very heavily affected, and their populations there are almost completely destroyed, as can be seen from the total absence, or almost total absence, of fruitbodies and mycorrhizas. In terms of 'strategies' this means that the K-selected fungi, characteristic of undisturbed, mature stands, are apparently more vulnerable than the r-selected fungi from young stands. These r-selected fungi can be seen as 'pioneers', that is, fungi that are able to colonize disturbed habitats. The implications for the nutrition of the trees are at present not fully known (Dighton & Mason, 1985, Dighton in press).

4. DISCUSSION

An important question in air pollution effects studies on trees has always been, whether the direct or the indirect effects are predominant in causing the reduced vitality and reduced growth of the trees and the ultimate die-back of the forests. In essence it is probably not one of these causes, but the varying combination of both types of effects, occurring simultaneously or sequentially, which is responsible for the phenomenon. It is quite clear that a complicated multi-causal complex is involved in the process of deterioration of a forest in reality. Direct effects on the above ground parts of trees are caused by several phytotoxic air pollutants, mostly in combinations, influenced by varying environmental conditions, determined by climatic and edaphic factors among other things. Indirect effects are evoked by atmospheric deposition of pollutants on the soil, causing changes in this important medium of the trees which may be detrimental. In this way direct effects may also be influenced by indirect ones, and vice versa. It is very difficult to say which are the most important, and it is most appropriate and useful to study all these effects together.
In below ground and rhizosphere processes it is useless to distinguish between direct and indirect effects, as is usually done in air pollution studies on above ground parts of organisms. Gaseous pollutants, when brought in direct contact, have great effects on the functioning of mycorrhizas: both O_3 and SO_2 severely reduced the oxygen consumption of excised mycorrhizal and non-mycorrhizal root fragments, and of mycorrhizal fungi in pure culture. It is, however, unlikely that fine roots, mycorrhizas, fungal hyphae (mycorrhizal and saprophytical) and soil microorganisms in natural soil systems will be exposed to these toxic gases. All gases will be readily absorbed into solution on the surface of soil particles. That is why the section on below ground effects speaks of "indirect effects" only, but apart from soil mediated, these indirect effects can also be tree-mediated.
An integrating approach has been applied as far as possible in the research on the thematic topic of biological and physiological effects of air pollution on Douglas fir, but there are still some possibilities unexplored till now. The combination of direct and indirect effects has only been studied in projects 107 and 108 A on the mycorrhiza of Douglas fir trees, influenced by O_3 in the air and NH_4^+ in the soil, but should also be investigated in relation to above ground phenomena. For example, the interaction of atmospheric uptake of SO_2 by

the needles and the uptake of the indirectly presented NH_4^+ by the roots could have been studied with respect to the disturbance of the pH-buffering capacity of the trees (project 115). This might explain among other things the very drastic effects seen in areas in the Netherlands with relatively high concentrations of SO_2 and NH_3, and the effects of fumigations with SO_2 + NH_3 in other projects of the DPPA (124/125), because NH_3 may have caused higher concentrations of NH_4^+ in the soil and in the trees.

Comparing the results of the research projects with the original objectives, it is clear that not enough variation in exposure levels has been applied to be able to determine the total quantitative relationships between air pollution and effects. A complete series of several fumigation experiments with varying exposure concentrations and exposure periods in different environmental conditions have not been carried out, because of lack of facilities and manpower. So, total no-adverse effect lines for a specific pollutant in relation to a special sensitive object have not been determined. This is a general feature in air pollution effect studies, and it should be changed so as to be more helpful for the sustaining of realistic policy measures. Also, to produce better simulation models, experiments with a series of exposure intensities should be performed.

An important gap is also present in the sense of experiments with more realistic combinations of several air pollutants as they occur in the field. Combination effects of several components will have to be known, to set the limits for the occurrence of these components in the air. Answers to the question which component is the most harmful for the forest stands have not been given by the projects. This is also very difficult in the light of the possible combination effects, which may be additive but also more or less than additive.

A lot of information has been produced by the research projects, especially on the mechanisms of action of air pollutants on plants/trees. The scientific merits of this research can be found in the new ideas to combine separate knowledge into the modelling of the processes involved in possible forest decline. The negative effects of air pollutants on the photosynthesis and growth of Douglas fir trees may be predicted by the use of these mechanistic simulation models. These models might be improved when also combination effects of several air pollutants and of air pollutants and other stress factors are included.

Comparison of the tree growth characteristics and the needle wax structure of Douglas fir trees from the two field research sites of the DPPA has not revealed a possible difference in effects of atmospheric deposition on both sites. It is likely that the air pollution at the two sites did not differ sufficiently to cause different effects. But it is also possible that the level of air pollution at both sites was too low to produce any effects in this sense. This can only be proven by comparison of the trees in filtered and non-filtered ambient air. Manipulation of the air environment in the Douglas fir stands to a well-known (measured) level of air pollution by artificial fumigations would have been of great help in looking for no-effect levels. At the moment there are mainly results of indoor fumigations of juvenile trees to show the potential effects of the pollutants.

Studying the growth of trees of two plots with slightly different loads of air pollution, De Kort (project 111-1) did not find any differences. Her observations did not, however, include trees from relatively unpolluted areas in the North of the Netherlands, nor from the heavily NH_3 polluted areas in the South of the country. As far as we are aware such data are not available. Jansen & De Vries (project 25) did not include growth when they studied the vitality of Douglas fir in plots in the northern, central and southern parts of the country, but their impression was that stands between 20 and 40 years old did not differ very much in growth. In particular, the growth in these medium-age stands from the southern region did not seem to be very bad (Jansen, own observations). This may, however, be different for other tree species, such as Scots pine, but also for older Douglas fir trees.

A reduction in photosynthesis, as found by Van Hove and Van Kooten when fumigating trees with SO_2 and O_3, could possibly occur also in field situations. Such a reduction in photosynthesis did not apparently influence the tree growth. This could only be so when normally, that is in an unpolluted situation, a great surplus of assimilates existed, and this surplus of energy was wasted by a 'waste respiration'. Waste respiration is indeed a well-known phenomenon in plants.

At least a part of this 'waste' could have taken place in the fine roots, in the form of respiration and exudation, thereby supplying rhizosphere microorganisms and mycorrhizal fungi. Indeed a large reduction of root respiration was found by Gorissen (project 107) when fumigating with O_3 and SO_2. It is therefore perfectly possible that also the dependent organisms receive less energy in an air polluted situation. This can explain, at least partly, the observed decline and shift in species composition of mycorrhizal fungi and rhizosphere microorganisms. However, quantitative data on photosynthesis and carbon transport through the plant are necessary for understanding the relation between the tree and its symbionts.

That tree growth was not reduced, as said above, may be a consequence of the large nitrogen input in forests. In most of the Dutch forests, nitrogen was always limiting. A small amount of nitrogen input may result in enhanced growth, even when the photosynthesis is reduced by atmospheric pollutants. As a consequence of this enhanced growth again less assimilates will be 'wasted' in the tree, and again less energy will be available for mycorrhizal fungi and rhizosphere microorganisms. Again, quantitative data are lacking, but both the nitrogen input and the reduction of the flora of mycorrhizal fungi started in the fifties in the southern part of the Netherlands.

Nitrogen input mainly takes places as ammonium input. This ammonium is rapidly nitrificated in the soil by soil microorganisms, and according to Van Breemen & Van Dijk (1988) the nitrification rates are sufficiently high to transform all atmospherical deposited ammonium to nitrate. Douglas fir prefers ammonium over nitrate (Gijsman project 83, Kamminga-Van Wijk project 108 B), but ammonium feeding has negative consequences for tree growth. However, ammonium is less available in relatively dry soil and most of the Dutch Douglas fir stands are situated on dry soils. It seems likely that with strong nitrification and in such dry conditions the trees will mainly feed on nitrate, especially in the growing season. In wetter seasons, autumn and winter, ammonium uptake may be larger than nitrate uptake. Possible negative effects are therefore to be expected in these seasons. We do not know whether such negative effects do really occur in Douglas fir trees. Needle yellowing in late winter and early spring is, however, a well known phenomenon in Scots pine, and could be related to this higher ammonium uptake in winter.

It seems therefore not unlikely that atmospheric pollutants have reduced tree growth less than was expected compared with air pollution levels in other countries, due to the relatively high deposition of ammonium followed by quick nitrification and in combination with a low availability of ammonium due to the relative drought of our forest soils. If this is true, growth reduction can be expected when ammonium is more readily available, that is at moister sites with excessive input of ammonium, or when the nitrification is insufficient.

The combination of an increasing concentration of ammonium and a low pH appeared to be also detrimental for mycorrhizal development. The lower capacity to colonize young, growing root tips may be the consequence of the reduced lateral growth rate of mycorrhizal fungi when pH is not optimal. Although some work has been done on the effects on tree growth of having a mycorrhizal symbiosis (project 108 B), most of the physiology of the symbiosis is still unknown.

In the light of the good tree growth as observed by De Kort (project 111-1) and the relatively poor mycorrhizal development in those stands as observed by Jansen & De Vries (project 108 A), it is questionable whether trees, of that age, really need mycorrhizas for their growth. In juvenile trees it is, however, evident that some strains of mycorrhizal fungi give very strong growth effects. This effect is generally thought to be the result of the production of plant growth hormones by the mycorrhizal fungus. Whether such effects still play a role in older trees is unknown, as are the effects of air pollutants on plant hormone production and functioning.

The results of the experiments on the effect of aluminium on mycorrhizal fungi and mycorrhizas are complicated. Jongbloed (project 108 C) found stronger effects on pure cultures in a liquid than in a solid medium. This may probably explain the high sensitivity of the fungus to aluminium, found in the hydroponic systems (Kamminga-Van Wijk, project 108 B), and the much lower sensitivity in the pot experiments (Jansen, project 108 A). Jongbloed studied only three isolates of mycorrhizal fungi, Kamminga-Van Wijk only one and Jansen's pot experiment had an uncontrolled mix of various species. If results from these experiments are contradictory with results reported elswhere, this can possibly be explained by the large inter- and intraspecific variation within mycorrhizal fungi. Negative effects of a high aluminium concentration seem a little reduced by high phosphate concentrations. In natural situations phosphate is usually available in low concentrations; effects of aluminium could possibly be a bit less detrimental on phosphate rich soils. In situations with a low pH where phosphate is less available or where a dense mat of grasses, as Deschampsia flexuosa in combination with their VA-mycorrhiza, is competing successfully for the available phosphate, a strong negative effect of aluminium on ectomycorrhizas may be expected.

5. CONCLUSIONS

- Short-term and long-term effects of SO_2 and O_3 in fumigation experiments have been shown to be responsible for disturbances of the metabolism of Douglas fir trees, leading to reduced growth. The information gathered about the uptake and effects of SO_2 and O_3 in needles has shown that these air pollutants may reduce the photosynthesis, and that recovery from this reduction is possible. The use of the quantitative information on exposure-response relations in building mechanistic simulation models may help to predict what is the ultimate effect on the growth of a forest stand.

- From the boundary layer resistance, stomatal conductance, and NH_3 concentration close to the needle surface, the NH_3 transfer into needles of Douglas fir trees could be calculated. Besides, a large quantity of NH_3 was found to be adsorbed to twig and needle surfaces.

- The specially adapted chlorophyll fluorescence measuring technique was of great use for the assessment of long-term effects of air pollutants on the photosynthesis of Douglas fir needles. NH_3, SO_2, NO_2 and some combinations (NH_3 + NO_2 and SO_2 + NO_2) have been shown to cause clear inhibition of the maximal electron transport rate (Jmax) in the photosystem II, but this reduction in Jmax could be compensated for by production of extra chlorophyll in the fumigations with NO_2, NH_3 and NO_2 + NH_3.

- Concentrations of NH_3, NO_2, SO_2 and O_3, normally occurring in the Netherlands, may not be high enough to cause drastic effects on the trees per se, but combinations with direct and indirect effects of other pollutants and other abiotic and biotic stresses may still be threatening the vitality of the trees. Combination effects of several air pollutants, other abiotic stresses and diseases and pests, however, have not been studied sufficiently in this research to be able to draw conclusions on their possible importance in the process of reduction in the vitality and growth of Douglas fir. But it is expected that all types of stress factors are involved in the integrated combination effects to be found in the forests.

- Xylem sap transport velocity may be investigated by the method of Heat Pulse Velocity, giving only relative values. Much variation in the results, specially between plots, was found and could not be related to air pollution effects. The method will only be useful for studying effects on patterns of xylem sap transport in trees.

- Studies on the year ring increment, wood production and water transport capacity of the Douglas fir trees at the two field study sites of the Dutch Priority Programme on Acidification revealed no significant differences between the trees of the two sites. No recent decline in radial, axial or volume growth was shown and a normal amount of sapwood was present, with only slightly sub-maximal moisture content.

- From the same two sites epicuticular needle wax morphology of Douglas fir, studied by Scanning Electron Microscopy, did not show any difference. Fumigations in open-top chambers with different concentrations of NH_3 over a period of 5 weeks did not affect the wax morphology in current year needles, but one-year-old needles exposed to these artificial fumigations from bud break onwards showed a severe degradation of the crystalline wax.

- Nitrogen is the main factor influencing processes in the rhizosphere. Douglas fir trees preferentially absorb nitrogen in ammonium form. Ammonium absorption in the absence of nitrate absorption has a lowered rhizosphere pH and a lowered internal pH as consequence. The lower rhizosphere pH will cause a changed morphology of the fine roots, and possibly a reduced water and nutrient uptake. A lowered internal pH, and probably also a high internal ammonium concentration, are symptoms of physiological disorders, leading to very poor growth and high mortality.

- Addition of ammonium sulphate reduced the size of the bacterial rhizosphere populations, reduced the mycorrhizal development, reduced the uptake of P, Mg, Ca, Fe and Zu, and increased the uptake of N. The combination of ozone and ammonium sulphate did not reduce the bacterial rhizosphere population but did decrease the mycorrhizal development. This effect has to be attributed to ammonium sulphate rather than to ozone.

- Ozone appeared to disturb the transport of photosynthates within the tree. Most of the assimilated $^{14}CO_2$ was retained in the needles and transport to other parts was apparently blocked. Reduced supply of photosynthates to roots may well explain the changes in rhizosphere bacteria and mycorrhizas, but detailed, quantitative data are still lacking.

- Although mycorrhizal fungi can use ammonium and seem not to react strongly to high ammonium concentrations, such a high ammonium concentration in combination with a low pH reduces the lateral growth of the fungi. This results in a lowered capacity to form mycorrhizas, but the consequence for the nutrition of the trees is unknown. In conclusion from the good growth and the poor mycorrhizal development of those trees, it is questionable whether trees of that age really need mycorrhizas.

- It is possible that no tree growth reduction occurred in the Netherlands, due to relatively high ammonium deposition, followed by nitrification and in combination with low ammonium availability due to relative drought of the forest soils.

6. REFERENCES

Anonymous, 1989. De vitaliteit van het Nederlandse bos 7. Directie Bos- en Landschapsbouw, Ministerie van Landbouw en Visserij. Rapport 1989-16, 26 pp.

Bjorkman, E., 1949. The ecological significance of the ecotrophic mycorrhizal association in forest tree. Svensk. Bot. Tidskr. 43: 223-262

Dighton, J., in press. Aquisition of nutrients from organic resources by mycorrhizal autotrophic plants. Experientia

Dighton, J. & Mason, P.A., 1985. Mycorrhizal dynamics during forest tree development. In: D. Moore, L.A. Casselton, D.A. Wood & J.C. Frankland (Eds), Developmental Biology of Higher Fungi. Univ. Press, pp. 117-139

Fogel, R. & Hunt, G., 1979. Fungal and arboreal biomass in a western Oregon Douglas fir ecosystem: distribution patterns and turnover. Can. J. For. Res. 9: 245-256

Genty, B., Briantais, J.M. & Baker, N.R., 1989. The relationship between the quantum yield of photosynthetic electron transport and quenching of chlorophyll fluorescence. Biochim. Biophys. Acta 990: 87-92

Hiltner, L., 1904. Uber neuere Erfahrungen und Probleme auf dem Gebiet der Boden-Bakteriologie und unter besonderer Berucksichtigung der Grundungung und Brache. Arb. Deut. Landw. Ges. 98: 59-78

Jansen, A.E., in press. The mycorrhizal status of Douglas fir in the Netherlands: its relation with stand age, regional factors, atmospheric pollutants and tree vitality. Agric. Ecosyst. Environm.

Kropff, M.J., 1989. Quantification of SO_2 effects on physiological processes, plant growth and crop production. Ph.D. thesis, Agricultural University, Wageningen, 201 pp., ISBN 90-9002942-7

Posthumus, A.C., 1991. Effects of air pollution on plants and vegetations. In: J. Rozema & J.A.C. Verkleij (Eds), Ecological Responses to Environmental Stresses. Kluwer Academic Press, Dordrecht, 191-198

Schreiber, U., Schliwa, U. & Bilger, W., 1986. Continuous recording of photochemical and non-photochemical fluorescence quenching with a new type of modulation fluorometer. Photosynth. Res. 10: 51-62

Slankis, V., 1951. Uber den Einfluss van ß-Indolylessigsaure und andere Wuchsstoffen auf das Wachstum von Kieferwurzel. Symb. Bot. Upsaliensis XI

Van Breemen, N. & Van Dijk, H.F.G., 1988. Ecosystem effects of atmospheric deposition of nitrogen in the Netherlands. Environ. Pollut. 54: 249-274

Van Hove, L.W.A., Tonk, W.J.M., Pieters, G.A., Adema, E.H. & Vredenberg, W.J., 1988. A leaf chamber for measuring the uptake of pollutant gases at low concentration by leaves, transpiration and carbon dioxide assimilation. Atm. Env. 22 (11): 2515-2523

Van Hove, L.W.A., Van Kooten, O., Adema, E.H., Vredenberg, W.J. & Pieters, G.A., 1989. Physiological effects of long-term exposure to low and moderate concentrations of atmospheric NH_3 on poplar leaves. Plant Cell Env. 12: 899-908

Van Kooten, O. & Van Hove, L.W.A., 1988. Fluorescence as a means of diagnosing the effect of pollutant induced stress in plants. In: P. Mathy (Ed.), Air Pollution and Ecosystems. D. Reidel Pub. Co. pp. 596-601

Visser, H., Noppert, F., Van Wakeren, H. & Vaessen, J., 1989. Xylem sap velocity in relation to weather and air pollution. IAWA Bull. n.s. 10 (4): 427-439

Visser, H. & Maessen, P.P.Th.M., 1990. Responses of trees to weather variations and air pollution: a tree-ring based approach. KEMA/De Dorschkamp, report 50385-MOF 90-3394/NPZR 73-7

INTEGRATED EFFECTS (FORESTS)

G.M.J.Mohren[1)]

1) Research Institute for Forestry and Landscape Planning "De Dorschkamp", Wageningen

A. PREFACE

As part of the Dutch Priority Programme on Acidification, groups of project reports with either methodological or thematic affinity are integrated in thematic reports that are used for further integration as part of the final evaluation procedure. This thematic report considers the effects of air pollution, atmospheric deposition and soil acidification on forests in the Netherlands. The complex nature of the problems involved in analyzing effects of acidification called for a large variety of research projects within the scope of the Priority Programme. Due to the multidisciplinary nature of the programme and the need for integration, thematic reports cannot be strictly separated. Hence, this report and the thematic reports on soil acidification, on biological and physiological effects, and on atmospheric deposition show some overlap.

The present report chiefly consists of a summary of the findings of the ACIFORN (ACIdification of FORests in the Netherlands) monitoring projects. The conclusions presented are the result of the joint interpretation of the material available up to date (May 1990), and should be considered a shared responsibility of the ACIFORN participants involved. Parts of this report were written or commented upon by A.W. Boxman, W. de Vries, P.W. Evers, A.J. Gijsman, P. Hofschreuder, I.T.M. Jorritsma, G.M.J. Mohren, A.F.M. Olsthoorn, W.L.M. Smeets, M.P. van der Maas, J.R. van de Veen, J. van den Burg, H.F.G. van Dijk, and A.W.M. Vermetten.

B. SUMMARY

This report summarizes the main findings from the ACIFORN monitoring projects, and from a number of related projects carried out as part of the Dutch Priority Programme on Acidification. These results are used to evaluate hypotheses on possible cause, mechanisms and effects of forest decline under conditions in the Netherlands. The hypotheses relate to traditional factors, to effects of ozone, to combined effects of SO_2, NO_x, and NH_3, to soil acidification, and to excess nitrogen deposition. With the exception of ozone, concentrations of air pollutants at the ACIFORN sites in 1988 and 1989 were rather low with two-year averages of 10 $\mu g\ m^{-3}$ for SO_2, 25 $\mu g\ m^{-3}$ for NO_2, 5 $\mu g\ m^{-3}$ for NO, and 5 $\mu g\ m^{-3}$ for NH_3. These concentrations are unlikely to have resulted in major short-term effects on tree physiology. Two-year average concentrations of ozone however, amounted to 50 $\mu g\ m^{-3}$, with hourly concentrations frequently reaching levels high enough to be damaging to plants (>200 $\mu g\ m^{-3}$). Long-term effects of low exposure to air pollutants may very well result in changes in assimilate distribution and decreased productivity, but these could not be established within the time-span of the present investigations.

At both sites, soil acidification mainly resulting from deposition of NH_3 and SO_2, has advanced to the point where more than 50% of the CEC is occupied by aluminium and base cation complexation is less than 5% in the mineral soil. Of the external proton input at Speuld 18% originates from NO_x, 49% from SO_2 and 33% from NH_3. In Kootwijk this amounts to 21% NO_x, 55% SO_2 and 24% NH_3. At both sites 83 and 79% of the proton production is estimated to be buffered by dissolution of aluminium hydroxides. Soil acidification amounts to 4.40 and 3.74 kmol $ha^{-1}y^{-1}$ for Speuld and Kootwijk. 50 to 60% of the nitrogen input is leached from the root zone as NO_3^-, accompanied by Ca^{2+}, K^+ and Mg^{2+}; both ecosystems still accumulate nitrogen. $[NH_4^+]/[K^+]$ and $[NH_4^+]/[Mg^{2+}]$ ratios are critical (>5) to a depth of about 10 cm below the soil surface; below 10 cm, ammonium concentrations decrease due to nitrification. At the ACIFORN sites, $[Al^{3+}]/[Ca^{2+}]$ ratios exceed the critical value of 1 below 20 cm from the soil surface. Nitrogen deposition, excluding gaseous uptake by the foliage, amounts to 47 and 50 kg $ha^{-1}yr^{-1}$ for the monitoring sites. Nitrogen deposition is mainly in the form of the NH_4^+, being a major source of protons leading to soil acidification due to nitrification. As a result of NH_4^+ preference for uptake, this deposition of NH_4^+ is considered potentially dangerous as its uptake leads to additional rhizosphere acidification and enhances nutrient disorder.
It is concluded that nitrogen deposition is the main factor governing present day large-scale changes in Dutch forest ecosystems. Nitrogen deposition has an immediate effect by causing nutrient imbalances of trees and stands, and is a major factor determining soil acidification. As a result of this soil acidification, leaching of cations and nitrate occurs, aluminium concentrations are increased, and root uptake capacity of water and minerals is reduced, thereby enhancing already existing nutrient imbalances.
Through changes in the root system, soil acidification makes the stands more vulnerable to drought and nutrient shortage. As a consequence of nutrient imbalances, especially related to increased uptake of NH_4^+ instead of NO_3^-, it is expected that the trees become more vulnerable to direct effects of pollutants due to limitations in metabolic buffering and detoxification capacity. With ongoing soil acidification a change from aluminium to iron buffering is expected on dry sandy soils within periods of 30-200 years from now, depending on soil conditions and acid deposition. This change is accompanied by a drop in pH of 1 unit, which is expected to severely affect tree growth.

1. INTRODUCTION

1.1 Aims and structure of the report

This thematic report contains the main results from the ACIFORN experiments and monitoring projects, and from a number of related projects carried out as part of the Dutch

Priority Programme on Acidification. As far as possible at this stage, the report aims at an integrated assessment of the effects of air pollution and soil acidification on Dutch forests based on the information available at present.

Most of the results discussed here refer to the Douglas fir stands that were used within ACIFORN; an attempt will be made to determine the general applicability of the ACIFORN results with regard to other sites within the Netherlands. Also, the key findings from ACIFORN will be compared to research results from other investigations published recently.

Despite a considerable research effort in recent years, experimental evidence and understanding of the impact of air pollutants and acidic deposition on trees and forests is still fragmentary, with large gaps in knowledge, conflicting results, incomplete data sets, etc. The sometimes highly complex nature of the chain of events possibly leading to forest decline renders it impossible to establish the validity of simple and clear causal mechanisms. The necessary differentiation with respect to site and species complicates an integrated assessment, especially when most information is available for one species, and for a restricted range of sites and pollution conditions only.

To discuss these matters, a dialectic approach is adapted here, by starting with a number of hypotheses that highlight a particular cause-and-effect scenario. After describing the stand structure and environmental conditions of the ACIFORN sites, these hypotheses are discussed in relation to the research carried out within the Dutch Priority Programme. Finally, the hypotheses will be compared, all evidence reconsidered, and a most likely cause-effect scenario for Dutch conditions will be proposed. The validity of the conclusions with regard to other sites and other species will also be discussed.

1.2 ACIFORN and the Priority Programme on Acidification

ACIFORN was established to analyze and quantify the effects of air pollution and soil acidification, among other stress factors, on forest growth and vitality. The majority of the monitoring work was carried out in two Douglas fir stands, and most of the discussion here will be limited to these stands. During the first phase of the Priority Programme (1985-1988), the main emphasis was on establishment of the field sites and on monitoring of ongoing processes. This included investigations into forest hydrology, soil physics and soil chemistry, air pollution, tree physiology, stand biometry, and root dynamics. During the second phase of the programme (1988-1990), emphasis shifted from monitoring to analysis, quantification, and integration of the results. Additional experimental work was started in this phase, and a number of modelling projects were included to combine the results of the experiments and monitoring work.

As most monitoring programmes were carried on into late 1989 or even early 1990, data interpretation and joint synthesis was limited up to now. It is expected that more results will emerge in the near future.

An intensive monitoring programme can have a twofold aim: to describe and analyse internal ecosystem structure from a functional point of view, and to provide a comparative analysis of comparable monitoring carried out under different conditions. Within ACIFORN, the main emphasis has been on the first aim: the description and analysis of internal system structure and the accompanying functional relationships. Hence the ACIFORN results alone cannot be used to compare processes in stands at different stages of decline; they merely serve to increase the understanding of forest functioning under ambient levels of air pollution and soil acidification. This understanding, which applies to the site conditions at the ACIFORN locations Speuld and Kootwijk, can be used to define scenarios of decline. These scenario's can then be evaluated against the background of data on soil and canopy processes collected elsewhere, e.g. in regional analysis of soil acidification and atmospheric deposition. Thus, through understanding of underlying principles, through comparison of relevant sites and species, the most likely pathways of the disturbances and their effects can be determined.

1.3 Materials used in writing the report

The basic material for this report comprises the final reports of the individual ACIFORN and related projects. In addition, a number of recently published overview papers were used to define the current hypotheses on forest decline, and to compare the main conclusions.

The projects within the Dutch Priority Programme considered in this report consist of:

Project: Title:

(* indicates that the project was part of the ACIFORN monitoring group)

Project:	Title:
P18:	The indirect effects of acid deposition on the vitality of the Dutch forests: Forest stand analysis
P83:	Nitrogen uptake and rhizosphere pH in Douglas fir
P100*:	Quantification of the effect of adverse soil conditions on the growth of Douglas fir and Scots pine by manipulating the water and nutrient supply in an existing stand

P101*: Air pollution in forest canopies

P102.1*: Monitoring of soil chemical parameters under Douglas fir and heather

P102.2*: Monitoring of soil physical parameters of two Douglas fir stands and heather

P103*: Root development of trees in environments affected by air pollution

P104.1*: Measurement and modelling of canopy water storage during and after rain, dew and fog

P105*: Research on the ecophysiology of conifers in forests in relation to air pollution and acid deposition on the soil

P107: Effects of air pollutants on the rhizosphere of Douglas fir

P111: Tree ring analysis and sapwood proportion in Douglas fir in relation to air pollution and acid deposition.

P112*: Coupling of soil, plant, and atmospheric processes and their disturbances through air pollution and soil acidification in a physiological model for Douglas fir stands.

P113: Data derivation, evaluation and application of a regional soil acidification model

P114.2: Dose-effect relationships for forest stands, to be applied in the Dutch Acidification System (DAS)

P115: Model development for the uptake of air pollutants and the effects on the physiology of Douglas needles in relation to drought

P116: Measuring sapstream velocities in Douglas fir

P118*: Experimental research on the effects of a decrease in deposition and improvements in the mineral balance on the vitality of forest in the Netherlands

In the following text, the outcome of these projects will be referred to by means of the relevant project number, e.g. P101 for the meteorological data in the project entitled "air pollution in forest canopies".

A list of the reports (or draft versions thereof) used in writing this summary report is given in the reference list.

2. PRESENT HYPOTHESE ON FOREST DECLINE

2.1 General aspects

The complex nature of forest decline at different sites within Europe, manifested by combinations of symptoms, cannot be explained by a single causal mechanism. At present, a number of hypothetical mechanisms have been proposed to account for the development of the various types of decline, each with their own symptoms and potential causes (Cowling et al. 1988; Blank et al., 1988; Klein & Perkins, 1988; Krahl-Urban et al., 1988; Mathy, 1988; Pitelka & Raynal, 1989; Prinz, 1987; Roberts et al., 1989; Schulze et al., 1989a; Van Breemen & van Dijk, 1988; Wellburn, 1989).

These hypotheses differ in their emphasis on the main causal factor, and can be used to discern the magnitude and relevance of the different types of disturbances. At present, although they cannot be used to accurately predict future developments, they indicate possible risks under various pollution scenarios. When considering the individual hypotheses, it should be kept in mind that none of the single causal mechanisms described is likely to explain all the decline phenomena reported from various countries or regions, although phenomena such as increased foliage shedding and foliage discoloration (yellowing) may appear to be ubiquitous.

2.2 Hypotheses

2.2.1 Traditional factors

The major requirements for plant growth such as photosynthetic active radiation, temperature, carbon dioxide, water and nutrients are regarded as the "traditional" factors influencing growth. Radiation, temperature, carbon dioxide and to some extent water are determined by climate; water and nutrient availability for plant growth are determined mainly by site conditions. According to this hypothesis traditional factors are the major cause of present forest decline. It suggests that the vulnerability of large parts of the forests in western Europe to climatic influences related with site and stand conditions, in combination with weather conditions in recent years, explains to a large extent the occurrence of forest decline.

In industrialized and densely populated areas in particular, forests are usually restricted to soils that are considered not profitable for agricultural purposes. These sites very often suffer from a limited supply of both water and nutrients. Site conditions such as limited water availability may have drastic effects on tree growth and needle loss during dry years such as 1976 and 1982/1983. Combinations of dry years with insect outbreaks could lead to forest damage that may take several years to recover. The type of nutrient deficiency depends on the soil type, the hydrological regime, and the atmospheric input. The latter is closely related to the proximity of the sea in combination with prevailing wind directions. In recent years, atmospheric input has changed considerably as a result of pollution from industrial sources, traffic, and agricultural activities.

In western Europe, forests are predominantly located on poor and dry sandy soil of pleistocene origin and consist of coniferous species in planted stands, especially in the lowlands, whereas stands originating from natural regeneration are more common in mountainous areas. In plantation forestry, tree breeding and provenance selection have been instrumental in improving the genetic material aimed at decreasing the risk of damage by frost, pests and diseases, and increasing yield. As part of the tree improvement programme, tree species and provenances have been planted outside the range of their natural habitat, for example in the case of Corsican pine (Pinus nigra var. maritima) in the Netherlands and silver fir (Abies alba) in parts of Central Europe. Increased vulnerability to disturbances due to incomplete adaptation, may have led to decline phenomena such as yellowing and premature shedding of foliage during years with extreme weather conditions. According to this hypothesis, damage will be concentrated in planted stands of exotic species in particular. The damage is assumed to be closely related to climatic extremes, with poor stand management acting as a predisposing factor.

2.2.2 Ozone and complex high-elevation disease

Ozone is a strong oxidant, highly reactive, and capable of disturbing a range of physiological processes by reacting with organic compounds. Leaves exposed to high doses of ozone display typical mottle symptoms, with decreased photosynthesis and increased dark respiration (Berry & Hepting, 1964; Miller & Parmeter, 1965; Skeffington & Roberts, 1985). Combinations of exposure to high concentrations of ozone and wet acidic deposition lead to increased nutrient leaching and partial breakdown of chlorophyll. Consequently, photosynthetic production is decreased and root growth retarded, making the stand more vulnerable to drought. According to this hypotheses the combination of ozone and acidic deposition leads to increased nutrient deficiency, decreased physiological activity, increased sensitivity to frost, and eventually to tree mortality and forest decline, especially at high elevations (Prinz, 1987).

2.2.3 Direct effects of SO_2, O_3, NO_x and NH_3 (multiple stress)

Gaseous air pollutants such as SO_2, NO_x, O_3, NH_3, and a number of other air pollutants may be harmful to plants by influencing physiological processes, or by causing leaf damage (Guderian, 1985). Air pollutants, notably SO_2, had been proven to cause damage to plants and forests as early as the end of the previous century (Wislicenus, 1908-1916). The most severe damage occurred near point sources and in industrial areas, and consisted of acute and visible damage to foliage. Oxidants other than O_3 have been shown to cause severe damage, but as concentrations of these pollutants are very low under the site conditions discussed here, they will not be dealt with any further (P101).

Air pollutants may affect physiological processes such as photosynthesis, respiration, stomatal regulation and phloem loading. SO_2 influences the metabolism of both broadleaved and coniferous trees, resulting in decreased photosynthesis and increased ageing of leaves and needles. Acute and visible damage under high pollutant exposure appears to be related to the capacity for metabolic detoxification and buffering. The uptake of gaseous air pollutants is mainly through the stomata; the stomatal opening together with ambient concentration difference determining the uptake rate (Unsworth & Black, 1981). Thus, CO_2-assimilation, transpiration, and uptake of gaseous air pollutants are strongly connected.

Under the "multiple stress" hypothesis it is assumed that exposure to combinations of air pollutants at low concentrations enhances the effects of the individual pollutants. Furthermore, preconditioning of plants in periods of drought or due to nutrient imbalances or deficiencies increases the plant's susceptibility to mixtures of air pollutants at low concentrations. Also, exposure to air pollutants such as ozone may render trees more susceptible to frost damage. Although the mechanisms underlying the final effect of combined exposure have yet to be revealed (McLaughlin, 1985), the results comprise increased ageing and shedding of foliage, decreased photosynthetic production and translocation of assimilates, and decreased root growth. The multiple stress hypothesis is sometimes expanded to include all factors that may influence growth (traditional factors, diseases, air pollutants, etc., see Schütt & Cowling, 1985).

2.2.4 Soil acidification, aluminium toxicity and nutrient deficiency

During the late seventies, development of forest damage was predicted due to soil acidification and the increase of aluminium in the soil solution. This conclusion was derived by Ulrich and co-workers, based on long-term research studies carried out in the Solling area near Göttingen (Ulrich, 1983 a & b). According to this hypothesis, soil acidification occurring naturally under coniferous forests is enhanced by deposition of acid and acidifying components from the atmosphere, leading to destabilization of forest nutrient cycling, and ultimately to deterioration and decline of entire stands. The main pollutants causing acidic deposition are sulphur dioxide (SO_2), nitrogen oxides (NO_x) and ammonia (NH_3). Occurrence of SO_2 and NO_x is closely related to industrial activity and road traffic; NH_3 originates mainly from intensive agriculture.

The mechanisms underlying forest decline mediated by soil acidification include various buffering mechanisms such as weathering of carbonates, silicates and aluminium hydroxides, and base cation exchange. Depletion of these buffers leads to a decrease in soil pH, a decrease in base cation availability, and an increase in aluminium concentrations in the soil solution (P113, De Vries & Kros, 1989; De Vries et al., 1989; Mulder et al., 1987; van Breemen et al., 1983 & 1984).

According to the hypothesis, ongoing soil acidification decreases the availability of the base cations required for tree nutrition (e.g. calcium, potassium, magnesium) and limits root development due to a high aluminium concentration in the soil solution (Murach, 1984). Also, in acid soils (pH < 4.1) competition for uptake sites may occur between Al^{3+} and Ca^{2+}, indicated by the $[Al^{3+}]/[Ca^{2+}]$ ratio. Consequently, nutrient deficiencies occur and the forest becomes vulnerable to drought. Decreased root uptake of base cations in this situation is due to decreased availability in the soil, and to restricted root uptake capacity in the presence of high aluminium concentrations (Foy et al., 1978; Keltjens & van Loenen, 1989).

In addition to damage due to changes in the soil, soil acidification and root damage as a result of aluminium are considered predisposing factors that increase susceptibility of a stand to gaseous air pollutants and extreme climatic conditions (Ulrich, 1983a).

2.2.5 Excess nitrogen deposition

A few years ago, Swedish and Dutch scientists (Nihlgård, 1985; Roelofs et al., 1985) postulated nitrogen to be a contributing factor to forest damage. At many sites within Europe it is evident that nitrogen is limiting growth. This situation is gradually changing due to increased deposition of nitrogen. Apart from changing tree nutrition, nitrogen deposition contributes to soil acidification (van Breemen et al., 1982), and changes biological activity in the soil (Boxman & Roelofs, 1988).

The hypothesis postulates that increased nitrogen deposition has primarily a fertilization effect leading to enhanced growth. Nitrogen can be taken up by plants as nitrate (NO_3^-) or ammonium (NH_4^+). When nitrogen is deposited as ammonium, base cation uptake may be hampered because trees prefer the uptake of ammonium to nitrate, and at the root surface competition may occur between ammonium and base cations (indicated by the ratios $[NH_4^+]/[K^+]$ and $[NH_4^+]/[Mg^{2+}]$). Thus, increased nitrogen availability may lead to nutrient imbalances and deficiencies of base cations such as magnesium or potassium (Roelofs et al., 1985; Schulze, 1989). Moreover, nitrogen availability may influence the distribution of assimilates within the tree, leading to a decrease in assimilate products allocated to the roots (Mansfield, 1988). These phenomena together with soil acidification, lead to increased leaching and decreased uptake of base cations. Again, decreased root growth may render the stands more susceptible to drought.

Nitrification of ammonium causes additional soil acidification, leading to the same type of disturbances as described in the previous section. The rate of ammonium input together with the rate of nitrification determines whether nitrate or ammonium uptake dominates under particular conditions, and the extent of soil and rhizosphere acidification caused by ammonium deposition and uptake.

High nitrogen content of the foliage may cause increased susceptibility to pests and diseases (de Kam et al., 1989) and frost (Aronsson, 1980). In addition, the uptake of nitrogen as ammonium could result in changes in the ionic balance of the plant causing a decrease of metabolic buffering and detoxification capacity with respect to gaseous uptake of pollutants (P83; Kropff, 1989).

2.3 Criteria for testing the hypotheses

The validity of any of the hypotheses cannot be established from the results of the ACIFORN monitoring research alone, but based on the processes described and documented in the ACIFORN results, the most likely chain of events to apply to the stands under study can be determined. The set of hypothesis listed above refers to different mechanisms that may explain the main symptoms of forest decline (premature foliage shedding and foliage discoloration) and the gradual changes within the forest ecosystems that are taking place, such as soil acidification.

Forest health surveys are dependent on comparative visual assessment of trees at standardized sampling plots, and on classification of symptoms in a number of damage classes. Damage classification heavily relies on the percentage of foliage loss compared with that of a healthy tree under the same growing conditions. Forest decline has mainly been interpreted in terms of needle discoloration, needle loss and crown thinning, reduced stand growth, and increased tree mortality. Phenomena such as needle loss, decreased growth, and tree mortality are mostly atypical and do not in themselves indicate which causal mechanisms underlie the overall result. This complicates the interpretation of disturbances of "vitality" in terms of physiological, hydrological and geochemical processes.

By means of intensive monitoring of selected ecosystems however, individual causal factors and processes relevant to conditions pertaining to the Netherlands can be evaluated. Using qualitative as well as quantitative ecosystem analysis, the relative importance of individual disturbances can then be evaluated by considering a range of tree and stand characteristics closely related to different physiological, hydrological and geochemical processes. In this way results from monitoring projects such as carried out within ACIFORN in combination with related experimental research can be used to determine the most likely chain of processes and events possibly leading to forest changes and forest decline.

The tree, stand and site characteristics that are used in the following sections of this report to evaluate the hypotheses described above, consist of:

SITE CONDITIONS

- climatic situation and weather conditions in recent years at the ACIFORN sites
- deposition rates of gaseous air pollutants and acidifying components
- concentrations of gaseous air pollutants within the canopy space

TREE STATUS

- stand structure and total biomass
- nutrient status of the trees
- biochemical components indicating disturbances in physiological processes

TREE AND STAND GROWTH

- photosynthesis
- transpiration
- respiration
- dry weight increase and stem volume increment

SOIL CHEMICAL STATUS

- acidity (pH)
- organic matter content and cation exchange capacity
- element ratios in the soil solution ($[Al^{3+}]/[Ca^{2+}]$, $[NH_4^+]/[K^+]$, $[NH_4^+]/[Mg^{2+}]$, indicating competition for uptake sites at the root surface)

RATE OF CHANGE IN SOIL CONDITIONS

- rate of soil acidification

- rates of deposition and leaching
- element budgets

In the following paragraphs, a general description of site and stand conditions at the ACIFORN sites will be given first. Next, the possible impact of gaseous air pollutants, through direct effects on tree physiology, will be discussed with respect to the hypotheses listed previously. Indirect effects of air pollution and acid deposition will be discussed in Section 5. Sections 6 and 7 contain an attempt at an integrated quantitative analysis to determine the relative importance of the various direct and indirect effects.

3. THE ACIFORN SITES SPEULD AND KOOTWIJK

In 1986 two forest stands were carefully selected as being suitable for an intensive study of forest growth and the influence of air pollution. A number of requirements had to be met for selection. The stands had to consist of Douglas fir of coastal provenance (USA), aged between 30 and 50 years, and with approximately the same stand treatment. To be able to study air flow and deposition rates over the forest, the stand needed to be part of a closed forest canopy covering an area of at least 600 m in all directions. The soil requirements had to be a profile without groundwater influence, not disturbed by digging and not too heterogeneous.

The ACIFORN sites finally selected are located in the Veluwe area, in the Centre of the Netherlands, at a distance of approximately 50 km from Wageningen. One of the stands (Kootwijk) is located on cover sand with some influence from previous agricultural activity before stand establishment. The soil is horizontally homogeneous and can be classified as a leptic podzol, well drained with a water table at 5 m below the soil surface. The other stand (Speuld) is located on preglacial sand deposits, which is more heterogeneous, especially in the subsoil where bands of loam and clay occur and can be classified as an orthic podzol/luvisol. The most homogeneous part within this stand was chosen for the monitoring projects. In Table 1 a summary shows the soil profiles of both sites. For further details on soil characteristics see P102.1 and P102.2.

Table 1. Summary descriptions of soil conditions at both ACIFORN sites Speuld and Kootwijk (taken from the descriptions of the sub-plots for root measurements). CEC and pH-KCl are indicated per horizon, double numbers indicate gradual differences within horizon (data from Tiktak et al., 1988)

	Speuld	Kootwijk
Location:	52°15'N, 5°41'E	52°11'N, 5°46'E
Soil profile:	Ortic podzol/ortic luvisol	Leptic podzol
Parent material:	Preglacial sand partly loamy/gravely	Eolian sand
Groundwater:	at 40 meter	at 5 meter
Profile description (simplified):		
Organic horizon:	-4 - 0 cm	-4 - 0 cm
Mineral horizons:	A1, 0 - 7 B2, 7 - 18 B3, 18 - 58 C, 58 - 125+	Ap, 0 - 37 B1s, 37 - 65 B2s, 65 - 90 C1, 90 - 150+
Cation exchange capacity:		
(mmolc/kg, LiEDTA)	A1: 75 B2: 35 B3: 30 to 10 C: 10	Ap: 50 to 20 B1s: 10 B2s: 5 C1: 5
pH-KCl:	Humus: 3.1 A1: 3.2 B2: 3.5 B3: 4.1 C: 4.3	Humus: 2.9 Ap: 3.0 to 3.9 B1s: 3.9 to 4.4 B2s: 4.5 C1: 4.6

3.1 Climate and weather conditions in 1988 and 1989

The geographical location of the site is 52.1 degrees northern latitude and 5.4 degrees eastern longitude. Altitude is 50 m at the Speuld site and 25 m at the Kootwijk site. In Table 2 the total amounts of precipitation, interception, transpiration, and total incoming radiation for both sites in 1988 and 1989 are given (hydrological data from P102.2). Figure 1 provides an overview of daily values for minimum and maximum temperature, global radiation, precipitation, wind speed, and humidity (data from P101 and P102.2).

Table 2. Precipitation, interception, transpiration, and total incoming radiation at the ACIFORN sites Speuld and Kootwijk during 1988, and 1989 (P101, P102.2). The uncertainty in the measurements of precipitation and radiation are 5 % or less (P101/P102.2)

	Speuld		Kootwijk	
	1988	1989	1988	1989
precipitation (mm)	1006	801	999	769
days with rain	225	248	225	248
Interception (mm)	495	350	438	282
Transpiration (mm)	316	380	285	340
Total radiation* (GJ m-2 yr-1)	3.15	3.64	3.15	3.64

* Only data for Speuld available

3.2 Air pollution, atmospheric deposition and soil input

Air pollution and meteorological conditions were monitored above and within the forest canopy (P101) to define the exposure of the forest to climate variables and air pollutants. The estimation of deposition was emphasized in the second phase of the National Programme, resulting in the study of flux-profile relationships, additional pollution and micro-meteorological measurements and more elaborate sampling and calculation procedures.

Results of the air pollution monitoring program are available for 1988 and 1989. Monthly averaged values of measured concentrations above the canopy for the Speuld location are provided in Table 3. Table 4 shows the monthly maximum values above the canopy for the same site. To indicate the occurrence of short periods of high O_3, Table 5 supplies a frequency distribution of half-hour values for 1988 and 1989 for the Speuld site. Figure 2 gives an overview of daily average values for SO_2, NO_x, O_3 and NH_3 for the same site. Data for the Kootwijk site will not become available until summer 1990. For the calculations presented later on in this report, air pollution data from the Speuld site were therefore used for the Kootwijk site as well.

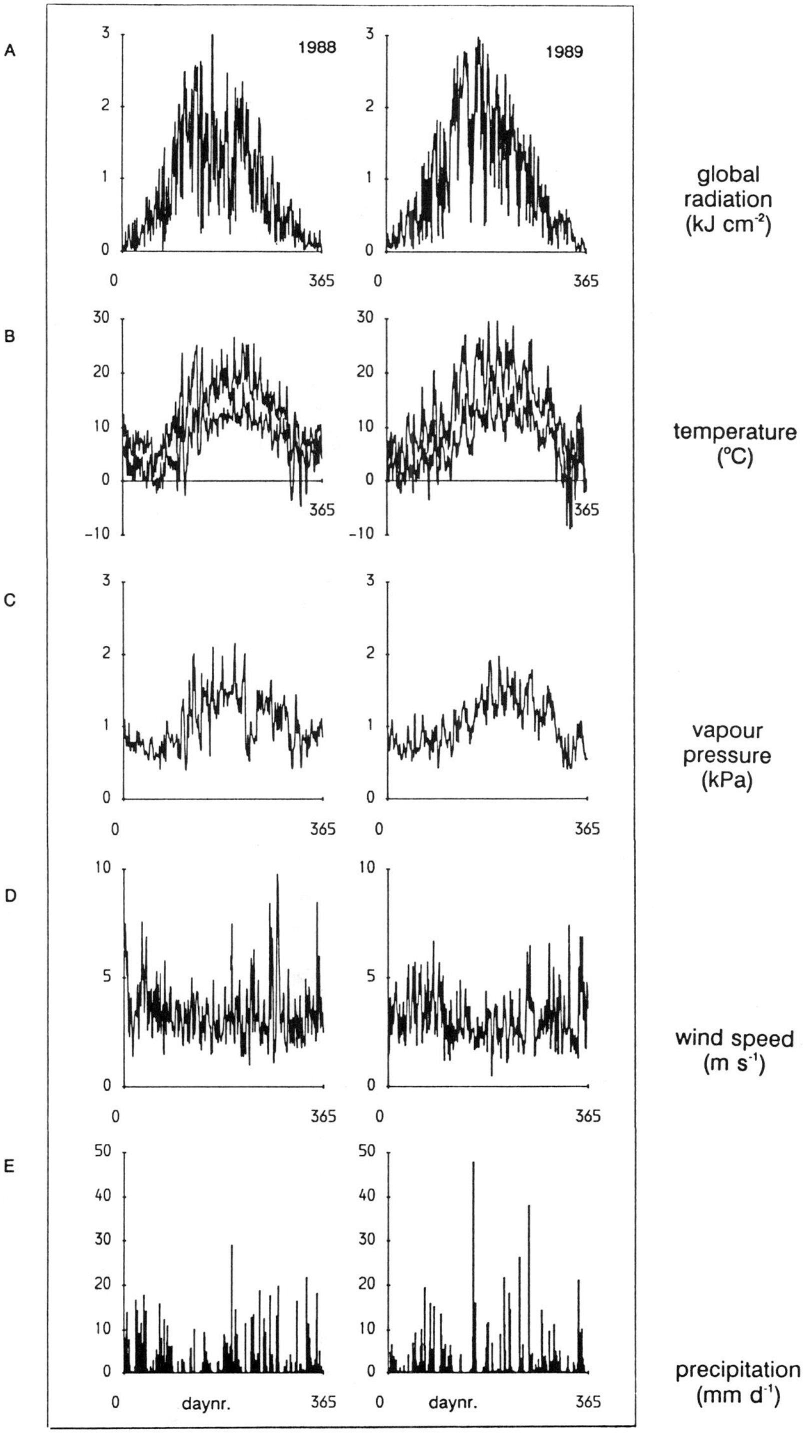

<u>Fig. 1.</u> Daily values for A: global radiation (kJ cm^{-2}); B: minimum and maximum temperature (degrees Celcius); C: humidity (kPa); D: wind speed (m s^{-1}) and E: precipitation (mm) for the Speuld location (P101) in 1988 and 1989

Table 3. Monthly average values of concentrations (in ppb*) of CO_2, NO_x, O_3 and NH_3 for Speuld (h = 30 m). The uncertainty in the monthly average values is 5 % or less (P101)

	CO_2 (ppm)	NO_x (ppb)	NO (ppb)	NO_2 (ppb)	SO_2 (ppb)	O_3 (ppb)	NH_3 (ppb)	O_x (ppb)
Jan 88	331.4	24.1	8.6	15.5	4.1	15.8	2.4	30.3
Feb 88	344.8	14.6	1.6	13.0	4.2	23.2	2.7	36.1
Mar 88	343.5	15.2	1.1	14.2	3.7	26.8	2.9	40.4
Apr 88	327.6	12.7	1.1	11.5	4.4	36.4	11.4	47.9
May 88	345.3	9.1	1.4	7.7	5.0	45.2	8.3	53.3
Jun 88	339.6	5.5	0.7	4.8	1.5	35.3	7.5	39.9
Jul 88	340.2	10.0	1.5	8.4	2.4	24.8	7.5	33.3
Aug 88	339.2	12.9	1.7	11.3	2.4	29.3	10.1	40.0
Sep 88	341.5	12.9	1.6	11.3	2.5	27.2	6.9	39.2
Oct 88	348.7	24.1	7.5	16.6	5.7	14.6	4.5	31.3
Nov 88	354.2	30.6	12.2	18.4	3.4	13.4	4.1	31.8
Dec 88	336.9	22.8	7.7	15.1	4.0	16.5	5.4	31.4
1988	341.1	16.2	3.9	12.3	3.6	25.7	6.1	37.9
Jan 89	357.0	39.4	18.2	21.2	4.7	6.1	9.0	27.3
Feb 89	342.8	29.9	10.0	19.9	5.4	16.6	9.2	36.5
Mar 89	348.0	19.6	2.6	16.9	5.5	25.9	6.4	42.4
Apr 89	345.0	10.3	1.0	9.4	3.4	36.2	7.5	45.5
May 89	335.2	9.5	0.5	8.9	4.5	51.2	8.5	60.1
Jun 89	326.0	8.6	0.7	7.9	2.4	46.0	7.7	53.9
Jul 89	341.0	9.0	0.6	8.4	3.1	41.3	12.3	49.9
Aug 89	334.4	12.8	1.4	11.4	2.8	33.8	10.8	45.2
Sep 89	342.5	15.0	2.2	12.8	4.2	27.6	11.1	40.4
Oct 89	352.5	20.7	5.4	15.3	3.5	19.2	8.0	33.9
Nov 89	365.3	35.0	14.0	21.0	6.3	8.7	6.0	29.7
Dec 89**	358.7	18.9	6.1	12.8	2.1	16.4	8.8	29.1
1989	345.7	19.1	5.2	13.8	4.0	27.4	8.8	41.2

* To convert concentration in ppb to $\mu g\ m^{-3}$ at 20°C and 1013 hPa multiply by 1.829 (CO_2), 1.247 (NO), 1.912 (NO_2), 2.661 (SO_2), 1.995 (O_3), and 0.707 (NH_3).

** Less than 50% of data available.

Table 4. Monthly average maximum values of concentrations of CO_2, NO_x, O_3 and NH_3 for Speuld (h = 30 m, data in ppb, see footnote Table 1) (P101)

	CO_2 (ppm)	NO_x (ppb)	NO (ppb)	NO_2 (ppb)	SO_2 (ppb)	O_3 (ppb)	NH_3 (ppb)	O_x (ppb)
Jan 88	384.9	129.1	109.6	41.8	21.8	45.2	11.9	48.8
Feb 88	405.0	93.1	60.7	42.2	30.1	44.3	18.6	49.1
Mar 88	389.5	69.2	30.8	45.5	40.0	52.5	10.2	59.0
Apr 88	365.7	55.4	26.4	47.9	26.7	88.2	59.2	97.2
May 88	388.0	46.1	14.0	34.1	27.1	97.1	69.7	111.3
Jun 88	394.5	37.2	9.2	36.2	20.7	89.6	25.1	94.8
Jul 88	404.9	44.5	26.1	26.5	20.2	73.9	46.9	75.2
Aug 88	421.5	44.4	21.0	38.3	17.9	122.1	56.9	131.4
Sep 88	421.5	57.2	30.5	38.3	19.0	74.8	28.3	92.5
Oct 88	412.8	117.2	88.4	50.0	32.2	52.0	28.8	126.1
Nov 88	417.9	133.4	101.5	41.5	17.2	46.9	32.0	49.5
Dec 88	375.8	154.1	127.1	38.2	35.1	48.4	25.4	51.2
max 1988	421.5	154.1	127.1	50.0	40.0	122.1	69.7	131.4
Jan 89	414.1	143.9	104.6	43.3	23.9	43.6	37.5	50.5
Feb 89	388.6	126.1	81.0	59.0	35.4	46.3	54.2	58.9
Mar 89	400.6	93.2	57.5	59.1	37.8	59.2	44.8	75.3
Apr 89	392.0	66.8	43.3	43.8	23.0	73.3	40.1	81.9
May 89	377.0	61.8	24.5	37.3	26.3	122.9	36.4	128.1
Jun 89	390.0	45.1	9.3	44.7	18.6	105.4	43.3	111.1
Jul 89	401.3	50.8	11.9	43.7	31.8	116.7	48.5	122.2
Aug 89	406.5	58.5	24.8	58.4	25.8	152.6	48.2	161.9
Sep 89	403.3	83.1	41.8	48.0	29.6	104.2	72.6	116.5
Oct 89	402.8	105.4	65.8	56.8	15.6	48.0	36.9	69.7
Nov 89	419.5	146.7	99.6	54.0	35.1	37.0	22.3	55.2
Dec 89	407.8	117.5	102.4	34.6	14.5	38.1	27.2	40.5
max 1989	419.5	146.7	104.6	59.1	37.8	152.6	72.6	161.9

Table 5. Number of half-hourly concentrations of O_3 above a certain concentration level Cx. The uncertainty in the half-hourly concentrations is estimated to be between 5-10 % (P101)

Cx(μg/m³)	50	100	150	200	250	Maximum conc.
Jan 88	331					90
Feb 88	637					89
Mar 88	788	12				105
Apr 88	1069	255	29			176
May 88	1180	466	117			194
Jun 88	1094	148	19			179
Jul 88	413	36				148
Aug 88	746	174	51	14		244
Sep 88	587	48				150
Oct 88	333	1				104
Nov 88	319					94
Dec 88	300					97
tot 1988	7797	1140	216	14		
Jan 89	72					87
Feb 89	387					93
Mar 89	778	33				118
Apr 89	1141	247				146
May 89	1306	692	148	46		246
Jun 89	1084	439	162	24		211
Jul 89	1065	397	115	25		233
Aug 89	924	263	108	42	6	305
Sep 89	794	124	45	2		208
Oct 89	544					96
Nov 89	61					74
Dec 89	172					76
tot 1989	8328	2195	578	139	6	

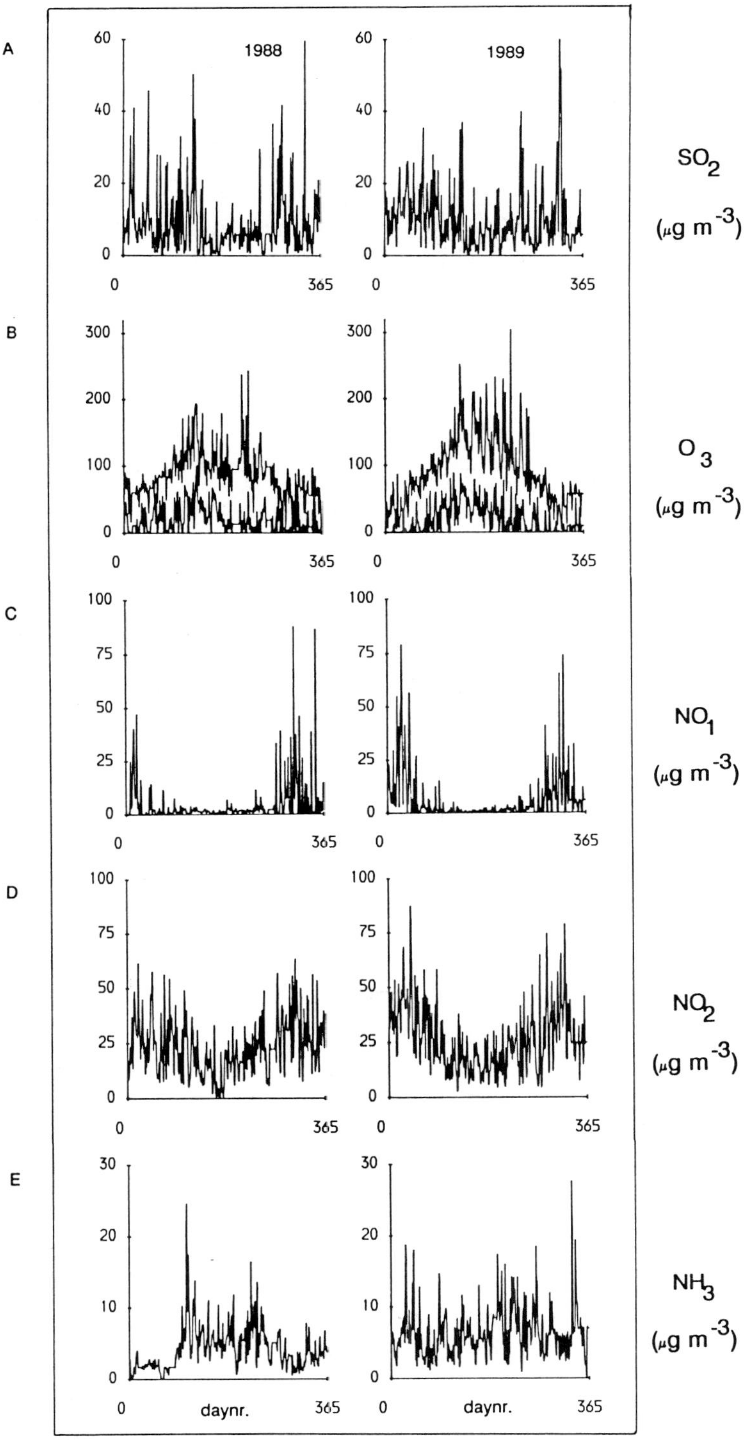

Fig. 2. Daily average values for A: SO_2; B: O_3; C: NO; D: NO_2, and E: NH_3 (all in µg m^{-3}) for Speuld in 1988 and 1989 (P101)

The values for SO_2, NO_x and O_3 are comparable with results from neighbouring stations from the National Air Quality Monitoring Network (P101). Data from this network can therefore be used to indicate the exposure of the forest to air pollutants for the Speuld site, remote from sources of air pollution. In the case of ozone, a slight difference may occur in summertime. A comparison is not possible for NH_3 because of the lack of data from other locations. NH_3 concentration fields are expected to be rather inhomogeneous and it is not feasible to determine the representativeness of the nitrogen input of the ACIFORN sites for the entire situation in the Netherlands. The considerable differences in throughfall data for different sites (Houdijk, 1990) imply that specific evaluation is needed. When available data on throughfall are analyzed, it appears that nitrogen deposition at the research sites is rather low in comparison with deposition rates in the eastern and southern part of the country. Local deposition (and hence concentration) of NH_3 largely depends on agricultural livestock density, which has increased steadily from around 1950 onwards (Asman, 1987).

SO_2 concentrations have decreased since the mid-seventies (Erisman et al., 1989). SO_2 and NO_x concentrations are strongly related to industrial activity and traffic (the diurnal course of NO_x shows a distinct peak value during morning rush hours, with a second somewhat lower peak during late afternoon), and have been increasing steadily from the fifties onwards. Incidental high peak values for NO occur during winter time, when strong inversions may develop under conditions of stable high atmospheric pressure and absence of wind. Apart from this, NO episodes can be attributed to the influence of traffic sources; soil emissions may play an important role in other periods with high concentrations (P101). O_3 concentrations are related to incoming solar radiation, and may build up on clear days. Examples can be seen in the Speuld data for May and August 1989. Average O_3 concentrations have been increasing gradually since the fifties.

The uptake of pollutants by the trees, the deposition of pollutants on the canopy, and the subsequent downwash in the form of throughfall water can be estimated by micro-meteorological methods combined with modelling or by calculations based on the difference between throughfall and rainfall composition. Deposition estimates based on micro-meteorological methods are not available as yet; Chapters 5 and 6 will deal with these approaches in greater detail. Analysis of data based on rainfall and throughfall is provided in Table 6. It should be kept in mind, that this type of balance modelling is prone to distortion due to leaching of elements and the underlying assumptions necessary to make the calculations (see footnotes accompanying Table 6). Assuming uniform deposition velocities over all comparable Dutch forests, and taking the concentrations for SO_2, NO_x and O_3 above the forest into account, one may expect the deposition at the sites to be representative of a Douglas fir forest in an extended forested area. Adjustment will have to be made for other tree species, small forests and forest edges (see thematic report on deposition).

able 6. Estimated mean bulk deposition, canopy interactions (uptake and leaching), and soil input for Speuld and Kootwijk in mol (+/-) ha-1yr-1 (mean values per hectare per year derived from period of March 1, 1987 until March 1, 1990; calculation of mean after calculation of annual fluxes). Data from P102.1 and P102.2*. For more detail on calculation procedures and interpretation, see P102.1

euld
D/WD)Na = 0.74**

	Na	K	Ca	Mg	Al	Fe	Mn	Cl	Cl (HCl)	H_2PO_4	H	NH_4	NO_3	SO_4	SO_4(sae) SO_4/Na=0.12	A-	SiO_2
roughfall	994	484	423	336	18	29	60	1384	0	7	74	2204	756	2017	119	339	21
lk open field	572	28	166	158	4	18	4	742	0	3	264	802	396	727	69	80	3
y deposition	422	20	121	116	3	14	3	549	94	3	-95	2002	395	1418	51	96	2
nopy leaching	0	436	135	62	11	-3	53	0	0	1	-95	-600	-35	-128	0	162	15
et deposition	524	20	134	142	2	1	2	689	0	1	294	685	360	697	0	57	4
y deposition	470	28	153	132	5	15	4	694	0	5	-124	2118	431	1567	0	104	2
tal deposition	994	48	287	274	7	17	6	1384	0	5	170	2803	792	2264	0	161	6
deposition	3595																
uptake foliage	635																

otwijk
D/WD)Na = 0.52**

	Na	K	Ca	Mg	Al	Fe	Mn	Cl	Cl (HCl)	H_2PO_4	H	NH_4	NO_3	SO_4	SO_4(sae) SO_4/Na=0.12	A-	SiO_2
roughfall	789	409	325	278	10	16	33	1091	0	12	55	2115	705	1833	95	296	18
lk open field	518	24	154	145	3	12	4	681	0	2	228	798	370	734	62	38	4
y deposition	271	12	82	76	1	6	3	364	46	2	-106	1799	382	1266	32	53	2
nopy leaching	0	373	89	57	6	-2	27	0	0	8	-67	-482	-46	-167	0	206	11

Kootwijk
(DD/WD)Na = 0.52**

	Na	K	Ca	Mg	Al	Fe	Mn	Cl	Cl (HCl)	H_2PO_4	H	NH_4	NO_3	SO_4	SO_4(sae) SO_4/Na=0.12	A-	SiO_2
Wet deposition	471	17	118	130	3	0	2	615	0	1	260	689	343	691	0	40	1
Dry deposition	318	20	119	91	2	6	4	476	0	3	-138	1908	408	1404	0	39	5
Total deposition	789	37	237	221	5	6	7	1091	0	4	122	2597	751	2095	0	79	6
N deposition	3349																
N uptake foliage	528																

* The calculation procedure contains a number of assumptions, that have to be known to be able to evaluate the data. A complete discussion can be found in P102.1; here, only the assumptions are listed:

A. Canopy leaching of Na is negligible, and dry deposition of Na is estimated as the difference between throughfall and bulk open field data. The ratio of Na bulk open field and dry deposition was used to calculate the dry deposition of aerosolic components Ca, Mg, K, Al, Fe, Mn, Cl and P. The main assumption is that the ratios of Na with these elements in bulk open field and dry deposition are equal.

B. The calculated canopy leaching of Cl is considered to be dry deposition of HCl, as leaching of Cl is unlikely.

C. Canopy leaching of K, Ca, Mg, Mn, Al and Fe is assumed to be caused by exchange with H^+ and NH_4^+. The uptake rate of H^+ and NH_4^+ is assumed to be equal to the mean of the ratio's of H^+ and NH_4^+ in throughfall and bulk open field. The rest term of H^+ and NH_4^+ is dry deposition.

D. Dry deposition of H^+ and NH_4^+ is balanced by dry deposition of Cl^-, NO_3^-, and SO_4^{2-}. The ratio of NO_3^- and SO_4^{2-} in dry deposition is assumed to be the same as the ratio of the difference between throughfall and bulk open field of NO_3^- and SO_4^{2-} after correction for seasalt-SO_4 (seasalt SO_4/Na = 0.12). The rest term for these components is canopy uptake.

E. Dry deposition and canopy leaching of weak acids are equal to the difference between the sum of cation and anions (on an equivalence basis).

F. Dry deposition and canopy leaching of Si can be calculated in the same way as for Ca, Mg, etc.

G. Real wet and dry deposition can be calculated by subtracting the difference between bulk open field, and adding wet-only data (total deposition remaining the same).

** (DD/WD)Na represents the ratio of Potassium in open field dry deposition over bulk open field deposition.

3.3 Biomass and stand structure

Table 7 provides a summary of the stand structure at both the ACIFORN monitoring sites, and Table 8 gives measured stem volume increments over 1988 and 1989 (data from P100, P103, P105). Stand structure in both stands is in agreement with yield table guidelines (LaBastide & Faber, 1972) and does not show any abnormalities. Volume increment as shown in Table 8 is difficult to estimate on small plots, and both sites may have some variation in local increment rates. Stem volume increments measured by permanent tapes at breast height of some 200 trees however, are provided in Table 8.

Table 8. Measured increment rates in Speuld and Kootwijk during 1988 and 1989. Data refer to measurements carried out at the subplot used for soil chemistry and soil hydrology monitoring. DBH increment was measured using permanent girthing tapes with an accuracy of 0.2 mm. Height increment was derived from height measurements using a telescopic measuring rod, which was read from the equipment towers, using binoculars. To obtain volume increment, tree volume was estimated using Schumacher & Hall's volume equation using genuine coefficients for Douglas fir in the Netherlands (LaBastide & Faber, 1972; Dik, 1984). Due to considerable heterogeneity within the stands at both locations, increment rates may vary over short distances, causing some discrepancy in growth rates as reported in separate project reports (P100, P103, P105)

	diameter increment (cm yr^{-1})		basal area increment ($m^2ha^{-1}yr^{-1}$)		height increment (m yr^{-1})		stem volume increment ($m^3ha^{-1}yr^{-1}$)	
	1988	1989	1988	1989	1988	1989	1988	1989
Speuld	0.77	0.71	2.41	2.31	0.87	0.75	31.6	30.8
Kootwijk	0.52	0.61	1.57	1.75	0.50	0.82	19.2	24.6

According to Table 8 it appears that increment rates in Speuld are somewhat higher than those in Kootwijk. This is confirmed in an analysis of ring width pattern (P111) although differences appeared to be less than those derived from diameter increments in 1988 and 1989. This discrepancy can be explained from differences in soil moisture availability between Speuld and Kootwijk (P102.2): Storage capacity within the rooted zone is somewhat higher in Speuld, hence water shortage in dry years such as 1989 will obviously be less than in Kootwijk.

Increment in both stands is higher than anticipated from the yield tables (expected increment would be around 20-25 $m^3ha^{-1}yr^{-1}$, with the higher value applying to Speuld, and the lower value to Kootwijk (LaBastide & Faber, 1972).

Table 7. Stand structure and biomass data for Speuld and Kootwijk. Data on stand structure refer to the situation at January 1, 1988. For more detail on structural analysis, together with a discussion of error-margins see P105

	age (yr)	basal area (m^2 ha^{-1})	av. height (m)	av. dbh (cd)	stem dens (ha^{-1})	stem volume (m^3 ha^{-1})	LAI* (m^2 m^{-2})	foliage-age distribution 1/2/3/4/5(%)	branch biom. (Mg ha^{-1})	fine roots (Mg ha^{-1})	coarse roots (Mg ha^{-1})
Speuld	29	33.4	18.5	21.9	886	296	11.2	36/29/21/10/4	24.2	2.99	31.0
Kootwijk	38	28.7	18.0	19.2	992	244	7.8	41/30/20/7/2	17.3	4.33	24.8

* The values given for LAI are derived from destructive sampling of branches in the upper, middle and lower parts of the crown. Branch samples were analysed with regards to branching structure, and average number of needles per branch-order. The variation in this sampling scheme is less than 10 %, but the accuracy cannot be established. Incidental LAI measurements using a direct beam sensor indicated a LAI of about 5.5 for both stands, whereas measurements with the LiCor-6200 canopy analyzer resulted in values for LAI of 7-8 for both stands. The beam sensor is likely to underestimate LAI in a coniferous canopy by a margin of up to 30 % (R.E. McMurtrie & R. Lang, Canberra, Australia, personal communication to G.M.J. Mohren, August 1989), which indicates a value of around 7-8. Average needle litter loss over the three-year monitoring period was estimated to be about 2.9 Mg $ha^{-1}yr^{-1}$ for Speuld, and about 2.4 Mg $ha^{-1}yr^{-1}$ for Kootwijk (P102.1); at an average needle life-span of 3.5 years and surface/dry weight ratio (SLA) of 0.5 ha Mg^{-1}, this gives an estimate for LAI of 5.1 for Speuld and 4.2 for Kootwijk. The reason for the discrepancy in LAI estimates is not clear; however, it is clear that the canopy is completely closed, and that LAI is sufficient to intercept over 95 % of incoming radiation.

3.4 Nutrient status

Table 9 provides element content of current year needles at the end of the year 1988 for nitrogen, phosphorus, potassium, calcium, magnesium and sulphur, together with estimates of the total amounts accumulated in the living biomass (data from P100, P105, P112).

Table 9. Nutrient concentrations of current year needles at the end of the year (as % of dry weight), and estimated total nutrient content (in kg per hectare) for Speuld and Kootwijk in 1988 and 1989. See P105 for complete data sets and extended discussions

	Speuld		Kootwijk	
	1988	1989	1988	1989
Nitrogen				
concentration	1.80	1.85	1.75	1.80
amount in biomass	550	550	510	510
Phosphorus				
concentration	0.11	0.12	0.12	0.12
amount in biomass	53.5	54.0	45.4	46.0
Potassium				
concentration	0.60	0.65	0.55	0.65
amount in biomass	200	200	171	175
Calcium				
concentration	0.24	0.26	0.20	0.20
amount in biomass	100	100	87.8	88.8
Magnesium				
concentration	0.10	0.11	0.09	0.12
amount in biomass	29	30	26.3	27.5
Sulphur				
concentration	0.20	0.20	0.20	0.20
amount in biomass	70	70	60.0	59.0

The data on nutrient content of the needles for both stands show no obvious deficiencies. Nitrogen content of 1.7-1.8 % appears to be within the optimal range, and no luxurious consumption and accumulation has taken place sofar. Phosphorus is rather low, and should be considered insufficient according to van den Burg (1988). Potassium is low in both stands, and tends to be insufficient in Kootwijk.

4. DIRECT EFFECTS OF GASEOUS AIR POLLUTANTS

4.1 Exposure and deposition at the ACIFORN sites

A distinction should be made between exposure (immission), dose (uptake) and deposition in the evaluation of direct effects of air pollutants on forest trees. The term deposition should be used for the total deposition onto the forest ecosystem; it is usually estimated by means of micro-meteorological methods above the canopy. Total deposition can be divided into dry deposition of aerosols and gases, and wet deposition including deposition of fog and dew (occult deposition). Another distinction can be made by looking at the destination of the deposition fluxes. Wet deposition will be partly intercepted by the canopy and partly reach the soil directly. Dry deposition consists of a direct flux to the soil, of stomatal uptake, and of adsorption onto (wet or dry) plant surfaces. For highly soluble gases the total deposition flux onto a wet canopy may be much larger than stomatal uptake. Evaporation of water and exchange of ions with the plant tissue may change the composition of intercepted wet deposition, which will finally reach the forest soil as throughfall water. Interrelationships between deposition, throughfall, and canopy exchange will be discussed in Chapter 5. Appendix A lists conversion factors to convert from ppb to $\mu g\ m^{-3}$ and vice versa.

To a large extent, the forest microclimate determines exposure; during daytime the turbulent exchange processes are very effective and concentrations of pollutants will be about equal inside and above the canopy (P101). In the case of coniferous trees, uptake is mainly governed by the stomatal resistance, which is large compared to atmospheric transport resistances. Thus, concentration gradients downwards from above the canopy are small. Obviously, the topography of the site can be important in determining leaf exposure in relation to concentrations above the canopy, but this can be ignored in the case of the ACIFORN stands as both sites are level. Laboratory studies on the subject of direct effects on trees often describe an exposure to pollutants in terms of average concentrations during the experiment, but in reality pollutant concentrations may vary considerably during the year and during the day. In case this variation is correlated with variation in stomatal opening, the use of average values may very well give misleading results. This is particularly important for ozone as ozone concentrations build up during the day, and reach a peak value during the afternoon.

Sofar only data from Speuld are available for both the ACIFORN sites, hence the discussion here will be limited to this set of data; preliminary analysis shows that differences between air pollution at Kootwijk and at Speuld are small. The air pollution concentration measurements at Speuld show that pollutant gradients over the canopy are indeed relatively small during the day. Monthly averages of the measurements at 30 m compare favourably with observations at nearby stations of the national air pollution monitoring network. If absolute values from Speuld for SO_2, NO_x, O_3 and NH_3 are examined, only one-hour average concentrations of O_3 reached levels possibly resulting in direct, short term visual damage ($\pm$ 200 μm^{-3}). This has been described in controlled experiments, for agricultural indicator species as well as for young trees (Posthumus et al., 1988). The frequency of occurrence of high concentration levels was greatest in 1989 (Table 5). Examples: levels above 100 μg m^{-3} (half-hour concentrations) occurred 990 times during the afternoon in 1988, and 1883 times in 1989. Levels above 200 μg m^{-3} occurred frequently in 1989 (70 hours).

The other important component under Dutch conditions, NH_3, demonstrated the highest concentrations during summer, amounting to annual averages of 4.3 and 6.2 μg m^{-3} for 1988 and 1989 at the Speuld site. These values are moderate for Dutch conditions; concentrations in the southern part of the Netherlands (the Peel-area) and nearer to agricultural areas are somewhat higher.

Compared to previous years, SO_2 concentrations were very low in 1988 and 1989; with annual averages of 9.6 and 10.6 μg m^{-3} respectively. All half-hour values were less than 100 μg m^{-3}.

4.2 Uptake of pollutants and their physiological effects

Exposure to air pollutants may affect the leaf surface, thereby altering the exchange of gases between the foliage and its surroundings, and they may lead to changes in physiological processes, via uptake through the stomata and subsequent dissolution in the interior of the leaf. Damage to the cuticle may result in increased transpiration, and increased uptake of pollutants, other than via the stomata. Moreover, additional leaching and uptake of minerals through the leaf surface can also be expected (Riederer, 1989). Although damage to the cuticle has been shown to occur under conditions of high pollutant load, especially ozone (Guderian, 1985), this appears to be of minor importance under conditions at the ACIFORN sites (Thysse & Baas, 1990). Under ambient conditions in the Netherlands, stomatal uptake of air pollution is the major pathway by which pollutants reach the leaf interior (Unsworth & Black, 1981; Freer-Smith & Dobson, 1989).

Physiological effects are related to the uptake of gaseous pollutants by the foliage and to biochemical reactions of the tree to that particular dose. These reactions can take the form of a short-term response such as decreased photosynthesis or increased respiration, and they may have long-term effects such as increased ageing or decreased phloem loading and altered assimilate distribution. Quite often the changes induced by the uptake of pollutants appear to be reversible for instance when photosynthesis is decreased due to slow detoxification of sulphite that results from uptake of SO_2 (Kropff, 1989; Lange et al., 1989), and removal of the pollutant load allows the metabolism to recover afterwards. In the case of visible injury and increased foliage loss, the damage is irreversible.

4.2.1 Short-term effects on tree physiology

The effects of short-term exposure on tree physiology arise from changes in physiological processes such as photosynthesis, transpiration, respiration and translocation. This may result in a decrease in assimilation of carbon dioxide, increased water use, or a decrease in the net amount of assimilates exported out of the leaf to other biomass components such as stems and roots.

Uptake of SO_2 may influence photosynthesis, by means of competition between SO_2 and CO_2 for the carboxylating enzyme. The magnitude of the reduction in photosynthesis depends on the rate of SO_2 uptake and the rate of detoxification via sulphite to sulphate. Total SO_2 uptake at Speuld and Kootwijk is rather limited due to low concentrations (annual averages of 9.6 and 10.6 $\mu g\ m^{-3}$ for 1988 and 1989, respectively), and high stomatal resistance of 300-500 $s\ m^{-1}$ at maximum photosynthesis (P115). The total amount of SO_2 taken up through the stomata at the ACIFORN sites was estimated to be about 2-3 kg $ha^{-1}yr^{-1}$ (Table 10). Using the model developed by Kropff (1989) this resulted in a negligible reduction in photosynthesis (Table 11). The uptake calculations are based on stomatal resistance determined by photosynthesis (using a ratio between internal and external CO_2-concentration of 0.6, P115), vapour pressure deficit (P115; Mohren, 1987; van Hove, 1989), and water potential of the foliage. The model used (P112, Mohren, 1987) applies to an entire stand of Douglas fir, and uses a multi-layer approach to estimate canopy photosynthesis from photosynthesis rates of individual foliage layers (Spitters, 1986; Spitters et al., 1986; Goudriaan, 1986; Spitters et al., 1989). Model calculations of total transpiration agree with estimated transpiration rates derived by monitoring the soil moisture (P102.2), indicating that stomatal resistance was accurately simulated.

Acute injury due to uptake of SO_2 may be due to a disturbance of the intracellular pH regulation or result from an accumulation of toxic anions (Kropff, 1989). At preconditioning external factors such as low temperature and high humidity, relatively high uptake rates (due to high humidity) combined with low detoxification capacity (due to low temperatures) may lead to internal acidification and high intracellular concentrations of SO_3^{2-} and HSO_3^-. These conditions did occur frequently at the ACIFORN sites during winter periods when some physiological activity may result in the opening of stomata and uptake of SO_2.

Table 10. Annual uptake of air pollutants in kg per hectare, as calculated by the model (P112/114.2) for Speuld and Kootwijk during 1988 and 1989 (using air pollution data from Speuld for both sites). To estimate uptake of S from SO_2, N from NO_x, and N from NH_3, multiply by 0.5, 0.3, and 0.82 respectively

	Speuld		Kootwijk	
	1988	1989	1988	1989
SO_2	3.26	3.58	2.54	2.57
NO_x	12.50	14.22	9.77	10.45
O_3	33.18	34.97	25.79	24.37
NH_3	3.45	4.67	2.69	3.29
uptake rates in case of non-limiting water supply:				
SO_2	3.40	4.08	2.65	3.25
NO_x	12.91	15.59	10.11	12.47
O_3	35.01	41.69	27.27	33.38
NH_3	3.66	5.34	2.85	4.29
uptake rates in case of non-limiting water supply, using maximum ozone concentrations:				
SO_2	3.40	4.08	2.65	3.25
NO_x	12.90	15.59	10.11	12.46
O_3	43.42	51.88	33.84	41.55
NH_3	3.66	5.34	2.85	4.29

Table 11. Calculated growth rates as % of potential growth rate (between brackets: calculated dry weight in Mg ha^{-1}, and stem volume increment in $m^3ha^{-1}yr^{-1}$ derived from the calculated dry weight increment) in Speuld and Kootwijk during 1988 and 1989, using the simulation model FORGRO (P112). For this calculation, no effects of nutrition are taken into account

	Speuld		Kootwijk	
	1988	1989	1988	1989
potential growth	100.0 (23.0/31.0)	100.0 (25.8/34.0)	100.0 (22.3/31.6)	100.0 (25.1/34.8)
growth reduced by water availability only % of potential	94.0	84.6	91.2	72.4
growth reduced by direct effects of pollutants only % of potential	99.9	99.9	99.9	99.9
growth reduced by water and direct effects of pollutants % of potential	93.9	84.5	91.1	72.3

The effects of O_3 exposure of up to one month on gas exchange of O_3 were described in some projects (P115) but have not yet been evaluated for the ACIFORN field situations. Under field conditions, high stomatal resistance of Douglas fir foliage (300 - 500 s m^{-1}) is an important threshold for O_3 to enter and cause physiological effects at ambient levels. In Douglas fir in controlled environments, O_3 concentrations of 400-800 $\mu g\ m^{-3}$ caused a proportional decrease in photosynthesis and stomatal conductance, and an increase in dark respiration. Exposure to O_3 may lead to a new stable physiological state at a lower efficiency that can be maintained for prolonged periods; in this case, the damage due to O_3 seems reversible. To some extent, O_3 may influence water relations through needle conductance (Freer-Smith & Dobson, 1989). After full needle development, the normal decrease in conductance during ageing of the new needles is not observed after O_3 fumigation, indicating that water use efficiency of fumigated needles is decreased. As stomatal behaviour remained largely intact, ozone probably interferes with the normal development of the cuticle or the epicuticular wax layer; in controlled experiments O_3 was associated with additional cuticular erosion (Cape, 1988). In addition to this, episodic exposure to very high concentrations of O_3 may lead to visible injury (leaf mottle). Occurrence of this type of injury, possibly leading to premature shedding and a decrease in total leaf area, may occur at the ACIFORN sites (see e.g. the episodes with high concentrations in May and August 1989).

In addition to an influence on photosynthesis and transpiration, O_3 appears to cause a delay of phloem loading after short exposures to realistic concentrations (100-200 $\mu g\ m^{-3}$) in controlled experiments (P107; Gorissen & van Veen 1988). Again, the trees seemed to be able to recover when exposure to O_3 was stopped. Inhibition of translocation at the expense of the rooting system also occurred after a month fumigation with 53 $\mu g\ m^{-3}$ SO_2.

In the Dutch situation, much attention has been focused on NH_3. In controlled experiments, the effects of NH_3 on leaf physiology were studied (van der Eerden, 1982; van Hove, 1989). The flux into the leaves is linear to the external concentration up to extremely high concentrations. NH_3 taken up by the foliage is efficiently assimilated when photosynthesis occurs. Uptake of NH_3 did not inhibit photosynthesis, but slightly stimulated it (van Hove, 1989). Low levels of SO_2 may counteract the stimulatory influence of NH_3.

The magnitude of direct short-term effects on forest growth in Speuld and Kootwijk can be assessed from the data gathered as part of the Dutch priority programme. With the possible exception of O_3, given the low levels of air pollution exposure in the stands of the fairly exposed Veluwe area, together with the high stomatal resistance of Douglas fir, it would appear that short-term effects are limited in scope.

4.2.2 Long-term effects

Long-term effects are less well defined than short-term effects, and are more difficult to monitor in the field, or to analyse experimentally under laboratory conditions. They either refer to changes in slow processes such as ageing and assimilate allocation, or relate to gradual accumulation of toxic substances that influence metabolic processes. Evidence suggests that long-term exposure to low concentrations may lead to accelerated ageing and leaf shedding (Zajaczkowska et al., 1981), thereby either shortening the growing season in annual crops and deciduous trees, or decreasing total leaf area in evergreen conifers. From analyses of both short-term and long-term effects in Vicia plants exposed to SO_2, Kropff (1989) concludes that short-term direct effects have a rather limited influence on productivity compared to the long-term effect of yellowing and early leaf fall.

Metabolic responses to long term exposure, irrespective of the plant species, are hardly understood so far. As long as the mechanism is not understood, it is difficult to define damage and measure its magnitude (Lange et al, 1989). Several explanations have been suggested e.g. for needle yellowing, which can be caused by ageing, water stress, nutrient deficit etc. In relation to air pollution, the mechanism behind needle yellowing can be a long-term effect such as a reduced pH buffer capacity following pollutant uptake (Kropff, 1989) or a magnesium deficit resulting from long-term soil acidification.

A hypothetical mechanism for damage from long-term direct effects of SO_2 assumes subtle injury to lead to increased ageing and leaf loss at low concentrations. Under these conditions, immediate effects on photosynthesis are minimal, possibly excluding situations with low temperature and high humidity when stomata are open and metabolic activity is restricted (P115, Freer-Smith & Mansfield, 1987). Empirical evidence indicates that even at rather low levels of SO_2 at 40-60 $\mu g\ m^{-3}$ the average life-span of leaves is decreased and early leaf shedding occurs (Kropff, 1989). At both the ACIFORN sites it seems unlikely that this phenomenon significantly influences productivity at this moment as Leaf Area Index (LAI) is high (Table 7). However, it may very well be important in stands with low LAI (such as Scots pine stands, P100), or in deciduous trees.

Changes in assimilate partitioning cannot be established from field data alone, but additional laboratory experiments (P107, Gorissen & van Veen, 1988) indicate that fine root growth may be decreased during exposure to O_3, making the stands more vulnerable to drought. It is suggested that reduced availability of assimilates elsewhere in the plant may result from temporary disturbances in phloem loading under exposure to O_3.

If the assimilate distribution within the plant changes to such an extent that root growth is permanently decreased and total root density is reduced, it may limit the uptake of water and nutrients. In that case, the stand becomes more susceptible to drought and nutrient shortage.

Some nutrient supply results from gaseous uptake of NO_2, NH_3 and SO_2. NH_3 and NO_2, after reduction to NH_3, can be incorporated in proteins and amino-acids. SO_2 is oxidized to SO_4^{2-}, and will have a fertilizing effect when the supply of sulphur in the soil is low. From Tables 10 and 12, it can be derived that gaseous uptake under conditions of high photosynthesis may supply some 10 % of the total amount of sulphur and some 5-10 % of the total amount of nitrogen required for growth. When no shortage of either nitrogen or sulphur exists, the gaseous uptake of N and S probably further contributes to nutrient imbalances already existing within the foliage (N and S uptake as discussed here refers to stomatal uptake only; additional cuticular uptake of N as NH_4^+ probably occurs in canopy interactions during canopy wetness, which will be discussed in Section 5.1).

Table 12. Calculated nutrient uptake rates (in kg per hectare), corresponding to water limited growth, accounting for different redistribution of N, P, K, Ca, Mg and S before litter loss (P112/P114.2)

	Speuld		Kootwijk	
	1988	1989	1988	1989
Nitrogen	89.	95.	80.	80.
Phosphorus	5.1	5.3	4.8	4.7
Potassium	33.2	32.8	31.0	29.7
Calcium	31.9	32.8	30.0	28.9
Magnesium	5.6	5.5	5.0	4.8
Sulphur	9.0	9.5	9.0	8.9

At the present stage of evaluation, no results are available yet from additional correlative analysis between the air pollution data sets and the dynamics of the structural elements such as root and branch dynamics at a detailed level, stem growth, and root-shoot ratios in the ACIFORN stands. These structural characteristics may be reflected in the progress of crown decline which is not detectable in the short term, yet may accumulate in the long term. The same will be done for ethylene synthesis, as biochemical parameters determined at the ACIFORN sites. The production of this stress marker differs widely between genotypes and between needle age classes (P105).

4.3 Conclusions

Most ACIFORN data on direct effects of air pollutants obtained up to now were gathered from controlled experiments. No confirmations of these results could be made as yet from the field data sets. This implies that conclusions regarding performance of adult trees are restricted. Neither from the controlled experiments carried out sofar, nor from the field observations, can conclusions regarding long-term effect caused by air pollution be drawn. This can only be achieved in a combined analysis consisting of experimental work and model analysis. Gaseous air pollutants at high concentrations have been shown to affect photosynthesis, respiration, photosynthate translocation from the needles and stomatal behaviour. Under current ambient conditions at Speuld, this seems unlikely to affect stand growth significantly. These conclusions with regard to the field situation at the ACIFORN sites are based on monitoring of growth and development over a limited number of years only; no extreme climatic events occurred over this period. As far as structural morphogenesis of the stands is concerned, sofar no conclusions can be drawn with regard to long-term effects.

Conclusions on both short-term and long-term direct effects only have limited scope when based on observational studies and short-term monitoring only, without the understanding of the underlying physiological processes that are involved. This implies that predictions of risk of future damage to forest trees inevitably include large uncertainties.

In the hypotheses on forest decline outlined in Section 2.2, direct effects are considered to be among the main disturbances in the hypothesis regarding impact of ozone and complex high-elevation decline, and in the hypothesis based on multiple stress. The present data suggest the possibility of both short term and long term effects under field conditions in the Netherlands.

Direct effects of ozone undoubtedly occur during episodes of high exposure, resulting in decreased photosynthesis, increased maintenance respiration, and decreased allocation of photosynthates to the roots. These episodes occur at times when physiological activity is high (during the growing season), implying that stomatal resistance is relatively low and O_3 can be readily taken up by the foliage. However, the combination of ozone together with acidic fog only rarely occurs, and it seems unlikely that this plays a major role under Dutch conditions. Nevertheless, ozone alone must be considered a primary risk factor with possibly acute damage and visible injury resulting from episodic exposure to concentrations that exceed damage thresholds.

When direct short-term effects are considered, the multiple stress hypothesis seems unlikely to be relevant under known conditions in the Netherlands, due to the low concentrations of SO_2. It should be noted however, that combinations of low temperatures, high humidity, and episodic or local high exposure are likely to result in temporarily decreased photosynthesis. Due to decreased oxidation of toxic anions such as sulphite and bisulphite, combined with decreased buffering capacity when metabolic activity is low, these conditions may lead to visible injury and foliage shedding near local sources of SO_2.

The magnitude of long-term effects of exposure to a combination of gaseous pollutants is unclear. Preliminary evidence indicates that increased ageing may lead to early foliage loss. In the case of conifer stands with a high leaf area index this may lead to only marginal effects, but in stands with a low leaf area index, and in deciduous forests, a major effect on growth could occur (Mohren & Bartelink, 1990).

In Douglas fir possible indirect long-term effects of gaseous air pollutants are most likely to be associated with decreased root growth, making the stands more susceptible to drought. When Douglas fir is planted on sites with an intermediate moisture supply, this indirect effect will not necessarily lead to a growth decline in periods with average moisture availability, but to a more pronounced growth decline under extreme drought.

With regard to the hypotheses on major causes of forest decline, only the hypothesis on multiple stress due to combined exposure to ozone, sulphur dioxide, nitrogen oxides and ammonia contains assumptions on long-term effects of the gaseous uptake of pollutants. In the Netherlands, the relevance of these long-term effects cannot be evaluated yet, and should not be ruled out.

5. ATMOSPHERIC INPUT AND SOIL ACIDIFICATION

5.1 Atmospheric input, canopy interactions and soil input

Atmospheric input (total deposition) of acidic substances may consist of wet and dry deposition. Part of both wet and dry deposition is intercepted by the canopy and may undergo transformations before it reaches the soil surface. Moreover, interactions with the intercepting surface (foliage or branch area) may take place, so that deposited material is taken up by the trees, and tissue components may leach. Both deposition and canopy interactions depend on the microclimate within the canopy space and on surface properties of foliage and woody material. Leaf wetness is particularly important in this respect, as the liquid on the outside of the foliage comprises the medium for chemical reactions of depositing components. Detailed analysis of canopy wetness and the dynamics of intercepted water at Kootwijk enables the prediction of conditions for canopy interactions at present (P104.1). In the near future, this will be used to develop models of canopy interactions and the relationship between deposition and throughfall.

Due to these canopy interactions, chemical composition of throughfall beneath the canopy may be different from the total deposition above the canopy. As the interactions that take place in the canopy space are largely unknown, generalization of deposition/throughfall relationship measured at the ACIFORN sites is restricted. In Section 6.2 methods are described to quantify some of the partial fluxes, and the results are compared.

The formation of ammonium sulphate from the interacting dry deposition of ammonia and sulphur dioxide during the time the canopy is wet, is an important canopy interaction under conditions with both NH_3 and SO_2 deposition, as in the Netherlands. SO_2 is oxidized to SO_4^{2-}, with NH_3 acting as the proton acceptor, thereby enabling the dissolution of SO_2 into the water film to continue (van Breemen et al., 1982). The ammonium sulphate formed in this way is washed off the canopy with intercepted precipitation when this exceeds canopy storage capacity, and is deposited on the forest floor with throughfall.

At the leaf or needle surface, H^+ and NH_4^+ are apparently taken up through the cuticle, in an exchange with K^+, Mg^{2+} and Ca^{2+} that are subsequently leached from the foliage, and washed off with rain to give additional K^+, Mg^{2+} and Ca^{2+} in throughfall. Leaching has been demonstrated many times, and additional leaching resulting from acidic precipitation is assumed (see review by Parker, 1983). Table 6 (P102.1) gives the estimated average annual amounts of leaching at the ACIFORN sites Speuld and Kootwijk. For potassium, total leaching estimated for Speuld amounts to 17 kg $ha^{-1}yr^{-1}$, and for Kootwijk 14.6 kg $ha^{-1}yr^{-1}$ out of a total throughfall of 16-20 kg $ha^{-1}yr^{-1}$. This means that leaching is considerably larger in the case of potassium than total deposition above the canopy (10-15 times as much), and that it amounts to some 10 % of the total amount of potassium in the biomass. These values agree with literature data supplied by Parker (1983) and Fiedler et al. (1973), and do not appear to be excessive. For calcium and magnesium the estimated deposition rates are higher, 5.7 and 4.7 kg $ha^{-1}yr^{-1}$ for calcium in Speuld and Kootwijk, and 3.3 and 2.7 kg $ha^{-1}yr^{-1}$ for magnesium in Speuld and Kootwijk, respectively. Leaching of calcium is about the same as bulk deposition, whereas leaching of magnesium is about half to one-third of bulk deposition. The uncertainty in the estimated leaching is small for potassium, but rather large for calcium and magnesium (P102.1).

Foliar uptake of nitrogen as NH_4^+ is estimated to be on average 8.4 kg $ha^{-1}yr^{-1}$ in Speuld and 6.8 kg $ha^{-1}yr^{-1}$ in Kootwijk. Comparison of these data (Table 6, P102.1) with literature values (Parker, 1983) does not indicate a strong increase in canopy exchange through the cuticle, and confirms the notion that cuticular damage plays only a minor role, if any, in both ACIFORN stands.

5.2 Soil acidification: processes occurring in Speuld and Kootwijk

Acidification processes can be described with the help of the proton budgets of the compartments of the ecosystem studied. Soil acidification, defined as the decrease of the Acid Neutralizing Capacity (ANC; Van Breemen et al., 1983), is the net effect of cation and anion precipitation and dissolution reactions.

Proton budgets for several compartments of the sites Speuld and Kootwijk are shown in Table 13. Proton production in the tree canopy takes place through NH_4^+ uptake, weak acid dissociation and net input of protons. Cation leaching (K^+, Ca^{2+}, Mg^{2+}, Mn^{2+}) and anion uptake (SO_4^{2-} and NO_3^-) accounts for the proton consumption. The main source and location of proton production is nitrification in the litter layer and in the first 10 cm of the profile. Buffering is highest in these layers too. At Kootwijk H^+-buffering is even complete at 10 cm depth but at Speuld significant H^+-buffering takes place at a depth of between 10 and 40 cm. This is also illustrated by the H-fluxes (Table 14). Given the higher acid neutralizing capacity (ANC) of the loamier soil at Speuld, this buffering profile suggests that historically, atmospheric deposition of N and/or S and proton production and buffering has been higher at Speuld than at Kootwijk. Higher deposition can be explained by higher interception by a larger canopy which can be expected on a more fertile soil with a higher water holding capacity. Furthermore the Speuld area is one of the oldest forest areas in the Netherlands, which means that historically speaking the interception capacity has been high for many years. Higher deposition at Speuld in recent years has been suggested by Van der Maas (P102.1). Higher deposition also leads to a higher degree of eutrophication with N through which less N can be immobilized. As a result NO_3^--leaching fluxes will be higher. Indeed it is found that NO_3^--fluxes at 90 cm depth are higher at Speuld (see Table 14). It is therefore suggested that the higher NO_3^--fluxes that are found at Speuld might result from higher deposition and/or lower immobilization of N (P102.1).

Table 13. Proton budgets for the tree canopy and five soil layers (litter layer (LL) plus 0 to 10 cm depth, 10 to 20, 20 to 40, 40 to 60 and 60 to 90 cm depth) at Speuld and Kootwijk. Results are mean values for the period 1 March 1987 to 1 March 1990, in kmol(+/-) $ha^{-1}yr^{-1}$ (P102.1)

Speuld	H production:						
	H-in	WA[1)]	N	Bio[2)]	Cp[3)]	Alp[4)]	Ad[5)]
Canopy	0.17	0.18	0.56				
LL-10	0.07		3.71	0.12			0.22
10-20	1.40	0.00	0.73	0.30	0.09		
20-40	1.03			0.47	0.04		0.02
40-60	0.28			0.15	0.02		0.15
60-90	0.18			0.06			0.75
	H consumption:						
	Cd[6)]	Ald[7)]	Ap[8)]	Bio	N	WA	H-out
Canopy				0.84			0.07
LL-10	0.63	1.90				0.13	1.40
10-20		1.27	0.28				1.03
20-40		1.00			0.33	0.04	0.28
40-60		0.26			0.00	0.06	0.18
60-90	0.22	0.13			0.44	0.02	0.21
Kootwijk	H production:						
	H-in	WA	N	Bio	Cp	Alp	Ad
Canopy	0.12	0.22	0.44				
LL-10	0.06		3.64	0.46			
10-20	0.78			0.14	0.13		0.01
20-40	0.52			0.20	0.33	0.22	
40-60	0.16		0.39	0.10			0.40
60-90	0.14	0.01	0.07	0.03		0.05	
	H consumption:						
	Cd	Ald	Ap	Bio	N	WA	H-out
Canopy				0.72			0.06
LL-10	0.39	2.41	0.42			0.16	0.78
10-20		0.24			0.30	0.01	0.52
20-40			0.34		0.72	0.05	0.16
40-60	0.27	0.66				0.00	0.14
60-90	0.12		0.03				0.15

1) WA : Weak Acids
2) Bio: Biological processes for all nutrients exclusive N (uptake, leaching, mineralisation)
3) Cp : Cation precipitation
4) Alp: Al-precipitation
5) Ad : Anion dissolution
6) Cd : Cation dissolution (exclusive of Al)
7) Ald: Al-dissolution
8) Ap : Anion precipitation

Table 14. Three-year mean atmospheric deposition (AD), throughfall (T) and drainage fluxes of water (mm) and solutes at 10, 20, 40, 60 and 90 cm depth for Speuld and Kootwijk after Cl-calibration (kmol(+/) ha^{-1} yr^{-1}; periods: 3 March 1987 - 1 March 1988, 1 March 1988 - 28 February 1989, and 28 February 1989 - 27 February 1990) (P102.1)

Speuld:	H^+	K^+	Na^+	Ca^{2+}	Mg^{2+}	Al^{3+}	Fe^{2+}	Mn^{2+}	NH_4^+
AD	0.17	0.05	0.99	0.29	0.27	0.01	0.02	0.01	2.80
T	0.07	0.42	0.99	0.42	0.34	0.02	0.03	0.06	2.20
10	1.40	0.35	0.98	0.84	0.60	1.92	0.09	0.14	0.99
20	1.03	0.25	0.90	0.70	0.51	3.18	0.07	0.16	0.23
40	0.28	0.11	0.97	0.34	0.47	4.18	0.01	0.15	0.04
60	0.18	0.06	1.05	0.19	0.38	4.44	0.00	0.18	0.00
90	0.22	0.13	1.03	0.34	0.34	4.57	0.00	0.19	0.01

	Cl^-	NO_3^-	SO_4^{2-}	$H_2PO_4^-$	WA[1)]	SiO_2
AD	1.38	0.79	2.26	0.01	0.16	0.01
T	1.38	0.76	2.14	0.01	0.35	0.02
10	1.38	3.25	2.51	0.01	0.21	0.92
20	1.38	3.22	2.22	0.00	0.22	0.85
40	1.38	2.69	2.22	0.00	0.18	0.85
60	1.38	2.65	2.36	0.00	0.12	0.59
90	1.38	2.22	3.11	0.00	0.10	0.76

Kootwijk	H^+	K^+	Na^+	Ca^{2+}	Mg^{2+}	Al^{3+}	Fe^{2+}	Mn^{2+}	NH_4^+
AD	0.12	0.04	0.79	0.24	0.22	0.01	0.01	0.01	2.60
T	0.06	0.41	0.79	0.33	0.28	0.01	0.02	0.03	2.12
10	0.78	0.03	0.83	0.48	0.47	2.43	0.04	0.06	0.29
20	0.52	0.02	0.83	0.34	0.37	2.66	0.03	0.05	0.23
40	0.16	0.01	0.70	0.10	0.23	2.44	0.01	0.05	0.00
60	0.14	0.01	0.73	0.22	0.24	3.10	0.00	0.07	0.00
90	0.15	0.02	0.79	0.19	0.28	3.05	0.01	0.07	0.00

	Cl^-	NO_3^-	SO_4^{2-}	$H_2PO_4^-$	WA	SiO_2
AD	1.09	0.75	2.10	0.00	0.08	0.01
T	1.09	0.71	1.93	0.01	0.30	0.02
10	1.09	2.52	1.65	0.01	0.14	1.01
20	1.09	2.16	1.66	0.00	0.13	0.78
40	1.09	1.21	1.32	0.00	0.08	0.50
60	1.09	1.60	1.72	0.00	0.08	0.48
90	1.09	1.67	1.69	0.00	0.09	0.50

[1)] WA: Weak Acids (HCO_3^- and organic acids)

Proton production through biocycling as presented in Table 13 (net result of mineralisation and uptake of cations and anions, disregarding uptake of NH_4^+ and NO_3^-) is relatively low in comparison with nitrate production and leaching. Production through biocycling for

Kootwijk is relatively high in the litter layer and the first 10 cm because of the assumed high uptake of K^+ in this layer. Weak acid dissociation and association are unimportant processes with respect to proton production or consumption at the low soil pH-values involved. Proton production through anion dissolution seems to be relevant at 40 to 60 cm at Kootwijk and at 60 to 90 cm at Speuld, but the significance of these values is not certain. Proton production through cation (inclusive of Al) precipitation is low. Proton consumption (buffering) mainly takes place through Al-dissolution while other processes (cation-dissolution, anion precipitation and N-immobilisation and/or uptake) are less important.

The flux data are given in Table 14 (102.1). NH_4^+-fluxes strongly decrease in the litter layer and the first 10 cm of the profile. NO_3^--fluxes strongly increased at these depths followed by a slight decrease further down the profile. Apparently the ecosystem is incapable of using all incoming N; 50 % (Kootwijk) to 60 % (Speuld) is leached below the root zone. Besides NO_3^- the nutrients Ca^{2+}, Mg^{2+} and K^+ are leached from the root zone too. In general atmospheric input of these components are shown to be equal to the leaching at 90 cm. The leaching values of K^+ at Speuld tend to be higher than the atmospheric input. High concentrations of Al^{3+} and H^+ (caused by high acidic atmospheric input), resulting in high saturation of the exchange complex with these components (Tiktak et al, 1988), reduce the possibility for the adsorption of Ca^{2+}, Mg^{2+} and K^+, and enhances their leaching. Phosphorus is not leached from the ecosystems but its availability for plants may be reduced because of formation of Al-P precipitates.

The ecosystem budgets for Speuld and Kootwijk (see Table 15) show that external input and atmospheric deposition are the main source of H^+ at both sites. Of the total external proton input at Speuld, it is calculated that 18 % originates from NO_x, 49 % from SO_2 (after SO_4-aerosol correction as 0.12·Na, see P102.1), and 33 % from NH_3. The percentages for Kootwijk are 21% NO_x, 55% SO_2 and 24% NH_3.

At Speuld and Kootwijk, respectively 83 and 79 % of the proton production is calculated to be buffered by dissolution of Al. Soil acidification amounts to 4.40 for Speuld and 3.74 $kmol.ha^{-1}.y^{-1}$ for Kootwijk (P102.1, see also thematic report by van Breemen and Verstraten).

Table 15. Proton budgets for the ecosystems Speuld and Kootwijk. Results are calculated with mean values from the period 1 March 1987 to 1 March 1990, in kmol (+/-) $ha^{-1}yr^{-1}$ (P102.1)

	H sources					
	external		internal			
	H	N	N	Bio[1)]	WA[2)]	Ad[3)]
Speuld	0.17	4.22	0.00	0.27		0.86
Kootwijk	0.12	3.51	0.00	0.21	0.01	

	H sinks internal				
	Cd[4)]	Ald[5)]	Ap[6)]	WA	H
Speuld	0.70	4.56		0.07	0.21
Kootwijk	0.32	3.04	0.38		0.15

1) Bio: Biological processes for all nutrients exclusive N (uptake, leaching, mineralisation)
2) WA : Weak Acids
3) Ad : Anion dissolution
4) Cd : Cation dissolution (exclusive of Al)
5) Ald: Al-dissolution
6) Ap : Anion precipitation

5.3 Soil acidification: effects on tree growth

Soil acidification may have three major consequences for tree growth:

1) base cations at the exchange complex, such as magnesium, potassium and calcium, are replaced by H^+ and Al^{3+}, whereby the cations, if not taken up by the trees, leach to the deep underground in periods of downward water flux; on poor sites, this leaching constitutes a severe loss of essential tree nutrients;
2) competition at the sites of uptake of NH_4^+ or Al^{3+} with cations such as magnesium, potassium and calcium, leading to a decreased uptake of nutrients by the fine roots, even when these nutrients are present in sufficient amounts in the bulk soil solution (Roelofs et al., 1985);
3) restricted root growth or a shift of fine root growth to the upper horizons of the profile, caused by chemical changes in the soil (increased [Al^{3+}]), leading to decreased uptake capacity for water and nutrients (Ulrich, 1983b; Murach, 1984)

ad 1)

Leaching of base cations is a common feature of soil acidification under circumstances where there is a precipitation surplus. This leads to increased shortage of mineral elements for tree nutrition in the case of soils already low in base cation supply, as is the case for most of the Dutch forest area on dry sandy soils. The occupation of the exchange complex with Al^{3+} and H^{+} also limits the buffering capacity of the complex for cations released from decomposing litter, thereby decoupling the mineral cycling to the extent that nutrients produced in decomposition are lost.

ad 2)

Keltjens & van Ulden (1987) and Keltjens and van Loenen (1989) show that base cation uptake per unit of root length is hampered in the presence of NH_4^+ and Al^{3+}-ions. This is supported by other greenhouse experiments (Boxman & Roelofs, 1988) and correlative field inventories (Boxman & van Dijk, 1988). Aluminium leads to a reduction in Ca^{2+} and Mg^{2+} uptake, which can be evaluated from the ratio between concentrations of aluminium and calcium ($[Al^{3+}]/[Ca^{2+}]$ ratio, Ulrich 1983b, see next Section). Thus, when soil acidification caused by nitrification of ammonium leads to an increase in free aluminum, and when at the same time NH_4^+ from atmospheric deposition is present, it can be expected that the base cation uptake is reduced in situations in which availability of base cations would remain unchanged. Combined with soil acidification and leaching of cations, it can be expected that both availability and uptake capacity will be decreased, the combination reinforcing the decrease in cation content of the plant tissue.

ad 3)

To enable the study of the effects of soil acidification on root growth, data are needed on root density in different soil horizons, on effects of pH and nutrient concentrations on root growth, on toxic effects of the soil solution, etc. Data on root density and root dynamics are scarce; in Speuld and Kootwijk, the average root density is around 0.5 cm^{-3}, which is low compared to agricultural crops (de Willigen & van Noordwijk, 1987). Although possible toxic effects of pH and $[Al^{3+}]$ on root growth can be established in greenhouse trials, the translation of results to mature trees under field conditions is difficult. The effects of simulated acid deposition were studied on seedlings of Douglas fir as part of P103 and P107. The results showed that in terms of effects of ammonium and toxic minerals on root length, effects on root systems of full grown trees can be estimated using simulation models in order to predict fine root densities in dependence of soil chemical conditions. As it turns out, specific root length (SRL, root length per unit of fine root dry weight) decreases with increasing % Al^{3+} at the exchange complex. Thus, it is expected that when the soil is within the aluminium buffering range, soil acidification leads to lower root density provided fine root mass remains the same.

5.4 Root activity in relation to soil conditions

Because concentrations of nutrients or toxic minerals at the root surface can be quite different from those in the bulk soil solution, the processes in the rhizosphere make it difficult to relate bulk soil parameters to consequences for root growth. Nevertheless, using the information available at present, some consequences for tree growth can be indicated.

In laboratory experiments carried out as part of P103, root development was severely inhibited at high ammonium supply. At high levels of ammonium the specific root length (SRL) was low compared to situations with low ammonium supply. Earlier investigations by Van den Driessche (1978) showed that with Douglas fir seedlings grown at pH 4 in a sand culture, SRL was larger with NO_3^- nutrition compared with NH_4^+ nutrition.

Unfavourable soil conditions for root growth in forest soil have been described by many authors, e.g. Rost-Siebert (1985), Ulrich & Pankrath (1983) and Ulrich et al. (1984). Concentrations of aluminium in special nutrient solutions lethal to different tree species are described by Eldhuset et al. (1987). In the field situation, Al^{3+} concentrations are usually lower than the lethal levels. However, fine root growth can be disturbed at sub-lethal levels. This appears as a decrease in root length (Keltjens & Van Loenen, 1989). Fine root dry weight is less affected. The specific root length is a simple morphological parameter: the ratio of fine root length to fine root biomass. It can be used to describe the influence of soil chemical conditions on the structure of the root system, and the influence of ammonium and aluminium on SRL via interactions in the rhizosphere as well (P103).

The concentration of nutrients and toxic minerals in the soil solution is variable due to changes in water content. Fine root growth is limited at low water content. Therefore, the SRL of the roots present at a specific time is the result of growth in sublethal concentrations in the period before that time. The SRL is the main parameter by which adverse soil chemical conditions can be established. However, the SRL is also dependent on soil bulk density because of mechanical resistance. The relationships of the SRL to soil chemical parameters have to be described per horizon or per soil layer.

It can be expected that the present levels of Al^{3+} have a negative influence on root length development under field conditions, as demonstrated in the SRL values in relation to bulk soil parameters (P103). At the same time, occurrence of relatively high concentrations of NH_4^+ in the soil solution of the top soil layer can be expected to influence uptake of Mg^{2+} and K^+. Thus, even if fine root biomass itself is not influenced by soil chemical conditions, the decrease in fine root length leads to a lower uptake capacity of the trees for water and nutrients, and uptake of cations in the top soil layer may be hampered through rhizosphere acidification due to high ammonium concentrations.

5.5 Criteria for evaluation of soil acidity

Soil acidity and soil acidification are aspects of soil chemistry that may interfere with tree growth in many ways, and no single soil parameter can be used to characterize this complex of interactions. As mentioned in Section 5.4, the main aspects involved in soils of pH around 4 (the Aluminium buffering range) are cation availability at the exchange complex and in the soil solution, Al^{3+} at the exchange complex and in the soil solution, and ammonium and nitrate concentrations in the soil solution. Also, proton activity (pH per se) is important when considering uptake of base cations such as K^+. Aluminium toxicity to plants is well know and extensively documented (see e.g. reviews by Cronan et al., 1990; Foy et al., 1978; Roy et al., 1988; Schaedle et al., 1989; van den Burg, 1990). Cronan et al. (1990), in a summary of the results from the ALBIOS (ALuminum in the BIOSphere) project state that "...Al toxicity varies across the landscape and is most likely under the following conditions: in forests containing sensitive or moderately sensitive tree species; in situations where much of the fine root biomass is concentrated in mineral horizons with <10-15% base saturation; in ecosystems characterized by high inputs of acidic deposition and slow soil sulphate adsorption; in forest stands where calcium, magnesium, or phosphorus are in marginal supply; and in forests experiencing drought stress and increased dependence on deeper rooting...". Virtually all of these conditions apply to the majority of the Dutch forest sites on sandy soils.

A wide range of Al toxicity thresholds has been reported in the literature, usually based on seedling experiments using solution cultures or greenhouse trails. The sensitivity of tree species may vary widely, increasing from relatively insensitive species such as birch, Scots pine, red oak and larch to moderately sensitive species such as Douglas fir and European beech, and to sensitive species such as Norway spruce (Schaedle et al., 1989; P113). Tree sensitivity depends on solution pH, on chemical speciation of the aluminium, on calcium concentration, on the form in which inorganic nitrogen is taken up (NO_3^- or NH_4^+), on mycorrhizal infection, on soil moisture, and on tree nutrient status. To some extent an antagonism appears to exist between aluminium and calcium, in which high concentrations of calcium appear to counteract the toxicity of aluminium. This antagonism was described by Rost-Siebert (1983) for seedlings of Norway spruce and European beech, and by Smit et al. (1987a) for Douglas fir. Results of these studies indicate that in acid forest soils ($pH(H_2O) \leq 4$), the ratio $[Al^{3+}]/[Ca^{2+}]$ can be used to describe the degree of aluminium toxicity. Based on the results of Rost-Siebert (1983), Ulrich (1983) proposes critical $[Al^{3+}]/[Ca^{2+}]$ ratios of 1.0 for Norway spruce, and 10 for birch. According to Ulrich (1983), major toxicity effects in Norway spruce might occur at a $[Al^{3+}]/[Ca^{2+}]$ ratio of 5.

Results of a correlative field study in the Netherlands (Roelofs et al., 1985) indicate a critical ratio above which aluminium may start to influence tree growth of black pine (Pinus nigra), comparable to the values for Norway spruce mentioned by Ulrich (1983).

Aluminium concentration and calcium and magnesium nutrition may interfere in many ways, and an empirical dimensionless number such as the $[Al^{3+}]/[Ca^{2+}]$ ratio is inadequate to describe the interactions entirely (Keltjens, 1990) and should be handled with care. From the literature data available at present, it appears however that the $[Al^{3+}]/[Ca^{2+}]$ ratio combines some of the major aspects governing possible aluminium toxicity possibly occurring at $pH(H_2O) \leq 4$. At both ACIFORN sites, the $[Al^{3+}]/[Ca^{2+}]$ ratio increases from about 3 at a depth of 10 cm, to a very critical value of about 5 at a depth of 20 cm (P102.1). This implies that most likely a very critical situation exists with regard to aluminium toxicity and calcium and magnesium uptake below 20 cm.

5.6 Consequences of nitrogen deposition on soil acidification and tree nutrition

From the preceding sections it has become evident that nitrogen deposition as NH_4^+ comprises a major contribution to soil acidification under Dutch conditions. High deposition rates of ammonium lead to acidification of forest soils, provided that nitrification takes place (Van Breemen et al, 1982) and may result in nutrient imbalances of forest stands (Roelofs et al, 1985; Mohren et al., 1986; Van den Burg, 1990). Unbalanced nutrition with nitrogen may have adverse effects on susceptibility of trees to diseases (De Kam et al., 1989) and frost resistance (Aronsson, 1980). The experiments carried out as part of P103 showed that high levels of ammonium deposition may have strong negative effects on root growth under field conditions. As a result, root densities will decrease and root uptake capacity will be reduced. At the same time, shoot growth may be stimulated due to the effect of increased nitrogen availability on shoot/root ratios, especially on sites where nitrogen had previously been deficient. Both phenomena will make the stands more vulnerable to drought, and may lead to nutrient imbalances and deficiencies of phosphorus, potassium, calcium and magnesium (Oren & Schulze, 1989).

The root uptake of nutrients, especially phosphorus, may be further reduced if mycorrhizal activity decreases as a result of increased nitrogen input (see thematic report on biological and physiological effects). Mycorrhizal activity is also expected to decrease due to increased concentrations of free aluminium in the soil solution.

Most atmospheric input consists of ammonium, which is converted into nitrate by nitrification. Two extremes can be distinguished in forest ecosystems: a situation with high nitrification, in which all ammonium input is converted to nitrate and N is taken up mainly as NO_3^-, with accompanying acidification due to proton production in the nitrification process, and a situation in which nitrification is inhibited, or insufficient to cope with ammonium inputs (Roelofs et al., 1988). In the latter case, nitrogen is taken up by the trees as ammonium, acidifying the rhizosphere through excretion of protons (P83). Also, differences in ammonium contribution to total nitrogen uptake affect the carboxylate concentration within the plant, thereby possibly affecting metabolic activity and the capacity to detoxify pollutants taken up directly through the stomata. H^+ produced during assimilation of NH_4^+ may contribute to internal acidification and rhizosphere acidification. Three situations may be distinguished (P83): if more than 65 % of the N taken up is absorbed as nitrate, the carboxylate concentration in the plant is at the appropriate level, and the rhizosphere is alkalized. This is a favourable situation under acid soil conditions. When nitrate accounts for between about 20 and 65 % of the total N uptake, the carboxylate content in the plant is still sufficient. But, with decreasing nitrate contribution in favour of uptake of ammonium, the roots excrete increasing amounts of protons, leading to acidification of the rhizosphere. When less than about 20 % of the amount of N taken up is in the nitrate form, the carboxylate concentration in the plant decreases leading to physiological disorders and rhizosphere acidification enhanced by proton excretion. It is thought that the latter situation leads to very poor growth and increased tree mortality.

Most of the processes leading to soil acidification are fairly well understood and can be quantified. Nitrification however, cannot be predicted with enough accuracy for future circumstances. Nitrification is a key process, making nitrate available for uptake, and thus resulting in higher pH values in the rhizosphere (van Breemen et al., 1987, P83).

5.7 Conclusions

Pertaining to the ACIFORN sites Speuld and Kootwijk:

Soil acidification, mainly as a result of deposition of NH_3 and SO_2 (as $(NH_4)_2SO_4$ in throughfall), has advanced to the point where more than half of the CEC is occupied by aluminium; base cation complexation is less than 5 % in the mineral soil, which according to Ulrich (1983 a & b) should be considered as a "worst case"; leaching of NO_3^-, accompanied by K^+ and Ca^{2+} from the rooted profile downwards occurs at both sites.

Calculations of annual fluxes showed high fluxes of NO_3^- at a depth of 90 cm: 50 to 60 % of the atmospheric N-input is leached from the root zone. Leaching of Ca^{2+}, Mg^{2+} and K^+ generally equalled atmospheric input. Base cations are not being retained by adsorption because the adsorption complex is saturated with Al^{3+} and H^+ as a result of the relative high concentration of these components. H^+-Buffering at Kootwijk is completed at a depth of 10 cm, while at Speuld significant H^+-buffering still takes place at 10 to 40 cm depth.

Proton budget calculations show that 95 % of the proton production is caused by external input. Of the total external proton input at Speuld, 18 % originates from NO_x, 49 % from SO_2 (after SO_4-aerosol correction for seaspray, see P102.1), and 33 % from NH_3. The percentages for Kootwijk are 21% NO_x, 55% SO_2 and 24% NH_3. About 80 % of the protons produced are consumed or buffered by Al dissolution. Buffering results in a depletion of the acid neutralizing capacity (ANC) of 4.40 and 3.74 kmol(+/-) $ha^{-1}yr^{-1}$ at Speuld and Kootwijk respectively.

Both nitrogen deposition and soil acidification have increased tree growth, to the extent that both Speuld and Kootwijk show growth rates according to the highest yield class, as indicated in current yield tables.

Considerable leaching of K^+ from foliage occurs at both sites, resulting from exchange of K^+ against NH_4^+ and H^+ at the surface of the foliage, possibly also from exchange against Na^+ near the seashore. This leaching does not appear to be excessive, but may cause additional potassium leakage from the ecosystem, as buffering capacity via the exchange complex in the soil is limited.

The fine root densities in the ACIFORN stands are low. This limits water and nutrient uptake (especially phosphorus). In Kootwijk the amount of fine roots is somewhat larger than in Speuld. Both stands are deficient in phosphorus, and tend to have a rather low potassium content (especially Kootwijk). Phosphorus deficiency in Kootwijk is less severe compared to Speuld, probably due to previous agricultural land-use.

Soil pH at both sites is expected to decrease within the next 30-50 years when the pool of aluminium hydroxide in the soil is depleted. Afterwards, a new stable situation will occur in the iron buffering range, with less Al in the soil solution and at the exchange complex. and increased proton activity. The consequences of a drop in pH on tree growth are unknown, but would be expected to be detrimental.

<u>Regarding effects of soil acidification on tree growth in general:</u>
Nitrogen deposition on poor sandy sites has improved nitrogen nutrition, but has progressed to the extent that on dry sites nitrogen is available in excess of demand (P18). At the same time, deficiencies of phosphorus, potassium, calcium and magnesium have been induced by removing limitations to tree growth due to nitrogen availability. Concomitantly, nitrogen deposition as NH_3 together with SO_2, after transformation to $(NH_4)_2SO_4$, is a

major cause of soil acidification.

Initially, soil acidification may enhance tree growth by increasing nutrient availability through protonation of the exchange complex, whereby base cations enter the soil solution. But then deficiencies may develop as nutrients leach as a result of higher concentrations in the bulk solution (provided a precipitation surplus exists), and subsequently minerals leak from the vegetation-uptake/litter-decomposition cycle.

With soil acidification taking place within the buffering range of aluminium, by dissolution of aluminium hydroxide, the concentration of free aluminium in the soil solution has been increasing in recent decades. The specific root length (SRL) of fine roots depends on soil conditions, and decreases with increasing [Al^{3+}], resulting in a decrease in root density with ongoing soil acidification. Due to low fine root densities, uptake of water and nutrients will be hampered when root density is decreased even further. It can therefore be expected that forest stands on dry sites will become more vulnerable to drought and nutrient deficiencies.

Douglas fir roots prefer the uptake of ammonium as a source of nitrogen, even at low pH. During ammonium uptake the pH of the rhizosphere decreases, and aluminium activity in the rhizosphere increases, thus stimulating aluminium toxicity. At the same time, ammonium uptake may interfere with pH buffering within the plant, and render trees more susceptible to internal acidification as a result of the uptake of gaseous air pollutants.

In view of the hypotheses outlined in Section 2.2, both the hypothesis on soil acidification and aluminium toxicity, and the hypothesis on nitrogen deposition and eutrophication seem to be highly pertinent to conditions in the Netherlands. Whether soil acidification and aluminium toxicity are the major mechanisms governing changes in site conditions for forest growth, or whether eutrophication by means of excess nitrogen deposition is considered to be the overall dominating factor, depends on the site conditions and on the criteria used. At the more fertile sites such as in Speuld and Kootwijk, soil acidification as such may not have had immediate negative effects on tree growth through aluminium toxicity so far. Nevertheless, soil acidification and its accompanying processes, such as loss of nutrients through leaching to the deep underground, at the rate at which it proceeds at present, should be considered highly restrictive for future forest development.

In the present situation, effects due to excess nitrogen deposition and altered conditions for tree nutrition dominate the changes in forest growth and development. Taking into account the major contribution to soil acidification by nitrogen deposition, it would seem appropriate to justify the hypothesis on excess nitrogen deposition as the main component of phenomena and processes underlying present day changes in Dutch forests.

6. QUANTIFICATION AND MODELLING

6.1 Potential growth rates and reductions due to traditional factors

To achieve a quantitative estimate of the magnitude of the reduction of forest growth due to traditional factors versus the reduction due to air pollution and soil acidification, a carbon-balance model of forest growth has been extended to include effects of air pollution and soil acidification (P112/114.2). Table 11 contains the results of this model for 1988 and 1989 at both Speuld and Kootwijk. The simulation results presented in Table 11 only refer to water and direct effects of air pollutants; no effects of nutrient limitations on growth were considered. The simulation results in Table 11 indicate the relative contribution of the various growth limitations on stand growth. As expected, direct short-term effects of air pollutants, notably SO_2 and O_3 are virtually absent in Speuld and Kootwijk. Again, it should be noted that the effects of both SO_2 and O_3 relate to short-term influences on photosynthesis and respiration. No long-term effects have been incorporated in the model, as no data were available to calibrate a dose-response relationship.

The calculated potential growth rates are somewhat less than the observed growth rates in Speuld in 1988 and 1989; the model results appear to underestimate the maximum growth rates. This may be due to several reasons, for a detailed discussion of which, see P112. As it is difficult to evaluate model performance by comparing model results with growth rates of only a few years, growth rates as simulated with the model were not adapted to improve the fit with the measurements. Here, the growth rates are used in a relative way, by comparing the reduction in growth rate due to water limitations and that due to direct effects of air pollution on photosynthesis and respiration.

As a result of limited water supply in the rooted soil profile, growth reductions in the order of 5 and 10 % occurred in 1988 in Speuld and Kootwijk respectively. In 1989, the reduction in growth rate appeared to be around 15 and 28 % for Speuld and Kootwijk respectively. This is compatible with the reduction in transpiration when expressed in terms of potential transpiration calculated by means of the Makkink equation (P102.2). This yielded reductions of 5 and 10 % for 1988, and 20 and 24 % for 1989 for Speuld and Kootwijk respectively. Note that in the carbon-balance model, transpiration is estimated using a Penman-Monteith balance equation, taking into account radiation, vapour pressure deficit, and canopy (stomatal) resistance, whereas the potential evaporation estimate according to Makkink only considers radiation and a calibration factor that depends on vegetation type (P102.2).

When only short-term direct effects of air pollutants on photosynthesis and respiration are studied, the resulting growth reduction appears to be negligible for pollution conditions such as determined in Speuld and Kootwijk in 1988 and 1989. Quantification of the short-term direct effect of uptake of air pollutants was taken from climate room trial (P115). The calculations presented in Table 11 assume no additional stomatal opening at low temperatures and high humidity. Sensitivity analysis has shown that although incorporation of these phenomena would add to the amount of pollutants taken up through the stomata, the additional effect on total annual photosynthesis and growth was minimal (P112). Note that the results of Table 11 do not contain any possible long-term effects. The analysis of combined effects of water limitations and direct short-term effects of pollutants, assuming them to be cumulative under the conditions in 1988 and 1989, shows that the resulting growth reductions were dominated by the effect of water shortage (Table 11).

To evaluate the possible magnitude of a long-term effect of exposure and uptake at low concentrations, model calculations were carried out with an empirical exposure index based on concentration (in $\mu g\ m^{-3}$) and time of exposure (in hours) for ozone (P112/P114.2, see also Reich, 1987; Lee et al., 1988; Posthumus et al., 1989). For reasons of convenience, the dose is simply expressed as $\mu g\ h\ m^{-3}$. The long-term dose-response relationship used, consisted of a threshold dose below which no additional foliage loss occurs of 100 $\mu g\ h\ m^{-3}$, a critical level of 400 $\mu g\ h\ m^{-3}$ at which additional foliage loss amounts to 10% of the normal foliage loss, and a doubling level of 400 $\mu g\ h\ m^{-3}$ at which foliage loss is doubled (with linear interpolation between these levels and linear extrapolation beyond 800 $\mu g\ h\ m^{-3}$). The total accumulating dose is estimated for each foliage age class separately, using 245 days in the year the foliage is formed, 365 days in all other years, and using an exposure time of 12 hours per day. For realistic concentrations of 50 $\mu g\ m^{-3}$ (25 ppb), this means that the critical level of 10% additional foliage loss will be reached after about two years of exposure. Threshold and critical levels for this type of dose estimation, as reported in the literature, vary widely (see e.g. Miller & Parmeter, 1965; Reich, 1987; Lee et al., 1988; Posthumus et al., 1989). The values used here are not unrealistic, but the results should be interpreted with caution. Therefore, Table 16 contains an indication for the magnitude of the effect only. The results as presented in Table 16 indicate that long-term effects of ozone may very well be of importance under conditions in The Netherlands, and should be further investigated. When expressed in terms of percentage reduction of dry matter increment, the calculations as presented here would indicate a loss of some 3-5% in case of medium to high LAI under ambient ozone. At lower LAI, these tentative results indicate a somewhat larger effect, possibly around 10-20% loss. These values are in agreement with estimates reported earlier (Van der Eerden et al., 1988; Tonneijck, 1989).

Table 16. Estimated magnitude of growth reductions, relative to potential growth rate, using input data for the Speuld site, at different extosure to SO_2 and O_3, and for stands with different initial Leaf Area Index, using the simulation model FORGRO (P112). For SO_2, only short-term direct effects have been taken into account, according to data from P115 and Kropff (1989). For O_3, both short-term direct effects (data from P115), as well as possible long-term effect are taken into account as indicated in the text. In case of ozone, simulation runs covered periods of five years. (--): no effects; (-): virtually no effects; (-/+): some reduction, limited effect (0-3%); (+): significant reduction of growth (at least 5-10%); (++): severe growth reductions (>20%), with likely tree mortality and stand decline

SO_2 (μg m^{-3})	25	50	100	300
LAI = 7.5	-	-/+	+	
O_3 (μg m^{-3})	25	50	100	300
LAI = 7.5	--	-	-	++
LAI = 4.5	-	-/+	+	++
LAI = 3.0	-/+	+	++	++

When the soil acidification model RESAM (P113) was used in combination with the carbon-balance model to simulate nutrient uptake, the nutrient supply as estimated for Speuld and Kootwijk was sufficient to maintain the nutrient concentrations in the biomass at the present level. For this purpose, nutrient uptake was estimated from nutrient demand by the plant and nutrient supply expressed in terms of concentrations in the bulk soil solution, according to De Willigen & van Noordwijk (1987). In their approach, both mass flow and diffusion are accounted for. An increase in nutrient content was simulated for calcium and magnesium. As no sign of an increase in uptake was found in the field plots the simulation result for calcium and magnesium would seem to be unrealistic. This apparent overestimation of nutrient availability in the combined model requires further attention in future research. Nutrient uptake appeared to be rather sensitive to soil moisture content (Gijsman, 1990). Results of the simulations of nutrient availability up to now, however, indicate that at Speuld and Kootwijk no extreme nutrient deficiencies due to soil acidification occur at present, with the possible exception of phosphorus, which was omitted in the soil models.

6.2 Deposition and uptake of gaseous air pollutants

To be able to quantify the direct and indirect effects of air pollution on forest growth the atmospheric input, canopy interactions, uptake of pollutants by the foliage and the flux to the soil have to be determined. As these fluxes and processes are related, the best approach would be to construct a mass balance for important pollutants. The biomass of the trees, the litter layer and the soil interact as important buffers. Slight changes in the buffering capacity alter the fluxes to a large extent. A reliable balance is obtained only by long term monitoring; model calculations must rely on the best estimate of fluxes and processes. The ways in which these estimates are obtained is only mentioned here; comparison of the different estimation procedures raises the question as to how consistent the different approaches are, which is treated in more detail elsewhere (thematic report on deposition by van Aalst et al.).

Total deposition for the ACIFORN sites can be calculated in several different ways:

- as the sum of the wet and dry deposition measured locally. Wet deposition is measured using bulk and wet-only rainfall collectors. Dry deposition is calculated from the gradients of wind speed, temperature and gaseous pollutants above the canopy. Data will not be available before the end of June, 1990.

- from emission fields and a transport/deposition model (see integrated report on atmospheric input). Hourly average concentrations and local parametrisations of deposition velocities and canopy resistances are used. The calculated fluxes are representative for a larger region (1 to 5 km grids).

- from measured throughfall and rainfall composition at the sites, using a balance model and some assumptions about processes in the canopy space. This method generates values for leaching and canopy uptake too, but with large uncertainties (see Table 6)

Canopy interactions are complex. They consist of the fluxes from the atmosphere to the canopy, the absorption in surface water layers and adsorption on dry leaves, chemical transformations on the leaf and the uptake through the stomata and cuticle. The uptake by the canopy and leaching of elements from the foliage are vital. An approach to calculate these quantities is described by Bredemeier (1988) and used for the Speuld data (P102.1).

Another way of obtaining the amount of pollutants taken up by the canopy is based on a the forest growth model (P112). This model only considers the stomatal uptake. As the resistance of the cuticle is high compared to the stomatal resistance, cuticular uptake is considered to be of little importance (Johanson, 1987; Riederer, 1989).

Table 10 gives the calculated stomatal uptake of air pollutants. Uptake rates are based on the model developed by Kropff (1989); stomatal resistance is derived from the combined effect of photosynthesis, needle water potential, and vapour pressure deficit of the surrounding air (P112, for details). As stomatal resistance is used in the calculation of transpiration as well, and simulated transpiration rates are comparable with transpiration estimated from field monitoring of soil water content (P102.2), it is likely that the pathway resistance to pollutant uptake is described with sufficient accuracy.

From Table 10 it can be seen that total stomatal uptake of pollutants is rather small; as a result, the short-term direct effects are very limited in scope, especially when expressed on an annual basis (Table 11 & 16). However, the possibility of acute damage under episodic high concentrations remains, especially when buffering capacity of the trees decreases, e.g. due to ammonium nutrition (P83, P115). Gaseous uptake of NO_2 may have the same effect (Kropff, 1989; Lange et al., 1989). It should be noted that Table 10 does not contain estimates of dry deposition, as dry deposition consists of more than stomatal uptake alone.

6.3 Conclusions on the relative importance of various growth disturbances for conditions in the Netherlands

Overall reduction in forest growth, resulting from air pollution and acid deposition, can be due to disturbances such as reduced primary production, an increase in the cost of repairs, or changes in assimilate allocation, and combinations thereof. It may result from a reduction in CO_2-assimilation or increased ageing and turnover of the foliage and root biomass. These in turn can be caused by either direct damage, by nutrient and water stress or by combinations of these. Nutrient stress caused by soil acidification should be considered as an indirect effect of air pollution.

At both the ACIFORN sites, the major growth influencing factors that determine the actual growth levels appear to be the availability of water and mineral nutrients. No extreme nutrient deficiencies occur on either site, and growth rates are high. Nutrient availability interferes with water availability to the extent that enough water must be available to facilitate nutrient uptake from the soil. In dry years, limitations in water supply overrule nutrients as the main growth limiting factor. Root density in both stands is low, and a further decrease, either through a shift in root/shoot ratio due to increased nitrogen availability, or through soil acidification and aluminium toxicity, will increase the effect of low water and nutrient availability.

The direct effect of gaseous air pollution appears to be of minor importance in reducing growth. Rather, an initial fertilization effect of gaseous uptake of NH_3 is expected. On less productive sites, this may lead to an increasing nitrogen content during foliage ageing (P18), clearly indicating nitrogen surplus. Should uptake of gaseous pollutants lead to an increase in ageing, as part of a chronic, long-term direct effect, this is unlikely to result in growth reductions, for the ACIFORN sites as LAI is high, and a small decrease will not show in total light absorption (Table 16, P112/P114.2; Mohren & Bartelink, 1990). At lower LAI, an effect of long-term exposure to ozone seems likely in sensitive species.

It is possible that some acute damage due to episodic high air pollution may have occurred during the monitoring period (1988 and 1989), but it seems unlikely that this could have caused foliage shedding to such an extent that growth was retarded.

Soil acidification in Speuld and Kootwijk has not yet proceeded to the point where large disturbances and nutrient imbalances result. Considering the soil conditions in both ACIFORN sites, it appears likely that this is not a representative situation for the entire forested area in the Netherlands (P113). On sites of lower productivity, but with equal deposition, soil acidification, aluminium toxicity, and leaching of cations will probably have resulted in more pronounced nutrient imbalances and growth reductions.

7. MOST LIKELY CAUSES OF FOREST DECLINE IN THE NETHERLANDS

7.1 Main additional disturbances relating to growth

Additional disturbances of forest growth under Dutch circumstances as a result of air pollution and atmospheric deposition appear to encompass three major aspects: firstly, an immediate effect of high nitrogen deposition resulting in widespread nutrient imbalances of trees and stands (P105, Roelofs et al., 1985; Mohren et al., 1986; Van Breemen & van Dijk, 1988; Van den Burg, 1990); secondly, soil acidification due to acidic deposition mainly as ammonium sulphate, resulting in leaching of cations and nitrate, disturbance of root uptake of water and minerals and, in the long run, exhaustion of the aluminium hydroxide buffer, resulting in further pH decrease in the top soil (P102.1, P113; De Vries & Kros, 1989; Van Breemen et al., 1987; Mulder et al., 1988); thirdly, increased risk of visible damage due to uptake of SO_2, NO_x, NH_3 and O_3 through internal acidification and decreased metabolic buffering capacity, caused by unbalanced nutrient uptake in combination with ammonium dominance in nitrogen uptake (P83; Kropff, 1989). Also, an increase of the risk of damage due to insects and pathogens can be expected with nitrogen deposition (De Kam et al., 1989), and present ozone levels point to possible long-term effects on foliage ageing.

Reconsidering the hypotheses on forest disturbance and possible forest decline outlined in Section 2.2, it would seem that the hypothesis on excess nitrogen deposition points to the most relevant complex of processes that cause forest changes under Dutch conditions. Nitrogen deposition leads to nutrient imbalance and is a major component of atmospheric deposition leading to soil acidification (van Breemen et al., 1982; van Breemen & van Dijk, 1988). The hypothesis on soil acidification and aluminium toxicity as major mechanisms causing forest decline is relevant in conditions in the Netherlands, in that soil acidification takes place in most forest soils, and aluminium concentration in the soil solution is relatively high due to the dissolution of aluminium hydroxides. However, at present the major consequence of this seems to be the reinforcement of nutrient imbalances induced by nitrogen deposition, by means of a further decrease in root density, instead of forest decline. This nutrient imbalance may lead to symptoms of forest decline (Oren & Schulze, 1989), and may act as a predisposing factor for impacts of drought and pests and pathogens. Within a broader timescale, however, soil acidification, leading to exhaustion of the aluminium buffering capacity of the soil within decades (De Vries & Kros, 1989), is the dominating factor that alters site conditions to such an extent, that conditions for tree growth deteriorate as a result of a true increase in acidity. As soil acidification at present is strongly related to nitrogen deposition, the causal factor remains the same in both cases, but the hypotheses suggest different mechanisms, both relevant but in different timescales. Soil acidification proceeding at the present rate should be considered as being detrimental in the long run. This means that at present, gradual large-scale changes are taking place in the majority of forest ecosystems in the Netherlands, ultimately resulting in severe restrictions for forest growth and development.

When contemplating the impact of nitrogen deposition and soil acidification, damage due to direct effects resulting from uptake of gaseous air pollutants by the foliage cannot be ignored. Although average concentrations are low and the total effect on photosynthesis and growth is negligible under prevailing conditions, considerable damage due to visible injury should not be ruled out. This visible damage may result from long-term exposure and from episodic high exposure, especially during winter months, or in spring when new foliage develops and O_3 levels may exceed critical levels during periods of several days or weeks (P101, see also Figure 2).

The conclusions outlined above are based on results of the research projects considered in this thematic report, and have been confirmed by others (e.g. van Breemen & van Dijk, 1988). In recent discussions on major causes of forest decline in Europe, the "nitrogen issue" has received considerable attention, and the present results confirm a large number of suggestions pertaining to possible effects of nitrogen deposition (see e.g. Brown et al., 1988; Cowling et al., 1988; Oren & Schulze, 1989; Schulze, 1989; Schulze et al., 1989a; Skeffington & Wilson, 1988; van den Burg, 1990).

The three aspects listed above will be discussed in more detail, taking into account possible combined effects, and considering the rate of change and the time span involved in the development of damage symptoms.

7.1.1 Nitrogen deposition

Nitrogen deposition in forests in the Netherlands is of the order of 40 - 80 kg $ha^{-1}yr^{-1}$; total nitrogen input at the ACIFORN sites amounts to 47 kg $ha^{-1}yr^{-1}$ for Kootwijk and 50 kg $ha^{-1}yr^{-1}$ for Speuld (P102.1). Both sites show accumulation of nitrogen over the monitoring period (P102.1); apparently the ecosystem is not saturated with nitrogen. This also follows from the nitrogen content of the foliage (1.7 - 1.8 % of dry weight), which is within the optimum range, and below the maximum value (P105; Table 9, van den Burg, 1988 & 1990).

The effect of increased nitrogen availability on tree growth depends on site conditions such as availability of soil moisture, and availability of other nutrients. Many sites occupied by forests consist of poor, well drained sandy soils, in which decomposition of soil organic matter together with atmospheric input is the only source of nutrients for plant growth. Under these conditions, nitrogen has been limiting growth, and an initial fertilizing effect of nitrogen deposition seems likely. As a result, nutrient imbalances have developed that are related to phosphorus, magnesium and potassium deficiencies (Roelofs et al., 1985; Mohren et al., 1986; van den Burg, 1990). Ultimately, the ecosystem will become nitrogen saturated, in which situation the input of nitrogen will be balanced by leaching of either NH4+ or NO_3- out of the root zone.

As long as nitrification proceeds at a rate that enables the trees to take up nitrate rather than ammonium, there is no imminent risk of metabolic disorder. In this case, soil acidification occurs due to proton production in nitrification, and subsequent protonation of the exchange complex and weathering of aluminium hydroxides neutralizes the acidity. Base cations will be leached in parallel with NO_3^-, and free aluminium in the soil solution may restrict root growth (Mulder et al., 1988). When nitrification is insufficient to cope with the NH4+ load, ammonium will be taken up instead of nitrate, leading to additional acidification of the rhizosphere (P83). As a result, uptake of base cations such as calcium and magnesium may be restricted and metabolic disorder may occur due to decreased carboxylate levels within the trees (Kropff, 1989). In the presence of ammonium, with high concentrations of NH_4^+ in the soil solution, the uptake of potassium and magnesium may decrease as a result of competition of cations at the root surface. Ratio's of NH_4^+/K^+ and NH_4^+/Mg^{2+} are used to indicate possible critical conditions. At Speuld and Kootwijk, NH_4^+/K^+ and NH_4^+/Mg^{2+} ratio's are critical to a depth of about 10 cm below the soil surface; below 10 cm, ammonium concentrations decrease due to nitrification (P102.1).

A likely mechanism causing additional reduction in the uptake of water and nutrients is a change in root/shoot ratio towards increased above-ground growth as a result of increased nitrogen availability (Mansfield, 1988). Both changes in root/shoot ratio and decreased root elongation due to toxic aluminium concentrations in the soil serve to decrease uptake capacity of the root system. The extent to which these changes occur is not clear. Data on assimilate allocation and root/shoot ratio's are scarce and not easily collected under field conditions; laboratory experiments show a clear effect of aluminium on root density, but this is difficult to detect in field situations due to soil heterogeneity (P103). However, both changes can be considered as additional disturbances that do not counteract but exacerbate nutrient imbalances.

7.1.2 Soil acidification from atmospheric input

The participation of the different buffering mechanisms in the neutralization of protons depends on soil acidity and on the degree of base saturation of the exchange complex. At the ACIFORN locations, dissolution of aluminium hydroxide is the major buffering mechanism. Under acid conditions as in Speuld and Kootwijk, cation exchange has proceeded to the point where occupation of the exchange complex by base cations is below 5 %, which can be regarded as the lowest possible value (P102.1). This implies that the active buffering mechanism for acid neutralization consists entirely of dissolution of aluminium hydroxide. Such is the case for the majority of the Dutch forest sites on dry sandy soils.

Another outcome is that tree nutrition in situations where weathering of clay minerals is limited, becomes entirely dependant on atmospheric deposition and decomposition of organic material in the soil. This influences nutrient availability and, on sites where net precipitation (throughfall) exceeds root uptake, nutrient losses by means of leaching to the underground result. Analysis of interception, infiltration and soil water movement clearly shows that this is the case both in Speuld and in Kootwijk (102.2, P104).

Uptake of cations by roots may decrease due to limited nutrient availability in combination with high concentrations of aluminium in the soil solution. The ratio between calcium and aluminium is used as an indicator (Ulrich, 1983b). At the ACIFORN sites, the Ca/Al ratio approaches a critical value (according to Ulrich, 1983b) at 20 cm below the soil surface and further downwards.

In Speuld and Kootwijk, as a result of soil acidification due to input of ammonium sulphate, aluminium concentrations in the soil are increased due to weathering of aluminium hydroxide (P102.1, P113). With ongoing soil acidification, the aluminium hydroxide pool will become exhausted starting with the top soil layer. Should soil acidification proceed to the point at which a transition from aluminium to iron buffering occurs, acidity (pH) will drop significantly, leading to completely altered conditions for plant growth. It would appear likely that such a transition will have far reaching consequences for the existing forest stand. At present, however, nutrient imbalances, decreased buffering, unfavourable Ca/Al ratio's and increased leaching of cations from the system dominate the changes occurring in the soil compartment.

Due to increased free aluminium concentrations in the soil solution, mycorrhizal activity is expected to decrease (see thematic report on biological and physiological effects), possibly resulting in a decrease in phosphate uptake.

7.1.3 Direct effects of air pollutants

Two-year average concentrations at 30 m height at the ACIFORN sites were about 50 µg m^{-3} for O_3, 10 µg m^{-3} for SO_2, 25 µg m^{-3} for NO_2, 5 µg m^{-3} for NO, and 5 µg m^{-3} for NH_3, with hourly concentrations of O_3 frequently reaching levels high enough to be damaging to plants (> 200 µg m^{-3}) (P101). Total uptake of gaseous air pollutants in conifers is limited due to low stomatal conductance, and present concentrations of SO_2, NO_x, and NH_3 as measured at Speuld are unlikely to result in major direct short-term effects on physiological processes. Due to a combination of predisposing factors such as low temperature, high humidity, and NH_4^+ nutrition, disturbances in metabolic buffering of protons produced in the uptake of SO_2 and NO_2 may occur, resulting in increased ageing and a decrease of LAI. The relevance of this phenomenon under Dutch conditions is not clear, and requires further investigation. Model calculations indicate that a long-term effect on foliage longevity is likely to result in growth decrease in coniferous stands of low LAI (<5), and in deciduous trees.

With regard to the spatial variability of SO_2, NO_x and NH_3 concentrations, it seems likely that the ACIFORN sites represent a rather "clean" situation, and that local occurrence of higher concentrations of SO_2, NO_x and NH_3 near point sources may very well lead to direct short-term disturbances, especially in combination with decreased metabolic buffering capacity due to the predisposing factors mentioned above.

The annual average concentration of ozone is about equal to the critical level presumably required to prevent serious damage (50 µg m^{-3}, see Schneider & Bresser, 1988), and frequently exceeds the threshold above which visible damage may occur (200 µg m^{-3},

Schneider & Bresser, 1988). Although no direct short-term effects on net photosynthesis or dark respiration were found in fumigation experiments with up to 400 μg m^{-3} O_3 (mainly due to the high stomatal resistance, see P115), changes in assimilate allocation were found (P107), and recent experimental results indicate that at a concentration level of 75 μg m^{-3} total biomass growth may be reduced by 10-15 % (Edwards et al., 1990). The results from project P115 were obtained using light levels of 200 W m^{-2}, which is low compared to field conditions. It is not unlikely that under conditions of high radiation interactions occur such as between uptake of O_3 and photo-inhibition, and between uptake of O_3 and nutrient shortages that may lead to decreased capacity to recover from ozone damage. Considering this, together with the possible long-term effect and the episodic high concentrations, the likely influence of ozone on assimilate allocation, and the possible long-term effects of low doses of ozone on foliage ageing, it must be concluded that ozone at concentrations as measured at the ACIFORN sites is bound to have an influence on growth. In addition, there is a considerable risk of acute injury due to ozone during episodes with high concentrations, such as in 1989.

7.2 Aspects and rate of forest decline

In the previous sections, it was concluded that nitrogen deposition, atmospheric input and soil acidification, and exposure to ozone are the main aspects causing changes in forest growth and development at the ACIFORN sites and in large parts of the forested area in the Netherlands. The changes taking place resulting from these external influences differ in impact, magnitude, rate of change, and the extent to which detrimental effects are reversible.

The rate of change in forest growth due to exposure to ozone will be in equilibrium with changes in ozone concentrations, except for visible injury. This implies that when ozone exposure increases, concomitant growth reductions will increase as well. At present, ozone concentrations increase by 1-2 % per year (Ashmore et al., 1985), and the negative effect of ozone on biomass increment is expected to increase at the same relative rate. Direct, short-term effects on process rates such as photosynthesis and respiration are very often fully reversible and the system usually responds rather quickly, in terms of days, hours, or even minutes. This implies that detrimental effects due to exposure to gaseous air pollutants at the level of the physiological processes involved may be removed simply by decreasing exposure. Visible injury is often irreversible, and a decrease in LAI due to visible injury may take 1-2 years to recover.

The rate of soil acidification with respect to deposition and leaching can be expressed in several ways, in relation to acid neutralizing capacity, or with respect to acidity (pH) of the soil solution. Change in acid neutralizing capacity is the best absolute measure of the rate of soil acidification (van Breemen et al., 1983), but regarding conditions for tree growth, it is desirable to evaluate the rate of change in cation availability, aluminium concentration, and pH as well (P83, P103, P113). The rate of change in these soil conditions depends largely on weathering of primary minerals, acidic deposition, the vegetation cycle, and buffering capacity. Most dry sandy forest soils in the Netherlands are within the aluminium buffering range (P102.1, P113), with a pH-value of around 4. At the present rate of deposition, a transition is expected to occur from the aluminium buffering range to the iron buffering range within decades; this transition will be accompanied by a decrease in pH of about 0.5 to 1.0 unit. This change is expected to occur within 30-50 years on sites that are poor in aluminium hydroxides, such as dry leptic podzols, and within 50-200 years on sites rich in aluminium hydroxides (P113). To a large extent, these changes in soil conditions are irreversible, as base cations are leached from the rooted soil profile and aluminium hydroxides dissolve in the top layers of the profile and precipitate in lower layers, resulting in a permanent change in the rooted profile.

The rate of accumulation of nitrogen in forest ecosystems with an input of some 50 kg $ha^{-1}yr^{-1}$ is considerable, when compared to the total amount of nitrogen present within the system (3.000 - 10.000 kg ha^{-1}, with some 300-700 kg ha^{-1} within the living biomass). As nitrogen is deposited in a form that can readily be taken up by the trees (NH_4^+ or, after nitrification, NO_3^-) a quick response can be expected following nitrogen deposition. Due to immobilization of nitrogen in organic material, nitrogen availability is not immediately increased to the extent at which saturation within the trees occurs. Also, at Speuld and Kootwijk the total N-demand by the growing stand is high, equalling about 100 kg $ha^{-1}yr^{-1}$. This is due to the high growth rate, and apparently the demand cannot be met by the supply from the soil (P105, Gijsman, 1990). The analysis of the ACIFORN sites demonstrates that at present the nitrogen content in the needles at both sites is below the maximum reported for Douglas fir, and that both sites still accumulate nitrogen. Nitrogen saturation may take several decades to occur, depending on the amount of organic matter in the soil and on the growth rate of the trees. At sites with high nitrogen deposition loads (up to 100-200 kg $ha^{-1}yr^{-1}$) and low growth rates e.g. due to other nutrient deficiencies or water shortage, a saturated situation already exists (P18, Boxman & van Dijk, 1988).

Nitrogen is rather mobile within the ecosystem and a reduction in deposition will lead to changes in soil conditions rather quickly. Preliminary results from ongoing experiments already indicate that removal of deposition may result in decreased nitrogen content of needles within one year (P118).

7.3 Generalization

7.3.1 Species, sites and silvicultural treatment

The majority of the forested area in the Netherlands consists of even-aged coniferous stands consisting of pine (45 % of total, predominantly Scots pine), larch (6 % of total), Douglas fir (5 %), and spruce (5 %, predominantly Norway spruce). Broadleaved forests consist mainly of oak (18 %, predominantly pedunculate oak), poplar and willow species (7.5 %), birch (5 %), beech (3 %), and other species (5.5 %). Most coniferous forests are situated on rather poor and dry sandy soils, low in exchangeable base cations and low in Acid Neutralizing Capacity (P18, P113). Compared to the ACIFORN sites, the majority of the forest sites are of an inferior quality with regard to nutrient and water availability, and acid neutralizing capacity. This implies that the situation as described for the ACIFORN sites concerning effects of impact of air pollution and atmospheric deposition will generally be less favourable at other forest sites in the Netherlands.

To generalize, attention has to be paid to stand and species characteristics that determine deposition and uptake of pollutants, and availability of nutrients and water in the soil. So far, the analysis of effects of air pollution and soil acidification for tree growth has been concentrated on Douglas fir, which from the statistics given above, appears to be of minor importance in Dutch forestry. However, due to the fundamental nature of most of the research dealing with underlying physiological and biochemical interactions, this does not restrict the applicability of the results for trees and stands in general.

In the case of Douglas fir, possible long-term effects of air pollutants are most likely to be associated with decreased root growth, making the stands more susceptible to extreme drought, and premature foliage shedding. Douglas fir planted on sites with intermediate moisture supply will not necessarily exhibit growth decline in periods with average moisture availability, instead there may be a more severe growth decline under extreme drought. This phenomenon is expected to be even more pronounced on drier sites such as the eolian sands covered by pine forests. Also, in the case of pine forests with low LAI, premature shedding of foliage will quickly lead to growth decline, whereas in Douglas fir stands with high LAI, as in Speuld and Kootwijk, somewhat increased foliage shedding will not immediately show as a reduction of primary production.

Many forest soils in the Netherlands - especially in the case of pine stands - are low in nutrient supply as there has been virtually no fertilization in the past. Other stands (Norway spruce, Japanese larch, Douglas fir) were fertilized with phosphorus, which improved soil P status. Nitrogen fertilization was considered to be uneconomical and fertilization with phosphorus and potassium produced only limited results due to the shortage of nitrogen.

Incidental fertilization using magnesium or copper was applied where severe deficiencies occurred. In general, nitrogen was limiting on most sites in the past. As nitrogen limitations have been eliminated due to atmospheric input in recent decades, the phosphorus and base cation deficiencies reappear, and fertilization is reconsidered (van den Burg, 1989, 1990). When considering fertilization in these situations, to prevent damage symptoms originating from nutrient imbalances, attention has to be paid to the buffering capacity of the rooted soil. The decrease in buffering capacity due to acidification, resulting in occupation of the exchange complex mainly by Al^{3+} and H^+, might be expected to decrease recovery and effectiveness of fertilizer applications. In the case of low initial cation exchange capacity, as on poor sandy soils, this process is less important. Nutrient availability is rather high in both Kootwijk and Speuld, which is reflected in the high productivity at both sites. Nutrient deficiencies with respect to phosphorus and potassium can be expected to be more severe on most of the other sites on sandy soils, and have been found in inventories that were carried out recently (P18, Boxman & van Dijk, 1988; Houdijk, 1990).

Following the planting of first generation pine stands on former drift sands or heathlands, site improvement due to accumulation of organic matter has enabled the establishment of more demanding species such as Douglas fir stands with a higher productivity than the original pines. This indicates the importance of soil organic matter content, and the build up of soil organic matter during primary succession. The importance of decomposition of soil organic matter as a source of nutrients for tree growth increases with soil acidification and atmospheric deposition of nitrogen as the demand for cations may increase, whereas the buffering of base cations by means of the exchange complex is limited due to protonation and occupation by aluminium.

Current forest practice in most of the stands consists of repeated thinning, thereby selecting the best trees to remain in the stand. Most coniferous stands have been planted. At present the emphasis has shifted to natural regeneration whenever possible. Due to rather intensive management, including repeated thinnings, mortality as a natural phenomenon in developing stands is poorly understood. Also, following predisposing decline, mortality very often results from secondary attacks by insects and pathogens. As a result, it is difficult, if not impossible, to establish a relationship between disturbances due to air pollution and soil acidification, and tree and stand mortality. Further analysis will be necessary in order to establish increased mortality risks related to, for example, nutrient imbalances or ongoing soil acidification.

7.3.2 Growth limitations

In the Netherlands, the length of the main growing season is between 150 - 250 days, with budflush occurring between late April and mid May. Due to mild winter conditions, evergreen conifers in the Netherlands may photosynthesize and transpire during the winter season as well, suggesting a potential net gain in carbon assimilation during all the winter months.

A major growth limitation results from water shortage on dry sites during the summer. Potential evapotranspiration during the growing season may amount to 300 - 400 mm. On average, rainfall is of the same order of magnitude, but 30 - 50 % is lost due to interception of precipitation by the crown canopy. In order to prevent water shortage occurring, even under conditions of average precipitation, the moisture holding capacity of the rooted soil should be around 200 mm. This only occurs on the best sites available for forest growth.

Due to the limited cation exchange capacity of the sandy soils, the soil nutrient status is rather poor in large parts of the forested area. It seems plausible that most sites were limited in nitrogen during the first half of the century, up to 1960. Fertilizer experiments in the late fifties demonstrated growth response to nitrogen fertilization on sites where phosphorus, potassium and magnesium supply was sufficient. Although fertilization with elements other than nitrogen was shown to eliminate deficiencies, the increment in growth was limited due to limited nitrogen supply (Blok et al., 1975).

Stem volume increment rates, when averaged over the total forested area over the recent decades, amounted to some 4-6 $m^3ha^{-1}yr^{-1}$. For optimal sites, current yield tables indicate attainable growth rates around 15-20 $m^3ha^{-1}yr^{-1}$ for Scots pine, and 25-30 $m^3ha^{-1}yr^{-1}$ for Norway spruce and Douglas fir (LaBastide & Faber, 1972). These values apply to an average of the best growing stands in the data base underlying the yield tables. True optimal growth can be expected to be somewhat higher, as clearly demonstrated by the results of P100 where an optimal nutrition and irrigation treatment according to the Ingestad technique was applied, and in Kootwijk after one year of treatment, volume growth increased to over 40 $m^3ha^{-1}yr^{-1}$ for Douglas fir.

7.4 Conclusions

Nitrogen deposition is the key factor determining the changes taking place in forest ecosystems in the Netherlands. Several decades of increased nitrogen deposition have led to the elimination of nitrogen deficiencies, increased growth, and subsequent nutrient imbalances due to deficiency of magnesium, potassium and phosphorus. Concurrently, the nitrogen input has caused considerable soil acidification, together with the deposition of SO_2 and NO_x. The interplay between increased nitrogen availability and soil acidification, leading to decreased cation availability, increased aluminium concentrations in the rooted soil profile and decreased root uptake of water and minerals, rendered the stands more susceptible to drought and attacks by pests and pathogens. Figure 3 gives a simplified overview of the chain of events leading to forest changes and, eventually, to forest decline, due to excess nitrogen deposition.

Nutrient imbalances resulting from nitrogen deposition have increased vulnerability to frost, to insects and pathogen damage, and susceptibility to gaseous air pollutants. Preliminary results from ongoing research indicate that this situation may change rather quickly with decreased nitrogen deposition. The elimination of nitrogen input could lead to decreased nitrogen content and restoration of nutrient balances within a few years after removal of the excess deposition (P118).

The ACIFORN investigations refer to highly productive conditions at sites favourable for tree growth. Thus, the results can be expected to indicate a lower range of damage. Leaching of base cations on sites with lower CEC but otherwise comparable deposition rates, would be larger, leading to accelerated mineral deficiencies and growth disturbances.

The situation in Speuld and Kootwijk with near optimal canopy assimilation and high productivity shows maximum stomatal uptake of gaseous air pollutants; the absence of short-term effects at ambient concentrations in these stands indicate that short-term effects of gaseous uptake of air pollutants is of minor importance compared to changes due to nitrogen deposition and soil acidification. This does not apply to ozone, which can be expected to influence growth on most sites as its concentration regularly exceeds base damage thresholds, and long-term effects of ozone are likely. It should be noted that exposure thresholds should be interpreted with care, as actual uptake depends also on stomatal resistance. As a result, coniferous trees are less sensitive than agricultural crops, and also pollutant uptake per unit of leaf area may be higher in deciduous trees.
The high value of LAI at both Speuld and Kootwijk results in low sensitivity of stand productivity to increased ageing of the foliage. Combined with the relatively short period of intensive monitoring so far, no long-term effects on growth have been found. Differences in LAI have to be taken into account when extrapolating the ACIFORN results to other situations. Stands of Scots pine for example, generally have lower LAI, to the extent that a reduction in leaf area will lead to a diminished light interception and hence to decreased growth (Mohren & Bartelink, 1990). In this case, a small increase in ageing may lead to significant growth decline. Should nutrient imbalances have the same effect as long-term exposure to air pollutants, as suggested by Oren & Schulze (1989), this difference between high LAI stands, such as the Douglas fir stands in Speuld and Kootwijk, and low LAI stands, such as Scots pine stands on poor sites, will be even more pronounced.

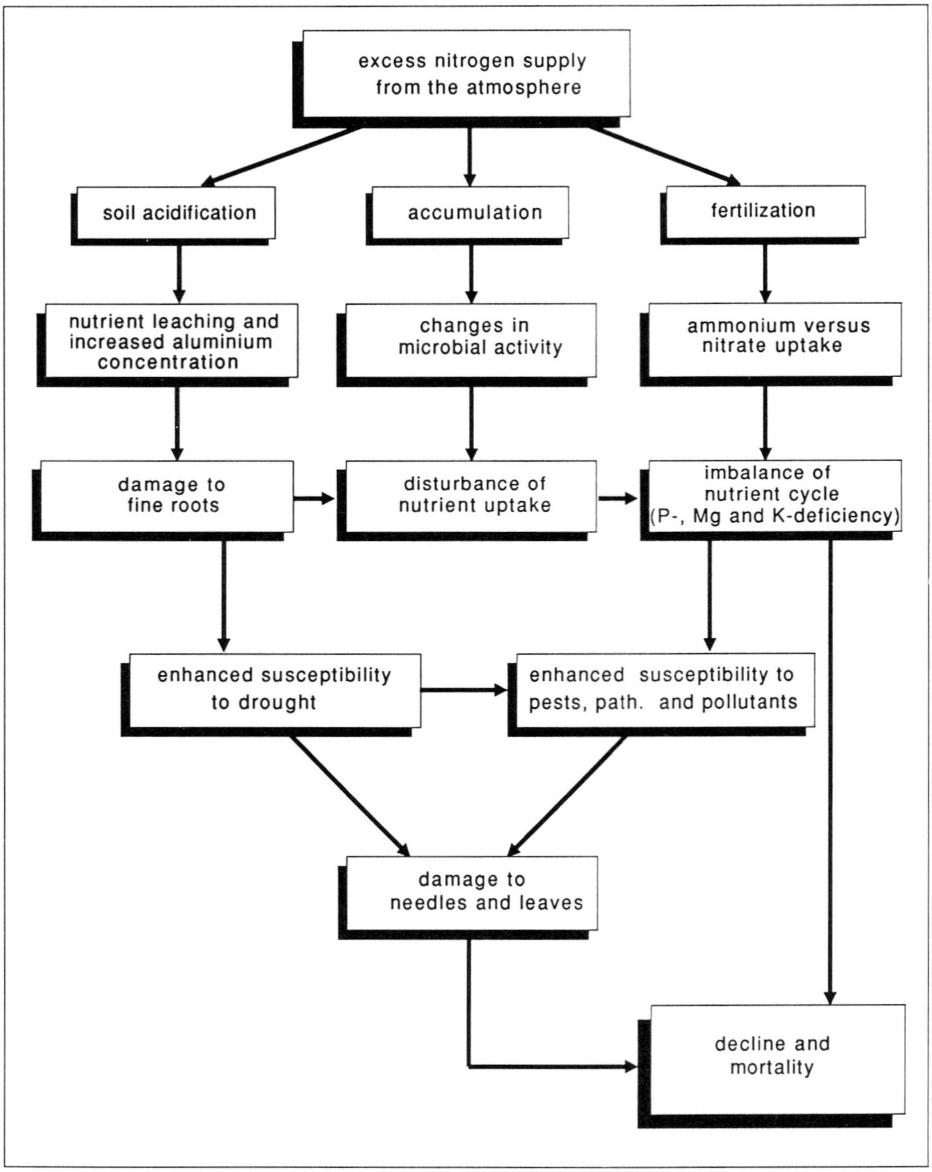

Fig. 3. Simplified overview of major disturbances leading to forest decline in the Netherlands: causes, pathways, consequences (adapted from Cowling et al., 1988)

8. SYNTHESIS, CONCLUSIONS AND PERSPECTIVE

8.1 Magnitude of forest damage in the Netherlands

At present, the majority of forests in the Netherlands have been under the influence of the atmospheric deposition of NH_3, SO_2, NO_x and O_3, and concomitant soil acidification. Most forest sites are, and have been in the past, vulnerable to nutrient deficiencies and drought (P102.2), and have only limited acid neutralizing capacity, making them vulnerable to acidification (P113). The more fertile forest sites away from local emission sources of nitrogen are capable of accumulating nitrogen; forests on poor sites in agricultural areas are expected to be near nitrogen saturation.

As a result of widespread high nitrogen deposition, nutrient imbalances have developed, and forest ecosystems are in transition from a nitrogen deficiency state to a nitrogen saturated state. At the same time, severe soil acidification has been progressing on non-calcareous soil to such an extent, that the cation exchange complex is almost exclusively occupied by protons and aluminium (P102.1, P113). The majority of the forest sites situated on dry sandy soils at present are within the aluminium buffering range, where the amount of aluminium hydroxide determines the acid neutralizing capacity. Under these conditions, the concentration of free aluminium in the top soil layers is high, leading to a decrease in root density (P103), and possibly diminishing mycorrhizal activity (P107). Base cation supply to the trees entirely depends on weathering, atmospheric deposition, and decomposition of organic material (P102.1, P113). In sandy soils, base weathering is limited, and part of base cation input from background atmospheric deposition is lost, due to limited buffering capacity of the exchange complex and subsequent leaching out of the rooted soil profile.

Deposition rates of ammonia may be higher near local sources of emission, resulting in NH_4^+/K^+ and NH_4^+/Mg^{2+} ratios that further decrease root uptake of K^+ and Mg^{2+} (P118). Rhizosphere acidification increases when nitrogen is taken up as NH_4^+ rather than NO_3^-, enhancing the toxic effects of aluminium on base cation uptake (P83, P103).

Concentration levels of ozone have reached values that could very likely result in damage to the interior of the foliage, necessitating additional photosynthesis products for repair (P101, P107, P115). Average levels of air pollution measured at the ACIFORN sites correspond well with results from the national monitoring network (P101); the exposure to ozone and the expected damage apply to most of the forested area (P101).

Water availability is low, even on favourable sites such as in Speuld and Kootwijk, leading to growth reductions in dry years (P102.2). Should nitrogen availability and soil acidification result in a decrease in root density, vulnerability to drought and nutrient deficiencies will increase.

8.2 Possible future developments

A transition from aluminium-nitrate dominated systems to ammonium dominated systems may occur in large areas of Dutch forests as a result of ongoing nitrogen deposition. This may lead to increased deficiencies of magnesium and calcium due to aluminium toxicity near the root tip and to changes in internal buffering of acidity making the trees more vulnerable to direct effects of air pollutants (P83, Kropff, 1989). At the same time, ongoing soil acidification runs parallel to cation leaching and increased aluminium concentrations due to dissolution of aluminium hydroxides. All these processes point to a deterioration of nutrient supply with ongoing nitrogen deposition and soil acidification. The rate at which the transition proceeds is dependent on the acid neutralizing capacity within the aluminium buffer range, the rate of ammonia deposition, the rate of nitrification in the top soil, and the hydrological regime.

The exhaustion of aluminium hydroxide in the top soil will reduce pH until a new stable situation has been established. Although it is virtually impossible to estimate root activity in this new state of equilibrium, additional deficiencies of nutrients such as phosphate are likely (Foy et al., 1978). The time span in which this change may take place can be derived from available soil acidification models, and appears to be between 30 - 100 years for poor sandy soils (P113).

8.3 Main remaining uncertainties

The occurrence of nitrification and hence the form in which nitrogen is taken up by the trees is crucial to cation uptake. An increase in NH_4^+-uptake at the expense of NO_3^- may gradually lead to rhizosphere acidification and increased impact of aluminium on cation uptake. Combined with a decreased metabolic buffering capacity of the foliage, this may very well prove to be a critical chain of events leading to forest decline under conditions of high ammonia input and episodic high concentrations of sulphur dioxide. Yet the main factors governing the rate of nitrification are only barely understood. Nitrification depends to some extent on pH and soil moisture status. If further soil acidification leading to exhaustion of the aluminium buffer, results in a drop in acidity, nitrification could be decisive in preventing initiation of a chain of events leading to severe nutrient disorder. In order to investigate this, it should be determined whether an increase in NH_4^+ uptake at the expense of NO_3^- uptake is to be expected with ongoing nitrogen deposition. Furthermore, the ensuing consequences for whole plant ionic balance and the metabolic buffering capacity of the foliage should be analysed quantitatively. It has been shown in seedling trials that severe injury leading to mortality may result from disturbances in whole-plant ionic balance in the case of ammonium feeding, and the relevance of this phenomenon under field conditions has to be determined.

With the exception of ozone, direct short-term effects due to uptake of other gaseous air pollutants appear to be of minor importance. Yet, long-term chronic effects of uptake of air pollutants occurring in low concentrations are expected to be more important, although the effects are not known, and the mechanisms not understood at all. Considering the possible interplay with nutrient status and the metabolic detoxification capacity derived from it, whereby the vulnerability to long-term effects may increase with increasing nutrient imbalance, this aspect warrants further attention. At present the available models, concerning both stand growth and soil acidification, appear to be suitable for performing such an analysis.

Canopy interactions will have to be better understood with regard to emission and regional deposition of air pollutants to enable evaluation of forest damage at large. Bulk deposition above the canopy differs considerably from throughfall and soil input below the canopy. Processes taking place at the surface of the leaves and exchange of ions through the epidermis and stomata determine the size of the canopy sink as well as leaching of minerals from the foliage, and are crucial in the calculation of proton budgets for the entire ecosystem. However, the critical factors determining these canopy interactions are largely unknown.

So far, results have been expressed in terms of growth and primary productivity, or in terms of changes in conditions for growth and productivity. To be able to link model predictions and scenario analysis to current practice in inventories of forest decline and health, the results have to be translated in terms of forest health and vitality, thereby accounting for aspects of mortality related to nutrient deficiency and drought. Analysis of mortality in undisturbed situations and under pressure from atmospheric deposition and soil acidification requires risk assessment, where uncertainties in dose-effect relationships can be translated into risks for damage. The relative importance of increased risks due to air pollution and soil acidification should thus be assessed by describing mortality in relation to site and stand conditions, especially concerning nutrient status.

9. REFERENCES

9.1 References (Dutch Priority Programme on Acidification)

P18: The indirect effects of acid deposition on the vitality of the Dutch forests: Forest stand analysis (Oterdoom/de Vries)

P83: The effect of high ammonium input to the soil on ionic uptake balance and rhizosphere pH of Douglas fir (Gijsman)

P100: Manipulation of the water and nutrient supply in two forest ecosystems in the Netherlands (de Visser)

P101: Air pollution in forest canopies (Vermetten et al.)

P102.1: Hydrochemistry of two Douglas fir ecosystems in the Veluwe, the Netherlands (van der Maas)

P102.2: Soil hydrological system characterization of the two ACIFORN stands using monitoring data and the soil hydrological model "SWIF" (Tiktak & Bouten)

P103: Root research on Douglas fir in the ACIFORN-project: Influence of soil acidification on fine root growth (Olsthoorn)

P104.1: Measurement and modelling of canopy water storage during and after rain, dew and fog (ACIFORN) (Bouten et al.)

P105: Impact of air pollution on ecophysiological relations in 2 Douglas fir stands in the Netherlands (Evers et al.)

P107: Effects of air pollutants on the rhizosphere of Douglas fir (Gorissen)

P111: Wood production, stem growth and water transport capacity of Douglas fir in the ACIFORN stands in Kootwijk and Garderen (de Kort)

P112: Quantifying the effects of air pollution and acid deposition on two Douglas fir stands in the Netherlands (Mohren et al.)

P113: Data derivation, evaluation and application of a regional soil acidification model (de Vries)

P114.2: Dose-effect relationships for forest stands, to be applied in the Dutch Acidification System (DAS) (Mohren et al.)

P115: Model development for the uptake of air pollutants and the effects on the physiology of Douglas needles in relation to drought (Smeets et al.)

P116: Measuring sapstream velocities in Douglas fir (Noppert)

P118: Effects of a decrease in atmospheric deposition on the mineral balance and vitality of the Dutch forest (van Dijk et al.)

9.2 References (International literature)

Aronsson, A. 1980. Frost hardiness in Scots pine (Pinus sylvestris L.). II Hardiness during winter and spring in young trees of different mineral nutrient status. Studia Forestalia Suecica, no 155: 1-27

Ashmore, M., N. Bell & J. Rutter 1985. The role of ozone in forest damage in West Germany. Ambio 14: 81-87

Asman W.A. 1987. Atmospheric behaviour of ammonia and ammonium. Ph.D.Thesis, Wageningen Agricultural University

Berry, C.R. & G.H. Hepting 1964. Injury to eastern white pine by unidentified atmospheric constituents. Forest Science 10: 2-13

Blank, L.W., T.M. Roberts & R.A. Skeffington 1988. New perspectives on forest decline. Nature 336: 27-30

Blok, H., van den Burg, J., van Goor, C.P., Jager, K. & L. Oldenkamp, 1975. Bemesting en minerale voeding van Douglas-cultures. Rijksinstituut voor onderzoek in de bos- en landschapsbouw "De Dorschkamp", Wageningen, Intern Rapport, nr. 69, 147 + 186 pp.

Boxman, A.W. & J.M. Roelofs 1988. Some effects of nitrate versus ammonium nutrition on the nutrient fluxes in Pinus sylvestris seedlings. Effects of mycorrhizal infection. Canadian Journal of Botany 66: 1091-1097

Boxman, A.W. & H.F.G. van Dijk 1988. Het effect van landbouw ammonium deposites op bos- en heidevegetaties (in Dutch). Vakgroep Aquatische Oecologie en Biogeologie, Katholieke Universiteit Nijmegen, Nijmegen, the Netherlands, 96 pp.

Bredemeier, M. 1988. Forest canopy transformation of atmospheric deposition. Water, Air and Soil Pollution 40: 121-138

Brown, K.A., P.H. Freer-Smith, G.D. Howells, R.A. Skeffington & R.B. Wilson 1988. Rapporteurs' report on discussions at the workshop on excess nitrogen deposition, Leatherhead, September 1987. Environmental Pollution 54: 285-295

Cape, J.N. 1988. Air pollutant effects on conifer leaf surfaces. In: J.N. Cape & P. Mathy (eds.): Scientific basis of forest decline symptomatology. Proceedings of a workshop jointly organized by the Commission of the European Communities, and the Institute of Terrestrial Ecology, Edinburgh, Scotland, 21-24 March 1988. CEC/DG-XII Air Pollution Report Series nr. 15

Cowling, E., B. Krahl-Urban & C. Schimansky 1988. Hypotheses to explain forest decline. In: Krahl-Urban, B., H.W. Papke, K. Peters & C. Schimansky (eds.): Forest decline: Cause-effect research in the United States of North America and Federal Republic of Germany. Jülich Nuclear Research Centre, Jülich, FRG, p. 120-125

Cronan, C.S., R. April, R.J. Bartlett, P.R. Bloom, C.T. Driscoll, S.A. Gherini, G.S. Henderson, J.D. Joslin, J.M. Kelly, R.M. Newton, R.A. Parnell, H.H. Patterson, D.J. Raynal, M. Schaedle, C.L. Schofield, E.I. Sucoff, H.B. Tepper & F.C. Thornton 1989. Aluminum toxicity in forests exposed to acidic deposition: the ALBIOS results. Water, Air and Soil Pollution 48: 181-192

De Kam, M., C.M. Versteegen, B.C. van Dam, J. van den Burg & D.C. van der Werf 1989. Effect of nitrogen and potassium fertilization on the development of Sphaeropsis sapinea in Pinus nigra. Acta Botanica Neerlandica 38: 354

De Vries, W. & J. Kros 1989. The long-term impact of acid deposition on the Aluminum chemistry of an acid forest soil. In: J. Kämäri, D.F. Brakke, A. Jenkins, S.A. Norton & R.F. Wright (eds.): Regional acidification models: Geographic extent and time development. Springer Verlag, Berlin, p. 113-128

De Vries, W., M. Pösch & J. Kämäri, 1989. Simulation of the long-term soil response to acid deposition in various buffer ranges. Water, Air, and Soil Pollution, 48: 349-390

De Willigen, P. & M. van Noordwijk 1987. Roots, plant production and nutrient use efficiency. Ph.D. thesis, Wageningen Agricultural University, the Netherlands

Dik, E.J., 1984. Estimating the wood volumes of standing trees in forestry practice. Research Institute for Forestry and Landscape Planning "De Dorschkamp", Wageningen, Extensive Report 19, nr.1, 114 pp.

Edwards, N.T., G.E. Taylor, M.B. Adams, G.L. Simmons & J.M. Kelly 1990. Ozone, acidic rain and soil magnesium effects on growth and foliar pigments of Pinus taeda L. Tree Physiology 6: 95-104

Eldhuset, T., A. Göransson & T. Ingestad 1987. Aluminum toxicity in forest tree seedlings. In: T.C. Hutchinson & K.M. Meema (eds.): Effects of atmospheric pollutants on forests wetlands and agricultural ecosystems. Springer Verlag, Berlin, p. 401-409

Erisman J.W., F.A.A.M. de Leeuw & R.M. van Aalst 1989. Deposition of the most acidifying components in the lower atmosphere. Atmospheric Environment 22: 1153-1160

Fiedler, H.J., W. Nebe & F. Hoffmann 1973. Forstliche Pflanzenernährung und Düngung. Gustav Fischer Verlag, Stuttgart, 481 pp.

Foy, C.D., R.L. Chaney & M.C. White 1978. The physiology of metal toxicity in plants. Annual Review of Plant Physiology 29: 511-566

Freer-Smith, P.H. & M.C. Dobson 1989. Ozone flux to Picea sitchensis (Bong) Carr and Picea abies (L) Karst during short episodes and the effects of these on transpiration and photosynthesis. Environmental Pollution 59: 161-176

Freer-Smith, P.H. & T.A. Mansfield 1987. The combined effects of low temperature and SO_2 + NO_2 pollution on the new season's growth and water relations of Picea sitchensis. New Phytologist 106: 237-250

Gijsman, A.J. 1990. Model calculations on ammonium and nitrate uptake by a mature Douglas-fir stand from a soil with high ammonium input (in preparation)

Gorissen, A. & J.A. van Veen 1988. Temporary disturbance of translocation of assimilates in Douglas firs caused by low levels of ozone and sulphur dioxide. Plant Physiology, 88: 559-563

Goudriaan, J. 1986. A simple and fast numerical method for the computation of daily totals of crop photosynthesis. Agricultural and Forest Meteorology 38: 249-254

Guderian, R. (ed.), 1985. Air pollution by photochemical oxidants. Formation, transport, control, and effects on plants. Springer-Verlag, Berlin, 346 pp.

Houdijk, A.L.F.M. 1990. Effecten van zwavel- en stikstofdepositie op bos- en heidevegetaties, Eindrapport VROM-proj. 64.10.22.00, KU-Nijmegen, vakgroep Aquatische Oecologie en Biogeologie

Johanson, C. 1987. Pine forest: negligible sink for atmospheric NOx in rural Sweden. Tellus 39b: 426-438

Keltjens, W.G., 1990. Effects of aluminum on growth and nutrient status of Douglas-fir seedlings grown in culture solution. Tree Physiology 6: 165-175

Keltjens, W.G. & E. van Loenen 1989. Effects of aluminium and mineral nutrition on growth and chemical composition of hydroponically grown seedlings of five different forest tree species. Plant and Soil, 119: 39-50

Keltjens, W.G. & P.S.R. van Ulden 1987. Effects of Al on nitrogen (NH_4^+ and NO_3^-) uptake, nitrate reductase activity and proton release in two sorghum cultivars differing in Al tolerance. Plant and Soil 104: 227-234

Klein, R.M. & T.D. Perkins 1988. Primary and secondary causes and consequences of contemporary forest decline. Botanical Review, 54: 1-43

Krahl-Urban, K., H.W. Papke, K. Peters & C. Schimansky (eds.) 1988. Forest decline: Cause-effect research in the United States of North America and Federal Republic of Germany. Jülich Nuclear Research Centre, Jülich, FRG

Kropff, M.J. 1989. Quantification of SO_2 effects on physiological processes, plant growth and crop production. Ph.D. thesis, Wageningen Agricultural University, Wageningen, 201 pp.

LaBastide, J.G.A. & P.J. Faber 1972. Revised yield tables for six tree species in the Netherlands. Stichting Bosbouwproefstation "De Dorschkamp", Wageningen, Uitvoerig Verslag Band 11, nr 1

Lange, O.L., U. Heber, E.-D. Schulze & H. Ziegler 1989. Atmospheric pollutants and plant metabolism. In: E.-D. Schulze, O.L. Lange & R. Oren (eds.): Forest decline and air pollution. Ecological Studies 77, Springer Verlag, Berlin, pp. 238-276

Lee, E.H., D.T. Tingey & W.E. Hogsett 1988. Evaluation of ozone exposure indices in exposure-response modelling. Environmental Pollution 53: 43-62

Mansfield, T.A. 1988. Factors determining root:shoot partitioning. In: J.N. Cape & P. Mathy (eds.): Scientific basis of forest decline symptomatology. Proceedings of a workshop jointly organized by the Commission of the European Communities, and the Institute of Terrestrial Ecology, Edinburgh, Scotland, 21-24 March 1988. CEC/DG-XII Air Pollution Report Series nr. 15

Mathy, P. 1988. Air pollution and ecosystems. Proceedings of an International Symposium held in Grenoble, France, 18-22 May 1987. D. Reidel Publishing Company, Dordrecht

McLaughlin, S.B., 1985. Effects of air pollution on forests: a critical review. Journal of the Air Pollution Control Association 35: 516-534

Miller, P.R. & J.R. Parmeter 1965. Effects of sustained low concentration ozone fumigation on Ponderosa pine. Phytopath 55: 1068

Mohren, G.M.J. 1987. Simulation of forest growth, applied to Douglas fir stands in the Netherlands. Ph.D. thesis, Wageningen Agricultural University, Wageningen, 184 pp.

Mohren, G.M.J. & H.H. Bartelink 1990. Modelling the effects of needle mortality rate and needle area distribution on dry matter production of Douglas fir. Netherlands Journal of Agricultural Science, 38: 53-66

Mohren, G.M.J., van den Burg, J. & F.W. Burger 1986. Phosphorus deficiency induced by nitrogen input in Douglas fir stands in the Netherlands. Plant and Soil, 95: 191-200

Mulder, J., van Grinsven, J.J.M. and N. van Breemen 1987. Impacts of acid atmospheric deposition on woodland soils in the Netherlands III. Aluminum chemistry. Soil Science Society of America Journal, 51: 1640-1646

Murach, D. 1984. Die Reaktion der Feinwurzeln von Fichten (Picea abies, Karst.) auf zunehmende Bodenversauerung. Göttinger Bodenkundliche Berichte 141: 1-126

Nihlgård, B. 1985. The ammonia hypothesis - an additional explanation of the forest dieback in Europe. Ambio 14: 2-8

Oren, R. & E.-D. Schulze 1989. Nutritional disharmony and forest decline: a conceptual model. In: E.-D. Schulze, O.L. Lange & R. Oren (eds.): Forest decline and air pollution. Ecological Studies 77, Springer Verlag, Berlin, pp. 425-443

Parker, G.G. 1983. Throughfall and stemflow in the forest nutrient cycle. Advances in Ecological Research 13: 58-133

Pitelka, L.F. & D.J. Raynal 1989. Forest decline and acidic deposition. Ecology, 70: 2-10.
Posthumus, A.C., A.E.G. Tonneijck & L.J. van der Eerden 1989. Exposure-effect relationships for plants in relation to several air pollutants. In: L.J. Brasser & W.C. Mulder (eds.): Man and his Ecosystem. Proceedings of the 8th World Clean Air Congress 1989, The Hague, the Netherlands, 11-15 September 1989, Volume 2, Elsevier Science Publishers, Amsterdam, the Netherlands, pp. 13-18

Prinz, B. 1987. Causes of forest damage in Europe. Major hypotheses and factors. Environment, 29: 11-15 & 32-37

Riederer, M. 1989. The cuticle of conifers: structure, composition and transport properties. In: E.-D. Schulze, O.L. Lange & R. Oren (eds.): Forest decline and air pollution. Ecological Studies 77, Springer Verlag, Berlin, pp. 157-192

Reich, P.B., 1987. Quantifying plant response to ozone: a unifying theory. Tree Physiology 3: 63-91

Roberts, T.M., R.A. Skeffington & L.W. Blank 1989. Causes of type 1 spruce decline in Europe. Forestry 62: 179-222

Roelofs, J.G.M., A.J. Kempers, A.L.F. Houdijk & J. Jansen 1985. The effect of airborne ammonium sulphate on Pinus nigra var. maritima in the Netherlands. Plant and Soil 84: 45-56

Roelofs, J.G.M., A.W. Boxman, H.F.G. van Dijk & A.L.F. Houdijk 1988. Nutrient fluxes in canopies and roots on coniferous trees as affected by nitrogen-enriched air-pollution. In: J. Bervaes, P. Mathy & P. Evers (eds.). Relationships between above and below ground influences of air pollutants on forest trees. Commission of the European Communities, Air Pollution Research Report 16, p. 205-221

Roy, A.K., A. Sharma & G. Talukder 1988. Some aspects of aluminium toxicity in plants. Botanical Review 54: 145-178

Rost-Siebert, K. 1983. Aluminium-Toxizität und -Toleranz an Keimpflanzen von Fichte (Picea abies Karst.) und Buche (Fagus sylvatica L.). Allgemeine Forstzeitschrift 38: 686-689

Rost-Siebert, K. 1985. Untersuchungen zur H^+- und Al-Ionentoxicität an Keimpflanzen von Fichte (Picea abies, Karst.) und Buche (Fagus sylvatica, L.) in Lösungskultur. Berichte des Forschungszentrums Waldökosysteme/Waldsterben, Universität Göttingen, FRG, 12: 1-219

Schaedle, M., F.C. Thornton, D.J. Raynal & H.B. Tepper 1989. Response of tree seedlings to aluminum. Tree Physiology 5: 337-356

Schneider, T. & A.H.M. Bresser, 1988. Evaluatierapport Verzuring. Dutch Priority Programme on Acidification, Report nr. 00-06. RIVM Bilthoven, the Netherlands

Schulze, E.-D. 1989. Air pollution and forest decline in a spruce (Picea abies) forest. Science, 244: 776-783

Schulze, E.-D., W. de Vries, M. Hauhs, K. Rosén, L. Rasmussen, C.-O. Tamm & J. Nilsson, 1989a. Critical loads for nitrogen deposition on forest ecosystems. Water, Air and Soil Pollution, 48: 451-456

Schulze, E.-D., O.L. Lange & R. Oren (eds.) 1989b. Forest decline and air pollution. A study of spruce (Picea abies) on acid soils. Ecological Studies 77, Springer Verlag, Berlin, 475 pp.

Schütt, P. & E.B. Cowling 1985. Waldsterben, a general decline of forests in Central Europe: Symptoms, development and possible causes. Plant Disease 69: 548-558

Skeffington, R.A. & T.M. Roberts 1985. The effects of ozone and acid mist on Scots pine saplings. Oecologia 65: 201-206

Skeffington, R.A. & E.J. Wilson 1988. Excess nitrogen deposition: issues for consideration. Environmental Pollution, 54: 159-184

Smit, H.P., N. van Breemen & W.G. Keltjens 1987a. Effects of soil acidity on Douglas fir seedlings 1. Rooting characteristics of natural regeneration of Douglas fir in strongly acid forest soils. Netherlands Journal of Agricultural Science 35: 533-536

Smit, H.P., W.G. Keltjens & N. van Breemen 1987b. Effects of soil acidity on Douglas fir seedlings 2. The role of pH, aluminium concnetration and nitrogen nutrition (pot experiment). Netherlands Journal of Agricultural Science 35: 537-540

Spitters, C.J.T. 1986. Separating the diffuse and direct component of global radiation and its implications for modelling canopy photosynthesis II: Calculation of canopy photosynthesis. Agricultural and Forest Meteorology, 38: 231-242

Spitters, C.J.T., van Keulen, H. & D.W.G. van Kraalingen 1989. A simple and universal crop growth simulator: SUCROS87. In: R. Rabbinge, S.R. Ward and H.H. van Laar (eds.): Simulation and systems management in crop protection. Simulation Monograph. Pudoc, Wageningen, p. 147-181

Spitters, C.J.T., Toussaint, H.A.J.M. & J. Goudriaan 1986. Separating the diffuse and direct component of global radiation and its implications for modelling canopy photosynthesis I. Components of incoming radiation. Agricultural and Forest Meteorology.
Tiktak, A., C.J.M. Konsten, R. van der Maas & W. Bouten, 1988. Soil chemistry and physics of two Douglas-fir stands affected by atmospheri deposition on the Veluwe, the Netherlands. Dutch Priority Programme on Acidification, Report nr. 03-01, 93 pp.

Thysse, G. & P. Baas 1990. Natural and NH3 induced variation in epicuticular needle wax morphology of Pseudotsuga menziessii (Mirb.) Franco. (in preparation)

Tonneijck, A.E.G. 1989. Evaluation of ozone effects on vegetation in the Netherlands. In: T. Schneider, S.D. Lee, G.J.R. Wolters & L.D. Grant (eds.): Atmospheric ozone research and its policy implications. Amsterdam, Elsevier, Studies in Environmental Science 35, p. 251-260

Ulrich, B. 1983a. A concept of forest ecosystem stability and of acid deposition as driving force for destabilization. In: B. Ulrich & J. Pankrath (eds.): Effects of accumulation of air pollutants in forest ecosystems. D. Reidel Publishing Company, Dordrecht, p. 1-32

Ulrich, B. 1983b. Soil acidity and its relations to acid deposition. In: B. Ulrich & J.

Pankrath (eds.): Effects of accumulation of air pollutants in forest ecosystems. D. Reidel Publishing Company, Dordrecht, p. 127-146

Ulrich, B., K.J. Meiwes, N. König & P.K. Khanna 1984. Untersuchsverfahren und Kriterien zur Bewertung der Versauerung und ihrer Folgen in Waldböden. Der Forst- und Holzwirt 11: 278-288

Ulrich, B. & J. Pankrath (eds.) 1983. Effects of accumulation of air pollutants in forest ecosystems. D. Reidel Publishing Company, Dordrecht, 389 pp.

Unsworth, M.H. & V.J. Black 1981. Stomatal responses to pollutants. In: P.G. Jarvis & T.A. Mansfield (eds.): Stomatal Physiology. Cambridge University Press, Cambridge, pp.187-203

Van Breemen, N., Burrough, P.A., Velthorst, E.J., van Dobben, H.F., de Wit, T., Ridder, T.B. and H.F.R. Reijnders. 1982. Soil acidification from atmospheric ammonium sulphate in forest canopy throughfall. Nature, 299: 548-550

Van Breemen, N., Driscoll, C.T. and J. Mulder 1984. Acidic deposition and internal proton sources in acidification of soils and water. Nature, 307: 599-604

Van Breemen, N., Mulder, J. and C.T. Driscoll 1983. Acidification and alkalinization of soils. Plant and Soil, 75: 283-308

Van Breemen, N., Mulder, J. and J.J.M. van Grinsven 1987. Impacts of acid atmospheric deposition on woodland soils in the Netherlands II. Nitrogen transformations. Soil Science Society of America Journal, 51: 1634-1640

Van Breemen, N. & H.F.G. van Dijk, 1988. Ecosystem effects of atmospheric deposition of nitrogen in the Netherlands. Environmental Pollution 54: 249-274

Van den Burg, J. 1988. Voorlopige criteria voor de beoordeling van de minerale voedingstoestand van naaldbomen op basis van de naaldsamenstelling in het najaar. Rijksinstituut voor Onderzoek in de Bos- en Landschapsbouw "De Dorschkamp", Report 522, Wageningen, 20 pp.

Van den Burg, J. 1989. Bemesting in de Nederlandse bosbouw. Meststoffen nr 1, p. 5-19

Van den Burg, J. 1990. De betekenis van de aluminiumconcentratie in het wortelmilieu voor de groei en de minerale voeding van enkele boomsoorten - Een literatuurstudie (in Dutch). "De Dorschkamp" Instituut voor Bosbouw en Groenbeheer, Wageningen, Rapport nr 580, 468 pp.

Van den Burg, J. 1990. Stickstoff- und Säuredeposition und die Nährstoffversorgung niederländischer Wälder auf pleistozänen Sandböden. Forst und Holz, in press

Van den Driessche, R. 1978. Response of Douglas-fir seedlings to nitrate and ammonium nitrogen at different levels of pH and iron supply. Plant and Soil 49: 607-623

Van der Eerden, L.J. 1982. Toxicity of ammonia to plants. Agriculture and Environment 7: 223-235

Van der Eerden, L.J., A.E.G. Tonneijck & J.H.M. Wijnands 1988. Crop loss due to air pollution in the Netherlands. Environmental Pollution 53: 365-376

Van Hove, L.W.A. 1989. The mechanism of NH_3 and SO_2 uptake by leaves and its physiological effects. Ph.D. thesis, Wageningen Agricultural University, Wageningen, 131 pp.

Wellburn, A. 1988. Air pollution and acid rain. Longman, Harlow, England, 274 pp.

Wislicenus, H. (ed.) 1908-1916. Waldsterben im 19. Jahrhundert. Sammlung von Abhandlungen über Abgase und Rauchschäden. Berlin, Paul Parey. Reprint VDI-Verlag Düsseldorf, 1985

Zajaczkowska, J., A. Lotocki, H. Morteczka & D. Witkowska. 1981. Ageing of assimilatory organs of conifers under natural and polluted environmental conditions. Polish Ecological Studies 7: 401-413

Appendix A: Conversion of concentration units at 20°C and 1013 hPa. 1 ppb corresponds to x µg m^{-3}; 1 µg m^{-3} corresponds to y ppb. x = M/24.055 (P101)

Component	Mol. weight M	x	y
CO_2	44	1.829	0.547
NO	30	1.247	0.802
NO_2	46	1.912	0.523
SO_2	64	2.661	0.376
O_3	48	1.995	0.501
NH_3	17	0.707	1.415

INTEGRATED EFFECTS (LOW VEGETATION)

H.F.van Dobben[1]

[1] Research Institute for Nature Management, Leersum

1. INTRODUCTION

According to commonly used definitions, heathland is a vegetation dominated by evergreen dwarfshrubs, with little or no shrubs or trees, and with a well developed moss layer (de Smidt 1975). In the western European lowland naturally occurring heathland is confined to a narrow coastal zone. Inland heath is man-made, although in some localities it may have been in existence for over 500 years. In the inland heath development of forest was prevented by mowing, sheep grazing and sod cutting. Up to the beginning of this century it was common practice to transport nutrients from heathland to the agricultural land surrounding the villages by a combination of sheep grazing and sod cutting. Sheep were kept overnight in stables with sods spread over the floor; the mixture of sods and dung was used as manure for the agricultural land (de Smidt 1982). After the invention of fertilizer this practice lost its economical significance, and the area of heathland rapidly declined (Table 1.1), most of it being reclaimed for agriculture or forestry. Presently most of the remaining heathland is located in nature reserves.

Table 1.1 Approximate total surface area of Dutch heathland in the course of time (after de Smidt 1984 and Van der Zande et al. 1988)

1800	800 000 ha
1833	600 000 ha
1907	450 000 ha
1940	100 000 ha
1970	61 000 ha
1983	42 000 ha

From a syntaxonomic point of view, heathland may belong to two classes: Oxycocco-Sphagnetea and Nardo-Callunetea (Westhoff & den Held 1969, de Smidt 1982). The former is characterized by the occurrence of Erica tetralix and Sphagnum spp.; wet heathlands, which are influenced by groundwater at least part of the year, and transitions between heathland and bog belong to this class. The latter class is characterized by the occurrence of Calluna vulgaris and Potentilla erecta (among others); dry heathland and transitions between heathland and grassland belong to it. Within the Nardo-Callunetea three alliances are distinguished: Empetrion nigri (coastal heath, characterized by the combination of Empetrum nigrum and Salix repens), Calluno-Genistion pilosi (dry inland heath, characterized by Calluna vulgaris) and Violion caninae (syn. Nardo-Galion; transition between heathland and grassland, characterized by among others Nardus stricta, Antennaria dioica, Arnica montana and Viola canina).

On a quantitative basis the Calluno-Genistion is by far the most important heathland type in the Netherlands. In many places, however, heathland is in a process of transformation into grassland or open forest. Table 1.2 gives the approximate surface area of the various types and their degradation stages estimated by Van der Zande et al. (1988). The Violion caninae is quantitatively unimportant and no estimates of its surface area exist. This type is usually found in close connection with the Calluno-Genistion, its occurrence being limited to patches with a somewhat higher soil pH and nutrient availability, for instance along paths or near sheep stables. In spite of their small surface area, Violion caninae patches can be very rich in species, and are therefore considered valuable with respect to nature conservation (Van der Zande et al. 1988).

Table 1.2 Approximate surface area of heathland types in the Netherlands in 1983 (after Van der Zande et al. 1988)

coastal heath	2 000 ha
wet heath	1 000 ha
heath with 20-60% tree cover	8 000 ha
heath with >75% grass cover	8 000 ha
dry heath with <20% trees and <75% grass	23 000 ha
	----------+
TOTAL	42 000 ha

The previously mentioned vegetation types rapidly declined during this century, especially after c. 1970 (Quene'-Boterenbrood 1988, Van der Zande et al. 1988). In the wet heathland, Erica tetralix is being replaced by Molinia caerulea, in the dry heathland Calluna vulgaris is being replaced by Molinia caerulea and Deschampsia flexuosa, and the Violion caninae species are disappearing altogether. Although many possible causes for the decline, like improper management, lowering of the groundwater table, and over-recreation are reported, most authors agree that air pollution and resulting acidification and eutrophication are key factors. A correlation between the SO_2 concentration and the decline of species in the Violion caninae was reported (Van Dam et al. 1986), the replacement of Erica by Molinia was shown to be dependent upon nitrogen availability (Berendse & Aerts 1984), and nitrogen enrichment is also a probable cause for the replacement of Calluna by Deschampsia (Heil & Diemont 1983, Berdowski 1987, van Ree & Meertens 1989).

The present study focused on the decline of the Calluno-Genistion and the Violion caninae. Throughout the project a distinction was made between 'dominant' species (Calluna and grasses) and Violion caninae species. Within the Violion caninae a distinction was made between 'sensitive' species, for which a significant correlation was shown between decline and SO_2 concentration by van Dam et al. (1986), and 'insensitive' species for which no such correlation was shown (Table 1.3).

Table 1.3 Species selected for detailed study

dominant species	Calluna vulgaris (L.) Hull Molinia caerulea (L.) Moench Deschampsia flexuosa (L.) Trin
Violion caninae species	
'sensitive'	Arnica Montana (L.) Antennaria dioica (L.) Gaertn. Viola canina (L.)
'insensitive'	Hieracium pilosella (L.) Agrostis capillaris (L.) Nardus stricta (L.) Gentiana pneumonanthe (L.)

Two hypotheses regarding the decline of species were tested for both the dominant, sensitive and insensitive species:

a. SO_x is the master factor, either through SO_2 toxicity, or through soil acidification and resulting aluminium toxicity or inhibition of cation uptake;
b. NH_y is the master factor, either through NH_3 toxicity, or through eutrophication and resulting replacement of slow-growing species by fast-growing ones.

The main objective of the project was to make a quantification of the effects of SO_2, NH_3 and $(NH_4)_2SO_4$ on heathland that can be used in environmental policy, e.g. setting standards. No attempts were made towards an evaluation of heathland management (for this, see Van der Zande et al. 1988), nor was it attempted to make a comprehensive study of nutrient cycling (for this, see Gimingham 1972 and Aerts 1989). Effects on fauna were not included either. In detail the aims were as follows:

1. to quantify the extent to which replacement of Calluna by grasses had taken place, with a regional differentiation;
2. to quantify the atmospheric deposition of NH_y and the fate of nitrogen in the ecosystem;
3. to quantify the relation between NH_y deposition and replacement of Calluna by grasses;
4. to quantify the effect of gaseous SO_2 and NH_3 on Calluna, grasses, Violion caninae species and mosses;
5. to estimate the effects of both soil acidification and eutrophication on Violion caninae

species, and the relative importance of each of these processes;

6. to estimate the importance of the naturally occurring stress factors frost, drought and plagues for the transition of heathland into grassland.

1.1 Set-up of the project

The replacement of Calluna by grasses was quantified for the complete heathland area of the Netherlands, using satellite images (Van Kootwijk 1989, section 2). Data on the autecology of Arnica montana were obtained through a comparison of former and actual Arnica stand in the Netherlands, Germany and Denmark (Fennema 1990, section 5). All other field measurements and experiments were carried out on the location 'Asselse Heide' (52°13' N, 5°51' E). This is a dry heathland site that still had a relatively vital Calluna vegetation at the start of the project. Soil chemical properties are described by Van der Maas (1990), physical properties and hydrology are described by Bizot et al. (1990). Atmospheric deposition of NHy was measured by the gradient method (Duyzer et al. 1989) and the throughfall method (Bobbink et al. 1990; section 10.1). Nutrient leaching was measured using filter plate lysimeters (Van der Maas 1990; section 10.3). The fate of nitrogen in the ecosystem was evaluated through experiments with 15N-labelled $(NH_4)_2SO_4$ and literature data (sections 7 and 10).

The effects of atmospheric deposition on plant species was evaluated in a number of experiments with both monocultures (sections 3, 4, 5 and 6) and mixed cultures (sections 7 and 8) of dominant, sensitive and insensitive species (Van der Eerden et al. 1989 and 1990). The effects of atmospheric deposition in combination with other stress factors were investigated for Calluna (section 9). Methods ranged from completely controlled laboratory experiments to observations under field conditions, and included water culture experiments with pH, Al and nitrogen series, fumigation experiments with SO_2 and NH_3, experiments with $(NH_4)_2SO_4$ supplemented artificial rainwater (in the laboratory as well as in the field), and nitrogen fertilization of field plots. For the fumigation experiments both fully controlled Plant Growth Chambers (PGC), a semi-controlled Fumigation Greenhouse (FG), Open Top Chambers (OTC) and a Field Fumigation System (FFS) were used. A full description of the experimental facilities is given by Van der Eerden et al. (1990).

In principle, three approaches could be used for the integration of the results of the various experiments:

1. Nutrient balances. Elimination of heathland species by more efficient nitrogen-users can be expected if accumulation of nitrogen in the soil takes place. The accumulation

can be estimated as the difference between input by atmospheric deposition, and output by uptake in biomass, leaching to the groundwater and removal by grazing or sod-cutting. Because estimates were available for all output terms, a maximum value for the input could be derived.

2. Dynamic simulation of growth and competition. Competition can be simulated using the concepts developed by de Wit (1960). For the Ericion this was done by Berendse & Aerts (1984). Their model was adapted for the Calluno-Genistion by Conijn & Berendse (1991), while another model was developed by Heil (1991). However, the use of these models was outside the scope of this study.

3. Risk assessment based on a toxicological model. Harmful effects determined in laboratory experiments can be extrapolated to the field using safety factors (Suter et al. 1985). This can also be done for a group of species, even if not all species have been individually tested (van Straalen & Denneman 1989). In this approach results of fumigation experiments were used to determine exposure levels which protect a given percentage of a community's species against e.g. visible injury or a decrease in biomass production or competitive strength.

All experiments are described in detail in project reports to which references are given or, if these are not given, the full description is found in Van der Eerden et al. (1989 and 1990).

2. REGIONAL QUANTIFICATION OF HEATHLAND DEGRADATION

Various inventories of heathland vitality have been made in the past (Diemont et al. 1982, Van der Zande et al. 1988). These were carried out by means of simple field observation, for instance as a part of forest inventories or by means of a questionnaire among heathland managers. However, if such a study is to cover the country's entire heathland area it may be very time-consuming. Therefore we used satellite images in order to find a quicker method that probably yields more accurate estimates. In principle, the earth's reflection in various wavelength bands is determined by its vegetation cover. When an empirical relation between these two entities can be assessed, satellite images provide a quick way to estimate vegetation composition. Details on our method are given by Van Kootwijk (1989).

2.1 Outline of the method

A satellite image consists of a large number of picture elements ('pixels') that are characterized by their spectral reflectances, each corresponding to a small part of the earth's surface. A common procedure in vegetation mapping is to find a number of pixels, termed 'training pixels', that represent pre-defined vegetation types. For these training pixels the relation between vegetation type and spectral reflectance is empirically determined. Subsequently, all other pixels can be assigned to one of the vegetation types on the basis of their spectral reflectance. However, in our case this method could not be used, because the aim was not to map vegetation types but to make a quantitative estimate of heathland vitality. Therefore the pixels were not classified, but instead a relation was determined between spectral reflectance and cover percentages of heath, grass and bare soil. The very exact localization of the training pixels required for this method was achieved by a 'grid search' method (for details see Van Kootwijk & van der Voet 1989).

Two LANDSAT images from 1987 were used, with a pixel size of 25 * 25 m^2, each covering half of the country. Independent calibrations were carried out for both images. Ground truth was obtained by making vegetation descriptions along transects using a line intercept method. These transects were roughly localized in the satellite image, using roads, forest edges, pools etc. as landmarks, and multiple regressions were carried out of the cover percentages of the various species on the reflectances of the corresponding pixels. The mean of the residual variances was used as a measure for the overlap between the field transect and the image transect, and an iterative procedure was used to find the maximum overlap.

In the field 25 m wide transects with a total length of 5.7 km were sampled with three lines 12.5 m apart. These data were recalculated to cover percentages of 25 * 25 m^2 ground elements, using three cover types: 'heath', 'grass', and 'bare soil', according to Table 2.1. A distinction was made in six degradation stages, dependent on the relative presence of these three types (Table 2.2).

Table 2.1 Definition of cover types

cover type	field cover
'heath'	Calluna, Erica, Empetrum, dead heath
'grass'	Molinia, Deschampsia, Nardus, dead grass, other green vegetation
'bare soil'	bare soil

Table 2.2 Definition of degradation stages

Main classification		
no degradation	0- 30%	'grass' per pixel
intermediate degradation	31- 50%	'grass' per pixel
strong degradation	51- 70%	'grass' per pixel
full degradation	71-100%	'grass' per pixel
Additional classes		
vital heathland	71-100%	'heath' per pixel
no grass	0%	'grass' per pixel

2.2 Results

Three of the seven transects that were described in the field could be localized by means of the grid search method. For the other transects this was not possible, probably due to strong vegetation changes that had occurred between the image date and the field-work, e.g. as a result of mowing, burning etc.

Two criteria were used in the selection of the pixels to be analyzed: (1) visual inspection: only pixels in selected heathland areas were analyzed, and within these areas pixels containing roads, water, bare soil etc. were discarded; (2) pixels with reflectance values that strongly deviated from the training set were discarded.

Because two training sets were available for the northern image, cross-prediction could be used to estimate the error in the predicted cover values. This error appeared to be c. 15%, leading to a confidence interval of c. 30% (absolute values for each individual pixel). For the mean cover over a large area the confidence limit will of course be smaller (for details see Van Kootwijk & van der Voet 1989).

A total of 276 431 pixels were analyzed, corresponding to a surface area of 17 277 ha, which is only 41% of the country's total estimated heathland area (Table 1.2). This difference can be ascribed to the following causes:

1. coastal heath (2000 ha) and wet heath (1000 ha) were left out of consideration;

2. c. 1000 ha of heathland in the extreme East of the country were outside the two images used;

3. c. 8000 ha were covered with trees for 20-60%, which means that c. 3200 ha were not analyzed;

4. c. 2500 ha were in very small parcels (10 ha or less). Most of these were not analyzed;

5. only pixels that really contain heath or grass-heath have been analyzed. The estimates in Table 1.2 are based on the total area of heathland reserves, that may also contain isolated trees or small forest lots, pools, roads etc.;

6. pixels with a spectral reflectance strongly deviating from the training set were not analyzed.

After subtraction of the areas summed up in 1-4, 53% of the remaining heathland area has been analyzed. The remaining difference is largely due to 5., and for a very small part due to 6.

Table 2.3 gives the distribution of the pixels over the degradation classes given in Table 2.2. Of the total analyzed area, 56% is covered with 'heath', 39% with 'grass' and 5% with 'bare soil'. It may be assumed that heathland containing more than 50% grass is due to change into grassland within 3-5 years (Berdowski 1987). On the basis of this assumption, only 32% of the Dutch heathland can be said to be 'vital' (>70% heath), 34% is 'less vital' (<70% heath and/or 31-50% grass), and 34% is 'degraded heathland' (51-100% grass). Table 2.4 gives the same distribution for each province.

<u>Table 2.3</u> Percentage distribution of the pixels (n=276 431) over the degradation classes given in Table 3.5

Main classification	
no degradation	42%
intermediate degradation	24%
strong degradation	20%
full degradation	14%
	-------+
	100%
Additional classes	
vital heathland	32%
no grass	11%

Table 2.4. Distribution of the total number of analyzed pixels over the degradation classes, per province. F=Friesland, D=Drenthe, O=Overijssel, G=Gelderland, U=Utrecht, N=Noord-Holland, B=Brabant, L=Limburg

Main classification	F	D	O	G	U	N	B	L
no degradation	22	24	49	51	30	37	37	32
intermediate degradation	26	24	27	22	28	29	27	29
strong degradation	27	27	18	17	26	22	23	25
full degradation	24	25	6	10	16	12	14	14
Additional classes								
vital heathland	16	16	32	41	21	30	28	21
no grass	2	3	8	17	4	4	8	6
Percentage distribution of surface area								
'heath'	45	44	59	61	51	57	55	51
'grass'	51	51	34	33	45	40	41	44
'bare soil'	4	5	8	6	4	2	4	5
Total heathland area (ha)	269	3339	970	8690	297	475	2837	400

2.3 Discussion

Our results indicate that about one-third of the Dutch heathland is vital, one-third is endangered and one-third has changed into grassland. A direct comparison with similar estimates made by others cannot be made because of the differences in method and definitions used. Diemont et al. (1982) estimate that 20% of the total heathland is covered with more than 50% grass, on the basis of a questionnaire and local mapping studies, carried out between 1973 and 1981. This estimate also includes coastal heath and wet heath. As coastal heath is still relatively vital compared to inland heath, the figure for inland heath might be somewhat above 20%, which leads to the conclusion that the percentage of grass-heath has increased form c. 22% to c. 32% over a period of c. 10 years. Van der Zande et al. (1988) give an estimate of 18% of the dry heathland area covered with more than 75% grass, also on the basis of a questionnaire, probably carried out in 1986-87. Coastal heath, wet heath and heath with more than 20% tree cover were not included in this estimate, so it may be comparable to ours, and in fact there is a close agreement with the figure given in Table 2.3 (14% covered with >75% grass). It may be concluded that the Dutch heathland is in a process of rapid transition into grassland. As c. 34% of the heathland area is due to change into grassland in the next 3-5 years (Table 2.3), the rate of change is probably increasing with time.

A descriptive study like the one presented here does not allow conclusions on the cause of heathland degradation. At first sight there is no correlation between the spatial distribution of degradation and the spatial distribution of SO_2 (Anon. 1988) or NH_3 (Erisman 1989). The two northern provinces Friesland and Drenthe, which have a relatively low deposition of SO_2 and NH_3, even have the highest percentage of strongly degraded heathland. However, if NH_3 is an important causal factor, a much finer spatial resolution should be used to detect a correlation, because the spatial variation in emission is high and transport distances are small. Moreover, an attempt to explain local heathland degradation on the basis of atmospheric deposition should take the effect of management into account. An area in which sod-cutting has recently taken place on a large scale may contain a high percentage of vital heathland, even if deposition is high. Such a detailed mechanistic approach was outside the scope of this study.

3. EFFECTS OF SO_2, NH_3 AND $(NH_4)_2SO_4$ ON GERMINATION AND ESTABLISHMENT OF HEATHLAND SPECIES

Sod-cutting is generally considered to be the most effective management practice to maintain heathland or to restore grassland into heathland (De Smidt et al. 1984, Van der Zande et al. 1988). This measure not only removes excess nutrients, but also causes a complete regeneration of the vegetation from seeds. In the regeneration process Calluna has an advantage over the grasses in having a long-living seed bank (Thomson & Grime 1979; Heil 1984). However, fertilization with nitrogen compounds has been shown to negatively affect germination and growth of Calluna seedlings (Helsper et al. 1983). Air pollution might act in a comparable way, and therefore we studied the effects of SO_2, NH_3 and $(NH_4)_2SO_4$ on germination and establishment of Calluna and Arnica.

3.1 Calluna vulgaris

Four to six years old Calluna plants were potted in topsoil from the site of collection. After one year pots were chosen in which sufficient seedlings were growing. The adult plants were then cut off and pots were fumigated for eight months in PGCs. Treatments and results are summarized in Figure 3.1; details are given by Dueck (1990).

Both germination and survival of seedlings appeared to be highest in filtered air. There was no significant difference between ambient air and the NH_3 treatments, but SO_2 further reduced both germination and survival. High O_3 concentrations were reached during the experimental period (peak daytime means up to 120 $\mu g.m^{-3}$). As the filters are very efficient in removing O_3, the difference between filtered and ambient air might be due to O_3, although an effect of SO_2 cannot be ruled out. It may be concluded that germination of Calluna vulgaris is negatively affected by SO_2 and O_3, and unaffected by NH_3.

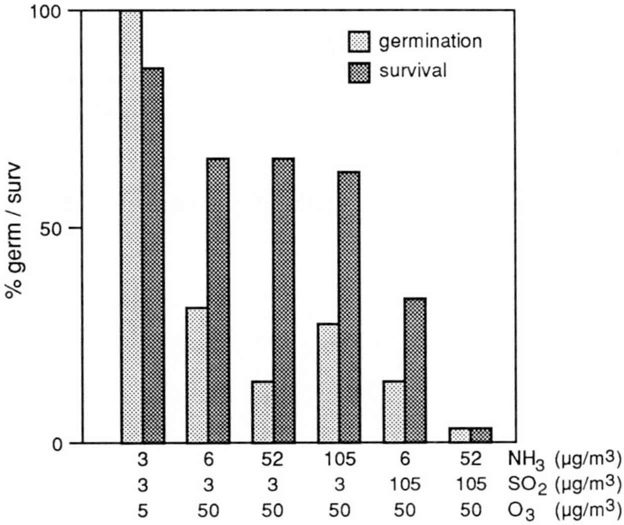

Fig. 3.1 Number of Calluna vulgaris seedlings newly germinated during fumigation (percentage of number in filtered air), and percentage of old seedlings surviving the fumigation

3.2 Arnica montana

Seeds of Arnica montana were spread on the surface of sandy soil (pH 7.0) and kept moist with tap water. They were fumigated with 0, 15, 60 and 240 $\mu g.m^{-3}$ NH_3 in a PGC during 35 days. At the end of this period the pH of the soil had significantly decreased to c. 6.4 at the highest NH_3 concentrations. In another experiment Arnica seeds were placed on top of a slice of oasis soaked with acetic acid buffer (pH 4.7). These were fumigated in an OTC with NH_3 (0, 53 and 105 $\mu g.m^{-3}$) and NH_3 + SO_2 (53 + 90 $\mu g.m^{-3}$) for 21 days. To determine the effect of $(NH_4)_2SO_4$ on germination, Arnica seeds were placed on filter paper moistened with a range of $(NH_4)_2SO_4$ concentrations (0, 25, 50, 100 and 400 $\mu mol.l^{-1}$). The effects of NH_3 and SO_2 on seedling survival were determined by putting one week old seedlings in sandy soil (pH 7.0) and fumigating them in the OTC with NH_3 (0, 25, 53 and 105 $\mu g.m^{-3}$) and NH_3 + SO_2 (53 + 90 $\mu g.m^{-3}$) for 16 days.

The results are summarized in Figure 3.2, details are given by Fennema (1990). Germination percentages at a low NH_3 concentration were significantly higher than both in 'clean' air or at higher NH_3 concentrations. The effect of additional SO_2 was not significant. Seedling survival followed a pattern comparable to that found for Calluna, with a decrease at high NH_3 concentrations and a strong additional reduction by SO_2. The effect of $(NH_4)_2SO_4$ on germination percentage was not significant. Apparently germination of Arnica is not directly negatively influenced by ambient NH_3 concentration in the Netherlands. But like Calluna, seedling survival may be negatively affected by ambient SO_2. Germination and establishment can also be influenced by vegetation changes induced by nitrogen deposition; this is discussed in 8.2. The stimulation of germination at low NH_3 concentration is apparently a direct effect of gaseous NH_3, as it is independent of pH and no effect was found for dissolved $(NH_4)_2SO_4$.

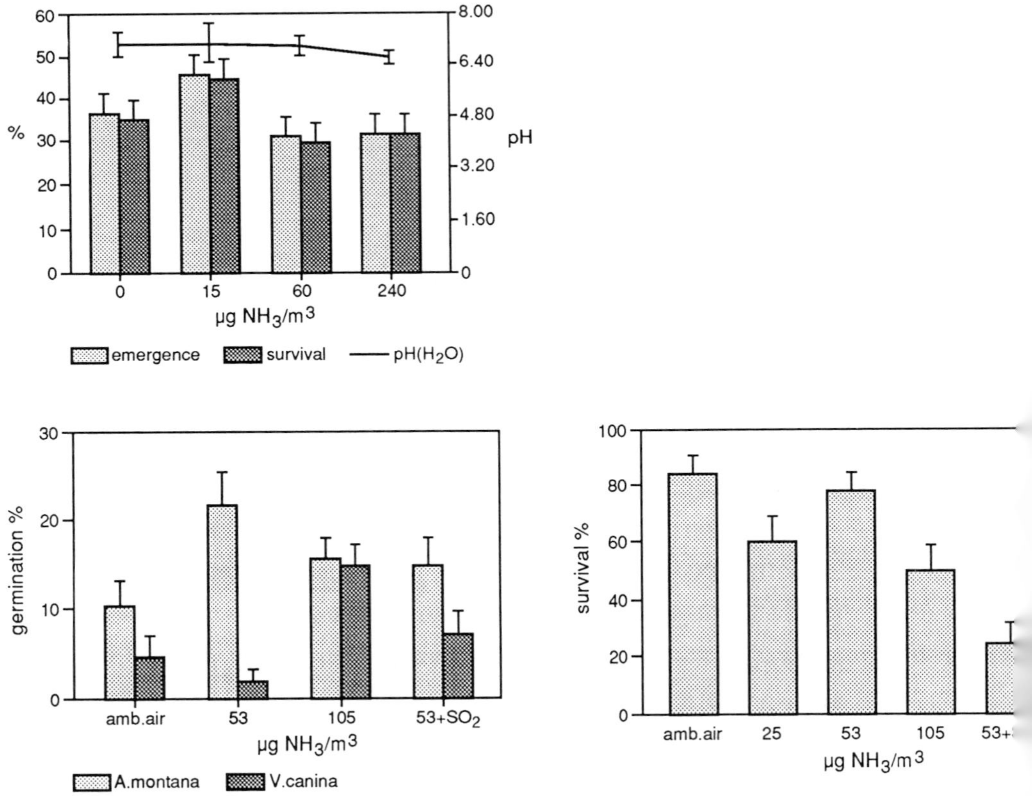

Fig. 3.2 Effects of fumigation with NH_3 on the emergence and survival of Arnica seedlings and the pH of the upper 0.5 cm soil layer (left), the cumulative germination percentage of Arnica and Viola seeds (lower left) and the survival of Arnica seedlings (lower right)

4. RESPONSE OF HEATHLAND SPECIES TO SO_2 AND NH_3

In order to quantify the effects of SO_2 and NH_3 on heathland species fumigation experiments were carried out in PGCs with a range of species, both 'dominant', 'sensitive' and 'insensitive' according to Table 1.2. As bryophytes have also been shown to be sensitive to air pollution (Greven 1989) a number of species from this group were included as well. In order to be able to extrapolate results obtained for a limited number of species to a complete ecosystem, a risk analysis technique was used.

4.1 Principles of risk analysis

Ecotoxicological risk analysis can be an important tool for estimating concentrations to protect sensitive species, but to date it is methodically poorly developed. The approach presented here has been derived from a method developed for water quality (Kooijman 1987) and soil quality (Van Straalen 1987; Van Straalen & Denneman 1989). Its principle is to determine the concentration of a toxic compound which does not negatively affect an ecosystem. An important assumption is that the protection of an ecosystem in effect means the protection of the individual species. In order to be able to protect the complete ecosystem, ideally all species must be tested to locate the most sensitive one. As this can only seldom be realized, a number of representative species from various 'niches' are chosen. These species are exposed to a range of concentrations, and a NOEC (no observable effect concentration) is determined. The NOEC is based on a sensitive parameter that is ecologically relevant for the particular species in its niche. For annual species, this parameter might be relative growth rate or reproductive capacity, while for perennials, it might be biomass increment, root:shoot ratio, competitive capacity or sensitivity to secondary stress factors. These factors often determine the abundance and are thus ecologically relevant. As the concentration to which a particular species is exposed increases, the parameter is negatively influenced, sometimes gradually, but often suddenly. The concentration at which the parameter becomes significantly different from the control treatment, is defined as the NOEC. NOECs of a number of species are assumed to be log-normally distributed, and a mean NOEC is calculated as the geometric mean of the NOECs of the tested species. The standard deviation of this mean value is used to derive the safety factor (T), which estimates a hazardous concentration for a certain proportion (∂) of the community. In this study ∂ was arbitrarily set to 0.05, or in other words, a value of T is calculated that aims at protection of 95% of the species. According to Kooijman (1987), T can be calculated as:

$$T = \exp\left[(3 s_m d_m/\pi^2) * \ln(1-\partial)/\partial)\right]$$

with:

s_m standard deviation of expected NOEC

d_m factor dependent on number of tested species, calculated according to Kooijman (1987)

∂ fraction of species that is not protected by the calculated safety factor

Dividing the (geometric) mean NOEC by T yields a concentration at which 100(1-∂)% of the community's species are protected.

4.2 SO_2

Table 4.1 shows the results of the fumigation experiments carried out with SO_2. Details of this experiment and its evaluation are given by Dueck et al. (1990). The effect of SO_2 on the grasses was generally small. All species reacted to SO_2 with a decrease of the root:shoot ratio, mostly due to a depression of root growth. Agrostis was stimulated in both root and shoot growth up to 66 $\mu g.m^{-3}$. Of the dicotyledonous species, Calluna was very tolerant: even at 520 $\mu g.m^{-3}$ no significant effects were found. The other species showed more pronounced effects, generally a decrease in biomass production with increasing SO_2 concentration, the roots more so than the shoots. For the bryophytes only visual damage (leaf tip necrosis) was assessed. In this group effects were found for all species, sometimes at unexpectedly low exposure levels (14 d 32 $\mu g.m^{-3}$).

The effect-threshold values given in Table 4.1 are the lowest SO_2 concentrations at which any of the growth parameters (altered root: shoot ratio, decreased root or shoot biomass, or leaf injury) were significantly different from the control. For the bryophytes the exposure period lasted only 18 days, and for this group threshold values were estimated for a 42 day period using the empirical relation $C * T^n = K$, where C and T are threshold value and time respectively, while n and K are species-specific constants.

The calculated threshold values indicate that effects of gaseous SO_2 on the dominant heathland species are unlikely in the Netherlands. Threshold values for these species are far above the current peak SO_2 values. For the Violion caninae species and the bryophytes, however, threshold values are in the same order of magnitude as the ambient values, at least for a number of species. In view of the higher SO_2 concentrations in the past (Van Dobben 1990), gaseous SO_2 is possibly a causal factor in the decline of the Violion caninae.

Table 4.1 Effect threshold values for SO_2 (in μg.m-3, derived by linear interpolation), for heathland species after 42 days exposure to 0, 32, 66, 130, 260 and 520 μg.m-3 SO_2. Effect criteria are: 1) altered root:shoot ratio; 2) decreased root or shoot dry weight; 3) leaf injury. All parameters significantly influenced at $p<0.05$

species	criterion	effect threshold
A) Dominant species		
Calluna vulgaris	2, 3	750
Molinia caerulea	1	300
Deschampsia flexuosa	1, 2, 3	750
B) Violion caninae species		
Arnica montana	1, 3	300
Antennaria dioica	2	75
Agrostis capillaris	1	400
Nardus stricta	1	200
Hieracium pilosella	3	50
C) Bryophytes		
Hypnum jutlandicum Holm. et Warncke	3	21
Dicranum polysetum Sw.	3	19
Pleurozium schreberi (Brid.) Mitt.	3	40
Campylopus flexuosus (Hedw.) Brid.	3	48
Dicranella heteromalla (Hedw.) Schimp.	3	144
Plagiothecium curvifolium Schlieph. ex Limpr.	3	220

Although our experiments confirm the results of Van Dam et al. (1986) in a general way (sensitivity of Violion caninae species to SO_2), there are discrepancies for individual species: Arnica and Nardus were found to be sensitive by Van Dam and insensitive in our study, while for Hieracium the reverse is true. Only Antennaria was found to be sensitive in both studies. A possible explanation for these differences is that in a short-term experiment effects that come about through soil acidification are not taken into account. Besides, in our experiments the first indications of adverse effects were scored, whereas Van Dam et al. evaluated a final response, i.e. the disappearance of a species. The mean NOEC (129 μg.m-3 SO_2) was divided by the safety factor T (16.02), which resulted in a mean effect threshold of 8 μg.m-3 SO_2 to protect 95% of the species in a heathland vegetation from chronic injury. This value is just below the ambient mean concentration over the Netherlands in 1989 (10 μg.m-3). Because the NOECs were derived on the basis of a 42 day fumigation, the long-term value might still be somewhat lower.

4.3 NH_3

The application of the risk analysis method to NH_3 requires an extra assumption. Except for bryophytes (Table 4.2) no direct toxic effects of NH_3 were observed. Initially, fumigation with NH_3 resulted in growth stimulation for all species that were tested. Yet NH_3 may disturb the metabolism of species that are naturally adapted to nutrient-poor conditions, e.g. by increasing their sensitivity to secondary stress factors like frost and drought, sometimes resulting in death following even mild winters (see section 9). Although the relation between growth stimulation and negative effects was not quantified, risk analysis was also carried out for NH_3, based on the assumption that for heathland species a specific growth stimulation eventually results in a negative effect.

Table 4.2 Exposure time (in day) and NH_3 concentration (in$\mu g.m^{-3}$) causing leaf chlorosis in bryophyte species

species	exposure time	concentration
Campylopus flexuosus (Hedw.) Brid.	14	105
Rhacomytrium lanuginosum (Hedw.) Brid.	23	25
Hypnum jutlandicum Holm. et Warncke	11	105
Pleurozium schreberi (Brid.) Mitt.	90	100

Plants were fumigated with NH_3, and the most sensitive growth parameters per species were chosen as for SO_2. A relationship between NH_3 concentration and growth was then found by linear regression. The NH_3 concentrations corresponding to growth stimulations of 10%, 25% and 50% are given in Table 4.3. These NH_3 concentrations were then incorporated into the equation yielding the safety factor (T), based on the assumption that a growth stimulation will eventually lead to detrimental effects. Because the growth stimulation which results in a negative effect is unknown, three degrees of growth stimulation were chosen. Table 4.4 indicates which proportion of the community can be protected, given that a particular growth stimulation results in negative effects for sensitive species.

Table 4.3 NH_3 concentrations (in $\mu g.m^{-3}$, derived by linear interpolation) at which 10%, 25% and 50% growth stimulation (heathland species), occurred after 90 days exposure to 25, 53 and 105 $\mu g.m^{-3}$ NH_3

species	proportional growth stimulation 10%	25%	50%
A) Dominant species			
Calluna vulgaris	8.2	15.9	41.0
B) Violion caninae species			
Agrostis capillaris	6.3	8.6	13.4
Antennaria dioica	8.5	17.5	48.6
Potentilla erecta	11.0	31.9	145.4
Viola canina	7.0	10.9	20.6

Table 4.4 Concentrations of NH_3 ($\mu g.m^{-3}$) which protect various proportions of the heathland species against negative effects, assuming the occurrence of such effects at growth stimulations of 10%, 25% or 50%

proportion of the species to be protected	proportional growth stimulation 10%	25%	50%
95%	1.7	3.3	3.7
90%	2.9	5.1	6.3
85%	4.0	6.8	8.9
80%	5.1	8.4	11.5

5. RESPONSE OF VIOLION CANINAE SPECIES TO SOIL FACTORS

Recent reviews on the distribution of heathland types in the Netherlands suggest a more rapid decline of the Violion caninae compared to the Calluno-Genistion (Quene'-Boterenbrood 1988, Van der Zande et al. 1988). In many cases Violion caninae species had disappeared before heathland changed into grassland (Berendse 1988, Houdijk 1990). Both SO_2 and NH_3 have been suggested as possible causes (Van Dam et al. 1986; Quene'-Boterenbrood 1988). However, information on the mechanism leading to this decline is lacking. Possible mechanisms include

(a) direct effects of gaseous compounds;

(b) soil acidification and direct effects of pH on nutrient availability or nutrient uptake;

(c) toxic Al levels in the soil resulting from acidification;

(d) toxic NO_3^- of NH_4^+ levels resulting from nitrogen deposition or soil acidification;

(e) outcrowding by more efficient nitrogen users at high N levels.

In order to gain more information on the importance of SO_2 and NH_3 and the role of soil factors, a number of field and laboratory studies were carried out on the ecology and physiology of Violion caninae species.

5.1 Field studies

Soil analysis of former and actual sites of Violion caninae species may yield more insight into the cause of the decline. Two studies of this type were performed, one including a wide range of heathland species and the other concentrating on Arnica montana.

5.1.1 Soil chemistry of rare heathland species in the Netherlands

Soil samples were taken from sites with one of the species summed up in Table 5.1. These species partly belong to the Violion caninae, partly to other, closely related alliances which are indicated in the table. The soils were grouped according to their chemical characteristics using PCA, and the correlation between the presence of the species and the first three PCA axes were determined (Houdijk 1990; Table 5.1).

The rare species appeared to occur on soils that have a systematically higher pH and higher NO_3^- content than the mean value over a large number of heathland sites. The soil NH_4^+ content in the 'rare species sites' was usually higher than in mean heathland sites, but lower than in Molinia or Deschampsia sites. CEC was usually low (up to 50 μmolc.kg^{-1} dw), 80% of the adsorption complex was made up by Al and Ca. On the basis of the PCA analysis (including all macronutrients) the species could be grouped as follows:

1. Lycopodium clavatum and Rhynchospora fusca: these species occurred on soils comparable to the mean heathland soil, with a pH around 4.1, high Al and low base cation contents.

2. Gentiana pneumonanthe, Narthecium ossifragum and Genista pilosa: these species occurred on soils with a pH somewhat higher than in mean heathland soil (c. 4.4), but with a high Al and low base cation content.

Table 5.1 Mean soil pH-H_2O values, and soil NH_4^+ and NO_3^- contents (in µmol.kg^{-1} dw) of sites with 'rare' heathland species. Letters indicate the alliances in which the species occur: Vc = Violion caninae, E = Ericion tetralicis (wet heath), GC = Calluno-Genistion pilosae (dry heath), TA = Thero-Airion (sandy places in dry heath). n=10 for rare species, n=20 for dominant species. Modified, after Houdijk (1990)

name	all.	pH	NH_4	NO_3
Pedicularis sylvatica L.	Vc	4.9	78	17
Lycopodium inundatum L.	E	4.6	68	11
Rhynchospora fusca (L.) Ait.f.	E	4.2	130	28
Dactylorhiza maculata (L.) Soo	Vc	4.8	121	49
Gentiana pneumonanthe L.	Vc	4.4	215	56
Narthecium ossifragum (L.) Huds.	E	4.4	220	37
Thymus serpyllum L.	TA	5.2	58	16
Polygala serpyllifolia Hose	Vc	4.6	169	21
Arnica montana L.	Vc	4.5	154	20
Lycopodium clavatum L.	GC	4.1	58	28
Genista pilosa L.	GC	4.3	199	14
Genista tinctoria L.	GC	5.4	41	20
Erica tetralix L.	E	4.1	55	0
Calluna vulgaris (L.) Hull	GC	4.1	84	1
Molinia caerulea (L.) Moench		4.2	248	17
Deschampsia flexuosa (L.) Trin.		4.1	429	29

3. Pedicularis sylvatica, Lycopodium inundatum, Orchis maculata, Thymus serpyllum, Polygala vulgaris, Arnica montana and Genista tinctoria: these species occurred on soils with a higher pH (c. 4.5-5.0), lower Al and higher base cation content than mean heathland soil.

All Violion caninae species except one (Gentiana pneumonathe) are in the third group. The preference of this alliance for somewhat 'richer' sites in heathland, reported in the earlier ecological literature (see e.g. Westhoff & den Held 1969) is confirmed by these results. It may therefore be tentatively concluded that soil acidification and resulting replacement of adsorbed base cations by Al is an important factor in the decline of the 'rare' heathland species, especially those from the Violion caninae. The dominant heathland species, and also the Ericetum species Lycopodium clavatum and Rhynchospora fusca have a preference for sites with a lower soil pH, and will therefore be less affected by a general decrease in soil pH.

The role of nitrogen is not quite clear. As the rare species sites have a higher soil nitrogen content than the heathland sites, total nitrogen does not seem to be an important factor in the decline. The low NO_3- content in the heathland sites can probably be explained from a slow mineralization of Calluna and Erica litter (see 10).

5.1.2 Soil chemistry of former and actual Arnica montana sites in the Netherlands, Germany and Denmark

Sites with Arnica montana were selected in the Netherlands, Northern Germany and Denmark. In addition, sites were selected in the Netherlands where Arnica had disappeared during the past decade. In spring and summer 1989, soils of these sites were sampled at two depths. Plots measuring 0.25 m^2 were harvested in July 1989, divided into Arnica and 'bulk' (non-Arnica) vegetation, dried, weighed and analyzed.

Detailed results are given by Fennema (1990). Actual Arnica stands have significantly higher pH values and base cation contents than former Arnica stands. The relation between the occurrence of Arnica and soil chemistry is given in Table 5.2. The occurrence of Arnica is significantly related to soil pH, and total Na, K, Mg and exchangeable Ca content, but not to total N, NH_4 or NO_3, and neither to Al. Table 5.2 applies to the overall comparison of former and actual Arnica stands, but comparable results are obtained if only the Dutch stands are included. Actual Arnica stands in Germany and Denmark tend to have higher base cation contents than those in the Netherlands, but there were no significant differences in pH.

Analysis of plant material showed that Dutch Arnica had a significantly higher nitrogen content and lower Ca and Mg contents than German or Danish Arnica; the standing crop of Arnica was significantly higher in Germany and Denmark. In the actual Arnica stands the bulk vegetation had a significantly lower N content and higher K, Ca and Mg contents than in the former Arnica stands. The above-ground biomass of the bulk vegetation was not different between the former and actual Arnica stands, however.

These results confirm the results of the previous paragraph. Sites with a low pH or a low base cation content are apparently unsuitable as habitats for Arnica. In the Netherlands, where SO_2 and NH_3 concentrations are high compared to Northern Germany and Denmark, Al and H may have replaced base cations due to soil acidification, which in turn led to the disappearance of Arnica.

Table 5.2 Significant ($p<0.05$) regression coefficients for the logit regression of the occurrence of Arnica on soil chemistry
Model function:

$$y = [EXP\ (a+bx)] / [1 + EXP\ (a+bx)]$$

in which a and b are regression coefficients, x can be any soil parameter, and y is the probability of occurrence of Arnica. P50 indicates values at which the predicted probability for the occurrence of Arnica is 50%. (pH-CaCl2; Na, K and Mg total contents in $mg.kg^{-1}$ soil; Ca exchangeable contents in $mmol.100g^{-1}$ soil; - = relation not significant)

	Spring 0 - 5 cm			Summer 0 - 5 cm		
	a	b	P_{50}	a	b	P_{50}
pH	-	-	-	-11.52	3.43	3.36
Na	-1.695	0.1539	11.01	-1.203	0.1065	11.30
K	-1.31	0.0272	48.16	-	-	-
Mg	1.108	0.0548	20.22	-1.147	0.0462	24.83
Ca	-1.485	5.1	0.29	-9.10	2.138	4.26
	Spring 5 - 10 cm	Summer 5 - 10 cm				
	a	b	P_{50}	a	b	P_{50}
Na	-2.139	0.411	5.20	-2.129	0.396	5.38
K	-1.76	0.0903	19.49	-	-	-
Mg	-0.663	0.0947	7.00	-0.4	0.0705	5.67
Ca	-0.182	6.66	0.03	-	-	-

Like the previous paragraph this study does not permit definitive conclusions on the role of nitrogen. Soil N apparently does not have a direct effect on Arnica. Although the nitrogen content of the bulk vegetation in former Arnica stands was higher than in actual Arnica stands, there was no difference in the bulk standing crop. Therefore the decline of Arnica is probably not due to competition; this will be further discussed in 8.1.

5.2 Laboratory studies

Increased dissolution of organically bound aluminium caused by soil acidification leads to high Al^{3+} concentrations in the soil solution (Van Dobben et al. 1991). Absorbed base cations are replaced by aluminium ions, and high aluminium concentrations relative to base cation concentrations are probably toxic to plants (Hüttermann 1985). The Al/Ca ratio is often used to estimate the probability of toxic effects; in a number of studies an effect threshold value of 1 is reported for tree saplings (see Boxman & van Dijk 1988). In order to evaluate and quantify the effects pH and aluminium on heathland species, and to determine whether an interaction with nitrogen exists, we carried out a number of pot and hydroculture experiments.

5.2.1 Effects of NH_4/NO_3 and Al/Ca ratios on Thymus serpyllum

In a pot experiment described by Houdijk (1990), Thymus serpyllum survived a treatment with artificial rainwater enriched with $(NH_4)_2SO_4$ to simulate a deposition of 140 kg N $ha^{-1}.y^{-1}$, on a soil buffered with $CaSO_4$ and Na_2CO_3. This again confirms that high nitrogen concentrations are not toxic by themselves. To assess the effects of Al and nitrogen form, uptake experiments were performed with roots of Thymus. Branches were cultured on a modified Hoagland solution until sufficient roots had developed, and then transferred to solutions with different Al/Ca and NH_4/NO_3 ratios. Nutrient uptake was then measured after eight hours. Tables 5.3 and 5.4 give the results of two of these experiments, which are described in full by Houdijk (1990). The uptake of potassium was inhibited at a Al/Ca ratio of 1, and excretion of potassium, magnesium and calcium increased at higher Al/Ca ratios. This seems to confirm earlier results with tree saplings, but it is not clear whether these results can be extrapolated to the field as Al/Ca ratios up to 5 were found in sites with Thymus (Houdijk 1990).

Table 5.3 Cumulative uptake (+) and release (-) of nutrients by roots of Thymus serpyllum during eight hours incubation in solutions with different Al/Ca ratios. Concentrations in $\mu mol.l^{-1}$, uptake in $\mu mol.g^{-1}$ root dw. From Houdijk (1990)

incubation soln.		K	Mg	Ca	Al
Al 0	Ca 100	5.2	-0.7	-2.2	0
Al 100	Ca 100	-4.7	-1.2	-1.6	-1.6
Al 500	Ca 100	-7.0	-1.1	-3.4	-14.2
Al 1000	Ca 100	-4.5	-4.3	-4.1	-27.9

Table 5.4 Cumulative uptake (+) and release (-) of nutrients by roots of Thymus serpyllum during eight hours incubation in solutions with different NH_4/NO_3 ratios. Concentrations in $\mu mol.l^{-1}$, uptake in $\mu mol.g^{-1}$ root dw. From Houdijk (1990)

incubation soln.		NH_4	NO_3	K	Mg	Ca
NH_4 0	NO_3 100	0.1	11.5	1.0	-0.2	0.9
NH_4 100	NO_3 100	5.2	13.8	1.1	-1.2	-0.8
NH_4 100	NO_3 0	5.5	0.3	-0.4	-1.2	-2.2

Base cation uptake was also inhibited by high NH_4/NO_3 ratios (Table 5.4); Thymus had a strong preference for nitrogen uptake in the form of NO_3. Here again the extrapolation to the field is difficult. In soil solutions of acidic sites in the Netherlands the NH_4/NO_3 ratio is always well above 1 in the topsoil (Van Dobben et al. 1991, Van der Maas 1990). The studies described in 5.1 showed that 'rare heathland species' have a preference for sites with a relatively high base cation availability. Therefore NH_3 deposition might have a double action, on the one hand causing soil acidification and thereby cation leaching, on the other hand causing a high NH_4/NO_3 ratio and thereby inhibiting cation uptake.

5.2.2 Effects of pH on heathland species

A comparison was made of root growth and survival of 'dominant', 'sensitive' and 'insensitive' species (Table 1.3) placed on hydroculture solutions with pH values, ranging from 2.5 to 6.5 in 0.5 unit intervals (adjusted by adding NaOH or H_2SO_4), in a greenhouse during four weeks. Root growth was used as a response variable, except for Deschampsia and Arnica, where many new roots developed from the stem base; for these species response was measured as survival rate. Results are given in Table 5.5. The dominant species appeared to have a lower pH optimum than the Violion caninae species, which is in agreement with findings of the field studies reported in 5.1. Also the 'sensitive' species tend to have a higher pH optimum than the 'insensitive' species, but without a clear-cut separation.

Table 5.5 Summary of ANOVA analysis regarding pH range of heathland species on hydroculture

species	pH opt	range with no sign. diff.	effect parameter
Molinia	3.5	3.0-3.7	root growth
Deschampsia	4.0	3.0-4.5	survival
Agrostis	4.5	3.8-4.8	root growth
Arnica	4.2	3.7-4.7	survival
Antennaria	6.0	4.0-6.5	root growth
Viola	4.7	4.5-6.5	root growth
Hieracium	4.5	4.0-6.5	root growth
Gentiana	4.0	4.0-4.8	root growth

5.2.3 Effects of aluminium on Violion caninae species

Hydroculture experiments were carried out with Arnica, Viola and Antennaria ('sensitive') and Agrostis ('insensitive'). Plants were cultivated from seeds, and grown in a greenhouse on solutions containing a range of Al concentrations during 4 weeks. Plant fresh weight and root length were measured at the start and end of the experiment; Table 5.6 gives the results. A significant decrease in growth rate was found at 11 ppm for Antennaria, which was the most sensitive of the tested species. Al stimulated growth of Arnica and Viola at low concentrations, but at high concentrations the effect was small or not significant. It can be concluded that the species studied are probably not affected by Al in the field situation, as soil solutions of acidic sites usually have Al concentrations in the range 2-15 ppm in the subsoil and lower in the topsoil (Van Dobben 1991; Asselse Heide: mean 2.2 ppm at 30 cm depth and much lower at 5 and 20 cm; Van der Maas 1990). Also at Al/Ca ratios far above 1 no significant inhibition of total growth or root growth took place (Al/Ca in soil solution at Asselse Heide: 0.25 at 5 cm depth and 3 at 30 cm; Van der Maas 1990). In another experiment Arnica was grown on a range of NH_4 and Al concentrations. No significant interaction between NH_4 and Al was found. The results of these experiments confirm the conclusions drawn by Pegtel (1987) on the basis of comparable experiments.

Table 5.6 Relative growth rate values (RGR) of heathland species on solutions with different Al concentrations. Different letters indicate significant differences. RGR was calculated as

(ln FWt2 - ln FWt1) * 1000 / (t2 - t1)

in which FW is fresh weight and t is time in days

Al concentration (ppm) Al/Ca (mol/mol) species	0 0	5.4 2	11 4	22 8	44 16
Antennaria	96a	76ab	36bc	26c	
Arnica	238a	336b	270a	274a	
Viola	57a	138b	172b	91b	
Agrostis		229a			217a

It is not easy to make a general conclusion on the effect of aluminium. Firstly the extrapolation of hydroculture experiments to the field is difficult. In soil solutions aluminium may be organically complex and therefore less toxic. Secondly, a wide variety of toxic Al levels is reported from different experiments (compare this experiment, Houdijk 1990, Boxman & van Dijk 1988), and probably large differences in sensitivity to aluminium exist between species. However, aluminium toxicity is probably unimportant for the decline of the Violion caninae.

6. RESPONSE OF DOMINANT SPECIES TO NITROGEN IN MONOCULTURES

In nutrient-poor conditions nitrogen is the most important factor that limits plant growth, and even small additions of nitrogen will stimulate growth. However, if differences exist between species in their ability to take up, store and use nitrogen, their competitive ability will be affected differently and ultimately vegetation changes will occur. Such a mechanism was shown for wet heath by Berendse & Aerts (1984). The uptake of nitrogen may take place in three ways:

1. above-ground uptake of gaseous NH_3;
2. above-ground uptake of NH_4^+ dissolved in rainwater;
3. below-ground uptake of NH_4^+ from soil solution.

In any case an increase in atmospheric NH_3 concentration will result in an increase in nitrogen uptake, either directly from the atmosphere or after dissolution in rainwater or soil solution. However, the rate of change in plant nitrogen content (and thereby plant growth) after a change in atmospheric NH_3 concentration is strongly governed by the uptake mechanism: it will be delayed by absorption in the litter and subsequent mineralization if below-ground uptake predominates, and fast if above-ground uptake predominates. Therefore nitrogen uptake and resulting growth stimulation was studied for Calluna vulgaris, Molinia caerulea and Deschampsia flexuosa in reaction to NH_3 fumigation, application of $(NH_4)_2SO_4$ supplemented rainwater and fertilization with $(NH_4)_2SO_4$.

6.1 Effects of gaseous NH_3

Calluna vulgaris, Molinia caerulea and Deschampsia flexuosa plants were potted in heathland soil (pH 4.5) and fumigated in a FG with filtered air (<10 $\mu g.m^{-3}$ NH_3) or air with 108 $\mu g.m^{-3}$ NH_3 for 38 weeks. Results for Calluna and Deschampsia are given in Figure 6.1. Calluna appeared to accumulate N in the leaves immediately, but growth reaction was much slower, starting after 12 weeks, but exponential up to the end of the experiment. The root:shoot ratio strongly decreased in the treated plants, indicating that only shoot growth is stimulated by NH_3. In contrast to Calluna, growth of Molinia was immediately stimulated by NH_3, but levelled off at the end of the experiment, probably due to senescence. There appeared to be some N accumulation in the leaves, but a strong accumulation in the basal internode and roots. Root:shoot ratio was unaffected by the treatments. The response of Deschampsia was intermediate between Calluna and Molinia.

There was a lag in growth response, and the biomass increase was mainly, but not completely due to shoot growth. The N-content of the leaves was higher in fumigated than in control plants, but decreased with time, even in fumigated plants.

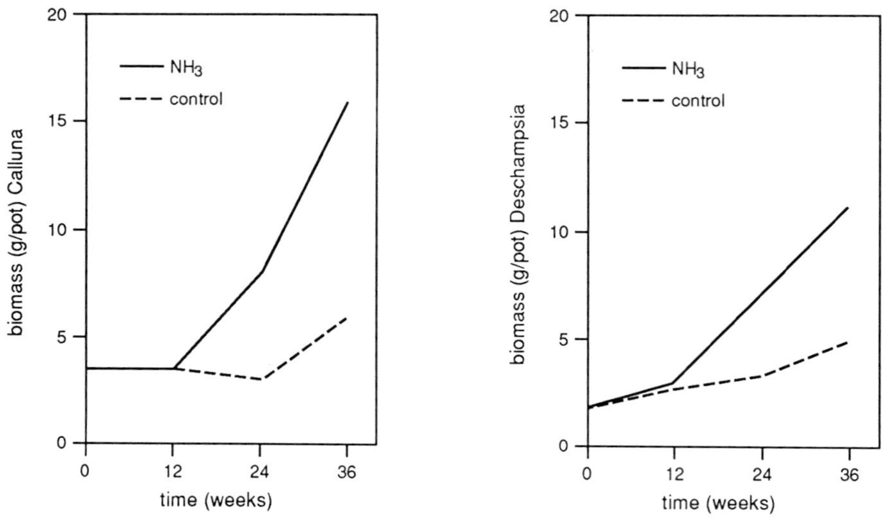

Fig. 6.1 Biomass increase of Calluna (left) and Deschampsia (right) fumigated with 108 $\mu g.m^{-3}$ NH_3

These experiments show that all three species can use atmospheric NH_3 as an additional source for nitrogen. The extra nitrogen is partly stored and partly used for growth. A total mass of nitrogen per plant can be calculated as the product of total biomass and N-content, and this value is a measure for 'potential growth', i.e. growth if nitrogen were the only limiting factor. In fact there are also other limiting factors, and storage of nitrogen takes place to a certain degree. Table 6.1 gives the slope parameters for the regression of total-N on time. The strongest increase of total-N on fumigation is found for Calluna, about twice that of the grasses.

Table 6.1 Parameter estimates for the regression of N-total (standardized to N-total = 1 at t = 0) on time (in weeks)

species	control	NH_3	NH_3:control
Calluna vulgaris	0.026	0.160	6.1
Molinia caerulea	0.075	0.235	3.1
Deschampsia flexuosa	0.032	0.111	3.5

Actual growth differs from potential growth in several respects. In Calluna, growth was exponential after a very slow start, in Molinia growth became more or less linear throughout the experiment, while Deschampsia showed intermediate behaviour. Furthermore, Molinia was able to distribute growth over both shoot and root while in Calluna and to a lesser extent Deschampsia NH_3 mainly stimulated shoot growth. Another difference between Molinia and Calluna is that NH_3 caused an increase in N-content in both root and shoot of Molinia but only in the leaves of Calluna. Deschampsia was again intermediate. It is also important that Calluna does not allocate the nitrogen taken up from the atmosphere to its roots, as the grasses do.

These results indicate that although Calluna can make an efficient use of atmospheric nitrogen, it may become more sensitive to above-ground injury by herbivores, frost etc., than the grasses. In Calluna, nitrogen only stimulates shoot growth and all nitrogen taken up is stored above-ground. Therefore re-growth of Calluna after injury will start only slowly. On the other hand the grasses use the extra nitrogen for both root and shoot growth, and store surplus nitrogen partly in their roots. Therefore re-growth after injury can be faster in the grasses.

6.2 Above-ground nitrogen uptake

The uptake of different forms of nitrogen was studied in an experiment in which Calluna vulgaris, Molinia caerulea and Deschampsia flexuosa were exposed to four levels of NH_3 (10, 60, 170 and 350 $\mu g.m^{-3}$), and three levels of $(NH_4)_2SO_4$ (20, 300 and 1000 $\mu mol.l^{-1}$ dissolved in artificial rainwater), during a three-month period, in a FG.

At the end of the experiment the NH_4^+ and NO_3^- of the soil appeared to increase with both increasing NH_3 and increasing $(NH_4)_2SO_4$ concentration. NO_3^- in the soil also increased with increasing NH_3, probably due to nitrification, in spite of the low pH which had remained constant at 4.3 throughout the experiment (see also section 10). No relation was found between the NO_3^- concentration in the soil and the $(NH_4)_2SO_4$ concentration applied. This suggests that while NH_3 stimulates nitrification, $(NH_4)_2SO_4$ does not, but an alternative explanation is that the plants prefer uptake of NO_3^- over uptake of NH_4^+ and that a high above-ground uptake of NH_3 reduces below-ground uptake of NO_3^-.

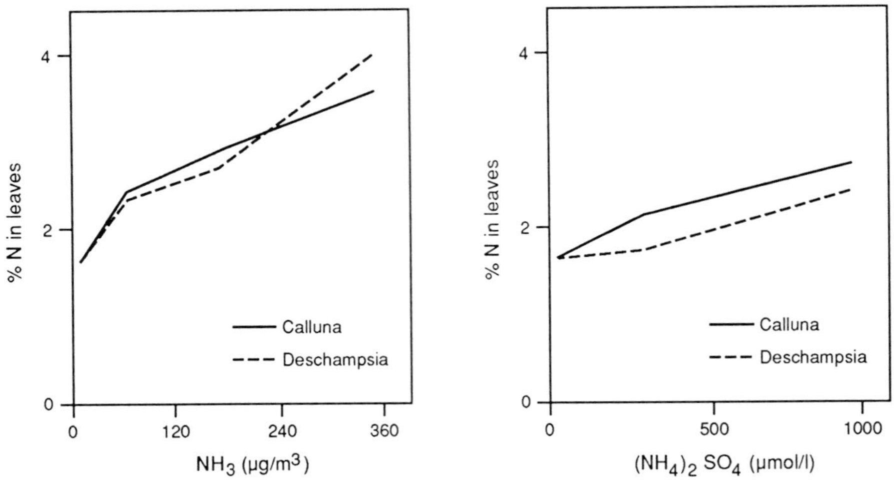

Fig. 6.2 Nitrogen content of the leaves of Calluna and Deschampsia after three months exposure to NH_3 (left) and $(NH_4)_2SO_4$ (right)

Although rather high concentrations of NH_3 and $(NH_4)_2SO_4$ were used, no clear treatment effects on either leaf necrosis or biomass production were found. The tested species appear to be insensitive as the highest NH_3 concentration is well above the threshold for leaf injury of sensitive species (Van der Eerden 1982). Both NH_3 and $(NH_4)_2SO_4$ were taken up by all three species. There appears to be a linear relationship between NH_3 concentration and nitrogen content of the leaves for all species (Figure 6.2). The relationship with $(NH_4)_2SO_4$ is also similar for the three species, but here the uptake levels off at the highest concentration. Linear interpolation to ambient values (10 $\mu g.m^{-3}$ NH_3, 150 $\mu mol.l^{-1}$ $(NH_4)_2SO_4$) suggests an equal effect of both nitrogen sources, but because wind speed in this experiment was low, the boundary layer resistance is high and uptake of NH_3 under field conditions may therefore be underestimated.

6.3 Below-ground nitrogen uptake

To investigate the possibility of below-ground uptake of NH_3 applied above-ground (after dissolution and possibly nitrification), an additional experiment was performed in which Calluna and Deschampsia were potted in heathland soil with and without additional fertilization with $(NH_4)_2SO_4$ (300 mg N kg^{-1} dw) and fumigated with 105 $\mu g.m^{-3}$ NH_3 for four months. Before the start of the experiment half of the pots were covered with activated charcoal in order to prevent NH_3 reaching the soil.

Fumigation raised the total nitrogen amount by a of factor 1.4 in Calluna and 1.8 in Deschampsia. In charcoal-covered soil these factors were 1.3 and 1.5, respectively. In unfumigated plants fertilization raised the total nitrogen amount by a factor of 1.1 in both species, whereas the effect of fertilization was not significant in fumigated plants.

These results indicate that both Calluna and Deschampsia take up NH_3 more rapidly above-ground. A smaller amount of nitrogen is probably taken up by the root system after deposition. However, extrapolation of these results to the field situation is difficult. On the one hand, the NH_3 concentration applied in the experiments is much higher than the ambient concentration, but on the other hand wind speed, and therefore deposition, was low compared to the field situation.

7. RESPONSE OF DOMINANT SPECIES TO SO_2, NH_3 AND $(NH_4)_2SO_4$ IN MIXED CULTURES

The results given in the previous paragraph suggest a more efficient use of extra nitrogen by the grasses, which would ultimately lead to the elimination of Calluna. However, it is not clear to what extent the results of these experiments can be extrapolated to field situations. Moreover, the effect of SO_2 on competition is not known. In order to be able to make a better extrapolation of the laboratory experiments to the field, a number of experiments were made to simulate competition between Calluna and Deschampsia under field conditions at different levels of nitrogen deposition and sulphur dioxide concentration.

7.1 Application of artificial rain supplemented with $(NH_4)_2SO_4$ under a roof

7.1.1 Experiments with potted plants

One-year-old Calluna and Deschampsia were planted in 50 l containers filled with heathland soil (pH 4.5, N 0.085% dw), in such a way that 1/3 of the surface was covered with Calluna, 1/3 was covered with Deschampsia and 1/3 was bare. In half the containers the soil was fertilized with $(NH_4)_2SO_4$ to an N content of 0.137%. The containers were placed outside under a plastic roof, which eliminated all wet deposition and a small part of dry deposition, altogether c. 50% of total deposition (see 10.1). Under this roof artificial rain with $(NH_4)_2SO_4$ concentrations of 20, 100, 200 and 400 $\mu mol.l^{-1}$ was applied weekly. The experiment lasted one year.

After a year both fertilization and nitrogen addition in rainwater had significantly stimulated N uptake and growth in Deschampsia, but for Calluna no such stimulation was found. Because in monoculture experiments a higher nitrogen availability resulted in a higher nitrogen uptake and a higher biomass production of both species, there is apparently an effect of competition. At the end of the experiment the Deschampsia plants were higher than the Calluna plants, and light interception had probably prevented utilization of the extra nitrogen by Calluna.

To further quantify this effect, parameters were estimated for the regression of biomass production on nitrogen addition with rainwater in the unfertilized soil. For the total biomass (Calluna + Deschampsia) the regression equation runs:

$$\text{biomass} = 21.4 + 0.021\ N \qquad (1)$$

in which biomass is recalculated to $t.ha^{-1}$, and N is recalculated to $kg.ha^{-1}.y^{-1}$; all parameters are significant at the 1% level; 30% variance is accounted for. The effect of nitrogen addition with rainwater on total biomass appears to be small but significant: an extra deposition of 40 $kg.ha.^{-1}.y^{-1}$ causes an increase in the standing crop with 4% compared to a situation without extra nitrogen. However, the experiment lasted only one year and the nitrogen can partly be stored in plant tissue or litter and in the long run cause a stronger increase in standing crop.

To estimate the effect of nitrogen deposition on the transition of heathland into grassland a parameter %G was defined as the percentage contribution of grass to total biomass:

$$\%G = D*100 / (C+D) \qquad (2)$$

in which C is Calluna biomass and D is Deschampsia biomass. The equation for the regression of this parameter on nitrogen deposition runs:

$$\%G = 64.1 + 0.21\ N \qquad (3)$$

(all parameters significant at the 1% level; 23% variance accounted for). Substitution of the values 0 and 19 (the estimated wet nitrogen deposition on the Asselse Heide) for N gives the values of c. 64% and c. 68% for %G. Therefore it may be concluded that

(a) the effects of ambient wet deposition levels on the balance between Calluna and Deschampsia is small but significant. However, a longer exposure period or interaction with an additional stress facter might have resulted in a stronger effect;

(b) even a complete elimination of wet deposition of nitrogen is not sufficient to prevent outcrowding of Calluna by Deschampsia in the long run: on the basis of equation (3) %G is expected to increase from c. 50% at the start of the experiment to c. 64% at the end at 0 $\mu mol.l^{-1}$ $(NH_4)_2SO_4$, and a longer exposure period would probably have resulted in a still higher %G. Apparently, dry deposition also has to be reduced to accomplish a dominance of Calluna over Deschampsia.

7.1.2 Field experiment

The effect of different $(NH_4)_2SO_4$ deposition levels on an existing vegetation was assessed through a field experiment on the Asselse Heide. Three plastic roofs were built over the vegetation, and under each roof six plots were made measuring 1.8 * 1.8 m^2. 50% of the Calluna plant in the plots were cut off at soil level, and in each plot cores with Deschampsia were planted between the Calluna, resulting in a 10% grass cover. Artificial rainwater with $(NH_4)_2SO_4$ added in concentrations of 15, 30, 50, 120, 240 and 480 $\mu mol.l^{-1}$ was sprayed over the plots once every 14 days. 2% of the added $(NH_4)_2SO_4$ was labelled with 15N. The experiment lasted 1.5 years (two growing seasons). Soil solution was sampled occasionally using vacuum ceramic cups at 5 and 20 cm depth.

At the end of the experiment Calluna was still strongly dominating at all $(NH_4)_2SO_4$ levels, but the ratio between Deschampsia and Calluna (expressed as %G) was significantly influenced by $(NH_4)_2SO_4$, with a minimum of 7.1 at 25-50 $\mu mol.l^{-1}$ $(NH_4)_2SO_4$, and increasing to 11.1 at the highest $(NH_4)_2SO_4$ level (Figure 7.1). If %G is calculated on the basis of N amount instead of biomass, the minimum at 30 $\mu mol.l^{-1}$ is more pronounced, indicating that at a low $(NH_4)_2SO_4$ level Calluna accumulated more nitrogen than Deschampsia without utilizing it for growth. The minimum in %G indicates that below 60 $\mu mol.l^{-1}$ Calluna made the most efficient use of the available nitrogen, while above 30 $\mu mol.l^{-1}$ Deschampsia was more efficient. This result is in contrast with the previous paragraph, where Deschampsia became dominant at all $(NH_4)_2SO_4$ levels. The explanation for this difference can be found in the initial density of Deschampsia relative to Calluna. In the experiment in 7.1.1 this density was high, resulting in light interception by Deschampsia, while in this experiment Deschampsia had low initial density and light interception by Calluna kept growth of Deschampsia at a low level. Comparable results were obtained by Aerts (1989) for mixtures of Calluna and Molinia.

The pH of soil solutions at the end of the experiment had significantly decreased from c. 4.0 to c. 3.5 at the highest $(NH_4)_2SO_4$ levels. This decrease may be due to (a) nitrification, or (b) replacement of H^+ by NH_4^+ at the absorption complex. NO_3^- concentration in the soil solution was low, which may be due to plant uptake (see also Section 10).

Figure 7.2 shows the fraction of total nitrogen in Calluna and Deschampsia taken up from the artificial rainwater, calculated on the basis of ^{15}N analyses, after 14 and 18 months. The uptake was higher in Deschampsia, although the uptake in the last 4 months relative to the first 14 months was higher in Calluna than in Deschampsia. This confirms earlier findings that the reaction of Calluna to nitrogen starts only slowly (see 6.1).

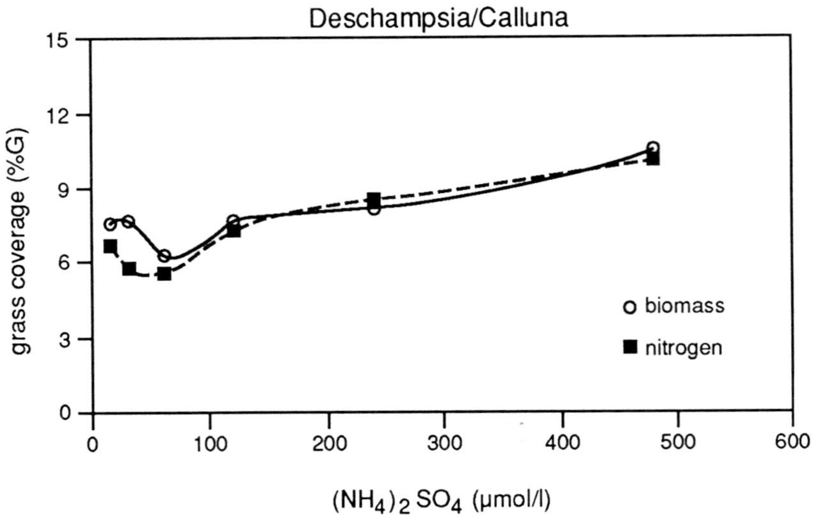

Fig. 7.1 Ratio of Deschampsia biomass relative to Calluna biomass, expressed as
$\%G = 100 * Desch / (Call + Desch)$
after 18 months as a function of $(NH_4)_2SO_4$ concentration in artificial rainwater

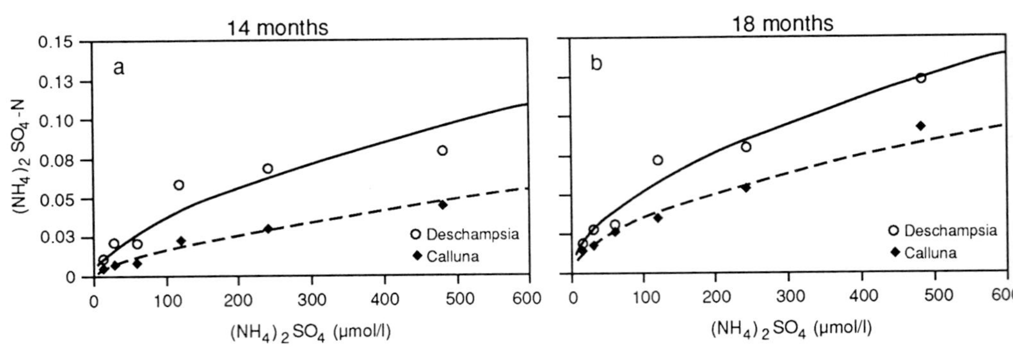

Fig. 7.2 Accumulation of 15N labelled nitrogen (as percentage of total N) in leaves of Calluna and Deschampsia, in a mixed vegetation treated for 14 (left) and 18 months (right) with artificial rain containing $(NH_4)_2SO_4$

The fraction of $(NH_4)_2SO_4$-nitrogen in litter was somewhat lower than the fraction in the leaves, while the fractions in humus and mineral soil were much lower (litter:humus:soil = c. 8:2:1). Apparently the litter has a large capacity to accumulate nitrogen. In humus and soil the fraction of $(NH_4)_2SO_4$-N almost linearly increased with $(NH_4)_2SO_4$ concentration, but in the litter layer the increase levelled off at high $(NH_4)_2SO_4$ concentrations. This may be explained as a dilution effect at high nitrogen availability, causing a stimulation of plant growth and thereby litter production. The fraction of $(NH_4)_2SO_4$-N in the foliage also levelled off at a high concentration (Figure 7.2). However, total uptake of labelled nitrogen appeared to be linearly related to $(NH_4)_2SO_4$ concentration. The non-linearity indicates a loss of nitrogen that is subject to further study. An important conclusion to be drawn from this experiment is that a relatively fast flow of the nitrogen into the litter compartment takes place.

7.2 Effects of SO_2 and drought on competition

To assess the effect of SO_2 on competition mature (c. 3-5 years) and young (<2 years) Calluna plants were potted in various densities, alone or together with Deschampsia or Molinia plants, and exposed to 90 $\mu g.m^{-3}$ SO_2 in a FFS for one year (1988/89, with a very mild winter in the middle of the exposure period). The control plants received the ambient SO_2 concentration at Wageningen, c. 10 $\mu g.m^{-3}$. There were significant effects of competition and plant density on the growth of Calluna, but the effect of SO_2 was not significant. The interactions were not significant either, although there were trends towards less growth inhibition by SO_2 at higher densities for old Calluna, and growth stimulation of Calluna by SO_2 in mixed cultures. It may therefore be concluded that there are no consistent effects of SO_2 on competition of the dominant species. If such effects exist they tend to benefit old Calluna and Calluna in competition with grasses, so that SO_2 without additional stress factors will probably not promote the transition of heathland into grassland.

An extremely dry period in spring 1989 offered the possibility to study the interaction between SO_2 and drought stress. Leaf damage was estimated in eight arbitrary classes ranging from 0 (no visual damage) to 7 (more than 90% of the leaf surface necrotic). Figure 7.3 gives the results, showing a strong increase of leaf damage in fumigated plants, especially in monoculture.

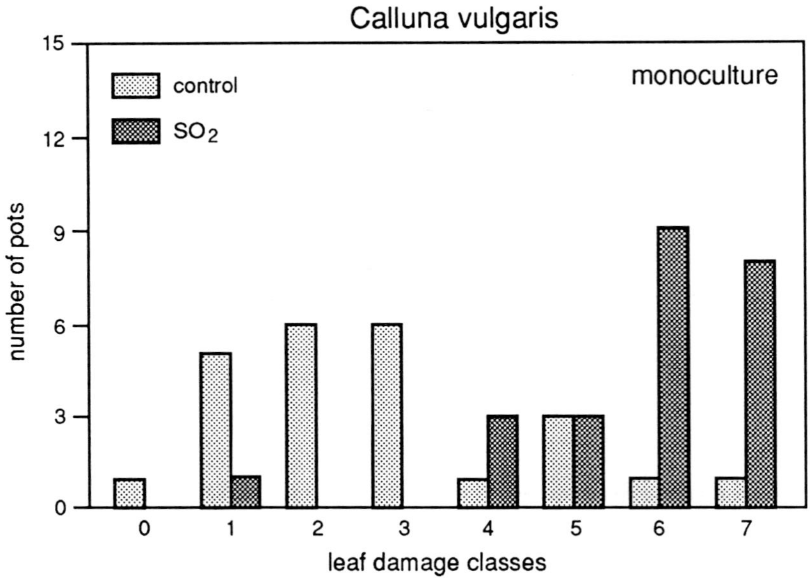

Fig. 7.3 Leaf damage of Calluna following drought and exposure to 10 (control) and 90 µg.m^{-3} SO_2 in monoculture. Damage is estimated in eight arbitrary classes ranging from 0 (no visible damage) to 7 (>90% necrosis)

7.3 Effects of NH_3 and drought on competition

The experiments described in 7.1 support the hypothesis that competition between Calluna and grasses is influenced by nitrogen. In these experiments nitrogen availability was varied through the application of different $(NH_4)_2SO_4$ concentrations in artificial rainwater. However, in section 6 evidence was found for a stronger uptake of gaseous NH_3 than of dissolved $(NH_4)_2SO_4$. Therefore another experiment was carried out in which mixed cultures of Calluna with Molinia and Deschampsia were fumigated with NH_3.

Calluna plants were potted alone or together with Deschampsia or Molinia in various densities. Monocultures of the grasses were also included. They were exposed to filtered air (c. 3 µg.m^{-3} NH_3), ambient air (c. 6 µg.m^{-3} NH_3), or ambient air supplemented with 53 or 105 µg.m^{-3} NH_3, in OTCs for two years.

All three species were significantly stimulated in growth by NH_3; no differences in growth were observed between filtered and ambient air. NH_3 had the strongest effect on shoot growth of Molinia (increased by a factor of 3-4), the effect was less in Deschampsia (increased by a factor of 2-3) and Calluna (increased by a factor of 2). In Figure 7.4 the competition of Calluna with Molinia or Deschampsia is expressed as relative yield (growth with a competing grass relative to growth in a monoculture). In filtered and ambient air, it made no difference whether 2 or 4 grass tussocks competed with Calluna. When exposed to NH_3, however, 4 tussocks of either grass species inhibited Calluna more than 2 tussocks did. The exception, Calluna with Molinia exposed to 53 $\mu g.m^{-3}$ NH_3 may be accounted for by damage to that treatment group by drought in spring 1989.

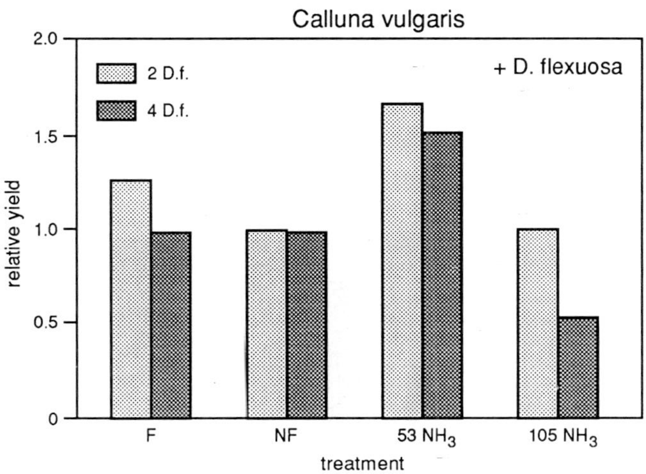

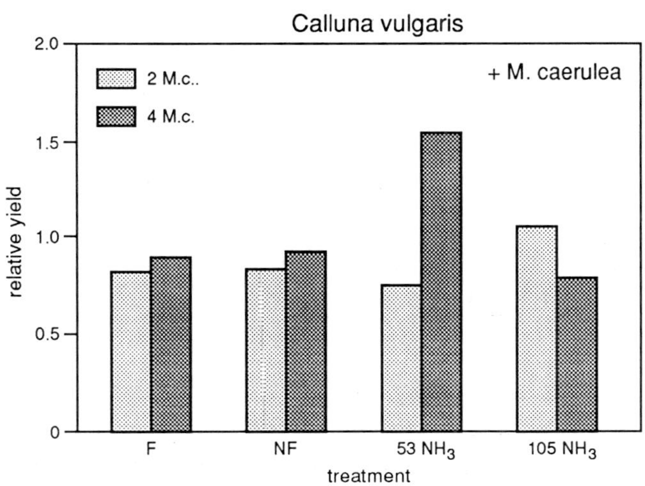

Fig. 7.4 Biomass production of Calluna in competition with Molinia (2 or 4 plants) or Deschampsia (2 or 4 plants), grown in filtered air (F), ambient air (NF) and air supplemented with 53 and 103 $\mu g.m^{-3}$ NH_3

In general, growth and competitive ability of Calluna was highest at 53 $\mu g.m^{-3}$ NH_3, while at 105 $\mu g.m^{-3}$ NH_3 it was equal to or lower than that in ambient air. This is probably also a result of drought damage, which was the strongest in the highest NH_3 treatment. Figure 7.5 shows the drought damage (estimated as in 7.2) in relation to NH_3 concentration. Drought damage strongly increased at high NH_3 concentrations, as it did at high SO_2 concentrations.

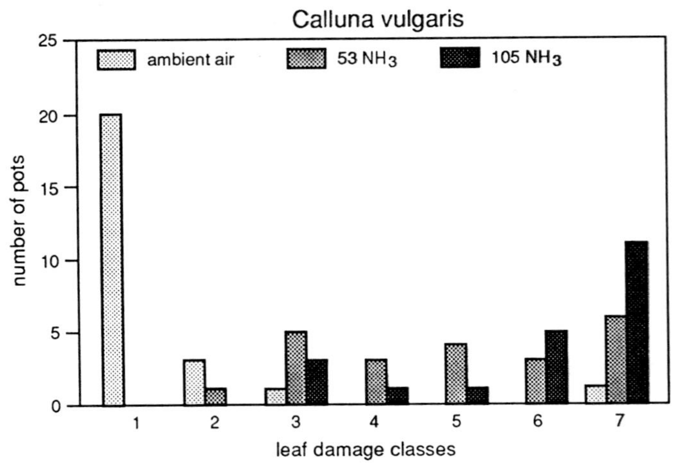

Fig. 7.5 Leaf damage of Calluna exposed to NH_3 during the drought period in spring 1989. Damage classes as in Figure 7.3, treatments as in Figure 7.4

Although the interpretation of this experiment is somewhat hampered by the drought period, the results seem to confirm the results of earlier experiments, in that the combination of a a high nitrogen availability and competition caused a growth reduction in Calluna.

8. RESPONSE OF VIOLION CANINAE SPECIES TO SO_2, NH_3 AND $(NH_4)_2SO_4$ IN MIXED CULTURES

In sections 4 and 5 evidence was found for the sensitivity of Violion caninae species to soil acidification, but there were no apparent effects of high nitrogen levels. However, all experiments were monoculture experiments, and in mixed cultures outcrowding of dicotyledonous species by grasses might occur at high nitrogen levels, analogous to the mechanism found for Calluna and Deschampsia. Therefore experiments were set up with Arnica and Agrostis in mixed culture at different nitrogen levels and sulphur dioxide concentrations. In this case Agrostis was chosen as a competitor rather than Deschampsia or Molinia because it commonly occurs in Violion caninae habitats, and its pH range is closer to that of Arnica than the pH range of Deschampsia or Molinia (Table 5.5).

8.1 Effects of $(NH_4)_2SO_4$ on potted Arnica / Agrostis mixtures

Arnica seedlings were potted in heathland soil (pH 4.4) together with two densities of Agrostis, covering 20% ('mix low') or 80% ('mix high') of the pot area. The pots were placed in the garden and were fertilized with 10, 50 and 90 kg N $ha^{-1}.y^{-1}$ as monthly gifts of $(NH_4)_2SO_4$ dissolved in water. Flowering and seed production of Arnica were recorded, and seed was collected during two years. At the end of the second year both species were harvested. The collected seed was put on moist filter paper at room temperature, and germination was recorded during 35 days.

Detailed results are given by Fennema (1990). The growth of Arnica in a monoculture reached an optimum at 50 kg N $ha^{-1}.y^{-1}$. In mixed culture, however, growth of Arnica was negatively correlated to nitrogen deposition, while growth of Agrostis was positively correlated (Figure 8.1). The total N amount in Arnica was positively correlated to N deposition in a monoculture, but negatively in a mixed culture. In addition, $(NH_4)_2SO_4$ and competition had a strong effect on flowering and seed production of Arnica. In the monoculture flowering was stimulated by $(NH_4)_2SO_4$, but in the mixed culture flowering was inhibited by $(NH_4)_2SO_4$. In many cases the inhibition was even complete (Figure 8.2a). $(NH_4)_2SO_4$ and competition had an analogous effect on seed production, germination rate and germination percentage (Figure 8.2b).

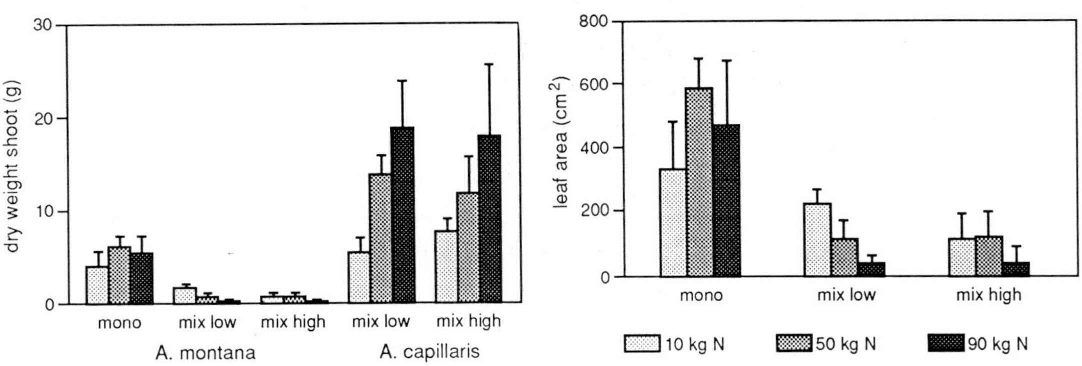

Fig. 8.1 Effects of additional $(NH_4)_2SO_4$ deposition in artificial rainwater, and competition with Agrostis on shoot dry weight (left) and leaf area (right) of Arnica. Mono = monoculture, mix low = initial density of Agrostis 20% pot area, mix high = initial density of Agrostis 80% pot area

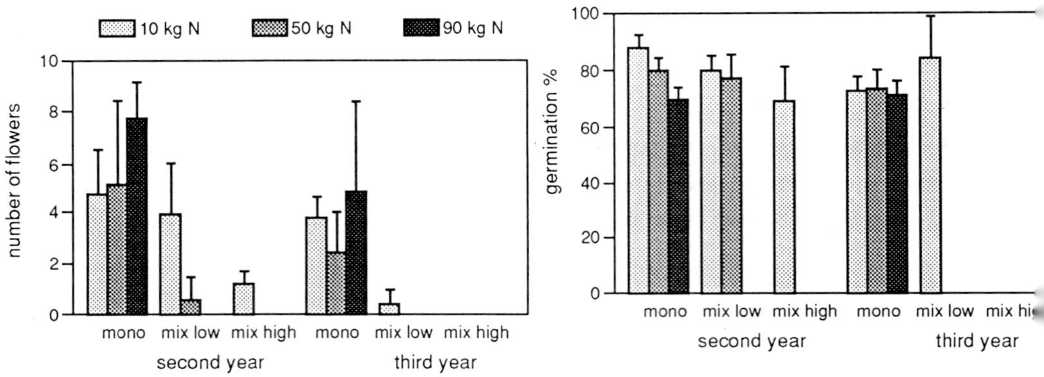

Fig. 8.2 Effects of $(NH_4)_2SO_4$ deposition in artificial rainwater and competition with Agrostis on flower production (left) and cumulative germination percentage of the harvested seed (right) of Arnica

These results indicate that if nitrogen availability is high, Agrostis is a much stronger competitor than Arnica. Because the effects on flowering and seed production were strongest, reproduction will probably be inhibited at nitrogen deposition levels that still permit the occurrence of Arnica.

8.2 Effects of vegetation density on germination and establishment of Arnica

As Arnica is easily crowded out by grasses at a high nitrogen availability, germination and establishment of this species can probably only take place in micro-habitats that provide sufficient open space. To test this hypothesis a field experiment was carried out on four locations, two in the Netherlands and two in Denmark. All locations belonged to the Violion caninae and were mown once in two years. In each location five plots were randomly selected, and the following subplots were made: (i) bare soil, (ii) artificial moss patch (cover 100%), (iii) artificial litter layer (thickness 0.5 cm) and (iv) undisturbed vegetation (cover 85%). In each subplot Arnica seeds were sown, and seedling emergence was recorded. Results are given in Figure 8.3, for details see Fennema (1990). Both emergence and survival appeared to be significantly lower in undisturbed (=dense) vegetation. Other treatment effects were not significant, neither was the difference between the Netherlands and Denmark. The observation that Arnica cannot establish in a dense vegetation is in accordance with e.g. Grubb (1977).

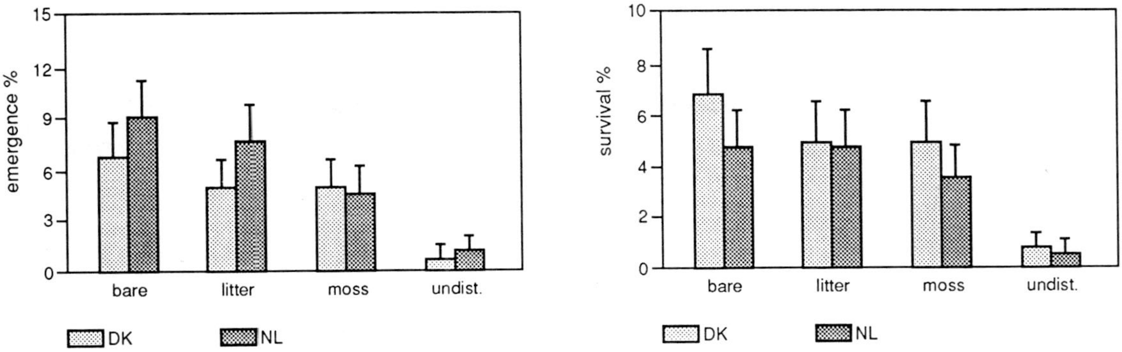

Figure 8.3 Effects of micro-habitat and locality (NL = the Netherlands, DK = Denmark) on seedling emergence (left) and survival (right) of Arnica

As Arnica seed requires no dormancy breaking and has a relatively high germination percentage (Pegtel 1988 and Figure 3.2), it needs long-distance dispersal mechanisms. Although no specific dispersal experiments were performed, seedlings were found at a maximum of 6 m from the parent plants both in the field and in the experimental garden. Therefore Arnica can probably only establish in an existing vegetation if (a) micro-habitats with a low density of phanerogamic species are available, and (b) a reproducing population is near. As nitrogen deposition both leads to a denser vegetation of competitors, and has a negative effect on seed production, it will probably hamper establishment of Arnica.

8.3 Effect of SO_2 on competition of Violion caninae species

Because Arnica proved to be sensitive to competition with Agrostis, and grasses generally are less sensitive to SO_2 than dicotyledonous species, SO_2 is likely to affect competition between Violion caninae species and grasses. To test this hypothesis an experiment was performed in which mixed cultures of Violion caninae species and either Agrostis or Nardus stricta were exposed to SO_2.

Arnica montana, Hieracium pilosella and Gentiana pneumonanthe (4 weeks old) were potted in heathland topsoil together with small Agrostis or Nardus plants. In addition, monocultures of the grasses and the dicotyledonous species were potted. The pots were placed in the FFS and exposed to ambient air (c. 10 $\mu g.m^{-3}$ SO_2), or air with 90 $\mu g.m^{-3}$ SO_2. After one year the plants were harvested.

Shoot growth of the grasses appeared to be unaffected by SO_2, but shoot growth of the dicotyledonous species was reduced, up to 40% for Arnica. Root growth was unaffected by SO_2. The effect of SO_2 on relative yield of mixed cultures of Gentiana and Arnica is shown in Figure 8.4. Growth of Gentiana was much more inhibited by competition with either grass than by SO_2. For this species no interaction was found between competition and exposure to SO_2. In Arnica there was a strong interaction: relative yield was depressed by c. 40% in the mixed cultures when exposed to SO_2. Agrostis proved to be a much stronger competitor than Nardus.

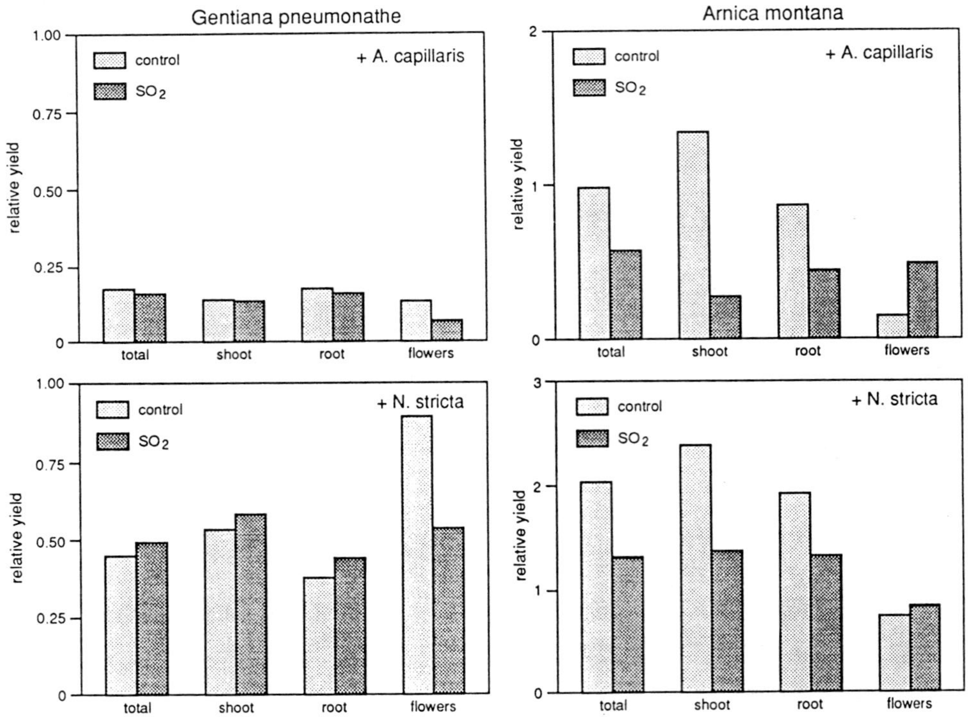

Figure 8.4 Effect of SO_2 (10 / 90 $\mu g.m^{-3}$) and competition (Agrostis / Nardus) on the biomass production of Arnica and Gentiana, relative to biomass production in monoculture

The results of these experiments confirm the earlier hypotheses that (a) Arnica is more sensitive to SO_2 than Gentiana (Table 1.3), and (b) effects of SO_2 on Violion caninae species may come about through changes in competitive strength (section 4).

8.4 Effect of combinations of SO_2 and NH_3 on competition of Violion caninae species

In the previous paragraphs both $(NH_4)_2SO_4$ and SO_2 appeared to affect competition between Violion caninae species and grasses. To investigate a possible interaction between nitrogen and SO_2 an experiment was performed in which mixed cultures of Arnica and Viola with Agrostis were fumigated with SO_2, NH_3 and their combination.

Arnica and Viola were potted in heathland topsoil, together with small Agrostis plants. Monocultures of Agrostis, Arnica and Viola were also potted. The pots were exposed to ambient air, and ambient air supplemented with 53 $\mu g.m^{-3}$ NH_3, 90 $\mu g.m^{-3}$ SO_2 and the combination in OTCs. After a year the plants were harvested.

In monocultures NH_3 generally appeared to stimulate both shoot and root growth; the strongest stimulation was found for Agrostis. On the other hand SO_2 generally inhibited growth, although in none of the species a very strong inhibition was found. Figure 8.5 gives the results of the experiments with mixed cultures. Here a strong growth inhibition was found for Arnica, especially in the presence of NH_3, which agrees with results obtained in other experiments (8.1, 8.2). Viola proved to be a stronger competitor than Arnica; its growth reduction in mixed culture in ambient air was negligible, and a slight growth reduction was found in the presence of SO_2 or NH_3. No indications were found for an interaction of SO_2, NH_3 and competition. The growth of Arnica in a mixed culture in the presence of SO_2 and NH_3 was about equal to the growth of the control plants, while the growth of Viola in SO_2 + NH_3 even exceeded that of the control plants.

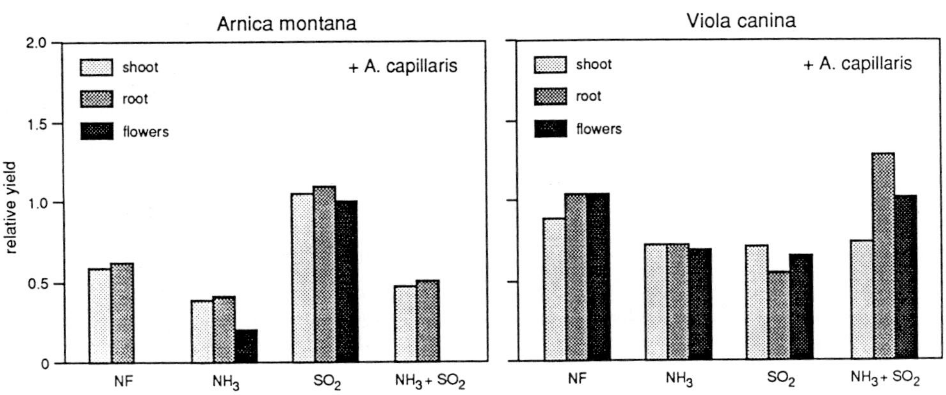

Figure 8.5 Effects of ambient air (NF), 53 $\mu g.m^{-3}$ NH_3, 90 $\mu g.m^{-3}$ SO_2 and the combination on the biomass production in mixed cultures of Arnica and Viola with Agrostis relative to monocultures

9. MULTIPLE STRESS EFFECTS ON CALLUNA VULGARIS

The experiments with mixed cultures showed that a closed canopy of Calluna can compete well with grasses, even under high levels of nitrogen deposition. Therefore the transition of heathland into grassland must be initiated by some secondary stress factor. The most obvious of these stress factors are frost, drought and plagues of the heather beetle Lochmaea suturalis Thomson. To get some insight in the probability of occurrence and the effects of these factors, and their interaction with the primary stress factors SO_2, NH_3 and $(NH_4)_2SO_4$, some additional measurements were made in the experiments described in sections 7.1, 8.2 and 8.3.

9.1 Frost sensitivity

Frost sensitivity was estimated as electrolyte leakage from leaves (attached to the shoot) in demi-water after short-term exposure to low temperatures. These measurements were carried out with plants taken from the experiments described in 7.1.1, 8.2 and 8.3, and an additional field experiment in which Calluna plots were fertilized with 10, 50 and 90 kg N $ha^{-1}.y^{-1}$. Results are summarized in Table 9.1. Significant interactions were found between frost sensitivity and treatments with both SO_2, NH_3 and $(NH_4)_2SO_4$, but there was no consistency in the season in which the effects occurred. Therefore both delayed winter hardening and premature dormancy breaking may take place. The effects of NH_3 and $(NH_4)_2SO_4$ seem to be most relevant, effects of SO_2 were only significant at temperatures that rarely occur in the Netherlands.

Table 9.1 Interaction between frost sensitivity of Calluna and treatments with SO_2, NH_3 and $(NH_4)_2SO_4$. Treatments lasted c. 1-2 years. Figures are highest temperatures (in ^oC) at which a significant ($p<0.05$) interaction occurs, n.s. = no significant interaction at -26^oC or above

treatment	measurement period		
	February	September	November
SO_2 (90 $\mu g.m^{-3}$)	-26	n.s.	n.s.
NH_3 (75 $\mu g.m^{-3}$)	-12	n.s.	n.s
NH_4 rain exp. (90 kg N $ha^{-1}.y^{-1}$)	n.s.	n.s.	n.s.
NH_4 fert. exp. (90 kg N $ha^{-1}.y^{-1}$)	n.s.	n.s.	-10

9.2 Drought sensitivity

Drought sensitivity was measured by (a) visual assessment of leaf damage after the drought period of spring 1989 in the experiments described in 7.2 and 7.3 (results given in these sections) and in the additional fertilization experiment (result given in Table 9.2); and (b) pressure-bomb measurements, in which a gradually increasing air pressure is applied to a shoot section, and the xylem water potential is assumed to be the negative value of the pressure at which exudation starts. These measurements were carried out in the artificial raining experiment (section 7.1.2) during the drought period, just before the application of rainwater (i.e., after two weeks drought), and in an additional experiment in a PGC after two weeks of fumigation with 100 $\mu g.m^{-3}$ NH_3 followed by 10 day drought + NH_3. Results are summarized in Table 9.3. Both SO_2, NH_3 and $(NH_4)_2SO_4$ appeared to have significant effects on drought sensitivity. Although relatively high concentrations were used in these experiments, the effects are so strong that relevant effects are likely to occur at ambient concentrations.

Table 9.2 Effects of the drought period in spring 1989 on leaf damage in fertilized Calluna plots. (means of 16 replicates; all differences are significant)

fertilization ($kg\ N\ ha^{-1}.y^{-1}$)	leaf damage (%)
10	17
50	54
90	90

Table 9.3 Interaction between drought sensitivity of Calluna and treatments with SO_2, NH_3 and $(NH_4)_2SO_4$. Treatments lasted c. 1-2 years except for NH_3 (2 weeks, see text). * = significant ($p<0.05$) interaction, n.d. = not determined

Treatment	method	
	leaf damage	xylem water potential
SO_2 (90 $\mu g.m^{-3}$)	*	n.d.
NH_3 (100 $\mu g.m^{-3}$)	n.d.	*
NH_4 (50 $\mu mol.l-1$)	*	*

9.3 Heather beetle plagues

Although the effects of nitrogen on the heather beetle was not an original aim of the project, an outbreak of a heather beetle plague on the Asselse Heide in August 1989 provided an opportunity for some additional experiments.

(1) Lochmaea larvae were collected from all plots of the artificial raining experiment, and numbers of larvae in the second and third stage of development were counted. There appeared to be a trend towards a higher percentage of third-stage larvae in treatments

with high $(NH_4)_2SO_4$ dosages.

(2) Lochmaea larvae were put on Calluna branches taken from plants that had been fumigated with NH_3 for 12 months in an OTC. After a week the larvae were counted and weighed. Both growth and development (expressed as the percentage of larvae in the third stage) significantly increased with increasing NH_3 concentration applied to the Calluna before the start of the experiment.

These results are in agreement with Heil & Brunsting (1985), who found both a faster development and a higher adult weight of Lochmaea fed on N-fertilized Calluna compared to unfertilized Calluna. Here again above-ground uptake of gaseous NH_3 appears to have an effect comparable to that of below-ground uptake of NH_4^+.

10. NITROGEN DYNAMICS OF HEATHLAND

Literature data and the experiments with mixed cultures described in section 7 showed that nitrogen availability is a crucial factor in determining the transition of heathland into grassland. In principle, two sources of nitrogen exist: (a) atmospheric deposition, and (b) mineralization of organic matter. In heathland, where legumes are absent or scarce, nitrogen fixation is probably negligible. In a growing ecosystem nitrogen is stored in both living biomass and dead organic matter. If organic matter accumulates, its importance as a potential source for nitrogen will increase with time. However, not all deposited nitrogen is necessarily stored in the system; excess nitrogen that is not taken up may be leached as NH_4^+ or NO_3^-. In the following section the flow and storage of nitrogen in heathland ecosystems will be discussed.

10.1 Deposition

Deposition of nitrogen on the Asselse Heide was measured by two independent methods:

(1) micrometeorology: From August 1987 through July 1988 the NH_3 concentration was measured by the denuder method together with relevant meteorological parameters according to a measurement strategy aimed to derive the yearly averaged deposition flux. Micrometeorological characteristics of the site were investigated in two campaigns. During these campaigns direct dry deposition measurements of NH_3, NO_x and SO_2 were carried out. Using the results of these campaigns and concentration fields derived from the National Monitoring Network, the deposition of NO_x, SO_2 and the wet deposition of NH_3 were calculated by the inference method. The dry deposition of NH_3 was calculated from the denuder measurements. This

study, which is described in detail by Duyzer et al. (1989) yielded a mean NH_3 concentration of 2.2 $\mu g.m^{-3}$ for the period August 1987 through July 1988, and an estimated nitrogen deposition of 41 kg N $ha^{-1}.y^{-1}$ in 1987 (wet 18, dry 23; NH_x 21, NO_x 20 kg N $ha^{-1}.y^{-1}$).

(2) throughflow deposition: throughfall and stemflow (=throughflow) were continuously sampled below the Calluna vegetation using specially designed samplers. The difference in deposition between throughflow and bulk deposition (=deposition in open rainwater collectors) can be used to estimate dry deposition. This method is described in detail by Bobbink et al. (1990), its outline is as follows: dry deposition is equal to throughflow deposition minus bulk deposition, minus plant leaching or plus plant uptake. Comparison of throughflow under living Calluna and under polythene model Calluna showed that (a) uptake or leaching of sulphate and probably also nitrate is negligible, (b) co-deposition of SO_2 and NH_3 takes place, so that their dry deposition is about equal (on charge basis), and (c) a large part of dry deposited NH_x is taken up by living Calluna. Therefore dry deposition of sulphate and nitrate was estimated as throughflow minus bulk deposition, and dry deposition of NHx was assumed to be equal to dry deposition of SO_x. This yielded an estimated nitrogen deposition of 34 kg N $ha^{-1}.y^{-1}$ for the period October 1988 through July 1989 (wet 19, dry 15; NH_x 28, NO_x 6 kg N $ha^{-1}.y^{-1}$).

Although the two methods show a reasonable agreement as far as total nitrogen deposition is concerned, the estimated values for the dry deposition of NO_x are very different (micromet.: 15 kg N $ha^{-1}.y^{-1}$ (including deposition of HNO_3), throughflow: 0.8 kg N $ha^{-1}.y^{-1}$). Further research would be needed to explain these differences. Measurements in another Calluna site ('Elspeetse Veld') using the throughflow method yielded a value of 44 kg N $ha^{-1}.y^{-1}$. The difference between this site and the Asselse Heide can be accounted for by a higher dry deposition of NO_x, probably caused by a higher traffic intensity. In the following sections an estimated average deposition of 35 kg N $ha^{-1}.y^{-1}$ will be used. Above-ground uptake of nitrogen estimated by the throughflow method amounts to c. 13 kg N $ha^{-1}.y^{-1}$. Dry deposition of SO_x is estimated as 16 kg S $ha^{-1}.y^{-1}$ by the throughflow method, and 20 kg S $ha^{-1}.y^{-1}$ by the micrometeorological method.

Measurements of dry deposition under the plastic roof described in section 7, using the throughfall method, showed that dry deposition of nitrogen under the roof is almost equal to dry deposition in the open field (13 kg N $ha^{-1}.y^{-1}$; NH_x 12, NOx 1 kg N $ha^{-1}.y^{-1}$; Bobbink et al. 1990).

10.2 Nitrogen storage in biomass and litter

Berendse (1988) gives estimates of total nitrogen storage in humus + litter + living biomass in various heathland ecosystems. Regression of the total nitrogen amount on the number of years after sod-cutting gives an estimated annual nitrogen storage of c. 33 kg N $ha^{-1}.y^{-1}$. Although this value agrees well with the nitrogen deposition estimated in the previous section, it may be an under-estimate as the highest total nitrogen amount reported by Berendse (c. 1000 kg N ha^{-1}) is well below the mean value for total nitrogen in humus + litter given by De Boer (1989) (mean over 12 Calluna stands: 1870 kg N ha^{-1}). Anyhow a considerable accumulation of nitrogen takes place, and nitrogen losses are probably small. In dry heathland, where aerobic soil conditions prevail, denitrification is probably negligible, and nitrogen leaching was found to be negligible in two heathland sites (see 10.3).

Heil (1984) estimated a biomass production of young (<10 y) Calluna of c. 2.4 $t.ha^{-1}.y^{-1}$, and a total nitrogen storage in living biomass of c. 15-20 kg N $ha^{-1}.y^{-1}$. Because above-ground uptake of dry deposited nitrogen was estimated as 13 kg N $ha^{-1}.y^{-1}$, Calluna can satisfy a large part of its nitrogen requirement through above-ground uptake, which is in agreement with earlier findings. From data on litter and humus accumulation given by Berendse (1988) and a mean nitrogen content in soil organic matter of c. 1.2% (De Boer 1989), a mean nitrogen flow into litter + humus of c. 11 kg N $ha^{-1}.y^{-1}$ can be derived. Therefore nitrogen is probably about equally divided over dead and living biomass in young Calluna stands, but in older stands nitrogen storage in living biomass levels off, while litter production increases (Berendse 1988). This means that in mature heathland mineralization becomes increasingly important as a source of plant-available nitrogen.

De Boer (1989) measured nitrogen mineralization of undisturbed heathland soil cores after 100 days incubation at 20°C, and found values of c. 250 kg N $ha^{-1}.y^{-1}$. Actual values will be far lower, however. Berendse (1988) reported values of c. 10 kg N $ha^{-1}.y^{-1}$ for young Calluna stands, c. 60 kg N $ha^{-1}.y^{-1}$ for old Calluna stands and up to 150 kg N $ha^{-1}.y^{-1}$ for Molinia and Erica stands (see also Berendse et al. 1987). In a stable situation, nitrogen mineralization cannot exceed nitrogen deposition, but vegetation changes (e.g. increasing dominance of grasses) can probably induce changes in the mineralization rate. As ericaceous species usually have a high lignin content, mineralization of their litter might be slower than mineralization of grass litter. Such a difference would explain the higher nitrate content found in soils under Molinia and Deschampsia dominated vegetation compared to soil under

Calluna or Erica (de Boer 1989, Houdijk 1990, Table 5.1). Literature data are not quite consistent on this point, but there are indications for a more rapid mineralization of grass litter compared to ericaceous litter. In a litterbag experiment Berendse (1986) found a weight loss of Molinia litter that far exceeded the weight loss of Erica litter, but in an in-situ incubation experiment equal nitrogen mineralization rates were found for Molinia and Erica litter (Berendse et al. 1987). De Boer (1989), in a laboratory incubation experiment at 20°C found equal nitrogen mineralization rates for Calluna, Erica and Molinia litter, but a c. 30% higher value for Deschampsia. M.van Vuuren in a litterbag experiment found mineralization rates in the order Calluna $\leq$ Molinia << Deschampsia (see Van der Eerden et al. 1990).

De Boer (1989) studied nitrification in heathland dominated by either Calluna, Erica, Deschampsia or Molinia, and found that in these soils nitrification commonly occurs. Nitrification in heathland is probably only limited by NH_4^+ availability and oxygen supply. Nitrifying bacteria extracted from heathland soils showed an exponential growth even down to pH 3.5. Although the assumption of a complete nitrification commonly used in calculations of acid load is generally realistic for heathland soils, ammonia cannot be considered an acidifying component in Calluna or Erica dominated sites as all deposited NH_4 is taken up by the vegetation.

10.3 Nutrient leaching

Soil solutions were collected at Asselse Heide using vacuum ceramic cup lysimeters at 5, 20 and 30 cm depth. The complex hydrology caused by the iron pan at 35 cm depth did not allow an estimation of nutrient fluxes, however. Therefore only preliminary conclusions are possible on the basis of concentration data (Van der Maas 1990).

Very low NH_4^+ concentrations were found at 30 cm depth, with a sudden rise after the Lochmaea outbreak in August 1989. This indicates a nearly complete uptake of nitrogen by a healthy vegetation, and a reduced uptake and/or increased mineralization as a result of damage by Lochmaea. Also potassium concentrations strongly increased after the outbreak, probably due to increased plant leaching. Nitrification is probably not very strong as NH_4^+ concentration at 30 cm depth far exceeded NO_3^- concentration. The low nitrogen leaching under healthy vegetation is in agreement with findings of Van Dobben et al. (1990).

11. DISCUSSION

To evaluate the effects of SO_2, NH_3 and $(NH_4)_2SO_4$ on heathland, two complementary approaches can be used, one based on the effects on individual species, aiming at critical levels (concentrations), and the other based on the effects on the ecosystem (including soil and dead organic matter), aiming at critical loads (depositions). Because effects of NH_3 and $(NH_4)_2SO_4$ come about through the same mechanism (higher nitrogen availability), these two pollutants will be taken together, and first the effects of sulphur and nitrogen compounds will be evaluated using both approaches. Finally, the relative importance of S and N compounds will be estimated.

11.1 SO_2

For SO_2 the critical level approach yielded clear results: the dominant species are rather insensitive to SO_2 and are probably only affected in the seedling stage, while the Violion caninae species and bryophytes are much more sensitive and effects at ambient concentrations are possible. A critical level of c. 8 $\mu g.m^{-3}$ was calculated, to protect 95% of the species from adverse effects, which may be visible injury, growth reduction or a decreased root:shoot ratio. However, it is difficult to determine the relevance of these effects in the field situation. Competition is probably very sensitive to the growth parameters on which the critical levels for the individual species are based. Therefore the most likely effects in the field are changes in the abundance of sensitive species that are growing together with less sensitive species. An additional uncertainty in the extrapolation to the field situation is due to the fact that the critical levels for the individual species were determined in fumigation experiments that do not take into account the temporal variation in pollution level and other environmental factors. At present it is not possible to quantify these uncertainties.

Deposition of SO_2 will lead to acidification and a higher sulphate availability. Sulphur deficiency does not occur in Dutch agricultural soils (Mes & Smilde 1970) and an additional experiment showed no effect of sulphate on Agrostis. Therefore effects of sulphate probably do not occur, and the ecosystem effects of SO_2 are acidification effects. If nitrification is complete and all depositied nitrogen is leached to the groundwater (which may be the case in degraded, grass-dominated heathland), total acid load equals the potential acid load which can be calculated by adding sea-salt corrected SO_x, NO_x and NH_y deposition together (on an charge basis). For the Asselse Heide this amounts to c. 4000 mol H^+ $ha^{-1}.y^{-1}$. If alle deposited nitrogen is taken up by the vegetation (which is probably the case in vital, Calluna or Erica dominated heathland), total acid load is the sum of sea-salt corrected SO_x and NO_x deposition, for the Asselse Heide c. 2000 mol H^+ $ha^{-1}.y^{-1}$. A given acid load will cause a decrease in soil pH at a rate which depends on nitrification, uptake of nitrogen in biomass and the buffer capacity of the soil. Acidification effects on dominant species are not expected at pH values down to pH c. 3.0 (which are not likely to occur), but a decrease in soil pH below c. 4.5 will lead to a gradual disappearance of Violion caninae species. In principle pH changes due to the deposition of acidifying compounds can be simulated by model calculations, although this was not attempted here. Violion caninae species usually occur in sites with soil properties different from the mean heathland soil (higher pH, CEC and base cation content), and characterization of these soils would require extra study. Although quantification is not possible, both field and laboratory studies have made it highly probable that the deposition of acidifying compounds has strongly contributed to the decline of the Violion caninae species. The mechanism behind the sensitivity of these species to soil acidity is not quite clear, however. Aluminium toxicity probably does not occur, but a lower cation availability combined with an inhibition of cation uptake (possibly also due to a high NH_4/NO_3 ratio) may be key factors.

11.2 NH_3 and $(NH_4)_2SO_4$

Above-ground nitrogen uptake proved to be very efficient and linearly related to concentration (at least for NH_3), in a range from zero up to 400 $\mu g.m^{-3}$ NH_3 or 200 $\mu mol.l^{-1}$ $(NH_4)_2SO_4$. At the ambient NH_3 concentration in the Netherlands, Calluna probably satisfies most of its nitrogen requirement through above-ground uptake. Only when a thick litter layer has been formed mineralization of organic nitrogen may become the dominant N source. The effect of nitrogen on a monoculture is always a growth stimulation, and as nitrogen is the most important growth-limiting factor in heathland ecosystems (Berendse 1988), even the smallest addition of nitrogen will have a stimulating effect. This means that no critical level exists for growth stimulation by NH_3.

The secondary effects of the growth stimulation can be divided into two groups: physiological effects and ecological effects. The physiological effects are metabolic or morphogenetic changes that for species adapted to nutrient-poor conditions will generally be harmful. Such effects might be (1) a lower root:shoot ratio, resulting in enhanced drought sensitivity, (2) a lower level of carbohydrate reserves, resulting in an enhanced frost sensitivity, or (3) a higher amino acid or protein content, resulting in a higher incidence of plagues. Experiments showed that all these effects are probably important in the field situation, but because the mechanism behind them was not quantified it is not possible to give exact concentrations at which such effects would become detectable. Most probably these concentrations are between 1 and 10 $\mu g.m^{-3}$ NH_3.

Ecological effects of nitrogen compounds come about through competition; an increase in nitrogen availability will in the long run lead to a dominance of the species that most efficiently use the extra nitrogen for their growth. In heathland ecosystems the grasses are the most efficient N-users. However, outcrowding of Calluna by grasses will not occur if it forms a closed canopy, even at nitrogen deposition levels far above the ambient level in the Netherlands; light interception will then prevent a strong growth of the grasses. Only if gaps are formed in the Calluna canopy, grasses will start to expand, and Calluna will eventually be outcrowded. The opening of the Calluna canopy can be caused by either frost, drought or Lochmaea plagues. Experiments showed that the critical nitrogen deposition level for the outcrowding of an open Calluna canopy by grasses must be below 20 kg N $ha^{-1}.y^{-1}$; a tentative quantification using model calculations by Heil (1991) leads to a value of c. 10-15 kg N $ha^{-1}.y^{-1}$ for the Calluna / Deschampsia system (Figure 11.1). The model developed by Berendse et al. (1988) for the Erica / Molinia system gives a critical load of c. 20 kg N $ha^{-1}.y^{-1}$.

Because inland heath is a man-made ecosystem, critical loads should be evaluated together with management practices. Even when nitrogen deposition is low, accumulated organic matter will become an increasingly important source for nitrogen in the absence of management, and thus favour the growth of grasses and the outbreak of Lochmaea plagues. Once the transition of heathland into grassland starts it may be a self-accelerating process, because decomposition of grass litter is faster than decomposition of Calluna litter. Lochmaea attacks not only kill Calluna, but also produce large amounts of easily degradable organic matter (excrements, dead larvae, plant particles).

It is clear that not only a decrease in nitrogen deposition, but also management aiming at the removal of nitrogen will decrease the probability of transition into grassland. Both grazing, mowing and sod-cutting could be used to remove nitrogen. However, grazing is not very effective (Berendse 1988), although it may favour a diversity in soil characteristics and thus promote the occurrence of Violion caninae species. After mowing regeneration of Calluna older than c. 15 years is very slow (Berdowski & Siepel 1988) and even slower at high N levels, while grasses probably regenerate faster. Thus the only effective measure is sod-cutting, which will remove c. 800 kg N ha^{-1} (Van der Zande et al. 1988). After sod-cutting Calluna will regenerate from seed. A higher critical load can be allowed if sod-cutting takes place more frequently, although a high frequency would result in a reduced floristic and faunistic diversity. A tentative quantification for the Erica / Molinia system using Berendse's (1988) model gives a critical load of c. 5-10 kg N $ha^{-1}.y^{-1}$ if sod-cutting takes place once every 50 years. At a sod-cutting frequency of 20 y^{-1} a critical load of 20 kg N $ha^{-1}.y^{-1}$ is permissible. Grazing and/or a still higher cutting frequency (up to 10 y^{-1}) would permit a deposition of c. 30 kg N $ha^{-1}.y^{-1}$.

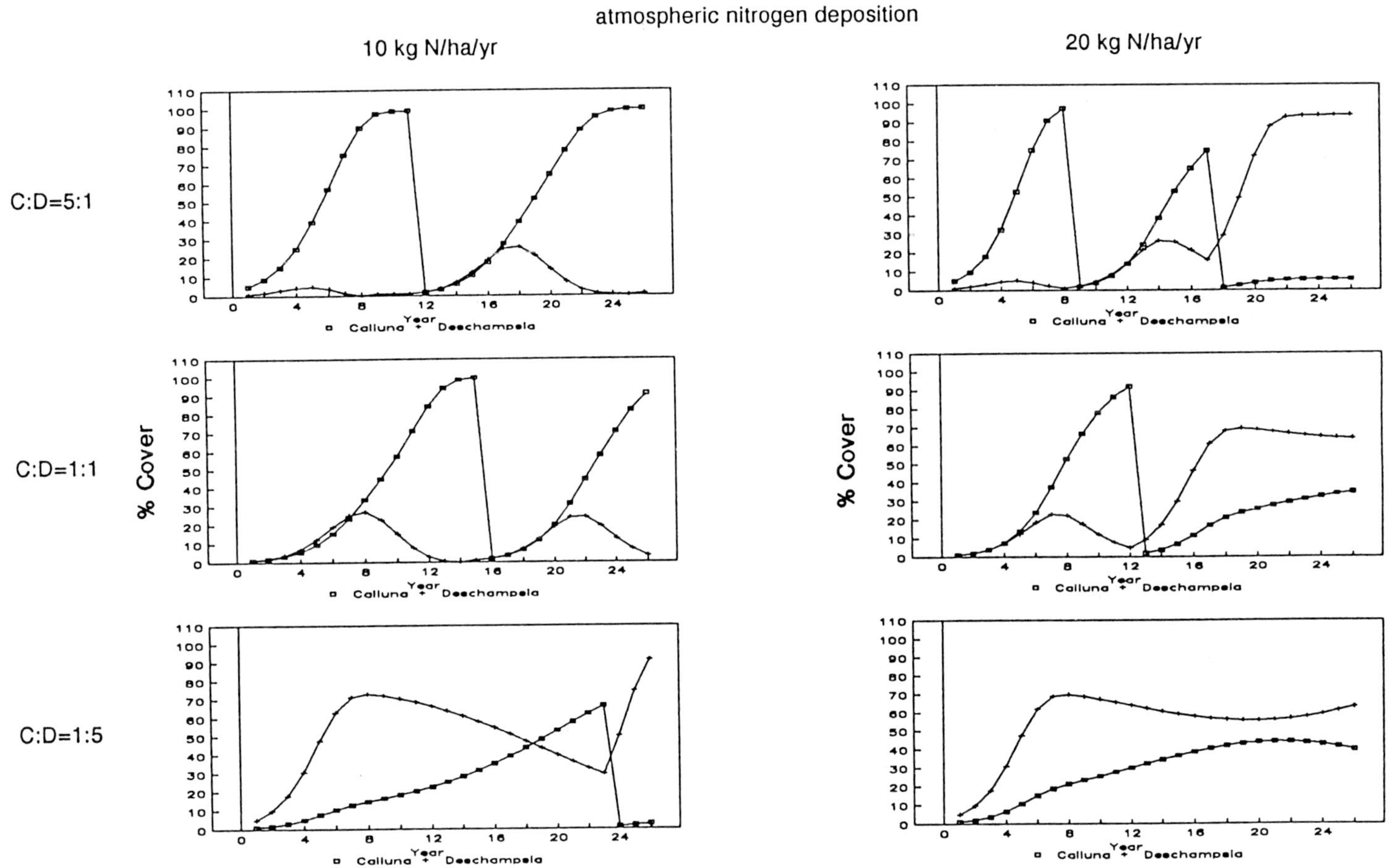

Fig. 11.1 Results of simulation model by G.W. Heil on Calluna and Deschampsia. C:D is ratio between Calluna and Deschampsia at the beginning of the simulation. Lochmaea plagues are randomly generated, with a probability dependent upon nitrogen availability

High nitrogen loads may also lead to the outcrowding of Violion caninae species by grasses. Fertilization with 50 kg N $ha^{-1}.y^{-1}$ had a strong negative effect on growth, flowering and seed production of Arnica in mixed cultures with Agrostis, and would in the long run probably lead to a complete outcrowding of Arnica. However, experiments with Viola showed that this species is a stronger competitor than Arnica. Furthermore, Arnica performed well in competition with Agrostis after fertilization with 10 kg N $ha^{-1}.y^{-1}$. As the fertilization experiments were made in the open air, the N doses were additional over the ambient deposition levels, and therefore ambient levels are probably not high enough to cause a complete outcrowding, although effects on reproduction are possible. This hypothesis is supported by the fact that the occurrence of Arnica in the field had no correlation with soil nitrogen content. It should be stressed, however, that these effects were only comprehensively studied for Arnica, while a limited number of experiments was made for other species. Therefore no definitive conclusions on the effects of nitrogen deposition on the Violion caninae are possible, but probably nitrogen deposition has not strongly contributed to its decline.

11.3 Evaluation

To compare the effects of SO_2, NH_3 and $(NH_4)_2SO_4$ a distinction should be made in the effects of gaseous SO_2 (detrimental to Violion caninae species), the effects of deposition of nitrogen compounds (which promote the transition of heathland into grassland), and the effects of soil acidification (caused by deposition of SO_2, NH_3 and NO_x; detrimental to Violion caninae species). Once heathland has become grassland the Violion caninae species will disappear, but these species may have disappeared before grass became dominant. As SO_2 concentrations have decreased with time (Van Dobben 1990) and concentrations of nitrogen compounds have increased with time (Asman et al. 1987), it may be tentatively concluded that in the past SO_2 has been an important factor for the decline of the Violion caninae, while at present NH_3 is the most important factor, causing a complete degradation of the heathland ecosystem.

A decrease in nitrogen deposition may lead to a relatively fast recovery of heathlands if the right management (sod-cutting) takes place. But because above-ground uptake is an important source for nitrogen, a decrease in NH_3 concentration will probably immediately retard the process of heathland degradation, even without an intensive management, at least in young heathland.

12. SUMMARY AND CONCLUSIONS

1. About one-third of the Dutch heathland area is still vital, one-third is due to become grassland in the next three to five years, and one third has completely changed into grassland.

2. Nitrogen deposition on Dutch heathland is c. 35-40 kg N $ha^{-1}.y^{-1}$.

3. Adverse effects of SO_2 on more than 5% of the heathland species can be expected at concentrations above a critical level of 8 $\mu g.m^{-3}$. Effects on dominant species (Calluna and grasses) will probably not occur at the current SO_2 levels in the Netherlands.

4. Above-ground nitrogen uptake as NH_3 or $(NH_4)_2SO_4$ is very efficient and linearly related to concentration in a range from zero up to 400 $\mu g.m^{-3}$ NH_3 or 200 µmol.l-1 $(NH_4)_2SO_4$.

5. On the individual plant level, nitrogen addition (as NH_3 or $(NH_4)_2SO_4$) causes growth stimulation even at low dosages. However, metabolic or morphogenetic changes may occur in heathland species that make them more sensitive to frost, drought and plagues. A critical level for NH_3 cannot be exactly defined but is probably in the range 5-10 $\mu g.m^{-3}$.

6. On the ecosystem level, nitrogen addition ultimately leads to the elimination of slow-growing species by fast-growing species, but Calluna will not be outcrowded by grasses at nitrogen deposition levels up to 150 kg N $ha^{-1}.y^{-1}$ if its canopy is closed.

7. Opening of a Calluna canopy can be caused by the secondary stress factors frost, drought or plagues of the heather beetle (see 5).

8. The critical nitrogen load for the outcrowding of an open Calluna canopy by grasses is c. 10-15 kg N $ha^{-1}.y^{-1}$.

9. The decline of the Violion caninae species is probably due to direct effects of gaseous SO_2 (see 3) and/or soil acidification. The probability of occurence of Violion caninae species decreases below soil pH 4.5. This value could not be translated into a critical load.

10. Outcrowding of Violion caninae by grasses can also take place, but is probably only important at nitrogen deposition levels above the ambient level in the Netherlands. Effects on reproduction and establishment are possible at ambient level, however.

13. REFERENCES

Aerts, R. 1989. Plant strategies and nutrient cycling in heathland ecosystems. Diss. Utrecht, 203 pp.

Anonymous. 1988. Luchtkwaliteit. Jaarverslag 1987. Report RIVM 228702009 Bilthoven, 175 pp.

Asman, W.A.H., Drukker, B. & Janssen, A.J. 1987. Estimated historical concentrations and depositions of ammonia and ammonium in Europe and their origin (1870-1980). Report IMOU R-87-2, Utrecht

Berdowski, J.J.M. & Siepel, H. 1988. Vegetative regeneration of Calluna vulgaris at different ages and fertilizer levels. Biol. Conserv. 46, 85-93

Berdowski, J.J.M. 1987. The catastrophic death of Calluna vulgaris in Dutch heathlands. Diss. Utrecht, 135 pp.

Berendse, F. & Aerts, R. 1984. Competition between Erica tetralix L. and Molinia caerulea (L.) Moench as affected by the availability of nutrients. Acta Oecol./Oecol. Plant. 5, 3-14

Berendse, F. 1986. Stikstofmineralisatie en strooiselproduktie in voedselarme ecosystemen. Vakbl. Biol. 66, 430-433

Berendse, F. 1988. De nutrie"ntenbalans van droge zandgrondvegetaties in verband met de eutrofiëring via de lucht I: een simulatiemodel als hulpmiddel bij het beheer van vochtige heidevelden. Report Agrobiological Research Centre, Wageningen, 51 pp.

Berendse, F., Beltman, B., Bobbink, R., Kwant, R., & Schmitz, M. 1987. Primary production and nutrient availability in wet heathland ecosystems. Acta Oecologica/ Oecologia Plantarum 8, 265-279

Bizot, N., Tiktak, A. & Bouten, W. 1991. Soil hydrological system characterization of a heather ecosystem. Report Laboratory of Physical Geography and Soil Science, University of Amsterdam, in press

Bobbink, R., Heil, G.W., & Raessen, M.B.A.G. 1990. Atmospheric deposition and canopy exchange in heathland ecosystems. Report, Department of Plant ecology and Evolutionary Biology, University of Utrecht, 80 pp.

Boxman, A.W. & van Dijk, H.F.G. 1988. Het effects van landbouw ammonium deposities op bos- en heidevegetaties. Report Department of Aquatic ecology en Biogeology, Catholic University Nijmegen

Brunsting, A.H.M. & Heil, G.W. 1985. The role of nutrients in the interactions between a herbivorous beetle and some competing plant species in heathlands. Oikos 44, 23-26

Conijn, J.G. & Berendse, F. 1991. De simulatie van de concurrentie tussen Calluna en Molinia in droge heidevelden. Report RIN, Arnhem, in prep.

De Boer, W. 1989. Nitrification in Dutch heathland soils. Diss. Wageningen, 96 pp.

De Smidt, J.Th. 1982. De Nederlandse heidevegetaties. Wet. Med. KNNV 144, 87 pp.

De Smidt, J.Th. 1975. Nederlandse heidevegetaties. Diss. Utrecht, 99 pp. + ann.

De Smidt, J.Th., Berdowski, J.J.M, Brunsting, A.M.H., Heil, G.W. & Zeilinga, R. 1984. Hedendaags heidebeheer. Natuur & Techniek 52, 590-709

De Wit, C.T. 1960. On competition. Agricultural Research Report 66, 1-82

Diemont, W.H., Blanckenborg, F.G. & Kampf, H. 1982. Blij op de hei? Innovaties in heidebeheer. Report RIN, Arnhem, 135 pp.

Dueck, Th. A. 1990. Effect of ammonia and sulphur dioxide on the survival and growth of Calluna vulgaris (L.) Hull seedlings. Funct. Ecol. 4, 109-116

Dueck, Th.A., van der Eerden, L.J. & Berdowski, J.J.M. 1990. Effects of SO_2 on heathland species in the Netherlands. Funct. Ecol., in press

Duyzer, J.H., Verhagen, H.L.M. & Erisman, J.-W. 1989. De depositie van verzurende stoffen op de Asselse Heide. Report TNO R 89/29, Delft, 42 pp. + ann.

Erisman, J.-W. 1989. Ammonia emissions in the Netherlands in 1987 and 1988. Report RIVM 228471006, Bilthoven, 66 pp.

Fennema, F. 1990. Effects of exposure to atmospheric SO_2, NH_3 and $(NH_4)_2SO_4$ on survival and extinction of Arnica montana L. and Viola canina L. Report RIN 90/14, Arnhem, 61 pp.

Gimingham, C.H. 1972. Ecology of heathlands. Chapman & Hill, London

Greven, H.C. 1989. Gevoeligheid van mossen voor zwaveldioxide. Buxbaumiella 22, 34-35

Grubb, P.J. 1977. The maintenance of species-richness in plant communities: the importance of the regeneration niche. Biol. Rev. 52, 107-145

Heil, G.W. & Diemont, W.H. 1983. Raised nutrient levels change heathland into grassland. Vegetatio 53, 113-120

Heil, G.W. 1984. Nutrients and the species composition of heathland. Diss. Utrecht, 139 pp.

Heil, G.W. 1991. CALLUNA: effects of acidification on dry heathland. In: G.J. Heij & T. Schneider (eds.): Acidification research in the Netherlands. Elsevier, Amsterdam, in press

Helsper, H.P.G., Glenn-Lewin, D., & Werger, M.J.A. 1983. Early regeneration of Calluna heathland under various fertilization treatments. Oecologia (Berlin) 58, 208-214

Houdijk, A.L.F.M. 1990. Effecten van zwavel- en stikstofdepositie op bos- en heidevegetaties. Report Department of Aquatic Ecology en Biogeology, Catholic University of Nijmegen, 124 pp.

Hüttermann, A. 1985. The effects of acid deposition on the physiology of the forest ecosystem. Experientia 41, 584-590

Kooijman, S.A.L.M. 1987. A safety factor for LC50 values allowing for differences in sensitivity among species. Water Research 21, 269-276

Mes, A.E.R & Smilde, K.W. 1970. Sulphur deficiency in the Netherlands. Sulphur Inst. J. 6, 16-18

Pegtel, D.M. 1987. Effects of ionic Al in culture solutions on the growth of Arnica montana L. and Deschampsia flexuosa (L.) Trin. Plant and Soil 102, 85-92

Pegtel, D.M. 1988. Germination in declining and common herbaceous plant populations co-occurring in an acid-peaty heathland. Acta Bot. Neerl. 37, 215-223

Quene'-Boterenbrood, A.J. 1988. Veranderingen in de flora van 17 overwegend droge natuurgebieden met verschillende ammoniakemissies in Nederland. Report SBB 1988-11, Utrecht, 243 pp.

Suter II, G.W., Barnthouse, L.W., Breck, J.E., Gardner, R.H. & O'Neill, R.V. 1985. Extrapolating from the laboratory to the field: how uncertain are you? IN: R.D. Cardwell, R. Purdy & R.C. Bahner (eds.): Aquatic toxicology and Hazard Assessment. American Society for Testing and Materials, Philadelphia, 400-413

Thomson, K. & Grime, J.P. 1979. Seasonal variation in the seed banks of herbaceous species in ten contrasting habitats. J. Ecol. 67, 893-921

Van Dam, D., van Dobben, H.F., ter Braak, C.J.F. & de Wit, T. 1986. Air pollution as a possible cause for the decline of some phanerogamic species in the Netherlands. Vegetatio 65, 47-52

Van Dijk, H.F.G. & Roelofs, J.G.M. 1987. Effects of excessive ammonium deposition on the nutritional status and condition of pine needles. Physiol. Plantarum 73, 494-501

Van Dobben, H.F. 1990. Effecten van luchtverontreiniging op korstmossen, resultaten van meerjarige studies. De Levende Natuur 91, 179-183

Van Dobben, H.F., Mulder, J., van Dam, H. & Houweling, H. 1991. The impact of acid atmospheric deposition on the biogeochemistry of moorland pools and surrounding terrestrial vegetation. Agricultural Research Report, in press

Van Kootwijk, E.J. & van der Voet, H. 1989. De kartering van heidevergrassing in Nederland met de Landsat Thematic Mapper satellietbeelden. Report RIN 89/2, Arnhem

Van Kootwijk, E.J. 1989. Inventarisatie van de vergrassing van de Nederlandse heide. Report RIN 89/1, Arnhem, 49 pp.

Van Ree, P.J. & Meertens, M.H. 1989. Verarming van de Veluwse heide in relatie met ammoniak-depositie. Report Province Gelderland, Arnhem, 49 pp.

Van Straalen, N.M. & Denneman, C.A.J. 1989. Ecotoxicological evaluation of soil quality criteria. Ecotox. and Env. Safety 18, 241-251

Van Straalen, N.M. 1987. Stofgehalten in de bodem - (geen) effect op bodemdieren. In; Bodemkwaliteit, Proc. Symp. Ede, pp. 75-84

Van der Eerden, L.J., Dueck, Th.A., Elderson, J., Van Dobben, H.F., Berdowski, J.J.M, & Latuhihin, M. 1989. Effects of SO_2, NH_3 and $(NH_4)_2SO_4$ deposition on terrestrial semi-natural vegetation on nutrient-poor sandy soils. Report IPO/RIN, Wageningen/Arnhem, 169 pp.

Van der Eerden, L.J., Dueck, Th.A., Elderson, J., Van Dobben, H.F., Berdowski, J.J.M, & Latuhihin, M. 1990. Effects of NH_3 and $(NH_4)_2SO_4$ deposition on terrestrial semi-natural vegetation on nutrient-poor soils. Report IPO/RIN, Wageningen /Arnhem, in press

Van der Eerden, L.J.M. 1982. Toxicity of ammonia to plants. Agric. Environ. 7, 223-235

Van der Maas, R. 1990. Hydrochemistry of two Douglas fir ecosystems and a heather ecosystem in the Veluwe, the Netherlands. Report Department of Soil Science and Geology, Agricultural University of Wageningen, in press

Van der Zande et al. ('Werkgroep Heidebehoud en Heidebeheer'). 1988. De heide heeft toekomst! Report, 135 pp. + ann.

Westhoff, V., & den Held, A.J. 1969. Plantengemeenschappen in Nederland. Thieme, Zutphen, 324 pp.

INTEGRATED MODELLING

T.N.Olsthoorn[1)]

1) National Institute of Public Health and Environmental Protection, Bilthoven

PREFACE

This report has been written in co-operation with the people listed below, involved in the development of the different modules of DAS. To a large extent they provided parts of the text directly or partially and their reports have been used.

A.H. Bakema (RIVM), F. Berendse (CABO), J.J.M.Berdowski (RIN), K.F. de Boer (RIVM), J.-W. Erisman (RIVM), L.J. van der Eerden (IPO), J.F. Feenstra (IvM), H.J.M. van Grinsven (RIVM), C. van Heerden (RIN), J.A. van Jaarsveld (RIVM), J. Kros (WSC), F.A.A.M. de Leeuw (RIVM), J.G. van Minnen (RIVM), G.M.J. Mohren (Dorschkamp), A.A. Olsthoorn (IvM), R. Thomas (RIVM), A.E.J. Tonneijk (IPO), W. de Vries (WSC), J.H.M. Wijnands (LEI) and F.G. Wortelboer (RIVM)

1. OVERVIEW OF THE DUTCH ACIDIFICATION RESEARCH PROGRAMME

Early 1983 the Dutch parliament questioned the Minister of Housing, Physical Planning and the Environment (VROM) on the subject of acidification.

Early 1984 together with the minister of Agriculture & Fisheries (L&V), he presented the general outline of the acidification policy and the research needed to obtain a better understanding of the phenomena, in order to provide for effective measures. At about the same time the Minister of VROM established a special Programme on Acidification, covering a number of research projects. Also at about this time, the Minister of Economic Affairs (EZ) commissioned a programme from the Energy Research Foundation (ECN), a project on Air Pollution, Forest Die Back and Acidification.

In mid 1984 a ministerial memorandum laid down the way in which the research was to be organized. It also pointed out the issues that needed additional research. This resulted in the Dutch Priority Programme on Acidification. These special issues are:

* Dose-effect relationships (forests, cultivated crops and semi-natural vegetation i.e.heathlands)
* NH_3-emission reductions, transport and transformation
* Effectiveness of control measures

In September 1984 a steering committee was set up. For early 1985 it decided on an Additional Acidification Research Programme focusing on the issues mentioned. The programme is financed by the Ministry of VROM, the Ministry Agriculture & Fisheries, and the Ministry of Economic Affairs, together with the refineries and electricity producing companies. The various projects are co-ordinated by a project team headed by RIVM, the National Institute for Public Health and Environmental Protection.

The main question to be answered is: "How will the environment benefit from the proposed measures?" The programme is aimed at scientific coherence in order to answer political questions that will be raised once the abatement measures and their accompanying costs will be presented. While in operation the programme should provide the policy makers with information whenever possible. For this purpose evaluation reports have been issued (Schneider & Bresser, 1988).

In order to arrive at a coherent view of the acidification in the Netherlands and to be able to decide on the best set of abatement measures, the programme should lead to a model covering the relevant factors. The on-going research carried out by many institutions simultaneously is directed towards insight in phenomena, parameter values, relationships between phenomena, the gathering of information, the interpretation of historical data, the realization of detailed models to explain observations and to forecast developments and the generation of scenarios for all relevant fields. Finally the results will flow directly or indirectly into the Dutch Acidification Systems Model (DAS), a regionalized high-level systems model.

The remainder of this paper focuses on the acidification model. It will be less concerned with the details of the various research projects, of which there are about 60, carried out by about 20 different institutions and universities.

2. DUTCH ACIDIFICATION SYSTEMS MODEL (DAS)

As has been pointed out before, the results of the various projects within the research programme are to contribute to a model capable of estimating and assessing in a consistent and coherent way the results and effectiveness of abatement measures proposed by politicians and perhaps others. As far as possible the effectiveness should be specified in terms of predicted effects.

The model must clearly contain the causality chain, i.e. emissions, depositions/ concentrations and effects on several receptor systems. It must do so in a regionalized fashion.

Packages of abatement measures specified over time, together with local and foreign autonomous economic and technical developments should yield the emission scenarios which are to be the driving force of the model. Fed into the air module these produce depositions and air-borne concentrations. They are in turn used by a set of effect modules to calculate effects on several relevant receptors. These effects may be used (i.e. interpreted) to alter scenarios in order to obtain a more optimal final result.

2.1 Important choices for the model and status quo

The DAS model was developed in 1985 (Figure 1).

The subdivision of the Netherlands on which it was based, was a compromise between detail, the availability of regional data and the accuracy with which an air-transport model could produce local time-averaged concentration and deposition figures.

For the Netherlands the compromise came out at areas of about $60x60km^2$. This resulted in a subdivision of the country into 20 areas (Figure 2). Each area is the combination of one or more so-called COROP areas that divide the Netherlands into 40 well-defined economic zones. The 20 areas used within DAS have become widely known as "Dutch acidification areas". They are treated both as emission and as receptor areas.

DAS focuses on effects within the Netherlands only. Hence, outside the Netherlands, a subdivision of Europe was needed for emission aggregation only. This was done as coarsely as possible, keeping in mind the relative contribution of the individual areas to the total deposition and concentration in the Netherlands. Until recently the rest of Europe was divided into 13 areas, but it was found necessary to refine the subdivision further to 19 areas (Figure 3).

In this way DAS is regionalized into 39 emission areas, 20 of which lie in the Netherlands. Those 20 have also been chosen as the receptor areas. These areas have been the basis for aggregation of emissions as well as for the aggregation of effects. Inventories have been carried out with this subdivision in mind.

The congruence between the Dutch emission and receptor areas has been chosen for practical reasons only. Different subdivisions may be better suited for the assessment of some of the effects. However, data collected for the model now fits within this spatial framework.

The time-span for the model is about a century. 1950 is taken as a starting point for the data collected in order to obtain a long-enough history to test the applicability of the model.

Dutch Acidification Systems Model

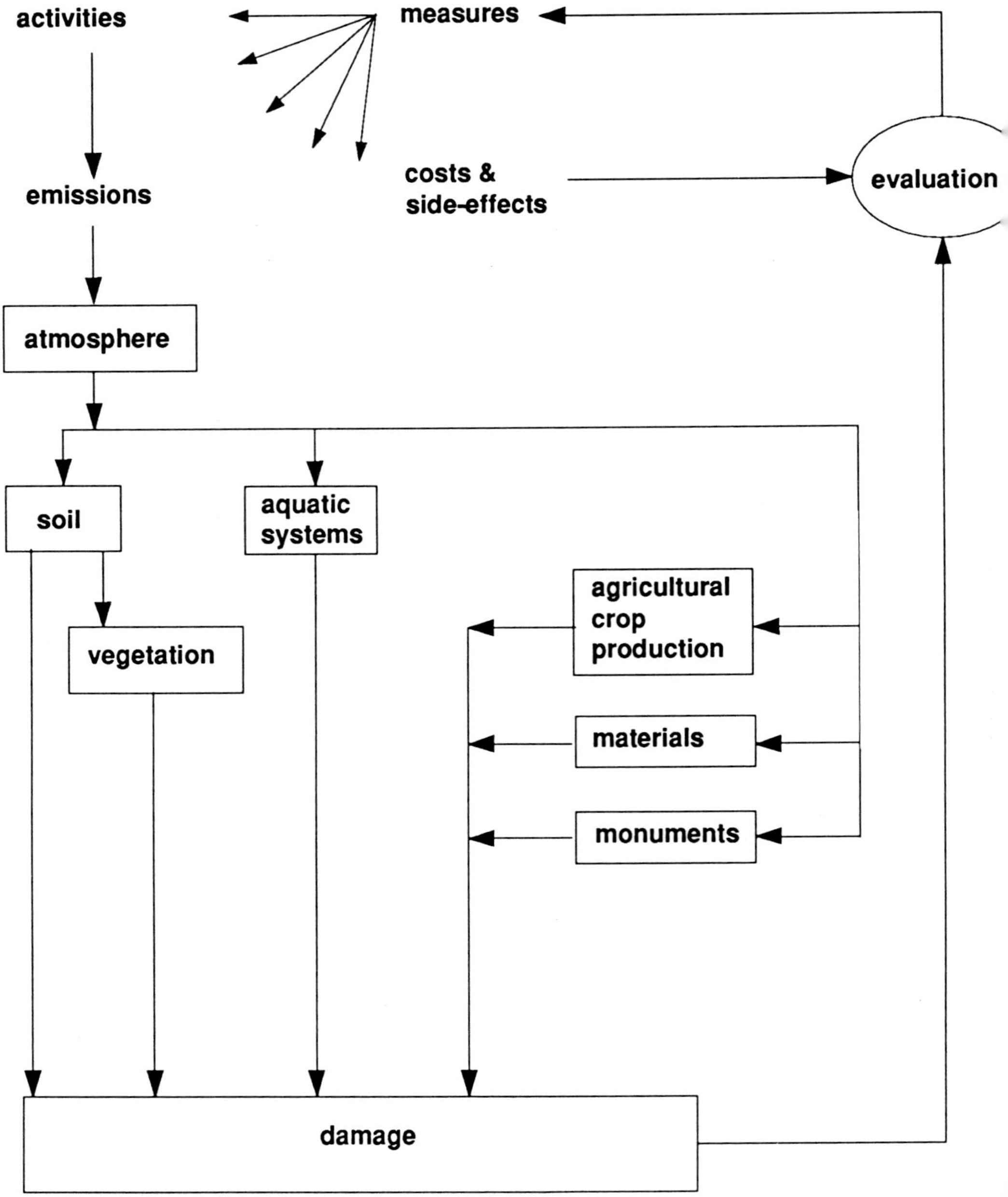

Fig. 1. Basic (conceptual) layout of DAS

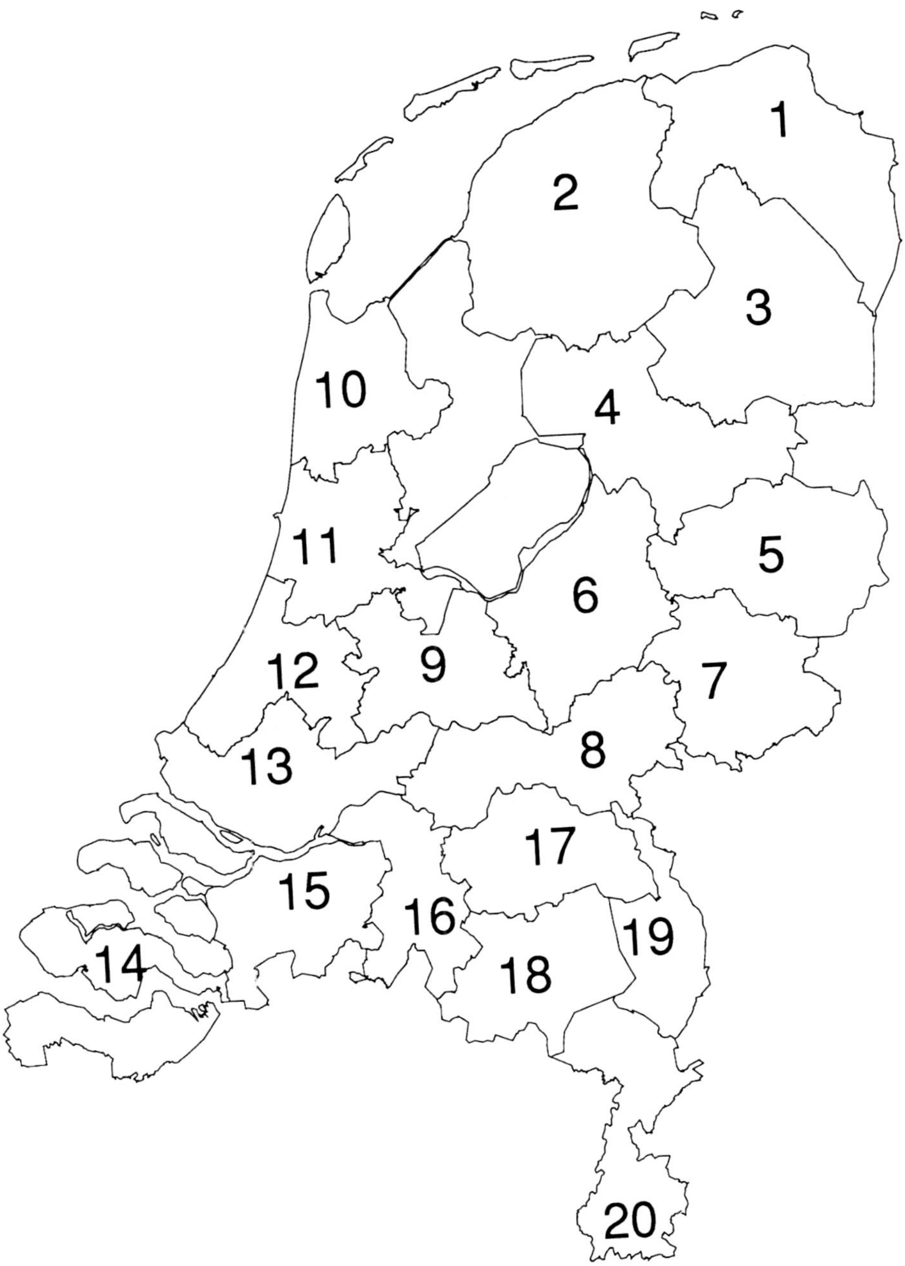

<u>Fig. 2.</u> The 20 Dutch acidification areas used by DAS

Fig. 3. The 19 European emission areas outside the Netherlands used by DAS

Time steps were set at one year, because for long-term forecasting, shorter time steps do not make sense. Some of the detailed submodules use a smaller time step internally for numerical reasons.

Included air-pollution compounds are the acidification agents NH_3, NO_x, and SO_2. VOC is included as a precursor of O_3, which is used by some of the effect modules as a damaging agent. Since DAS has no module for O_3, VOC is not used.

The effect modules include sensitive materials, monuments, crops, soils, forests (Douglas fir), heathlands and aquatic systems. Damage to public health is not taken into account nor are cultural assets like old books and paintings.

2.2 History of DAS

The Dutch Priority Programme on Acidification started in 1984. In 1985 a first integrated model was constructed along the above-mentioned lines for 33 emission areas and 20 receptor areas. It was implemented in the simulation language DYNAMO on an IBM mainframe (Waltmans, 1985, Koster & Waltmans, 1986) using a 4-year time step. The choice of the DYNAMO simulation language was based on the dynamic character of the processes and their different time characteristics.

The model at that time calculated among other things the damage to crops, the average pH in soils, the Al/Ca ratio and leaf weight of forests. At that time, the modules were in an experimental stage and the data were not of high quality. Much input from the various partners and projects was needed. Nevertheless, the model proved the feasibility of an integrated approach.

The development of the DYNAMO-based model was terminated due to software problems and associated mainframe computation costs. There has been an attempt to translate the program into Pascal, but that has not been successful because of the limitations of the available compiler.

The reworked version of the current DAS was implemented by RIVM within two months early 1987, using the C programming language (Kernighan and Ritchie, 1978, 1988) on a VAX under UNIX. It was far from complete. It included the causality chain from sectoral emissions down to depositions and provided a framework to which effect modules could be connected.

A number of the effect modules have since then been developed independently by participating institutes (Winand Staring Centre, The Research Institute for Forestry and Urban Ecology, the Dutch Research Institute for Nature Management and RIVM itself) in part using other programming languages such as FORTRAN and C++. DAS thus consists of a heterogeneous set of models, called modules.

Beside DAS some PC-based models have been developed as an aid to policy-makers (Olsthoorn and De Leeuw, 1988, Olsthoorn, 1988). The aim of these models is to provide tools that are handy to use, also outside the institute and that, though simplified and steady-state, can give answers to many questions in a fast and easy way while lending themselves to rapid analysis of alternative abatement options.

Much effort was spent on the development of various effect modules by scientists at RIVM and other institutes. In spite of the original goals some of these effect modules have grown out into stand-alone computation intensive modules that are difficult to integrate into DAS in a regionalized fashion.

Since September 1989 the set of individual modules has been put together and the documentations written in the form of Module Specifications (Bakema et al., 1990). In the meantime development continues and uncertainty analysis is being carried out on the most complex modules (RESAM, see 3.6, SOILVEG, see 3.7 and AQUACID, see 3.10).

3. DESCRIPTION OF THE DAS-MODULES

DAS consists of an emission module, an air module and a series of effect modules (see Figure 1). This chapter describes these modules in a more general way, leaving out smaller modules that are mainly of interest with respect to the implementation and actual usage. They are included in chapter 4 which describes the implementation. For more detailed information of the individual modules the reader should refer to the separate module specifications (Bakema et al., 1990) or, for background information, to the original documentation of the developers of the submodules (see references in the section that describes the specific module).

3.1 Emissions and abatement measures module

Detailed information on the emission module of DAS can be found in Thomas et al., 1988. In their report historic emissions for the period 1950-1980 and scenarios covering the period 1980-2030 are described. Emissions from the Netherlands, emissions from the surrounding countries (FRG, Belgium, France and UK) and those from other European countries further away from the Netherlands are derived separately.

Historic emissions for the Netherlands and abroad have been derived from various reports. These data are fixed and available to DAS from a file.

To generate scenarios for the Netherlands, RIVM Laboratory for Waste Materials and Emissions (LAE) has an Environmental Information and Planning System (RIM) available. This is an interactive computer system for evaluation of scenarios in terms of emissions, their abatement and associated costs (Thomas et al. 1989). RIM is surrounded by its own set of models for specific sectors.

With RIM emissions scenarios can be generated for the compounds SO_2 and NO_x per economic sector for the Netherlands as a whole, for a number of reference years, starting in 1980. RIM also calculates the costs of the abatement measures that are included in the emission scenarios.

Inputs are user-defined scenarios derived from demographic, economic and energy projections and selected abatement measures (such as catalytic converters, flue-gas treatment and application of clean technology). Also, activity levels can be changed by structural measures such as energy conservation and fuel switches.

To obtain the necessary broad acceptance of the RIM-projections, data for the base year and the future development are derived from official sources. As such the Central Planning Bureau (CPB) produces economic forecasts, the Energy Study Centre (ESC) provides energy-use forecasts, the Central Bureau of Statistics (CBS) publishes base-year data on emissions and figures on production, traffic, agriculture and demography, while agricultural data are obtained from the Agricultural Economic Institute (LEI).

To obtain emissions per DAS area in the Netherlands, the national emissions obtained from RIM are disaggregated within DAS, based on the emission distribution per compound and sector in 1980, which is known from the Dutch Emission Inventory System (ER) (TNO, 1985). DAS can accommodate such disaggregation vectors for any year if these become available.

Emission scenarios for the surrounding countries FRG, Belgium, France and the UK are produced by the European Model for Acidification (EMAC), developed by the Agricultural University of Wageningen (van Ierland and Oskam, 1988). EMAC calculates emissions per sector for SO_2 and NO_x for the areas in DAS that surround the Netherlands for a number of reference years, starting in 1980. For the period 1980-2000 these scenarios are based on the "Energy 2000 study" (Guilmot et al., 1986). Assumptions made for the period

2000-2030 can be found in Thomas et al., 1988, p 53.

The sectors used by DAS are shown in the table below. They differ somewhat from those available in EMAC and RIM.

The effective height of emission sources plays an important role in air-transport and deposition. Hence, the air-transport module of DAS (see section 3.2) needs this information. It uses transfer matrices for the generic stack heights "high", "medium" and "low". Each economic sector is put into one of these categories as shown in Table 3.1.1.

<u>Table 3.1.1</u> Economic sectors used in DAS, derived from EMAC and RIM models together with allocated generic stack heights

Economic sector	EMAC	RIM	SO_2			NO_x			NH_3	
			High	Med	Low	High	Med	Low	Low	Med
Agriculture	+	+			+			+	+	
Domestic		+							+	
Industrial NH_3-emis.		+							+	+
Other high sources	+	+		+			+			
Power plants	+	+	+			+				
Primary Metal	+	+		+			+			
Refineries	+	+	+			+				
Chemical industry		+		+			+			
Fertilizer industry		+		+			+			
Other low sources	+				+			+		
Other traffic	+				+			+		
Private cars	+				+			+		
Trucks / busses		+			+			+		
Ships	+				+			+		

For European regions that are not neighbouring the Netherlands, scenarios for SO_2 and NO_x for the same reference years are derived from literature and the statements made by the individual countries on the expected emissions by the end of the century. For the period 2000-2030 relative changes have been assumed equal to those of the countries in the EMAC model. The emission scenarios for these countries are generated for SO_2 and NO_x for low and high stacks respectively, without a sectoral subdivision.

As a first basic approximation NH_3 emissions for countries outside the Netherlands have been assumed constant for the entire period 1980-2030. However, alternative scenarios can be generated and used by DAS.

Though the calculation of the emissions and costs involve merely the multiplication of scenario-dependent activity levels with the associated emission factors and measure-specific cost factors, the entire system has become a complex dedicated database application due to the multitude of possible choices. The same holds for the EMAC model. For this reason RIM and EMAC have not been integrated with DAS. Instead scenarios are generated separately and DAS uses the generated emission data as its starting point. However, DAS does provide means to alter these emissions for what-if analyses (see INTERACT modules, chapter 4).

More information can be obtained from Dr. R. Thomas, RIVM, Laboratory for Waste Materials and Emissions (LAE), P.O. Box 1, 3720 BA Bilthoven, the Netherlands, +31-30-749111.

3.2 The air-transport and transformation module (SRM)

The air-transport and transformation module of DAS, called SRM (from "Source-Receptor Matrices") focuses on SO_2, NO_x and NH_3 and their derivatives. In the Netherlands each of them counts for at least 20% of the total deposition of what is called potential acid (i.e. the sum of SO_x, NO_x and NH_x as expressed in equivalents per ha per year). Together, these compounds cause over 90% of the (potential) acid depositions.

Additional sources of acid deposition, like HCl, are only of local importance, mainly around waste incinerators, but their relative importance may grow if deposition could be reduced to levels as low as about 1400 mol H per ha per year from the current values of about 5000 mol H per ha per year.

The depositions and the concentrations of these compounds and their derivatives are calculated by means of transfer matrices (sometimes called source-receptor matrices). These matrices have been generated by RIVM Air Research Laboratory (LLO) by means of their long-range transport model, called TREND (Van Jaarsveld, 1988).

The TREND model may be characterized as a statistical long-range Lagrangian model. The model includes transport, dispersion, chemical transformation and deposition of pollutants. Long-term averaged values of concentration and deposition are calculated as weighted averages over 72 classes (12 wind and 6 stability/mixing height classes). Dry deposition parameters are based on analysis of data of the National Air Quality Monitoring Network (Elskamp, 1989), on results obtained within the framework of the National Programme on Acidification and on literature. Wet deposition losses are calculated taking into account the rain probability and the mean duration of precipitation events (both depend on wind direction) and assuming Poisson statistics for precipitation events. The chemical transformations are described by linear, first-order rate coefficients which are, however, different for each of the meteo-classes (e.g. dependent on solar radiation, temperature etc.). This means that there is no interaction between SO_2, NO_x and NH_3.

In this way we have matrices for the following combinations:

Table 3.2.1 Available matrices for DAS.

Emission	NH_3	NO_x	SO_2
Stack	low medium	low medium high	low medium high
Concentrations	NH_3 NH_4	NO_x NO_3	SO_2 SO_4
Dry depositions	NH_x	NO_x	SO_x
Wet depositions	NH_x	NO_x	SO_x

Matrices for dry deposition of NH_4 and SO_4 are not available as yet. At the moment a fixed fraction of total dry deposition of NH_x and SO_x must therefore be used.

Further information may be obtained from Ing. J.A. van Jaarsveld and Dr.F.A.A.M. de Leeuw, RIVM, Air Research Laboratory (LLO), P.O.Box 1, 3720 BA Bilthoven, the Netherlands, Tel: +31-30-749111.

3.3 Filtering factors

Using TREND and the transfer matrices, derived from TREND, deposition are obtained for a Dutch landscape with an average texture (mixing length). Receptor area specific correction factors (also called filtering factors) for forest, heathland and water bodies are provided by RIVM Air-Research Laboratory (Erisman et al., 1987, 1990). Correction is applied to deal with different texture, but as yet, no correction can be applied based on the properties of the vegetation itself.

Further information may be obtained from Ing. J.-W. Erisman, RIVM, Air Research Laboratory (LLO), P.O.Box 1, 3720 BA Bilthoven, the Netherlands, Tel: +31-030-749111

3.4 Ozone scenarios

O_3 at ground level is an important factor in the damage to vegetation and possibly to human health as well. O_3-concentration data are needed for the crop-growth season (May-Sept) as daily averages (10.00-17.00h). According to the research into the effects of O_3, peak values are significant for damage to human health, while average values during the growing season seem to correlate better with damage to crops (Eerden et al, 1988).

O_3 is not emitted, but formed in the atmosphere by chemical processes. In this context only tropospheric ozone (i.e. in the lowest 10-12 km of the atmosphere) is considered.

The long-term averaged ozone concentrations at ground level are strongly related to the concentrations in the free troposphere. In the free troposphere the oxidation of the less reactive CO and CH_4 in the presence of NO_x is mainly responsible for the formation of O_3. At ground level, in the planetary boundary layer, the more reactive volatile organic compounds (VOC) play a key role. Until now CH_4 and CO have not been considered within the acidification programme. No emission scenarios for CO and CH_4 are available at the moment. These emission scenarios will be developed in the global biosphere programme that has started recently within RIVM. Both measurements and model studies indicate a positive trend in tropospheric O_3 concentrations.

The ozone formation is described by a complex non-linear chemistry. The formation rate depends on a large number of parameters: meteorological conditions (e.g. solar radiation, temperature, relative humidity, etc.), concentration and composition of the VOC mixture, NO_x concentration. In contrast to the source-receptor relations obtained for the acidifying components, there is no direct, "simple-to-use" correlation between emissions of NO_x and VOC on the one hand and O_3 concentrations on the other. Moreover, emissions on a European and global scale have to be considered, as the formation of O_3 takes place on a timescale varying from hours (in the planetary boundary layer by oxidation of reactive VOC) to years (in the troposphere by oxidation of CO and CH_4).

Until now the modelling of the O_3 concentrations has been too complicated for easy use in an integrated acidification model such as DAS. Also, the complex O_3 model of RIVM Air-Research Laboratory is still under development and needs improvement before it can be included in the DAS-framework. In the mean time DAS will use O_3 scenarios instead of a proper module. These scenarios will be provided by the RIVM Air Research Laboratory.

Further information may be obtained from Dr. F.A.A.M. de Leeuw, RIVM, Air Research

Laboratory (LLO), P.O.Box 1, 3720 BA Bilthoven, the Netherlands, Tel: +31-30-749111.

3.5 Effect modules

DAS contains a soil acidification module (RESAM, section 3.6) and a module in which soil and vegetation are dynamically connected (SOILVEG, section 3.7). It also contains a module to calculate the impact of N-deposition (which is two thirds of total acid deposition in the Netherlands) on the competition between two wet-heathland plant species (ERICA, section 3.9). The development of a module for the competition between dry-heathland species (80-90% of total heathland in the Netherlands) is still in the definition phase. A more general module to describe growth of forests is also under development and will be available in 1990 (FORGRO, section 3.8). The aquatic systems module simulates shallow acid-sensitive lakes (AQUACID, section 3.10). Furthermore three simple modules based on dose-effect relationships are available to calculate damage to crops (AGRIPROD, section 3.11), materials (MATERIALS, section 3.12) and monuments (MONUMENTS, section 3.13) respectively. The modules for soil, soil and vegetation and aquatic systems are complicated, computation intensive modules and therefore impose limitations on the number of scenarios that can be evaluated on a regionalized scale.

3.6 The soil acidification module (RESAM)

The long-term and large-scale soil response to acid deposition will be analysed by RESAM, a process oriented Regional Soil Acidification Model, developed by the Soil Survey Institute of Winand Staring Centre in Wageningen, with technical input from RIVM. RESAM predicts annual-averaged fluxes and concentrations of the major chemical species in soils of characteristic forest ecosystems in the Netherlands. It thus serves to evaluate the long-term effects of deposition on the chemistry of forest soil and soil water. A detailed description is given by its developers, De Vries et al. 1988, De Vries and Kros, 1989.

RESAM is based on the conceptual relationship between forest nutrient cycling and soil acidification. Nutrient cycling strongly affects model prediction in the uppermost soil horizons. This is important, since RESAM has been coupled to various vegetation models (see SOILVEG section 3.7 and FORGRO, section 3.8).

RESAM output is confined to major ions in soil water relevant to the growth and vitality of forests and to groundwater quality. Ions include the macro nutrients NH_4 NO_3, SO_4, Ca, Mg, K, H, Al, Na, Cl, HCO_3 and RCOO. Input includes atmospheric deposition of SO_2, NO_x, NH_x, base cations, chloride and infiltration of water.

RESAM simulates major biogeochemical processes occurring in the forest canopy, the litter layer and the mineral horizons. The latter take care of vertical heterogeneity. The lower boundary of the soil compartment is the mean lowest water-table. Processes include canopy interactions, element cycling, nitrogen transformations, geochemical weathering and cation exchange reactions (see table 3.6.1).

Table 3.6.1 Processes included in RESAM, compounds involved and process formulation

Process	Compounds involved	Process formulation
Leaf uptake	NH_3, NO_x, SO_2	linear function of dry deposition
Leaf exudation	Ca, Mg, K	first order reaction
Litter fall	N, Ca, Mg, K, S	first order reaction
Root decay	N, Ca, Mg, K, S	first order reaction
Mineralization	N, Ca, Mg, K, S	first order reaction
Root uptake	N, Ca, Mg, K, S	several options*
Nitrification	NH_4, NO_3	first order reaction
Weathering of silicates	Al, Ca, Mg, K, Na	first order reaction
Weath./precip. of Al-hydroxides	Al	first order reaction
Cation exchange	H, Al, Ca, Mg, K, Na, NH_4	Gaines-Thomas equations
Sorption of sulphate	SO_4	Langmuir equation
Dissociation of CO_2	HCO_3	CO_2 equilibrium
Protonation of RCOO	RCOO	first order reaction

* See text.

Hydrological data, such as infiltration, evaporation and soil-moisture content per soil layer are derived from a hydrological model SWATRE (Belmans et al., 1983).

While RESAM calculates the annual average fluxes and concentrations of the major elements in characteristic forest soils, it does not simulate growth, but instead uses a fixed, but vegetation dependent nutrient cycle with a net uptake that is either constant or a forcing function of time.

From the air module of DAS (section 3.2) year-averaged depositions are taken for SO_2, NO_x and NH_x. Direct oxidation of SO_2 to SO_4, of NO_x to NO_3, and of NH_3 to NH_4 in deposited compounds is assumed. H deposition is then calculated to balance the charge in the total deposition. Base cations Ca, Mg, K and Na, (and also Cl) are included as area-specific depositions, averaged over the period 1978-1985, taken from the national network on rain water quality (Buijsman et al. 1988, Elskamp, 1989).

Within the canopy, foliar uptake of S and N from SO_2 and NO_x is taken as a fraction of dry deposition. Uptake of NH_3 (as NH_4) relates to exudation. Exudation from leaves is modelled for the elements Ca, Mg, K and Na. The rate depends on their concentration in the leaf and the leaf's biomass.

All biological processes except nitrification and denitrification are assumed to take place in the litter layer. Even though root uptake is distributed across the different soil layers, root decay is allocated to the litter layer.

The organic material in the litter layer is assumed to originate from fallen leaves and decaying roots only. Living biomass of leaves and roots is assumed to be in steady state. Fixed rates are used for leaf-fall and the mortality of the roots. The nutrient concentrations of N, S, Ca, Mg and K in leaves and roots are used to calculate nutrient fluxes by litter fall.

Mineralization is modelled in the litter layer only. The mineralization fraction for either fresh and old litter depends on pH and the groundwater level. For roots, mineralization equals root death. Part of the N from mineralization becomes immobilized by micro-organisms. This fraction depends on the N-content of the substrate and microbes.

The following processes occur in every soil horizon underneath the litter layer:

Protonation, which is proportional to RCOO--concentration with a rate-coefficient depending on pH.

Root uptake balances the gains and losses of the vegetation. These are the incoming foliar uptake and the outgoing fluxes via litter fall, root death, foliar exudation and net uptake. A fixed ratio for NH_4/NO_3 is used. Nutrient concentrations within the vegetation are constant in time. The uptake is distributed across the soil layers according to the water-uptake pattern. The total root uptake is either fixed but based on a given growth (stems, branches) assuming a stationary element cycle on an annual basis, or is variable determined by the water uptake and nutrient concentration per soil layer.

Nitrification and denitrification are modelled as a first-order reaction of NH_4 and NO_3 concentrations respectively. The rates depend on pH and the groundwater level.

Weathering of Ca, Mg, K and Na from primary minerals is a first-order reaction of their concentrations in the solid phase. That of Al is a fixed fraction of the total weathering, depending on the type of base cation.

Weathering of carbonate is described as a first-order reaction. Weathering fluxes depend linearly on carbonate amount and the degree of undersaturation of the solution with respect to $CaCO_3$.

The module has two alternatives for the calculation of the dissolution of aluminium hydroxide. It can use a first-order reaction rate that depends on the Al-amount and the degree of undersaturation with respect to gibbsite. Alternatively it can use Elovich equations together with undersaturation with gibbsite.

Cation exchange is described by Gaines-Thomas equations for Al, H, Ca, Mg, K, Na and NH_4. These are solved with respect to mass and charge balance.

Solute transport in the soil is downward only. It is calculated explicitly as the product of concentrations and percolation flow. This flow equals precipitation surplus minus the sum of root-water uptake in the overlaying and current layers.

An uncertainty analysis based on Monte Carlo analysis using PRISM software (Janssen et al., 1990) has been carried out on RESAM and is published separately (Kros et al. 1990).

In order to regionalize RESAM in the DAS context, a separate module has been developed (see section 4, SELINP/GENINP) to generate RESAM inputs for all of the receptor areas and all of the soil-vegetation combinations directly from the available database. The list of vegetations include Scots pine, black pine, Douglas fir, Japanese larch, Norway spruce, oak, beach, heathland and grassland. The list of soils include fine-textured and coarse-textured leptic podzols, fine-textured gleyic-podzols, fine-textured humic podzol, fine-textured humic gleysol, fine-textured calcaric fluviosol and, finally, fine- and coarse-textured albic arenosols.

More information can be obtained from: Ir. W. de Vries, and Ir. J. Kros, Winand Staring Centre, P.O. Box 125, 6700 AC Wageningen, the Netherlands, Tel. +31-8370-19100.

3.7 The soil and vegetation module (SOILVEG)

The soil and vegetation module, SOILVEG, integrates soil and vegetation processes for douglas-fir stands on sandy soils. The module has been developed by the Dutch Research Institute for Nature Management (RIN) with input from RIVM (Berdowski et al., 1990). SOILVEG implements a version of RESAM for its soil processes (See section 3.6). Testing and implementation was done in co-operation with RIVM.

SOILVEG simulates growth of Douglas-fir stands. For a given soil-type it dynamically calculates the chemical composition of soil and soil water, the nutrient status of plant parts and tree growth. It produces these figures for the situation each year for March as well as for the year-averaged situation.

In SOILVEG, plant growth may be hampered by stresses due to pH, Al, SO_2 and O_3 and nutrient shortages.

SOILVEG doesn't consider PO_4, undergrowth and fructification and uses year-averaged input data.

Biotic functions of the module are leaf uptake, leaf exudation, leaf and branch fall, stem and root death, thinning, mineralization of organic matter, root uptake, nutrient transport within the plants and the soil, C-balance, plant growth, nitrification, denitrification and reallocation of nutrients.

Abiotic functions are protonation of organic acids, weathering, cation sorption and transport. The same soil formulation of the processes is used as in RESAM (see section 3.6). Compared to RESAM, SOILVEG lacks cation exchange and carbonate-weathering. Like RESAM, SOILVEG receives precalculated soil-water contents and infiltration information from the model SWATRE (Belmans et al., 1983).

From the air module of DAS (section 3.2) year-averaged depositions are taken for SO_2, NO_x and NH_x, as well as concentrations for SO_2 and NH_3. Season averaged O_3 concentrations are supplied separately. The same is valid for the depositions of chloride and the base cations Ca, Mg, K and Na. They are taken from the national monitoring rain water quality network and have been averaged over the period 1978-1985. Direct oxidation of SO_2 to SO_4, of NO_x to NO_3, and of NH_3 to NH_4 in deposited compounds is assumed. H deposition is then calculated to balance ionic charge of total deposition.

Uptake of gaseous NH_3 is modelled as a fraction of NH_3 concentration. SO_2 and O_3 are not taken up but can hamper photosynthesis. Leaf uptake of deposited nutrients and components (SO_4, NH_4, NO_x, Ca, Mg, K, Na, Cl and H) is a fraction of deposition flux. However, the uptake fraction for Ca, Mg, K, Na, Cl and H is set to zero, due to lack of information.

Leaf exudation is also normally set to zero but can be calculated for N, Na, Ca, Mg, and K as a fraction times the leaf concentration. H is exudated to maintain the charge balance in the leaves.

Leaves (needles) of different age have their own specific fall fraction. This fraction depends on N-concentrations and shortages of K and Mg in the leaves. The concentration in fallen leaves differs from that in living leaves because of reallocation of nutrients, for which a reallocation factor is used. For N this factor depends on the N-concentration in the leaves.

Branches fall according to a fixed fall fraction. Reallocation is also applied, using the same mechanism as for leaves using specific coefficients.

Death of woody and fine roots is calculated using a sinusoidically varying death rate during the year. Reallocation of nutrients for roots is done in the same way as for leaves but using other coefficients.

Stem death is a linear function of total stem mass. Stems are not mineralized. Stems are also removed by a prescribed thinning regime. This thinning only removes stems and living branches. Leaves and roots remain in the system.

The soil part of SOILVEG lacks the cation exchange and carbonate weathering processes. SOILVEG assumes a constant water content and assumes that infiltration is the upper limit for transpiration on a monthly basis.

Root uptake of nutrients is the product of a selective uptake fraction, water uptake and the nutrient concentration in the soil solution. These uptake fractions determine nutrient preferences and depend on the season (sine function) and on tree age. The sum of the water uptake in all layers equals transpiration. Leaf amount influences transpiration; however, there is no feed-back to soil moisture. Hence, the module cannot simulate specific situations that may occur during a year such as upward soil moisture flow and drought stresses.

Root uptake is linearly dependent on pH and exponentially on Al concentration. Below a certain threshold it also depends linearly on available fine roots.

CH_2O formed by photosynthesis is allocated to plant compartments using allocation fractions. For nutrients are adjusted to stimulate growth of specific plant parts and for ageing. Reallocation of nutrients in leaves, branches and fine roots occur just before they die.

Maximum photosynthesis follows a sine during the year. Maximum photosynthesis is reduced below a threshold leaf area index (logarithmically) and by SO_2 and O_3 concentrations. Available C in specific plant parts depends furthermore on respiration (dependent on N, Mg, Ca and K conc. in the plant) and the plant-part specific allocation factors.

Potential photosynthesis minus maintenance respiration determines potential growth. If one of the nutrients reaches its minimum plant-tissue growth concentration then actual growth is limited by the available nutrients, otherwise it is limited by available CH_2O.

Year-averaged SO_2 concentrations and yearly peak concentration for SO_2 reduce photosynthesis in a logarithmic way. Season averaged O_3 concentrations do so in a linear way.

Uptake of nutrients is reduced linearly by low pH values and exponentially by high Al solution concentrations. Furthermore, Al reduces growth of fine roots according to an exponential relationship.

Mineralization of dead leaves and dead roots is considered for N. It follows a sinusoid during the year. Ca, Mg and K are released instantaneously. Fresh litter decays as a fraction of yearly leaf fall. For old leaf-litter a first-order decay function is used, depending on the mean groundwater level, pH and N content. The latter is used to calculate the immobilization. Fine roots mineralize according to a first-order function of dead-roots amount. Mineralization of woody roots is neglected.

Nitrification depends linearly on NH_4 in the soil layers, the mean groundwater level and pH.

Denitrification is proportional to the NO_3 concentration and dependent on pH and the highest groundwater level during the year.

Protonation of $RCOO^-$ is related to pH which reduces the protonation rate. $RCOO^-$ stems from mineralization of organic materials (leaves and roots). Its total flux equals the sum of those of NH_4, Ca, Mg and K minus that of NO_3.

Weathering of Ca, Mg, K and primary Al is assumed to be congruent. It depends linearly on mineral pools (silicate and aluminium oxide). Therefore Al is released automatically from silicate weathering. Al oxide is the second source of Al in the module. The weathering rate is dependent on H and on undersaturation with respect to gibbsite.

Linear sorption is considered for Ca, Mg and K. Redistribution of these components is calculated after all other processes have occurred.

Convective transport is driven by water infiltration minus root-water uptake in the layers. Transpiration equals total water uptake and depends on the leaf-area index using an empirical relation. The water carries along SO_4, NO_3, NH_4, Ca, Mg, K, Na, Cl, RCOO and H. From these pH and Ca/Al- and NH_4/K-ratios are calculated for each layer.

An uncertainty analysis has been carried out on SOILVEG and is published separately (Van Grinsven et al., 1990).

In order to regionalize the use of SOILVEG in DAS, input files have been generated for Douglas fir stands on 4 types of soil (coarse-textured leptic podzol, fine textured gleyic podzol, fine-textured humic podzol and coarse-textured albic arenosol) in the 20 receptor areas. These can be combined with deposition/concentration scenarios.

More information can be obtained from Dr. J.J.M. Berdowski and Ir. C. van Heerden, Dutch Research Institute for Nature Management (RIN), P.O. Box 9201, 6800 HB Arnhem, the Netherlands, +31-85-452991.

3.8 Forests module (FORGRO)

The Research Institute for Forestry and Urban Ecology (De Dorschkamp) is developing a forest growth model of which a simplified version is to be included in DAS. FORGRO was developed from existing simulation models and is based on the simulation approach developed by de Wit and co-workers, mainly at the Department of Theoretical Production Ecology of Wageningen Agricultural University in collaboration with a number of agricultural research institutes in Wageningen. More extended documentation can be found in the simulation monographs edited by Penning de Vries and Van Laar (1982), Van Keulen and Wolf (1986) and Rabbinge et al. (1989).

A detailed description of the model itself can be found in a series of reports containing technical model documentation (Jorritsma et al., 1990, Mohren, 1990, Mohren et al., 1990).

The general forest growth module was developed to analyze the relationships between environmental variables and growth of a forest stand. The module is based on a carbon-balance approach for whole trees and stands and simulates stand dynamics with time steps of one day. The module is based on the underlying physical, chemical and physiological processes. So far, the main emphasis has been on Douglas fir. At this stage of module development, it cannot be applied to deciduous trees.

To enable the analysis of nutrient cycling and the effect of soil acidification on growth, the entire carbon-balance module has been coupled to the soil acidification module RESAM (see section 3.6). To estimate root uptake from demand by the growing forest and supply as estimated using RESAM, a root uptake submodel developed by De Willigen and Van Noordwijk (1987) is used.

FORGRO itself is primarily meant for research purposes. It's too complicated and computation intensive to be used as a module of DAS in a regionalized fashion. Therefore a simplified version will be derived from it after uncertainty analysis has shown the most important variables. FORGRO's input needs are such that data for the simplified version can be collected on a regional scale. Delivery of this version is foreseen by the middle of 1990.

More information can be obtained from Dr. G.M.J. Mohren, Research Institute for Forestry (De Dorschkamp), P.O. Box 73, 6700 AA Wageningen, the Netherlands, Tel. +31-8370-95322.

3.9 Heather module (ERICA)

A module for the assessment of acid deposition on semi-natural vegetation (dry heathlands) was to become part of DAS. At the moment a module for wet heathlands (called ERICA in the DAS-context) is available. It has been developed under the name NUCOMP (Nutrient Competition) by Berendse (1987) of the Centre for Agrobiological Research (CABO), Wageningen and ported to DAS by RIVM.

The wet heathland for which the above module has been developed only makes up 10 to 20% of the total heathland area in the Netherlands. Therefore another module was to be developed by the Dutch Research Institute for Nature Management (RIN). Because of the extra time involved with the development of SOILVEG, this heathland module was still in its definition phase by May 1990.

ERICA simulates the competition between two species, Erica and Molinia, found on wet heathlands. In ERICA these species depend on N-input and light only. N-input comes from deposition, sheep and mineralization of litter. On the other hand nitrogen is removed by sheep, by nitrification and by mechanical means. The soil compartment is not simulated. Therefore acidification of heathland soils as such cannot be accounted for by this module. Acid by itself does not play a role in this module, the effects are due to eutrophication by N-input. However, N makes up about two thirds of the total potential acid input. Hence, this module produces useful information on acid-related changes of the vegetation of these heathlands under Dutch circumstances.

More information can be obtained from Dr. J.J.M. Berdowski and Ir. C. van Heerden, Dutch Research Institute for Nature Management (RIN), P.O. Box 9201, 6800 HB Arnhem, the Netherlands, +31-85-452991 and Dr.Ir. F. Berendse, Centre for Agrobiological Research (CABO), P.O. Box 14, 6700 AA Wageningen, the Netherlands, +31-8370-19017.

3.10 Acid-sensitive surface waters (AQUACID)

The module for aquatic systems, AQUACID, has been developed by RIVM (Wortelboer, 1990, 1991). It envisages explaining the phenomena observed in shallow acid-sensitive lakes by simulating the change in vegetation in these lakes due to acidification.

Acid-sensitive surface waters in the Netherlands are restricted to isolated heathland lakes. These waters do not have a catchment area and no surface-water inflows and outflows. These heathland lakes are found in sandy soils with low calcium contents. Low amounts of nutrients and organic material used to be the normal situation. They are therefore weakly buffered and acid-sensitive.

The surface area of these lakes ranges from 1 to 100 ha, with an average depth of 1 m. The regional groundwater level is much deeper then their bottom, but the water is kept in place locally due to impermeable iron-ore layers. The only supply of water stems from precipitation. Losses of water are due to evaporation and underground seepage across rims of the ore plates. Atmospheric deposition determines the nutrient input of these lakes.

Small brooks that originate from higher sandy areas are acid-sensitive as well. They have not been considered, since they are normally better buffered than the heathland lakes, so acid effects will show up later, after 50 years for instance. The acidification of these brooks will be reflected by the acidification of their watersheds, which could be derived using a soil acidification model like RESAM (see section 3.6).

The heathland lakes formerly had a very low production of biomass. Deposition only brought minor amounts of nutrients. Vegetation was dominated by submerged plants, mainly Littorella uniflora.

Research by the Catholic University of Nijmegen (Roelofs, 1983) and the Research Institute for Nature Management (Van Dam et al., 1981, Van Dam 1987) has shown that both vegetation and diatoms are subject to change. Moreover the heathland lakes have become more acid and show an increase in the amount of nutrients. Also, an accumulation of organic matter on the sediments has been observed. The new vegetation has become dominated by Juncus bulbosus.

Both situations are characterized in Table 3.10.1.

Table 3.10.1 Characteristics of non-acidified and acidified heathland lakes

Parameter	Original heathland lakes	Acidified heathland lakes
Dominant plants	Littorella uniflora	Juncus bulbosus
pH	6.5	3.9
CO_2 ($\mu mol.l^{-1}$)	40	150
NH_4^+ ($\mu mol.l^{-1}$)	5	40
Sediment	sandy, oxygenated	organic, reduced

Research focuses on the mechanisms behind these phenomena, but until now there has been no satisfactory explanation for the observed biological succession.

AQUACID aims to explain the observed phenomena on the time scale and with the speed found in practice, based on present chemical and physiological knowledge.

The process of acidification is seen as a balance (whether or not disturbed) between acidifying and alkalizing processes. In-lake acidity and alkalinity producing processes are regarded as the most important processes with respect to the ability of these aquatic systems to recover from acidification.

Current acidification models do not show very much detail (if they have it at all) in formulations concerning the in-lake production of alkalinity, and the role of submerged macrophytes therein. Besides, the role of a watershed is negligible in the case of the heathland lakes in the Netherlands, while many models focus on this aspect in much detail.

The module starts with the following assumptions:

* The aquatic systems are completely isolated with respect to surface water, groundwater

and surface runoff (no catchment area).

* Atmospheric input and the restricted input of allochtonous organic matter are the only sources of nutrients.
* Only submerged growing macrophyte species are considered.
* The reduction of water surface by lateral growth of shore plants (which appears to be enhanced by the higher nutrient levels) is not considered.

The module simulates the competition between the above mentioned submerged macrophyte species Littorella uniflora and Juncus bulbosus, which are characteristic of the possible states of these aquatic systems.

The modelling is based upon the ecophysiology of the species themselves, which differ greatly (Table 3.10.2):

Table 3.10.2 Characteristics of the modelled macrophyte species

Littorella uniflora	Juncus bulbosus
Low productivity	High productivity
Abstracts CO_2 from sediment	Abstracts CO_2 from the water
Organic material is broken down relatively well	Organic material breaks off slowly, and tends to accumulate
Only NO_3^- is taken up as a nutrient	Takes up NO_3^- as well as NH_4^+

Productivity produces acidity, while decomposition produces alkalinity. So, with a shift in species, the characteristics of the system under study may change completely with on-going acidification.

The module now contains one water layer and one soil layer. Chemical equilibrium is preserved in all layers (Table 3.10.3).

More information can be obtained from Drs. F.G. Wortelboer, RIVM/LWD, P.O. Box 1, 3720 BA Bilthoven, the Netherlands, Tel: +31-30-749111.

3.11 Agricultural production module (AGRIPROD)

This module, called AGRIPROD, has been implemented by RIVM, based on the research by the Research Institute for Plant Protection in Wageningen, the Netherlands (IPO) and the Agricultural Economics Research Institute in The Hague, the Netherlands (LEI).

Table 3.10.3 Processes included in the model, together with the components involved

Process	Compounds involved
Primary Production	Biomass of Littorella uniflora and Juncus bulbosus
Uptake/release by macrophytes	O_2, NO_3^-, NH_4^+, SO_4^{2-}, CO_2, H^+
Mineralization	O_2, NO_3^-, NH_4^+, SO_4^{2-}, CH_4, CO_2, H_2S, H^+
Oxidation	NH_4^+, H_2S, CH_4, H^+
Diffusion	All dissolved components
Exchange with atmosphere	CO_2, O_2 and CH_4
Dry deposition	SO_2, SO_4^{2-}, HNO_3, NO_3^-, NH_4^+
Wet deposition	SO_4^{2-}, NO_3^-, NH_4^+, H^+, OH^-
Acid-base reactions	H^+, OH^- CO_2-buffer system
Input of allochtonous organic matter	

AGRIPROD uses dose-effect relations that directly link the decrease of crop production to average concentration of O_3 and SO_2 in the air. O_3 is the most important, while other agents are irrelevant in the Netherlands on a regional scale. For SO_2, year-averaged concentrations are used. For O_3, season-averaged concentrations are used (May-Sep, 10.00-17.00h).

Damage is specified as a loss percentage of the total crop volume. The damage relations are linear and non-linear regression equations that link damage to concentrations. There is an equation for O_3 and one for SO_2, while there are coefficients for each crop or group of related crops.

For each acidification area the yearly production of the crops has been estimated. With the actual concentrations and the damage-regression equations the total reduction percentage of the agricultural production per area is calculated in terms of volume as well as money. For crops that suffer more from damage because of the concurrent loss of appearance, a 'severity' factor is applied on top of the volume loss. No market mechanisms are considered; the reduction of agricultural profit is based on the market price for each crop in an undamaged state. Prices may be updated for each year. Price scenarios for investigations into the future are not available. It is assumed that relative crop prices are stable and thus the relative agricultural losses can be calculated as a percentage of the total production.

Further information may be obtained from: Drs. A.E.G. Tonneijk, Ir L.J. van der Eerden IPO, P.O.Box 9060, 6700 GW Wageningen, the Netherlands, +31-8370-19151. and Ir. J.H.M. Wijnands, LEI, P.O.Box 29703, 2502 LS The Hague, the Netherlands, +31-70-3614161.

Implementation: Drs. F.G.Wortelboer, RIVM, P.O.Box , 3720 BA Bilthoven, the Netherlands, +31-30-749111.

3.12 Materials module (MATERIALS)

The module MATERIALS has been developed by RIVM, based on a study by the Free University of Amsterdam - Institute for Environmental Studies, (IVM). (See Gosseling et al., 1990). Its structure matches that of AGRIPROD (See section 3.11).

MATERIALS uses direct dose-effect relations between the year-averaged SO_2-air concentration and the damage.

Damage may be due to two factors: the loss of life expectancy for the materials and the increase of maintenance (mainly painting).

Materials considered are (Table 3.12.1):

Table 3.12.1 Kind of material damage considered

Material	life expectancy	maintenance
Duplex steel		*
Zinked steel		*
Painted steel		*
Zinked objects	*	
Zinc	*	

These materials are found in various kinds of objects like bridges, containers, balconies, etc. The "stock at risk" for these object categories has been estimated for each of the acidification areas in terms of total exposed area for duplex, zinked and painted steel and in terms of money value for zinked objects and zinc.

For each of these categories the normal life expectancy and specific maintenance-cost figures are provided. Based on these values and the SO_2 concentration total damage to these materials can be calculated.

There are no scenarios available for the future use of the various materials in the different areas. This use may very well be influenced by the damage itself, in that better protection is applied to new constructions beforehand. This accounts for the applied construction methods and materials. It then becomes very difficult to estimate damage per sé. The question can be raised as to whether the use of materials that are better resistant to acidification can be considered a kind of damage or whether this should be seen as caused by technological development.

More information can be obtained from Drs. J.F. Feenstra and Dr. A.A. Olsthoorn, IVM/VU (Institute for Environmental Studies), Free University of Amsterdam, De Boelelaan 1115, Provisorium III, 1091 HV Amsterdam, the Netherlands, Tel: +31-20-5483827.

3.13 Cultural assets (MONUMENTS)

The module MONUMENTS has been developed by RIVM, based on a study by the University of Amsterdam - Institute for Environmental Studies (IVM) (See Gosseling et al., 1990). Its structure matches that of AGRIPROD (See section 3.11).

The study for damage to cultural assets due to acidification has been limited to the damage to monuments that have natural-stone parts that are continuously exposed to outside air. Damage to other cultural assets such as paper, paintings, textiles etc. is not considered.

The approach taken stems from a study of Cambridge Decision Analysts, (1987). The gradual erosion of natural stone is taken as the measure of damage. Total loss is assumed as soon as one cm of thickness of natural stone has disappeared. Statues will then be completely formless whereas in larger monuments there will be spots that are much more severely damaged than this one cm might suggest.

It has been impossible to produce validated dose-effect relations for these monuments. From German research (Luckat, 1981) an empirical dose-effect relation between dry SO_x deposition (no concentrations) and the dissolution of a much used Baumberger Stone was taken for the estimation. The time it takes to render monuments worthless, can then be calculated from the actual environmental conditions.

Even with monuments it proved very difficult to estimate the stock at risk. The number of monuments is known, but the amount of natural stone at risk and the part of it that is exposed to outside air are not. Moreover it seems impossible to estimate the value of monuments. Therefore, only the number of monuments is used to estimate the damage.

The estimated damage to the monuments in the Netherlands is thus expressed as the reduced lifetime [years/yr] of the monuments in an area. These figures may be multiplied by the number of monuments to obtain a certain total for the damage.

More information can be obtained from Drs. J.F.Feenstra and Dr. A.A. Olsthoorn, IVM/VU (Institute for Environmental Studies), Free University of Amsterdam, De Boelelaan 1115, Provisorium III, 1091 HV Amsterdam, the Netherlands, Tel: +31-20-5483827.

4. IMPLEMENTATION

DAS consists of modules (see Figure 4) implemented as programs that can run independently. The series of modules characterize the causality chain. Feedback is therefore limited to within the individual modules. For instance there is no feed-back possible between effect modules and the air-transport module. Many of the modules are implemented as filters; they accept input and produce output that can be directly read by following modules.

The use of separate programs as modules guarantees easy distributed development of the "overall model" as well as reusability of the software. In this way the modules can be combined in several ways and can be exchanged without programming. New modules can be added into the system. Moreover, the modules can be used on their own, for other purposes and also for testing and uncertainty analysis. The main disadvantage of this approach is speed. But ease of maintenance and extendability is believed to be more important.

Most of the modules run under the UNIX operating system, on various computers within RIVM. RESAM and FORGRO also run in their host institutes under another operating system. The INTERACT modules (see below) are spreadsheet-based applications. They reside on a PC. Communication between modules is done via RIVM's local area network.

Emission scenarios for the Netherlands are generated in a separate database system with its own surrounding submodels, the so-called Dutch Environmental Information and Planning System (RIM) of the Laboratory of Waste Removal and Emissions (LAE). RIM delivers emissions per compound and per sector for the Netherlands as a whole for a number of reference years, starting in 1980. It also calculates costs of the abatement measures associated with each scenario.

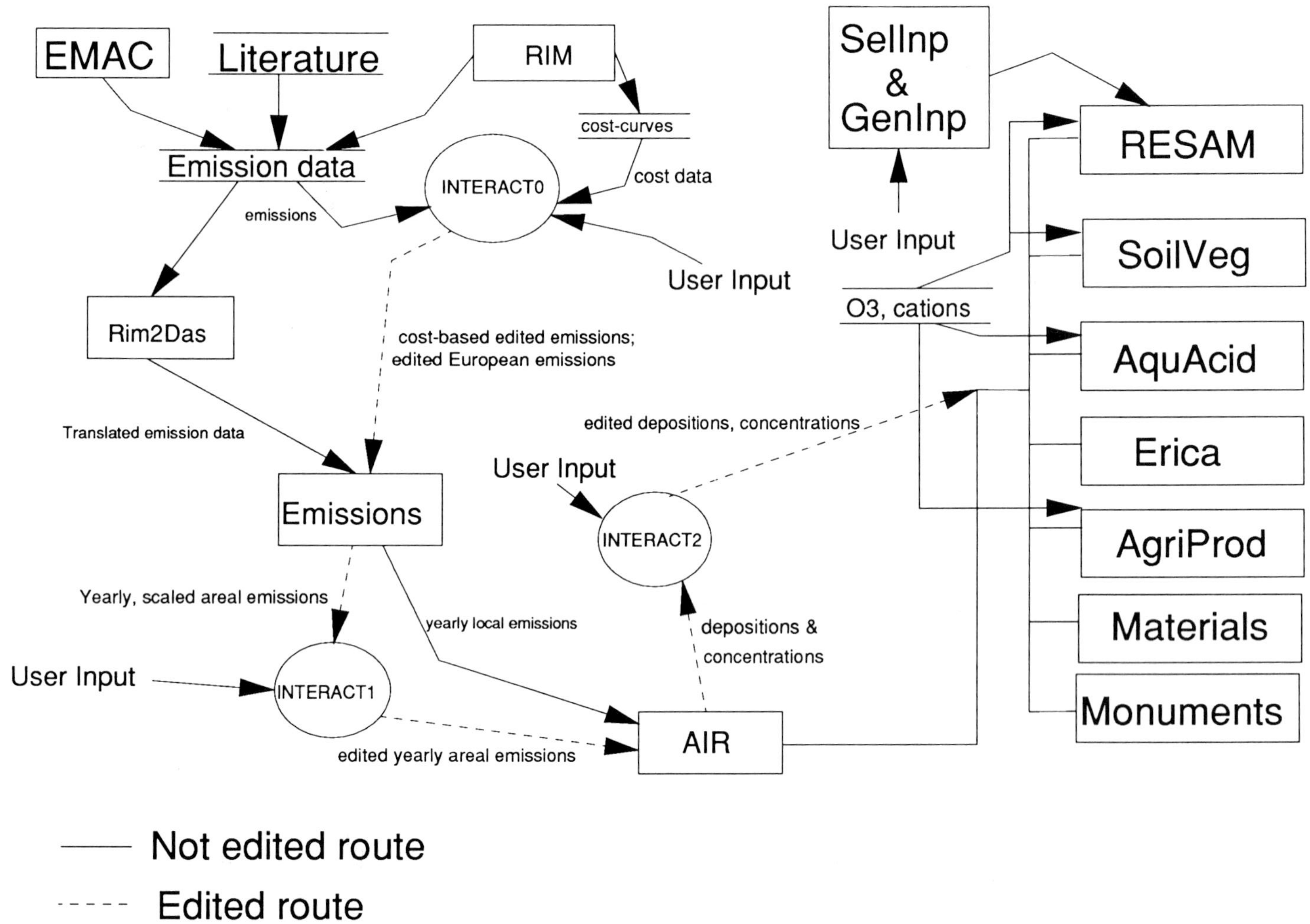

Fig. 4. Implementation overview of DAS, actual modules

Emission scenarios for surrounding countries are produced by the Agricultural University of Wageningen using the EMAC model. As with RIM, EMAC delivers emissions per sector and compound for the areas in DAS that surround the Netherlands for a number of reference years, starting in 1980.

For other European countries scenarios are built by RIVM by hand for the same reference years, using literature and information from international committees. These emissions are given per compound for low and high stacks respectively, without a sectoral subdivision.

These three emission sets are fed into the DAS module RIM2DAS. RIM2DAS translates specific codes of RIM and EMAC into those used in DAS and adds generic stack heights ("low", "medium" or "high") to sectors, to yield data in the following format:

Year	Sector	Compound	Emission	Unit
1980	XYrefineries	SO_2.high	14300	ton/yr

where XY is omitted for the Netherlands and replaced by 2 characters that identify a European area.

Instead of passing the three sets of sectoral emission data directly to the module RIM2DAS, you may want to load them into the spreadsheet-based module INTERACT0. INTERACT0 produces the same output as RIM2DAS, but lets you alter the sectoral emission values based on cost-abatement curves that it abstracts from the RIM-system. Using the cost-abatement curves you may allocate money to the abatement of a specific compound in a specific sector and year, relative to the current scenario. Using this information INTERACT0 alters the basic sectoral emission scenarios. It further acts in the same way as RIM2DAS and produces the same output format. INTERACT0 is a spreadsheet-based PC-application but delivers its output via the network to the host computer where it can be used by other modules.

The output of RIM2DAS or INTERACT0 is fed into the module called EMISSIONS. EMISSIONS interprets these records. It applies sector-specific and time-specific distribution-vectors to spread the Dutch national emissions over the Dutch areas. It then adds them up to produce emissions per compound, per stack height and per area. It also adds the sectoral foreign emissions in each European area together per compound and generic stack height ("high", "medium" or "low"). It then fills in the gaps by interpolation between reference years used by RIM and EMAC and sorts the output. It optionally scales data to other units. The output is a file with emissions per compound, per stack height and per area in the following format:

Year	Area#	Compound	Emission	Unit
1980	1	SO_2.high	42320	ton/yr
1980	3	NO_x.low	33000	ton/yr

etc.

This file can be directly fed into the air and transport module SRM. Before applying SRM, you may want to change the generated emissions using the module INTERACT1. Using this spreadsheet-based module you can set reduction factors for chosen compound/stack-heights ("high", "medium" or "low") in chosen areas at chosen times. The changes remain valid until new values are encountered in the data file. Instead of the reduction factors, absolute emissions may be specified in INTERACT1.

The emission file can now be processed by SRM to produce year-averaged concentrations and depositions. It only contains compound/stack height and area references and starts in 1980. Historic data are available from 1950 in the same format. They are added to the other data and the entire file is fed into SRM.

SRM starts by reading a set of source-receptor matrices (transfer matrices), one matrix for each specific input and output compound. SRM then continues by reading and interpreting the records of its emissions input. It produces year-averaged concentrations and depositions in the 20 Dutch receptor areas over the whole period for which there is input and for the set of compounds for which there are matrices and input.

The output of SRM has the following format:

Year	Area#	Compound	Dep/Conc	Unit
1980	1	SO_x.dry	1500	mol/ha/yr
1980	3	NO_x	4	$\mu g/m^3$

etc.

This format is compatible with those mentioned before. The unit string e.g. "mol/ha/yr", "$\mu g/m^3$", etc., together with a compound-string extension that can be omitted or set to "dry" or "wet" uniquely identifies the meaning of the values and serves to differentiate between deposition and concentrations.

This output of SRM can be used by all of the effect modules. Each interprets these records.

Before passing it on to the effect modules, you may want to alter the deposition and concentration output of SRM using the module INTERACT2. INTERACT2 allows you to change specific compounds (concentrations or depositions) in chosen areas and for chosen years. You can do so either by specifying relative factors or by typing in absolute values. This spreadsheet-based module produces a file that contains the changes to be made to the output of SRM. It is put automatically from the PC onto the host machine, where another program takes it up and brings about the changes to the original SRM output.

The effect modules RESAM, SOILVEG (FORGRO) and AQUACID also need base-cation deposition, chloride deposition and O_3 concentrations. Currently, DAS has no modules for these compounds. It uses prescribed scenarios instead. These are merged into the output of SRM to produce a complete deposition/concentration file that can be used by all of the effect modules.

The effect modules interpret the deposition and concentration records themselves, picking up compounds needed and ignoring the rest. The effect modules may be run independently.

In order to regionalize RESAM, a twin-module called SELINP/GENINP has been developed to select sets of combinations of vegetation and soil types that exist in the receptor areas. Thus the same vegetation and soil type combination may exist in different areas which in turn have a different deposition load. After selection of the combinations by SELINP, GENINP generates the proper input files for RESAM. It uses the database that contains the properties of the vegetation and soil types and their occurrence in the different areas. A regionalized scenario result is then obtained by running RESAM for each area with the area- and scenario-specific depositions and concentrations for all the area-specific combinations of vegetation and soil types. Clearly, execution of the computation intensive RESAM module for each of the maximum set of some 1500 vegetation/soil/area-combinations is a big task and can only be done for a limited set of scenarios.

Regionalization of SOILVEG is done in a simpler way. Only four soils and one vegetation type (Douglas fir) are discriminated. The required specific datafiles (parameters and initialization files) have been prepared in advance. They are combined with the occurrence of the soil types in each area to calculate a regionalized measure of potential forest damage. Of course runs can be restricted to only those soil-area combinations on which Douglas fir is really present. A maximum of 80 runs is needed to obtain complete information on potential Douglas fir damage on the 4 soils in the 20 Dutch acidification areas.

Regionalization of AQUACID is done by putting a standardized heathland lake in the areas in which these lakes are present in reality. Area-specific deposition factors, calculated using the actual sites of the lakes within the areas, are used to calculate the actual deposition in each area. 16 runs are needed to obtain complete regionalized information on the potential damage and behaviour of these lakes.

Regionalization with ERICA, the heathland module, is done by putting a standardized heathland in the different areas to obtain a measure of potential behaviour, or they are combined with those areas in which this vegetation is present in reality. 20 runs are needed to obtain complete regionalized information on the potential damage and behaviour of these heathlands.

The other modules AGRIPROD, MATERIALS and MONUMENTS are simple dose-effect relationships. Those modules have been regionalized internally and always produce answers for all of the receptor areas.

5. ADDITIONAL PC-BASED MODELS

Two non-dynamic PC-based models have been made to accompany DAS. They are used to more rapidly provide answers to questions that do not require a dynamic model and for which DAS itself is too computation intensive. These PC-based models are thus suitable for screening sets of scenarios in advance. They also make some features of DAS portable.

5.1 Prolog based policy model to calculate acid deposition on PC (ACID.EXE)

With this steady state model (Olsthoorn, 1988) emission reductions may be specified for any combination of economic sectors, sector activities, emitted compounds and emission areas. The model provides for this on various levels of aggregation up to the European scale. The model assumes that acidification policies may address sectors, activities, compounds and areas in any combined way and on different levels of aggregation. Then a policy will, in general, be partially effective only, because an economic sector will reduce its emission no further than the maximum of the set of heterogeneous requirements it is subjected to.

You can let the model calculate the total emission in acid equivalents for the specified combination and aggregation levels. At the same time, you can have it calculate the resulting average acid deposition for the Netherlands. You can thus calculate the combined effect of several sets of simultaneously active policy measures directed against acid deposition.

5.2 Spreadsheet based deposition model for PC (1985.WK1)

A steady state model has been constructed on a spreadsheet, using the same matrices as DAS. It serves to rapidly calculate the deposition onto the 20 Dutch receptor areas caused by the emission of SO_2, NO_x and NH_3 by economic sectors in the Netherlands and Europe (Olsthoorn and De Leeuw, 1988). It calculates the origin of the depositions and so can be used to devise abatement measures. The graphical capabilities of the spreadsheet are used to present the various results.

6. ABBREVIATIONS

1985.WK1	Accompanying spreadsheet based PC-model to DAS, developed by RIVM.
ACID.EXE	Accompanying PC-model to DAS, developed by RIVM.
AGRIPROD	Submodule of DAS, develeped by IPO, LEI and RIVM, to calculate damage to agricultural crops.
AQUACID	Submodule of DAS for aquatic systems, developed by RIVM. It simulates the acidification process in acid-sensitive heathland lakes.
CABO	The Centre for Agrobiological Research, Wageningen.
COROP	Division of the Netherlands into 40 economic zones, originated from CBS.
CPB	Central Planning Bureau, The Hague.
CBS	Central Bureau of Statistics, Rijswijk.
DAS	Dutch Acidification Systems Model.
De Dorschkamp	The Research Institute for Forestry and Urban Ecology, Wageningen.
ECN	Energy Research Foundation, Petten.
EMAC	A model for the generation of emission scenarios for acidifying compounds (SO_2, NO_x and VOC) in surrounding countries, developed bu LUW. Used to generate emission scenarios for DAS.
EMISSIONS	Submodule of DAS, developed by RIVM, to distribute national emissions across Dutch acidification areas.
ER	The Dutch Emission Inventory System
ERICA	Submodule of DAS for effects on wet heathlands, developed under the name NUCOMP by Berendse (1987) CABO. It simulates the competition between two wet-heathland plant species due to N input.
ESC	The Energy Study Centre, Petten.
EZ	Ministry of Economic Affairs, The Hague.
FORGRO	A general forest growth model under development by De Dorschkamp, to become a submodule of DAS.

GENINP	A submodule of DAS, developed by RIVM, to generate input files for RESAM.
INTERACT	The modules INTERACT0..INTERACT2 are submodules of DAS, developed by RIVM, that allow the user to change scenario data interactively.
IPO	The Research Institute for Plant Protection, Wageningen.
IvM	Free University of Amsterdam - Institute for Environmental Studies, Amsterdam.
LAE	RIVM Laboratory for Waste Materials and Emissions, Bilthoven.
L&V	Ministry of Agriculture & Fisheries, The Hague.
LEI	The Agricultural Economics Research Institute, The Hague.
LLO	RIVM Air Research Laboratory, Bilthoven.
LUW	Agricultural University of Wageningen, Wageningen.
MATERIALS	Submodule of DAS, developed by IvM and RIVM, to calculate the damage to materials.
MONUMENTS	Submodule of DAS, developed by RIVM and IvM, for the damage to monuments.
PC	Personal computer.
RESAM	Soil acidification module of DAS, developed by the Soil Survey Institute of Winand Staring Centre.
RIM	Environmental Information and Planning System of RIVM, used by DAS as scenario module for Dutch emissions.
RIM2DAS	Submodule of DAS, developed by RIVM, connecting RIM to DAS.
RIN	The Dutch Research Institute for Nature Management, Arnhem.
RIVM	National Institute of Public Health and Environmental Protection, Bilthoven.
SELINP	Submodule of DAS, developed by RIVM. It selects combinations of vegetations and soils to be included in regionalized use of RESAM. Input of SELINP is fed into GENINP.
SOILVEG	The soil and vegetation module of DAS, developed by RIN. It simulates growth of Douglas-fir stands and uses soil process formulations of RESAM.
SRM	The air-transport and transformation module, developed by RIVM. SRM uses source receptor matrices for the components SO_2, NO_x and NH_3.
TREND	Lagrangian long-range transport model of LLO.
UNIX	Computer operating system.
VOC	Voltile organic compounds.

VROM Ministry of Housing, Physical and Natural Environment, The Hague.

WSC Winand Staring Centre (including Soil Survey Institute), Wageningen.

6. REFERENCES

Coördinatiegroep Systeemonderzoek Verzuring, 1986. Onderzoek naar de effectiviteit van maatregelen ter voorkoming en bestrijding van de effecten van zure regen. ("Research on the effectiveness of the measures to provide and abate effects due to acid rain", In Dutch). RIVM, Verzuringsonderoek rapport 69-1, Bilthoven

Alcamo, J., M. Amann, J.-P. Hettelingh, M. Holmberg, L. Hordijk, J. Kämäri, L. Kauppi, P. Kauppi, G. Kornai & A. Mäkelä, 1987. Acidification in Europe; a simulation model for evaluating control strategies, AMBIO Vol 16, 5:232-245

Alcamo, J., R. Shaw, L. Hordijk (Eds.), 1990. The RAINS model of acidification; science and strategies in Europe, Kluwer Academic Publishers, Dordrecht, Boston, London

Baedecker P.A. et al., 1989. Effects of acidic deposition on materials, state of science/technology, NAPAP

Bakema A.H., K.F.de Boer, G.W.Bultman, J.J.M.van Grinsven, C.van Heerden, R.M.Kok, J.Kros, J.G.van Minnen, G.M.J.Mohren, T.N.Olsthoorn, W.de Vries & F.G.Wortelboer, 1990. Dutch Priority Programme on Acidification. Dutch Acidification Systems Model - Specifications. RIVM, Report nr. 114.1-01, Bilthoven

Belmans, C., J.H.G. Wesselingh & R.A. Feddes, 1983. Simulation of the water balance of a cropped soil: SWATRE. J. Hydrol. 63:271-271

Berdowski, J.J.M., C. van Heerden & J.G. van Minnen, 1990. SOILVEG: a model for effects of acid atmospheric deposition on soil and forest. Part A: Model structure, vegetation processes and preliminary results. Part B: Soil processes. Dutch Research Institute for Nature Management, Arnhem, in prep.

Berendse, F., 1988. Een simulatiemodel als hulpmiddel bij het beheer van vochtige heidevelden (A simulation model as a tool to maintain wet heathlands). CABO, Wageningen, Report 743

Buijsman, E. & W.F. Blom, 1989. Luchtkwaliteit 1988 (Air quality, 1988). RIVM, Bilthoven, Report 228702015, 200pp.

Cambridge Descision Analists Ltd., 1987. A Decision Analysis of Costs and Benefits of Acid Deposition and Photochemical Oxidants, Control Policies in the European Community. 3 Volumes: Vol 1: Analytical structure, data input and supporting information. Vol 2: Analysis results. Vol 3: Policy document. EEC, study contract No B6612/02/85/AD. DGXI and VROM, Brussels and Leidschendam, Sept 1987

Dam, H. van, G. Suurmond & C.J.F. ter Braak, 1981. Impact of acidification on diatoms and chemistry of Dutch moorland pools. Hydrobiologia, 83:425-459

Dam, H. van, 1987. Monitoring of chemistry, macrophytes and diatoms in acidifying moorland pools. Dutch Research Institute for Nature Management (RIN), Report 87/19, Arnhem, 115pp.

Derwent, R.G, O.Hov, W.A.H. Saman, J.A. van Jaarsveld & F.A.A.M. de Leeuw, 1989. An intercomparison of long-term atmospheric transport models; the budgets of acidifying species for the Netherlands. Atmospheric Environment Vol 23, No.9:1893-1909

Eerden, L.J. van der, A.E.G. Tonneijk & J.J.M. Wijnands, 1988. Crop Loss Due to Air Pollution in the Netherlands. Research Institute for Plant Protection (IPO) & Agricultural and Economics Institute (LEI), Wageningen & The Hague. Also in Environm. Pollut. 53(1988)365-376

Elskamp H.J. (Ed.), 1989. National air-quality monitoring network. Technical Descriptions. RIVM, Bilthoven, Report 228702017, 172pp.

Erisman, J.W., F.A.A.M. de Leeuw & R.M. van Aalst, 1987. Depositie van de voor verzuring in Nederland belangrijkste componenten in de jaren 1980-1986 (Deposition of the most important acidifying components in the Netherlands in 1980-1986). RIVM, Bilthoven, Report 228473001, 57pp.

Florax, J.P.G.G.M., P.J.Kowalik & G.M.J. Mohren 1990. Water relations in Douglas-fir stands: A modelling approach to tree water status and transpiration. Research Institute for Forestry and Urban Ecology "De Dorschkamp", in prep.

Gosseling, H.J., A.A.Olsthoorn & J.F.Feenstra, 1990. Schade aan materialen door verzuring. ("Damage to materials by acidification, Stock at Risk and Damage Functions"). IVM-VU report, Nr. W-170, Jan 1989, revised April 1990, 31pp.

Grinsven, J.J.M. van, P.H.M. Janssen, J.G. van Minnen, C. van Heerden, J.J.M. Berdowski & R. Sanders, 1990. SOILVEG: A model for effects of acid atmospheric deposition on soil and forest. Uncertainty analysis. Jointly by RIVM and Winand Staring Centre, Bilthoven and Wageningen, 1990, in prep.

Guilmot, J. et al., 1986. Energy 2000, a reference projection and alternative outlooks for the European Community and the world to the year 2000. Cambridge University Press

Hettelingh, J.-P., 1989. Uncertainty in modelling regional environmental systems: The generalization of a watershed acidification model for predicting broad-scale effects. PhD Thesis, Free University of Amsterdam, available as IIASA research report RR 90-3, Laxenburg, Austria, 1990

Hove, L.W.A. van, 1989. The mechanism of NH_3 and SO_2 uptake by leaves and its physiological effects. Ph.D. thesis, Wageningen Agricultural University, Wageningen 1989, 131 pp.

Ierland, E.C. van & E. Oskam, 1988. A regional sectoral scenario model for SO_2 and NO_x emissions in European Countries. In Environmental Policy in a market economy. Pudoc, Wageningen

Jaarsveld, J.A. van, R.M. van Aalst & D. Onderdelinden, 1986. Depositions of metals from the atmosphere into the North Sea: Model Calculations. RIVM, Bilthoven, Report, 842015002

Jaarsveld, J.A. van, 1988. Modelling of NOy budgets over the Netherlands. Paper presented at ECE-WMO Workshop on Modelling, Transformation Processes and Transport of Air Pollution with special Reference to Nitrogen Oxides, Potsdam (DDR), 21-24 March, 1988

Jaarsveld, J.A. van, 1989. A model approach for assessing transport and deposition of acidifying components on different spatial scales. WMO, special environment report No. 17 "Changing composition of the troposphere". WMO, Geneva, Report 724

Jaarsveld, J.A. van, 1990. A quantitative model analysis of year to year changes in concentration and deposition. Paper presented at the NATO/CCMS 18th ITM on air pollution, modelling and its application, 13-17 May 1990, Vancouver, Canada

Janssen, P.H.M., W. Slob & J. Rotmans, 1990. Gevoeligheidsanalyse en onzekerheidsanalyse; een inventarisatie van ideeân, methoden en technieken uit de literatuur. (Sensitivity analysis, uncertainty analysis; a literature inventory of ideas, methods and techniques. In Dutch). RIVM, Bilthoven, report 958805001

Jorritsma, I.T.M., G.M.J. Mohren & J.R. van de Veen 1990. Effects of air pollution and soil acidification in a general model of forest growth: FORGRO 3.3. Research Institute for Forestry and Urban Ecology "De Dorschkamp", in prep.

Kernighan, B.W. & D.M. Ritchie, 1988. The C programming Language. Prentice Hall, Software series, ISBN 0-13-110370-9

Keulen, H. van & J. Wolf (eds.) 1986.Modelling of agricultural production: weather, soils and crops. Simulation Monograph. Pudoc, Wageningen, 478 pp.

Koster, P.K. & M. Waltmans, 1986. Integrated acidification model, 1st interim report (In Dutch). RIVM, Bilthoven, Systeemonderzoek Verzuring, Project 89

Kros, J., P.H.M. Janssen, W. de Vries & C.I. Bak, 1990. Gebruik van onzekerheidsanalyse bij modelberekeningen. Een toepassing op het regionale bodemverzuringsmodel RESAM. (Use of uncertainty analysis for model calculations applied on RESAM, in Dutch). Winand Staring Centre, Wageningen, Report 65, 125 pp.

Kros, J., W. de Vries, P.H.M. Janssen & C.I. Bak, 1990. The uncertainty of forecasting regional trends of forest-soil acidification. Paper to be presented on the Int. Conf. on Acid Deposition, Glasgow, Sept. 1990

Luckat, S., 1981. Quantitative Untersuchung des Einflusses von Luftverunreinigungen bei der Zerstörung von Naturstein. (Influence of air pollution on destruction of natural stone, in German). Umweltbundesamt, Berlin, Forschungsbericht 106 08 003/02

Mohren, G.M.J., 1987. Simulation of forest growth, applied to Douglas fir stands in the Netherlands. Ph.D.-thesis, Wageningen, Agricultural University, Wageningen 1987, 184 pp.

Mohren, G.M.J. (ed.), 1990. Water relations and nutrient demands in a general forest growth model: FORGRO 3.1. Research Institute for Forestry and Urban Ecology "De Dorschkamp", in prep.

Mohren, G.M.J. & H. Bartelink, 1990. Modelling the effects of needle mortality rate and needle area distribution on dry matter growth of Douglas-fir in the Netherlands. Netherlands Journal of Agricultural Science 38: 53-66

Mohren, G.M.J., I.T.M. Jorritsma, J.P.G.G.M. Florax, H.H.Bartelink & J.R.van de Veen 1990a. FORGRO 3.0: A basic forest growth model. Model documentation and listing. Research Institute for Forestry and Urban Ecology "De Dorschkamp", Report nr. 524 (in prep.)

Mohren, G.M.J., I.T.M. Jorritsma, J.R. van de Veen & J.P.G.G.M. Florax, 1990b. Quantification of the effects of air pollution and acid deposition on two Douglas fir stands in the Netherlands. Research Institute for Forestry and Urban Ecology "De Dorschkamp", in prep.

Mohren, G.M.J. & R. Rabbinge 1990.Growth-influencing factors in dynamic models of tree growth. Proceedings of the conference on "Forest growth: Process Modelling of Responses to Environmental Stress". April 19-22, 1988, Gulf Shores, Alabama, U.S.A. (in press)

Mohren, G.M.J., J.R. van de Veen, A. Olsthoorn & I.T.M. Jorritsma 1990c. Root growth and uptake of water and nutrients in a general forest growth model: FORGRO 3.2. Research Institute for Forestry and Urban Ecology "De Dorschkamp", in prep.

Olsthoorn, T.N., 1988. Integrated environmental models for management and policy makers (In Dutch). RIVM, Report 758805003, Bilthoven, 17pp.

Olsthoorn, T.N. & F.A.A.M. de Leeuw, 1988. Calculation of acid deposition in the Netherlands based on transfer matrices. (In Dutch). RIVM, Bilthoven, Report 758805005, 37pp + diskette

Olsthoorn, T.N., 1988. Interactive policy model to experiment with acidifying emissions and depositions (In Dutch). RIVM, Bilthoven, Report 758805002, 23pp + diskette

Penning de Vries, F.W.T. & H.H. van Laar (eds.), 1982.Simulation of plant growth and crop production. Simulation Monograph. Pudoc, Wageningen, 320 pp.

Rabbinge, R., S.R. Ward & H.H. van Laar (eds.), 1989. Simulation and systems management in crop protection. Simulation Monograph. Pudoc, Wageningen, 420 pp.

Roelofs, J.G.M., 1983. Impact of acidification and eutrophication on macrophyte communities in soft waters in the Netherlands. I. Field Observations. Aquatic Botany 17:139-155

Schneider, T. & A.H.M. Bresser, 1986. Dutch priority Program on Acidification; report no. 00-01, Programme And Projects. RIVM, Bilthoven

Schneider, T. & A.H.M. Bresser, 1988. Dutch priority Programme on Acidification. RIVM, Evaluation Report on Acidification (In Dutch). RIVM, report no. 00-06, Bilthoven

Schneider, T. & A.H.M. Bresser, 1988. Evaluation of Phase 1 of the Dutch priority Programme on Acidification. Summary presented at the Third meeting of Acidification Research Co-ordinators (MARC III), Bracebridge, Canada

Shriner, D.S., W.W. Heck, S.B. McLaughlin, D.W. Johnson, J.D. Joslin & C.E. Peterson, 1990. Response of vegetation to atmospheric deposition and air pollution. NAPAP, Report 18, ORNL/ESD Publication No. 3452

Thomas, R., W.G. van Arkel, H.-P. Baars, E.C. van Ierland, K.F. de Boer, E. Buijsman, T.J.H.M. Hutten & R.J. Swart, 1988. Emission of SO_2, NO_x, VOX and NH_3 in the Netherlands and Europe in the period 1950-2030. The Emissions Module the Dutch Acidification Simulation Model. RIVM, report 758472002, Bilthoven, 118pp.

Thomas, R., R.J.M. Maas, R.J. Swart & R.M. Kok, 1989. Environmental Information and Planning System "RIM". In Carlsson et al. (Eds) Risì Int. Conf. on Env. Models Emissions and Consequences, Risì, Denmark, 22-25 May 1989, Elsevier, Amsterdam

TNO, 1985. Gegevens Emissieregistratie 1e en 2e ronde (Data from the Emission Inventory System, 1st and 2nd inventory.) TNO, Delft

Vries, W. de, M.J.P.H. Waltmans, R. van Versendaal & J.J.M. van Grinsven, 1988. Aanpak, structuur en voorlopige procesbeschrijving van een bodemverzuringsmodel voor toepassing op regionale schaal (Approach, structure and current process formulation of a model to simulate soil acidification on a regional scale, in Dutch). Staring Centre, Report 2014, Wageningen

Vries, W. de & H. Kros., 1989. De lange termijn effecten van verschillende depositiescenario's op representatieve bosbodems in Nederland (Long-term effects of various depositon scenarios on representative forest soils in the Netherlands, In Dutch). Winand Staring Centre, Report 30, Wageningen

Vries, W. de, M. Posch & J. Kämäri, 1989. Simulation of the long-term soil response in various buffer ranges. Water, Air and Soil Pollution 48:349-390

Vries, W. de & J. Kros, 1990. Long-term impact of acid deposition on the aluminium chemistry of an acid forest soil. In J. Kämäri, D.F. Brakke, A. Jenkins, S.A. Norton & R.F. Wright, Eds., 1988. Regional acidification models geographic extent and time development. Springer Verlag, Berlin, Heidelberg, New-York, 1990, p113-118

Vries, W. de & J. Kros, 1990. Assessment of critical loads and the impact of deposition scenarios by steady state and dynamic soil acidification models. Winand Staring Centre, Wageningen, in prep.

Waltmans, M., 1985. Prototype of a simulation model directed to the effectiveness of policy measures against acidification (In Dutch), RIVM, Bilthoven

Willigen, P. de & M. van Noordwijk, 1987. Roots, Plant Production and Nutrient Use Efficiency. Thesis, Wageningen Agricultural University, Wageningen, 282 pp.

Wortelboer, F.G., 1990. A model of the competition between two macrophyte species in acidifying shallow soft-water lakes in the Netherlands. Hydrobiol. Bull. 24, 91-107

Wortelboer, F.G., 1991. AQUACID: Acidification model of shallow soft-water lakes in the Netherlands. RIVM, Bilthoven, in prep.

ASSESSMENT OF CRITICAL LOADS AND THE IMPACT OF DEPOSITION SCENARIOS BY STEADY STATE AND DYNAMIC SOIL ACIDIFICATION MODELS

W. de Vries[1)]
J. Kros[1)]

1) The Winand Staring Centre, Wageningen

ABSTRACT

An overview is given of steady state and dynamic soil acidification models, together with applications on a local and national scale. The application of a simple steady state model shows that the average critical load for potential acidity on non-calcareous sandy forest soils in the Netherlands is approximately 1400 mol_c ha^{-1} yr^{-1}. This is based on critical values for the Al concentration and molar Al/Ca ratio in the soil solution. A further decrease in soil solution pH, caused by depletion of Al-hydroxides, is also avoided at this level. Deposition reductions generally lead to a fast improvement of the soil solution chemistry. Results of dynamic model simulations show that deposition reductions up to 1400 mol_c ha^{-1} yr^{-1} as an average for Dutch forest before 2010 cause a substantial decrease in the exceedance of critical Al concentrations and Al/Ca ratios in forest topsoils (from about 80% to 20%).

PREFACE

This report summarizes the results of the model-oriented acidification research at the Winand Staring Centre for Integrated Land, Soil and Water Research during the period 1985-1990. This research was part of the first and second phase of the Dutch Priority Programme on Acidification and was carried out in the context of project 7113: "Development, data derivation, evaluation and application of a regional soil acidification model".

Originally, project 7113 only aimed at the development and application of a dynamic soil model (RESAM) within an overall Dutch Acidification Simulation model (DAS). The ultimate aim was the assessment of long-term impacts of various emission-deposition scenarios on Dutch forest soils in order to assist policy makers in the choice of optimal abatement strategies. However, within the context of these projects, simple steady state models (START, MACAL) were also developed to assess long-term critical loads of nitrogen and sulphur, as this became an important political issue during the research period. Furthermore, research was broadened to an international scope by developing a relatively simple dynamic soil model (SMART) within the overall Regional Acidification and Information Simulation model (RAINS) developed at the International Institute for Applied System Analysis (IIASA). This model will be applied on a European scale.

This report summarizes all model-related research results of the Winand Staring Centre up to December 1990, both with respect to critical loads and to long-term impacts of emission-deposition scenarios with an emphasis on national applications. The methods, data, literature sources etc. that have been used to obtain these results are described in separate reports and articles. This report can also be used as a guideline for reading these reports and articles.

In the coming period 1991-1993, the research will be continued within the context of two new research projects with the common title: "Assessment and mapping of critical acid loads on forest ecosystems and evaluation of abatement strategies". One project (7160) aims at application on a National scale and the other (7156) at application on a European scale.

SUMMARY

An overview is given of steady state and dynamic soil acidification models together with applications on a national and European scale. The steady state models (SMB, START and MACAL) have been developed for the assessment of critical loads for nitrogen and sulphur, whereas the dynamic models (SMART and RESAM) have been developed to gain insight in the long-term impact of deposition scenarios.

SMB (Simple Mass Balance model), START (Steady State SMART model) and SMART (Simulation Model for Acidifications Regional Trends) are one-layer models, which are specifically developed for application on a European scale, whereas MACAL (Model to Assess a Critical Acid Load) and RESAM (Regional Soil Acidification Model) are multi-layer models, developed for application in the Netherlands. Processes included in all models are base cation weathering, net uptake of nitrogen and base cations, Al dissolution from hydroxides and element leaching from the rootzone. Multi-layer models also include nutrient cycling by litterfall, root decay, mineralization and uptake, whereas dynamic models include the effects of cation exchange and sulphate adsorption/desorption.

Average critical loads for the sum of sulphur and nitrogen on non-calcareous forest soils, as derived by the application of the steady state SMB model, vary between 1300-1900 $mol_c\ ha^{-1}\ yr^{-1}$. For groundwater and surface water, the values are lower, i.e. 400-700 $mol_c\ ha^{-1}\ yr^{-1}$. Although the uncertainty in critical load values can be large, this shows that substantial emission reductions are needed.

Dynamic model simulations of the long-term soil response to current deposition rates show that the pH of calcareous dune soils with a low carbonate content will drop significantly in the coming decades unless major emission/deposition reductions are implemented. This is due to the low buffering capacity of dune soils by cation exchange and Al dissolution. Simulations about the long-term impact on non-calcareous soils show a further decrease in the pH in the topsoil due to depletion of Al-hydroxides.

Dynamic simulations of the impact of deposition changes show that deposition reductions generally lead to a fast improvement of the soil solution chemistry. This includes an increase in pH and a decrease in Al, SO_4 and NO_3 concentration and Al/Ca and NH_4/K ratio. However, for the NO_3 concentration and NH_4/K ratio there is a clear time lag between the reduction of deposition and concentration, which is mainly due to N mobilization from the litter layer.

An analysis of the uncertainty in long-term model predictions in the Netherlands shows that deposition of SO_2, NO_x and NH_3 and parameters that determine nitrogen and aluminium dynamics, play an important role in this uncertainty. On a European scale, the uncertainty in CEC and base saturation is also (very) important.

Predicted median values of element concentrations in the soil solution of non-calcareous sandy forest soils in 1990 generally show a reasonable agreement with measured median values for 150 forest stands in the same year. However, there is a clear underprediction of the base cations Ca and K, especially in the subsoil which might be due to the long-term effect of liming and fertilization.

Results of critical load calculations for all Dutch forest-soil combinations show that the median values are nearly all in a range between 1000-1500 $mol_c\ ha^{-1}\ yr^{-1}$ for non-calcareous sandy soils, between 1500-2000 $mol_c\ ha^{-1}\ yr^{-1}$ for loess soils and above 3000 $mol_c\ ha^{-1}\ yr^{-1}$ for clay soils, peat soils and calcareous soils.

Results of dynamic model simulations for the Netherlands as a whole, show that deposition reductions up to 2500 $mol_c\ ha^{-1}\ yr^{-1}$, which is a target for the year 2000, reduces the exceedance of a critical Al concentration of 0.2 $mol_c\ m^{-3}$ and a critical Al/Ca ratio of 1.0 in forest topsoils from about 80% to 50%. Reductions up to 1400 mol_c as an average for the Netherlands, which are aimed before 2010, cause a further decrease in the exceedance of these criteria to 20%. Further reductions up to 1400 $mol_c\ ha^{-1}\ yr^{-1}$ as an average for forest areas only, lead to a negligible exceedance of these criteria. Depletion of Al hydroxides is also avoided at this level. These simulations also show that the influence of the tree species on the soil solution chemistry is more important than the influence of the deposition area.

1. INTRODUCTION

Apart from direct effects, resulting in visual damage of the forest canopy, the deposition of SO_2, NO_x and NH_3 affects forest vitality by indirect, soil-mediated effects on the roots. The most notable effects are mobilization of Al (acidification) and accumulation of NH_4

(eutrophication), causing inhibition of base cation (Ca, MG and K) uptake by roots, due to unfavourable molar ratios of Al and NH_4 to these base cations. Additional indirect effects of nitrogen include changes in vegetation due to a high nitrogen supply, increased susceptibility to frost and fungal diseases related to high leaf N contents and increased nitrate leaching to the groundwater (De Vries, 1988; De Vries and Gregor, 1990; De Vries, 1991a).

Information on the long-term effects of acid deposition on soils is very important for the formulation of policies for emission reductions. In this respect, models provide an important tool to derive critical acid loads and to assist decision makers in evaluating the effectiveness of abatement strategies.

In order to analyse environmental impacts from various emission scenarios, the integrated Dutch Acidification Simulation model (DAS) has been developed. This overall model gives a quantitative description of the linkages between emissions, deposition and environmental impacts such as soil acidification and effects on terrestrial and aquatic ecosystems. A similar overall model called RAINS (Regional Acidification Information and Simulation model) has been developed at the International Institute for Applied Systems Analysis (IIASA) for application on a European scale. Within the DAS and RAINS model, the soil acidification models form an important link between atmospheric deposition and effects on forest, groundwater and surface water.

This report summarizes the structure and application of the various steady state (SMB, START and MACAL) and dynamic soil acidification models (SMART and RESAM) that have been developed in this context. Chapter 2 gives an overview of the developed models. A summary of simulation results is given in chapter 3. Chapter 4 contains a short discussion about the uncertainties related to the modelpredictions and chapter 5 gives a state of the art description of regional model applications. In each chapter the articles and/or reports dealing with the subject in more detail are referred to. Special emphasis is given to the models and results on the national scale. The report ends with conclusions and recommendations (chapter 6).

2. MODELS

2.1 Introduction

A clear distinction has to be made between steady state and dynamic soil acidification models. Steady state models are particularly useful to derive critical loads for S and N (acid). These models, which only include processes that influence acid production and consumption during infinite time, such as weathering and net uptake, directly predict chemical values for relevant ions in the soil solution (e.g. Al, Ca, NH_4, etc.) in an equilibrium situation. Dynamic soil acidification models are particularly useful to predict the time period before a critical chemical value has been reached. These models also include processes that influence the acid production and consumption on a finite time scale, such as cation exchange, nitrogen mineralization/immobilization and sulphate adsorption/desorption.

Summarizing, steady state models are useful to determine the final emission amounts based on an acceptable acidification status in an eventually reached steady state situation, whereas, dynamic models are necessary to determine an optimal emission scenario, based on the temporal evolution of the acidification status.

Here, we describe the structure of the various steady state - and dynamic soil acidification models that have been developed for application on a National and European scale.

2.2 Steady state models

2.2.1 Basic principles

An overview of the methodology to derive critical loads, including the description of various steady state models, is given in Sverdrup et al. (1990) and De Vries (1991b). De Vries (1991b) focuses on the steady state models that have been developed at the Winand Staring Centre, i.e. simple mass balance models for deriving critical acid -, nitrogen - and ammonium loads, a one-layer soil model excluding nutrient cycling (START) and a multi-layer (variable depth) soil model (MACAL) including nutrient cycling.

In all steady state models, the derivation of critical loads is based on the implicit assumption that dynamic processes such as cation exchange, adsorption/desorption of sulphate and ammonium and mineralization/immobilization dynamics of nitrogen, sulphur and base cations are unimportant for the assessment of a long-term critical load.

Further assumptions in all steady state models are:
- Negligible nitrogen fixation;
- Negligible net uptake, reduction and precipitation of sulphate
  (sulphate is assumed to behave as a tracer).

A justification of these assumptions is given in De Vries and Gregor (1990) and De Vries (1991a, b).

2.2.2 Simple mass balance models

2.2.2.1Acidification model

The simplest steady state model developed, to derive critical loads for potential acidity is a one-layer simple mass balance model (SMB). This model ignores the effect of nutrient cycling by litterfall, mineralization and root uptake on the soil solution composition in the rootzone. It includes only net uptake of nitrogen and base cations. Additional to the assumptions given in par. 2.2.1, NH_4 leaching is assumed negligible which implies a complete NH_4 uptake and/or nitrification in the rootzone.

The critical load for potential acidity is calculated as (De Vries, 1991a, b):

$$CL\ (AC_{pot}) = BC^{*}_{dd} - BC_{gu} + BC_{we} + N_{gu} + N_{de} + N_{im}(crit) - Alk_{le}(crit) \quad (1)$$

where $CL(AC_{pot})$ stands for the critical load of potential acidity, BC^{*}_{dd} is the dry deposition of base cations not balanced by chloride, BC_{gu} and N_{gu} is the growth uptake (net uptake needed for forest growth) of base cations and nitrogen respectively, BC_{we} is the base cation weathering, N_{de} is the denitrification flux, $N_{im}(crit)$ is the long-term critical nitrogen immobilization, and Alk_{le} (crit) is the critical leaching flux of alkalinity. The element fluxes in equation (1) are all given in $mol_c\ ha^{-1}\ yr^{-1}$.

In words, equation (1) states that the critical load for potential acidity is equal to the net neutralizing input of base cations (BC^{*}_{dd}-BC_{gu}+BC_{we}) plus the net uptake and long-term immobilization of nitrogen minus a critical alkalinity leaching.

The alkalinity [Alk] can be defined as (De Vries, 1991a):

$$[Alk] = [HCO_3] + [RCOO] - [H] - [Al] \quad (2)$$

where [] denotes the concentration in $mol_c\ m^{-3}$.

In forest soils, the critical pH is such (about 3.7) that $[HCO_3]$ can be neglected. [RCOO] can be ignored, by assuming that it is completely bound to Al and by taking [Al] equal to the in-organic Al-concentration, which is toxic to roots (De Vries, 1988; De Vries and Gregor, 1990; De Vries 1991a, b).

For forest soils, the critical alkalinity leaching flux thus simplifies to:

$$Alk_{le}(crit) = - H_{le}(crit) - Al_{le}(crit) \tag{3}$$

There are three options for the calculation of the critical Al leaching flux depending upon the criterion used, i.e. (De Vries 1991a, b): (1) a criterion for the Al concentration in the rootzone (equation 4), (2) a criterion for the or molar Al/Ca ratio in the rootzone (equation 5) and (3) a negligible depletion of Al-hydroxides (equation 6):

$$Al_{le}(crit) = FW.[Al](crit) \tag{4}$$

$$Al_{le}(crit) = RAlCa(crit) \,.\, (BC^{*}_{dt} + BC_{we} - BC_{gu}) \tag{5}$$

$$Al_{le}(crit) = r \,.\, BC_{we} \tag{6}$$

where [Al](crit) is a critical Al concentration in $mol_c\ m^{-3}$, RAlCa(crit) is a critical molar Al/Ca ratio in $mol_c\ mol_c^{-1}$ and r is a stoichiometric ratio of Al to BC weathering. These options affect the critical load value (see par. 3.1).

The critical H leaching flux is calculated as:

$$H_{le}(crit) = FW.[H](crit) \tag{7}$$

where FW is the water flux in $m^3\ ha^{-1}\ yr^{-1}$.

The critical H concentration is related to the critical Al concentration according to (Sverdrup et al., 1990; De Vries, 1991a, b):

$$[H](crit) = ([Al](crit) / K_{gibb})^{1/3} \tag{8}$$

where K_{gibb} is the gibbsite equilibrium constant in $(mol_c\ m^{-3})^{-2}$.

The value of the critical Al concentration is determined by the critical Al leaching flux divided by the water flux. Although this steady state model has been developed for application on forest soils, it can also be used to derive critical loads for groundwater and surface water. The only difference in the application of equation (1) for the various receptors is a change in system boundaries, i.e. the rootzone for forest soils, the unsaturated zone (depth to groundwater extraction) for groundwater and a catchment for surface water (Sverdrup et al., 1990; De Vries, 1991a, b). This influences the areal weathering rate (in $mol_c\ ha^{-1}\ yr^{-1}$) by a difference in the considered depth of the soil profile. Apart from weathering there is also a difference in critical alkalinity leaching for the various receptors. This is due to different criteria for the critical alkalinity value

In forest soils, the critical alkalinity is negative. It thus leads to an increase in the critical load value (equation 1) by allowing a critical rate of Al buffering. Contrary to forest soils, the critical alkalinity value for groundwater and surface water is positive since the critical pH for these systems is above 5. This causes a strong increase in [HCO_3] that is related to [H] and the partial CO_2 pressure (pCO_2 in mbar)according to:

$$[HCO_3] = KCO_2 \, . \, pCO_2 \, / \, [H] \tag{9}$$

where KCO_2 is the dissociation constant for CO_2 in $(mol_c \, m^{-3})^2 \, mbar^{-1}$.

2.2.2.2Eutrophication models

Independent from acidification, an upper limit is set on the nitrogen deposition by the eutrophication aspect. This load can be calculated as (Schulze et al., 1989; De Vries, 1991a, b):

$$CL(N_{dt}) = N_{gu} + N_{de} + N_{im}(crit) + NO_{3,le}(crit) \tag{10}$$

The critical nitrate leaching rate ($NO_{3,le}$(crit)) can be calculated by multiplying the water flux with a critical nitrate concentration [NO_3](crit).

It is also possible to derive a critical load for NH_3-N deposition. The accumulation of ammonium in the soil, induced by the deposition of NH_3, appears to inhibit the growth of ectomycorrhizae, which play an important role in the nutrient uptake by many coniferous trees. Imbalanced nutrient concentrations in the soil solution thus cause potassium and magnesium deficiencies, resulting in chlorotic yellow brown needles. An indication of nutrient imbalance is an increased molar NH_4/K ratio.

A simple model to calculate a critical NH_4 deposition is:

$$CL(NH_{4,dt}) = RNH_4K(crit) \, . \, (K_{tf} + K_{lf}) - N_{lf}.(1-f_{ni}) \, / \, (1-f_{ni}) \tag{11}$$

where RNH_4K(crit) is a critical molar NH_4/K ratio and where the subscripts tf stand for throughfall, lf for litterfall and ni for nitrification.

This simple model is based on the folowing assumptions (De Vries, 1991a):
- throughfall and total deposition of NH_4 are equal;
- mineralization of NH_4 and K is equal to the N and K input by litterfall;
- root uptake does not affect the molar NH_4/K ratio;
- weathering of K is negligible.

2.2.3 The one-layer model START

The steady state one-layer model START, has been developed specifically for the assessment and mapping of critical loads on a European scale. It calculates the critical load of both total acid and nitrogen in an indirect way. This is done by comparing the predicted alkalinity and nitrate concentration at a given deposition rate with their critical concentration, thus computing a critical load exceedance (De Vries, 1991b). The model also accounts for the possibility of incomplete nitrification in the rootzone using a linear relationship between the NH_4 input and the nitrification rate.

START is a stationary variant of the dynamic one-layer model SMART (see par. 2.3.2) and can be chosen as an option within this model. It includes the major elements in a forest soil i.e. H, Al, BC, NH_4, NO_3 SO_4 and HCO_3. Again BC is a lumped expression for the sum of Ca, Mg, K and Na corrected for Cl. The model consists of a set of mass balance equations, which describe the soil input-output relationships for the cations (BC, NH_4) and the strong acid anions (SO_4, NO_3), and a set of equilibrium equations, which describe the equilibrium soil processes (for H, Al and HCO_3).

For X = BC, NH_4, NO_3 and SO_4, the concentration is calculated according to:

$$[X] = (X_{dt} + X_{int}) / FW \qquad (12)$$

where the subscript int refers to the interaction in the soil, which equals weathering minus uptake for BC, uptake for NH_4 and NO_3 and zero for SO_4. The concentration of [H], [Al] and [HCO_3] is determined by the equilibrium equations (8) and (9) and the concentration of the other ions by the charge balance principle

$$[H] + [Al] + [BC] + [NH_4] = [NO_3] + [SO_4] + [HCO_3] \qquad (13)$$

Combination of the equations (8), (9), (12) and (13) leads to one equation with one unknown, i.e. [H] which is solved iteratively.

2.2.4 The multi-layer model MACAL

The multi-layer model MACAL (Model to Assess a Critical Load) has been developed specifically for the assessment and mapping of critical loads for the Netherlands. Contrary to the one-layer model START, it includes the impact of nutrient cycling and water uptake on the soil solution composition. Furthermore, the model incorporates canopy interactions i.e. foliar uptake and foliar exudation by using a first order reaction (foliar exudation) and a linear relationship with deposition (foliar uptake). Ions included in MACAL are similar to START except that base cations are not lumped, i.e. H, Al, Ca, Mg, K, Na, NH_4, NO_3, SO_4, Cl and HCO_3.

MACAL describes a steady state situation, implying that mineralization equals litterfall and that the total uptake equals litterfall plus foliar exudation minus foliar uptake (maintenance uptake) plus a net uptake from the soil. In the original MACAL model (De Vries, 1988), predictions are made for compartments of 10 cm thickness up to a depth of 80 cm by describing weathering and uptake as a function of depth. In a later version, described in De Vries (1991b) and De Vries et al. (1991b) predictions can be made for any given depth.

Major differences between the two MACAL versions are that the original version (1) does not account for internal proton production and consumption due to nutrient cycling and (2) assumes that the proton load is completely neutralized by weathering and Al dissolution, thus neglecting leaching of H (De Vries, 1988).

Similar to START, the last MACAL version is based on the charge balance principle (equation 13) and concentrations of Ca, Mg, K, Na, NH_4, NO_3, SO_4 and Cl are determined by deposition and soil interactions (equation 12) while concentrations of H, Al and HCO_3 are calculated from the gibbsite equilibrium (equation 8) and the CO_2 equilibrium (equation 9). The gibbsite equilibrium constant is described as a function of depth. The various soil interactions, i.e. water and nutrient uptake, nitrification and weathering are also described as a function of depth. When this depth is taken as equal to the rootzone, the net effect of nutrient cycling is nil and MACAL predictions are identical to those of START (De Vries, 1991b; De Vries et al., 1991b).

2.3 Dynamic models

2.3.1 Basic principles

Similar to the steady state models, both a dynamic one-layer and multi-layer model has been developed at the Winand Staring Centre, in order to predict important soil solution parameters such as pH, Al concentration, molar Al/Ca ratio, molar NH_4/K ratio and NO_3 concentration as a function of time. The one-layer model called SMART (Simulation Model for Acidification Regional Trends) is the dynamic variant of START, while the multi-layer model called RESAM (Regional Soil Acidification Model) is the dynamic variant of MACAL. The major difference between the steady state and dynamic models is the inclusion of cation exchange and sulphate adsorption in SMART and RESAM. Furthermore, RESAM also includes mineralization (in MACAL equal to litterfall), denitrification and organic acid protonation.

Apart from denitrification that is included in RESAM, both models are also based on the assumptions regarding nitrogen and sulphur stated in par. 2.1.1. Other assumptions are:
- the soil consists of (a) homogeneous compartment(s) of constant density;
- the element input mixes completely within the (a) soil compartment(s).

2.3.2 The one-layer model SMART

The dynamic one-layer model SMART has been developed in a joint cooperation with the International Institute of Applied Systems Analysis (IIASA) and the Water and Environment Research Institute in Finland (De Vries et al., 1989a, b).

SMART has been developed especially to acquire insight into the impacts of different emission scenarios on forest soils in Europe. Consequently the model will be applied on a European scale within the overall framework of RAINS (Regional Acidification Information and Simulation model). Moreover, national applications of SMART are planned for Finland as part of the Finnish Integration Acidification model HAKOMA. Apart from effects on forest soils, SMART can also be used to predict the effects of acid deposition on surface waters in a dynamic way. In this context the model has been coupled to a lake model and applications are underway for analyzing lake water acidification on a large regional scale in Scandinavia.

As with the steady state models the model structure of SMART is based on the charge balance principle. A relation diagram is shown in Figure 1.

State variables depict the quantities of chemical constituents in minerals (carbonates, silicates and hydroxides) and on the exchange complex, as well as the ion concentrations in the soil solution. Rate variables depict the processes that influence state variables. This includes the net input of elements (deposition minus net uptake and net immobilization) and water (precipitation minus interception and minus evapotranspiration), as well as various neutralizing reactions, i.e. the dissolution (weathering) of carbonates, silicates and/or Al hydroxides, and cation exchange.

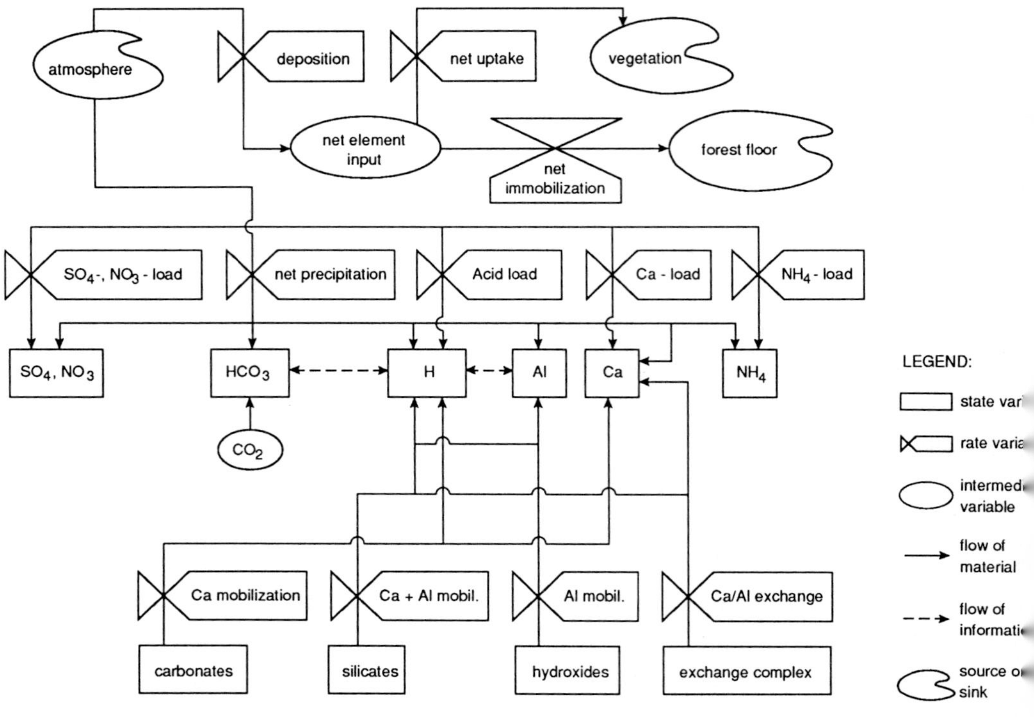

Fig. 1. Relation diagram of the SMART model

Ions included in SMART are similar to START, i.e. H, Al, BC, NH_4, NO_3, SO_4 and HCO_3. Comparable to START (see par 2.1.3), atmospheric deposition of SO_2, NO_x, NH_3 and BC, net uptake of N and BC and net release of BC by silicate weathering are described by zero order reactions and are required as input to the model. Weathering of Al hydroxides and dissociation of CO_2 are described by the same equations given before (equations (5) and (9) respectively).

The major difference between SMART and START is the inclusion of cation exchange between H, Al and BC using Gaines Thomas equations. Cation exchange may play an important role in neutralizing the acid input during a limited time period, e.g. in loamy soils with a high CEC and a high base saturation. SMART also includes an equilibrium equation for carbonate weathering. However, the effect of this process can be included in START, by using a high weathering rate for calcareous soils. At present SMART also includes sulphate adsorption using a Langmuir equation. A complete overview of the various process formulations in SMART including the initialization procedure and the solution methods are described in De Vries et al. (1989b).

2.3.3 The multi-layer model RESAM

The dynamic multi-layer model RESAM has been developed in a joint cooperation with the National Institute of Public Health and Environmental Protection (RIVM) in the Netherlands to evaluate the impact of various abatement strategies on forest soil acidification in the Netherlands (De Vries, 1987a; De Vries et al., 1988; De Vries and Kros, 1989a, b). The model includes the same elements as MACAL (H, Al, Ca, Mg, K, Na, NH_4, NO_3, SO_4, Cl and HCO_3) as well as organic anions (RCOO).

The model structure of RESAM is based on the relationship between soil acidification and forest element cycling (De Vries and Breeuwsma, 1987). A relation diagram of the model is shown in Figure 2.

RESAM simulates the major biogeochemical processes occurring in forest canopy, litter layer and mineral soil horizons i.e. (1) foliar uptake and foliar exudation; (2) litterfall and root decay; (3) mineralization; (4) root uptake; (5) nitrification and denitrification; (6) protonation of organic anions; (7) carbonate weathering; (8) weathering of primary minerals containing aluminium and base cations; (9) aluminium hydroxide dissolution/ precipitation; (10) cation exchange of hydrogen, aluminium, base cations and ammonium; (11) sulphate adsorption and (12) dissolution/speciation of inorganic carbon.

As in MACAL, foliar exudation and litterfall are described by first order reactions. Foliar uptake is considered as a fraction of the dry atmospheric deposition and root uptake is equal to the sum of litterfall, foliar exudation and root decay minus foliar uptake plus a given net growth. Net growth is either described by a logistic function or as a constant increase. Root uptake per soil layer is assumed to be proportional to the transpiration per soil layer. Root decay, nitrification, denitrification, protonation of organic anions and mineral weathering are described by first order reactions.

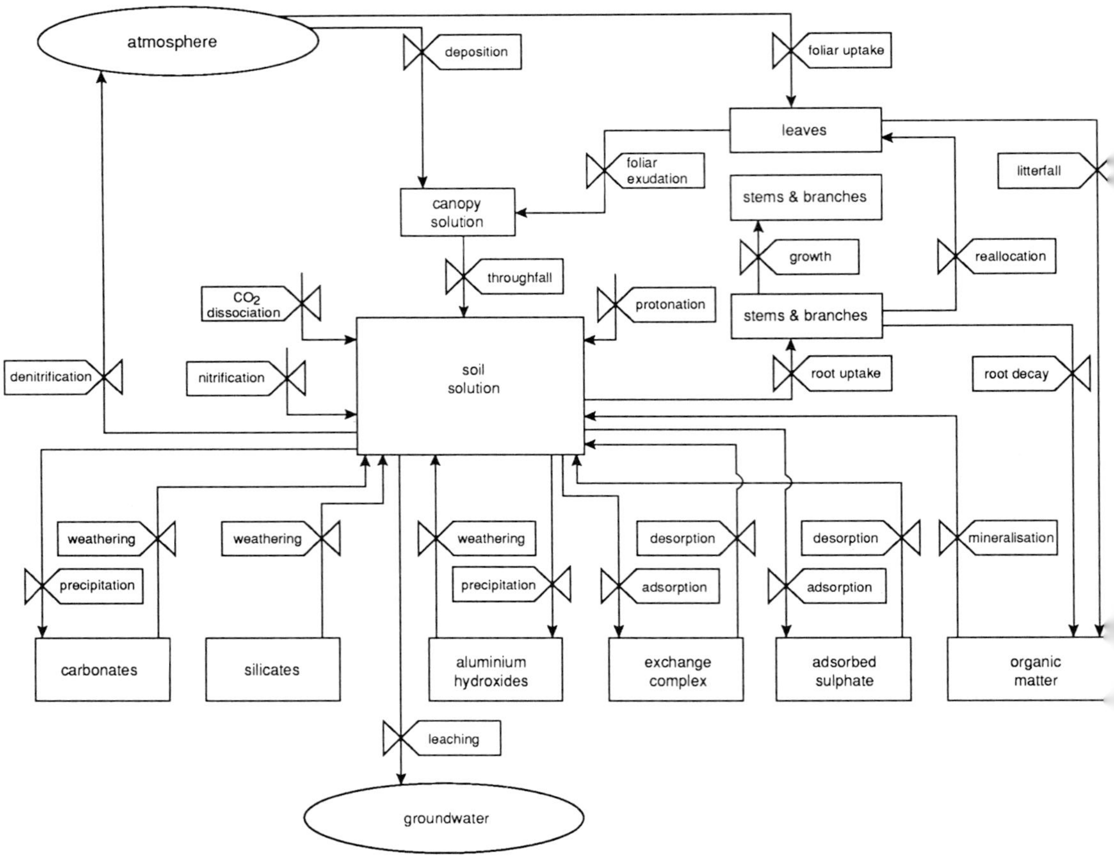

Fig. 2. Relation diagram of the RESAM model.

As in SMART, cation exchange and sulphate sorption are treated as equilibrium reactions, using Gaines Thomas equations and a Langmuir isotherm respectively and CO_2 dissolution is computed from equation (9). However, unlike SMART the dissolution of calcium and aluminium from carbonates and hydroxides respectively, is described as first order reactions, which are rate-limited by the degree of undersaturation. If supersaturation occurs the calcium or aluminium concentration is set to equilibrium. Unlike SMART, RESAM thus consists not only of a set of mass balance equations and equilibrium equations but also includes rate-limited equations. A complete overview of the model structure and process formulations in RESAM is given in De Vries and Kros (1989a).

2.4 Summary

A summarizing overview of the processes and process formulations included in the various steady state and dynamic models is given in Table 1.

3. MODEL PREDICTIONS

3.1 Critical loads

An overview of the assessment of critical nitrogen and potential acid loads on forest ecosystems in the Netherlands is given in De Vries (1988, 1989). A more general overview including groundwater and surface water is given in De Vries and Gregor(1990) and De Vries (1991a). De Vries and Gregor (1990) focus on sensitive ecosystems in Europe as a whole, whereas De Vries (1991a) focuses on the Netherlands only.

3.1.1 Nitrogen

An overview of average critical nitrogen loads derived for the Netherlands is given in Table 2 (compare De Vries, 1991a). The values for forests and heathlands are applicable to non-calcareous sandy soils with deep groundwater tables (> 2 m). Most Dutch forests and heathlands (> 80%) on these soils occur (De Vries et al., 1989c). The values for surface waters relate to hydrologically isolated shallow softwater systems.

Regarding vegetation changes in forests and heathlands, the first value is based on equation (10). We used an average net nitrogen uptake of 300, 500 and 400 $mol_c\ ha^{-1}\ yr^{-1}$ for coniferous forest, deciduous forest and heathland respectively. For all ecosystems, denitrification and long-term nitrogen immobilization was considered negligible, whereas the natural nitrate leaching rate has been taken at 100 $mol_c\ ha^{-1}\ yr^{-1}$. The second value is derived from experimental data about loads above which vegetational changes do occur. For surface waters the value is only based on experimental data (De Vries, 1991a).

The critical loads for frost damage and fungal diseases on coniferous forests are related to a critical nitrogen content in needles of 2.0% (De Vries, 1991a). Values are all based on literature data. The first value is based on results of a dynamic soil vegetation model and the second value on empirical data about the relationship between nitrogen and deposition and nitrogen content in needles.

The values related to nutrient imbalances refer to NH_3-N only and are also based on both experimental data (first value) and a simple model, i.e. equation (11) (second value) using a critical NH_4/K ratio of 5. The model derived value (1500 $mol_c\ ha^{-1}\ yr^{-1}$) is based on a N/K ratio of 5 in needles and an average K input by throughfall of 300 $mol_c\ ha^{-1}\ yr^{-1}$ and a nitrification fraction of 0 (worst-case).

Table 1. Processes an process formulations included in START, MACAL, SMART and RESAM

Processes	START	MACAL	SMART	RESAM
Hydrological processes:				
	Precipitation surplus	Variable flow with depth	Precipitation surplus	Hydrologic submodel
Biogeochemical processes:				
Foliar uptake	-	Proportional to dry deposition		-Proportional to dry deposition
Foliar exudation	-	Proportional to H and NH_4 deposition		-Proportional to H and NH_4 deposition
Litterfall	-	First order reaction		-First order reaction
Root decay	-	-	-	First order reaction
Mineralization/ immobilization	Zero order reaction[1)]	Zero order reaction[1)]	Zero order reaction[1)]	First order reaction
Net uptake	Constant growth	Constant growth	Constant growth	- Constant growth - Logistic growth
Maintenance uptake	-	Forcing function[2)]	-	Forcing function[2)]
Nitrification	Proportional to NH_4-deposition	Proportional to NH_4-deposition	Proportional to NH_4-deposition	First order
Denitrification	Proportional to wet NO_3-input	Proportional to wet NO_3-input	Proportional to wet NO_3-input	First order reaction
Geochemical processes:				
CO_2 dissociation	Equilibrium	Equilibrium	Equilibrium	Equilibrium
RCOO protonation	-	-	-	First order reaction
Carbonate weathering	-	-	Equilibrium	First order reaction
Silicate weathering	Zero order reaction	Zero order reaction	Zero order reaction	First order reaction[3)]
Al-hydroxide weathering	Gibbsite Equilibrium	Gibbsite Equilibrium	Gibbsite Equilibrium	- First order reaction - Elovich equation
Cation exchange	-	-	Gaines Thomas equation[4)]	Gaines Thomas equation[4)]
Sulphate adsorption	-	-	Langmuir equation	Langmuir equation

1) START, MACAL and SMART only include long-term net nitrogen immobilization.

2) In MACAL the maintenance uptake equals litterfall plus foliar exudation minus foliar uptake and in RESAM it is the sum of litterfall, root decay and foliar exudation minus foliar uptake.

3) In RESAM there is also the option to include a dependence of pH on the weathering rate.

4) In SMART cation exchange is limited to H, Al and BC whereas it includes H, Al, Ca, Mg, K, Na and NH_4 in RESAM.

Table 2. Average critical nitrogen loads ($mol_c\ ha^{-1}\ yr^{-1}$) for terrestrial and aquatic ecosystems in the Netherlands

Effects	Coniferous forests	Deciduous forests	Heathlands	Surface waters
Vegetation changes	400-1400	600-1400	500-1400	1400
Frost damage/Fungal diseases	1500-3000	-	-	-
Nutrient imbalances[1)]	800-1250	-	-	-
Nitrate leaching	900-1500	1700-2900	2000-3600	-

1) Refers to NH_3-N only

The two values for nitrate leaching are derived by using equation (10) again, using critical nitrate concentrations of 25 mg l^{-1} (0.4 $mol_c\ m^{-3}$) and 50 mg l^{-1} (0.8 $mol_c\ m^{-3}$) respectively, which are the target value and the standard value for drinking water in the Netherlands. For the precipitation surplus annual average values of 150, 300 and 400 mm yr^{-1} were used for coniferous forest, deciduous forests and heathlands respectively.

From Table 2 it can be concluded that most effects of nitrogen will be prevented at loads below 1000 $mol_c\ ha^{-1}\ yr^{-1}$. At this load vegetation changes might still occur in the long term. However, this effect will also be slowed down considerably compared with the present nitrogen input in the Netherlands, which is equal to 3000 $mol_c\ ha^{-1}\ yr^{-1}$ on an average up to 10,000 $mol_c\ ha^{-1}\ yr^{-1}$ in areas with intensive animal husbandry.

3.1.2 Potential acidity

An overview of the average critical acid loads derived for forests and surface waters in the Netherlands is given in Table 3 (compare De Vries, 1991a). For heathlands a critical acid load has not been determined since the effect of Al on heather is unclear. Similar to nitrogen, the values are applicable for non-calcareous well drained sandy (forest) soils and hydrologically isolated poorly buffered pools.

Table 3. Average critical loads for potential acidity ($mol_c\ ha^{-1}\ yr^{-1}$) for terrestrial and aquatic ecosystems in the Netherlands

Effects	Criteria	Coniferous forests	Deciduous forests	Surface waters
Root damage	Al $<0.2\ mol_c\ m^{-3}$	1100	1700	-
	Al/Ca $<1.0\ mol\ mol^{-1}$	1400	1400	-
Aluminium depletion	$\Delta Al(OH)_3 = 0$	1200	1500	-
Aluminium leaching[1)]	Al $<0.02\ mol_c\ m^{-3}$	500	300	-
Fish dieback	Al $<0.003\ mol_c\ m^{-3}$	-	-	400

1) The values are related to a depth of 2 m.

Except for surface waters, all values have been derived by application of equation (1). For both soil and groundwater, the water flux has been taken to be equal to the precipitation surplus, i.e. the water draining from the rootzone. It should be noted that the precipitation surplus is not a very adequate value for the water flux in forest soils. The critical chemical levels for Al and Al/Ca are related to the topsoil (first 30 cm) where most of the fine roots occur. Consequently, it is better to use the annual water flux at this depth. However, this also affects the weathering rate and uptake rate of base cations and nitrogen (expressed in $mol_c\ ha^{-1}\ yr^{-1}$), thus requiring the use of multi-layer models (see also De Vries, 1991a, b).

The values in Table 3 regarding forests are all based on:

- a seasalt corrected base cation input by dry deposition of 300 $mol_c\ ha^{-1}\ yr^{-1}$ for coniferous forests and 150 $mol_c\ ha^{-1}\ yr^{-1}$ deciduous forests;
- a net uptake of base cations of 300 and 350 $mol_c\ ha^{-1}\ yr^{-1}$ for coniferous and deciduous forests respectively;
- a net uptake of nitrogen equal to 300 and 500 $mol_c\ ha^{-1}\ yr^{-1}$ for coniferous and deciduous forest respectively (see par. 3.1.1);
- an average base cation weathering rate of 200 $mol_c\ ha^{-1}\ yr^{-1}$ for a rootzone of 1 m;
- a critical H-leaching rate of 300 $mol_c\ ha^{-1}\ yr^{-1}$ for coniferous forests and 600 $mol_c\ ha^{-1}\ yr^{-1}$ for deciduous forests. This is calculated by multiplying a critical H concentration of 0.2 $mol_c\ m^{-3}$ (pH = 3.7) by the previously given precipitation surpluses of 150 mm yr^{-1} and 300 mm yr^{-1} for coniferous and deciduous forests (see equation 7; par. 2.2.2.1).

The critical Al leaching rate related to root damage is calculated by (1) multiplying the precipitation surplus by a critical Al concentration of 0.2 $mol_c\ m^{-3}$ (300 and 600 $mol_c\ ha^{-1}\ yr^{-1}$ for coniferous and deciduous forests; see also equation 6 par. 2.2.2.1) and (2) multiplying the base cation leaching rate by a critical equivalent Al/Ca ratio of 1.5 (600 and 300 $mol_c\ ha^{-1}\ yr^{-1}$ for coniferous and deciduous forests; see also equation 7 par. 2.2.2.1). The critical Al-leaching rate in relation to depletion of Al-hydroxides has been derived by multiplying the weathering rate by 2.0 (400 $mol_c\ ha^{-1}\ yr^{-1}$; see also equation 8 par. 2.2.2.1), which is the assumed stoichiometric equivalent ratio of Al to BC in the weathering of primary minerals (De Vries, 1991a).

Regarding root damage, similar values to those given in Table 3 have been derived before by application of the original MACAL model, using the same criteria for the top 30 cm (De Vries 1988). These critical loads are not higher because the original MACAL model neglects H leaching (see par. 2.2.3).

The value for groundwater has been derived by using the same values for nitrogen uptake and precipitation surplus. For the weathering rate a value of 400 mol_c ha^{-1} yr^{-1} was used, by multiplying the weathering rate of 200 mol_c ha^{-1} yr^{-1} m^{-1} by an average depth to phreatic surface of 2 m. The critical alkalinity value used equals 0.14 mol_c m^{-3} (De Vries, 1991a).

For surface water the value is based on empirical data. The critical load for surface water is much less than for forest soils, since the critical Al concentrations in surface water are very low (a positive critical alkalinity leaching term). Terrestrial vegetations can stand higher Al concentrations than fish and aquatic vegetations.

3.1.3 Policy implications

Comparison of Table 2 and 3 shows that critical acid loads are more stringent than critical nitrogen loads. This is especially true for groundwater (drinking water). From the viewpoint of eutrophication, the critical nitrate leaching to groundwater has been defined in relation to drinking water quality. However, from the viewpoint of acidification, the critical nitrate leaching level is determined by the inherent acidity production. Together with the acidity production caused by sulphate leaching, it should not exceed the critical acid load. When the upper limit for nitrogen deposition related to eutrophication is higher than the critical load for total acidity, the acidification limit overrides the eutrophication limit.

On the basis of the critical deposition levels given above a target acid deposition level of 1400 mol_c ha^{-1} yr^{-1} has been set for the Netherlands with a nitrogen input below 1000 mol_c ha^{-1} yr^{-1}. At this level the most serious effects will be prevented. The aim is to reach this load in the coming 20 years (2010) with an interim target load of 2400 mol_c ha^{-1} yr^{-1} in the year 2000. Considering present acid loads on forests, which vary between 4000-15000 mol_c ha^{-1} yr^{-1} (Kros et al., 1990; 1991), this implies considerable emission reductions up to 80%.

3.2 Long-term impacts

3.2.1 Soil response to present deposition rates

3.2.1.1 Calcareous soils

The behaviour of SMART in different buffer ranges between pH 7 and pH 3 has been evaluated by analyzing the response of an initially calcareous soil of 50 cm depth to a constant high acid load (5 $kmol_c$ ha^{-1} yr^{-1}) over a period of 500 yr. In calcareous soils weathering is fast and the pH remains high (near 7) until the carbonates are exhausted. Results indicate a time lag of about 100 yr for each percent $CaCO_3$ before the pH starts to drop. In non-calcareous soils the response in the range between pH 7 and 4 mainly depends on the initial amount of exchangeable base cations. A further decrease in pH to values near 3.0 occurs when the Al oxides and/or hydroxides are exhausted. The analyses show that this could occur in acid soils within several decades (De Vries et al., 1989a, b).

As an example Figure 3 shows the temporal pH trajectories of initially calcareous soils (0.5% $CaCO_3$) for different CEC values that are representative for a sandy soil (reference value), a loamy soil (upper value) and an extremely poor sandy soil (lower value), such as the dunes in the Netherlands (De Vries et al., 1989b).

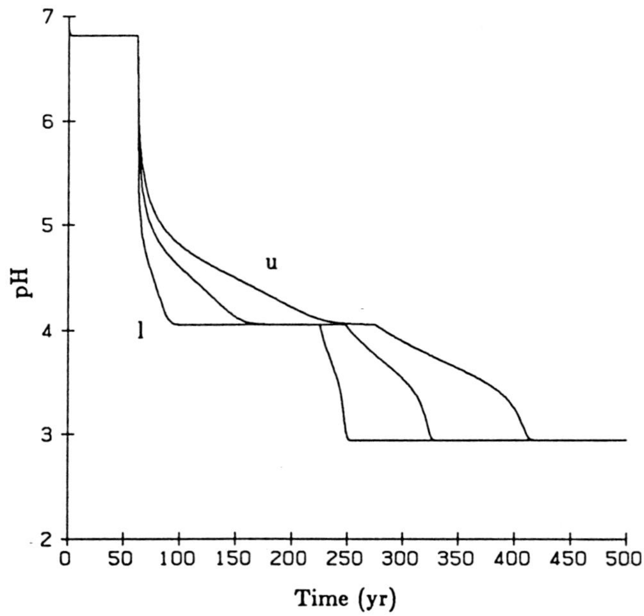

Fig. 3. Temporal development of the pH of an initially calcareous soil for varying CEC values in response to a constant high acid load

For roughly the first 50 years the soil stays in the carbonate buffer range and the pH remains high. However, as soon as the carbonates are exhausted, there is a sudden drop in pH at high acid loads (Figure 4), which becomes even more pronounced with decreasing CEC. This is obvious, because the H-production in non-calcarious soils is initially mainly neutralized by exchange between H and BC, and the buffer capacity of the soil related to this mechanism decreases strongly with decreasing CEC.

The extremely sharp drop in pH for the lower CEC value results from the almost negligible exchange buffer capacity in this poor sandy soil. This leads to an almost direct switch from the carbonate buffer range (pH = 6.8) to the Al buffer range (pH I 4.0), because base cation weathering is far too low to neutralize this high acid input. Actually, this predicted sudden drop in pH is somewhat overestimated, since carbonate weathering appears to become rate-limited at carbonate contents near 0.3% (see further).

An increase in CEC causes a more gradual decrease in pH, especially in the pH range 5.0 to 4.0. As long as the soil contains Al oxides and/or hydroxides, the pH remains above 4.0, but as soon as these minerals are exhausted, the pH drops further to values near 3.0. Again, this is illustrated most clearly for the poor sandy soil, where the pH drops suddenly, after the Al-hydroxide has been exhausted. In the other soils, the change is more gradual, because of exchange between H and Al. Apart from exchange, H buffering below pH 4.0 can also be the result of rate-limited dissolution of Al from oxides and/or hydroxides (De Vries and Kros, 1989a, b), but this is not considered in SMART.

The RESAM model has also been used to analyse the long-term (100 yr) impact of acid deposition on calcareous soils, i.e. a calcareous dune ecosystem near the coast of the Netherlands (De Vries et al., 1991a). The (potential) acid deposition rate has been taken equal to 3 $kmol_c\ ha^{-1}\ yr^{-1}$. More information on deposition values is given in Table 4 (par. 3.2.1.2). A steady state nutrient cycle has been assumed (mineralization equals litterfall and root decay) and sulphate adsorption/desorption has been ignored in the simulations.

Results from the simulations on the calcareous dune ecosystem (De Vries et al., 1991a) show that the impact of atmospheric deposition on calcareous soils becomes very important when the carbonate content has dropped below 0.3%. In this situation, natural acidification due to dissociation of CO_2 decreases strongly, since the pH drops below 6, and acid deposition becomes the dominating acidifying mechanism. Compared with the SMART simulation results for calcareous soils (Figure 3, par. 3.2.1.1), the pH drop is less sudden since carbonate weathering is described as a rate-limited reaction in RESAM. However, model predictions still indicate that in dune soils with low carbonate contents (<0.5%) the pH in the topsoil can drop from about 6.5 to 3.0 in several decades, because the weathering rate of silicates is very low and the buffercapacity of the exchange complex and of amorphous Al hydroxides is almost negligible.

In the Netherlands dune ecosystems are very valuable from a floristic viewpoint since about 64% of all plant species do grow in the dunes, especially in the moist valleys. Compared with forests and heathlands, the impact of acid deposition on soils and vegetations in coastal dunes, and the possible role of dune management practices in delaying acidifying effects have not yet received much attention. With some exceptions, attention has mainly been focused on the natural effects of decalcification on soils and vegetation. However, it can be concluded that paying more attention to the impact of atmospheric inputs on dune areas is absolutely justified, since the processes of acidification and eutrofication are an enormous potential threat for this ecosystem which is strongly determined by its mineral nutrient status (De Vries et al., 1991a).

3.2.1.2 Non-calcareous soils

The behaviour of RESAM has also been evaluated by simulating the long-term (100 yr) impact of acid deposition on a representative non-calcareous forest soil (De Vries and Kros, 1989a, b) and a non-calcareous dune soil (De Vries et al., 1991a). Again, a steady state nutrient cycle has been assumed and sulphate adsorption/desorption has been ignored (see par. 3.2.1.1).

The total deposition values used are given in Table 4. It represents the annual average deposition for the Netherlands (forest soil) and the average deposition near the coast (dune soils). The deposition rates for the dune vegetation have also been used to analyse the impact on a calcareous soil (see par. 3.2.1.1).

Table 4. Total deposition values ($kmol_c\ ha^{-1}\ yr^{-1}$) used for the simulation on representative forest and dune soils

Ecosystem	Deposition values ($kmol_c\ ha^{-1}\ yr^{-1}$)								
	H	NH_4	NO_3	SO_4	Ca	Mg	K	Na	Cl
Douglas stand	2.2	1.4	1.6	2.0	0.3	0.3	0.1	1.2	1.9
Dune vegetation	1.0	1.0	0.8	3.2	1.5	3.0	0.5	12.0	15.0

There are major differences in the deposition of sulphate, chloride and base cations due to the sea spray effect. Differences in nitrogen load are relatively low. However, on most forests the deposition of both NH_4 and SO_4 will be higher than the annual average values used here, due to forest filtering. The total acid load, which equals the sum of H plus twice the amount of NH_4, is 5 and 3 $kmol_c\ ha^{-1}\ yr^{-1}$ for the forest and dune soil respectively. For forest soils, this load is relatively low. Throughfall data, which include the forest filtering effect, indicate a range of 4-15 $kmol_c\ ha^{-1}\ yr^{-1}$ (Kros et al., 1990, 1991).

Results of the simulations on a forest soil (leptic podzol) below Douglas stands (De Vries and Kros, 1989a, b) indicate that, with the present high acidifying inputs in the Netherlands, the pool of readily dissoluble Al in amorphous hydroxides may be completely exhausted in the next decades. This is illustrated in Figure 4 which shows the initial amount of Al hydroxides in the first five layers (Figure 4a) and the changes in this amount during the simulation period for periods of 25 years (Figure 4b).

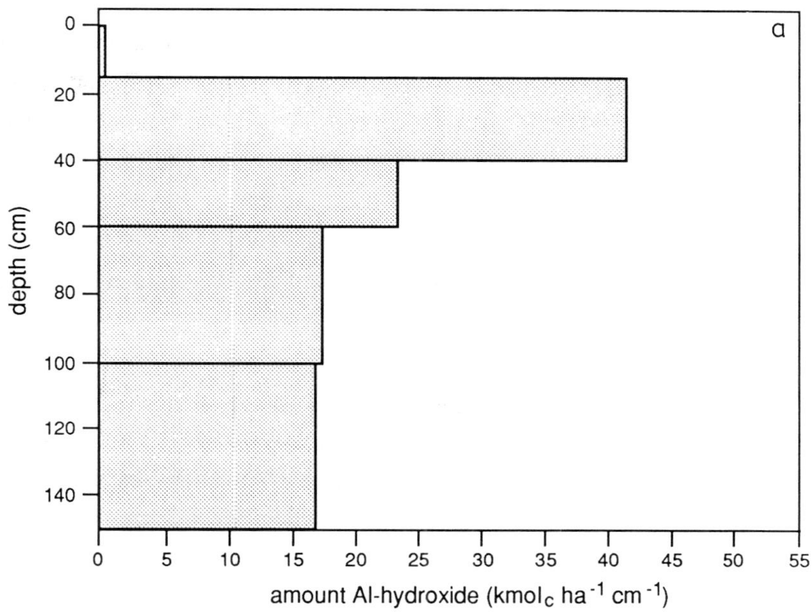

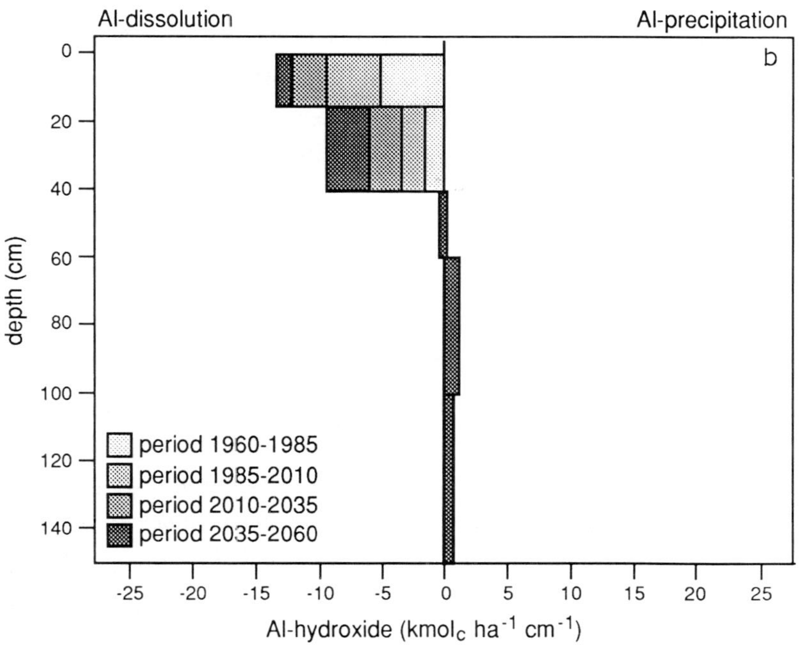

Fig. 4. Initial amount of Al hydroxide per soil layer (a) and changes during the simulation period (b) in an acid forest soil

The Al dissolution in layer 1 appears to decrease with time, while increasing in layer 2, due to a decrease in the amount of Al hydroxide in the first layer. Figure 4 shows that the amount in the first layer is nearly depleded within a period of 100 years. This indicates that the pool of weatherable Al in acid forest soils in the Netherlands is seriously limited. At a greater depth, there is some precipitation of Al hydroxides (layer 4 and 5), caused by the incongruent weathering of primary minerals. The rate-limited dissolution of Al leads to Al concentrations which are unsaturated with respect to gibbsite in the topsoil and saturated in the sub-soil. This situation is commonly found in acid forest soils in the Netherlands (Kleijn et al., 1987, 1989).

The decrease in Al dissolution causes lower concentrations and adsorbed amounts of Al in the topsoil, whereas it increases for H. This is illustrated in Figure 5.

Figure 5a shows that H and Al are the main cations at the adsorption complex. During the simulation period the adsorbed H fraction increases from approximately 0.2 to 0.5, while the Al fraction decreases from 0.7 to 0.4. The other exchangeable cations decrease only slightly. Due to the decreasing amount of amorphous Al hydroxide, which causes a lower H buffering rate, the H concentration in the first soil layer increases from 0.2 $mol_c\ m^{-3}$ (pH 3.7) to 0.4 $mol_c\ m^{-3}$ (pH 3.4) in 100 years. On the other hand the Al concentration decreases from 1.0 to 0.3 $mol_c\ m^{-3}$ (Figure 5b). The decrease in the sum of Al and H concentration is caused by an increasing NH_4 concentration and a decreasing NO_3 concentration, due to the assumed inhibition of nitrification at lower pH values. The base cation concentrations remain fairly constant during the simulation period, because the pH effect on base weathering was assumed to be negligible (not shown in Figure 5b).

The dramatic changes in pH and Al chemistry in the forest topsoil may have serious ecological consequences (De Vries and Kros, 1989b).

Simulations have also been made for a (dry) leptic podzol below an oak stand and for a (wet) humic gleysol below Douglas fir and oak stands respectively, to evaluate the influence of tree species and soil type (groundwater level) (De Vries and Kros, 1989a). Regarding tree species, results show lower concentration levels and a more uniform concentration profile below the oak stand compared with Douglas fir, due to higher values for the water flux and a more uniform water uptake pattern in the oak stand. Regarding soil type, predicted concentrations of Al and NO_3 were lower while NH_4 concentrations were higher in the humic gleysol compared with the leptic podzol, due to a decrease in nitrification and an increase in denitrification.

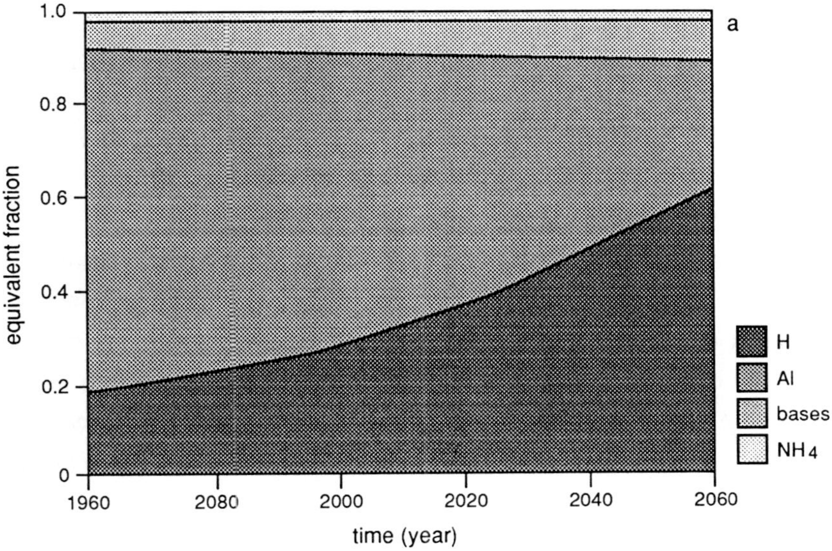

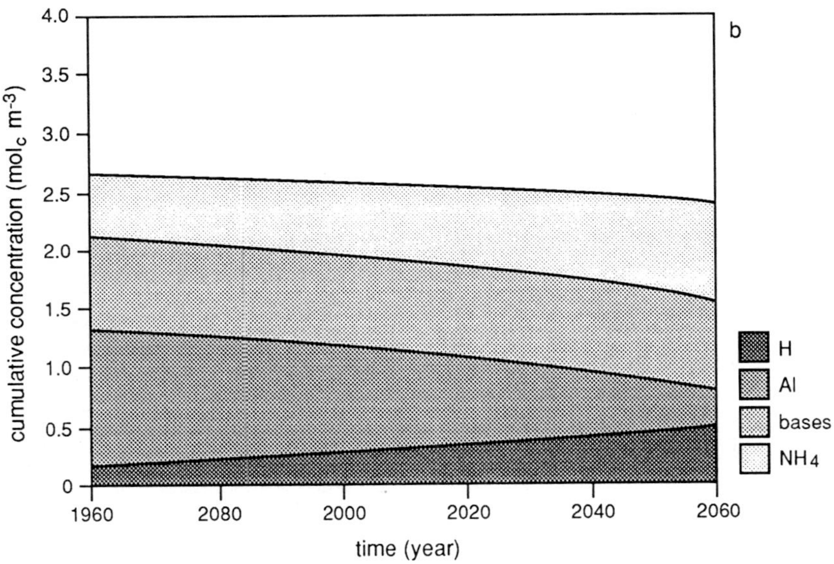

Fig. 5. Temporal development of cumulative equivalent fractions (a) and of cumulative concentrations of H, Al, base cations (Ca+Mg+K+Na) and NH_4 (b) in an acid forest topsoil in response to a constant high acid load

In non-calcareous dune soils, the depletion of amorphous Al-hydroxides is also very important, causing a pH drop from about 4.0 to 3.0 in several decades in deeper soil layers, which is due to very low content of Al hydroxides (De Vries et al., 1991a).

3.2.2 Soil response to deposition changes

The SMART model has also been used to analyse the impact of various deposition scenarios on non-calcareous soils for a time period of 100 yr (De Vries et al., 1989a, b). As an example, the influence of different deposition scenarios on the temporal evolution of the molar Al/BC ratio in a non-calcareous sandy soil is shown in Figure 6.

Figure 6a shows the time pattern of the total deposition level for each scenario. The scenarios start with background levels for SO_2, NO_x and NH_3 which increase to "European average" levels within 25 years and stay there for another 25 years. Within the next 25 years deposition levels are reduced by 0%, 30%, 70% and 90% for scenarios 1, 2, 3 and 4, respectively, and stay there for another 25 years. The background levels are based on literature information on SO_2 and NO_x and NH_3. The "European average" values are rough estimates based on average throughfall data from 51 sites in Europe. The various reductions are related to political goals and critical loads for various receptors. The 30% reduction refers to the (minimal) reduction aimed at by most (western) European countries by the year 1993. A 70% reduction is the political goal in the Netherlands for the year 2010 based on a critical load of 1400 mol_c ha^{-1} yr^{-1} for coniferous forests (De Vries 1988), and a 90% reduction might be the ultimate goal to protect the most sensitive surface waters (De Vries and Gregor, 1990; De Vries, 1991a).

Figure 6b shows the resulting time pattern of molar Al/BC ratios. The results of scenario 1 illustrate the considerable time lag between the period of acid deposition increase and molar Al/BC ratio increase (appoximately 25 year). The time period before the molar Al/BC ratio becomes critical (above 1.0) is approximately 35 year (see Figure 6b). This coincides with a base saturation of approximately 5%.

The deposition reductions between 50 and 75 year for scenarios 2, 3 and 4 almost directly influence the predictions of the molar Al/BC ratio. This is conceivable, because a decreased input of S and N directly influences the concentrations of SO_4 and NO_3 and, in turn, this directly influences the Al concentration (charge balance), because Al dissolution is the dominant buffer mechanism in acid sandy soils. Figure 6b shows that a deposition reduction of 30% still leads to an increase in the molar Al/BC ratio. A reduction of at least 70% (up to 1000 mol_c ha^{-1} yr^{-1}) is needed to arrive at a final molar ratio of 1.0, but the time taken to achieve this ratio is more than 100 yr (not shown in Figure 6b). However, one should be

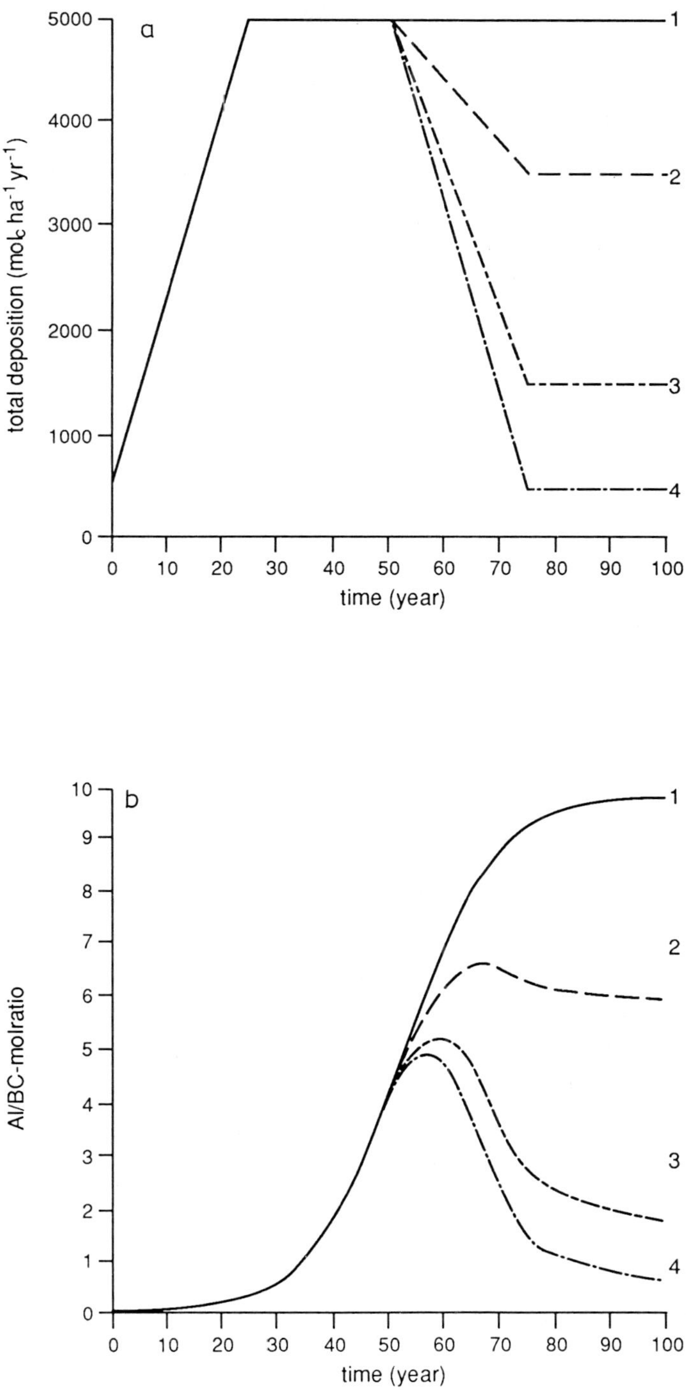

Fig. 6. Temporal trajectories of total deposition levels for four scenarios (a), and the resulting molar Al/BC ratio (b) in a non-calcareous sandy soil

aware that the nutrient cycle is not included in SMART. The molar Al/BC ratio is completely dependent on geochemical interactions. Consequently, the predicted molar ratios will be most reliable for subsoils (e.g. below B-horizons). In the topsoil the molar Al/BC ratio will be more favourable because of a net input of base cations by deposition and mineralization (De Vries et al., 1989a, b). The deposition level of 1000 to 1500 $mol_c\ ha^{-1}\ yr^{-1}$ is consistent with the critical load of 1400 $mol_c\ ha^{-1}\ yr^{-1}$, which has been derived for Dutch forest soils in relation to a critical molar Al/BC ratio of 1.0 (see par. 3.1.2).

As with SMART, the impact of various deposition scenarios on a non-calcareous forest soil (leptic podzol below Douglas) has also been evaluated with RESAM. The results lead to similar conclusions, i.e. deposition reductions almost directly reduce concentrations in SO_4, NO_3, and Al in various soil layers thus improving the soil water quality (De Vries and Kros, 1989a).

4. UNCERTAINTIES

4.1 Critical loads

The uncertainty in the average critical load values derived before can be large, and is mainly determined by the uncertainty in (1) critical chemical values set for the receptor, (2) model structure and (3) data (De Vries 1991a, b; Sverdrup et al., 1990).

4.1.1 Critical chemical values

Uncertainties in critical chemical values for a given receptor partly reflect our lack of knowledge regarding the effect of acid deposition and are partly due to a natural range in sensitivity. This uncertainty can be very large, especially for critical acid loads on forests (soils) since the range in Al toxicity appears to be very large for different tree species (De Vries, 1988; 1991a). A change in critical Al concentration directly influences the critical load by the critical alkalinity leaching term (par. 2.1.2.1). The importance can be illustrated as follows: taking an average precipitation surplus of 200 mm yr^{-1}, which is a reasonable value for the forests in the Netherlands and assuming a critical inorganic Al concentration of 2 mg l^{-1}, which corresponds to an alkalinity value of about -0.4 $mol_c\ m^{-3}$, leads to an critical alkalinity leaching of -800 $mol_c\ ha^{-1}\ yr^{-1}$. However, when an critical value of 5 mg l^{-1} is assumed, the critical alkalinity equals about -0.8 $mol_c\ m^{-3}$ and the leaching term becomes -1600 $mol_c\ ha^{-1}\ yr^{-1}$ thus increasing the critical load by 800 $mol_c\ ha^{-1}\ yr^{-1}$.

For groundwater and surface water the uncertainty in critical alkalinity leaching is much less, since the range in critical chemical levels for alkalinity is much lower. Like critical acid loads, the critical nitrogen load on forests related to frost damage and nutrient imbalances can also be large due to uncertain criteria, whereas the load related to nitrate leaching to groundwater is much more reliable taking 50 mg l^{-1} of NO_3 as an accepted value.

4.1.2 Model structure

Uncertainties in the modelstructure relate to the assumptions that have been made to simplify the "real world". Unlike the uncertainty in acceptable chemical values and data, it is rather difficult to quantify the uncertainty due to modelling assumptions. The underlying premise in an uncertainty analysis is that the model structure is correct or at least represents current knowledge adequately.

In this context it is important to note that use of a one-layer model will most likely cause an underprediction of critical acid loads. The annual average water flux at 30 cm depth is much higher than the precipitation surplus, thus affecting the critical acidity leaching. This difference is approximately 100-150 mm yr^{-1} which causes an increase in acidity leaching of 400-600 mol_c ha^{-1} yr^{-1}. The increase in critical load might be in the same order of magnitude, considering that the overall effect of the sum of weathering and uptake of nitrogen and base cations is small (cf De Vries, 1991a). This increase in critical load is illustrated by the results of the national application of RESAM (see par. 5.2.5). These results show that approximately 50% of the forest topsoils have Al contrations and molar Al/Ca ratios below critical levels at an average load of 2500 mol_c ha^{-1} yr^{-1} for the Netherlands. This can be regarded as an indication of the median critical acid load for forest soils.

A large source of uncertainty may also occur when deriving critical loads in specific forest areas due to the occurrence of N fixation, denitrification or a complex hydrology including seepage or surface runoff. However, the assumptions that these processes can be neglected (see par. 2.1.1) are reasonable for the receptors defined in paragraph 3.1.1 and 3.1.2, i.e. well drained non-calcareous sandy soils.

4.1.3 Data

Uncertainties in data are due to lack of knowledge, including measurement errors, and spatial variability. The values that have been used for weathering, uptake and precipitation surplus are (long-term) averages for coniferous and deciduous forests. The uncertainty for the critical load at a specific location may be in the order of 50%, due to spatial variability in these data. Results of a national application of the SMB model including 12 tree species and 18 non-calcareous sandy soils show a range in critical loads between approximately 500 and 2500 mol_c ha^{-1} yr^{-1} (see par. 5.1.4).

The uncertainties in critical loads for groundwater are also high because of uncertainties regarding weathering and denitrification in the saturated zone. Consequently, the phreatic level has been taken as the target for the assessment of critical loads both for nitrogen and potential acidity. At the depth of groundwater extraction (generally more than 10 m) critical loads may be much higher due to the occurence of these process in the saturated zone.

4.2 Long-term impacts

4.2.1 Introduction

As in critical load derivations, the uncertainty in the predicted long-term impact of acid deposition is due to the modelstructure and to data. Furthermore, the spatial and temporal aggregation also affects the output of dynamic models.

Unlike RESAM, the SMART model has not been subjected to a thorough uncertainty analysis. However, the sensitivity of the model output to various parameters has been evaluated. This showed that the CEC and base saturation are the parameters which most influence the model output (Al/BC and pH), when using a range of values that can be expected on a European scale (De Vries et al., 1989a, b).

Assuming that the structure (and implementation) of RESAM is correct, the uncertainty in model response to a given deposition scenario has been evaluated in relation to data uncertainty (including spatial variability) (Kros et al., 1990; 1991). Here, data refer to model inputs (the source terms to an ecosystem), model variables describing the initial state of an ecosystem and model parameters characterizing the rate of processes changing that state. However, in the following, inputs, parameters and variables are all referred to as parameters. The main aim was to aquire an insight into which additional data most improve the reliability of predictions, to have a guideline regarding data derivation for a regional application of RESAM. In this context we also investigated whether average input data, used in a regional application in order to limit the number of simulations, produce adequate average model outputs for a specific soil vegetation combination.

4.2.2 Methodology

The uncertainty analysis was performed by using Monte Carlo simulation techniques in combination with regression analysis. The uncertainty in model outputs was quantified by giving frequency distributions of input parameters instead of deterministic values. The resulting frequency distributions of the model output were analysed by regression analysis to evaluate the contribution of the uncertainty of various parameters to the model uncertainty.

The presented uncertainty analysis has been restricted to a leptic podzol with Douglas fir, subject to a reducing deposition scenario between 1987 and 2010. The used average deposition levels in 1987 for NH_4, NO_3 and SO_4 together with their uncertainties are given in Table 5.

Table 5. Frequency distributions of the total deposition of SO_4, NH_4 and NO_3 (mol_c ha^{-1} yr^{-1}) used in an uncertainty analysis with RESAM

Ion	Distribution type	Deposition (mol_c ha^{-1} yr^{-1}) Mean	Standard deviation	Minimum	Maximum
SO_4	normal	3740	1250	2340	5770
NH_4	normal	4195	1270	2475	6540
NO_3	normal	970	265	690	1570

The values in Table 5 are based on 27 throughfall sites in areas with intensive animal husbandry. The potential acid deposition, which is approximately equal to twice the NH_4 deposition, ranges between approximately 5000-13000 mol_c ha^{-1} yr^{-1}. Using all available throughfall data this range is even 4000-15000 mol_c ha^{-1} yr^{-1}.
Used reduction fractions for SO_4, NH_4 and NO_3 are 0.63, 0.58 and 0.40 for the period between 1987 and 2000 and 0.58, 0.40 and 0.29 between 2000 and 2010. These values are based on the intended emission reductions up to 2010. The deposition scenario for each Monte Carlo simulation was obtained by multiplying the randomly sampled initial deposition value for SO_4, NH_4 and NO_3 by the corresponding reduction factor.

The investigated output variables were pH, NO_3 concentration, molar Al/Ca ratio and molar NH_4/K ratio in the rootzone, which are generally used as indicators for soil acidification and for potential forest damage. As with the simulations described before in par. 3.2.1, we assumed a steady state nutrient cycle to restrict the number of parameters. For this soil profile (i.e. leptic podzol), with four distinguished layers (A0 (litter layer, 4 cm), A1 (15 cm), Bh (25 cm) and C (20 cm)), RESAM needs about 200 parameters. Ultimately approximately 75 parameters have been described by frequency distributions (uniform, normal or lognormal), including data regarding the average, standard deviation, minimum and maximum, while performing 250 Monte Carlo runs.

4.2.3 Uncertainty in modelpredictions

As an example, Figure 7 shows the development of the median, mean 97.5 and 2.5 percentile of the molar NH_4/K ratio in the topsoil (A1 horizon) as a function of time, together with a reference run in which average parameter values have been used. The molar NH_4/K ratio is important in describing possible nutrient imbalances in the soil by nitrogen deposition.

Figure 7 shows that the decreasing nitrogen deposition (mainly as NH_4) causes a decrease in the mean and standard deviation of the molar NH_4/K ratio. However, the variation coefficient (standard deviation divided by the mean value) slightly increases.

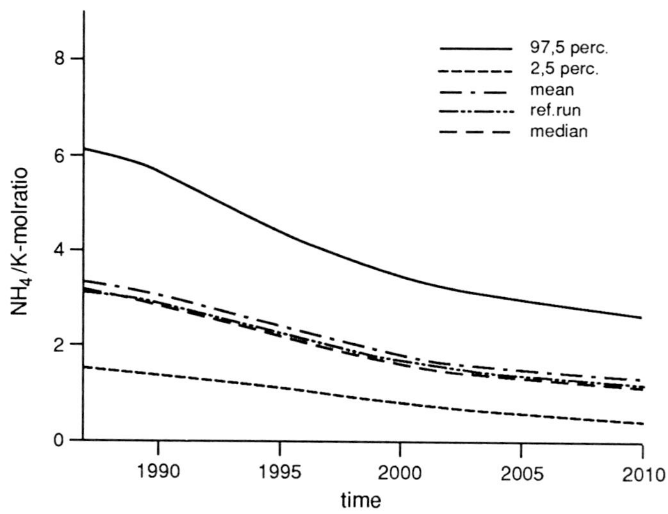

Fig. 7. Temporal development of the molar NH_4/K ratio in layer 1 of an acid forest soil response to a reducing acid load

Furthermore, the temporal development of the reference run appears to be very close to the mean value of the 250 simulation runs. It is even not distinguishable from the median value. This indicates an overall linear behaviour between model input parameters and model outputs, which was ascertained by statistical (linear regression) analysis. In this analysis the R^2 was more than 0.9.

Overall results for the various output parameters showed that the uncertainty (variation coefficient) is relatively high for the molar NH_4/K and Al/Ca ratio, whereas it is relatively low for the pH and the NO_3 concentration. Statistical analysis showed that the relation between the parameters and the model responses in most cases could be described by a linear regression model ($R^2 > 0.8$; Kros et al., 1990; 1991).

4.2.4 Contribution of modelparameters to the uncertainty in modelpredictions

The impact of the most important model parameters on the molar NH_4/K ratio, assessed by this analysis, is illustrated in Figure 8 by the temporal development of the root of the partial uncertainty contribution (RTU). RTU is a statistical entity relating model output to model input, accounting for possible correlations between modelinput parameters. An increase in RTU implies an increase in the uncertainty contribution of the considered model input to the considered model output (Kros et al., 1990; 1991).

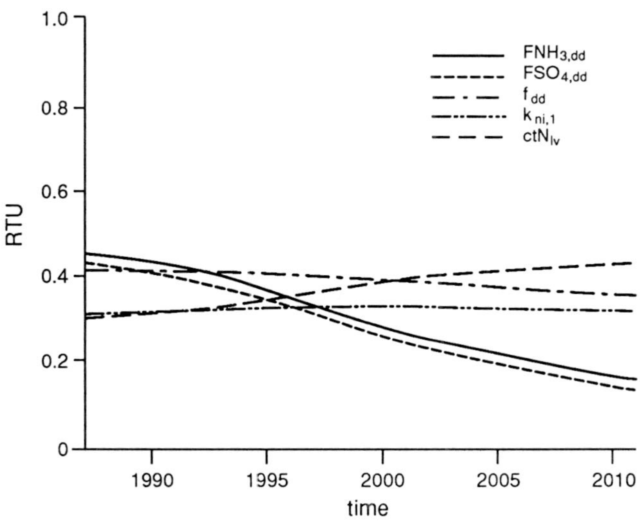

Fig. 8. Temporal development of the RTU between model parameters and the molar NH_4/K ratio in an acid forest topsoil

Figure 8 shows that the contribution of the various parameters to the uncertainty changes in time. However, during the simulation period the uncertainty in the molar NH_4/K ratio in the topsoil appears to be mainly determined by parameters influencing the NH_4 concentration, i.e. the dry deposition of NH_3 ($FNH_{3,dd}$), the nitrification rate constant in layer 1 ($k_{ni,1}$) and the nitrogen constant in leaves (ctN_{lv}), and by parameters influencing the K concentration, i.e. a dry deposition factor for base cations (f_{dd}), and to a lesser degree the wet deposition of K (FK_{dw}) and the potassium content in leaves (ctK_{lv}). The last two parameters are not shown in the Figure. According to Figure 8 the dry deposition of SO_2 ($FSO_{2,dd}$) also determines the uncertainty in the NH_4/K ratio, but this is due to a strong correlation between the dry deposition of NH_3 and SO_2, included in the analysis.

Overall results showed that the uncertainty contribution of the various parameters depended on the considered output variable, soil compartment and time. However, in most cases the uncertainty in the deposition of SO_2, NO_x and NH_3 and parameters determining the nitrogen and aluminium dynamics play an important role in the resulting uncertainty of the considered model output (Kros et al., 1990; 1991). A simple sensitivity analysis performed earlier (De Vries and Kros, 1989b) already showed that the changes in Al chemistry in the topsoil are strongly influenced by the parameters regulating nitrification and Al dissolution, because these processes mainly regulate H production and consumption, respectively. The relative unimportance of CEC and base saturation, compared with the SMART evaluation is due to the low values (and the small range) for the base saturation of Dutch forest soils.

5. REGIONAL ASSESSMENT

5.1 Critical loads

5.1.1 Introduction

In order to assess cost-effective emission reduction scenarios, it is necessary to aquire an insight into the regional variation of critical loads for various receptors. In this context, a mapping excercise has to be carried out by all European countries, the U.S. and Canada, before February 1991. The maps will be presented to the Executive Body (EB) for the Convention on Long Range Transboundary Air Pollution (LRTAP) of the United Nations-Economic Commision for Europe (UN-ECE), which has installed a Task Force on Mapping (TFM) in this respect.

A methodology for the assessment and mapping of critical loads and critical load exceedances has been given in Sverdrup et al. (1990); De Vries (1991b) and Hettelingh and De Vries (1990). It includes guidelines for critical chemical values, quantification of the receptor distribution, calculation methods (models), input data and mapping procedures. The overview given by Sverdrup et al. (1990) includes all relevant available models whereas De Vries (1991b) focuses on the models SMB, START and MACAL. Hettelingh and De Vries (1990) have summarized the various information sources in a separate mapping vademecum. This also includes guidelines for mapping critical concentration levels related to direct effects on crops and materials.

At present the SMB model has been applied to assess critical loads on forests, both on a National and a European scale. The START and MACAL model will be applied on a European and National scale respectively during the year 1991. Here, we summarize the quantification of the receptor (forest/soil) distribution and the data collection procedure for both applications. Furthermore, an example of the SMB model output is given.

5.1.2 Deposition areas and receptor distribution

Emissions and thereby depositions strongly vary in space. Consequently, deposition areas have to be defined, seeking for an optimum between the number of areas and the spatial variability within each area. The shape and spatial resolution of a deposition area directly affect the mapping procedure. Maps can either be displayed as polygons or as grids. In order to ensure uniform cartographic presentation, grid maps have been used for both the European and National application. For the European application, the so-called RAINS grid of 1.0° longitude x 0.5° latitude was used. The length of a grid element in the south-north direction is fixed to 56 km but the width in the east west direction varies between 38 to 91 km depending on the latitude (approximately 50 to 60 km in Central Europe). For the Netherlands a 10 x 10 km grid has been used because detailed information regarding tree species and soil types exists at this scale. The total number of grids containing forests equals 434.

For the European application, a distinction has been made between coniferous and deciduous trees. Although detailed information on the areal distribution of various tree species (Pine, Fir, Spruce, Oak, Beech and Birch) was available, this has not been used since the spatial allocation of individual tree species over soil types was not known. Soil types were distinguished based on the legend of the FAO-UNESCO Soil Map of the World (1974). The code of a mapping unit contains four items, the dominant soil unit, the associated soil unit, the dominant texture class and the dominant slope class. For our application the soils were distinguished on the basis of the dominant soil unit, texture class and slope class.

The areal distribution of soils was digitized by estimating the fraction of each mapping unit (soil type, texture class and slope class) within each grid square. The resolution of the FAO-UNESCO soil map was such that a grid contained one to seven mapping units, the mean number being 2.2. The total number of grids equals 2364. Similar to soils, the areal distribution of coniferous and deciduous forests was obtained by estimating the fraction of these forests for each grid square using aeronautic maps. The distribution of both soils and forests within a grid was not known. In estimating the spatial distribution we assumed that forests are not evenly distributed over all soil types, but instead, they are mainly located on areas with steep slopes and poor soils (low weathering rates and coarse texture) (De Vries et al., 1991d).

For the National application, a distinction has been made in twelve tree species and 23 soil types. Tree species included are Pinus Sylvestris (Scotch Pine), Pinus Nigra (Black Pine), Pseudotsuga Menziesii (Douglas fir), Picea Abies (Norway Spruce), Larix Leptolepis

(Japanese Larch), Quercus Robur (Oak), Fagus Sylvatica (Beech), Populus Spec (Poplar), Salix Spec (Willow), Betula Pendula (Birch), Fraxinus Nigra (Ash) and Alnus Glutinosa (Black Alder). Soil types were differentiated in 18 non-calcareous sandy soils (mainly podzolic soils), calcareous sandy soils, loess soils, non-calcareous clay soils, calcareous clay soils and peat soils on the basis of a recent 1 : 250 000 soil map of the Netherlands (De Vries et al., 1991b).

Information on the area (distribution) of each specific forest-soil combination in a grid was derived by overlaying the digitized forest- and soil database. This was done by a grid-overlay of the digitized 1 : 250 000 soil map with a spatial resolution of 100 m x 100 m and a data base with tree species information with a spatial resolution of 500 x 500 m for each 10 x 10 km grid (De Vries et al., 1991b).

5.1.3 Data collection

Data that are needed to map critical loads, are the deposition, weathering and uptake of elements and the water flux. MACAL also needs litterflux data. These data depend on location, forest type (tree species) and soil type as shown in Table 6.

Table 6. The influence of location, tree species and soil type on input data as considered in the National and European application (x = considered, - = not considered)

	Deposition	Weathering	Uptake	Water flux	Litter flux
Location	x	-	x[1)]	x	-
Tree species	x	-	x	x	x[2)]
Soil type	-	x	x[2)]	x[2)]	-

1) Not considered in the National application.
2) Not considered in the European application.

An overview of the collection of the data determining the various fluxes of elements and water is given below.

Deposition rates

The average deposition of SO_2, NO_x and NH_3 on each grid has been derived from atmospheric models, or source-receptor matrices based on such models, within the overall RAINS and DAS model. The bulk deposition of base cations (Ca, Mg, K and Na) and Cl has been derived from 22 weather stations in the Netherlands and 76 in Europe, using interpolation techniques to get values for each grid. The total deposition on each tree species has been estimated using filtering factors for SO_2, NO_x and NH_3 and dry deposition factors for base cations and chloride. These data, which are vegetation dependent, have been derived from available literature (De Vries et al, 1991c; 1991d).

Weathering rates

For Europe, weathering rates have been derived by using a relative simple pedo-transfer function with soil type and texture class (De Vries, 1991b; De Vries et al., 1991d). For the Netherlands data are derived from (1) information on base cation depletion rates in soil profiles (De Vries and Breeuwsma, 1986) and (2) column and batch experiments which have been conducted during five years on the most relevant soil types and soil horizons in the Netherlands (De Vries et al., 1991c).

Uptake rates and litterfall fluxes

Uptake rates are determined by forest growth and element contents in stems and branches, whereas element fluxes by litterfall are determined by the biomass of leaves, the turnover rate and the element content. For the Netherlands, forest growth estimates for all relevant combinations of forest and soil type are based on a literature inventory (De Vries et al., 1990b). For Europe, average soil independent growth rate estimates per grid have been used, which are based on a data compilation available at IIASA. Biomass and turnover rates of leaves (needles) and contents of the elements N, K, Ca and Mg in stems, branches and leaves are based on a literature survey for major tree species i.e. Pinus Silvestris (Scotch pine), Pinus Nigra (Black pine), Pseudotsuga Menziesii (Douglas fir), Picea Abies (Norway spruce), Larix Leptolepis (Japanese larch), Quercus Robur (Oak) and Fagus Sylvatica (Beech) (De Vries et al., 1990b).

Water fluxes

Water fluxes are determined by the precipitation rate minus the sum of interception, evaporation and transpiration (evapotranspiration). In MACAL, the transpiration is described as a function of depth by empirical functions (linear, quadractic, etc.). Precipitation estimates have been derived from 280 weatherstations in the Netherlands and 325 in Europe, using interpolation techniques to obtain values for each grid. Interception fractions, relating interception to precipitation, have been derived from literature data for all tree species considered. For Europe, an average value for the sum of evaporation and transpiration has been derived for each grid. Values are calculated with an empirical model as a function of precipitation, altitude and latitude (De Vries et al., 1991d). For the Netherlands data for evaporation and transpiration have been calculated for various combinations of tree species and soil type with a separate hydrological model called SWATRE (De Visser and De Vries, 1989).

5.1.4 Modelpredictions

Results of the application of the SMB model for the Netherlands are given in De Vries et al. (1991b). An example of model output is given in Figure 9.

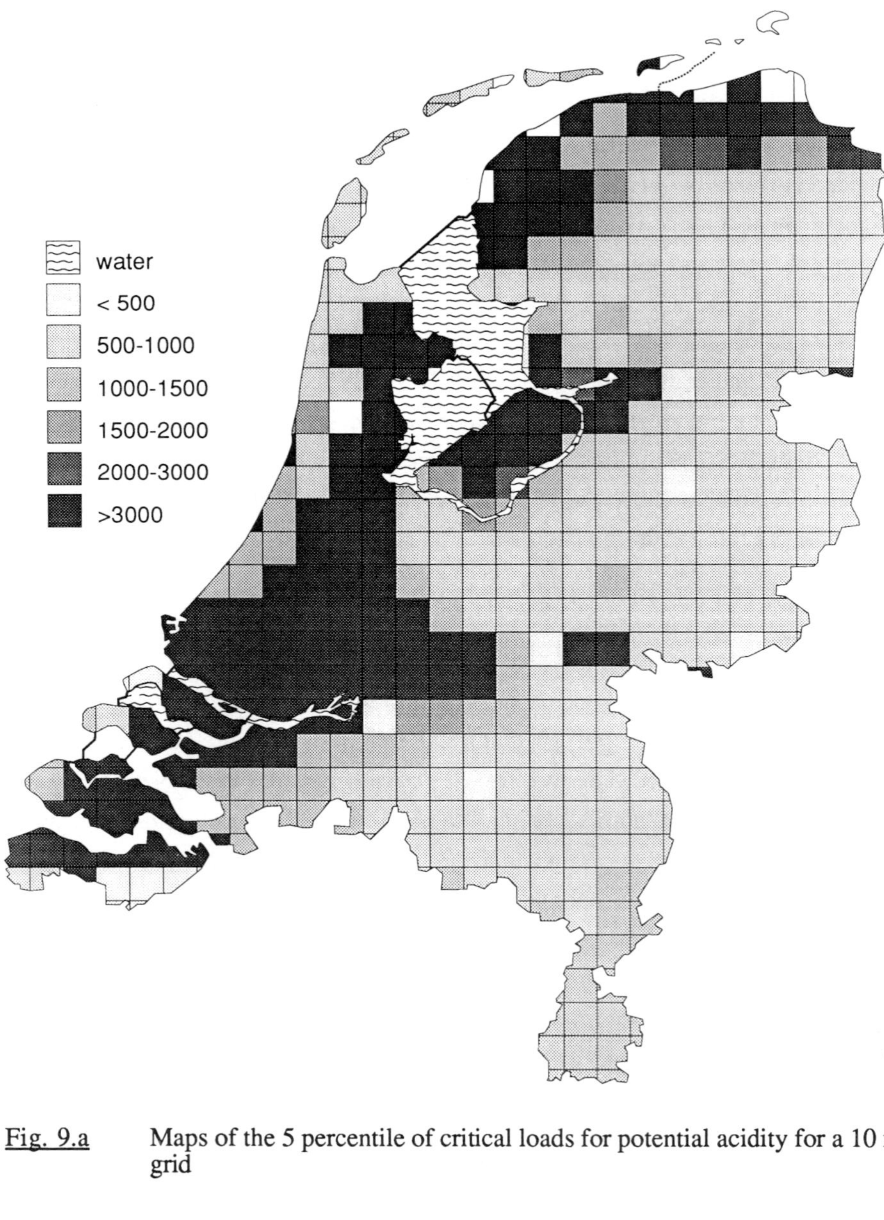

Fig. 9.a Maps of the 5 percentile of critical loads for potential acidity for a 10 x 10 km grid

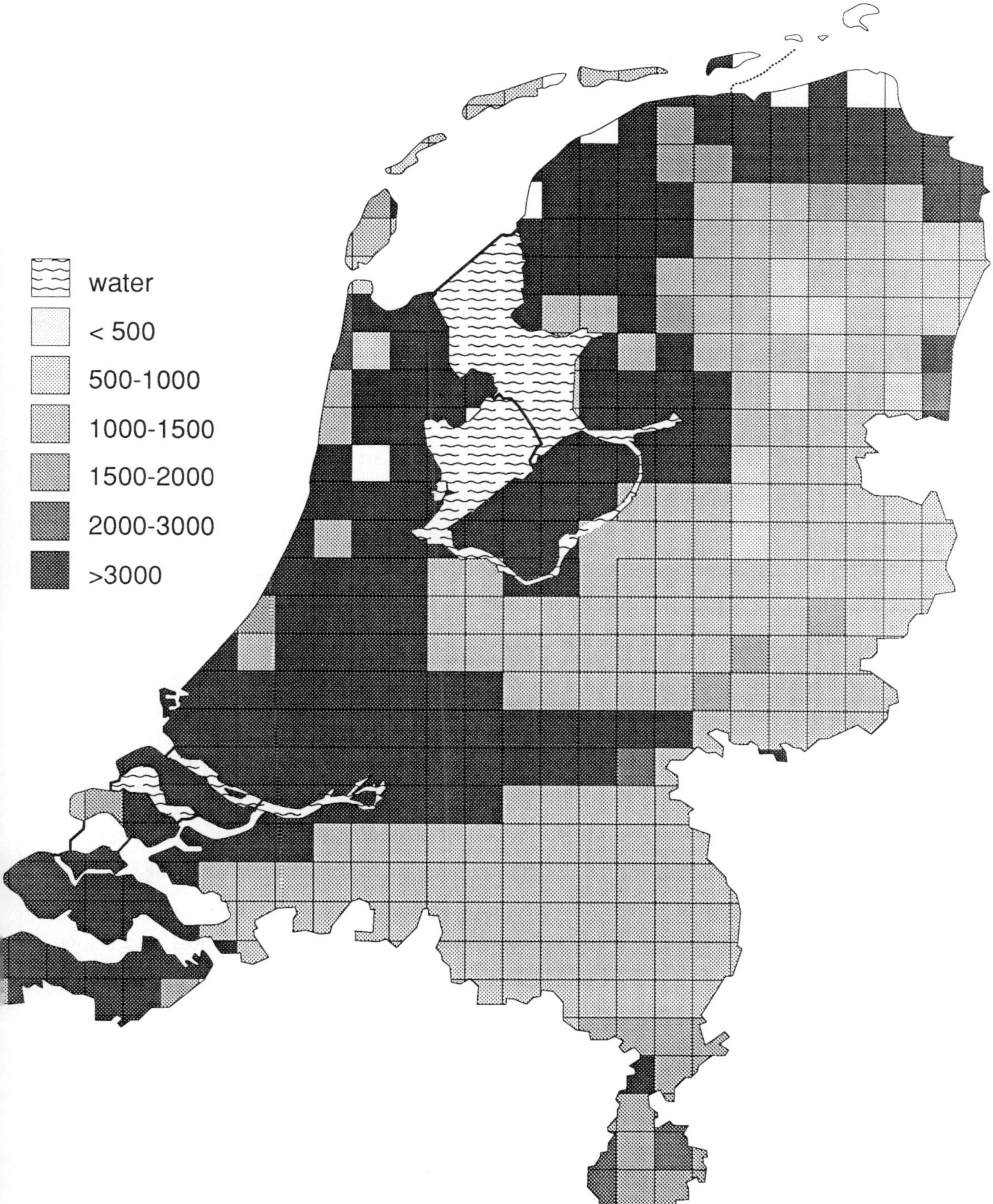

Fig. 9.b Maps of the 50 percentile of critical loads for potential acidity for a 10 x 10 km grid

Comparison of the critical load maps given in Figure 9 shows that the choice of the percentile hardly influences the area with a median critical load above 1500 mol_c ha^{-1} yr^{-1}. This area mainly includes the loess soils in the southern part of the Netherlands(Limburg) and the calcareous soils (sand and clay), clay soils and peat soils in the western and northern part of the country (Zeeland, Zuid- and Noord-Holland, Groningen, Friesland and Flevoland).

Median critical loads for loess soils in the southern part of the Netherlands are nearly all in a range between 1500 and 2000 mol_c ha^{-1} yr^{-1}. The major reason for these relatively high critical loads is the large input of base cations (mainly Ca) in this area, caused by the occurrence of limestone quarries (De Vries et al., 1991b).

Critical loads for calcareous soils, clay soils and peat soils are always above 3000 mol_c ha^{-1} yr^{-1}. For these soils a separate procedure has been used to calculate the critical loads. For calcareous soils the critical S load was based on the load associated with a critical annual level of SO_2 of 20 g m^{-3}. This was done since the base cation weathering rate of calcareous soils is such that it can neutralize any acid input without mobilizing aluminium. However, above a certain SO_2 level, direct effects may occur. The critical S load thus calculated is equal to 4100 mol_c ha^{-1} yr^{-1}. Similar to non-calcareous soils, the critical N load was based on the net uptake and immobilzation of nitrogen. For clay and peat soils, the critical N load was based on the load associated with a critical annual level of NO_2 of 30 g m^{-3}. This was done since the nitrogen input on these soils is almost completely compensated by denitrification. The critical N load thus calculated is equal to 3200 mol_c ha^{-1} yr^{-1}. Similar to non-calcareous sandy soils, the critical S load was based on the net input of base cations (dry deposition plus weathering minus uptake) minus the critical alkalinity leaching.

The choice of the percentile clearly influences the area with a median critical load below 1500 mol_c ha^{-1} yr^{-1}. This area mainly includes the non-calcareous sandy soils in the central and eastern part of the country (Drenthe, Overijssel, Gelderland, Utrecht and Noord-Brabant) where most of the Dutch forests occur. Most striking is the shift from 1000-1500 mol_c ha^{-1} as a median value to 500-1000 mol_c ha^{-1} yr^{-1} as a 5 percentile value.

The overall median value for the critical load of potential acidity on non-calcareous sandy soils is approximately 1200 mol_c ha^{-1} yr^{-1}. This is in the lower range of the average critical load values for forests given before in Table 3 (see par. 3.1.2). As an example, the range in critical loads on coniferous forests is shown in Table 7.

Table 7. The range in model input and model output for coniferous forests on non-calcareous sandy soils

Parameter	Min	5%	50%	95%	Max
BC^*_{dd}	188	228	330	716	1370
BC_{we}	120	180	270	580	580
BC_{gu}	187	187	292	428	654
N_{gu}	197	197	310	612	612
Ac_{le}	103	110	448	742	920
$Cl(Ac_{pot})$	541	650	1140	1658	2346

The critical acidity leaching is calculated as the minimum value of the three options mentioned in par. 2.2.2.1 (equation 4, 5 and 6).

It should be noted that use of a 5 percentile critical load value based on a simple mass balance model is a very strict criterium. Use of this model implies that the critical Al concentration and Al/Ca ratio is applied at the bottom of the rootzone. Values for these parameters in the forest topsoil are generally lower because of nutrient (Ca) cycling, transpiration and Al mobilization with depth (see also par. 4.1.2). Applications with the MACAL model show that critical loads based on critical values at a depth of 30 cm are significantly higher. The increase in critical load, if Al concentrations and Al/Ca ratios in the topsoil are considered, is also illustrated by the results of the RESAM application (par. 5.2.5). Results of this multi-layer model show that approximately 50% of the forests on non-calcareous sandy soils is below critical soil solution levels in 2000 at an average load of 2500 $mol_c\ ha^{-1}\ yr^{-1}$. This can be regarded as an indication of the median critical acid load considering forest topsoils.

5.2 Long-term impacts

5.2.1 Introduction

The methodology for mapping the output of the dynamic models SMART and RESAM at a specific future time period is similar to mapping critical loads. It also requires information on the area of forest-soil combinations in each deposition (receptor) area together with deposition data and data related to soils, forests, and the combination of both. Here we describe the quantification of the receptor distribution and the data collection for the regional application of SMART and RESAM. Furthermore, examples are given of the national application of RESAM, together with results of a regional model validation.

5.2.2 Deposition areas and receptor distribution

As with START, the RAINS grid system is used for a preliminary regional application of

SMART (De Vries et al., 1991d). The procedure used for assessing the receptor distribution within each RAINS grid is described before (par. 5.1.2).

RESAM has not been applied on a grid basis. Here, the shape and spatial resolution of the deposition areas has been delineated by 20 areas with relevant statistical information on emissions (De Vries, 1990; De Vries et al., 1991c). An overview of the deposition areas is given in Figure 10.

In order to limit both data acquisition and computation time, the regional application of RESAM has been restricted to tree species and soil types (receptors) of major importance. Forests are represented by seven important tree species i.e. Pinus Sylvestris (Scotch Pine), Pinus Nigra (Black Pine), Pseudotsuga Menziesii (Douglas Fir), Picea Abies (Norway Spruce), Larix Leptolepis (Japanese Larch), Quercus Robur (Oak) and Fagus Sylvatica (Beech).

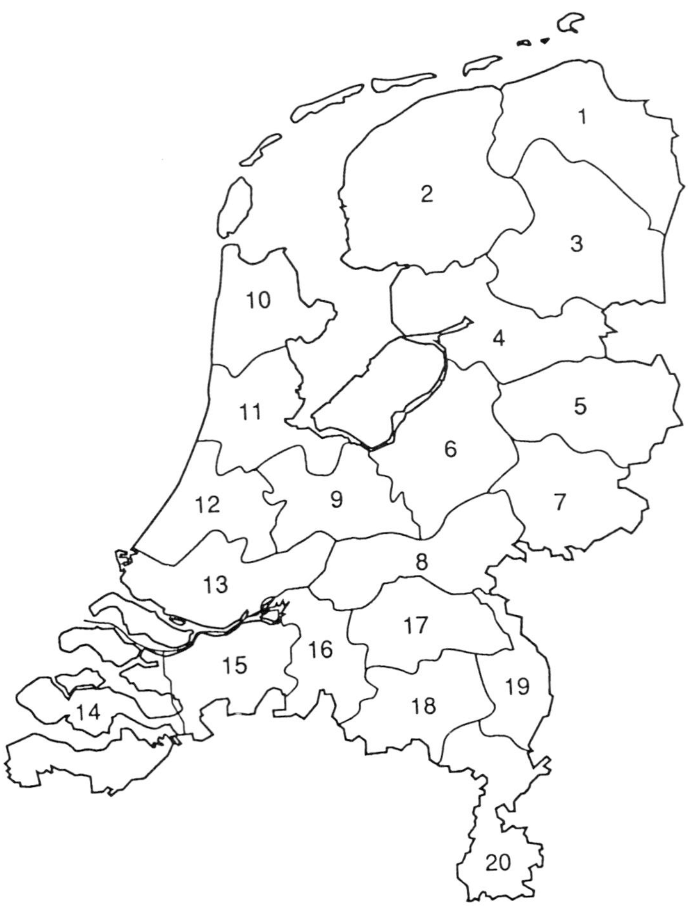

Fig. 10. The 20 deposition areas considered in the regional application of RESAM

Forest soils are confined to acid sandy soils, as these cover 80 to 90% of the Dutch forest area. Furthermore, these soils are sensitive to acidification. The soils are characterized by 14 representative profiles and cover a broad range in soil properties, such as the organic matter content, texture and groundwater level, which influence major hydrological, biochemical and geochemical processes in the soil. The vertical heterogeneity is taken into account by differentiating between soil layers (horizons). An overview of the designation and thickness of the horizons in the various soil profiles is given in De Visser and De Vries (1989).

The forest/soil combinations that have been included comprise nearly 65% of the total Dutch forest area of which more than 50% is covered by Pinus Sylvestris (Scotch Pine). The remaining 35% comprises tree species such as Populus Spec (Poplar) and Betula Pendula (Birch) (approximately 20%) and soil types such as calcareous sandy soils, clay soils, loess soils and peat soils (approximately 15%). The spatial distribution (area) of the forest-soil combinations considered in each of the 20 receptor areas has been assessed by a grid-overlay procedure as described in par. 5.1.2.

5.2.3 Data collection

The various data described before in par. 5.1.3, i.e. deposition data, weathering rates, growth (and turnover) rates, element contents in tree compartments and water fluxes in soil layers, also are needed for SMART and RESAM. Actually, the data acquisition described in par. 5.1.3 was orginally carried out for the application of these models.

Additional data for SMART compared with START are bulk density, cation exchange capacity, base saturation and cation exchange constants for the various soil types. The bulk density and cation exchange capacity has been related to the organic matter content, whereas the base saturation has been related to the texture class, using so-called pedo-transfer functions. Organic matter content and texture class have been derived from the FAO Soil Map of Europe. Cation exchange constants are based on literature data (De Vries et al., 1991d).

Additional data for RESAM to those required by MACAL include the element amounts in litter, primary minerals, hydroxides and on the adsorption complex together with rate and equilibrium constants influencing the mobilization rate of the modelled soil processes, i.e. nitrification, denitrification, protonation, base cation weathering, Al dissolution and cation exchange (De Vries, 1991c).

The amount in primary minerals is based on a total analysis of the considered soil types and soil horizons, whereas the adsorbed amount of cations is based on field data (Kleijn et al.,

1989). All other soil data are derived from a soil information system or related to basic soil characteristics, available in the system. Relations (pedo-transfer functions) that have been used are the derivation of bulk density and cation exchange capacity from the organic matter content and the derivation of the sulphate sorption capacity from the amount of amorphous Al hydroxides, using regression equations. A discussion of the use of the pedo-transfer function approach in parameterizing regional water quality models, together with application examples, has also been given by De Vries (1987b; 1990) and De Vries et al. (1989d).

Nitrification and protonation rate constants are derived by calibration, whereas the denitrification rate is based on literature information. Constants determining the Al dissolution rate are based on one-year batch experiments that have been conducted for nearly all soil types and soil horizons included in the regional application. Cation exchange constants are based on the simultaneous measurement of elements on the adsorption complex and in the soil solution for the most important soil types i.e. all Podzols and the Albic Arenosol (De Vries et al., 1989c).

5.2.4 Model validation

In order to gain insight in the reliability of the model predictions, model results of the soil solution chemistry in 1990 were compared with soil solution measurements in 150 forest stands during the period March to May in the same year. The tree species included in the field survey are similar to those included in the simulations.

A comparison of median values of important soil solution parameters is given in Table 8. The pH, Al concentration, molar Al/Ca ratio and molar NH_4/K ratio in the topsoil (top 20 to 30 cm) and pH, Al and NO_3 concentration are important indicators of forest stress, whereas pH, Al and NO_3 concentration in the subsoil are important indicators of potential groundwater pollution. For most of these parameters, critical concentration levels have been defined (see par. 3.1.1 and 3.1.2). The SO_4 concentration has been added to acquire an insight into the relative contribution of S and N in soil acidification.

The agreement between model results and field data is good (difference < 10%) for pH and for SO_4, reasonable (difference between 10-30%) for the Al/Ca ratio and NO_3 concentration and poor (difference > 30%) for the NH_4/K ratio and Al concentration.

Table 8. Median values of important soil solution parameters as measured in the field and predicted by RESAM

Parameter	Topsoil		Subsoil	
	measured	predicted	measured	predicted
pH (-)	3.6	3.7	3.9	3.8
Al ($mol_c\ m^{-3}$)	0.7	0.4	0.6	1.2
Al/Ca (-)	1.3	1.7	-	-
NH_4/K (-)	1.7	2.7	-	-
NO_3 ($mol_c\ m^{-3}$)	-	-	0.5	0.7
SO_4 ($mol_c\ m^{-3}$)	-	-	1.1	1.0

It should be kept in mind that one measurement in early spring can give rise to both lower and higher concentrations compared with the flux weighted anual average value depending upon the difference in actual 1990- and 30 year averaged meteorologic data. Comparison between model results and field data for the tracers Na and Cl in both topsoil and subsoil shows that the model results are always (slightly) lower, especially in the topsoil. In the subsoil, comparison is better. This is partly an explanation for the underprediction of Al concentrations in the topsoil. The overprediction of Al in the subsoil can partly be explained by an overestimation of the NO_3 and SO_4 concentration which influences the Al mobilization. The remaining difference is most probably due to the long-term effect of liming (Ca) and fertilization (mainly K) and/or a higher base cation input from the atmosphere which will cause an increase in base cation concentration and a decrease in Al concentration. This also explains the overestimation of the molar Al/Ca ratio and molar NH_4/K ratio by RESAM. More detailed information on the regional validation of RESAM is given in De Vries et al. (1991c).

5.2.5 Model predictions

The response of Dutch forest soils to deposition reductions has been evaluated for three emission-deposition scenarios during the period 1990-2050. The deposition reduction between 1990 and 2000 is based on expected abatement strategies causing emission reductions (scenario NMP+), whereas the reductions between 2000 and 2050 are based on various deposition aims.

Values for the average deposition levels on the Netherlands during this period are given in Table 9. Average deposition levels on forests are 20% higher.

Table 9. Average acid deposition levels on the Netherlands during the period 1990-2050 for three scenarios

Scenario	Deposition (mol_c ha^{-1} yr^{-1}) 1990	2000	2010	2050
1	4360	2200	2200	2200
2	4360	2200	1400	1230
3	4360	2200	1230	700

Results of the long term impacts of these scenarios are given in De Vries et al. (1991c). As an example trends in median values of the Al concentration and Al/Ca ratio in the topsoil are given in Figure 11.

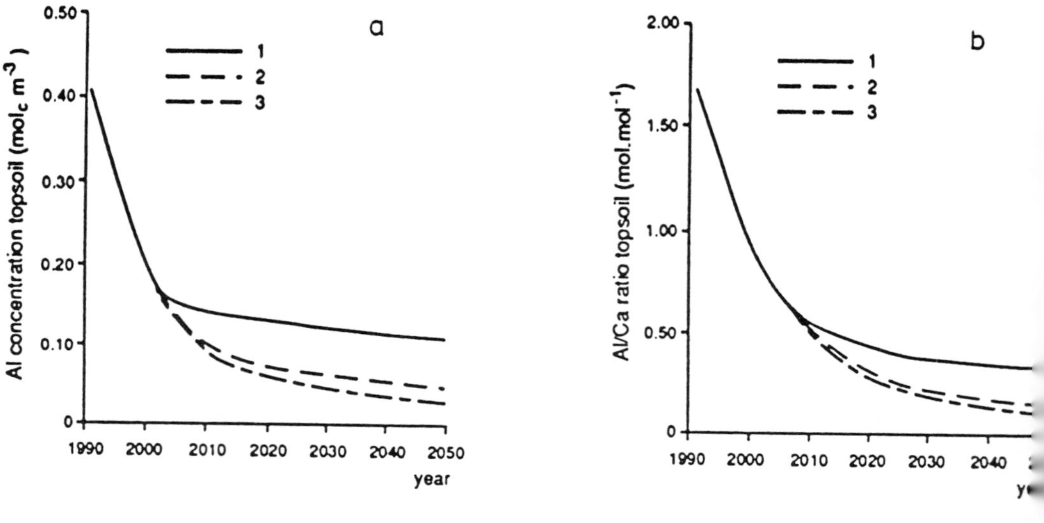

Fig. 11. Future trends in median values of the Al concentration (a) and the molar Al/Ca ratio in the topsoil (b) in response to three deposition scenarios

Between 1990 and 2000 there is no difference in trends for the three scenarios because the deposition values are similar. In this period the average deposition in the Netherlands drops from approximate 4400 to 2200 $mol_c\ ha^{-1}\ yr^{-1}$. In response to this deposition reduction there is a considerable decrease in the Al concentration (Fig. 11a) and Al/Ca ratio (Fig. 11b) in the topsoil. Both medians drop below a critical value of 0.2 $mol_c\ m^{-3}$ and 1.0 mol mol^{-1} respectively before the year 2000. Between 2000 and 2050 there is still a small decrease in Al concentration and Al/Ca ratio for scenario 1 although the deposition level remains the same. For the scenarios 2 and 3 there is still a substantial decrease in median values especially up to 2020.

The difference between scenario 2 and 3 is much smaller than between scenario 1 and 2. This is consistent with the difference in deposition values in 2050, i.e. 2200, 1230 and 700 $mol_c\ ha^{-1}\ yr^{-1}$ respectively.

Trends in the area exceeding critical levels for Al and Al/Ca in the topsoil are given in Figure 12.

The figure shows that the area exceeding a critical Al concentration and Al/Ca ratio of 0.2 $mol_c\ m^{-3}$ and 1.0 mol mol^{-1} respectively is approximately 75% and 65% in 1990. In the year 2000, the percentage of forest soils exceeding both values is still approximately 40% and 30% respectively. In the year 2050 this area is negligible for the scenarios 2 and 3. For scenario 1 it remains approximately 25% and 5% for the Al concentration and Al/Ca ratio respectively.

It is important to stress that a deposition reduction according to scenario 2 is enough to avoid exceedances in Al concentration or Al/Ca ratio in forest topsoils. The Al hydroxide depletion also stops for this scenario between 2020 and 2030 (not presented here). The average deposition level at this time is close to 1400 $mol_c\ ha^{-1}\ yr^{-1}$ which is the average critical load derived for the effects of aluminium on forests (see par. 3.1.2).

The response of "N related parameters", such as the molar NH_4/K ratio and NO_3 concentration, to a deposition reduction is small compared with the "Al related parameters". This is mainly due to N mobilization from litter, which in turn is caused by a decrease in the N content of leaves in response to decreased N deposition. This causes a time lag between the reduction in N deposition and N (NH_4, NO_3) concentration.

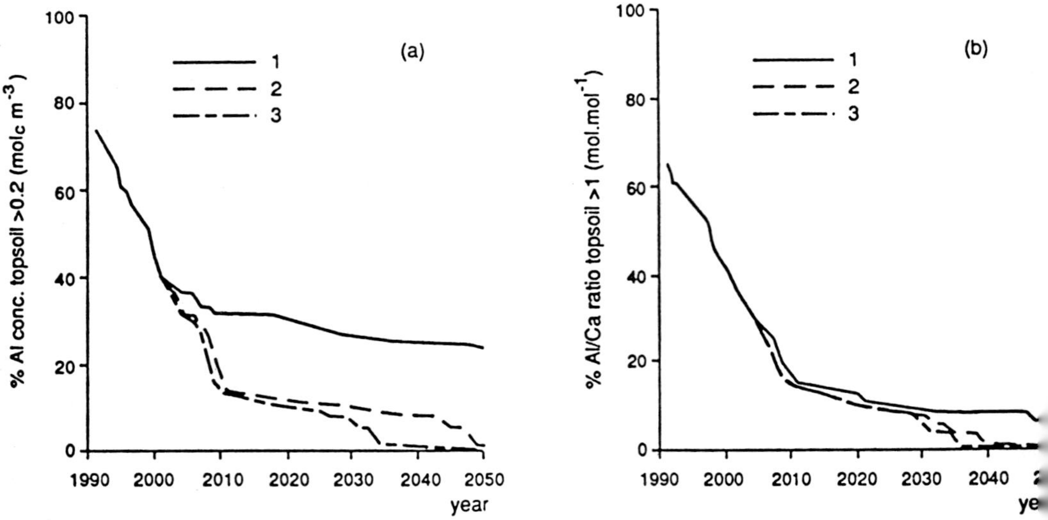

Fig. 12. The percentage of forest soils exceeding a critical Al concentration of 0.2 mol_c m^{-3} (a) and a critical molar Al/Ca ratio of 1.0 in the topsoil (b) in response to three deposition scenarios

Tree species appeared to have a larger influence on the soil solution chemistry than deposition areas. This is illustrated in Table 10 for three different tree species, i.e. Douglas Fir, Scotch Pine and Oak and for two deposition areas, i.e. 3 (Drenthe) and 18 (Eindhoven). Both areas have a relatively high proportion of forests but the average acid load in area 18 is much higher (5900 mol_c ha^{-1} yr^{-1}) than in area 3 (4000 mol_c ha^{-1} yr^{-1}).

Table 10. Influence of tree species and deposition area on median values of the Al concentration and molar Al/Ca ratio in forest topsoils in 1990, 2010 and 2050 in response to scenario 3

Tree species	Deposition area	Al concentration			Molar Al/Ca ratio		
		1990	2010	2050	1990	2010	2050
Douglas fir	all[1)]	0.8	0.2	0.0	3.7	1.2	0.2
Scotch pine	all	0.6	0.1	0.0	3.1	1.0	0.2
Oak	all	0.3	0.1	0.0	1.5	0.4	0.1
All[2)]	3	0.4	0.1	0.0	3.1	1.1	0.3
All	18	0.7	0.1	0.0	4.3	1.6	0.3

1) Deposition area 1 to 20.
2) Scotch Pine, Black Pine, Douglas fir, Norway Spruce, Japanese Larch, Oak and Beech.

The decrease in concentrations and ratios according to Douglas fir > Scotch Pine > Oak is mainly due to a decrease in evapotranspiration. The differences between the tree species are in agreement with the field data described in par. 5.2.4. The influence of tree species and deposition areas decreases with time, which corresponds to the decrease in deposition differences.

6. CONCLUSIONS AND RECOMMENDATIONS

The most important conclusions from this research are:

1 Average critical loads for potential acidity on well drained non-calcareous sandy forest soils in the Netherlands vary between 1100-1700 mol_c ha^{-1} yr^{-1}. For groundwaters and surface waters values are lower, i.e. 400-700 mol_c ha^{-1} yr^{-1}. This is much less than present inputs (4000-15000 mol_c ha^{-1} yr^{-1}). Although the uncertainty in critical load values can be large, this shows that substantial emission reductions are needed.

2 The pH of calcareous dune soils with a low carbonate content will drop significantly in the coming decades unless major emission/deposition reductions are implemented.

3 In the long run, the present high acid loads in the Netherlands will cause a depletion of the pool of readily dissoluble Al-hydroxides in non-calcareous forest and dune soils. This will cause a further decrease in the soil solution pH in the topsoil.

4 Deposition reductions generally lead to a fast improvement of the soil solution chemistry. This includes an increase in pH and a decrease in Al, SO_4 and NO_3 concentration and Al/Ca and NH_4/K ratio. However, for the NO_3 concentration and NH_4/K ratio there is a clear time lag between deposition- and concentration reduction which is mainly due to N mobilization from the litter layer.

5 Uncertainties in the deposition of SO_2, NO_x and NH_3 and in parameters determining nitrogen and aluminium dynamics, play an important role in the uncertainty of long-term model predictions in the Netherlands. On a European scale, the uncertainty in CEC and base saturation is also (very) important.

6 Predicted median values of element concentrations in the soil solution of non-calcareous sandy forest soils in 1990 generally show a reasonable agreement with measured median values for 150 forest stands in the same year. However, there is a clear underprediction of the base cations Ca and K, especially in the subsoil which might be due to the long-term effect of liming and fertilization.

7 Deposition reductions up to 1400 $mol_c\ ha^{-1}\ yr^{-1}$ as an average for the Netherlands before 2010 cause a substantial decrease in the exceedance of a critical Al concentration of 0.2 $mol_c\ m^{-3}$ and a critical molar Al/Ca ratio of 1.0 in the topsoil (from about 75% to less than 20%). Further reductions up to 1400 $mol_c\ ha^{-1}\ yr^{-1}$ as an average for forests areas only, lead to a negligible exceedance of these criteria. It is also enough to avoid depletion of Al hydroxides.

8 Results of critical load calculations for all Dutch forest-soil combinations show that the median values are nearly all in a range between 1000-1500 $mol_c\ ha^{-1}\ yr^{-1}$ for non-calcareous sandy soils, between 1500-2000 $mol_c\ ha^{-1}\ yr^{-1}$ for loess soils and above 3000 $mol_c\ ha^{-1}\ yr^{-1}$ for clay soils, peat soils and calcareous soils.

9 In the Netherlands, the influence of the tree species on the soil solution chemistry appears to be more important than the influence of the deposition area. However, the dependance of soil solution chemistry on tree species and deposition areas decreases in future predictions due to deposition reductions.

In order to aquire a better insight into (1) the long-term effects of Al depletion, (2) the long-term effects of acid deposition on a National and European scale and (3) the reliability of long-term predictions, additional research is needed with respect to:

1 The adsorption/desorption and precipitation/dissolution of phosphate in relation to Al- and Fe-buffering;
2 The acidification status (CEC, base saturation and soil solution composition) of all non-agricultural soils in the Netherlands and in Europe including loamy soils, clay soils and peat soils;
3 The uncertainty in regional model predictions due to the uncertainty in model structure and the variability in model parameters.

The above-mentioned aspects will be investigated in the period between 1990 and 1993 (in the context of the projects 7160 and 7156; see preface).

Ad 1 Adsorption/desorption kinetics and phosphate equilibria with Al and Fe will be implemented in RESAM using available information on the important reaction mechanisms. Data on rate and equilibrium constants for phosphate will be derived from the literature. Data on Al and Fe weathering influencing phosphate interaction processes, will be derived from laboratory experiments.

Ad 2 Data will be gathered in the Netherlands for loamy soils and peat soils regarding CEC, base saturation, soil solution composition, weathering rates and exchange constants. Model predictions with RESAM will be made for these soils in combination with sandy soils, to assess the long-term impact of acid deposition on the total Dutch forest area. The range in CEC and base saturation of European forests will be derived from available data in various European countries together with data becomming available through the International Cooperative Programme (ICP) on forests which is part of the Working Group on Effects resulting under the EB of LTRAP. The data will be stored in a data base.

Ad 3 For both RESAM and SMART, the uncertainty in regional model predictions will be assessed in relation to the spatial variability in model parameters, using the data described above. With respect to RESAM, the uncertainty in the model structure also will be quantified by comparing various process formulations in the model. Furthermore, the impact of temporal aggregation will be assessed by comparing long-term predictions, using annual average data (present situation), with predictions including in-year variability.

7. REFERENCES

Hettelingh, J.P. and W. de Vries, 1990. Mapping vademecum. Bilthoven, National Institute of Public Health and Evironmental Protection. Technical report

Kleijn, C.E. and W. de Vries, 1987. Characterizing soil moisture composition in forest soil in space and time. In W. van Duyvenbooden and H.G. van Waegeningh (Eds.): Vulnerability of Soil and Groundwater to Pollutants, Proc. Int. Conf. Noordwijk aan Zee, 1987, the Netherlands: 591-600

Kleijn, C.E., G. Zuidema en W. de Vries, 1989. De indirecte effecten van atmosferische depositie op de vitaliteit van Nederlandse bossen. Deel 2. Depositie, bodemeigenschappen en bodemvochtsamenstelling van acht Douglasopstanden. Wageningen, STIBOKA, Rapport nr. 2050

Kros, J., P. Janssen, W. de Vries en C. Bak, 1990. Het gebruik van onzekerheidsanalyse bij modelberekeningen: een toepassing op het regionale bodemverzuringsmodel RESAM. Wageningen, Staring Centrum, Rapport 65

Kros, J., W. de Vries, P. Janssen and C. Bak, 1991. The uncertainty in forecasting regional trends of forest soil acidification. Water, Air and Soil Pollution (Submitted)

Schulze, E.D., W. de Vries, M. Hauhs, K. Rosén, L. Rasmussen, C.O. Tamm and J. Nilsson, 1989. Critical loads for nitrogen deposition on forest ecosystems. Water, Air and Soil Pollution, 48: 451-456

Sverdrup, H., W. de Vries and A. Henriksen, 1990. Mapping critical loads. A guidance manual to criteria, calculation methods data collection and mapping. Background document for the Task Force on Mapping, Bad Harzburg, 22-23 May, 1990

Visser, P.H.B. de en W. de Vries, 1989. De gemiddeld jaarlijkse waterbalans van bos-, heide en grasland vegetaties. Wageningen, STIBOKA, Rapport nr. 2085

Vries, W. de and A. Breeuwsma, 1986. Relative importance of natural and anthropogenic proton sources in soils in the Netherlands. Water, Air and Soil Pollution, 28:173-184

Vries, W. de and A. Breeuwsma, 1987. The relation between soil acidification and element cycling. Water, Air and Soil Pollution, 35:293-310

Vries, W. de, 1987a. A conceptual model for analysing soil and groundwater acidification on a regional scale. Proc. Int. Symp. on Acidification and Water Pathways. Bolkesjo, 1987, Norway: 185-194

Vries, W. de, 1987b. The role of soil data in assessing the large scale input of atmospheric pollutants on groundwater quality. In W. van Duyvenbooden and H.G. van Waegeningh (Eds.): Vulnerability of Soil and Groundwater to Pollutants. Proc. Int. Conf. Noordwijk aan Zee, 1987, the Netherlands: 897-910

Vries, W. de, 1988. Critical deposition levels of nitrogen and sulphur on Dutch forest ecosystems. Water, Air and Soil Pollution, 42:221-239

Vries, W. de, M.J.P.H. Waltmans, R. van Versendaal en J.J.M. van Grinsven, 1988. Aanpak, structuur en voorlopige procesbeschrijving van een bodemverzuringsmodel voor toepassing op regionale schaal. Wageningen, STIBOKA, Rapport nr. 2014

Vries, W. de, 1989. Kritische depositieniveau's van stikstof en zwavel op Nederlandse bossen. Proc. Nationaal Symposium Verzuring, Ede, 15 juni 1989: 15-22

Vries, W. de en J. Kros, 1989a. De lange termijn effecten van verschillende depositiescenario's op de bodemvochtsamenstelling van representatieve bosecosystemen. Wageningen, Staring Centrum, Rapport 30

Vries, W. de and J. Kros, 1989b. The long term impact of acid deposition on the aluminium chemistry of an acid forest soil. In J. Kämäri, D.F. Brakke, A. Jenkins, S.A. Norton and R.F. Wright (Eds.): Regional Acidification Models. Geographic Extent and Time Development: 113-128

Vries, W. de, M. Posch and J. Kämäri, 1989a. Modelling time patterns of forest soil acidification for various deposition scenarios. In J. Kämäri, D.F. Brakke, A. Jenkins, S.A. Norton and R.F. Wright (Eds.): Regional Acidification Models. Geographic Extent and Time Development: 129-150

Vries, W. de, M. Posch and J. Kämäri, 1989b. Simulation of the long-term soil response to acid deposition in various buffer ranges. Water, Air and Soil Pollution, 48:349-390

Vries, W. de, A. Breeuwsma en F. de Vries, 1989c. Kwetsbaarheid van de Nederlandse bodem voor verzuring. Een voorlopige indicatie in het kader van de Richtlijn "Ammoniak en Veehouderij". Wageningen, Staring Centrum, Rapport 29

Vries, W. de, O.F. Schoumans, J.F. Kragt and A. Breeuwsma, 1989d. Use of models and soil survey information in assessing regional water quality. In G. Jousma, J. Bear, Y.Y. Haimes and F. Walter (Eds.): Groundwater contamination: Use of models in decision-making, Proc. Int. Conf. Amsterdam, 1987, the Netherlands: 419-432

Vries, W. de, 1990. Philosophy, structure and application methology of a soil acidification model for the Netherlands. In J. Kämäri (Ed.). Impact models to assess regional acidification: 3-21

Vries, W. de, A. Hol, S. Tjalma en J.C. Voogd, 1990. Voorraden en verblijftijden van elementen in een bosecosysteem: een literatuurstudie. Wageningen, Staring Centrum, Rapport 94

Vries, W. de and H.D. Gregor, 1990. Critical loads and critical levels for the environmental effects of air pollutants. In: M.J. Chadwick and M. Hutton (Eds.): Acid Depositions in Europe: Environmental effects, control strategies and policy options. Stockholm Environment Institute

Vries, W. de, 1991a. Assessment and policy implications of critical loads for nitrogen and sulphur in the Netherlands. Water, Air and Soil Pollution (submitted)

Vries, W. de, 1991b. Methodologies for the assessment and mapping of critical loads and the impact of abatement strategies on forest soils. Wageningen, The Winand Staring Center for Integrated Land, Soil and Water Research (in preparation)

Vries, W. de, J. Kros and J. Klijn, 1991a. Simulation of the long-term impact of atmospheric deposition on dune ecosystems in the Netherlands. Journal of Applied Ecology (in preparation)

Vries, W. de, J. Kros, R. Hootsmans, G.J. van Uffelen and J.C. Voogd, 1991b. Assessment and mapping of critical loads for nitrogen and sulphur on Dutch forest soils (in preparation)

Vries, W. de, J. Kros, C. van der Salm and J.C. Voogd, 1991c. The long-term impact of three emission-deposition scenarios on Dutch forest soils. Water, Air and Soil Pollution (in preparation)

Vries, W. de, M. Posch, J. Kämäri and W. Schöpp, 1991d. Long-term soil response to acidic deposition in Europe. Water, Air and Soil Pollution (in preparation)

ANNEX 2 REVIEW REPORT

1. INTRODUCTION

1. The Second Phase of the Dutch Priority Programme on Acidification was established in 1988 and asked to address three main questions:

* What is the quantitative contribution of O_3 SO_2, NO_x and NH_3 to direct effects on vegetation?
* What is the quantitative contribution of SO_2, NO_x and NH_3 to the indirect effects (via the soil) on vegetation?
* What is the effectiveness of abatement measures in terms of environmental effects?

This programme is now in its final phase and will terminate in 1990.

2. An independent Review Team (Annex 1) has been convened to provide a critical assessment of the Second Phase Programme in answering these questions and also to make recommendations on whether there is justification for future work in this area. In undertaking this task, therefore the Review Team has interpreted its remit to address the following issues:

- Has the Second Phase Programme been successful in providing answers to the three main questions, have the research groups focussed on the main issues and what is the quality of the research under review?
- If not, what further information is required to fill gaps in knowledge with some attempt to identify priorities for future work?
- How relevant is the Second Phase Programme to the ongoing assessment of critical loads/levels in the Netherlands?

Individual assessments have been made on the separate components of the Second Phase of the Dutch Priority Programme on Acidification and these have been brought together in a conclusion and recommendation section at the end of the report. The value of extending the programme into a third phase is also considered.

Finally, although outside the specific remit of the Review Team, all members felt that the many lessons learned through running an integrated research programme on acidification could well have relevance to future planning on climate change impact research. For this reason, the Review Team has identified a number of projects or research areas in the current programme which could be translated without difficulty into the field of climate change impacts research.

2. GENERAL COMMENTS

In addition to the detailed observations made by individual reviewers in the seven project area chapters, the Review Team discussed the Second Phase Programme in its entirety and has concluded that :

- Total nitrogen deposition and ammonia and its subsequent effects appears to have become the major focus for research and is now dominating the programme.
- The Review Team were impressed with the scientific standard of the Second Phase Programme. It agrees that the focus on ammonia and total nitrogen deposition is a positive step in the programme and meets concern over a major pollutant of specific importance in the Netherlands. However, it is the view of the Review Team that because of this, the Second Phase Programme as it is at present comprised will not provide the desired quantification of the three main questions. In general, there appears to be considerable coordination and integration in the Second Phase Programme, but the Review Team were concerned over an apparent lack of integration at this stage between e.g. the biological and physiological effects projects and the integrated effects on forests projects and urges more effort in these areas.
- In targetting sensitive areas e.g. forests, soils, heathlands and acid waters, the Second Phase Programme is ideally placed to provide data on critical loads and levels. The programme has been successful in developing and applying models to establish critical loads for nitrogen and acidifying compounds. These models need validation and the results from the programme (e.g. roof exclusion project, Speuld studies etc.) offer good opportunities for progress in this area. Some of the results on the effects of SO_2 and O_3 on natural vegetation will also contribute to the derivation of critical levels. However, the Review Team noted that these sensitive areas represent less than 20% of the total land mass of the Netherlands. For this reason, the Second Phase Programme alone cannot provide a comprehensive understanding of air pollution effects for the country as a whole but on the basis that protection for sensitive areas should more than adequately protect less senstitive areas, this is probably not necessary.
- The Review Team noted that in attempting to quantify the effects of air pollution, the Second Phase Programme has until now been somewhat introspective and appears to have ignored results from other relevant scientific projects taking place both within The Netherlands and elsewhere. By using data from outside the Second Phase Programme, a wider interpretation and quantification of effects could be possible.
- The Review Team recognises the potential of the integrated modelling project. In spite of the imperfections and the inevitable use of broad based assumptions in this work, the Review Team saw the modelling work as the unifying force in the programme to bring

together the outputs from all the projects. However, the Review Team felt it was important that full collaboration between project leaders and modellers should be encouraged if full use is to be made of this tool.

- The programme is solely air pollution orientated and appears to ignore the possible contributing role to damage from other stresses. The Review Team felt that to assign all the effects related in the Programme to air pollution might be misleading especially in relation to future decisions on emission control policies and that more attention should be paid to this issue.
- The programme as it stands is not organised to provide answers on the effectiveness of abatement measures in terms of environmental effects. Indeed the cessation of long term monitoring and concentration of short term research projects in study areas could specifically preclude this.
- The Review Team identified several areas requiring further resolution and has recommended the continuation of further work in these areas into a Third Phase Programme. The Review Team however, envisages such a programme being specifically targetted to these problems and would therefore probably require less resources than those devoted to the Second Phase Programme. Identification of the specific areas needing further investigation is made in the specific chapters.
- Several project areas have direct relevance to evaluating the possible impacts of climate change and these have been identified where possible throughout the report.
- The Review Team considered that the positive aspects of the Second Phase Programme far outweighed those of a negative nature. It also felt that if attention could be paid to the criticisms in this report in a Third Phase Programme, the combination of information from the First, Second and Third Phase Programmes would then provide a firmer basis for future policy decisions on air pollution emission control in the Netherlands.

3. EMISSIONS OF NH_3

3.1 Introduction

The role of NH_3 in soil acidification and other ecological effects has gained more attention in the last few years. It is estimated that about 80% of the NH_3 emissions are derived from animal husbandry all the year round while only 20% are from other sources. Therefore the researchers of the 2nd Phase Programme restricted their investigations to issues of livestock production in the report "Ammonia Emissions from Livestock Production".

This review is based on the thematic report presented on the 11th June. The programme of work at the start had a clear focus on the declared goals of the research plan to establish an "inventory of emission factors and the extrapolation to the national scale". These goals are addressed in 6 projects (nr. 130, 131, 132, 133, 190.2 and 200.1). A short description of 5 of these projects exist in the 2nd phase programme. Nr. 200.1 was not yet available. Because very little information was available about the NH_3 emission from the different systems of livestock production and manure handling systems, it was impossible to complete inventory. Therefore this part of the work was incorporated in the Research Programme for Animal Manure. Moreover the research on emissions from housing systems and on the prevention and abatement of NH_3 has been included in this programme. A second change in the working plan was made when investigations were started in poultry houses instead of cattle houses.

Inspite of these organizational problems the work on NH_3 emissions from animal production and on abatement measures of NH_3 emissions for housing is ahead of the rest of the world. The work on developing the inventory of emission factors and its extrapolation to the national scale is also progressing well. It is important to stress that this part of the programme is of basic importance to the whole because only these projects are emission related whereas the other parts are transport, input and effect related. It is also important to stress that NH_3 emission must be considered as a loss of nitrogen fertiliser. Therefore it is necessary to minimize NH_3 emissions not only for environmental reasons but also from economic points of view. Costs of mechanisation have to be taken in account.

3.2 Comments on quality of work

The work on abatement measures for NH_3 emission has been a central factor in the programme. It has shown that NH_3 release during and after manurespreading can be reduced by a variety of measures up to 90% . This is true for the application on arable land as well as on grassland. However, at least ten factors can influence the emission. The results obtained are in good agreement with investigations abroad.

A model was developed in the Programme to describe the NH_3 release after application of slurry on arable land. The model provides a useful insight in the factors which influence NH_3 losses. However, several key factors (wind speed, soil surface temperature and pH, rainfall, gas filled pore fraction of the soil) depend on weather conditions which cannot be predicted. Thus, it can hardly be used in practice. However, the results contribute significantly to the better understanding of NH_3 emissions under various conditions.

The results show that the NH_3 emission from grazing is considerably influenced by the nitrogen content of the grass which is taken up by the cows. The NH_3 emission from grazing contributes about 13% of the total emission in the Netherlands.

Much work was devoted to the emissions from housing systems. Valuable information is given on the amount of NH_3 emitted from layers as well as from broiler houses which has led to a considerable correction of the existing inventory data for the NH_3 emissions from poultry houses in the Netherlands. The values are in good agreement with German investigators for layers in battery keeping systems; disagreements exist for broilers. The investigations on reducing NH_3 emissions by almost two thirds from layers battery systems have been extremely valuable when forced drying of the manure is used. Investigations on boilers are not yet conclusive.

Research on pigs showed that to effectively reduce NH_3 emissions, urine should be removed from the floor of the system. The handling of the manure is more important than the amount of manure stored in the stable. This critical finding should influence the design of future manure removal regimes. There are indications that flushing can reduce ammonia release from buildings. Unfortunately the results of experiments with flushing systems are not included in the report.

The results on the research on cattle are still incomplete. However, the authors believe that the data are still sufficient to improve the emission factor for cattle from 8 to 6 kg NH_3 per animal per year.

Measurements of the NH_3 emission from noncovered slurry tanks are still incomplete and the results obtained from different estimation methods are inconclusive. Covering the slurry storage can reduce NH_3 release up to 90%.

Biofilters and also scrubbers can considerably reduce the NH_3 concentration in the exhaust air of forced ventilated animal houses. Considerable efforts are necessary to clean the filter bed and to work up the filter material several times a year. These findings are not fully supported by experience in Germany.

In summary, the original goals were generally reached in this stage of the programme. The results obtained significantly contribute to the understanding of NH_3 emissions under various conditions. The report gives an almost complete inventory of the factors and sources of NH_3 in animal production. However, the determination of quantitative relationships for some factors influencing the emission is still inadequate. Consequently, the calculation of the national NH_3 emission is only as good as the information on the different sources. These studies have accomplished the third objective.

The knowledge of the research group could be used to measure methane production from livestock production in respect to the climate change programme.

3.3 Weaknesses of the project

No information is given on the measuring methods for NH_3 which are used in the projects. The reliability of analyses cannot be estimated.

It is important to show that the results are valid under practical conditions.

It is less useful to start the work on NH_3 emissions from housing in poultry facilities because these emissions are of minor quantitative importance (about 10% of the total emission, only).

3.4 Gaps

More information is needed on the reduction of nitrogen input by animal feedstuffs and the use of additives in feed and slurry to prevent NH_3 release (bacterial/chemical, pH value). More detail should be provided on the different methods for measuring NH_3 emission in general. Attention must be given to the comparability and interpretation of results obtained with the different methods. Standardization of these methods should be considered.
The release of NH_3 from arable land where no manure is applied is not included in the research but in the case of the Netherlands this may not be significant.

3.5 Future work/priorities

Because of the important role of NH_4^+ in the soil acidification, reduction measures should have priority.
Continuation in improving the management for animal keeping and manure handling and spreading. Alternative approaches could be to decrease the number of animals per unit area or to lower the productivity per animal. This should be considered in relation to the present situation of overproduction on a large scale in animal production in Europe. However, this implies far reaching social and economic consequences for agriculture.
It is recommended to continue the work on the abatement of emissions after landspreading of manure, from buildings, stores and during grazing. Specifically:

- to include modelling of emissions from natural ventilated buildings (esp. cattle)
- modelling the emission of uncovered and covered stored manure at different dry matter, pH, temperatures and additives
- include investigations on reducing the nitrogen content in the feed
- producing guidelines for a good management practice for buildings and landspreading to avoid NH_3 release
- standardization of measurement procedures and sampling set-ups
- continuation of the emission inventory

The knowledge of the research group can be used to measure methane production from livestock production in relation to the climate change programme. While reducing NH_3 emissions from landspreading care should be taken to prevent other environmental problems like NO_3^- leaching or N_2O emissions.

4. ATMOSPHERIC INPUT FLUXES

4.1 Introduction and general comments

This review is based on the thematic report presented on the 11th June. Additional background papers have been consulted and references to some are made in specific comments to follow. It is recognized that many important details from recent field studies which have only been subject to a preliminary analysis are not included in the thematic report.

The presentation provided a broad overview of the approaches adopted, the main results and an unbiased interpretation of some contentious issues. The programme of work has a clear focus on the deposition of acidifying compounds on heathlands and forest (HNO_2, HNO_3, NH_3, NO_2, SO_2, SO_4^{2-}, NH_4^+, NO_3^-). The work has been approached in a energetic and professional manner and the initial objectives have to a large extent been successfully accomplished. Considerable attention has been given to intercomparison, quality assurance and quality control, and this has provided a sound basis for the interpretation provided. It is clear however, that this review has occured before sufficient time has been taken to provide full analysis and interpretation of the data and the preparation of appropriate peer-reviewed papers.

The Review Team noted that the concept of total acidifying input used throughout the programme implies a known chemical and biological fate of deposited NH_x and NO_x. It is the maximum acidifying input and may in some circumstances be larger than the net acidic input. There should therefore be a short section introducting the concept and briefly stating the strengths and weaknesses of this approach.

In the case of the atmosphere-surface exchange of gaseous ammonia the Dutch acidification research programme is considerably ahead of the rest of the world. The work on developing models to scale fluxes (and estimate the associate errors) for the landscape is also progressing well and complements the extensive national monitoring of gaseous and particulate pollutants.

4.2 Relevance of the deposition research to the major goals of phase 2

The work on deposition of gaseous ammonia deposition has been central to the phase 2 programme (projects 101, 128, 129, 190 and 191). It has shown that for the sites of interest for effects (heathland and forest) that ammonia is a major component of the "acidifying" input. For heathland, for which most of the data have been fully interpreted the major features of the dry deposition process have been demonstrated, and a sound basis for modelling inputs has been provided. Some technical problems remain, including the requirement for continuous monitoring of ammonia to enable inputs by dry deposition to be calculated.

A closely related development, for the monitoring of dry deposition (for SO_2 in the work reported) is a welcome innovation and shows considerable promise but reported uncertainties in the mean deposition velocities of $\pm$ 15% at the Zegveld site appear to be conservative. Detailed "campaign" comparisons between such simplified methods and more rigorous flux measurements are required for validation.

For measurements over forests there has been less time for detailed analysis, however the reported ammonia deposition fluxes are a notable achievement. Such measurements are subject to considerable difficulty, in particular the investigation of flux/gradient relationships at the experimental site and the satisfactory achievement of sensible heat fluxes provides the necessary basis for interpretation of the gradients in NH_3. The very extensive data set of concentrations of all gases contributing to the acidic input above this forest together with the very detailed description of canopy characteristics and physiology from the biological projects within the programme represent a excellent opportunity for collaboration between the different disciplines during the interpretation, synthesis and modelling phase. It is not clear from the report whether differences between the NH_3 deposition velocity estimates at this site from the campaign measurements and from the long term monitoring of concentration at two heights will be reconciled following the necessary additional analysis. This is an important comparison.

For NO and NO_2 exchange above vegetation the programme to date has not provided an understanding of the processes controlling the flux. In the absence of this understanding it is difficult to see how estimates of fluxes across the countryside can be calculated for the Netherlands with an uncertainty of only 36%! (table. 3.5). The work reported for NO and NO_2, however is in good agreement with recent results in other countries and international collaborative work by the Dutch groups is an important link with the wider scientific community. A better understanding of the variables controlling NO and NO_2 exchange above vegetation and soils should be possible within the next 2 years.

The throughfall and stemflow measurements form an important but poorly developed link between the atmospheric scientists in this programme and biological and ecological scientists. There are clearly larger gross fluxes of the major mobile ions below canopies of vegetation in throughfall for regions with large ambient concentrations of the gaseous precursors of the same ions (notably SO_4^{2-}, NO_3^- and NH_4^+) than at clean air locations. The difficulties in interpreting these throughfall chemistry measurements are recognized and discussed in the review documents provided and in the background papers.

The total deposition has been estimated using different methods. Results from the Speuld forest show rather good agreement between the methods based on throughfall measurements and estimates of total deposition from a model using measured concentrations of the gases and particle species. However, in figure 5.1 where results of total deposition estimates from a range of different sites in the Netherlands using the same two methods are compared, there are large discrepancies between the different estimates for total sulphur deposition. The large SO_4^{2-} in throughfall relative to estimated atmospheric deposition in particular is a feature of the results at many of the "most polluted" sites. It is not possible from the information presented to show which estimate is "least wrong". However, the apparent proportionally between the two estimates is suggestive of either a much larger leaf surface

SO_2 uptake than has been assumed in the dry deposition model, a large leaching term in the throughfall SO_4^{2-} (which may be proportional to total sulphur deposition) or larger particle deposition rates then has been assumed.

There are notable discrepancies between estimates of inputs at the Speuld site in the different documents provided and some evidence of inconsistencies in the units (moles ha^{-1}, equivalent ha^{-1}, and kg ha^{-1}). These presumably result from the lack of time for preparation of the documents but should be more consistent, or justified in future reports.

In this brief review of the work a detailed discussion of these important arguments concerning interpretation of different estimated inputs is not appropriate but the thematic report shows encouraging signs of a fairly recent joint interpretation of data from these two very different fields of work. From the analysis of the tracer studies at the Speuld forest site and integration of data from the range of groups who have been involved at the same site a more detailed interpretion of the net fluxes and their sources should be possible.

As the international debate between scientists involved in the interpretation of deposition estimates based on different methods is still very active the Review Team encourages the involvement of the Dutch workers with their new data to participate in this debate.

4.3 Generalizing fluxes across the countryside

The thematic report provides a detailed description of the methods used to extrapolate from the field measurements of deposition fluxes to the countryside. This is an essential part of the programme since it represents the application of the measurements made for practical tasks. Taking roughly a third of the report it is however a much larger proportion of the space than the work in this area represents within the programme. The approaches appear reasonable for atmospheric and boundary layer resistance terms r_a and r_b. The surface resistances at least for SO_2 could be improved by explicity including the stomatal resistance term, calculated from stomatal responses of major plant species and temperature and radiation. The extrapolation of NO and NO_2 fluxes to the countryside is a dubious procedure in the absence of good estimates of the canopy resistance r_c. The extrapolation of NH_3 fluxes to arable cropland areas is also uncertain. Our understanding of NH_3 exchange processes over arable land is poor (the heathland is important for effects and well understood for NH_3 deposition but only represents 1% of the countryside). These arguments are minor critisms of the overall programme, and the modelling of deposition to the countryside is important. However, this criticism should be seen as highlighting the uncertainties in these extrapolation exercises. The report provides a brave estimate of uncertainties in total deposition, and it is a welcome addition. The values within the table are less certain than they appear. It would be helpful to the reader if the authors were to provide a range of error estimates with best and with worse case assumptions (including for example in a "worst case", combinations of highly correlated errors).

Wet deposition merits brief comment since in this area the Netherlands operates a high quality network of precipitation collectors, and rigorous quality assurance/quality control protocols and within the input budgets this term and the associated uncertainties are well defined.

4.4 Dew and fog studies

The recognition that surface wetness is an important feature in modifying the surface resistance for SO_2 and NH_3 deposition on vegetations is widely recognised. The programme to investigate these effects is original and has provided valuable techniques and interesting data. It is not clear from the results presented that a adequate understanding of the sources of the large concentrations of major ions in these liquid films have been defined. Clearly there is more scope for interpretation and analysis of the results obtained but the importance of dew to annual inputs of NH_3 and SO_2 on different vegetation has yet to be demonstrated. The fog studies show large concentrations of major ions in fog deplets and these may be much more important as a mechanism producing "effects" on vegetation than as a contributor to annual inputs of pollutant to any ecosystem in the Netherlands. It would therefore seem appropriate to develop closer links between the groups engaged in characterizing fog droplet composition and groups who have the facilities to test the direct and indirect effects of the large concentrations of SO_4^{2-}, NO_3^-, NH_4^+ and H^+ on trees. As radiation fog is largely a winter phenomenon and is the mechanism producing most "fog" in the Netherlands this work could logically be directed towards effects on conifers and "winter injury" studies.

4.5 Gaps in knowledge

The report contains several brief sections on research requirements but no clear priorities within them. The short list below includes some of the suggested items in order of priority:

- The initial priority within the existing groups in this field is for a more detailed analysis and interpretation and publication of the existing experimental and monitoring data.

The priority experimental tasks have been grouped below into three projects each of which may be justified by more extensive arguments than the brief ones attached.

1. Dry deposition

The areas with most uncertainty for NH_3 deposition are the arable cropland districts, which are important in the national budget but less important for effects. The net exchange of nitrogen oxides NO, NO_2 across the country is still very uncertain and mechanisms regulating exchange poorly understood. The dry deposition "monitoring" facility should be subject to a comparison against detailed micrometeorological measurement systems in the campaign type of equipment.

The dry deposition of aerosols remain uncertain especially for forests isolated trees and hedgerows and should be the subject of a new field study.

2. Measurements of the sources of SO_4^{2-}, NO_3^- and NH_4^+ in throughfall and stemflow. The requirement here is for more intensive and mechanistic studies than the monitoring which characterizes most work to date. It is necessary to include both biological and atmospheric scientists both for the comparisons between methods and for the synthesis and interpretation of results.

3. The influence of combinations of trace gases on deposition processes. In particular the co-deposition phenomenon of SO_2 and NH_3 which appears central to the Dutch acidification issue. This process is poorly understood in field conditions and is intimately linked with leaf surface properties and the dew and wet surface studies. It should also be extended to other trace gases.
An important component of all three areas outlined above is the development of models of deposition to help integrate the results of the different studies and to provide fluxes over large areas and over long time intervals. Such work should be initiated within each of the above topics rather than be developed in isolation.

5. SOIL ACIDIFICATION / NITROGEN CYCLING

5.1 Introduction

The projects "Soil acidification and nitrogen cycling" and "Assessment of critical loads and the impact of deposition scenarios by steady-state and dynamic soil acidification models" have provided valuable information to help answer critical questions posed in the Second Phase of the Dutch Priority Programme on Acidification (DPPA). This research has provided quantitative information on indirect (soil mediated) effects of SO_2, NO_x and NH_3 inputs on vegetation (question 2), as well as the response of soil and soil solution chemistry to changes in atmospheric deposition (question 3). The overall quality of the research in this component of the DPPA is excellent and the investigators have strong international reputations.

5.2.1 Review of the project "Soil acidification and nitrogen cycling"

The approach used in this component of the DPPA was to develop ecosystem-level balances of major elements at 18 sites, primarily located in the central Netherlands. The sites selected for this investigation included different vegetation and soil types. The element balance research was supported by additional soil characterization, process-level studies and modelling activities at some or all of the study sites.

The element balance approach used in this investigation is the most appropriate method to address the critical research questions. The data obtained clearly demonstrate that rates of soil acidification are very high and this effect is largely due to atmospheric deposition of nitrogen and sulphur. Soil acidification is enhanced by the oxidation of NH_4^+ inputs by chemolithotrophic bacteria, resulting in elevated production of nitric acid. The input of acidity (sulphuric and nitric acids) coincides with leaching of Al and nutrient cations from these base-poor soils.

A strength of this effort is the apparent close cooperation with the "Integrated effects on forests" research activities. Indeed soil solution chemistry and hydrology from the ACIFORN sites was used to help interpret information collected on forest health and form the basis for conclusions that deleterious effects of acidic deposition on forest ecosystems are likely due to soil mediated processes (question 2).

Time series analyses were conducted using monitoring data in an attempt to determine the response of soil solution chemistry to abatement strategies (question 3). While results show that surface soil solutions are decreasing in pH and increasing in NO_3^- and Al concentrations, these data should be interpreted with caution. Other investigations have shown extended lengths of increasing or decreasing NO_3^- in drainage water with no statistically significant long-term trend over the entire period of record. The investigators did not discuss time-series analysis of SO_4^{2-} or the base cations. It is not clear whether other solutes were not examined or simply did not exhibit a trend. In any event, the period of study is probably too short to conduct a detailed trend analysis. An extension of the monitoring programme is necessary so that this assesment can continue. The investigators are also encouraged to expand the scope of the trend analysis by examination of temporal patterns in precipitation chemistry in collaboration with atmospheric deposition researchers.

There are some minor concerns with the material presented in this section of the DPPA. First, soil/soil solution data were generally presented as single values. It would be helpful to also provide the reader with information on the variability associated with sample collection (e.g. standard deviation, standard error, range). This would help put patterns in perspective and facilitate any uncertainty analysis that might be conducted with the data.

A second concern is associated with the modelling activities. While the investigators are to be commended for the development/application of a model (WATERSTOF) to help interprete their research results, it is not clear how this modelling effort is coordinated or integrated with the critical loads modelling using START/RESAM. Are these efforts complementory or in competition?

The investigators involved in this study have conducted good quality research that clearly demonstrates the magnitude of the acidification problem at the study sites. This work appears to be well integrated with other experimental research activities.

5.2.2 Review of the project "Assessment of critical loads and the impact of deposition scenarios by steady-state and dynamic models"

Modelling activities are an important component of integrated ecological studies. Models can be used to identify gaps in knowledge, test hypothesis, integrate components of a research programme and help direct future research. The modelling approach used in this investigation is sound. Model applications range from simple steady-state models to relatively complex dynamic simulation models. While steady-state models might be used to determine critical loads, they are probably not adequate to help integrate the process-level studies of the DPPA. The RESAM model is similar in structure to many of the comprehensive acidification models used today (e.g., MAGIC, ILWAS).

There is concern that steady-state models might not be appropriate for critical load calculations. If steady-state processes are the predominant factors regulating the acid-base status of ecosystems, then this approach might be appropriate. If dynamic processes are important (e.g., changes in exchangeable cation of biomass element pools) dynamic simulation models should be used.

Modelling activities to date have focused on conducting long-term simulations and determining critical loads for forest ecosystems. While these are important endproducts, the dynamic RESAM model should also be calibrated using data collected at the ACIFORN sites (e.g., Speuld and Kootwijk), as well as intensive study sites in other countries. Detailed biogeochemical information could be then used to conduct a rigorous uncertainty analysis and evaluate model performance. In addition, the model(s) could be verified using data obtained from the roof manipulation experiments. This activity would provide a sound test of the model structure, help integrate field activities and identify important processes that could be investigated experimentally. Modelling activities based on information collected at intensive field sites could also help evaluate the effects of abatement measures on forest soil/soil solution chemistry. Once the model is calibrated/verified at intensive study sites, it can be applied on a regional scale to determine critical loads with more confidence.

The results of the regional soil survey and subsequent model application are interesting activities and should prove very important. There have been few investigations on scaling plot/watershed models up to regional scales and determination of the uncertainties involved. This is an important activity and should be coordinated with other research groups. Many countries are currently trying to develop critical load maps and are experiencing problems in applying models at regional and/or national scales. It is encouraging that an international workshop on this subject is planned for the near future in the Netherlands.

5.3 Relevance to an integrated assessment of acidification in the Netherlands

The soil acidification/nitrogen cycling studies were established at several intensive sites within relatively a limited area in the Netherlands. As a result it is unclear whether the infomation collected at the ACIFORN sites can be regionalized or extrapolated to the country as a whole. In particular, atmospheric deposition measured at the ACIFORN sites is low relative to other areas in the country. Soil acidification documented at these sites may be less severe than in other areas of the Netherlands. Moreover, it is not evident if climatic, hydrologic, soil and vegetation characteristics at the intensive study sites are representative of the country. Efforts to map atmospheric deposition, precipitation inputs, runoff, soil physical and chemical characteristics as well as vegetation characteristics should help determine methods to regionalize the ACIFORN data. In particular, the ongoing soil/soil solution survey should help provide much needed regional information on the extent of soil/soil solution acidification and be important for critical loads modelling activities.

5.4 Gaps in knowledge on soil acidification/nitrogen cycling

While this investigation has provided definitive information on the nature and extent of soil acidification at the intensive study sites, several other important observations emerged. Most critical of these may be the marked depletion of soil Al pools. This effect is potentially significant because soil Al is an important pH buffer. If reactive Al is depleted, soil solution pH will be depressed even further. This suggests that a better understanding of the Fe buffering system in soil is needed.

In addition, during soil development Al is retained in the lower mineral soil and coincides with immobilization of naturally occuring organic acids. This mineral soil organic matter represents the predominant organic carbon pool in forest soils and is largely responsible for the cation exchange characteristics. Dissolution and transport of soil Al, due to atmospheric inputs of strong acids, has disrupted soil development and may represent irreversible damage to the forest ecosystem. If organic solutes are no longer immobilized within the profile or if deposited organic acids are remobilized with the dissolution of soil Al, cation exchange capacity may decrease and buffering of nutrient cations by the soil exchange complex be diminished. In addition, it is well established that complexation of organic matter by Al reduces mineralization rates. Therefore, changes in the soil Al concentration and speciation could have a marked effect on the dynamics of soil organic carbon. An examination of the relevant literature on tropical soils might provide some additional information on this process. The investigators did not discuss patterns in dissolved organic carbon (DOC) concentration/transport in soil solutions. Further analysis of these data may yield important information.

The investigators hypothesized that elevated soil solution Al may have altered phosphorus availability particularly through effects on mycorrhiza. Results from the research activities on "Integrated effects on forests" suggests that forest systems may be phosphorus growth limited. The effort to monitor phosphorus chemistry of the ACIFORN sites might be intensified to evaluate effects of elevated Al on the phosphorus cycle.

Clearly additional work is needed to assess the impact of Al on the nutrient (organic carbon, P) status of forest systems.
The investigators reported elevated leaching of NH_4^+ to groundwater. This is an unexpected finding and warrants additional study. It is not evident why nitrification is incomplete at these sites or what are the consequences of elevated concentrations of NH_4^+ in groundwater.

5.5 Relevance to critical loads assessment
This work is highly relevant to the ongoing critical loads assessment. The studies provide valuable information to determine critical chemical values of heathland and forest systems in the Netherlands. The DPPA is closely coupled with the Critical Loads Programme. The linkage between the DPPA and the Critical Loads Programme could be strengthed (as discussed previously) through application of the RESAM model to the ACIFORN sites.
The Dutch critical loads researchers are to be commended for using a range of modelling approaches (steady-state, dynamic) in their assessment and their efforts in sensitivity analysis. Moreover, they have been at the forefront of critical load acitivities in Europe. They should be encouraged to continue the exchange of models and data sets, and to initiate discussion on approaches the application of acidification models over regional scales.

5.6 Conclusions and recommendations
In summary, the work on soil acidification/nitrogen cycling is high quality and has produced definitive information on the response of soil to atmospheric deposition of strong acids in the Netherlands. However, additional effort should be made:
1. to apply the SMART/RESAM models at the ACIFORN sites, and
2. to establish an intensive study site in an area of maximum atmospheric deposition.
Moreover, the time frame of the monitoring effort in phase 2 of the Programme was too short to assess the response of soil solution chemistry to emission abatement strategies or changes in atmospheric deposition.

Based on the results obtained, future research activities should be considered.

1. The intensive monitoring studies should be extended to include ecosystem-level manipulations. This approach has already been initiated with the roof exclusion experiments. It is recommended that the RESAM model be applied at these and other manipulation sites. This could provide a rigorous verification of the model(s) and increase confidence in future critical load calculations. Model application should also help focus experimental work at the sites on important processes.

2. A few (e.g., 4-6) intensive sites should be established for long-term monitoring. Biogeochemical monitoring at the ACIFORN sites should be expanded to include measurement of trace gas (CO_2, N_2O, CH_4) fluxes, and changes in soil pools. Through data collected, it would be possible to assess the changes in soil, soil solution chemistry and vegetation that occur in response to emission abatement measures. In addition, an integrated long-term monitoring programme would have direct linkages with future global change research. Critical issues in global change research include storage/cycling of carbon, trace gas consumption/emissions, changes in hydrology and the effects of changes in atmospheric deposition on ecosystem processes. These could all be addressed in part through an integrated long-term monitoring programme. This intensive monitoring programme might also be coupled with extensive regional surveys so that observations can be regionalized. The biogeochemical modelling efforts initiated as part of the Programme (e.g., RESAM, START, SOILVEG, FORGRO) could also serve as a foundation for global change modelling initiatives. These models could be expanded to include those processes which might be affected by global change. However, as with acidification research, modelling activities associated with a global change research programme should be closely coupled with experimental/field research activities.

3. In addition, process-level studies should be initiated to further evaluate important findings of the soil acidification/nitrogen cycling research. These might include detailed investigations of:

1. the factors which regulate nitrogen mineralization, nitrification and immobilization,
2. the effects of depletion of soil Al pools,
3. the effects of elevated Al leaching on phosphorus availability on forest soils,
4. the effects of elevated Al leaching on pools and turnover of soil organic carbon,
5. the soil Fe buffering system, and
6. the implications of NH_4^+ leaching to groundwater,

6. BIOLOGICAL AND PHYSIOLOGICAL EFFECTS

This thematic report encompasses 11 projects which can be classified as follows:
- studies under controlled conditions on the impact of gaseous pollutants on the above-ground parts of the plants (projects 109, 110, 115)
- in-situ studies on mycorrhizae (project 108.1)
- studies under controlled conditions on the impact of nitrogen and acid deposition on the root system (projects 83, 108.1, 108.3)
- study of the combined effect of gaseous pollutants (O_3) on the above ground part and of

NH_4^+ on the root system (projects 107, 108.1)
- various studies on growth (111.1), on sap flow (116), on epicuticular wax (111.2).

These projects have provided some important information on the biological and physiological effects of O_3, SO_2, NO_x and NH_3. Studies of mechanisms and processes appear to have been preferably undertaken under controlled conditions. In comparison with other similar national programmes, this is one strong point of the Dutch programme. The studies focussed on rhizosphere, and in particular on the effect of "excessive" ammonium input, have provided many new results. The investigations into the effects of a gaseous pollutants in combination with NH_4^+ ions are original and fit well into the integrated approach. Crops are mentioned in one project only, but the Review Team noted from the earlier report that research already available on agricultural crops is thought to be sufficient to draw policy conclusions.

There is an unavoidable consequence of this relative predominance of fundamental work. The problem of extrapolation of the results to the actual status of the Dutch forests.

Concerning gaseous pollutants, the rapporteurs pointed out that the concentrations applied in fumigation experiments are significantly higher than ambient concentrations.

As for experiments on artificial medium, it is difficult to compare the NH_4^+, Al^{3+} etc. concentrations applied with actual conditions. In the pot experiments, such comparisons do not seem to have been attempted systematically. The investigations on the rhizosphere have answered some questions such as the relationship between NH_4^+ and NO_3^- uptake but have also raised many new questions. The complex relations between mycorrhizae development, root development, growth of the above-ground part, concentrations in base cations, Al^{3+}, NH_4^+ and pH are still poorly understood. In particular, the effect of Mg/Al and Ca/Al ratios on the development of mycorrhizae is not clearly defined.
As far as physiological and biological effects are concerned, the major gaps in knowledge remaining after the second phase of the DPPA are:

1. The long-term effect of ozone on forest trees, in combination with nutritional and water stresses
"Long-term" should here be measured in years. As physiological characteristics are different in young plants and adult trees, the results obtained on the former cannot readily be extrapolated to the latter.

2. The effect of nitrogen and acid deposition on rhizosphere
As already mentioned, many inconsistencies and gaps remain. As the process of interaction between phosphate and aluminium has not been clearly described; the collaboration of biogeochemists and microbiologists is advisable. In addition, the role of fungi microflora (excluding mycorrhizae) should not be underestimated. The pot experiments should take into account more carefully the diversity of soils.

3. The impact of nitrogen deposition on tree growth
In several European regions, a significant increase in growth for the past 10, 20 or 30 years has been found. According to authors not quoted in this report, a similar trend seems to have been observed in the Netherlands. It may be interesting to carry out a thorough dendroecological study on a greater number of plots characterized by contrasted nitrogen deposition levels and nutrient and water supply conditions.

4. The effect of SO_2 and of NH_4^+ uptake on the physiology of forest trees in winter
The authors of the thematic report consider this effect to be likely, but no conclusive experiment is provided.

The work undertaken during the second phase of the DPPA has shown clearly that gaseous pollutants (SO_2, NH_3, O_3) and acidic or acidifying deposition can damage trees. However, it does not allow definition of quantitative relations between the level of pollution and that of damage, either because the concentrations applied were significantly higher than ambient concentrations, or because the effect of the factors investigated (nitrogen deposition in particular) varies significantly with stand conditions.

It is recommended:
- attempts should be made to reach a better balance between research under controlled conditions and in situ studies, and to try and bridge the gap between these two approaches. In situ studies should not consist only in monitoring a small number of plots as so far undertaken. Manipulative experiments carried out on the ecosystem on the one hand (such as the current investigations with deposition filtering roofs), or field studies taking into account the diversity of ecological conditions are better adapted to evaluation of this issue.
- more fundamental research in certain fields, in particular in whole-tree physiology, by focussing studies on the uptake and transfer within the tree of nutrients and water should be carried out.

These two approaches are also interesting in the context of the study on a possible global climate change:

- the impact of climate (which should be considered as a potential cause of forest decline and not only a contributing factor) on the forest state of health will largely depend on site conditions and stand history
- the influence of the modification of parameters such as water, CO_2 and mineral nutrients on the physiological function of trees is still poorly known.

7. INTEGRATED EFFECTS (LOW VEGETATION)

7.1 Comments on quality of presentation and/or work

The standard of descriptive and experimental approaches of the scientists involved in this part of the Dutch Priority Programme on Acidification is one of the highest in the world. The second phase of the programme could however also build on experience and hypotheses developed in the first phase. There also appears to exist much published data on related topics that has been done outside of the Priority Programme, before or during the existence of the current Programme.

Integration is one of the most demanding tasks - not only intellectually but also in terms of resource committment, and it is time consuming. More integration could have been achieved had been more time and contacts been available to the rapporteurs and the Review Team. Possibly some effort could be made to remedy this situation before the final version of the reports is presented in October 1990.

The aim of the project report "was to make a quantification of the effects of air pollution on heathland, for guidance in determining environmental policy, e.g. in setting standards". Heathland management and nutrient cycling have been dealt with elsewhere but fauna has not been investigated. This appears to be a major omission (see 7.3.).

7.2 Relevance of current reports to giving an integrated picture of the present acidification situation

7.2.1 The extent to which Calluna in conservation areas is replaced by grasses has been quantified with a regional differentiation, from satellite pixel documents. The extent of the replacement appears to be more severe than expected with an acceleration of change in recent years. However as there are uncertainties in the satellite approach, the results should be checked against more extensive ground surveys. These could take into account the emission situation around a given heathland area, its soil conditions and its microhabitats and thereby provide a better estimate of the chances of regeneration by management of the sinks and/or sources of NH_x.

7.2.2 The atmospheric deposition of nitrogen compounds and the fate of N in the ecosystem have been considered elsewhere in this review.

7.2.3 The relationship between nitrogen deposition and replacement of Calluna by grasses has been quantified. The critical load based on modelling results appears to be 10 to 15 kg N/ha/year in "unmanaged" situations. Up to 30 kg N might be tolerated where grazing combined with a high frequency of sod cutting takes place. However, this model is not yet calibrated. Other approaches may also be used to establish quantitative values on critical loads for heathlands. These critical loads are considerably lower than the measured present deposition rates.

7.2.4 The risk analysis approach for effects of SO_2 and NH_3 on grasses and Calluna respresents an innovative and original development. However the underlying assumptions require experimental validation, particularly as the extrapolation of effects from short time fumigation at high concentrations to long term exposure to low concentrations assumes linear concentration/response relationships down to the ambient concentrations at sites in the Netherlands, that is not supported by the international literature on the subject.

7.2.5 Things are more complicated in the "real world" with several species competing with each other for light, water and nutrients. Obviously high levels of SO_2 may cause significant vitality reductions during seed production and germination, and nitrogen availability changes competition equilibria for established plants. But these processes apparently are not sufficiently well understood for understanding the dynamics of the Violion caninae, whose sites are characteristic microhabitats with open ground available for germination, slightly better nutrient supply (cations) and higher pH levels, more "continental" microclimate, etc. These sites are created de novo in $\pm$ short time intervals. The causes of these micro-site-dynamics is little understood but their importance should not be overlooked.

7.2.6 The importance of "secondary" stress factors, such as more susceptibility to drought, frost or insect plagues, due to the different consequences of better nitrogen supply, has been clearly demonstrated. Better quantification in this respect appears to be difficult and not really necessary for the purpose of the programme results. Nevertheless, what are the underlying adaptations of plants to nutrient poor, acidic conditions?

7.3 Identification of gaps in knowledge

7.3.1 Based on field experience the investigated plants have been classified in "dominant" and "Violion caninae" species. There are however "sensitive" and "insensitive" ones in the latter. These groups of species appear to be characterized by common "morphological"

features: perennial, "annual", and vegetatively colonizing life forms in each respective case. These species appear to be preadapted to the dynamics of their sites by these features. This raises the questions: - Who/what causes the dynamics of the microsite and/or the disturbance of the soils? - Under what conditions, and how effectively? - Who or what are the dispersers of the seeds and the pollinators of the flowers?
Possibly it will turn out that animals like small mammals, rabbits, even burrowing insects like ants, ant lions, earth bees, gryllidae are the key factors for understanding the system "Violion caninae", as they dig up "less acidified" soil to the surface. They provide climatically preferred microhabitats with little competition from other plants at least for some of the time. - The significant contribution of the heather beetle to the stability of the Calluna-system has only been found by chance.

7.3.2 There exists an impact of habitat change - as triggered by NH_3 - and other pollution - on animals of different classes and families. This should be documented and evaluated, as a great deal of the recreation value of heathland is dependent on the presence of animals. There are also independent reasons for dealing with animals per se. In the Netherlands, conditions to investigate the interrelationships of deposition, habitat change, plant and animal population reactions, are obviously better than in most European countries, as a huge amount of specialist knowledge has been accumulated. This knowledge could be organized and interpreted, using the modelling of the changing conditions for retrospective and actual comparisons and this would also have relevance for possible impact of climate change.

7.3.3 In ecosystems with acid soils, break down of litter is dominated by fungi activity, and to a certain degree mesofauna is involved in mechanically preparing the substrate. Apparently the mechanisms of litter production and accumulation are well understood. But it is not clear that this is true also for litter breakdown and mineralization.

7.3.4 The chapter on risk assessment indicates that SO_2 under present conditions in the Netherlands is of minor importance. But the proportion of growth stimulation via nitrogen input to ecosystems will have significant impact on the proportion of the species that can be protected from extinction. Who decides what proportion to chose? Is there any accepted sociological approach to deal with problems like these?

7.3.5 Does the efficient above-ground uptake of nitrogen in the living canopy, and the documented fast flow of N into the litter compartment, contribute to nitrogen accumulation in the litter? Under what conditions is it leached to the groundwater? What to do in long-term perspective with accumulating litter? Is it a significant sink for the CO_2 surplus? A modelling approach could help solve the problems.

7.4 Relevance of phase 2 programme to critical loads/levels assessment

Risk assessment based on modelling results produces a critical level of 8 μg SO_2/m^3 in the air necessary to protect "95% of the species" of heathland from adverse growth effects. But seedlings appear to be more sensitive, and the competition equilibria between species probably are even more so. Temporal variation in pollution levels and seasonality in vitality of the investigated plants was also not be taken into account. Fumigation experiments under "field conditions" taking propagation and competition into account should therefore be carried out.

As there exists significant N uptake via the canopy, no critical level can be defined for growth stimulation by NH_3/NH_4^+ as stimulation already works at very low concentrations. In the long run, by competition, fast growing plants are favoured over slow growing species, especially those, like grasses, that are capable to allocate their N resources effectively via internal cycling between roots and shoots. The consequences for accumulation and mineralization (conditions, effectiveness) still have to be understood. This is another opportunity for a modelling approach.

Drought, frost and plague sensitivity in heather plants rise significantly at NH_x levels of 1 to 10 μg/m^3 of air. From this viewpoint, microclimatic and edaphic preconditions of sites (top, slope, valley; dry, moist; exposition) should be considered in a landscape-ecological approach in order to augment the survival probability of heathland ecosystems at least in specific sites and to concentrate conservation efforts upon "optimal" places. Compared to the current deposition loads in the Netherlands , the critical load of 10 to 15 kg N/ha/yr for replacement of Calluna by Deschampsia (and for the survival of the Violion caninae-species) is obviously low. This calls for more investigations in the possibilities for reduction of NH_3 emission at source.

Management techniques could triple this critical load when sod cutting is done at intervals of much less than 20 years and grazing takes place simultaneously. "Heathland" managed like this is qualitatively different from that heathland known before and up to the early sixties.

The results of the "phase 2 projects" provide data on critical levels and critical loads. They are sufficient to deal with only one part of the relevant questions. In the future more emphasis should be put upon questions concerning the "production conditions" of emission, most of which are addressed in the chapter on emission of ammonia.

In a situation of overproductivity on a large scale it is not longer the ecosystems ("sinks") that have to be adapted or changed but the systems of production ("sources").

7.5 Major conclusions and recommendations arising from sections 7.2. to 7.4. and also any interface with possible climate change impact research programme in the future

Eutrophication via the air has caused many plant species to decline, some to disappear. A few species were favoured by the same conditions to the extent that they are now dominant

at places that formerly were unattractive or present where they were unable to survive before. This change in species composition is triggered mainly by "indirect" effects. Direct "toxic" effects of SO_2, NH_x, NO_x on plant tissues appear of less importance in the Netherlands.
There obviously are many more "indirect" effects involved. However the Dutch Priority Programme on Acidification addresses "indirect" effects via the soils as opposed to "direct" ones via the air. A differentiation of these two aspects is helpful for analysis, for synthesis however they need to be brought together.

Changes in plant species composition, successions that are triggered by nitrogen input up to excess nitrogen availability, force many other ecofactors to change. With better nitrogen supply:

* small scale spacial habitat heterogenity is monotonized and there is more synchronization of plant development
* less radiation reaches the ground, causing less temperature diversity and less temperature oscillations on a daily and yearly scale. Similar observations apply for moisture. So, microclimate is changing towards less "continental" conditions. As this is the case simultaneously on large areas, the same is expected on a landscape scale - at least during spring and summer.
* plants start to grow earlier, they grow faster, higher, larger, stay vital for longer time in fall, some even during winter. So they are longer available and better digestible for e.g. micro-organisms, caterpillars, deer, wintering geese and so inducing population reactions, with consequences back to the plants; species composition, stand structure etc.. For the same reasons, plants are more exposed to drought, frost and plagues, that are labilizing the stand continuities.
* building up more biomass, more water appears to be necessary for evapotranspiration, more cations are needed, causing more acidification in the soils. In case some sort of harvest or accumulation "exports" the biomass from the system. What to do with the surplus production?
* in spite of this, only a small fraction of the CO_2 exhaust from fossil fuel burning can be fixed by biomass growth in industrial countries.

A few of these aspects have been dealt with in the first and second phases of the Dutch Priority Programme on Acidification. More of them could be addressed in a Third Phase. Man as the triggering partner in these changing ecosystems should be considered to a greater extent - not only in the case of heathland (and forest), as changes in his basis of living are likely to induce reactions on economic and sociological levels, and not in every case of positive value.

8. INTEGRATED EFFECTS (FORESTS)

8.1 General comments

The presentation and thematic report were of a high standard. The combination of the FORGRO and RESAM models make a valuable tool and their operation to produce the predictions of the impacts of water shortage, gaseous air pollutants and nutrient inbalance (as presented in the thematic and monitoring report) is a considerable credit for the modelling group (projects 112, 113, 114.1, 114.2 and 115) and to those involved in the experimental work which has provided information for their operation. The two items "Soil acidification/nitrogen cycling" and "Integrated effects on forests" have been well integrated and those involved with both these areas have made significant progress. Internationally there have been few programmes in which experimental and monitoring work designed to elucidate mechanisms (project 83, 103, 105 and 107 and those reviewed under the theme "Biological and physiological effects") have been followed through with such thorough work at the stand level. This effort has clearly been justified by the results which were presented and the Priority Programme is an international authority on the function and growth of Douglas fir stands.

The Review Team were however concerned that work on forest effects has not advanced to the stage at which the objectives of the second phase have been met. The selection of two sites with only moderate nitrogen inputs and good forest growth, and the more advanced state of the monitoring projects than the pollution injection and other manipulation experiments are considered to be the main reasons for this.

The individual projects with a monitoring and quantification role at the Speuld and Kootwijk sites on both physical, chemical (projects 101, 102.1 and 102.2) and on biological aspects (projects 103, 104.1, 105, 111 and 116) have provided information critical for achieving the present state of knowledge. It is important that the output of FORGRO and RESAM are interpreted accurately and the thematic report should include additional text briefly explaning the models and their output more thoroughly. The values presented in table 15 must be interpreted sensibly relative to those of table 6. It is difficult in places in the thematic report to determine what new information from the Second Phase of the Dutch Priority Programme on Acidification has been made use of. Some of the manipulative studies (projects 100, 105, 107 and 118) appear not to have reached the point at which their conclusions could be used to the full in the theme synthesis, so that smaller scale fumigation work or published work from elsewhere were relied on in these areas. This must give some pointers as to where further effects can now be justified.

8.2 Detailed comments; soils

In spite of the selection of two sites at which N inputs are below the average value for the

Netherlands, the soil studies have shown that N and S inputs $(NH_4)_2SO_2$ in throughfall) are driving soil acidification at Speuld and Kootwijk. More than half the CEC is occupied by aluminium with base saturation being as low as 5% in the mineral soil. NH_4^+ and NO_3^- fluxes indicate that nitrification is occuring in the litter layer and top soil. Al/Ca are high in the soil solution, although Al/Ca and Al/Mg ratios were not presented in the thematic report or developed in relation to assessment of critical loads. The soil chemistry therefore seems clear and, although the further shift of pH and move into the iron buffering range must be considered, the weaker area of knowledge for the forest sites appears to be the biological consequences of the chemical changes. Work using ^{15}N could help to determine the N cycling in the forest-canopies and in quantifying the ecosystem N cycle. There remains uncertainty over the factors which control nitrification (specifically pH and soil moisture status). Work on the processes which control the amount of organic matter in soils (including decomposition rates) and on the factors which determine the related N transformations, including denitrification, should be considered.

8.3 Detailed comments; pot experiments

Seedling trials have shown that reliance on NH_4^+, as would presumably occur if nitrification was entirely inhibited, reduces mycorrhizal frequency in the rhizosphere and bacterial populations on the root surface, root growth is also reduced (smaller SRL) and plant ionic balance disturbed. The relationship between NH_4^+ or NO_3^- uptake and rhizosphere pH has been demonstrated in some elegant experiments. In some pot experiments NH_4^+ applications have led to seedling mortality. It is of considerable interest that the impact of ammonium deposition on the microbial community appears to have a larger influence on carbon allocation than the impact of short term O_3 fumigations (project 107).

8.4 Detailed comments; stands level studies

The two stands already show signs of altered nutrition with potassium and phosphorus deficiencies developing. With the suggestion from the soil models that exhaustion of aluminium hydroxide and of the Al bound organically will probably take 30 to 100 years, it is clear that investigations of the biological consequences of such a shift will have to be based on manipulation experiments within the forest stand, or on analytical work at different sites at which soil deterioriation has proceeded further. The Douglas fir stands seen in the Peel area, where current-year foliage is severely yellowing, argenine accumulation is evident in foliage and tree mortality is occuring, must be obvious candidates for this work. The thematic report paid little attention to the acute damage occuring in some areas of the Netherlands, this is a notable defeciency. This may purely reflect emphasis on Douglas fir at the ACIFORN sites. However better integration of the work being conducted at Speuld and

Kootwijk with the work on Pinus sylvestris in the Peel area is also to be encouraged. It now seems necessary to devote more research effort to the species and areas in which "effects" are clearly taking place. Examination of the biochemical route by which nitrogen detoxification occurs and of the role of iron in this appears a priority. It would be sensible to have such work associated with the continuing manipulation studies at the ACIFORN sites as well as in the areas of dramatic damage.

The association between excess nitrogen and increased susceptibility to fungal pathogens, as in the case of Sphaeropsis sapinea on Scots pine, does not appear to have been fully elucidated and this aspect was not covered in the integrated effects on forest. In contrast the experimental work on the effects of O_3 and possiby of SO_2 on phloem loading, photoassimilate distribution and therefore on root growth and root/shoot ratios was made use of.

Increased susceptibility of Douglas fir stands to drought emerging as the second most likely hypothesis for problems of forest growth after nitrogen effects and associated acidification of soils. Field verification of the operation of this damage sequence appears to rely solely on the observations of poor mycorrhizal associations and of poor root growth. This highlights the need for a better understanding of canopy effects and puts emphasis on these aspects of project 105 which still require attention. More research on plant water relations and the impacts of drought on growth and on nutrient pollutant responses are recommended, particularly in view of the importance of such interactions to climate change. The analysis of stand ecophysiology still needs to address transpiration, photosynthetic and respiration rates and how they are altered by severe nitrogen stress as well as the associated biochemical effects (argenine accumulation and phloem loading). The direct effects of O_3 and possibly of NH_3 resulting from long-term, low concentration exposures also remain an important gap in knowledge. Information on these effects from the cuvette studies in the Douglas fir stands now seems an important requirement for development of the stand growth model.

8.5 General conclusions

The comments made above make it clear that the experimental work on young trees, the stand level measurements which have been made at Speuld and Kootwijk and the modelling which has been conducted to synthesis these have contributed towards answering the first two questions addressed in the second phase. Quantitative assessments of the role of direct effects of O_3, SO_2, NO_x and NH_3 can be made with varying degrees of certainty. Separation of the contributions made by O_3, SO_2, NO_x and NH_3 will remain impossible until projects 100, 103 and 105 have been completed and some more sharply focused experiments conducted.

Overriding emphasis can be placed on the role of excess nitrogen operating indirectly through soil acidification and effects on mycorrhiza, root morphology and tree nutrient status. There hase been some excellent science done, but it is has not so far been sharply enough focussed on the three questions of the second phase of the Priority Programme. The significance of O_3 episodes and of interacting effects of NH_3, SO_2, NO_x, NH_4^+ and SO_4^{2-}, such as predisposition to fungal pathogens and to low temperature injury, still require attention. This puts a limit on the modelling work and on the confidence with which critical loads for forest ecosystems can be assessed.

The question of site numbers and location has been raised on a number of occasions, particularly with reference to the intensive work on forest stands. These issues can best be decided by a clear examination of the objectives of the research programme. If the objective is to characterize forest conditions relative to pollutant deposition, then more sites than one covering a range of pollutant climate and damage are required. An alternative approach is to manipulate the conditions at a single site. A full range of techniques can be adopted at the same site, resulting in high quality and well integrated work. This approach is dependent on accurately reproducing the range of conditions experienced across the country and therefore on a thourough monitoring network. This gives rise to the necessity of having two tiers of sites with specific, differing objectives.

8.6 Priorities for future work

A number of priorities for future work emerge.

Firstly, the need to continue analytical work at the stand level so as to fill the specific gaps identified above, in the report on "Biological and physiological effects" and to enable the modelling to be based on effects and parameters derived from large trees.

Secondly the manipulative experiments offer the opportunity to assess critical loads by the correction of nutrient inbalances (Ingestad treatment) and to examine rates of recovery in soils, tree nutrition and physiology which might occur with the abatement of NH_3 emissions. The manipulation studies may also be valuable in verification of the FORGRO model. These projects are at an early stage but the results presented to date are already of considerable value. Continuation of these experiments, full analysis and publication will be important, although the treatments in which the S and N content of throughfall is reduced to preindustrial values may be of more practical relevance for environmental protection and forestry than the optimum nutrition treatment.

Finally, because of the high quality of work which has been conducted at, or in association with the ACIFORN sites, an integrated, process based understanding of the function of Douglas fir stands has been produced. Some specific additional work has been identified above which will improve the use of this model for assessment of critical loads of acidic pollutants. Similarly the programme is excellently placed to address the impacts of rising CO_2 concentrations and climate change on forest stands.

9. INTEGRATED MODELLING

9.1 Review of the integrated modelling activities

While very difficult and ambitious, the integrated modelling effort (Dutch Acidification System (DAS) model) is an important activity. The DPPA is to be commended for initiating this effort. It appears that DAS has been primarily developed as a predictive model as well as to serve as an interface between scientists and policy makers/resource managers. While this is an important application, an equally important aspect of modelling is its use as an integration/research tool. For a study to be integrated, experimental researchers and modellers must work together. Modellers need information from experimental researchers to parameterize/calibrate the model. Experimentalists can also help improve process representation within the model and possibly simplify the model. Experimental researchers can be assisted by modellers in interpreting data and testing hypotheses. Moreover, modelling can help identify critical processes and help focus research activities. It appears that there is good interaction between modellers and experimentalists in the DPPA. The investigators are encouraged to use the DAS model to facilitate interaction between experimental researchers and modellers.

The individual submodels (e.g. RESAM, SoilVeg, Aqu Acid, Erica) should be calibrated, and verified if possible, with data collected from intensive study sites. The calibrated submodels should be critically reviewed by experimental researchers working on individual processes. This might be effectively accomplished through modelling workshops. In these workshops, the experimental researchers could suggest one or two hypothesis that can be tested with the model. They should also indicate how the model should be run to test the hypotheses of interest and what results they anticipate from the simulations. Following completion of these simulations by modelling groups, the results should be discussed with all researchers involved in the study. Workshops will facilitate communication between modellers and experimentalists.

Considerable effort has been made to make the DAS model user-friendly. This approach has the advantage that the model is available for many people to use as a research and predictive tool. Unfortunately, it has the disadvantage that people can use the model who don't fully understand its limitations or the assumptions used in its development. As a result, detailed sensitivity and uncertainty analyses should be conducted to help the user interprete the limitations of model predictions. It appears that this effort is underway. Every effort should be made to complete this exercise before the model is widely distributed.

9.2 Relevance of integrated modelling to an integrated evaluation of acidification in The Netherlands

The integrated modelling effort is the only hope of achieving an integrated assessment of effects of acidic deposition in the Netherlands. As a result, this component of the study is critical to the overall success of the DPPA. However, the integrated assessment will only be quantitative if modelling activities are accompanied by a rigorous calibration, verification and uncertainty analysis.

9.3 Relevance to critical loads/critical levels assessment

Again the integrated modelling work is probably the most effective way to determine critical levels/critical loads of air pollutants for the Netherlands. The determination of critical levels/critical loads should be preceded by: 1. calibration/verification of the individual submodels using data collected from intensive study sites (e.g. ACIFORN) and 2. conducting an uncertainty analysis on model parameters.

9.4 Conclusions and recommendations for future activities

In summary, the integrated modelling effort is a critical component of the DPPA. The most important aspect of this work might be as a tool to integrate the overall Programme. However, it appears that the individual submodels could be used more effectively as research tools. Close cooperation between the individual researchers and modellers could improve model development applications. The models should be used to test individual research hypotheses and suggest/direct future research activities. A more formal effort should be made to calibrate/verify the submodels using data from intensive study sites and an uncertainty analysis should be conducted.

Model development/application should be an ongoing activity. Many of the DAS submodels are directly relevant and could be expanded to address questions critical to global change. It seems that the DPPA is well positioned to move directly in global change research.

THEMATIC CONCLUSIONS

Emission of NH_3

The ammonia emission project meets the objective relating to abatement strategies. Reduction of NH_3 from animal husbandry is of basic importance for the abatement of the nitrogen mediated soil acidification, because livestock production contributes to about 80% to the NH_3 emission. Based on the present knowledge the most efficient reduction of the NH_3 release can be achieved by reducing the emission after spreading. Efficient measures to reduce NH_3 emission from buildings and stores are mostly still under progress.

Atmospheric input fluxes

The programme on inputs of ammonia by dry deposition on heathland and forest have convincingly demonstrated the processes regulating deposition and has provided the means to quantify input estimates. This is a notable achievement.

Ammonium inputs from the atmosphere to heathland and forest represent the dominant input of atmospheric nitrogen (most of this 60 - 80% as NH_3 deposition) and a major input of potential acidity (30 to 50% depending on location).

The main processes regulating NH_3 and SO_2 inputs to heathland and forest have been quantified and regional fluxes for those surfaces may be estimated with uncertainties of $\pm$30 to $\pm$50%.

Soil acidification / N cycling, biological / physiological effects, integrated effects

The experimental work on young trees, the stand level measurements which have been made at Speuld and Kootwijk and the modelling which has been conducted to synthesise these, have provided only partial answers to the questions placed in the Second Phase of the DPPA.

The direct effects of O_3 on trees cannot be ruled out on the basis of the information provided. Furthermore, the interacting effects of gaseous air pollutants with other abiotic (drought and frost) and biotic (fungal) agents have not been investigated thoroughly.

The DPPA research effort on soil acidification and nitrogen cycling has clearly demonstrated that acidification of soil solutions at the ACIFORN sites is due to atmospheric deposition of sulphur and nitrogen compounds. Therefore, any indirect (soil mediated) acidification effects on vegetation can be attributed to atmospheric deposition. The relationships between soil chemistry and tree nutritional status has been identified but not fully quantified. The biological effects of changes in soil chemistry, and particularly these on mycorrhizae also require further attention.

Preliminary results are available from the forest manipulation experiments which give early indications only of the probable effectiveness of abatement measures.

Integrated effects (low vegetation)

The projects on "integrated effects (low vegetation)" have concentrated on dry heathland habitats. The projects have been innovative and have provided a comprehensive treatment of the loss of plant species that formerly were characteristic for Dutch heathlands.

The rate of change from Calluna to grasses on protected "heathlands" is proceeding faster than expected and is shown to be due to nitrogen deposition, mainly as NH_3, as a background force. The underlying causes of micro-site-dynamics at the sites of the less competitive species still require better understanding.

Integrated modelling

Integrated modelling effort is a critical component of the DPPA. The most important aspect of this work will be as a tool to integrate the overall programme.

GENERAL CONCLUSIONS

1. The Review Team felt the true value of the work in the Second Phase Programme still has to be realised. This is due to the lack of time available for the project leaders to undertake a full and rigorous analysis of the data in their programmes. Consequently, much of the data analysis of the Second Phase Programme was not available for review.

2. The Review Team considered the overall quality of the research to be good. However some variety in quality was noted and this is commented on in the thematic reports. Some of the work is of world class quality . The research in the Netherlands is far ahead of other countries as regard emissions, depositions and effects of NH_3 and NH_4^+.

3. The Team considered the objectives of the Programme were too ambitious to be met in a short term research project of this nature.

4. Most of the scientific evidence presented in the Second Phase Programme indicates a need to address specific questions in a Third Phase Programme.

5. The Programme has been successful in developing and applying models to establish critical loads for total nitrogen deposition and acidifying compounds. These models, however, need validation. The results from the programme offer good opportunities in this area.

6. There are indications of a lack of integration of the different thematic reports. This results in part from a lack of time for synthesis and integration and is evidenced by contradicting conclusions within and between reports.

7. The research papers under review would have stronger impact if they were to be subject to peer review and publication in the scientific literature. The Review Team considered the Programme to be of sufficient value for publication in a single volume. This step in itself would assist in integration.

8. Research in the Netherlands should be further encouraged to play a major role in collaborative efforts to develop the critical loads/levels concept, both nationally and internationally. This is especially important in those areas where the Dutch research programme is strong particularly the study of pathways of ammonia and total nitrogen from emission and deposition to effects and integrated modelling.

GENERAL RECOMMENDATIONS

1. The Review Team strongly supports the continuation of research in specific areas in a Third Phase.

2. It is recommended that before any future research is undertaken full evaluation of data from the First and Second Phase Programmes should be undertaken and a series of well formulated and realistic objectives should be defined.

3. The Review Team recommended for consideration the inclusion of the following elements in the future research programme. Details regarding these are supplied in the specific conclusions section.

- full data evaluation of results from the First and Second Phase Programme
- the establishment of some form of long term environmental monitoring network to meet the requirements of the objective relating to abatement strategies
- the continuation of relevant projects in the Second Phase Programme to capitalise in the studies so far undertaken
- the continuation and establishment of manipulative field experiments to verify some of the preliminary conclusions reached in the Second Phase
- indepth research and integration of data relating to the history of the experimental sites and their ecological dynamics
- development and much wider use of integrated models throughout the work of the Dutch Priority Programme on Acidification. This aspect could make integrated modelling the core of the third phase.

REVIEW TEAM

REVIEWERS

C.T. Driscoll
Department of Civil and Environmental Engineering, Syracuse University
220 Hinds Hall
SYRACUSE, NEW YORK 13244-1190
USA
tel: 315-4432311 fax: 315-4434936

integrated modelling
soil acidification/N cycling

H.Ellenberg
FG Wildtierökologie und Jagd
Institut für Weltforstwirtschaft und Ökologie
Bundesforschungsanstalt für Forst- und Holzwirtschaft
Leuschnerstrasse 91
2050 HAMBURG-80
FRG
tel: 40-73962405 fax: 40-73962480

integrated effects (low-vegetations)

D. Fowler
Institute of Terrestrial Ecology
Bush Estate
PENICUICK, MIDDLOTHIAN EG250QB
UK
tel: 31-4454343 fax: 31-4453943

atmospheric input fluxes

P. Freer Smith
Forest Research Station, Alice Holt Lodge
FARNHAM, SURREY GU104LH
UK
tel.: 0420-22255 fax: 0420-23653

integrated effects (forests)
soil acidification/N cycling

J. Hartung
Institut fur Tierhygiene und Tierschutz
Tierartzliche Hochschule
Bunteweg 17P
D-3000 HANNOVER
FRG
tel: 511-85688312 fax: 511-8567685

emissions of NH3

G. Landmann
Inst.Nat.de la Recherche Agronomique
Centre de Recherches Forestieres Nancy
F-54280 CHAMPENOUX
France
tel: 83394077 fax: 83394069

biological and physiological effects

J. Nilsson
National Swedish Environmental
Protection Board
P.O.Box 1302
171 52 SOLNA
Sweden

general review

R.B. Wilson
UK Department of the Environment
Room B352, Romney House
43 Marsham Street
LONDON SW1P 3PY
UK
tel: 71-276-8316 fax: 71-2768501

chair

ANNEX 3 PROJECTS AND PUBLICATIONS FIRST AND SECOND PHASE DUTCH PRIORITY PROGRAMME ON ACIDIFICATION

Project no. **00**
Title The assessment of the dry deposition flux of NH_3 on the "Speulderbos" forest based on ECN data
Projectleader J.H.Duyzer
Address Organisation for Nature Scientific Research TNO Technology for Society
P.O.Box 217, 2600 AE DELFT
Tel. 015-696263

Project no. **00.1**
Title Uitbreiding van de nota "Evaluatie Verzuringsonderzoek"
Projectleader Drs.H.M.A.Jansen
Address Free University of Amsterdam, Institute for Environmental Studies
P.O.Box 7161, 1007 MC AMSTERDAM
Tel. 020-5483827

01. Jansen H.M.A., 1989
Verzuring: een overzicht van de situatie in enige Europese landen
Instituut voor Milieuvraagstukken, Vrije Universiteit Amsterdam

Project no. **01**
Title Field monitoring "Winterswijk": the role of acid atmospheric deposition in the biogeochemical balance of a forest ecosystem
Projectleader Prof.Dr.J.M.Verstraten
Address University of Amsterdam, Laboratory for Physical Geography and Soil Science
Dapperstraat 115, 1093 BS AMSTERDAM
Tel. 020-5257415

01. Bouten W., Schaap M.G., Bakker D.J. and Verstraten J.M., 1991
Modelling soil water dynamics in a forested ecosystem I: a site specific evaluation
submitted to Hydrological Processes
02. Bouten W. and Witter J.V., 1991
Modelling soil water dynamics in a forested ecosystem II: evaluation of spatial variation of soil profiles
submitted to Hydrological Processes
03. Tiktak A. and Bouten W., 1991
Modelling soil water dynamics in a forested ecosystem III: model description and

evaluation of discretisation
submitted to Hydrological Processes

Project no. **02**
Title Field monitoring for research on the role of acid atmospheric deposition in the biochemical balance of forest and heathland ecosystems
Projectleader Prof.Dr.Ir.N.van Breemen
Address Agricultural University, Department of Soil Sciences and Geology
P.O.Box 37, 6700 AA WAGENINGEN
Tel. 08370-82424

01. Breemen N.van, Visser P.H.B.de and Grinsven J.J.M.van, 1986
Nutrient and proton budgets in four soil-vegetation systems underlain by Pleistocene alluvial deposits
Journal of the Geological Society, London, 143: 659-666

02. Booltink H.W.G., Pape Th., and Breemen N.van, 1988
Field monitoring for research on the role of acid atmospheric deposition in the biogeochemical balance of forest and heathland systems
Department of Soil Science and Geology, Agricultural University Wageningen, The Netherlands
Dutch Priority Programme on Acidification, report nr. 02-01

03. Breemen N.van, Visser W.F.J. and Pape Th., 1988
Biochemistry of an oak-woodland ecosystem in The Netherlands affected by acid atmospheric deposition
Agricultural Research Reports 930, p. 97, Pudoc Wageningen

04. Breemen N.van and Dijk H.F.G.van, 1988
Ecosystem effects of atmospheric deposition of nitrogen in The Netherlands
Env.Pollution 54: 249-274

05. Breemen N.van, Boderie P.M.A. and Booltink H.W.G., 1989
Influence of airborne ammonium sulfate on soils of an oak woodland ecosystem in The Netherlands: Seasonal dynamics of solute fluxes
Acid Precipitation vol. 1, case studies, D.C.Adriano and M.H.Havas (eds.), Springer Verlag, p. 209-236

06. Pape Th., Breemen N.van, and Oeveren H.van, 1989
Calcium cycling in an oak-birch woodland on soils of varying $CaCO_3$ content
Plant and Soil 120, p. 253-261, Kluwer Academic Publishers

07. Velthorst E.J. and Breemen N.van, 1989
Changes in the composition of rainwater upon passage through the canopies of trees

and of ground vegetation in a Dutch oak-birch forest
Plant and Soil 119: p. 81-85

08. Breemen N.van, 1990
Deterioration of forest land as a result of atmospheric deposition in Europe: a review
7th North American Forest Soils Conference, Sustained productivity of forest soils, edited by S.P.Gessel, D.S.Lacate, G.F.Weetman and R.F.Powers; Forestry publications, Facultaty of Forestry, MacMillan Bldg., University of British Columbia, Vancouver, B.C., Canada, V6T 1W5, p. 40-48

Project no. **03**

Title Monitoring of soil chemical and soil physical parameters under Douglas fir and heather in support to the integrated field research in the Dutch Priority Programme on Acidification

Projectleader Prof.Dr.Ir.N.van Breemen

Address Agricultural University, Department of Soil Sciences and Geology
P.O.Box 37, 6700 AA WAGENINGEN

Tel. 08370-82424

01. Maas M.P.van der and Valent A.P.M., 1988
In situ conservation of throughfall samples
In: Monitoring air pollution and forest ecosystem research; proceedings of a workshop jointly organized by commissions of the EG and RIVM
Edited by: Bresser A.H.M. and Mathy P.; Air pollution report series of the Environmental Research Programme of the EC, pp. 137

02. Tiktak A., Konsten C.J.M., Maas M.P.van der and Bouten W., 1988
Soil chemistry and physics of two Douglas fir stands affected by acid atmospheric deposition on the Veluwe, The Netherlands
Dutch Priority Programme on Acidification, report nr. 03-01, 93 p.

03. Breemen N.van, 1990
Deterioration of forest land as a result of atmospheric deposition in Europe: a review
7th North American Forest Soils Conference, Sustained productivity of forest soils, edited by S.P.Gessel, D.S.Lacate, G.F.Weetman and R.F.Powers; Forestry publications, Facultaty of Forestry, MacMillan Bldg., University of British Columbia, Vancouver, B.C., Canada, V6T 1W5, p. 40-48

Project no. **04**
Title Nitrogen budget and conversions in the litter layer
Projectleader Prof.Dr.J.M.Verstraten
Address University of Amsterdam, Laboratory for Physical Geography and Soil Science
Tel. 020-923030

01. Bouten W., Duijsings J.J.H.M., Tietema A. and Verstraten J.M., 1986
The balance study, a method in forest ecological research (in Dutch, with English summary)
Ned.Bosbouwk.Tijdschrift 58 10: 256 - 261

02. Verstraten J.M., Bouten W., Duijsings J.J.H.M. and Tietema A., 1986
The biochemical balance study, a method in forest ecological research (in Dutch)
In: Bosecosysteem onderzoek in permanente proefperken; Verslag van een themadag op 13 juni 1986
Pudoc, Wageningen, p. 46 - 53

03. Verstraten J.M., Duijsings J.J.H.M., Bouten W. and Tietema A., 1986
Biogeochemical balance study of a forest ecosystem; Forest dynamics research in Western and Central Europe
Pudoc, Wageningen, p. 124 - 131

04. Tietema A. and Verstraten J.M., 1988
The nitrogen budget of an oak-beech forest ecosystem in the Netherlands in relation to atmospheric deposition
Dutch Priority Programme on Acidification, report no. 04-01

05. Verstraten J.M., Tietema A., Bouten W. and Dopheide J.C.R., 1988
Veldmonitoring voor het onderzoek naar de rol van atmosferische depositie in biogeochemische balans van eiken-beuken bosecosysteem nabij Winterswijk; Strooiselafbraak en de stikstofcyclus van een eiken-beuken bosecosysteem
Rapport FGBL-UvA, Amsterdam 17 pp.

06. Verstraten J.M., Tietema A. and Dopheide J.C.R., 1990
Bodemverzuring: principes en voorbeelden
Geogr.Tijdschrift 1989-4: 251 - 261

07. Tietema A., Duijsings J.J.H.M., Verstraten J.M. and J.W.Westerveld, 1990
Estimation of actual nitrification rates in a acid forest soil
In: Harrison A.F., Ineson P. and Hean O.W. (eds.), Nutrient cycling in terrestrial ecosystems. Field methods application and interpretation, 190 - 197, Elsevier Applied Science, London and New York

08. Verstraten J.M., Dopheide J.C.R., Duijsings J.J.H.M., Tietema A. and Bouten W.,

1990
The proton cycle of a deciduous forest eocystem in the Netherlands and implications for soil acidification
Plant and Soil, 127, 61 - 69

09. Tietema A., Bouten W. and Wartenburgh P.E., 1990
Nitrous oxidynamics in an acid forest soil in the Netherlands
Forest Ecology and Management
in press

10. Tietema A., Verstraten J.M. and Wijk A.J.van, 1990
The nitrogen cycle of an oak-beech forest ecosystem in the Netherlands at increased nitrogen deposition: 1. biochemical nitrogen transformations and solute fluxes
Biogeochemistry
submitted

11. Tietema A. and Verstraten J.M., 1990
The nitrogen cycle of an oak-beech forest ecosystem in the Netherlands at increased nitrogen deposition: 2. the nitrogen and proton budget
Biogeochemistry
submitted

<u>Project no.</u> **05**

<u>Title</u> A simulation model for interactions of acid atmospheric deposition with the unsaturated soil; model development and validation

<u>Projectleader</u> Ir.J.J.M.van Grinsven

<u>Address</u> Agricultural University, Department of Soil Sciences and Plant Nutrition
De Dreyen 3, 6703 BC WAGENINGEN

<u>Tel.</u> 08370-82531

01. Grinsven J.J.M.van, Mulder J. and Breemen N.van, 1984
Hydrochemical budgets of some Dutch woodland soils with high inputs of atmospheric acid deposition; Mobilization of aluminium
In: Eriksson, E. (ed.), Hydrochemical balances of freshwater systems, IAHS Publication no. 150: p. 237 - 247

02. Grinsven J.J.M.van, Kloeg G.D.R. and Riemsdijk W.H.van, 1986
Kinetics and mechanism of mineral dissolution in a soil and pH values below 4
Water, Air and Soil Pollut. 31: p. 981 - 990

03. Grinsven J.J.M.van, Haan F.A.M.de and Riemsdijk W.H.van, 1986
Effects of acidic deposition on soil and groundwater
In: T.Schneider (ed.), Acidification and its policy implications, Elsevier Sci.Publ.,

Amsterdam, The Netherlands

04. Grinsven J.J.M., Breemen N.van, Riemsdijk W.H.van and Mulder J., 1987
The sensitivity of acid forest soils to acid deposition
Proc.Int.Sympos. Acidification and Water Pathways, Bolkesjo 4-5 May, Norwegian National Committee for Hydrology (ISBN 82-554-0486-4), kp. 365 - 374

05. Grinsven J.J.M.van, Breemen N.van and Mulder J., 1987
Impacts of acid atmospheric deposition on woodland soils in The Netherlands: I. Calculation of hydrologic and chemical budgets
Soil Sci.Soc.Amer.J. 51: p. 1629 - 1634

06. Grinsven J.J.M.van, Booltink H.W.G., Dirksen C., Breemen N.van, Bongers N. and Waringa N., 1988
Automated in situ measurement of unsaturated soil water flux
Soil Sci.Soc.Amer.J.
in press

07. Grinsven J.J.M.van, 1988
Impact of acid atmospheric deposition on soils; Quantification of chemical and hydrologic processes. Thesis
16 september 1988, Wageningen; promotoren: Dr.Ir.N.van Breemen, Dr.Ir.F.A.M.de Haan, Co-promotor: Dr.Ir.W.H.van Riemsdijk, pg. 1-215

08. Grinsven J.J.M.van, Kros J., Breemen N.van, Riemsdijk W.H.van and Eek E.van, 1989
Simulated response of an acid forest soil to acid deposition and mitigation measures
Neth.Journal of Agricultural Science 37, pp. 279-299

Project no. **06**

Title A simulation model for interactions of acid atmospheric deposition with the unsaturated soil; determination of kinetic parameters

Projectleader Dr.Tj.Th.Lub

Address Energy Research Foundation in The Netherlands
P.O.Box 1, 1755 ZG PETTEN

Tel. 02246-4949

Project no. **07**

Title A simulation model for interactions of acid atmospheric deposition with the unsaturated soil; field validation of a soil acidification model with the St.Arnold lysimeter (BRD)

Projectleader Ir.J.J.M.van Grinsven

Address Agricultural University, Department of Soil Sciences and Plant Nutrition
De Dreyen 3, 6703 BC WAGENINGEN

Tel. 08370-82531

01. Grinsven J.J.M., Wesselink B.G., Schroeder M. and Breemen N.van
Soil acidification and solute budgets for forested lysimeters in Nordrhein Westfalen
Zeitschrift für Pflanzenernährung und Bodenkunde
accepted for publication

Project no. **08**

Title The aluminium budget of acidic soils under forest and heathland under the influence of acid rain

Projectleader Prof.Dr.Ir.N.van Breemen

Address Agricultural University, Department of Soil Sciences and Geology
P.O.Box 37, 6700 AA WAGENINGEN

Tel. 08370-82424

01. Mulder J., 1988
Impact of acid atmospheric deposition on soils: Field monitoring and aluminium chemistry. Thesis
16 September 1988, Wageningen; Promotor: Dr.Ir.N.van Breemen, pg. 1-163

02. Mulder J., Breemen N.van and Eijck H.C., 1989
Depletion of soil aluminium by acid deposition and implications for acid neutralization
Nature 337: 247-249

03. Mulder J., Breemen N.van and Driscoll, C.T., 1989
Aluminium chemistry of forest soils strongly affected by acid atmospheric deposition in The Netherlands and in Denmark
presented before the Division of Environmental Chemistry American Chemical Society, New Orleans LA, September 1987

04. Mulder J., Christophersen et al., 1990
Water flow paths and hydrochemical controls in the Birkenes catchment as inferred from a rainstorm high in seasalts
Water Resources, vol. 26, no. 4, pp. 611-62

05. Grinsven J.J.M. van., Breemen N. van and Mulder J., 1987
Impacts of Acid Atmospheric Deposition on Woodland Soils in the Netherlands: I Calculation of Hydrologic and Chemical Budgets
Soils Science Society of America Journal Vol. 51, no. 6, November-December 1987 pp. 1629-1646

06. Mulder J, Dobben H.F. van, Visser P.H.B. de, Booltink H.W.G. and Breemen N. van., 1987
Effect of vegetation cover (pine forest vs. no vegetation) on atmospheric deposition and soil acidification
Acidification and Water Pathways, Int. Symp. Bolkesjo, Norway, p. 79-90

07. Mulder J and Breemen N. van., 1987
Differences in aluminium mobilization in podsols in New Hampshire (USA) and in The Netherlands as a result of acid deposition
NATO ASI Series, Vol. G16, pp. 361-376

08. Smit H.P., Breemen N. van and Keltjens, W.G. van., 1987
Effects of soil acidity on Douglas fir seedlings. 1. Rooting characteristics of natural regeneration of Douglas fir in strongly acid forest soils
Netherlands Journal of Agricultural Science 35, 1987 pp. 533-536

09. Smit, H.P., Keltjens W.G. and Breemen N. van., 1987
Effects of soil acidity on Douglas fir seedlings. 2. The role of pH, aluminium concentration and nitrogen nutrition (pot experiment)
Netherlands Journal of Agricultural Science 35, 1987 pp. 537-540

Project no. **10**

Title The role of micro-organisms in the nitrogen budget of forest soils in relation to soil acidification

Projectleader Prof.Dr.A.J.B.Zehnder

Address Agricultural University, Department of Microbiology
H.van Suchtelenwaeg 4, 6703 CT WAGENINGEN

Tel. 08370-82105/83100

01. Stams A.J.M. and Lutke-Schipholt I.J., 1990
Nitrate accumulation in leaves of vegetation of a forested ecosystem receiving high amounts of atmospheric ammonium
Plant and Soil 125: 143-145

02. Stams A.J.M. and Marnette E.C.L., 1990
Investigation of nitrification in forest soils with soil percolation columns
Plant and Soils 125: 135-141

03. Stams A.J.M., Lutke-Schipholt I.J., Marnette E.C.L., Beemsterboer B. and Woittiez J.R.W., 1990
Conversion of ^{15}N ammonium in forest soils
Plant and Soil 125: 129-134

04. Stams A.J.M., Flameling E.M. and Marnette E.C.L., 1990
The importance of autotrophic vs. heterotrophic oxidation of atmospheric ammonium in forest ecosystems with acid soil
Fems Microbiol. Ecol., in press

05. Stams A.J.M., Booltink H.W.G., Lutke-Schipholt I.J., Beemsterboer B., Woittiez J.R.W. and van Breemen N., 1990
A field study on the fate of ^{15}N ammonium to demonstrate nitrification of atmospheric ammonium in an acid forest soil
Biogeochemistry, in press

Project no. **11**
Title The influence of acid deposition on microbial processes
Projectleader Dr.J.W.Vonk
Address Organisation for Applied Nature Scientific Research, Technology for Society, Technology for Society, Institute for Applied Chemistry
P.O.Box 108, 3700 AC ZEIST
Tel. 03404-55444

01. Vonk J.W., 1987
Soil acidification and microbial processes: The fate of inorganic nitrogen in acid heathland and forest soil
TNO report no. R87/331, 8 December 1987

02. Vonk J.W., Barug D. and Bosma T.N.P., 1988
Fate of mineral nitrogen in acid heathland and forest soils
In: P.Mathy (ed.), Air pollution and ecosystems; D.Reidel Publishing Comp.; Dordrecht, 1988 pg. 835-840

Project no. **12** (see also 18)
Title The indirect effects of acid deposition on the vitality of the Dutch forests; pedologic research
Projectleader Ir.W.de Vries
Address Soil Survey Institute (STIBOKA)
Marijkeweg 11, 6709 PE WAGENINGEN
Tel. 08370-74353

01. Oterdoom J.H., Burg J.van den and Vries W.de, 1987
Resultaten van een oriënterend onderzoek naar de minerale voedingstoestand en bodemchemische eigenschappen van acht Douglasopstanden met vitale en minder vitale bodem in Midden-Nederland, winter 1984/1985
Wageningen, Rijksinstituut voor Onderzoek in de Bos- en Landschapsbouw "De Dorschkamp", report nr. 470

02. Kleijn C.E. and Vries W.de, 1987
Characterizing soil moisture composition in forest soil in space and time
In: W. van Duijvenbooden and H.G. van Waegeningh (Eds.): Vulnerability of Soil and Groundwater to Pollutants. Proc. Int. Conf. Noordwijk aan Zee, 1987, The Netherlands: 591-600

03. Kleijn C.E., Oterdoom H., Vries W.de and Hendriks C., 1987
De indirecte effecten van atmosferische depositie op de vitaliteit van Nederlandse bossen. Deel 1: Beschrijving van de onderzoeksopzet. Stichting voor Bodemkartering, Wageningen, report nr. 2010

04. Kleijn C.E., Zuidema G. and Vries W.de, 1989
De indirecte effecten van atmosferische depositie op de vitaliteit van Nederlandse bossen. Deel 2. Depositie, bodemeigenschappen en bodemvochtsamenstelling van acht Douglasopstanden. Wageningen, Stichting voor Bodemkartering, report nr. 2050

05. Reurslag A., Zuidema G. and Vries W.de, 1990
De indirecte effecten van atmosferische depositie op de vitaliteit van Nederlandse bossen. Deel 3. Simulatie van de waterbalans van acht Douglas-opstanden. Wageningen, Staring Centrum, report 76

Project no.	**13**
Title	Development of a regional model for analysing the acidification of Dutch forest soils
Projectleader	Ir.W.de Vries
Address	Soil Survey Institute (STIBOKA) Marijkeweg 11, 6709 PE WAGENINGEN
Tel.	08370-74353

01. Hettelingh J.P. and Vries W.de, 1990
Mapping vademecum
Bilthoven, National Institute of Public Health and Environmental Protection
Technical report

02. Kros J., Janssen P., Vries W.de and Bak C., 1990
Het gebruik van onzekerheidsanalyse bij modelberekeningen: een toepassing op het

regionale bodemverzuringsmodel RESAM
Wageningen, Staring Centrum, report 65

03. Kros J., Vries W.de, Janssen P.and Bak C., 1991
The uncertainty in forecasting regional trends of forest soil acidification
Water, Air and Soil Pollution
submitted

04. Schulze E.D., Vries W.de, Hauhs M., Rosen K., Rasmussen L., Tamm C.O. and Nilsson J., 1989
Critical loads for nitrogen deposition on forest ecosystems
Water, Air and Soil Pollution, 48: 451-456

05. Sverdrup H., Vries W.de and Henriksen A., 1990
Mapping critical loads. A guidance manual to criteria, calculation methods, data collection and mapping. Background document for the Task Force on Mapping, Bad Harzburg, 22-23 May, 1990

06. Visser P.H.B. de and Vries W.de, 1989
De gemiddeld jaarlijkse waterbalans van bos-, heide en grasland vegetaties
Wageningen, STIBOKA, report nr. 2085

07. Vries W. de and Breeuwsma A., 1987
The relation between soil acidification and element cycling
Water, Air and Soil Pollution, 35:293-310

08. Vries W. de, 1987a
A conceptual model for analysing soil and groundwater acidification on a regional scale
In: Proc. Int. Symp. on Acidification and Water Pathways. Bolkesjo, 1987, Norway: 185-194

09. Vries W. de, 1987b
The role of soil data in assessing the large scale input of atmospheric pollutants on groundwater quality
In: W. van Duijvenbooden and H.G. van Waegeningh (Eds.): Vulnerability of Soil and Groundwater to Pollutants. Proc. Int. Conf. Noordwijk aan Zee, 1987, The Netherlands: 897-910

10. Vries W. de, 1988
Critical deposition levels of nitrogen and sulphur on Dutch forest ecosystems
Water, Air and Soil Pollution 42:221-239

11. Vries W. de, Waltmans M.J.P.H., Versendaal R.van and Grinsven J.J.M.van, 1988
Aanpak, structuur en voorlopige procesbeschrijving van een bodemverzuringsmodel voor toepassing op regionale schaal
Wageningen, STIBOKA, report nr. 2014

12. Vries W. de, 1989
Kritische depositieniveau's van stikstof en zwavel op Nederlandse bossen
In: Proc. Nationaal Symposium Verzuring, Ede, 15 juni 1989: 15-22

13. Vries W. de and Kros J., 1989a
De lange termijn effecten van verschillende depositiescenario's op de bodemvochtsamenstelling van representatieve bosecosystemen
Wageningen, Staring Centrum, report 30

14. Vries W. de and Kros J., 1989b
The long term impact of acid deposition on the aluminium chemistry of an acid forest soil
In: J. Kamari, D.F. Brakke, A. Jenkins, S.A. Norton and R.F. Wright (Eds.): Regional Acidification Models. Geographic Extent and Time Development: 113-128

15. Vries W. de, Posch M. and Kamari J., 1989a
Modelling time patterns of forest soil acidification for various deposition scenarios
In: J. Kamari, D.F. Brakke, A. Jenkins, S.A. Norton and R.F. Wright (Eds.): Regional Acidification Models. Geographic Extent and Time Development: 129-150

16. Vries W. de, Posch M. and Kamari J., 1989b
Simulation of the long-term soil response to acid deposition in various buffer ranges
Water, Air and Soil Pollution 48:349-390

17. Vries W. de, Breeuwsma A. and Vries F.de, 1989c
Kwetsbaarheid van de Nederlandse bodem voor verzuring. Een voorlopige indicatie in het kader van de Richtlijn "Ammoniak en Veehouderij"
Wageningen, Staring Centrum, report 29

18. Vries W. de, Schoumans O.F., Kragt J.F. and Breeuwsma A., 1989d
Use of models and soil survey information in assessing regional water quality
In: G. Jousma, J. Bear, Y.Y. Haimes and F. Walter (Eds.): Groundwater contamination: Use of models in decisionmaking, Proc. Int. Conf. Amsterdam, 1987, The Netherlands: 419-432

19. Vries W. de, 1990
Philosophy, structure and application methodology of a soil acidification model for the Netherlands
In: J. Kamari (Ed.). Impact models to assess regional acidification: 3-21

20. Vries W. de, Hol A., Tjalma S. and Voogd J.C., 1990
Voorraden en verblijftijden van elementen in een bosecosysteem: een literatuurstudie
Wageningen, Staring Centrum, report 94

21. Vries W. de and Gregor H.D., 1990
Critical loads and critical levels for the environmental effects of air pollutants
In: M.J.Chadwick and M. Hutton (Eds.): Acid Depositions in Europe: Environmental

effects, control strategies and policy options; Stockholm Environment Institute

22. Vries W. de, 1991a
Assessment and policy implications of critical loads for nitrogen and sulphur in the Netherlands
Water, Air and Soil Pollution
submitted

23. Vries W. de, 1991b
Methodologies for the assessment and mapping of critical loads and the impact of abatement strategies on forest soils
Wageningen, The Winand Staring Center for Integrated Land, Soil and Water Research. Report ..

24. Vries W. de, Kros J. and Klijn J., 1991
Simulation of the long-term impact of atmospheric deposition on dune ecosystems in the Netherlands
Journal of Applied Ecology
in prep.

25. Vries W. de, Kros J., Hootsmans R. and Uffelen G.J.van, 19991b
Assessment and mapping of critical loads for nitrogen and sulphur for Dutch forest soils
in prep.

26. Vries W. de, Kros J., Salm C.van der and Voogd J.C., 1991c
The long-term impact of three emission-deposition scenarios on Dutch forest soils
Water, Air and Soil Pollution
in prep.

27. Vries W. de, Posch M., Kamari J. and Schopp W., 1991d
Long-term soil response to acidic deposition in Europe
Water, Air and Soil Pollution
in prep.

<u>Project no.</u> **14**

<u>Title</u> Air pollution in forest; research on air pollution in forests in relation with physiological parameters of trees and acid deposition on the soil (monitoring of 2 Douglas fir stands)

<u>Projectleader</u> Ir.P.Hofschreuder

<u>Address</u> Agricultural University, Department of Air Pollution
P.O.Box 8129, 6700 EV WAGENINGEN

<u>Tel.</u> 08370-82684

01. Vermetten A.W.M., Hofschreuder P. and Harssema H., 1986
Dry deposition of gaseous pollutants in a Douglas fir forest
In: H.-W. Georgii (ed.), Atmospheric pollutants in forest areas, Proceedings symposium in Oberursel (B.R.D.), november 1985., p.1-11. Reidel Dordrecht

02. Evers P., Konsten C.J.M. and Vermetten A.W.M., 1988
Acidification research on Douglas fir forests in the Netherlands (ACIFORN-project)
In: P.Mathy (ed.): Air Pollution and Ecosystems, proceedings Symposium Grenoble, 18-22 May 1987, p.887. Reidel Dordrecht

03. Vermetten A.W.M., Hofschreuder P. and H.Harssema, 1988
Air pollution monitoring in a Douglas fir forest
In: P.Mathy(ed.): Air Pollution and Ecosystems, proceedings COST 612-Symposium in Grenoble, 18-22 May 1987, p.903-906. Reidel Dordrecht

04. Vermetten A.W.M., 1988
(in Dutch) Luchtverontreiniging in bossen (Air pollution in forest canopies)
NAPV 1st phase 1985 - 1987, Final report project 14. LUW, Dep. of Air Pollution, report R-306

05. Vermetten A.W.M. and Hofschreuder P., 1989
(in Dutch) Progress report over 1988, project 101.1 en 101.2.,Vakgroep LUW-Luvo, rapport R-332

06. Vermetten A.W.M. and, Hofschreuder P., 1989
Deposition of gaseous pollutants in a Douglas fir forest: First results of the ACIFORN-project
In: H.-W. Georgii (ed.), Mechanisms and Effects of Pollutant Transfer into Forests, Proceedings symposium in Oberursel (B.R.D.), 24-25 november 1988., p.61-68. Kluwer Academic Publishers, Dordrecht

07. Hofschreuder P. and Vermetten A.W.M., 1989
Sampling and analysis of air pollutants for exposition and deposition estimates in forests; Site description ACIFORN project
In: A.H.M. Bresser, P. Mathy(eds.): 'Air Pollution Monitoring and Forest Ecosystem Research', proceedings CEC-Workshop, Bilthoven, the Netherlands, 20-21 February 1989

08. Hofschreuder P. and , Vermetten A.W.M., 1989
The effects of pollutants in a Douglas fir forest
In: L.J. Brasser, W.C.Mulder:'Man and his Ecosystem'. Proceedings 8th World Clean Air Congress, 11-15th September 1989, The Hague, the Netherlands, Vol 2, p. 207-211. Elsevier, Amsterdam 1989

09. Vermetten A.W.M., Hofschreuder P. and Versluis A.H., 1990
Air pollution in forest canopies, final report project 14, 101.1, 102.1, Netherlands

Priority Programme on Acidification, report no. 101-09, Wageningen Agricultural University, Dept. of Air Pollution, report no. R-424, August 1990

10. Versluis A.H., Vermetten A.W.M. and Hofschreuder P., 1990
Air pollution in forest canopies: presentation of selected data, Netherlands Priority Programme on Acidification, report no. 101-10, Wageningen Agricultural University, Dept. of Air Pollution, report no. R-444, September 1990

11. Vermetten A.W.M., Hofschreuder P, Versluis A.H., Bij E.S.van der and Tongeren J.J.van, 1990
Luchtverontreiniging in bossen: meetopzet, kwaliteitscontrole en dataverwerking, Wageningen Agricultural University, Dept. of Air Pollution, Report R-445/NAPV report 101-11, September 1990

12. Versluis A.H., Vermetten A.W.M., Bouten W., Maas R.van der and Hofschreuder P., 1990
Canopy interactions in Douglas fir forest: Throughfall composition resulting from deposition under wet and dry conditions. Netherlands Priority Programme on Acidification, report no. 101-12, Wageningen Agricultural University, Dept. of Air Pollution, report no. R-446, October 1990

Project no. **14a**

Title Measurement and modelling of canopy water storage during and after rain, dew and fog (ACIFORN)

Projectleader Ir.W.Bouten

Address University of Amsterdam, Laboratory for Physical Geography and Soils Science
Dapperstraat 115, 1093 BS AMSTERDAM

Tel. 020-5257412

see project 104.1

Project no. **15** (see project 105)

Title Research on the oeco-physiology of conifers in forests in relation with air pollution and acid deposition on the soil

Projectleader Dr.P.Evers

Address Research Institute for Forestry and Urban Ecology "De Dorschkamp"
P.O.Box 23, 6700 AA WAGENINGEN

Tel. 08370-95111

01. Evers P.W., 1986
Fysiologische parameters van stress bij bomen
Gewasbescherming 1986, 17 (4), p. 110

Project no. **18** (see also 12)
Title The indirect effects of acid deposition on the vitality of the Dutch forest
Projectleader Ir.A.F.M.Olsthoorn
Address Research Institute for Forestry and Urban Ecology "De Dorschkamp"
P.O.Box 23, 6700 AA WAGENINGEN
Tel. 08370-95111

01. Kleijn C.E., Oterdoom J.H., Vries W.de and Hendriksen C., 1987
De indirecte effecten van atmosferische depositie op de vitaliteit van Nederlandse bossen; 1. Beschrijving van de onderzoeksopzet
Rapport STIBOKA, nr. 2010, Wageningen, 74 p.

02. Oterdoom J.H., Burg J.van den and Vries W.de, 1987
Resultaten van een oriënterend onderzoek naar de minerale voedingstoestand en de bodemchemische eigenschappen van acht douglasopstanden met vitale en minder vitale bomen in Midden-Nederland, winter 1984/1985
Rapport RBL De Dorschkamp, nr. 470, Wageningen, 47 p.

03. Oterdoom J.H., 1987
Mogelijke samenhang tussen N-voorziening en groei en gezondheid van bomen en opstanden
In: A.W.Boxman & J.F.M.Geelen (eds.) Effecten van NH_3 op organismen, Proceedings studiedag Nijmegen, dec. 1986, Laboratorium voor Aquatische Oecologie, KU, Toernooiveld, Nijmegen: 126-129

04. Oterdoom J.H., Postma R., Olsthoorn A.F.M. and Vries W. de
De indirecte gevolgen van atmosferische depositie op de vitaliteit van Nederlandse bossen, 5. Vitaliteitskenmerken, naaldsamenstelling fijne wortels en groei van acht douglas opstanden
Rapport Additioneel Programma Verzuringsonderzoek 18.02, RIVM, Bilthoven (in press)

05. Oterdoom J.H. and Burg J.van den
De indirecte gevolgen van atmosferische depositie op de vitaliteit van Nederlandse bossen, 4. De naaldsamenstelling van acht douglas opstanden van het Additioneel Programma Verzuringsonderzoek
Rapport Additioneel Programma Verzuringsonderzoek 18.01, RIVM, Bilthoven (in press)

06. Ancker J.A.M.van den, Evers P.W., Maessen P.P.Th.M., Oterdoom J.H. and Tweel P.A.van den, 1987
Inventarisatie van boomvitaliteit - een discussie
NBT 59 (12): 405 - 417
07. Oterdoom J.H., 1987
Mogelijke samenhang tussen N-voorziening en groei en gezondheid van 126 bomen en opstanden
In: A.W.Boxman and J.F.M.Geelen (red.); Proceedings BEL-studiedag "Effecten van NH_3 op organismen", gehouden op 12 december 1987
Nijmegen: Faculteit der Wiskunde en Natuurwetenschappen Katholieke Universiteit, 1987, pp. 126 - 129
08. Oterdoom J.H. and Burg J.van den, 1988
De indirecte effecten van atmosferische depositie op de vitaliteit van Nederlandse bossen; 4. De naaldsamenstelling van acht douglasopstanden van het additioneel programma verzuringsonderzoek (1984/85 tot en met 1986/87)
Report nr. 18.01, Additioneel Programma Verzuringsonderzoek
09. Oterdoom J.H., Vries W.de, Olsthoorn A.F.M. and Postma R., 1990
De indirecte effecten van atmosferische depositie op de vitaliteit van Nederlandse bossen; 5. Vitaliteit, naaldsamenstelling, groei en fijne wortels van acht douglas opstanden
Report nr. 18.02 (draft), Additioneel Programma Verzuringonderzoek

Project no.	**19**
Title	The effect of nitrogen deposition from agriculture on forest and heathland vegetation
Projectleader	J.G.M.Roelofs
Address	Catholic University of Nijmegen, Laboratory of Aquatic Ecology Toernooiveld 1, 6525 ED NIJMEGEN
Tel.	080-558833

01. Breemen N.van and Dijk H.F.G.van, 1988
Ecosystem effects of atmospheric deposition of nitrogen in The Netherlands
Env.Pollution 54: 249-274

Project no. **22**

Title Development of explanatory models for the relation between exposure to air pollution and the effect on growth, development and production of crop plants

Projectleader Dr.M.J.Kropff

Address Research Institute for Plant Protection (IPO)
P.O.Box 9060, 6700 GW WAGENINGEN

Tel. 08370-19151

Refereed Journal Papers

01. Kropff M.J., 1987
Physiological effects of sulphur dioxide, I. The effects of SO_2 on photosynthesis and stomatal regulation of Vicia faba
L.Plant, Cell and Environment 10. 753-760

02. Kropff M.J., 1989
Modelling short term effects of sulphur dioxide, I. A model for the flux of SO_2 into leaves and effects on leaf photosynthesis
Netherlands Journal of Plant Pathology 95, 195-213

03. Kropff M.J., 1989
Modelling short term effects of sulphur dioxide, II. Quantification of biochemical characteristics determining the effect of SO_2 on photosynthesis of leaves
Netherlands Journal of Plant Pathology 95, 214-224

04. Kropff M.J. and Goudriaan J., 1989
Modelling short term effects of sulphur dioxide, III. The effect of SO_2 on photosynthesis of leaf canopies
Netherlands Journal of Plant Pathology 95, 265-280

05. Kropff M.J., Mooi J., Goudriaan J., Smeets W., Leemans A., Kliffen C. and Zalm A.J.A.van der, 1989
The effect of long-term SO_2 exposure on a field crop of broad bean (Vicia faba L.), I. Depression of growth and yield
New Phytologist 113, 337-344

06. Kropff M.J., Mooi J., Goudriaan J., Smeets W., Leemans A. and Kliffen C., 1989
The effects of long-term SO_2 exposure on a field crop of broad bean (Vicia faba L.), II. Effects on growth components, leaf area development and elemental composition
New Phytologist 113, 345-351

07. Kropff M.J., 1990
Effects of long-term open-air fumigation with SO_2 on a field crop of broad bean (Vicia fa L.), III. Quantitative analysis of damage components

New Phytologist, 115, 357-365

08. Kropff M.J., Smeets W., Meijer E., Zalm A.J.A.van der and Bakx E.J., 1990
Effects of sulphur dioxide on photosynthesis: The role of temperature and humidity
Physiologia Plantarum, in press

09. Kropff M.J., 1990
Long term effects of sulphur dioxide on plants, SO_2 metabolism and regulation of intracellular pH
Plant and Soil, in press

Thesis

10. Kropff M.J., 1989
Quantification of SO_2 effects on physiological processes, plant growth and crop production
Ph.D.Thesis, Agricultural University, Wageningen, 201 pp. ISBN 90-9002942-7

Symposium Proceedings

11. Kropff M.J., Smeets W., Mooi J., Goudriaan J., Leemans A., 1989
Growth and production of faba bean crops exposed to SO_2 in the field: Experimental data analysed with a simulation model
In: Man and his ecosystem, L.J.Brasser and W.C.Mulder (eds.), Proceedings of the 8th World Clean Air Congress, The Hague, Holland, 11-15 Sept., 1989, pp. 29-34

12. Mohren G.M.J., Jorritsma I.T.M., Kropff M.J. and Vermetten A.M.W., 1990
Quantifying direct effects of air pollutants on forest growth
In: Conference abstracts Int.Symp.on Acidic Deposition, Glasgow, Sept. 1990, pp.90

13. Kropff M.J. and Smeets W.L.M., 1990
Quantitative analysis of the influence of temperature and humidity on photosynthetic depression by sulphur dioxide
In: Conference abstracts Int.Symp on Acidic Deposition, Glasgow, Sept. 1990, pp. 109

Other publications

14. Kropff M.J., Smeets W., Meijer E., Zalm A.J.A.van der, Kooijman A. and Leemans A., 1989
Uptake of SO_2 by leaf canopies and effects on growth and production: an integrated experimental and simulation study, LUW-LUVO,TPE/IPO report 121 pp., Wageningen

Project no. **25**

Title Qualitative and quantitative research on the relation between ectomycorrhiza of Pseudotsuga menziesii, vitality of the host and acid rain

Projectleader Dr.E.J.M.Arnolds

Address Agricultural University, Biological Station
Kampsweg 27, 9418 PD WIJSTER

Tel. 05936-441

01. Jansen A.E. and Vries F.W.de, 1987
Project qualitative and quantitative research on the relation between ectomycorrhiza of Pseudotsuga menziesii, vitality of the host and acid rain (1985-1986)
Report 25-01, Dutch Priority Programme on Acidification, RIVM, Bilthoven

02. Jansen A.E., 1988
Relation between mycorrhizas and fruitbodies and the influence of tree vitality in Douglas fir plantations in The Netherlands
In: A.E.Jansen, J.Dighton and A.H.M.Bresser (Eds.), Ectomycorrhiza and acid rain Proceedings of the workshop/expert meeting on Ectomycorrhiza, December 10-11, 1987, Berg en Dal, The Netherlands, pp. 68-76. Commission of the European Communities, Air Pollution Research Report 12, Bilthoven

03. Jansen A.E., 1988
Air pollution effects on ectomycorrhizal fungi
In: J.N.Cape and P.Mathy (eds.), Scientific basis of forest decline symptomatology Proceedings of workshop held 21-24 March 1988, Edinburgh (Scotland), pp. 182-189, Commission of the European Communities, Air Pollution Research Report 15, Brussels

04. Jansen A.E., 1988
Influence of acid rain on mycorrhizal fungi and mycorrhizas of Douglas fir in The Netherlands
In: P.Mathy (ed.), Air pollution and ecosystems, Proceedings of an International Symposium, Grenoble, 18-22 May, 1987, Reidel Dordrecht

05. Jansen A.E., 1988
Report on a Workshop on ectomycorrhiza held in Berg en Dal, The Netherlands, 10-11 Dec. 1987
In: J.Bervaes, P.Mathy and P.Evers (eds.), Relationships between above and below ground influences of air pollutants on forest trees, Commission of the European Communities, Air Pollution Research Report 16; 260-262

06. Dighton J., Jansen A.E. and Unestam T., 1988
Conclusions of the workshop "Ectomycorrhiza and acid rain"

In: A.E.Jansen, J.Dighton and A.H.M.Bresser (Eds.), Ectomycorrhiza and acid rain, Proceedings of the workshop/expert meeting on Ectomycorrhiza, Dec. 10-11, 1987, Berg en Dal, The Netherlands, pp. 68-76, Commission of the European Communities, Air Pollution Research Report 12, Bilthoven

07. Jansen A.E., Dighton J. and Bresser A.H.M. (Eds.), 1988
Ectomycorrhiza and acid rain, Proceedings of the workshop/expert meeting on Ectomycorrhiza, Dec. 10-11, 1987, Berg en Dal, The Netherlands, 194 pags., Commission of the European Communities, Air Pollution Research Report 12, Bilthoven

08. Jansen A.E. and Nie H.W.de, 1988
Relations between mycorrhiza and fruitbodies of mycorrhizal fungi in Douglas fir plantations in The Netherlands, Acta Bot.Neerl. 37: 243-249

09. Jansen A.E. and Vries F.W.de, 1988
Qualitative and quantitative research on the relation between ectomycorrhiza of Pseudotsuga menziesii, vitality of the host and acid rain, Report 25-02, Dutch Priority Programme on Acidification, RIVM and LUW, Wageningen

10. Eerden L.J.van der, Lekkerkerk L.J.A., Smulders S.M. and Jansen A.E., 1989
Effects of ozone and ammonia on Douglas fir (Pseudotsuga menziesii)
Proceedings IUFRO meeting Air Pollution and Forest Decline, Interlaken, Switzerland, Oct. 2-8, 1988 (4 pages)

11. Jansen A.E. and Vries F.W.de, 1989
Mycorrhizas on Douglas fir in The Netherlands
Agr.Ecosyst. Environ.: 197-200

12. Cudlin P., Jansen A.E. and Mejstrik V., 1990
Distinction of four Lactarius ectomycorrhizae on Pseudotsuga menziesii (Mirb.) Franco using epifluorescence microscopy of cross sections
In: A. Reisinger and A.Bresinsky (Eds.), Fourth International Mycological Congress, Abstracts: 73, Regensburg

13. Jansen A.E.
The mycorrhizal status of Douglas fir in The Netherlands: its relation with stand age, regional factors, atmospheric pollutants and tree vitality
Agric.Ecosyst.Environ. (in press)

see also project 108

Project no. **26**
Title Effects of acid rain on grasslands
Projectleader Dr.G.W.Heil*
Address University of Utrecht, Department of Plant Ecology
Lange Nieuwstraat 106, 3512 PN UTRECHT
Tel. 030-394515

* present affiliation: Resource Analysis, Zuiderstraat 110, 2611 SJ DELFT, tel. 015-122622

01. Heil G.W. and Diemont W.H., 1983
Raised nutrient levels change heathland into grassland
Vegetatio 53: 113-120

02. Diemont W.H. and Heil G.W., 1984
Some long term observations on cyclical and seral processes in Dutch heathlands
Biological Conservation 30 (3): 283-290

03. Brunsting A.M.H. and Heil G.W., 1984
Zure regen en insektenplagen
Lucht en Omgeving 1 (5): 149-152

04. Brunsting A.M.H. and Heil G.W., 1985
The role of nutrients in the interactions between a herbivorous beetle and some competing plant species in heathland
Oikos 44: 23-26

05. Heil G.W. and Dam D.van, 1986
Luchtverontreiniging en heidevegetaties, rapport vierde studiedag heidebeheer 1985, Ede, p. 131-136

06. Heil G.W. and Dam D.van, 1986
Vegetation structures and their roughness lengths with respect to atmospheric deposition
Proc.of the Seventh World Clean Air Congress, Sydney, Australia, 25-29 August 1986, vol. 5: 16-22

07. Heil G.W. and Dam D.van, 1986
Zure Regen: Wetenswaardigheden omtrent eigenschappen van vegetaties met betrekking tot invang van "zure regen"
Bulletin voor het onderwijs in de Biologie 17 (102): 155-158

08. Dam D.van, Dobben H.F.van, Braak C.F.J.ter and Wit T.de, 1986
Air pollution as a possible cause for the decline of some phanerogamic species in The Netherlands

Vegetatio 65: 47-52

09. Dam D.van and Heil G.W., 1986
Throughfall chemistry of herbaceous species growing on lysimeters
Proc.of the Seventh World Clean Air Congress, Sydney, Australia, vol. 5: 222-224

10. Heil G.W. and Dam D.van, 1986
Verzuring door natuurlijke processen en door zure regen
Bulletin voor het onderwijs in de Biologie 17 (102): 158-161

11. Bobbink R., Tooren B.F.van and Dam D.van, 1986
Effekten van luchtverontreiniging op kalkgraslandvegetaties
Natuurhistorisch maandblad 75 (12): 283-242

12. Heil G.W. and Bruggink M., 1987
Competition for nutrients between Calluna vulgaris (L.), Hull and Molinia caerulea (L.) Moench.
Oecologia (Berlin) pp. 105-108

13. Heil G.W., Dam D.van and Heijne B., 1987
Catch of atmospheric deposition in relation to vegetation structures of heathland
In: W.A.H.Asman and H.S.M.A.Diederen (eds.), Ammonia and acidification, Proceedings of the EURASAP, RIVM, Bilthoven, The Netherlands, 13-15 April 1987, pp. 107-123

14. Dam D.van, Heil G.W. and Heijne B., 1987
Acid rain in chalk grasslands; some data on ion concentrations in bulk precipitation, throughfall and soil solution
In: P.J.M.van der Aart (ed.), The Utrecht Plant Ecology News Report, pp. 135-140

15. Dam D.van, Heil G.W.and Heijne B., 1987
Throughfall chemistry of grassland vegetation: a new method with ion-exchange resins
Functional Ecology 1: 423-427

16. Heil G.W., Heijne B. and Dam D.van, 1987
Atmospheric deposition with respect to the structure of semi-natural grassland vegetation
Proceedings of the Symposium on effects of Air Pollution on Terrestrial and Aquatic Ecosystems, Grenoble 18-22 May 1987, pp. 530-536

17. Heil G.W., Werger M.J.A., Mol W.de, Dam D.van and Heijne B., 1988
Capture of atmospheric ammonium by grassland canopies
Science 238: 764-765

18. Roebertsen H., Heil G.W. and Bobbink R., 1988
Digital picture processing: a new method to analyse vegetational structures
Acta Botanica Neerlandica 37 (2): 187-192

19. Heil G.W., 1988
LAI. of grasslands and their roughness length
In: J.T.A.Verhoeven, G.W.Heil and M.J.A.Werger (eds.), Vegetation Structure in Relation to Carbon and Nutrient Economy, SPB Academic Publishing BV., The Hague, pp. 149-155

Project no. **28**

Title The history of the decline of the Litterellion under the influence of N enrichment and acidification

Projectleader J.G.M.Roelofs

Address Catholic University of Nijmegen, Laboratory of Aquatic Ecology
Toernooiveld 1, 6525 ED NIJMEGEN

Tel. 080-558833

01. Arts G.H.P., 1987
Geschiedenis van de verzuring van zwak gebufferde wateren in Nederland onder invloed van atmosferische depositie
Rapport 28-01 Dutch Priority Programme on Acidification; in opdracht van het Ministerie van VROM, 50 pp.

02. Arts G.H.P., 1988
Historical development and extent of acidification of shallow soft waters in the Netherlands
In: Mathy P. (ed.); Air pollution and ecosystems, Proc.Int.Symp.Grenoble 18-22 May 1987, p. 928-933

03. Arts G.H.P., 1990
Aquatic Bryophyta as indicators of water quality in shallow pools and lakes in the Netherlands
Ann.Bot.Fennici 27: 19-32

04. Arts G.H.P., 1990
Deterioration of atlantic soft-water systems and their flora, a historical account
Thesis Catholic University Nijmegen, Krips Repro Meppel, 197 pp.

05. Arts G.H.P. and Buskens R.F.M., 1989
Aanvoer van venvreemd water: een noodzaak?
In: Aanvoer van gebiedsvreemd water: omvang en effecten op oecosystemen (red. J.G.M.Roelofs)
Uitgave Vakgroep Aquatische Oecologie en Biogeologie, K.U.Nijmgen, p. 100-110

06. Arts G.H.P., Haan A.J.de, Siebum M.B. and Verheggen G.M., 1989
Extent and historical development of the decline of Dutch soft waters

Proc.K.Ned.Akad.Wetensch. C92: 281-295

07. Arts H.G.P. and Leuven R.S.E.W., 1988
Floristic changes in shallow soft waters in relation to underlying environmental factors
Freshwat.Biol. 20: 97-111

08. Arts G.H.P., Leuven R.S.E.W., Roelofs J.G.M., Schuurkes J.A.A.R., Smits H.A. and Tromp V.A., 1986
Verzuring en waterplantengemeenschappen: een historisch perspectief
In: Waterverzuring in Nederland en België; Oorzaken, effecten en beleid; Proc.Studiedag "Waterverzuring" K.U.Nijmegen, p. 103-115

09. Arts G.H.P., Schaminée J.H.J. and Munckhof P.J.J.van den, 1988
Human impact on origin, deterioration and maintenance of Litterelletaliacommunities
In: Proc. 5th Symposium on Synanthropic Flora and Vegetation (Chief Ed.M.Zaliberová), Martin, Czechoslovakia, 22-27 August 1988, p. 11-18

10. Arts G.H.P., Velde G.van der, Roelofs J.G.M. and Swaay C.A.M.van, 1990
Successional changes in the soft-water macrophyte vegetation of (sub)atlantic, sandy, lowland regions during this century
Freshwat. Biol. 24
in press

11. Leuven R.S.E.W., Kersten H.L.M., Schuurkes J.A.A.R., Roelofs J.G.M. and Arts G.H.P., 1986
Evidence for recent acidification of lentic soft waters in the Netherlands
Water, Air and Soil Pollut. 30: 387-392

12. Leuven R.S.E.W., Schuurkes J.A.A.R., Arts G.H.P. and Roelofs J.G.M., 1986
Oorzaken, omvang en effecten van waterverzuring in Nederland en België
In: Waterverzuring in Nederland en België; Oorzaken, effecten en beleid; Proc.Studiedag "Waterverzuring", K.U.Nijmegen, p. 87-102

<u>Project no.</u> **37**

<u>Title</u> Research on the relative contribution of nitrogen emission from agriculture to acid rain

<u>Projectleader</u> Drs.W.Bleuten

<u>Address</u> University of Utrecht, Geographical Institute
P.O.Box 80115, 3508 TC UTRECHT

<u>Tel.</u> 030-532749

01. Draaijers G.P.J., Ivens W.P.M.F. and Bleuten W., 1988
Atmospheric deposition in forest edges measured by monitoring canopy throughfall

Water, Air and Soil Pollution, 42, 129-136

02. Draaijers G.P.J., Ivens W.P.M.F., Bos M.M. and Bleuten W., 1989
The contribution of ammonia emissions from agriculture to the deposition of acidifying and eutrofying compounds onto forests
Environmental Pollution, 60, 55-66

03. Ivens W.P.M.F., Draaijers G.P.J., Bleuten W. and Bos M.M., 1989
The impact of air-borne ammonia from agricultural sources on fluxes of nitrogen and sulphur towards forest soils
Catena, 16, 535-544.

04. Draaijers G.P.J., 1989
Verzuring en eutrofiëring van Nederlandse bossen: bijdrage van de landbouw en gevolgen voor bodem en grondwater
Geografisch Tijdschrift, 4, 262-271

05. Bleuten W., Ertsen A.C.D. and Draaijers G.P.J., 1989
De gevolgen van versnippering voor de verzuring en de eutrofiëring van natuurgebieden in Nederland
Geografisch Tijdschrift, 4, 272-281

06. Ivens W.P.M.F., Kauppi P., Alcamo J. and Posch M., 1990
Empirical and model estimates of sulphur deposition onto European forests
Tellus 42B, 294-303

07. Draaijers G.P.J., Ivens W.P.M.F. and Bleuten W., 1987
The interaction of NH_3 and SO_2 in the process of dry deposition on plant surfaces
In: W.A.H. Asman and H.S.M.A. Diederen (Eds.), Ammonia and Acidification, pp 141-149. Proceedings of the Symposium of the European Association for the Science of Air Pollution (EUROSAP), Bilthoven 13-15 april 1987

08. Bleuten W., Ivens W.P.M.F. and Draaijers G.P.J., 1987
Throughfall sampling method for research on spatial variability of dry deposition in a forest stand and the effect of forest edges
In: Air Pollution Research Report 6, pp 107-115. Proceedings of a Workshop on Definition of European Pollution Climates and their Perception by Terrestrial Ecosystems, Bern 27-30 april 1987

09. Ivens W.P.M.F., Draaijers G.P.J. and Bleuten W., 1987
Atmospheric nitrogen deposition in a forest next to an intensively used agricultural area
In: P. Mathy (ed.), Air Pollution and Ecosystems, pp 536-541. Proceedings of a Symposium on effects of air pollution on terrestrial and aquatic ecosystems, Grenoble 18-22 mei 1987

10. Draaijers G.P.J., Ivens W.P.M.F. and Bleuten W., 1987

Measurements on the temporal variability of the atmospheric deposition in forests by sampling throughfall on an event basis
In: R. Perry, R.M. Harrison, J.N.B. Bell and J.N. Lester (Eds.), Acid Rain: Scientific and Technical Advances, pp 205-212. Proceedings of the International Acid Rain Conference, Lissabon 1-3 september 1987

11. Draaijers G.P.J. and Bleuten W., 1989
Spatial variability of the atmospheric deposition onto forests on a local and regional scale level
In: L.J. Brasser and W.C. Mulder (Eds.), Man and his Ecosystem, Vol. 2, 195-200. Proceedings of the 8th World Clean Air Congress, Den Haag 11-15 September 1989.

12. Ivens W.P.M.F., Klein Tank A., Kauppi P. and Alcamo J., 1989
Atmospheric deposition of sulphur, nitrogen and base cations onto European forests: observations and model calculations
In: J. Kamari, D.F. Brakke, A. Jenkins, S.A. Norton & R.F. Wright (eds), Regional Acidification models, Geographic Extent and Time development, Springer, Berlin, 103-112

13. Draaijers G.P.J., Meijers R., Bleuten W., Klein Tank A.M.G., 1989
Mogelijke aanpassingen van opstandstructuren ter vermindering van de atmosferische depositie
In: Proceedings van het symposium 'Vitaliteit 1989 - en wat kan de beheerder', Rhenen 22-23 November 1989

14. Draaijers G.P.J., Ek R.van and Meijers, R., 1990
The impact of forest stand structure on atmospheric deposition
Proceedings of the International Conference on Acidic Deposition: its nature and impact, Glasgow 16-21 September 1990

15. Ivens W.P.M.F., Draaijers G.P.J. and Bleuten W., 1986
Ruimtelijke en temporele variabiliteit van de atmosferische depositie in bossen
Vakgroep Fysische Geografie, R.U. Utrecht. AD 1986-1. Publ. reeks Nat. Progr. Zure Regen nr. 37-04 (22p)

16. Draaijers G.P.J., Ivens W.P.M.F. and Bleuten W., 1986
Droge depositie op bomen in relatie tot windrichting en mist
Vakgroep Fysische Geografie, R.U. Utrecht. AD 1986-2. Publ. reeks Nat. Progr. Zure Regen nr. 37-05 (22p)

17. Ivens W.P.M.F., Brinksma M.A., Draaijers G.P.J. and Bleuten W., 1987
Meting van atmosferische depositie in de boswachterij Kootwijk door opvang van doorval en stamafvoer
Vakgroep Fysische Geografie, R.U. Utrecht. AD 1987-1. Publ. reeks Nat. Progr. Zure Regen nr. 37-06 (20p)

18. Brinksma M.A., Ivens W.P.M.F., Draaijers G.P.J. and Bleuten W., 1987
Sequentiële bemonstering van doorvalwater als grondslag voor de berekening van de droge depositie op Douglas
Vakgroep Fysische Geografie, R.U. Utrecht. AD 1987-2. Publ. reeks Nat. Progr. Zure Regen nr. 37-07 (16p)

19. Draaijers G.P.J., Ivens W.P.M.F. and Bleuten W., 1987
Atmosferische depositie in bosranden met name met betrekking tot ammoniak en ammonium
Vakgroep Fysische Geografie, R.U. Utrecht. AD 1987-4. Publ. reeks Nat. Progr. Zure Regen nr. 37-08 (22p)

20. Draaijers G.P.J., Bleuten W., Ivens W.P.M.F., Asman W.A.H., Erisman J.W. and Pinksterboer E.F., 1987
Verspreiding en depositie van ammoniak en ammonium in bosranden. Vakgroep Fysische Geografie, R.U. Utrecht. AD 1987-5. Stichting voor Atmosferische Chemie R 1987-01 (12p)

21. Ivens W.P.M.F., Draaijers G.P.J., Bos M.M. and Bleuten W., 1988
Dutch forests as air pollutant sinks in agricultural areas: a case study in the central parts of the Netherlands on the spatial and temporal variability of atmospheric deposition to forests
Dept. of Physical Geography, State University of Utrecht, AD 1988-01. Dutch Priority Programme on Acidification, report 37-09 (43p)

22. Ivens W.P.M.F., Lovblad G., Westling O. and Kauppi P., 1989
Throughfall monitoring as a means of monitoring deposition to forest ecosystems - evaluation of European data
Dept. Of Physical Geography, University of Utrecht, report no. AD-1989-01. Swedish Environmental Research Institute (IVL), Goteborg, report no L46, 64p.

Project no. **38/39**
Title NH_3 emissions from soils and diffuse sources
Projectleader L.F.Heidema
Address Organisation for Applied Nature Scientific Research TNO Technology for Society
P.O.Box 217, 2600 AE DELFT
Tel. 015-696900

01. Heidema L.F., 1986
NH_3 emissions from soils and diffuse sources (in Dutch)
TNO report R 86/58

02. Hollander J.C.Th., 1989
Emissie van ammoniak uit dierlijke mest: emissie na toediening aan akkerbouwgrond en emissie uit een rundveestal
TNO report R 89/215

Project no. **47a**
Title Master project: air pollution research; phase 1: forest die-back and acidification; (I) modelling and evaluating studies
Projectleader Dr.J.Slanina
Address Energy Research Foundation ECN
P.O.Box 1, 1755 ZG PETTEN
Tel. 02246-4949

01. Janssen A.J., 1988
Wet deposition of air pollutants around coal-fired power plants with desulphurization installations
ECN contract report ECN-88-56 (in Dutch)
02. Janssen A.J., 1988
Some aspects of yearly variations in meteorology and air pollution
ECN report ECN-88-129 (in Dutch)
03. Slanina J., 1988
Report for project 4452: Modeling and evaluation
ECN contract report ECN-88-55 (in Dutch)

Project no. **47b**
Title Master project: air pollution research; phase 1: forest die-back and acidification; (II) emission research
Projectleader Ir.P.T.Alderliesten
Address Energy Research Foundation ECN
P.O.Box 1, 1755 ZG PETTEN
Tel. 02246-4949

01. Slanina J., 1988
Overview of deposition research at ECN for the period 1985 - 1988
ECN conctract report ECN-88-51 (in Dutch)
02. Mallant R.K.A.M., 1988
Overview of ECN research concerning fog and dew for the period 1985 - 1988
ECN contract report ECN-88-58 (in Dutch)

03. Klockow D., Niessner R., Malejczyk M., Kiendl H., Berg B.vom, Keuken M.P., Wayers-Ypelaan A. and Slanina J., 1989
Determination of nitric acid and ammonium nitrate by means of a computer controlled thermodenuder system
Atmospheric Environment 23, 1131 - 1138

04. Keuken M.P., Wayers-Ypelaan A., Mols J.J., Otjes R.P. and Slanina J., 1989
The determination of ammonia in ambient air by an automated thermodenuder system
Atmospheric Environment 23, 2177 - 2185

05. Keuken M.P., Schoonebeek C.A.M., Wensveen A.van and Slanina J., 1988
Simultaneous sampling of NH_3, HNO_3, HCl, SO_2 and H_2O_2 in ambient air by a wet annular denuder system
Atmospheric Environment 22, 2541 - 2548

06. Slanina J., Mols J.J. and Baard J.H., 1990
The influence of outliers on results of wet deposition measurements as a function of measurement strategy
Atmospheric Environment 24A, 1843 - 1860

07. Mallant R.K.A.M. and Arends B.G., 1989
Fog and dew, chemistry and effects on vegetation
In: Proc. of 8th World Clean Air Congress, 1989, Den Haag; ed.: L.J.Brasser, W.C.Mulder, vol. 2, p. 183 - 188

08. Arends B.G. and Eenkhoorn S., 1990
The influence of manganese leached from plants on dew chemistry
Environ.Technol. 11, 181 - 188

Project no. **47c**

Title: Master project: air pollution research; phase 1: forest die-back and acidification; (IIII) cloud, fog and dew research

Projectleader Drs.R.K.A.M.Mallant

Address: Energy Research Foundation ECN
P.O.Box 1, 1755 ZG PETTEN

Tel. 02246-4949

01. Mallant R.K.A.M., Slanina J., Masuch G. and Kettrup A., 1986
Experiments on H_2O_2 containing fog exposures on young trees
In: Proc.of the 2nd US-Dutch symposium: Aerosols, Research, Risk Assessment and Control Strategies, Williamsburg VA, May 1985, eds. Lee, Schneider, Grant, Verkerk, Lewis Publishers Inc. 1986, p. 901 - 980

02. Masuch G., Kettrup A., Mallant R.K.A.M. and Slanina J., 1985

Histological effects of H_2O_2 on the structure of beech leaves and spruce needles
In: Proc. of the International Workshop on Physiology and Biochemistry of Stressed Plants, May 20 - 21, 1985, GSF-Bericht 44/85 pp. 14 - 24

03. Masuch G., Kettrup A., Mallant R.K.A.M. and Slanina J., 1985
Wirkungen von wasserstoffperoxidhaltigem saurem Nebel auf die Laubblaetter junger Buchen (Fachus sylvatica L.)
In: Proc.of the VDI-Kolloqium Waldschaeden, Goslar, June 18 - 20, 1985, VDI-Berichte 560, VDI Verlag Düsseldorf FRG, 1985, pp. 761 - 776

04. Slanina J., Mallant R.K.A.M., Masuch G. and Kettrup A., 1985
The role of clouds, fog and dew in tree die-back
In: Proc.of the Cost Action 61 Abis meeting, Petten, the Netherlands, December 9 - 10, 1985

05. Masuch G., Kettrup A., Mallant R.K.A.M. and Slanina J., 1986
Effects of H_2O_2-containing acidic fog on young trees
Int.Journ.of Environmental Analytical Chemistry, vol. 27 pp. 183 - 213

06. Kettrup A.A., Kicinski H.G., Masuch G., Mallant R.K.A.M. and Slanina J., 1986
Hydrogen peroxide containing fog as air pollutant - causes and effects on spruce needles
In: Proc. of the 7th World Clean Air Congress, Sydney Australia, August 1986, vol. IV, pp. 228 - 253

07. Masuch G., Paul V.H. and Mallant R.K.A.M., 1986
Effects of acidic fog containing H_2O_2 on the sensitivity of agricultural crops to important fungal diseases
paper presented at the third Int.Conf. on Aerobiology, August 6-9, 1986

08. Mallant R.K.A.M., 1986
A fog chamber and wind tunnel facility for calibration of cloud water collectors
paper presented at the NATO advanced research workshop Acid deposition processes at high elevation sites, Edinburgh, September 1986

09. Slanina J. and Mallant R.K.A.M., 1986
The incorporation of pollutants in fog and dew
paper presented at the Kolloqium Aktuelle Aufgaben der Messtechnik in der Luftreinhaltung, Heidelberg, September 1986

10. Mallant R.K.A.M., 1986
The use of laboratory generated fog for testing fog water collectors, cloud chemistry experiments and effect studies
In: Aerosols, Proc. of the second internation aerosol conference, Berlin, September, 1986, pp. 71 - 74, Pergamon Press

11. Mallant R.K.A.M., Kos G.P.A. and Westen A.van, 1986

Chemistry and chemical composition of marine aerosols at the Dutch North Sea coast in relation to aerosol size
In: Aerosols, Proc. of the second international aerosol conference, Berlin, September 1986, pp. 49 - 52, Pergamon Press

12. Mallant R.K.A.M., 1986
Design and characterization of fog water collectors
paper presented at the workshop on Fog Chemistry, Frankfurt, December 1986

13. Mallant R.K.A.M., Slanina J., Masuch G. and Kettrup A., 1988
Effects of H_2O_2-containing acidic fog on young trees
In: Air pollution and ecosystems, ed. P.Mathy, Publ.D.Reidel, p. 306 - 311, 1988

14. Valente R.J., Mallant R.K.A.M., McLaren S.E., Schemenauer R.E. and Stogner R.W., 1989
Field intercomparison of ground-based cloud physics instruments at Whitetop Mountain, Virginia
Journal of Atmospheric and Oceanic Technology, vol. 6, pp. 396 - 406, 1989

15. Mallant R.K.A.M. and Arends B.G., 1989
Fog and dew, chemistry and effects on vegetation
In: Man and his ecosystem, L.J.Brasser and W.C.Mulder (eds.), Elseviers Science Publishers, Proceedings of the 8th World Clean Air Congress, 1989, The Hague, vol. 2, 1989

16. Mallant R.K.A.M. and Kos G.P.A., 1990
An optical device for the detection of clouds and fog
Aerosol Science and Technology, 13: 196 - 202, 1990

<u>Project no.</u> **47d**

<u>Title</u> Master project: air pollution research; phase 1: forest die-back and acidification; (IV) wet and dry deposition

<u>Projectleader</u> Dr.J.Slanina

<u>Address</u> Energy Research Foundation ECN
P.O.Box 1, 1755 ZG PETTEN

<u>Tel.</u> 02246-4949

Project no. **54**

Title Deposition of non -anthropogenic emissions

Projectleader Ir.H.S.M.A.Diederen

Address Organisation for Applied Nature Scientific Research TNO Technology for Society
P.O.Box 217, 2600 AE DELFT

Tel. 015-569330

01. Locht J.V. and Aalst R.M.van, 1988
Depositie van verzurende stoffen op de Nederlandse bodem van niet-antropogene herkomst
MT-TNO rapport R88/458

Project no. **62a**

Title Reduction of NH_3 losses from manure depots (stable, silo)

Projectleader Ing.W.Kroodsma

Address Institute of Agricultural Engineering IMAG
Mansholtlaan 10-12, 6708 PA WAGENINGEN

Tel. 08370-76300

01. IMAG publication
Ammoniakemissie uit rundveestallen
in prep.
02. IMAG publication
Ammoniakemissie uit mestvarkenstallen
in prep.
03. IMAG publication
Ammoniakemissie uit pluimveestallen en uit opslag van droge mest
in prep.

Project no. **62b**

Title Reduction of NH_3 losses when spreading manure

Projectleader Ing.A.Cappon

Address Institute of Agriculture Engineering IMAG
Mansholtlaan 10-12, 6708 PA WAGENINGEN

Tel. 08370-76300

01. Bruins M.A. and Cappon A., 1988
Ammoniakemissie tijdens het uitrijden van mest
IMAG nota 341 (HAB), February 1988

02. Bruins M.A., 1990
De ammoniakemissie tijdens en na het uitrijden van varkens-, runder- en kippemest
IMAG-nota P556, IMAG, Wageningen

03. Klarenbeek J.V. and Bruins M.A., 1988
Ammonia emissions from livestock buildings and slurry spreading in the Netherlands
In: Volatile emissions from livestock farming and sewage operations; eds. Nielsen V.C., Voorburg J.H. and l'Hermite P., Elsevier Applied Science, London-New York; ISBN 1-85166-227-8

04. Klarenbeek J.V., 1989
Ammoniakemissies bij het uitrijden van mest
In: Perspectieven voor de aanpak van de mest- en ammoniakproblematiek op bedrijfsniveau; eds. Jongebreur A.A. and Monteny G.J., Dienst Landbouwkundig Onderzoek, Wageningen

05. Klarenbeek J.V. and Bruins M.A., 1990
Ammonia emissions during and after land spreading of animal slurries
In: Odour and ammonia emissions from livestock productions; eds. Nielsen V.C., Pain B.F. and Hartung J., CEC Brussels
in press

06. Pain B.F. and Klarenbeek J.V., 1988
Anglo-Dutch experiments on odour and ammonia emissions from landspreading livestock wastes
Research report 88-2, IMAG, Wageningen

07. Pain B.F., Phillips V.R., Clarckson C.R. and Klarenbeek J.V., 1989
Loss of nitrogen through ammonia volatilisation during and following the application of pig or cattle to grassland
J.Sci.Food Agric. 47, 1 - 12

08. Phillips V.R., Pain B.F., Clarckson C.R. and Klarenbeek J.V., 1990
Studies on reducing the odour and ammonia emission during and after the land spreading of animal slurries
Farm Buildings & Engineering (7)2, ISSN 0265-5373

09. Locker D.R., Pain B.F. and Klarenbeek J.V., 1989
Ammonia emisisons from cattle, pig and poultry wastes applied to pasture
Environmental Pollution 56, 19 - 30

Project no. **63**

Title Abatement of NH_3 emission

Projectleader Ing.J.V.Klarenbeek

Address Institute of Agriculture Engineering IMAG
Mansholtlaan 10-12, 6708 PA WAGENINGEN

Tel. 08370-94911

01. Demmers T.G.M. and Scholtens R., 1989
Bestrijding van ammoniakemissies met behulp van biowassers
IMAG-nota 422 (HAB), IMAG, Wageningen
02. Demmers T.G.M., 1989
Bestrijding van ammoniakemissies met behulp van luchtwassers
IMAG-nota 423 (HAB), IMAG, Wageningen
03. Demmers T.G.M., 1989
Adsorption und Nitrifikation von Ammoniak im Biowäscher
VDI Berichte 735, pp. 147 - 160, VDI, Düsseldorg
04. Scholtens R., Demmers T.G.M. and Klarenbeek J.V., 1988
Bestrijding van ammoniakemissie
IMAG-nota 360 (HAB), IMAG, Wageningen
05. Scholtens R., Klarenbeek J.V. and Bruins M.A., 1988
Control of ammonia emissions with biofilters and bioscrubbers
In: Volatile emissions from livestock farming and sewage operations; eds. Nielsen V.C., Voorburg J.H. and l'Hermite P., Elsevier Applied Science, London-New York, ISBN 1-85166-227-8

Project no. **64**

Title Research on the possibilities to reduce ammonia emissions from grassland

Projectleader Drs.N.Vertregt

Address Centre for Agricultural Biological Research CABO
Bornsesteeg 65, 6708 PD WAGENINGEN

Tel. 08370-75700

01. Vertregt N. and Rutgers B., 1987
Ammoniak-emissie uit grasland
Dutch Priority Programme on Acidification, Report 64-1, 27 pp., CABO verslag 65
02. Vertregt N. and Rutgers B., 1988
Ammonia volatilization from grazed pastures
Dutch Priority Programme on Acidification, Report 64-2, 37 pp., CABO report 84

03. Vertregt N. and Rutgers B., 1988
Ammonia volatilization from urine patches in grassland
In: Volatile emissions from livestock farming and sewage operations (eds: V.C. Nielsen, J.H.Voorburg, P.L'Hermite), Elsevier Applied Science

04. Vertregt N and Rutgers B., 1987
Ammonia volatilization from urine patches in grassland
In: Animal manure from urine on Grassland and Fodder Crops (eds.: H.G.van der Meer et al.), Martinus Nijhoff Publishers, Dordrecht

Project no. **65**

Title NH_3 emission from the soil after animal manure has been brought in or up

Projectleader Drs.H.G.van Faassen

Address Institute for Soil Fertility IB
P.O.Box 30003, 9750 RA HAREN

Tel. 050-337777

01. Chardon W.J., Molen J.van der and Faassen H.G.van, 1990
Modelling ammonia emissions from arable land
Paper presented at the CEC Workshop on ammonia and odour emissions from livestock production, Silsoe, UK, 26-29 March 1990

02. Chardon W.J., Molen J.van der and Faassen H.G.van, 1990
Modelling ammonia emission from arable land
Poster presented at the 1990 annual meeting of the Amer.Soc.of Agronomy, San Antonio; Agronomy Abstracts 1990: 225

03. Faassen H.G.van, Chardon W.J., Bril J. and Vriesema R., 1990
Ammonia volatilization from arable land after surface application or incorporation of dairy cattle slurry
Final report project 132 (and 65), nota Inst.voor Bodemvruchtbaarheid, nr. 236

04. Molen J.van der, Bussink D.W., Vertregt N., Faassen H.G.van and Boer D.J.den, 1989
Ammonia volatilization from arable and grassland soils
In: K.Henriksen (ed.), Nitrogen in organic wastes applied to soils, Academic Press, London, 1989, p. 185-201

05. Molen J.van der, Faassen H.G.van, Leclerc, M.Y., Vriesema R. and Chardon W.J., 1990
Ammonia volatilization from arable land after application of cattle slurry; 1. Field estimates. Neth.J.Agr.Sci. 38: 145-158

06. Molen J.van der, Beljaars A.C.M., Chardon W.J., Jury W.A. and Faassen H.G.van,

1990
Ammonia volatilization from arable land after application of cattle slurry; 2. Derivation of a transfer model. Neth. J.Agr.Sci. 38: 239-254

Project no. **66**
Title Ammonia in The Netherlands: from emission to deposition
Projectleader Ing.J.H.Duyzer
Address Organisation for Nature Scientific Research TNO Technology for Society
P.O.Box 217, 2600 AE DELFT
Tel. 015-696900

01. Duyzer J.H. et al., 1987
Measurements of dry deposition velocities of NH_3 and NH_4^+ over natural terrains
MT-TNO Delft, report R 87/273
02. Duyzer J.H., Verhagen H.L.M. and Erisman J.W., 1989
De depositie van verzurende stoffen op de Asselsche Heide (The deposition of acidifying compounds on the "Asselsche Heide" (in Dutch)
MT-TNO Delft, report R 89/29
03. Duyzer J.H., Bouman A.M.H., Aalst R.M.van, Diederen H.S.M.A., 1987
MT-TNO publication P 87/34
04. Duyzer J.H. and Diederen H.S.M.A., 1989
Measurements of dry deposition velocities of NH_3 over heathland and forest
submitted to Journal of Geophysical Research
MT-TNO publication P 89/23
05. Duyzer J.H., 1990
Dry deposition of NH_3 over forest
In: Proceedings Conference on acidic deposition, its nature and impacts, Glasgow 16 - 21 September 1990
to be published

Project no. **67**

Title Measuring nitrogen losses at fertilization or manuring due to evaporation of NH_3

Projectleader Ir.D.J.den Boer

Address Dutch Fertilizer Institute/Research and Advisory Institute for Cattle Husbandry
Runderweg 6, 8219 PK LELYSTAD

Tel. 03200-22514

Reports

01. A84-024I, Meten van stikstofverliezen op grasland als gevolg van NH_3-vervluchtiging, 1986, Den Haag
02. A84.024.II, Bepaling van de NH_3-emissie bij beweiding in het kader van het N-stromenonderzoek, 1987, NMI, Den Haag
03. A.84.024-III, Bepaling van de NH_3-emissie bij beweiding in het kader van het N-stromenonderzoek in mei, juni, juli 1986, 1987, NMI, Den Haag
04. A87.024, NH_3-vervluchtiging bij beweiding (N-stromenonderzoek 1987), 1988, NMI, Den Haag
05. A88.024, Ammonia volatilization from a rotationally grazed sward, 1989, pp. 1-11, NMI, Den Haag
06. A88.024II, Resultaten van NH_3-emissiemetingen bij beweiding van 1986 tot en met 1988, NMI, Den Haag

Publications

07. Boer D.J.den, 1988
Ammoniakvervluchtiging bij beweiding
Jaarverslag PR 1987, pp. 47-50
08. Bussink D.J., 1989
Ammoniakvervluchtiging bij beweiding lager dan verwacht
Praktijkonderzoek, 2e jaargang nr. 5, pp. 2-4
09. Molen J.van der, Bussink D.W., Vertregt N., Faassen H.G.van and Boer D.J.den, 1989
Ammonia emission from arable and grassland soils
In: Proceedings of ISWA/DAKOFA/AUC specialized seminar Nitrogen in organic wastes applied to soils, Aalborg: Ammonia emission from arable and grassland soils, pp. 185-201
10. Bussink D.W., 1990
Ammonia volatilization from a rotationally grazed sward
In: R.Merckx, H.Vereecken en K.Vlassak, eds.; Proceedings Fertilization and the

environment, Leuven, pp. 305-313, Leuven, University Press

11. Jarvis S.C. and Bussink D.W., 1990
Ammoniakemissionen bei Weidehaltung
In: Ammoniak in der Umwelt, VDI/KTBL Symposiumsband, S.26.1-26.9, Braunschweig, KTBL Verlag in Münster-Hiltrup

Project no. **68**

Title Split applications of slurry on silage maize

Projectleader Ir.J.Schröder

Address Research Station for Arable Farming and Field Production of Vegetables (PAGV)
P.O.Box 340, 8200 AK LELYSTAD

Tel. 03200-22714

01. Schröder J., 1990
Stikstofverliezen bij de teelt van mais
Meststoffen 1/2, p. 25 - 32
NMI, Den Haag

02. Schröder J. and Lande Crember L.C.N.de la
Toedienen van drijfmest in mais (vervolgonderzoek 1985 - 1987)
PAGV-verslag nr. 85, 52 pp. + 43 bijlagen
PAGV, Lelystad

Project no. **69**

Title Effectiveness of measures to prevent and reduce the effects of acid rain

Projectleader Ir.A.H.M.Bresser

Address National Institute of Public Health and Environmental Protection (RIVM)
P.O.Box 1, 3720 BA BILTHOVEN

Tel. 030-742970

01. Bresser A.H.M., 1986
Effectiveness of measures to prevent and reduce the effects of acid rain (system study acidification
Summary report of phase 1 and 2

<u>Project no.</u> **70a**

<u>Title</u> Supplemental measurements

<u>Projectleader</u> Dr.J.Slanina

<u>Address</u> Energy Research Foundation (ECN)
P.O.Box 1, 1755 ZG PETTEN

<u>Tel.</u> 02246-4236

<u>Project no.</u> **70b**

<u>Title</u> Supplemental measurements

<u>Projectleader</u> Th.R.Thijsse

<u>Address</u> Organisation for Applied Nature Scientific Research TNO, Division of Technology for Society
P.O.Box 217, 2600 AE DELFT

<u>Tel.</u> 015-696900

01. Th.R.Thijsse, 1988
Supplemental hydrocarbon measurements at the location Kootwijk from 20 January till 19 February 1988
TNO report R 88/165

<u>Project no.</u> **71a**

<u>Title</u> Validation of methods and quality control for the Dutch Priority Programme; outdoor air quality measurements

<u>Projectleader</u> Ir.H.S.M.A.Diederen

<u>Address</u> Organisation for Applied Nature Scientific Research TNO Technology for Society
P.O.Box 217, 2600 AE DELFT

<u>Tel.</u> 015-569330

01. Diederen H.S.M.A., 1988
Vergelijkbaarheid van resultaten van analysemethoden toegepast voor onderzoek van natte depositie in het verzuringsonderzoek (Validation of methods and quality control for the Dutch Priority Programme on Acidification, air quality and deposition measurements (in Dutch)
TNO report R 88/439

Project no. **71b**

Title Quality assurance; soil measurements

Projectleader Dr.J.B.Somers

Address Lake IJssel Development Authority (RIJP)
P.O.Box 600, 8200 AP LELYSTAD

Tel. 03200-99111

Project no. **72**

Title The influence of a crop on withdrawal of pollutants from the atmosphere, especially ozone, in the course of the growing season

Projectleader Dr.Ir.M.Janssen-Jurkovicova

Address KEMA
P.O.box 9035, 6800 ET ARNHEM

Tel. 085-569111

01. Beld L.van de, Bruin W.J.de, Duuren H.van and Pul W.A.J.van, 1987
Concentratiemetingen van O_3 en CO_2 in en boven een gewasveld
KEMA-rapport 51514-MOL, 87-3111

02. Bruin W.J.de, Jaspers J.E., 1987
De ontwikkeling van een ozonmonitor voor eddy-correlatie metingen
KEMA-rapport 51514-MOL, 87-3157

03. Stolk A.P., 1989
Onderzoek naar het transport van ozon naar een maisgewas
stageverslag Vakgroep Luchthygiëne en -verontreiniging, LUW
KEMA-rapport 00609-MOZ 89-1034

04. Jong M.A.de, 1989
Analyse van de gemeten stomataire weerstanden van snijmais op het proefveld Sinderhoeve gedurende het seizoen 1987
KEMA-rapport 00606-MOB 89-3248

05. Rietbergen J.M., Römer F.G., Pul W.A.J.van and Stolk A.P., 1989
Bepaling van de depositie van verzurende stoffen in een maisveld door middel van doorvalmetingen; meetopzet en meetresultaten
KEMA-rapport 51514-MOC 89-3268

06. Jong M.A.de and Schoofs M.J., 1989
Analyse van de gemeten stomataire weerstanden van snijmais op proefveld Sinderhoeve gedurende het seizoen 1988
KEMA-rapport 51514-MOB 89-3323

07. Besteboer S.I., Jong M.A.de, Stortelder B.J.M., and Westen B.J., 1989

Evaluatie van de porometriemethode toegepast op mais
KEMA-rapport 82106-MOZ 89-3359

08. Stolk A.P., Pul W.A.J.van and Römer F.G., 1990
Doorvalwater onderzoek in een gewasveld
KEMA-rapport 51514 90-3406

09. J.Beck, 1988
Depositie van ozon op kale grond
Doctoraal verslag, Vakgroep Natuur en Weerkunde, LUW

10. Vermeulen A.T., 1989
Het meten van ozon-fluxen boven mais
Doctoraal verslag, Vakgroep Meteorologie, LUW

11. Kosters M., 1989
Eddy correlatiemetingen van ozonfluxen boven en in een maisgewas
Doctoraal verslag, Vakgroep Meteorologie, LUW

12. Pul W.A.J., Hakvoort H. and Jacobs A.F.G., 1991
Techniques used in the determination of the flux of ozone towards a maize crop
KEMA scientific and technical reports

13. Pul W.A.J.van and Jacobs A.F.G.
The conductance of a maize crop for ozone and water vapour under various environmental conditions; a comparison
Agricultural and Forest Meteorology
submitted

14. Aben J.M.M., 1989
Estimation of plant-related resistances determining the ozone flux to broad bean plants (Vicia faba far Metissa), after long-term exposure to ozone in an open top chamber
In: Air pollution and ecosystems (P.Mathy ed.), Proceedings of an international congress held in Grenoble, France, 18 - 22 May 1987, pp. 616 - 619, D.Reidel Publising Company, Dordrecht, The Netherlands

15. Aben J.M.M. and Janssen-Jurkovicova, 1989
Changes in photosynthesis and stomatal conductance of Vicia faba after short-term fumigation with ozone
In: Man and his ecocystem (L.J.Brasser and W.C.Mulder eds.), volume 2, Proceedings of the 8th World Clean Air Congress 1989, pp. 61 - 66, Elsevier, Amsterdam

16. Aben J.M.M., 1990
A system to determine whole-plant exchange rates of ozone, carbon dioxide and water vapour
KEMA Scientific and Technical Reports, 8

17. Aben J.M.M., Janssen-Jurkovicova M. and Adema E.H., 1990
Effects of low-level ozone exposure under ambient conditions on photosynthesis and stomatal control of Vicia faba L.
Plant, Cell and Environment, 13, 463 - 469

Project no. **73**
Title Analysis of the influence of the weather, air pollution and other external factors on tree growth
Projectleader Ir.H.Visser
Address KEMA
P.O.Box 9035, 6800 ET ARNHEM
Tel. 085-569111

01. Molenaar J. and Visser H., 1987
The Kalman filter in dendro-climatology
In: Proceedings ICIAM 87, Paris-La Villette (A.H.P. van der Burgh en R.M.M. Mattheij, eds.), pp. 203-214

02. Molenaar J. and Visser H., 1989
Fundamental aspects of the Kalman filter with examples regarding load forecasting and acid rain
KEMA Scientific & Technical Reports, vol. 7, no. 1, pp. 47-61

03. Vaessen R.J., 1988
Meteorological data for dendroclimatological applications in the Netherlands
KEMA-Dorschkamp rapport 50385-MOF 87-3165 NPZR 73-5

04. Visser H. 1986
Analysis of tree-ring data using the Kalman filter
IAWA Bulletin n.s., vol. 7(4), pp. 289-297

05. Visser H., 1989
Fir dying in the Bavarian forest and the role of SO_2 emissions: a dendroecological search for cause and effect relations
In: Proceedings International Congress on Forest Decline Research, Friedrichshafen, oktober 1989, vol. 1, pp. 31-32

06. Visser H. and Maessen P.P.Th.M., 1989
Responses of trees to weather variations and air pollution: tree-ring research in the Netherlands
In: Proceedings of the 8th World Clean Air Congress, September 1989 (L.J. Brasser and W.C. Mulder, eds.), vol. 2, pp. 287-292

07. Visser H. and Maessen P.P.Th.M., 1990

Responses of trees to weather variations and air pollution: a tree-ring based approach
KEMA-Dorschkamp rapport 50385-MOF 90-3394 NPZR 73-7

08. Visser H. and Molenaar J., 1986
Time-dependent responses of trees to weather variations: an application of the Kalman filter
KEMA-Dorschkamp rapport 50385-MOA 86-3041 NPZR 73-1

09. Visser H. and Molenaar J., 1987
Time-dependent responses of trees to weather variations: an application of the Kalman filter
In: Proceedings of the International Symposium on Ecological Aspects of Tree-Ring Analysis (G.C. Jacoby en J.W. Hornbeck, eds.), pp. 579-590

10. Visser H. and Molenaar J., 1988
Kalman filter analysis in dendroclimatology
Biometrics 44, pp. 929-940

11. Visser H. and Molenaar J., 1989
Detecting time-dependent climatic responses in tree rings using the Kalman filter
In: Methods of Dendrochronology (E.R. Cook en L.A. Kairiukstis, eds.), pp. 270-277

12. Visser H. and Molenaar J., 1990
Estimating trends in tree-ring data
Forest Science, vol. 36, no. 1, pp. 87-100

13. Visser H. and Molenaar J., 1991
Fir decline and the role of SO_2 emissions: a dendroecological search for cause and effect
Forest Science (accepted)

14. Visser H., Noppert F., Wakeren J.H.A. van, and Vaessen, R.J., 1989
Xylem sap velocity in relation to weather and air pollution
IAWA Bulletin n.s., vol. 10 (4), pp. 427-439

Project no. **74**

Title: Research on the chemical contents and the deposition of fog, dew (and rime) also on behalf of studies on exposure-effect relations

Projectleader Dr.F.G.Römer

Address: KEMA
P.O.box 9035, 6800 ET ARNHEM

Tel. 085-569111

01. Haas J.de, 1984
Ontwerp van een verschiltemperatuurmeter om de "stabiliteit" van de grenslaag te bepalen
KEMA stageverslag nr. MO-L 84-309

02. Janssen L.H.J.M. and Wakeren J.H.A. van, 1985
Deposities van dauw; voorstel voor een onderzoekstrategie
KEMA memorandum nr. VII 85-73 MOL

03. Steenkist R., 1985
Dauwvorming en zure regen (een eerste analyse)
KEMA report nr. III 9134-85 MOL

04. Römer F.G., Winkel B.H.te, Janssen L.H.J.M., Wakeren J.H.A. and Steenkist R., 1985
Additioneel Programma Verzuringsonderzoek, project 74: Onderzoek naar de chemische samenstelling en de afzetting van mist, dauw (en rijp); voortgangsverslag 1, KEMA notitite nr. 00574-MOL 85-1065

05. Galesloot L., 1985
Het thermisch gedrag van een gewasdeel en de bepaling van warmte-overdrachtscoëfficienten
KEMA stageverslag nr. MO-L-0574 85-297

06. Barendse L., 1985
Ontwikkeling van een temperatuurverschilmeter
KEMA memorandum nr. MO-L-0574 85-107

07. Piepers W., 1986
Onderzoek aan een heatfluxmeter; Uitbreiding: vochtmetingen
KEMA stageverslag

08. Steenkist R., 1986
Karakterisering van de meteorologische toestand in de atmosferische grenslaag
KEMA rapport nr. 50583-MOL 86-3105

09. Vlug N.H.G., 1986
Systeemontwerp van een data-acquisitiesysteem ten behoeve van de meetopstelling voor metingen aan dauw
KEMA rapport nr. 50583-APB 86-387

10. Wakeren J.H.A.van and Janssen L.H.J.M., 1986
Temperatuurvariaties van een kunstmatig oppervlak in de buitenlucht; een gevoeligheidsanalyse ten behoeve van metingen aan dauw
KEMA rapport nr. 50583-MOL 86-3060

11. Winkel B.H.te, 1986
Literatuuronderzoek (1) naar het reactieverloop van nitriet- en sulfietionen in

atmosferisch water
KEMA rapport nr. 50583-MOL 86-3036

12. Römer F.G., Winkel B.H.te, Janssen L.H.J.M., Wakeren J.H.A.van and Steenkist R., 1986
Additioneel Programma Verzuringsonderzoek, project 74: Onderzoek naar chemische samenstelling en de afzetting van mist, dauw (en rijp), voortgangsverslag 2
KEMA notitie nr. 00574-MOL 85-1639

13. Römer F.G., Winkel B.H.te, Janssen L.H.J.M., Wakeren J.H.A.van and Steenkist R., 1986
Additioneel Programma Verzuringsonderzoek, project 74: Onderzoek naar chemische samenstelling en de afzetting van mist, dauw (en rijp), voortgangsverslag 3
KEMA notitie nr. 00574-MOL 86-1034

14. Römer F.G. and Winkel B.H.te, 1987
Onderzoek naar de chemische samenstelling van dauw en mist in de periode 1985-1987
KEMA rapport nr. 50583-MOL 87-3170

15. Steenkist R., 1987
Bladtemperaturen en dauwvorming in een gewaslaag tijdens dauwvormingsomstandigheden
KEMA notitite nr. 00574-MOL 87-1392

16. Winkel B.H.te and Wakeren J.H.A.van, 1987
Beknopte beschrijving van metingen aan weeginrichting van de dauwplatenopstelling
KEMA notitie nr. 50583-MOC 87-2009

17. Römer F.G., Winkel B.H.te, Janssen L.H.J.M., Wakeren J.H.A.van and Steenkist R., 1987
Additioneel Programma Verzuringsonderzoek, project 74: Onderzoek naar chemische samenstelling en de afzetting van mist, dauw (en rijp), voortgangsverslag 4
KEMA notitie nr. 00574-MOL 87-1494

18. Römer F.G., Winkel B.H.te, Janssen L.H.J.M., Wakeren J.H.A.van and Steenkist R., 1987
Additioneel Programma Verzuringsonderzoek, project 74: Onderzoek naar chemische samenstelling en de afzetting van mist, dauw (en rijp), voortgangsverslag 5
KEMA notitie nr. 00574-MOL 87-2015

19. Römer F.G., Slangewal H.J. and Rietbergen J.M., 1987
De chemische samenstelling van regen te Arnhem, periode 1984-1986
KEMA rapport nr. 98371-MOZ 87-3146

20. Wakeren J.H.A.van, Janssen L.H.J.M. and Arjona F., 1988
IJking dauwtafel

KEMA rapport nr. 50583-MOF 88-3198

21. Steenkist R., 1988
Bladtemperatuur en dauwvorming in een gewaslaag tijdens dauwvormingsomstandigheden II
KEMA rapport nr. 50583-MOF 88-3184

22. Steenkist R. and Jacobs A., 1988
Test sonische anemometer
KEMA notitite nr. 50583-MOF 88-1267

23. Vaessen R.J., 1988
Modelberekeningen aan dauw gebruikmakend van synoptische uurlijkse waarnemingen vergeleken met metingen met een dauwautomaat over de periode 1985-1987
KEMA rapport nr. 50583-MOF 88-3221

23. Römer F.G., Janssen L.H.J.M., Wakeren J.H.A.van, Winkel B.H.te, Steenkist R. and Vaessen R.J., 1988
De depositie van waterdamp en de chemische samenstelling van dauw
Eindrapportage over de periode 1985-1988, KEMA rapport nr. 50583-MOC 88-3222

Project no. **77**

Title The influence of SO_2, NO_x, NH_3 and O_3 on photosynthesis

Projectleader Prof.Dr.W.J.Vredenberg

Address Agricultural University, Department of Plant Physiological Research
Gen.Foulkesweg 72, 6703 BW WAGENINGEN

Tel. 08370-82800

Publications

01. Kooten O.van and Hove L.W.A.van, 1988
Fluorescence as a means of diagnosing the effect of pollutant induced stress in plants
In: Air Pollution and Ecosystems (P.Mathy, ed.), D.Reidel Co., pp. 596 - 601

02. Kooten O.van, Hove L.W.A.van and Wijk K.J.van, 1988
The effect of long term exposition of poplars to low concentrations of SO_2 and NH_3
In: Applications of Chlorophyll Fluorescence (H.Lichtenthaler ed.), Kluwer Acad.Publ., Dordrecht, pp. 203 - 209

03. Rosema A., Gecchi G., Pantani L., Radicatti B., Romuli M., Mazzinghi P., Kooten O.van and Kliffen C., 1988
Air pollution effects on the fluorescence of Douglas fir and poplar
In: Applications of Chlorophyll Fluorescence (H.K.Lichtenthaler ed.), Kluwer Acad.Publ., Dordrecht pp. 307 - 317

04. Schapendonk A.H.C.M., Dolstra O. and Kooten O.van, 1988
The use of chlorophyll fluorescence as a screening method for cold tolerance in maize
Photosynth.Res. 20, 235 - 247

Abstracts

05. Kooten O.van, Lahey G. and Hove L.W.A.van, 1987
Fluorescence as an indicator of air pollutant effects on plants
Acta Bot.Neerl.suppl. 1, 36-2, 61

06. Kooten O.van and Schapendonk A.H.C.M., 1987
Fluorescence as a stress indicator in plant breeding
Acta Bot.Neerl. suppl. 1, 36-2, 62

07. Snel J.F.H., Kooten O.van and Vredenberg W.J., 1987
Characterization of chlorophyll a fluorescence response upon application of modulated illumination
Plant Physiol. 83s, 161

Project no. **78**

Title Uptake of gaseous air pollutants by leaves

Projectleader Prof.Dr.E.H.Adema

Address Agricultural University, Department of Air Pollution
P.O.Box 8129, 6700 EV WAGENINGEN

Tel. 08370-82684

01. Hove L.W.A.van, Koops A.J., Adema E.H., Vredenberg W.J. and Pieters G.A., 1987
Analysis of the uptake of atmospheric ammonia by leaves of Phaseolus vulgaris L.
Atmospheric Environment 21 (8): 1759 - 1763

02. Hove L.W.A.van, Adema E.H. and Vredenberg W.J., 1987
The uptake of atmospheric ammonia by leaves
Acta Botanica Neerlandica 36 (2), suppl. 1: 60

03. Hove L.W.A., 1987
De opname van atmosferisch ammoniak door bladeren
In: Effecten van NH_3 op organismen (eds. A.W.Boxman en J.F.M.Geelen): 35 - 44

04. Hove L.W.A.van, Tonk W.J.M., Pieters G.A., Adema E.H. and Vredenberg W.J., 1988
A leaf chamber for measuring the uptake of pollutant gases at low concentrations by leaves, transpiration and carbon dioxide assimilation
Atmospheric Environment 22 (11): 2515 - 2523

05. Hove L.W.A.van and Adema E.H., 1988

The uptake of atmospheric ammonia by leaves
In: Air pollution and ecosystems (ed. P.Mathy): 734 - 738, Reidel, Dordrecht

06. Hove L.W.A.van and Adema E.H., 1988
The uptake of NH_3 and SO_2 by leaves
Report project 78, Dutch Priority Programme on Acidification, 24 pp.

07. Booij C.G., Hove L.W.A.van and Adema E.H., 1987
Expositie-response-relatie modellen en onderzoek naar de opname van ammoniak door bladeren
Eindrapportage ERR-project (R-265), Vakgroep Luchthygiëne en -verontreiniging, LUW

08. Kooten O.van, Hove L.W.A.van, 1988
Fluorescence as a means of diagnosing the effect of pollutant induced stress in plants
In: Air pollution and ecosystems (ed. P.Mathy): 596 - 601, Reidel Dordrecht

09. Kooten O.van, Hove B.van and Wijk K.J.van, 1988
The effect of long term exposition of poplars to low concentrations of SO_2 and NH_3
In: Applications of Chlorophyll Fluorescence (ed. H.K.Lichtenthaler): 203 - 209, Kluwer, Dordrecht

see also projects 109 and 110

Project no. **79**
Title Acid rain: a biophysical analysis of some soil chemical factors
Projectleader Dr.H.B.A.Prins
Address University of Groningen, Department of Plant Physiology
P.O.Box 14, 9750 AA HAREN
Tel. 050-639111

see project 108

Project no. **80a**
Title Wood structure and quality in relation to declining tree vitality
Projectleader Dr.P.Baas
Address Rijksherbarium
Schelpenkade 6, P.O.Box 9514, 2300 RA LEIDEN
Tel. 071-130541

01. Baas P. and Bauch J., 1986
The effects of environmental pollution on wood structure and quality

Special issue IAWA Bull. n.s. 7 (4)

02. Baas P. and Kort I.de, 1986
Bossterfte en houtkwaliteit
Houtkwaliteit 39: 8-11

03. Kort I.de and Baas P., 1986
Diktegroei en houtstructuur van vitale en niet-vitale Douglas in de Peel
Nederlands Bosbouwtijdschrift 58: 52-57

04. Kort I.de and Baas P., 1986
Growth ring width and wood structure of vital and non-vital Douglas fir in the Netherlands
In: Proceedings of the 18th IUFRO World Congress 1986, Division 5: 539
(abstract poster)

05. Kort I.de, 1986
Wood structure and growth ring of vital and non-vital Douglas fir (Pseudotsuga menziesii) from a single stand in the Netherlands
IAWA Bull. n.s. 7: 309-318

06. Ancker J.A.M.van der, Kort I.de and Oudenaarden J.van, 1987
Vitaliteitsbeoordeling van Douglas aan de hand van luchtfotografie, kroonaanzicht en jaarringanalyse
Nederlands Bosbouwtijdschrift 59: 309-317

07. Kort I.de, 1989
Houtkwaliteit (nog) niet in gevaar door verzuring
Houtwereld 42: 21-23

08. Kort I.de, 1989
Growth ring patterns and wood anatomy of Douglas fir (Pseudotsuga menziesii (Mirb. Franco) and pendunculate oak (Qeurcus robur L.) in relation to vitality
Acta Bot.Neerl. 38: 97 (abstract paper)

09. Kort I.de, 1990
Wood production, stem growth and water transport capacity of Douglas fir in the ACIFORN stands in Kootwijk and Garderen
Report Rijksherbarium/Hortus Botanicus, Leiden, 20 p., 20 figs.

10. Kort I.de, 1990
Tracheid length in vital and non-vital Douglas fir (Pseudotsuga menziesii) in the Netherlands
IAWA Bull. n.s. 11: 203-209

11. Kort I.de, 1990
Radial increment and hydraulic fitness of Douglas fir
IAWA-IUFRO Wood Anatomy Symposium 1990, IAWA Bull. n.s. 11:128

(abstract paper)

12. Kort I.de, 1990/1991
Growth, wood anatomy and water transport capacity in Douglas fir (Pseudotsuga menziesii (Mirb.) Franco) in Kootwijk and Garderen
Acta Bot.Neerl. (in press, abstract paper)

Project no. **83**

Title The pH in the rhizosphere as a link between soil pH and the functioning of the root-system

Projectleader Dr.M.van Noordwijk

Address Institute for Soil Fertility, IB
P.O.Box 30003, 9750 RA HAREN

Tel. 050-337777

01. Gijsman A.J., 1989
PHGRAD, a simulation model for the description of pH gradients in the rhizosphere (in Dutch with English summary)
Dutch Priority Programme on Acidification, report 83-02
02. Gijsman A.J., 1990
Rhizosphere pH along different root zones of Douglas fir (Pseudotsuga menziesii), as affected by source of nitrogen
In: Van Beusichem M.L. (ed.) Plant Nutrition - Physiology and Applications, Kluwer Academic Publishers, Dordrecht, The Netherlands, pp. 45-51
Also: Plant and Soil 124: 161-167
03. Gijsman A.J., 1990
Nitrogen nutrition of Douglas fir (Pseudotsuga menziesii) on strongly acid sandy soil
I. Growth, nutrient uptake and ionic balance
Plant and Soil 126: 53-61
04. Gijsman A.J., 1990
Nitrogen nutrition of Douglas fir (Pseudotsuga menziesii) on strongly acid sandy soil
II. Proton excretion and rhizosphere pH
Plant and Soil 126: 63-70
05. Gijsman A.J., 1990
The effect of high ammonium input to the soil on ionic uptake balance and rhizosphere pH of Douglas fir
Final report of project 83 of the Dutch Priority Programme on Acidification; Institute for Soil Fertility Research (Haren), The Netherlands
06. Gijsman A.J. and Noordwijk M.van, 1990

Critical ammonium: nitrate uptake ratios for Douglas fir determining rhizosphere pH and tree mortality
Presented at the Second Int.Symp. on Plant-Soil Interactions at low pH, Beckley, USA (24-29 June 1990)

07. Gijsman A.J., 1990
Nitrogen nutrition and rhizosphere pH of Douglas fir
Ph.D.Thesis, Univ.of Groningen (The Netherlands)

08. Gijsman A.J.
Soil water content as a key factor determining the source of nitrogen (NH_4^+ or NO_3^-) absorbed by Douglas fir (Pseudotsuga menziesii) and the pattern of rhizosphere pH along its roots
submitted to Can.J.For.Res.

09. Gijsman A.J. and Willigen P.de, 1990
Modeling ammonium and nitrate uptake by a mature Douglas fir stand from a soil with high atmospheric NH_x input
Neth.J.Agric.Sci. (in press)

10. Gijsman A.J., Noordwijk M.van, Floris J. and Brouwer G., 1991
An inflatable minirhizotron system for root observations with improved soil/tube contact
submitted to Plant and Soil

Project no. **84**

Title Root development of trees in environments affected by air pollution

Projectleader Prof.Dr.Ir.R.A.A.Oldeman

Address Agricultural University, Department of Silviculture and Forest Ecology
P.O.Box 342, 6700 AA WAGENINGEN

Tel. 08370-84426

01. Olsthoorn, A.F.M. 1988
Monitoring of root growth
In: P. Mathy (ed.): Air pollution and ecosystems. Proceedings International Symposium Grenoble, France, May 1987. Commission of the European Communities, EUR 11244. Reidel Publishing Company, Dordrecht, Boston: 888-890

02. Olsthoorn, A.F.M. 1988
Root research on Douglas-fir with special attention to effects of acidification.
In: A.E. Jansen, Dighton, J. and Bresser, A.H.M. (eds.): Ectomycorrhiza and acid rain, Proceedings of the Workshop on Ectomycorrhiza/Expert meeting, December

10-11 1987, Berg en Dal, the Netherlands. Air Pollution Report no 12, Commission of the European Communities, Brussels, Eur 11543, p77-78

03. Olsthoorn, A.F.M. 1988
Wortelonderzoek aan bomen als onderdeel van het Nederlandse verzuringsonderzoek in Douglasbossen. Nederlands Bosbouwtijdschrift 60 (7): 222-224.

04. Olsthoorn, A.F.M.
Root research on Douglas-fir in the ACIFORN project: Influence of soil acidification on fine root growth. Report 103.01 Dutch Priority Programme on Acidification, RIVM, Bilthoven, The Netherlands, 1991
(in press)

05. Olsthoorn, A.F.M.
Fine root density and root biomass of two Douglas-fir stands on sandy soils in the Netherlands. I. Root biomass in early summer
Neth. J. Agric. Sc. 1991.
(in press)

06. Olsthoorn A.F.M., Baren B.van and Bosch, A.L. 1986
The perforated soil system or "Perforon", a versatile root observation method. Proceedings 18th IUFRO World Congress, Sept. 1986, Vol 2, II, page 849, poster 254

07. Olsthoorn A.F.M., Baren B.van and Hopman M., 1988
Ammonium influence on root growth and rhizosphere pH of Douglas-fir seedlings
Abstracts, Symposium International Society of Root Research "Plant roots and their environment", Uppsala, August 1988: part 2, poster 23

08. Olsthoorn A.F.M. and Florax J.P.G.G.M.
Reduction of forest transpiration by decreasing root densities as a result of soil acidification: model calculations. Proceedings E.C. Workshop "Above- and belowground interactions in forest trees in acidified soils", Simlångsdalen, Sweden, May 1990

09. Olsthoorn A.F.M., Keltjens W.G., Baren B.van and Hopman M.C.G.
Influence of ammonium on fine root development and rhizosphere pH of Douglas-fir seedlings in sand
Plant and Soil 1991

10. Olsthoorn A.F.M. and Tiktak A.
Fine root density and root biomass of two Douglas-fir stands on sandy soils in the Netherlands. II. Periodicity of fine root growth and estimation of belowground carbon allocation
Neth. J. Agric. Sc. 1991.

11. Tiktak A., Bouten W., Jans W.W.P. and Olsthoorn A.F.M.

Temporal dynamics of shoot extension and fine root activity in a Doulas fir forest in the Veluwe, The Netherlands. Draft Report Dutch Priority Programme on Acidification 203.9

12. Van der Maas M.P.van der, Belde J.J.M., Klap J. and Olsthoorn A.F.M. 1989
Dynamics of the potassium status of two Douglas-fir stands
Poster Abstract. International Congress on Forest Decline Reseach: State of Knowledge and Perspectives, Friedrichshafen, Germany, October 1989, Vol. 1: 484. Poster No 227

Project no. **86**
Title The influence of acid deposition on the microflora in the rhizosphere
Projectleader Dr.J.A.van Veen
Address Research Institute ITAL
P.O.Box 48, 6700 AA WAGENINGEN
Tel. 08370-91911

01. Gorissen A. and Veen J.A.van, 1988
Temporary disturbance of translocation of assimilates in Douglas firs caused by low levels of ozone and sulphur dioxide
Plant Physiol. 88: 559-563

02. Gorissen A., Schelling G.S. and Veen J.A.van, 1991
Concentration dependent effects of ozone on translocation of assimilates in Douglas fir
J.Environ.Q. (submitted)

03. Gorissen A., Joosten N.N., Smeulders S.M. and Veen J.A.van
Effects of ozone on carbon economy of Douglas fir; Part I. Juvenile Douglas firs grown at two soil moisture levels
J.Environ.Q. (submitted)

04. Smeulders S.M., Gorissen A., Joosten N.N. and Veen J.A.van
Effects of ozone on carbon economy of Douglas fir; Part II. Full-grown Douglas firs at field conditions
J.Environ.Q. (submitted)

05. Gorissen A., Jansen A.E. and Olsthoorn A.F.M.
Influence of two year application of ammonium sulphate on growth of juvenile Douglas fir and the rhizosphere microflora
Can.J.Bot. (submitted)

06. Gorissen A., Joosten N.N. and A.E.Jansen
Effects of ozone and ammonium sulphate on carbon partitioning to mycorrhizal roots

of juvenile Douglas firs
New Phytol. (submitted)

Project no. **87**
Title Biological aspects of the S flow in the soil in relation to the enhanced deposition of S combinations, especially SO_2
Projectleader Dr.Ir.P.Doelman
Address Research Institute for Nature Management
P.O.Box 9201, 6800 HB ARNHEM
Tel. 085-452991

Project no. **90a**
Title Effects of SO_x/NH_x on natural vegetation
Projectleader Drs.H.F.van Dobben
Address Research Institute for Nature Management
P.O.Box 46, 3956 ZR LEERSUM
Tel. 03434-52941

01. Berdowski J.J.M., Heerden C.van and Minnen J.G.van, 1991
SOILVEG part A: model structure, vegetation processes and preliminary results
Report RIN, Arnhem
in press
02. Conijn J.G. and Berendse F., 1991
De simulatie van de concurrentie tussen Calluna en Molinia in droge heidevelden
Report RIN, Arnhem
in prep.
03. Fennema F., 1990
Effects of exposure to atmospheric SO_2, NH_3 and $(NH_4)_2SO_4$ on survival and extinction of Arnica montana L. and Viola canina L.
Report RIN 90/14, Arnhem 61 pp.
04. Eerden L.J.van der, Dueck Th.A., Elderson J., Dobben H.F.van, Berdowski J.J.M. and Latuhihin M., 1989
Effects of SO_2, NH_3 and $(NH_4)_2SO_4$ deposition on terrestrial semi-natural vegetation on nutrient-poor sandy soils
Report IPO/RIN, Wageningen/Arnhem, 169 pp.
05. Eerden L.J.van der, Dueck Th.A., Elderson J., Dobben H.F.van, Berdowski J.J.M. and Latuhinin M., 1990
Effects of NH_3 and $(NH_4)_2SO_4$ deposition on terrestrial semi-natural vegetation on

nutrient-poor soils
Report IPO/RIN, Wageningen/Arnhem
in press

06. Kootwijk E.J.van and Voet H.van der, 1989
De kartering van heidevergrassing in Nederland met de Landsat Thematic Mapper satellietbeelden
Report RIN 89/2, Arnhem

07. Kootwijk E.J.van, 1989
Inventarisatie van de vergrassing van de Nederlandse heide
Report RIN 89/1, Arnhem, 49 pp.

Project no. **90b** (see 90a)
Title Effects of SO_x/NH_x on natural vegetation
Projectleader Ir.L.J.van der Eerden
Address Research Institute for Plant Protection
P.O.Box 9060, 6700 GW WAGENINGEN
Tel. 08370-19151

Project no. **90c**
Title The dry deposition of sulphur and nitrogen compounds on the "Asselsche Heide" heathland
Projectleader J.H.Duyzer
Address Organisation for Applied Nature Scientific Research TNO Technology for Society
P.O.Box 217, 2600 AE DELFT
Tel. 015-696900

Project no. **92**
Title Net acid load of the soil under heather and forest as a result of atmospheric deposition
Projectleader Ing.H.Loman
Address Institute for Soil Fertility (IB)
P.O.Box 30003, 9750 RA HAREN (Gr.)
Tel. 050-346541

01. Loman H., 1989
Netto zuurbelasting van de bodem door atmosferische depositie
Net acid load of the soil due to atmospheric deposition (in Dutch)

Inst.voor Bodemvruchtbaarheid, Report 207, 1989
02. Harmsen K., Loman H. and Neeteson J.J., 1990
A derivation of the Pierre-Sluijsmans equation used in the Netherlands to estimate the acidifying effect of fertilizers applied to agricultural soils
Fertilizer Research
in press

Project no. **96**
Title Survey of vitality of natural vegetation on dry, oligotrophic soil
Projectleader Dr.D.C.P.Thalen
Address Research Institute for Nature Management
P.O.Box 46, 3956 ZR LEERSUM
Tel. 03434-52941

Project no. **98**
Title Transport, deposition and transformation in clouds of acidifying substances
Projectleader Drs.W.A.H.Asman
Address University of Utrecht, Institute of Meterology and Oceanography
Princetonplein 5, 3584 CC UTRECHT
Tel. 030-533276

Project no. **99**
Title Measuring dry deposition of air pollutants
Projectleader Ing.J.H.Duyzer
Address Organisation for Applied Nature Scientific Research TNO Technology for Society
P.O.Box 217, 2600 AE DELFT
Tel. 015-696900

01. Duyzer J.H., and Bosveld F., 1986
Meting van droge depositie van luchtverontreiniging. Interim rapport.
Measurements of dry deposition of air pollution. Interim report (in Dutch).
MT-TNO Delft, report R 86/159
02. Duyzer J.H., Meijer G.M. and Aalst R.M.van 1986
Metingen van droge depositie van luchtverontreiniging
Measurements of dry deposition of air pollution (in Dutch)
MT-TNO Delft report R 86/290
03. Duyzer J.H. and Bosveld F.C., 1988

Measurements of dry deposition fluxes of O_3, NO_x, SO_2 and particles over grass/heathland vegetation and the influence of surface inhomogeneity
MT-TNO Delft, report R 88/111

04. Duyzer J.H., Meijer G.M. and Aalst R.M.van, 1983
Measurement of dry deposition velocities of NO, NO_2 and O_3 and the influence of chemical reactions
Atmospheric Environment Vol. 17 no. 10 p. 2117-2120.

05. Duyzer J.H., 1985
Measurements of dry deposition of gases and particulates in the Netherlands using eddy correlation and gradient methods. Workshop on Measurement of Atmospheric Acidity, COST 611 Working Group 1. XII/ENV/57/85. Montelibretti (Rome) Italië, 25-26 juni 1985
MT-TNO publication P 85/48

06. Duyzer J.H. and Aalst R.M.van, 1985
Problems in the assessment of dry deposition fluxes. Proceedings of Arbeitssitzung zur Bestimmung der Trockenen Deposition Atmosphärischer Spurenstoffen. GSF München, 9-10 sept 1985, BRD
MT-TNO publication P 85/61

07. Duyzer J.H., 1987
Factors affecting dry and wet deposition. Proceedings of the COST 612 Workshop: Definition of European pollution climates and their perception by terrestrial ecosystems, 28-30 april 1987, Bern, Zwitserland. Air pollution reseach report 6 EC no. EUR 11432 EN
MT-TNO publication P 87/46

08. J.H. Duyzer, 1989
Micrometeorological techniques for the measurement of trace gas exchange
In: Exchange of trace gases between terrestrial ecosystems and the atmosphere, M.O. Andreae, S.D. Schimel, editors. p. 189-207 John Wiley and Sons Ltd. D. Fowler

09. Duyzer J.H. et al., 1990
Measurements of NO_2 fluxes using the gradient method during the Halvergate experiment
Proceedings of the COST 611 Working Group 3 Workshop Madrid, 12-14 March 1990

10. Fowler D., Baldochi D. and Duyzer J.H., 1990
Defining impacts by dry and occult deposition. Conference on acidic deposition, its nature and impacts, Glasgow, 16-21 September 1990
to be published

Project no. **100**

Title Quantification of the effect of adverse soil conditions on growth of Douglas fir and Scots pine by manipulation of water and nutrient supply in an existing stand

Projectleader Prof.Dr.Ir.N.van Breemen

Address Agricultural University, Department of Soil Science and Geology
P.O.Box 37, 6700 AA WAGENINGEN

Tel. 08370-84145

01. Visser P.H.B.de and Breemen N.van, 1989
Optimal nutrition in two forest stands that are exposed to acid atmospheric deposition
Poster presentation, Proceedings Int.Plant Nutrition Colloqium in Wageningen, August 1989

02. Visser P.H.B.de and Breemen N.van, 1989
Eliminating unfavourable soil factors as an approach to quantify indirect effects of acid atmospheric deposition
Poster presentation, Int. Symposium on Forest Decline Research, Friedrichshafen, October 1989

03. Visser P.H.B.de and Breemen N.van, 1990
Optimal nutrition in two forest stands exposed to acid atmospheric deposition
Plant Nutrition - physiology and application, 69-72

Project no. **101.1**

Title Air pollution in a Douglas fir forest; management and infrastructure of 2 forest research stations (ACIFORN)

Projectleader Ir.P.Hofschreuder

Address Agricultural University, Department of Air Pollution
P.O.Box 8129, 6700 EV WAGENINGEN

Tel. 08370-82684

see project 14

Project no. **101.2**

Title Air pollution in a Douglas fir forest; air pollution research (ACIFORN)

Projectleader Ir.P.Hofschreuder

Address Agricultural University, Department of Air Pollution
P.O.Box 8129, 6700 EV WAGENINGEN

Tel. 08370-82684

see project 14

Project no. **102.1**

Title Monitoring of soil chemical parameters under Douglas fir and heather (ACIFORN)

Projectleader Prof.Dr.Ir.N.van Breemen

Address Agricultural University, Department of Soil Science and Geology
P.O.Box 37, 6700 AA WAGENINGEN

Tel. 08370-82424

01. Maas M.P. van der, 1990
Hydrochemistry of two Douglas fir ecosystems and a heather ecosystem in the Veluwe
Report 102.1-0.1, May 1990
02. Breemen N.van and Verstraten J.M., 1990
Soil acidification and N cycling: summary of research in the Dutch Priority Programme on Acidification
Theme Report

Project no. **102.2**

Title Monitoring of soil physical parameters on two Douglas fir stands (ACIFORN)

Projectleader Prof.Dr.J.M.Verstraten

Address University of Amsterdam, Laboratory of Physical Geography and Soil Science
Dapperstraat 115, 1093 BS AMSTERDAM

Tel. 020-5257451

01. Tiktak A., Bouten W. and Schaap M., 1990
SWIF: A simulation model of soil water flow in forested ecosystems
FGBL-rapport 44, 69 pp.
02. Tiktak A. and Bouten W., 1990

Soil water dynamics in a forest ecosystem; I: Model description
to be published

03. Tiktak A. and Bouten W., 1990
Modelling soil water dynamics in forested ecosystems: deterministic versus cascade models (ecological modelling)
to be published

04. Tiktak A. and Bouten W., 1990
Hydrological budgets
to be published

05. Maas R.van de, Breemen N.van and Tiktak A., 1990
Chemical budgets
to be published

06. Tiktak A. and Bouten W., 1990
Hydrological budgets
to be published

07. Breemen N.van and Verstraten J.M., 1990
Soil acidification and N cycling: summary of research in the Dutch Priority Programme on Acidification
Theme Report

08. Heimovaara T.J. and Bouten W., 1990
A computer-controlled 36-channel Time Domain Reflectometry system for monitoring soil water contents
Water Res.Res. 26 no. 10: 2311 - 2316

Project no. **103**

Title Root development of trees in environments affected by air pollution (ACIFORN)

Projectleader Prof.Dr.Ir.R.A.A.Oldeman

Address Wageningen Agricultural University, Department of Silviculture and Forest Ecology
P.O.Box 342, 6700 AH WAGENINGEN

Tel. 08370-84426

see project 84

Project no. **104.1**
Title Measurement and modelling of canopy water storage during and after rain, dew and fog (ACIFORN)
Projectleader Ir.W.Bouten
Address University of Amsterdam, Laboratory of Physical Geography and Soil Science
Dapperstraat 115, 1093 BS AMSTERDAM
Tel. 020-5257412

01. Bouten W., 1986
Microwave transmission as a measure of the amount of intercepted water on the crown of a coniferous vegetation
Rap. FGBL 27: 14 pp. (in Dutch)

02. Bouten W. and Water E.de, 1987
Microwave transmission, a new tool in forest hydrological research
Rap. FGBL 33: 16 pp. (in Dutch)

03. Bouten W., Schaap M. and Hakkaart P., 1989
Monitoring and modelling rainfall interception and canopy wetness
In: Symp. Monitoring air pollution and forest ecosystem research, p.149-152

04. Bouten W., Swart P.J.F. and Water E.de, 1991
Microwave transmission, a new tool in forest hydrological research
(in press, J.of Hydrology)

05. Bouten W. and Schaap M., 1990
Interception and drainage dynamics in a Douglas fir forest
to be published in J.of Hydrol.

06. Bouten W. and Renssen H., 1991
Modelling the vertical distribution of wet canopy evaporation
to be published

07. Bouten W., Schaap M. and Swart P.J.F., 1990
Monitoring and modelling canopy water storage amounts of a Douglas fir forest
Report 104.1 Dutch Priority Programme on Acidfication

08. Bouten W., Schaap M.G. and Snoeij P., 1991
Monitoring and modelling canopy water storage amounts of a Douglas fir forest
Rep. Dutch Priority Programme on Acidification 104.1-01
in press

Project no. **104.2**
Title Leaf wetness recording
Projectleader Ir.W.P.Mulder/Dr.K.Schurer
Address Technical and Physical Engineering Research Service
P.O.Box 356, 6700 AJ WAGENINGEN
Tel. 08370-19143

Project no. **105**
Title Research on the oeco-physiology of conifers in forests in relation with air pollution and acid deposition on the soil (ACIFORN)
Projectleader Dr.P.Evers
Address Research Institute for Forestry and Urban Ecology "De Dorschkamp"
P.O.Box 23, 6700 AA WAGENINGEN
Tel. 08370-95374

01. Ancker J.A.M.van de, Evers P.W., Maessen P.P.T.M., Oterdoom J.H. and Tweel P.A.van de, 1987
Inventarisatie van boomvitaliteit
Ned.Bosbouwtijdschrift 59 (12): 5-17
02. Burg J.van den, Evers P.W., Martakis G.F.P., Relou J.P.M. and Werf D.C.van der, 1987
De conditie van opstanden van grove den (Pinus sylvestris) and Corsicaanse den (Pinus nigra var.maritima) en de minerale voedingstoestand in de Peel en op de Veluwe, najaar 1986
Rapport De Dorschkamp
03. Relou J.P.M. and Evers P.W., 1988
Ethylene production induced by damage caused by Sphaeropsis sapinea in Scots pine in the Peel and the Veluwe
In: Relationships between above and below ground influences of air pollutants on forest trees, Bervaes, J.C.A.M., Mathy, P., Evers P.W. (eds.), CEC, Brussels, p. 196-205
04. Bervaes J.C.A.M., Mathy P. and Evers P.W., 1989
Whole tree physiology
CEC, Brussels, 198 pp.
05. Evers P.W., Konsten C.J.M. and Vermetten A.W.M., 1988
Acidification research on Douglas fir forests in the Netherlands (ACIFORN project)
In: Air pollution and ecosystems, Mathy, P. (ed.), Reidel Dordrecht, p. 887
06. Evers P.W., Maas P.W.van der and Belde J.J.M., 1990

Direct and indirect effects of acid deposition on growth and mineral status of Douglas fir
In: International congress on forest decline research, Ulrich, M. (ed.), p.232-235

07. Jans W.W.P., Evers P.W., Relou H. and Swart W., 1990
Biomass dynamics of the ACIFORN Douglas fir stands
Dorschkamp report, in press

08. Maas M.P.van der, Evers P.W. and Belde J.J.M., 1990
Dynamics of the potassium status of two Douglas fir stands
In: International congress on forest decline research, Ulrich, M. (ed.), p.227

09. Evers P.W., 1986
Analysis of direct and indirect effects of air pollutants on the physiology of forest trees
In: Andersson, F., Mathy, P. (eds.), Direct effects of dry and wet deposition on forest ecosystems - in particular canopy interactions, CEC, Brussels, p. 291-296

10. Evers P.W., Konsten C.J.M. and Vermetten A.W.M., 1988
Acidification research on Douglas fir forests in The Netherlands
In: Air pollution and ecosystems, Mathy, P. (eds.), Riedel, Dordrecht, p. 887

11. Evers P.W., Cortes P., Beek H.van de, Jans W., Donkers J., Belde J., Relou H. and Swart W., 1988
Influence of air pollution on tree physiology
In: Air pollution and ecosystems, Mathy, P. (eds.), Riedel, Dordrecht, p. 907-910

12. Evers P.W., Jans W.W.P. and Swart W.A.J.M., 1988
Monitoring the canopy structure and development of the ACIFORN Douglas fir stands
In: Scientific base of forest decline symptomatology, Cape, J.N., Mathy, P. (eds.) CEC, Brussels, P.273-282

13. Evers P.W., Maas M.P.van der and Belde J.J.M., 1990
Direct and indirect effects of acid deposition on growth and mineral status of Douglas fir
In: International congress on forest decline research, Ulrich, M. (ed.), p.232-235

14. Maas M.P.van der, Evers P.W. and Belde J.J.M., 1990
Dynamics of the potassium status of two Douglas fir stands
In: International congress on forest decline research, Ulrich, M. (ed.)

15. Ancker J.A.M.van den, 1986
Het proefperceel Ugchelen vanaf de luchtfoto; boomvitaliteit - boomhoogte en representativiteit
Rapport De Dorschkamp - Geosens B.V.

16. Burg J.van den, Evers P.W., Martakis G.F.P., Relou J.P.M. and Werf D.C.van

der, 1988
De conditie en de minerale-voedingstoestand van opstanden van Grove den (Pinus sylvestris) and Corsicaanse den (Pinus nigra var.maritima) in de Peel en op de zuidoostelijke Veluwe, najaar 1986
Rapport nr. 519, De Dorschkamp

17. Minnen J.G.van, 1988
Een oriënterende vergelijking van de waterrelaties van bomen, een onderzoek uitgevoerd in het kader van het ACIFORN-project
Verslag nr. 88-31, De Dorschkamp/LUW, Vakgroep Bosteelt en Bosoecologie

18. Cortes P.M., Beek J.van der and Blom J.H.G., 1988
Field gas exchange laboratory: technical report
Report nr. 511, De Dorschkamp

19. Relou J.P.M. and Evers P.W., 1987
Ethyleenproduktie als gevolg van aantasting door Sphaeropsis sapinea van Grove den in de Peel en op de Veluwe
Rapport nr. 494, De Dorschkamp

20. Ancker J.A.M.van den, 1987
Een vitaliteitsbeoordeling met behulp van kleuren - infrarood luchtfoto's (schaal 1:2500) van twee percelen Douglas
Rapport nr. 492, De Dorschkamp

21. Evers P.W. (red.), 1988
Koppeling van ecofysiologische parameters van luchtverontreinigingsinvloeden aan biometrie in de ACIFORN Douglasopstanden
Rapport nr. 512, De Dorschkamp

22. Evers P.W., 1990
Concept final project report: Impact of air pollution on ecophysiological relations in 2 Douglas fir stands in the Netherlands
Report De Dorschkamp

23. Ancker J.A.M.van den, Evers P.W., Maessen P.P.Th.M., Oterdoom J.H. and Tweel P.H.van den, 1987
Inventarisatie van boomvitaliteit, een discussie
Nederlands Bosbouwtijdschrift 1987, 59 (12): 405 - 418

Project no. **106.1**

Title Nitrification in acid soils: mechanisms and the influence of biotic and abiotic factors

Projectleader Prof.Dr.M.Verstraten

Address University of Amsterdam, Laboratory of Physical Geography and Soil Science
Dapperstraat 115, 1093 BS AMSTERDAM

Tel. 020-5257451

01. Boer W. de, Duyts H. and Laanbroek H.J., 1988
Autotrophic nitrification in a fertilized acid heath soil
Soil Biol. Biochem. 20: 845-850

02. Boer W.de, Duyts H. and Laanbroek H.J., 1989
Urea stimulated autotrophic nitrification in suspensions of fertilized, acid heath soil
Soil Biol. Biochem. 21: 349-354

03. Boer W.de, and Laanbroek H.J., 1989
Ureolytic nitrification at low pH by Nitrospira spec.
Arch Microbiol. 152: 178-181

04. Boer W.de, Klein Gunnewiek P.J.A.,Troelstra S.R., and Laanbroek H.J., 1989
Two types of chemolithotrophic nitrification in acid heathland humus
Plant & Soil 119: 229-235.

05. Troelstra S.R., Wagenaar R. and Boer W.de
Nitrification in Dutch heathland soils. I. Nitrate production in undisturbed soil cores
Plant and Soil, accepted

06. Boer W.de, Klein Gunnewiek P.J.A. and Troelstra S.R.
Nitrification in Dutch heathland soils. II. Characteristics of nitrate production
Plant and Soil, accepted

07. Boer W. de, 1989
Nitrification in Dutch heathland soils
PhD Thesis, Agricultural University, Wageningen, The Netherlands.

Published, accepted and submitted reports and papers

08. Tietema A. and Verstraten J.M., 1988
The nitrogen budget of an oak-beech forest ecosystem in The Netherlands in relation to atmospheric deposition
Dutch Priority Programme on Acidification, Report no. 04-01

09. Tietema A, Duijsings J.J.H.M., Verstraten J.M. and Westerveld J.W., 1990
Estimation of actual nitrification rates in an acid forest soil
In: Harrison, Ineson and Heal, Nutrient cycling in terrestrial ecosystems, field

methods, Application and Interpretation, Elsevier Applied Science

10. Emmer I.M. and Tietema A., 1990
Temperature-dependent nitrogen transformations in acid oak-beech forest litter in The Netherlands
Plant and Soil, 122, 193-196

11. Tietema A., Bouten W. and Wartenbergh P.E., 1990
Nitrous oxide dynamics in an acid forest soil in The Netherlands
Forest Ecology and Management (Submitted)

12. Tietema A., Verstraten J.J.M. and Wijk A.J.van., 1990
The nitrogen cycle of an oak-beech forest ecosystem in The Netherlands at increased nitrogen deposition: 1. Biochemical nitrogen transformations and solute fluxes
Biogeochemistry (submitted)

13. Tietema A. and Verstraten J.M., 1990
The nitrogen cycle of an oak-beech forest ecosystem in The Netherlands at increased nitrogen deposition: 2. The nitrogen and proton budget
Biogeochemistry (submitted)

Planned papers

14. Boer W.de, Tietema A., Klein Gunnewiek P.J.A. and Laanbroek H.J.
The chemolithotrophic ammonium oxidizing community in a nitrogen saturated acid forest soil in relation to pH dependent nitrifying acitivity
accepted by Soil Biology and Biochemistry

15. Tietema A., Boer W.de, Riemer L. and Verstraten J.M.
Vertical distribution of nitrate production in nitrogen saturated acid forest soils in relation to pH-dependent nitrifying activities
to be submitted to Soil Biology and Biochemistry

16. Tietema A. and Wessel W.W.
The relation between nitrogen transformations in acid forest litter from several forests in The Netherlands

17. Wessel W.W. and Tietema A.
Methodological aspects of the 15N pool dilution method in nitrogen cycling studies

18. Tietema A., Verstraten J.M. and Voorthuyzen I.M.
Mass loss and nutrient dynamics in decaying leaf and needle litter in several forests in The Netherlands

19. Tietema A., Verstraten, and L Riemer
Nitrogen cycling and soil acidification in several forest sites in The Netherlands in relation to different nitrification rates

20. Tietema A., Riemer L. and Verstraten J.M.

In situ nitrogen transformation rates in several forests in The Netherlands

21. Boer W.de and Tietema A., 1990
Nitrification on Dutch acid forest and heathland soils: mechanisms and effects of biotic and abiotic factors
Report project 106.1, Dutch Priority Programme on Acidification, RIVM, Bilthoven, the Netherlands

22. Boer W.de, Tietema A., Klein Gunnewiek P.J.A. and Laanbroek H.J., 1991
The chemolithotrophic ammonium-oxidizing community in a nitrogen saturated acid forest soil in relation to pH-dependent nitifying acitivity
Soil Biology and Biochemistry
submitted

Project no. **106.2**

Title Nitrification in acid soils: mechanisms and the influence of biotic and abiotic factors

Projectleader Dr.H.J.Laanbroek

Address Institute for Ecological Research
P.O.Box 440, 666 ZG HETEREN

Tel. 08306-23064

see project 106.1

Project no. **106.3**

Title Nitrification in acid soils: mechanisms and the influence of biotic and abiotic factors

Projectleader Prof.Dr.J.M.Verstraten

Address University of Amsterdam, Laboratory of Physical Geography and Soil Science
Dapperstraat 115, 1093 BS AMSTERDAM

Tel. 020-5257451

see project 106.1

Project no. **107**

Title Effects of air pollutants on the rhizosphere of Douglas firs

Projectleader Dr.J.A.van Veen

Address Research Institute ITAL
Keyenbergseweg 6, 6704 PJ WAGENINGEN

Tel. 08370-91911

see project 86

Project no. **108.1**

Title Acid rain and ectomycorrhiza of Pseudotsuga menziesii: an integration of laboratory and field research (a: field and greenhouse research)

Projectleader Dr.E.J.M.Arnolds

Address Agricultural University
Kampsweg 29, 9418 PD WIJSTER (Gr.)

Tel. 05936-441

01. Eerden L.J.van der, Lekkerkerk L.J.A., Smulders S.M. and Jansen A.E., 1989
Effects of ozone and ammonia on Douglas fir (Pseudotsuga menziesii)
IUFRO-proceedings

02. Jansen A.E., 1989
De eerste bijeenkomst van het Europees Comité voor de bescherming van paddestoelen
Coolia 32: 11-13

03. Moore L.M., Jansen A.E. and Griensven L.J.L.D.van, 1989
In vitro mycorrhization of Cantharellus cibarius
Acta Bot.Neerl. 38: 273-278

04. Jansen A.E. and Vries F.W.de, 1989
Mycorrhizas on Douglas fir in The Netherlands
In: Proceedings 2nd European Symposium on Mycorrhizae, held in August 14-20, 1988, Prague, Czechoslovakia, Agric.Ecosyst.Environ. 28: 197-200

05. Jansen A.E. (in press)
The mycorrhizal status of Douglas fir in The Netherlands: its relation with stand age, regional factors, atmospheric pollutants and tree vitality
Agric.Ecosyst. Environ.

06. Gorissen A., Jansen A.E. and Olsthoorn A.F.M. (in prep.)
Influence of ammoniumsulphate on growth of juvenile Douglas-fir, on mycorrhizal status and on bacteria in the rhizosphere

07. Jansen A.E., 1990
Conservation of fungi in Europe
The Mycolist 4: 83-85
08. Jansen A.E. (in press)
How Netherlands mycologists started worrying about decline of fungi
In: Conservation of fungi and other cryptogams in Europe, edited by the Lodz Society of Sciences and Arts
09. Jansen A.E., Jongbloed R. and Kamminga-van Wijk C.
Acid rain influences mycorrhizas of Douglas fir (poster)
In: Proceedings Int.Mycological Congress, Aug.-Sept. 1990, Regensburg

Project no. **108.2**
Title Acid rain and ectomycorrhiza of Pseudotsuga menziesii: an integration of laboratory and field research (b: laboratory reserach)
Projectleader Dr.H.B.A.Prins
Address University of Groningen, Biological Centre
P.O.Box 14, 9750 AA HAREN (Gr.)
Tel. 050-632304/632281

01. Jansen A.E., Kamminga-van Wijk C., Jongbloed R, 1990
Acid rain and ectomycorrhiza of Douglas fir
Proceedings 4th International Mycological Congress, Abstracts: 128, Regensburg, Germany 28th Aug. - 3rd Sept. 1990
02. Kamminga-van Wijk C, Bollen P., Roozendaal C.van and Prins H.B.A., 1990
Effects of NH_4^+, NO_3^- and Al^{3+} on mycorrhizal and nonmycorrhizal Pseudotsuga menziesii seedlings grown on hydroculture
Physiologica Plantarum, Abstracts of the 7th Congress of the Federation of European Societies of Plant Physiology, Umea, Sweden 79(2)
03. Kamminga-van Wijk C. and Prins H.B.A., 1987
Ectomycorrhizal effects on the physiology and growth of Pseudotsuga menziesii grown on hydroculture
In: A.E.Jansen, J.Dighton and A.H.M.Bresser (eds.) Ectomycorrhiza and acid rain; Proceedings of the workshop/expert meeting on ectomycorrhiza, Dec. 10-11, 1987, Berg en Dal, The Netherlands, Commission of the European Communities, Air Pollution Research Report 12, Bilthoven: 153-168
04. Kamminga-van Wijk C. and Prins H.B.A., 1988
Influence of acid rain effects on mycorrhizal and non-mycorrhizal Pseudotsuga menziesii seedlings grown on hydroculture

Abstracts of the 6th Congress of the Federation of European Societies of the 6th Congress of the Federation of European Societies of Plant Physiology, Split, Yugoslavia, 4-10 Sept.

05. Kamminga-van Wijk C. and Prins H.B.A., 1989
The influence of pH on ectomycorrhizal development of Pseudotsuga menziesii innoculated with Laccaria bicolor in hydroculture
Agriculture, Ecosystems and Environment, 28: 213-217

06. Jongbloed R.H. and Borst-Pauwels G.W.F.H., 1988
Effects of Al^{3+} and NH_4^+ on growth and uptake of K^+ and $H_2PO_4^-$ by three ectomycorrhizal fungi in pure culture
In: A.E.Jansen, J.Dighton and A.H.M.Bresser (eds.) Ectomycorrhiza and acid rain; Proceedings of the workshop/expert meeting on ectomycorrhiza, Dec. 10-11, 1987, Berg en Dal, The Netherlands, Commission of the European Communities, Air Pollution Research Report 12, Bilthoven: 47-53

07. Jongbloed R.H. and Borst-Pauwels G.W.F.H., 1989
Effects of ammonium and pH on growth and potassium uptake by the ectomycorrhizal fungus Laccaria bicolor in pure culture
Agriculture, Ecosystems and Environment, 28: 207-212

08. Jongbloed R.H. and Borst-Pauwels G.W.F.H., 1989
Differential response of some ectomycorrhizal fungi to cadmium in vitro
Acta Bot. Neerl. 39 (3), in press

09. Jongbloed R.H. and Borst-Pauwels G.W.F.H.
Effects of ammonium and pH on growth of some ectomycorrhizal fungi in vitro (submitted)

10. Jansen A.E., Kamminga-van Wijk C., Jongbloed R.H. and Vries F.W.de
Acid rain and ectomycorrhiza of Pseudotsuga menziesii, an integration of laboratory and field studies
in prep.

11. Gorissen A. and Jansen A.E.
Effects of ozone and ammonium sulphate on carbon partitioning to mycorrhizal roots of juvenile Douglas firs
in prep.

12. Gorissen A., Jansen A.E. and Olsthoorn A.F.M.
Effects of two years application of ammonium sulphate on growth of juvenile Douglas fir and the rhizosphere microflora
in prep.

Project no. **109**

Title Effect of air-borne pollutants on the phtotosynthesis of Douglas fir (Pseudotsuga menziesii)

Projectleader Prof.Dr.W.J.Vredenberg/Dr.M.E.Bossen

Address Agricultural University, Department of Plant Physiological Research
Gen.Foulkesweg 72, 6703 BW WAGENINGEN

Tel. 08370-82800/82808

01. Kooten O.van, and Hove L.W.A.van, 1988
Fluorescence as a means of diagnosing the effect of pollutant induced stress in plants
In: Air Pollution and Ecosystems (P.Mathy, ed.), D.Reidel Publ.Co., pp. 596-601

02. Kooten O.van, Hove L.W.A.van and Wijk K.-J.van, 1988
The effect of long term exposition of poplars to low concentrations of SO_2 and NH_3
In: Applications of Chlorophyll Fluorescence (H.K.Lichtenhaler ed.), Kluwer Acad. Publ., Dordrecht, pp. 203-209

03. Rosema A., Cecchi G., Pantani L., Radicatti B., Romuli M., Mazzinghi P., Kooten O.van and Kliffen C., 1988
Air pollution effects on the fluorescence of Douglas fir and poplar
In: Applications of Chlorophyll Fluorescence (H.K.Lichtenhaler ed.), Kluwer Acad. Publ., Dordrecht, pp. 307-317

04. Schapendonk A.H.C.M., Dolstra O. and Kooten O.van, 1988
The use of chlorophyll fluorescence as a screening method for cold tolerance in maize
Photosynth. Res. 20, 235-247

05. Kooten O.van, Hove L.W.A.van and Vredenberg W.J., 1989
The effect of prolonged exposure to air-borne pollutants on the photosynthesis of Douglas fir (Pseudotsuga menziesii) studied with in vivo chlorophyll fluorescence
In: Current Research in Photosynthesis, vol. IV (M.Baltscheffsky ed.) Kluwer, Dordrecht, Boston, London, pp. 19.611-614

06. Hove L.W.A.van, Kooten O.van, Adema E.H., Vredenberg W.J. and Pieters G.A., 1989
Physiological effects of long term exposure to low and moderate concentrations of atmospheric NH_3 on poplar leaves
Plant Cell Environ. 12, 899-908

07. Aben J.J.M. and Kooten O.van
Effects of prolonged fumigation with ozone on photosynthesis of Vicia faba L.
Z. Naturforsch. (submitted)

08. Kooten O.van, Meurs C. and Loon L.C., 1990
Photosynthetic electron transport in tobacco leaves infected with tobacco mosaic virus

Physiol.Plant (in press)

09. Kooten O.van and Snel J.F.A., 1990
Use of chlorophyll fluorescence nomenclature in plant stress physiology
Photosynth.Res., 25, 147 - 150

10. Hove L.W.A. van, Kooten O.van, Wijk K.-J. van, Vredenberg W.J., Adema E.H. and Pieters G.A.
Physiological effects of long-term exposure to low concentrations of SO_2 and NH_3 on poplar leaves (submitted)

11. Bossen M.E., Kooten O.van and Vredenberg W.J., 1990
Effects of air-borne pollutants on the photosynthesis of Douglas fir (Pseudotsuga menziesii)
Final report project 109, Dutch Priority Programme on Acidification, 15 p.

12. Snel J.F.H., Kooten O.van and Hove L.W.A. van, 1991
Assessments of stress in plants by analysis of photosynthetic performance
Tr.Analyt.Chem.
in press

13. Loon L.C.van, Kooten O.van, Linders E.G.A., Meurs C. and Wijdeveld M.M.G.
Recognition and disease development in the tobacco-TMV system
In: Recognition and response in plant-virus interaction, NATO ASI series H41 (R.S.S.Fraser ed.), Springer, Berlin, Heidelberg, pp. 311 - 328

<u>Abstracts</u>

14. Kooten O.van, Lahey G. and Hove L.W.A.van, 1987
Fluorescence as an indicator of air pollutant effects on plants
Acta Bot.Neerl.suppl. 1, 36-2, 61

15. Kooten O.van and Schapendonk A.H.C.M., 1987
Fluorescence as a stress indicator in plant breeding
Acta Bot.Neerl.suppl. 1, 36-2, 62

16. Snel J.F.H., Kooten O.van and Vredenberg W.J., 1987
Characterization of chlorophyll a fluorescence response upon application of modulated illumination
Plant Physiol. 83, 161

17. Kooten O.van, Hove L.W.A.van and Vredenberg W.J., 1989
The effect of prolonged exposure to air-borne pollutants on the photosynthesis of several plant species studied with in vivo chlorophyll fluorescence
Physiol. Plant, 76 (part 2), A53

18. Kooten O.van, Hove L.W.A.van, Mensink M.G.J. and Vredenberg W.J., 1989
The effect of prolonged exposure to air-borne pollutants on the photosynthesis of Douglas fir (Pseudotsuga menziesii) studied with in vivo chlorophyll fluorescence

In: Abstracts 2nd Joint Meeting Belgian and Dutch Soc.Plant Physiol., Antwerpen, Belgium, 49

19. Kooten O.van, Hove L.W.A.van, Mensink M.G.J. and Vredenberg W.J., 1990
Comparison of CO_2 assimilation rate with photosynthetic electron transport rates in needles of Douglas fir exposed to moderate concentrations of SO_2, NO_2 or NH_3
Plant Physiol., 935, 99

20. Hove L.W.A.van, Kooten O.van, Mensink M., Bossen M.E., Adema E.H. and Vredenberg W.J., 1990
The physiological effects of a long-term exposure to low concentrations of NH_3, SO_2 and NO_2 on Douglas fir (Pseudotsuga menziesii)
In: Proceedings International Conference on Acidic Deposition, Glasgow, Book of abstracts p. 537

Project no. **110**

Title The uptake of gaseous air pollutants by needles of the Douglas fir and effects thereof on physiological processes

Projectleader Prof.Dr.E.H.Adema

Address Agricultural University, Department of Air Pollution
P.O.Box 8129, 6700 EV WAGENINGEN

Tel. 08370-82684

01. Hove L.W.A.van, Koops A.J., Adema E.H., Vredenberg, W.J. and Pieters G.A., 1987
Analysis of the uptake of atmospheric ammonia by leaves of Phaseolus vulgaris L.
Atmospheric Environment 21 (8): 1759 - 1763

02. Hove L.W.A.van, Adema E.H. and Vredenberg W.J., 1987
The uptake of atmospheric ammonia by leaves
Acta Botanica Neerlandica 36(2), suppl.1 :60

03. Hove L.W.A. van, 1987
De opname van atmosferisch ammoniak door bladeren
In: Effecten van NH_3 op organismen (eds. A.W. Boxman en J.F.M. Geelen): 35-44

04. Hove L.W.A.van, Tonk W.J.M., Pieters G.A., Adema E.H. and Vredenberg W.J., 1988
A leaf chamber for measuring the uptake of pollutant gases at low concentrations by leaves, transpiration and carbon dioxide assimilation
Atmospheric Environment 22 (11): 2515-2523

05. Hove L.W.A.van and Adema E.H., 1988
The uptake of atmospheric ammonia by leaves

In: Air pollution and ecosystems (ed. P. Mathy): 734-738. Reidel, Dordrecht

06. Hove L.W.A.van and Adema E.H., 1988
The uptake of NH_3 and SO_2 by leaves
Report project 78. Dutch Priority Programme on Acidification. 24 pp.

07. Hove L.W.A.van, Adema E.H., Vredenberg W.J. and Pieters G.A., 1989
A study of the adsorption of NH_3 and SO_2 on leaf surfaces
Atmospheric Environment 23 (7): 1479-1486

08. Hove L.W.A.van, Kooten O.van, Adema E.H., Vredenberg W.J. and Pieters G.A., 1989
Physiological effects of long term exposure to low and moderate concentrations of atmospheric NH_3 on poplar leaves
Plant, Cell & Environment 12 (9): 899-908

09. Hove L.W.A.van, 1989
The mechanism of NH_3 and SO_2 uptake by leaves and its physiological effects
Thesis Agricultural University Wageningen, The Netherlands

10. Hove L.W.A.van, Tonk W.J.M. and Adema E.H., 1989)
A leaf chamber for measuring the uptake of pollutant gases at low concentrations by leaves, transpiration and carbon dioxide assimilation
(Reply on the comment of Allan H. Ledge)
Atmospheric Environment 23 (7): 1617-1618

11. Hove L.W.A.van, Vredenberg W.J. and Adema E.H., 1990
The effect of wind velocity, air temperature and humidity on NH_3 and SO_2 transfer into bean leaves (Phaseolus vulgaris L.)
Atmospheric Environment, 24A (5): 1263-1270

12. Hove, L.W.A.van, Kooten, O.van, Wijk K.J.van, Vredenberg W.J., Adema E.H. and Pieters G.A.
Physiological effects of long term exposure to low concentrations of SO_2 and NH_3 on poplar leaves (submitted for publication)

13. Booij C.G., Hove, L.W.A.van, Adema E.H., 1987
Expositie-respons-relatie modellen en onderzoek naar de opname van ammoniak door bladeren
Eindrapportage ERR-projekt (R-265, vakgroep Luchthygiëne en -verontreiniging, LUW)

14. Kooten O.van and Hove L.W.A.van, 1988
Fluorescence as a means of diagnosing the effect of pollutant induced stress in plants
In: Air pollution and ecosystems (ed. P. Mathy): 596-601, Reidel Dordrecht

15. Kooten O.van, Hove B.van and Wijk K.J.van, 1988
The effect of long term exposition of poplars to low concentrations of SO_2 and NH_3

In: Applications of Chlorophyll Fluorescence (ed. H.K. Lichtenthaler): 203-209, Kluwer Dordrecht

16. Kooten O.van, Hove L.W.A.van and Vredenberg W.J., 1989
The effect of prolonged exposure to air-borne pollutants on the photosynthesis of Douglas fir (Pseudotsuga menziesii) studied with in vivo chlorophyll fluorescence
Physiol.Plant. (part 2) A53 (abstract)

17. Bicanic D., Harren F., Reuss J., Woltering E., Snel J., Voesenek L.A.C.J., Zuidberg B., Jalink H., Bijnen F., Blom C.W.P.M., Sauren H., Kooijman M., Hove L.van, Tonk W. , 1989
Trace detection in agriculture and biology
In: Photoacoustic, Photothermal and Photochemical Processes in Gases (ed. P.Hess). pp. 213-245. Topics in Current Physics 46, Springer Verlag

18. Sauren H., Hove L.W.A.van, Tonk W.J.M., Jalink H. and Bicanic D., 1989
On the adsorption properties of ammonia to various surfaces
In: Monitoring of atmospheric pollutants by tunable diode laser spectroscopy, (eds. Grisar R., Schmidtke G., Tacke M. and Restelli, G.). pp. 196-201, Kluwer, Dordrecht, Boston, London

19. Sauren H., Akker, D.van der, Hove B.van, Bicanic D., Jalink H., Torfs P., Tonk W., Quist J., 1989
Experimental determination of the absolute adsorption of gases ammonia on metal, polymers and new coatings used as construction materials for photo-acoustic cell
In: Technical Digest 6th topical international meeting on photo-acoustic and photothermal phenomena, Johns Hopkins University Baltimore Maryland USA: pp.97-101, abstract

20. Hove L.W.A.van, Mensink M.G.J., Adema E.H. and Vredenberg W.J., 1990
The physiological effects of low concentrations of NH_3, SO_2 and NO_2 on Douglas fir (Pseudotsuga menziesii)
Final report project 110, Dutch Priority Programme on Acidification

21. Hove L.W.A.van, Kooten O.van, Mensink M.G.J., Bossen M.E., Adema E.H. and Vredenberg W.J., 1990
The physiological effects of a long term exposure to low concentrations of NH_3, SO_2 and NO_2 on Douglas fir (Pseudotsuga menziesii)
International Conference on Acidic Deposition; its Nature and Impacts, 16-21 September 1990, Glasgow, UK (abstract)

22. Kooten O.van, Laheij G. and Hove L.W.A.van, 1987
Fluorescence as an indicator of air pollutant effects on plants
Acta Bot.Neerl. suppl. 1; 36-2:61 (abstract)

23. Kooten O.van, Hove L.W.A.van, Mensink M.G.J.and Vredenberg W.J., 1990

Comparison of CO_2 assimilation rate with photosynthetic electron transport rates in needles of Douglas fir exposed to moderate concentrations of SO_2, NO_2 or NH_3
Plant Physiol.
in press

24. Kooten O.van, Hove L.W.A.van, Mensink M.G.J. and Vredenberg W.J., 1989
The effect of prolonged exposure to air-borne pollutants on the photosynthesis of Douglas fir (Pseudotsuga menziesii) studied with in vivo chlorophyll fluorescence
In: Abstracts 2nd Joint Meeting Belgium and Dutch Society Plant Physiol., Antwerpen, Belgium:49

25. Snel J.F.H., Kooten O.van and Hove L.W.A.van, 1991
Assessments of stress in plants by analysis of photosynthetic performance
Trends in analytical chemistry
in press

<u>Project no.</u> **111.1**
<u>Title</u> Tree ring analysis and sapwood proportion in Douglas fir in relation to air pollution and acid deposition
<u>Projectleader</u> Prof.Dr.P.Baas
<u>Address</u> University of Leiden, Rijksherbarium/Hortus Botanicus
P.O.box 9514, 2300 RA LEIDEN
<u>Tel.</u> 071-130541

see project 80a

<u>Project no.</u> **111.2**
<u>Title</u> Functional needle anatomy of Douglas fir in relation to air pollution and acid deposition
<u>Projectleader</u> Prof.Dr.P.Baas
<u>Address</u> University of Leiden, Rijksherbarium/Hortus Botanicus
P.O.Box 9514, 2300 RA LEIDEN
<u>Tel.</u> 071-130541

01. Thijsse G. and Baas P., 1990
Natural and NH_3 induced variation in epicuticular needle wax morphology of Pseudotsuga menziesii (Franco) Mirb. Trees 4 (or 5)
(in press)

Project no. **112**
Title Coupling of soil, plant and atmospheric processes and their disturbances through air pollution and acidification in physiological model for Douglas fir stands (ACIFORN)
Projectleader Dr.Ir.G.M.J.Mohren
Address Research Institute for Forestry and Urban Ecology "De Dorschkamp"
P.O.Box 23, 6700 AA WAGENINGEN
Tel. 08370-95321

01. Florax J.P.G.G.M., Kowalik P.J.K. and Mohren G.M.J., 1990
Water relations in Douglas-fir stands: A modelling approach to tree water status and transpiration
Research Institute for Forestry and Urban Ecology "De Dorschkamp", report nr. 592 (in press)

02. Jorritsma I.T.M., Mohren G.M.J. and Veen J.R.van de, 1990
FORGRO 3.3: A quantification of direct effects of air pollutants on forest growth, Model documentation and listing
Research Institute for Forestry and Urban Ecology "De Dorschkamp"
(in press)

03. Mohren G.M.J., 1988
Report on group discussions and final recommendations
In: J.Bervaes, P.Mathy and P.Evers (eds.): Relationships between above and below ground influences of air pollutants on forest trees, Commission of the European Communities, Air Pollution Research Report 16, p. 262-269

04. Mohren G.M.J. (ed.), 1990
Water relations and nutrient demands in a general forest growth model: FORGRO 3.1.
Research Institute for Forestry and Urban Ecology "De Dorschkamp"
(in prep.)

05. Mohren G.M.J. (ed.), 1990
Integrated effects of air pollution and soil acidification on forests
Summary report as part of the Dutch Priority Programme on Acidification, Wageningen/Bilthoven, approx. 75 pp.

06. Mohren G.M.J. and Bartelink H., 1990
Modelling the effects of needle mortality rate and needle area distribution on dry matter growth of Douglas-fir
Netherlands Journal of Agricultural Science, 38: 53-66

07. Mohren G.M.J., Jorritsma I.T.M., Florax J.P.G.G.M., Bartelink H.H. and Veen

J.R.van de, 1990
FORGRO 3.0: A basic forest growth model; Model documentation and listing Research Institute for Forestry and Urban Ecology "De Dorschkamp", report nr. 524 (in press)

08. Mohren G.M.J., Jorritsma I.T.M., Kropff M.J., Vermetten A.W.M. and Tiktak A., 1990
Quantifying direct effects of air pollution on forest growth
paper presented at the International Conference on Acidic Deposition, Glasgow, 16-21 September 1990

09. Mohren G.M.J. and Rabbinge R., 1988
Dynamic models of carbon water, and nutrient interactions in trees
In: J.Bervaes, P.Mathy and P.Evers (eds.): Relationships between above and below ground influences of air pollutants on forest trees, Commission of the European Communities, Air Pollution Research Report 16, p. 60-75

10. Mohren G.M.J. and Rabbinge R., 1990
Growth-influencing factors in dynamic models of tree growth
In: R.K.Dixon, R.S.Meldahl, G.A.Ruark and W.G.Warren (eds.): Process Modeling of Forest Growth Responses to Environmental Stress, Timber Press Inc., Portland, Oregon USA, p. 229-240

11. Veen J.R.van de, Mohren G.J.M., Olsthoorn A.F.M. and Jorritsma I.T.M., 1990
Effects of soil acidification, root growth and uptake of water and nutrients in a general forest growth model: FORGRO 3.2
Research Institute for Forestry and Urban Ecology "De Dorschkamp"
(report in prep.)

<u>Project no.</u> **113**
<u>Title</u> Dataderivation, evaluation and application of a regional soil acidification model
<u>Projectleader</u> Ir.W.de Vries
<u>Address</u> Soil Survey Institute
Marijkeweg 11, 6709 PE WAGENINGEN
<u>Tel.</u> 08370-74353

see project 13

Project no. **114.1**
Title System research acidification; Dutch Acidification Systems Model (DAS)
Projectleader Ir.T.N.Olsthoorn
Address National Institute of Public Health and Environmental Protection
P.O.Box 1, 3720 BA BILTHOVEN
Tel. 030-749111

01. Bakema A.H., Berendse F., Boer K.F.de, Bultman G.W., Grinsven J.J.M.van, Heerden C.van, Kok R.M., Kros J., Minnen J.G.van, Mohren G.M.J., Olsthoorn T.N., Vries W.de and Wortelboer F.G., 1990
Dutch Acidification Systems Model - Specifications
Dutch Priority Programme on Acidification, report nr. 114.1-01
02. Boer K.F., 1990
Spreadsheets and GIS in Integral Modeling
Paper presented at the Conference for Computer Science for Environmental Protection, Vienna 19-21 September
03. Boer K.F. and Padding P., 1990
GIS tools for Acidification Modeling
Paper presented at the EGIS conference, Amsterdam, 10-13 April 1990
04. Olsthoorn T.N.and Leeuw F.A.A.M.de, 1988
Berekening van de zure depositie op Nederland op basis van overdrachtsmatrices
RIVM rapport nr. 758805005
05. Gosseling H.J., Olsthoorn A.A. and Feenstra J.F., 1990
Schade aan materialen door verzuring
IVM rapport nr. W90-002
06. Olsthoorn T.N., 1988
Interactief beleidsmodel voor experimenteren met verzurende emissies en deposities
RIVM rapport nr. 758805002
07. Olsthoorn T.N., Jaarsveld J.A.van, Knoop J.M., Egmond N.D.van, Mülschlegel J.H.C. and Duijvenbooden W.van, 1990
Integrated modeling in The Netherlands
In: Fenham, J., H.Larsen, G.A.Mackenzie and B.Rasmussen, 1990, Environmental Models: Emissions and Consequences, Developments in Ecological Modeling 15, Elsevier, Amsterdam
08. Thomas R., Arkel W.G.van, Baars H.P., Ierland E.C.van, Boer K.F.de, Buijsman E., Hutten T.J.H.M. and Swart R.J., 1988
Emission of SO_2, NO_x, VOC and NH_3 in The Netherlands and Europe in the period 1950-2030

RIVM rapport nr. 758472002

09. Wortelboer F.G., 1990
A model of the competition between two macrophyte species in acidifying shallow soft-water lakes in The Netherlands
Hydrobiol.Bull., in press

10. Wortelboerg F.G., 1990
AQUACID: Acidification model of shallow soft-water lakes in The Netherlands
RIVM report, in prep.

Project no. **114.2**
Title Dose-effect relationship for forest stands, to be applied in the Dutch Acidification Systems model (DAS)
Projectleader Dr.Ir.G.M.J.Mohren
Address Research Institute for Forestry and Urban Ecology "De Dorschkamp"
P.O.Box 23, 6700 AA WAGENINGEN
Tel. 08370-95321

see project 112

Project no. **114.3**
Title Development of a predictive model on a regional scale of the effects of air pollution on heathlands on dry sandy soil
Projectleader Dr.J.J.M.Berdowski
Address Research Institute for Nature Management
P.O.Box 9201, 6800 HB ARNHEM
Tel. 085-452991

see project 90a

Project no. **115**
Title Model development for the uptake of air pollutants and the effects on the physiology of Douglas needles in relation to drought
Projectleader Dr.Ir.C.J.H.Booij
Address Research Institute for Plant Protection
P.O.Box 9060, 6700 GW WAGENINGEN
Tel. 08370-19151

01. Kropff M.J., 1987
Physiological effects of sulphur dioxide 1. The effect of SO_2 on photosynthesis and stomatal regulation of Vicia faba L.
Plant, Cell and Environment 10: 753-760.

02. Kropff M.J., 1989
Modelling short-term effects of sulphur dioxide.1. A model for the flux of SO_2 into leaves and effects on leaf photosynthesis
Netherlands Journal of Plant Pathology 95, 195-213

03. Kropff M.J., 1989
Modelling short-term effects of sulphur dioxide. 2. Quantification of biochemical characteristics determining the effect of SO_2 on photosynthesis of leaves
Netherlands Journal of Plant Pathology, 95, 214-224

04. Kropff M.J., Smeets W., Meijer E., Zalm A.J.A.van der and Bakx E.J.
Effects of sulphur dioxide on photosynthesis: The role of temperature and humidity
Submitted to Physiologia Plantarum

05. Kropff M.J. and Goudriaan J., 1989
Modelling short term effects of sulphur dioxide. 3. The effect of SO_2 on photosynthesis of leaf canopies
Netherlands Journal of Plant Pathology 95, 265-280

06. Kropff M.J., Mooi J., Goudriaan J., Smeets W., Leemans A., Kliffen C. and Zalm, A.J.A.van der, 1989
The effects of long-term open-air fumigation with SO_2 on a field crop of broad bean (Vicia faba L.). I. Depression of growth and yield
New Phytologist 113, 337-344.

07. Kropff M.J., Mooi J., Goudriaan J., Smeets W., Leemans A. and Kliffen C., 1989
The effects of long-term open air fumigation with SO_2 on a field crop of broad bean (Vicia faba L.). II. Effects on growth components, leaf area development and elemental composition
New Phytologist 113, 345-351

08. Kropff M.J., 1990
Effects of long-term open-air fumigation with SO_2 on a field crop of broad bean (Vicia faba L.) III. Quantitative analysis of damage components
New Phytologist, in press

09. Kropff M.J.
Long term effects of sulphur dioxide on plants, SO_2 metabolism and regulation of intracellular pH
Submitted to Plant and Soil

10. Kropff M.J., Smeets W., Mooi J., Goudriaan J. and Leemans A., 1989

Growth and production of faba bean crops exposed to SO2 in the field: Experimental data analysed with a simulation model
In: Man and his ecosystem, L.J.Brasser and W.C.Mulder (Eds.), Proceedings of the 8th World Clean Air Congress, The Hague, Holland, 11-15 sept. 1989, pp. 29-34

11. Kropff M.J., 1989
Quantification of SO_2 effects on physiological processes, plant growth and crop production
Ph.D.Thesis, Agricultural University, Wageningen, 201 pp.ISBN 90-9002942-7

12. Kropff M.J.,1988
Simulation analysis of the effects of SO_2 on crops
LUW-LUVO,TPE/IPO report, Wageningen, 79 pp.

13. Kropff M.J., Smeets W., Meijer E., Zalm A.van der, Kooijman A. and Leemans A., 1989
Uptake of SO_2 by leaf canopies and effects on growth and production: an integrated experimental and simulation study
LUW-LUVO,TPE/IPO report.121 pp., Wageningen

14. Smeets W.L.M., Kropff M.J.
Effects of long-term open-air fumigation with SO_2 on photosynthesis, growth and ionic composition of Douglas-fir seedlings (in prep.)

15. Smeets W.L.M., Kropff M.J. and Meijer E.
The effect of O_3 on photosynthesis and stomatal regulation of Douglas-fir in relation to waterstress (in prep.)

16. Smeets W.L.M., Kropff M.J.
Modelling effects of O_3 on leaf photosynthesis of Douglas-fir (in prep.)

17. Smeets W.L.M., Kropff M.J.
Single and combined effects of long-term fumigation with O_3 and SO_2 on photosynthesis and growth in broad bean and Douglas-fir (in prep.)

Project no. **116**

Title Measuring sapstream velocities in the Douglas fir

Projectleader Drs.H.A.Jenner

Address Joint Laboratories and Consulting Services of the Dutch Electricity Supply Companies N.V. KEMA
P.O.Box 9035, 6800 ET ARNHEM

Tel. 085-563008

01. Visser H., Noppert F., Wakeren H.van and Vaessen J., 1989
Xylem sap velocity in relation to weather and air pollution

IAWA bulletin n.s., Vol.10 (4), 1989: 427-439

Project no. **117**

Title Research into the deposition flux of water vapour and the chemical composition of dew, field measurements and modelling

Projectleader Dr.F.G.Römer

Address Joint Laboratories and Consulting Services of the Dutch Electricity Supply Companies N.V. KEMA
P.O.Box 9035, 6800 ET ARNHEM

Tel. 085-562599

01. Beljaars A.C.M., Holtslag A.A.M. and Westrhenen, R.M., 1989
Description of a software library for the calculation of surface fluxes
De Bilt, KNMI. Technical Report TR-112

02. Berkowicz R. and Prahm L.P., 1982
Evaluation of the profile method for estimation of surface fluxes of momentum and heat
Atm. Env. 16:12, 2809-2819

03. Businger J.A., 1971
Flux profiel relationships in the atmospheric surface layer
J. Atm. Sc., 28, 181-189

04. Businger J.A., et al., 1986
Evaluation of the accuracy with which dry deposition can be measured with correct micrometeorological methods
J. Clim. and Appl. Met., 25, 1100-1124

05. Hicks B.B., 1986
Measuring dry deposition: a re-assesment of the state of the art. Bound.-Lay Met., 30, 75-90

06. Hofschreuder P., Vermetten A.W.M., 1989
The effects of pollutants in a Douglas fir forest
In: Man and his Ecosystem, Amsterdam, Elsevier; Proceedings 8th World Clean Air Congress 1989

07. Keuken M.P., 1989
The determination of acid-deposition-related compounds in the lower atmosphere
Petten, ECN, Report 215

08. Vermetten A.W.M., Hofschreuder P. and Harssema, H., 1985
Deposition of gaseous pollutants in a Douglas fir forest
Contribution to the Colloquium "Deposition and interception of atmospheric

Pollutants in Forests", Oberursel (BRD), November 6/7 1985

09. Vermetten A.W.M., 1990
Mondelinge mededeling
List of publications, Project 117, contribution of ECN

10. Arends B.G., Eenkhoorn S
The influence of manganese on the oxidation of sulfite in dew water
In: Mechanisms and Effects of Pollutant-Transfer into Forests, ed. H.-W. Georgii, 1989, 231-238, Kluwer Acad. Publ.

11. Mallant R.KA.M., Arends B.G.
Fog and dew chemistry and effects on vegetation
In Proc of 8th World Clean Air Congres 1989, the Hague, eds. L.J. Brasser, W.C. Mulder, vol. 2, 183-188

12. Arends B.G., Wensveen E.van, R.K.A.M.Mallant
The oxidation of sulfite by nitrogen dioxide studied in a fog chamber
In: The Role of Clouds in Atmospheric Chemistry and Global Climate, 1989, 93-96

13. Mallant R.K.A.M., Arends B.G.
Simulation of fog in plant exposure experiments
Proc.Conf.Environmental Research with Plant Chambers, Neuherberg, 9-11 october 1989 (in press)

14. Masuch G., Kicinski H.G., Kettrup A., Slanina S., Mallant R.K.A.M., Arends B.G.
Effects of sulfite-containing fog on young Douglas firs
Proc.Conf.Environmental Research with Plant Chambers, Neuherberg, 9-11 october 1989 (in press)

15. Arends B.G., Eenkhoorn S.
The influence of manganese leached from plants on dew chemistry
Environ. Technol. Letters (in press)

16. Winkel B.H.te, Römer F.G., 1989a
Onderzoek naar de chemische samenstelling van dauw en mist. Overzicht metingen 1988
KEMA- report nr. 50583-MOC 89-3264.

17. Wakeren J.H.A.van and Janssen L.H.J.M., 1989b
Statusrapport dauwtafel augustus 1989. Een uitgewerkte meting van de dauwtafel met een foutenanalyse
KEMA report nr. 50583-MOF 89-3311.

18. Overdijk E.P, 1989c
Een PSI-simulatieprogramma voor het temperatuurverloop van een gewas tijdens dauwvormingsomstandigheden

KEMA report nr. 50583-MOF 89-1325.
19. Steenkist R., 1990
Dauwvorming en concentraties in een gewaslaag tijdens dauwvormings-omstandigheden III
KEMA report nr.50583-MOF 89-3304.
20. Römer F.G., Winkel B.H. te, 1989
The deposition flux of reactive substances to wet surfaces
Proceedings of the 8th World Clean Air Congress, The Hague, The Netherlands, 11-15 September 1989, Volume 3, Amsterdam, Elsevier Science Publishers B.V., pp. 451-456
21. Römer F.G., Winkel B.H. te, Janssen L.H.J.M., 1989
The deposition of acidifying components to wet surfaces. Occurrence and chemical composition of dew
Proceedings of the Fifth European Symposium on Physico-chemical behaviour of Atmospheric Pollutants, Varese, Italy, 25-28 September 1989.
22. Janssen L.H.J.M., Römer F.G., 1990
The frequency and duration of dew occurrence over a year. Model results compared with measurements
Submitted for publication in Tellus B.

Planned reports

23. Onderzoek naar de chemische samenstelling van dauw en mist. Overzicht metingen 1989
24. Onderzoek naar de chemische samenstelling van mist in de periode 1987-1989
25. Modellering van de chemische samenstelling van dauw
26. Bepaling van de depositieflux van waterdamp op een bos
27. Research into the chemical composition of dew and the deposition flux of water vapour, field measurements and modeling (evaluation of integrated results)
28. Dauwvorming in een hoog gewas

Project no. **118**

Title: Experimental research on the effects of a decrease in deposition and improvements in the minerals balance on the vitality forests in The Netherlands

Projectleader: J.G.M.Roelofs
Address: Catholic University
Toernooiveld, 6525 ED NIJMEGEN
Tel. 080-652860

01. Arts G.H.P., 1987
Geschiedenis van de verzuring van zwak gebufferde wateren in Nederland onder invloed van atmosferische depositie
Report 28-01 Dutch Priority Programme on Acidification

02. Arts G.H.P., 1988
Historical development and extent of acidification of shallow soft waters in The Netherlands, Lab.of Aquatic Ecology, Catholic University Nijmegen, The Netherlands
In: Proceedings Int.Symp.: Air pollution and Ecosystems, Grenoble, 18-22 May 1987, p. 928-933

03. Arts G.H.P., Schaminée and Munckhof P.J.J., 1988
Human impact on origin, deterioration and maintenance of Litterelletalia-communities
In: Proceedings Symposium Synanthropic Flora and Vegetation V, Martin, Czechoslovakia, p. 11-18

04. Arts G.H.P. and Leuven R.S.E.W., 1988
Floristic changes in shallow soft waters in relation to underlying environmental factors
Freshwat.Biol., 20: 97-111

05. Arts G.H.P., 1988
Waterverzuring in Overijssel; Effecten van verzurende en bemestende depositie op zwak gebufferde wateren in de provincie Overijssel in historisch en toekomstig perspectief
Aquat.Oec.Nijmegen, in opdracht van de Provincie Overijssel, 1988, 59 p. + 4 bijl.

06. Arts G.H.P., Haan A.J.de, Siebum M.B. and Verheggen G.M., 1989
Extent and historical development of the decline of Dutch soft waters
In: Proceedings Kon.Ned.Akademie van Wetenschappen, C 92(3): 281-295

07. Boxman A.W. and Roelofs J.G.M., 1986
Some physiological effects of NH_4^+ and Al^{3+} on pine forest ecosystems
In: Proceedings Int.Symp. "Neue Ursachenhypothesen", UBA, Berlin, 407-414

08. Boxman A.W., Sinke R.J. and Roelofs J.G.M., 1987
Effects of NH_4^+ on the growth and K^+ (^{86}Rb) uptake of various ectomycorrhizal fungi in pure culture
Water, Air and Soil Pollution, 31: 517-522

09. Boxman A.W., Dijk H.F.G.van and Roelofs J.G.M., 1987
Some effects of ammonium sulphate deposition on pine and deciduous forests in The Netherlands
In: Acid Rain: Scientific and Technical Advances, R.Perry, R.M.Harrison, J.N.B.Bell and J.N.Lester (eds.), p. 680-687

10. Boxman A.W., Dijk H.F.G.van and Roelofs J.G.M., 1987
Effecten van ammonium op de groei en fosfatase activiteit van geïsoleerde mycorrhiza- en saprofytische schimmels op de nutriënten-opnames door dennekiemlingen
In: A.W.Boxman en J.F.M.Geelen (eds.); Acute en chronische effecten van NH_3 (en NH_4) op levende organismen; Katholieke Universiteit Nijmegen, p. 115-125

11. Boxman A.W. and Geelen J.F.M. (eds.), 1987
Acute en chronische effecten van NH_3 (en NH_4^+) op levende organismen
In: Proceedings van de BEL-studiedag "Effecten van NH_3 op organismen", 12 december 1986, Laboratorium voor Aquatische Oecologie, Katholieke Universiteit Nijmegen, 133 pp.

12. Boxman A.W., Dijk H.F.G.van, Houdijk A.L.F.M. and Roelofs J.G.M., 1988
Critical loads for nitrogen - with special emphasis on ammonium;
In: Critical loads for nitrogen and sulphur I.Nilsson and P.Grennfelt (eds.), Workshop
Report Skokloster, Sweden, p. 295-322

13. Boxman A.W. and Roelofs J.G.M., 1988
Some effects of nitrate versus ammonium nutrition on the nutrient fluxes in Pinus sylvestris seedlings; Effects of mycorrhizal infection, Can.J.Bot., 66: 1091-1097

14. Boxman A.W. and Dijk H.F.G.van, 1988
Het effect van landbouw ammonium op bos- en heidevegetaties, Department of Aquatic Ecology and Biogeology and Ministry of Housing, Physical Planning and Environment, Report project 19, Dutch Priority Programme on Acidification, p. 96

15. Dijk H.F.G.van and Roelofs J.G.M., 1986
Effects of airborne ammonium on the nutritional status and condition of pine needles
In: Air pollution research report 4. Direct effects of dry and wet deposition on forest ecosystems - in particular canopy interactions, p. 40-50; Commission of the European Communities, Proceedings Workshop Lökeberg 19-23 Oct. 1986

16. Dijk H.F.G.van, Boxman A.W. and Roelofs J.G.M. 1987
De effecten van ammoniumdepositie op de voedingsstatus en conditie van dennenaalden
In: A.W.Boxman and J.F.M.Geelen (red.), Acute en chronische effecten van NH_3 (en NH_4^+) op levende organismen, Katholieke Universiteit Nijmegen, p. 105-114

17. Dijk H.F.G.van and Roelofs J.G.M., 1988
Effects of excessive ammonium deposition on the nutritional status and condition of pine needles
Physiol.Plant., 73: 494-501

18. Dijk H.F.G.van, Creemers R.C.M., Rijniers J.P.L.W.M. and Roelofs J.G.M.,

1989
Impact of artificial ammonium-enriched rainwater on soils and young coniferous trees in a greenhouse; I. Effects on the soils, Environ.Poll., 62: 317-336

19. Dijk H.F.G.van, Louw M.H.J.de, Roelofs J.G.M. and Verburgh J.J., 1990
Impact of artificial ammonium-enriched rainwater on soils and young coniferous trees in a greenhouse, II. Effects on the trees, Environ.Poll., 63: 41-59

20. Houdijk A.L.F.M., 1990
Effecten van zwavel- en stikstofdepositie op bos- en heidevegetaties, Department of Aquatic Ecology and Biogeology and Ministry of Housing, Physical Planning and Environment; Dutch Priority Programme on Acidificaiton, p. 124

21. Roelofs J.G.M., Kempers A.J., Houdijk A.L.F.M. and Jansen J., 1985
The effect of air-borne ammoniumsulphate on Pinus nigra var. maritima in The Netherlands; Plant and Soil. 84: 45-56

22. Roelofs J.G.M. and Boxman A.W., 1986
The effect of air-borne ammoniumsulphate deposition on pine forests; Proc. Int.Symp. "Neue Ursachenhypothesen", UBA, Berlin, 415-422

23. Roelofs J.G.M. and Dijk H.F.G.van, 1986
The effect of airborne ammonium deposition on canopy ion-exchange in coniferous trees
In: Air pollution research report 4. Direct effects of dry and wet deposition on forest ecosystems - in particular canopy interactions pp. 34-39; Commission of the European Communities; Proceedings Workshop Lökeberg, 19-23 Oktober 1986

24. Roelofs J.G.M., 1986
The effect of air-borne sulphur and nitrogen deposition on aquatic and terrestrial heathland vegetation
Experientia, 42: 372-377

25. Roelofs J.G.M., 1987
Acid rain and the flora of The Netherlands
In: Proc. F.O.E. conference: The effects of acid rain and air polltuion in Northern Europe, London, January 1987: 55-68, ISBN O 905966 49X

26. Roelofs J.G.M., Boxman A.W. and Dijk H.F.G.van, 1987
Effects of airborne ammonium on natural vegetation and forests
In: Asman W.A.H. and H.S.M.A.Diederen (eds.), Proceedings Eurosap Symposium on ammonia and acidification, Bilthoven, p. 266-276

27. Roelofs J.G.M., Boxman A.W. and Dijk H.F.G.van, 1987
Effecten van ammonium op bos en heidevegetaties
In: A.W.Boxman and J.F.M.Geelen (eds.), Acute en chronische effecten van NH_3 (en NH_4^+) op levende organismen, Katholieke Universiteit Nijmegen, p. 96-104

28. Roelofs J.G.M., Boxman A.W. and Dijk H.F.G.van, 1988
Effects of airborne ammonium on natural vegetation and forests
In: P.Mathy (ed.), Air pollution and ecosystems, D.Reidel Publishing Company, p.876-880
29. Roelofs J.G.M., Boxman A.W., Dijk H.F.G.van and Houdijk A.L.F.M., 1988
Nutrient fluxes in canopies and roots of coniferous trees as affected by nitrogen-enriched air pollution, CEC Air Pollution Research Report 16: 205-221
30. Roelofs J.G.M., Boxman A.W, and Dijk H.F.G.van, 1989
Effects of airborne ammonium on natural vegetation and forests, N.N.A.Ber., 2(1): 38-41

<u>Project no.</u> **119**
<u>Title</u> The role of combination stress, canopy exchange versus nutrient stress in the soil, on the plant species composition of heathlands
<u>Projectleader</u> Dr.G.W.Heil
<u>Address</u> University of Utrecht, Department of Plant Ecology
Lange Nieuwstraat 106, 3512 PN UTRECHT
<u>Tel.</u> 030-394515

01. Bobbink R., Bik L. and Willems J.H., 1988
Effects of nitrogen fertilization on vegetation structure and dominance of Brachypodium pinnatum (L.) Beauv. in chalk grasslands
Acta Bot. Neerl., 37, 231-242
02. Bobbink R. and Willems J.H., 1988
Effects of management and nutrient availability on vegetation structure of chalk grassland
In: During, H.J., Werger, M.J.A. & Willems, J.H. (eds.), Diversity and pattern in plant communities. SPB Academic Publishing, The Hague, pp. 183-193
03. Bobbink R., 1988
De toename van Gevinde kortsteel in Zuidlimburgse kalkgraslanden. Oorzaak - Gevolg - Toekomstig beheer
Publ.Natuurhist. Gen. Limburg, 37(2), 72 pp.
04. Heil G.W., Werger M.J.A., Mol, W.de, Dam, D.van and & Heijne B., 1988
Capture of atmospheric ammonium by grassland canopy
Science 239, 764-765
05. Heil G.W., 1988
LAI of grasslands and their roughness lengths
In: Verhoeven, J.T.A., Heil, G.W. & Werger, M.J.A. (eds.), Vegetation structure in

relation to carbon and nutrient economy. SPB Academic Publishing, The Hague, 149-155

06. Roebertsen H., Heil G.W. and Bobbink R., 1988
Digital picture processing: a new method to analyse vegetation structure
Acta Bot. Neerl., 37, 187-192

07. Bobbink R., Dubbelden K.den and Willems J.H., 1989
Seasonal dynamics of phytomass and nutrients in chalk grassland
Oikos, 55, 216-224

08. Bobbink R., 1989
Brachypodium pinnatum and the species diversity in chalk grassland
Ph.D. Thesis, Utrecht

09. Bobbink R., Heil G.W., Dam D.van and Heijne B., 1989
Canopy-exchange processes and deposition of sulphur and nitrogen in heathland
Abstracts International Congress on Forest Decline Research: State of Knowledge and Perspectives, Friedrichshafen, 125-126

10. Heil G.W., Bobbink R., Dam D.van and Heijne, B., 1989
LAI of grasslands: a measure for ammonium deposition from polluted air
Proc. 8th World Clean Air Congress 1989, The Hague, Vol. 3, 491- 495

11. Heijne B., Dam D. van, Heil G.W. and Bobbink R., 1989
The influence of the "acid rain" component ammonium sulphate on Vesicular Arbuscular Mycorrhiza
Proc. 8th World Clean Air Congress 1989, The Hague, Vol. 2, 257-261

12. Dam D.van, Bobbink R., Heil G.W. and Heijne B., 1989
Nitrogen and sulphur cycling in chalk grassland; the influence of acid rain
Proc. 8th World Clean Air Congress 1989, The Hague, Vol. 2, 201- 206

13. Heijne B., Heil G.W. and Dam D.van, 1989
Relations between acid rain and vesicular arbuscular mycorrhiza
Agriculture, Ecosystems and Environment, 29, 187-192

14. Bobbink R., 1990
Effects of nutrient enrichment in Dutch chalk grassland
J. Applied Ecol., vol. 27 (in press)

15. Dubbelden K.C.den and Leeflang L., 1990
Invloed van luchtverontreiniging op natuurterreinen in Limburg; degradatie van natuurwaarden door verzuring en vermesting en bestrijding van de gevolgen door effectgerichte maatregelen
Vakgroep Botanische Oecologie en Evolutiebiologie, Rijksuniversiteit Utrecht

16. Tooren B.van, Dam D.van and During H.J., 1990
The relative importance of precipitation and soil as sources of nutrients for

Calliergonella cuspidata (Hedw.) Loeske in chalk grassland
Funct. Ecol. 4, 101-107

17. Dam D.van, Heil G.W., Heijne B. and Bobbink R., 1990
Atmospheric deposition and sulphur cycling in chalk grassland: a mechanistic model simulating field observations
Biogeochem., 9, 19-38
18. Dam D. van, 1990
Atmospheric deposition and nutrient cycling in chalk grassland
Ph.D. Thesis, Utrecht
19. Bobbink R., Heil G.W. and Raessen M.B.A.G., 1990
Atmospheric deposition and canopy exchange in heathland ecosystems
Department of Plant Ecology and Evolutionary Biology, University of Utrecht, 80pp.
20. Bobbink R., Heil G.W. and Scheffers M., 1990
Depositie van NOx op wegbermvegetaties, Vakgroep Botanische Oecologie en Evolutiebiologie, Rijksuniversiteit Utrecht / Rijkswaterstaat, Delft, 64 pp.
21. Dam D.van, Heil G.W., Bobbink R. and Heijne B.
Atmospheric deposition to grassland canopies: lysimeter budgets discriminating between interception deposition, mineral weathering and mineralization
accepted, Air, Water and Soil Pollution
22. Heijne B., Hofstra J.J., Heil G.W., Dam D.van and Bobbink R. (subm.)
Effects of atmospheric ammonium sulphate on VAM infection of five heathland species
23. Dam D.van, Heil G.W., Bobbink R. and Heijne B. (subm)
Throughfall below grassland canopies: a comparison between conventional and ion-exchange methods
24. Dam D.van, Heil G.W., Bobbink R. and Heijne B. (subm.)
Impact of atmospheric deposition on nutrient cycling in chalk grassland: throughfall, canopy exchange, nitrogen turnover and input/output budgets

<u>Project no.</u> **120**

<u>Title</u> Additional measurements (ACIFORN)

<u>Projectleader</u> Th.R.Thijsse

<u>Address</u> Organization for Applied Scientific Research TNO, Division of Technology for Society
P.O.Box 217, 2600 AD DELFT

<u>Tel.</u> 015-696290

01. Thijsse Th.R.
Supplemental hydrocarbon measurements at the location Kootwijk (in Dutch)
in press

Project no. **122**
Title Field and laboratory study on effects of acidification on heathland species
Projectleader Drs.H.F.van Dobben
Address Research Institute for Nature Management
Broekhuizerlaan 2, 3956 NS LEERSUM
Tel. 085-452991

01. Dam D.van, Dobben H.F.van, Braak C.F.J.ter and Wit T.de, 1986
Air pollution as a possible cause for the decline of some phanerogamic species in the Netherlands
Vegetatio, 65: 47-52
02. Fennema F., 1990
Effects of exposure to atmospheric SO_2, NH_3 and $(NH_4)_2SO_4$ on survival and extinction of Arnica montana L. and Viola canina L
RIN report 90/14, Arnhem
03. Fennema F.
Effects of NH_3 deposition on the extinction of Arnica montana L.
(in prep.)
04. Fennema F.
Effect of sulphur dioxide and ammonia on the reproduction success of Arnica montana L.
(in prep.)

Project no. **123**
Title Descriptive diagnostic modelling of the effects of acidification on the natural values of the forest in The Netherlands (DAS)
Projectleader G.van Wirdum
Address Research Institute for Nature Management
P.O.Box 46, 3956 ZR LEERSUM
Tel. 03434-52941

Project no. **124**

Title Effects of ammonia and ammonium sulphate on natural vegetation growing on dry nutrient poor soils

Projectleader H.F.van Dobben

Address Research Institute for Nature Management
P.O.Box 46, 3956 ZR LEERSUM

Tel. 03434-52941

01. Ballach H.J., Dueck Th.A. and Eerden L.J.M.van der, 1987
Einfluss von Schwefeld oxid auf die Blattpigmente von Potentilla erecta (L.)Raeusch.
STAUB 47, 267-271

02. Berdowski J.J.M., 1991
Growth and reproduction of Arnica montana L. influenced by ammonium sulphate
in prep.

03. Berdowski J.J.M., 1991
Photosynthesis, root respiration and carbon balance of Calluna vulgaris (L.) Hull, influenced by SO_2, NH_3 and ammonium sulphate
in prep.

04. Dueck Th.A., Eerden L.J.M.van der and Berdowski J.J.M., 1987
Toxische en eutrofiërende effecten van luchtverontreiniging op planten
In: Verslag zesde studiedag Heidebeheer: "Heidebeheer en Luchtverontreiniging"

05. Dueck Th.A., 1989
Air pollution and reproductive processes in natural plant species
Ecological responses to environmental stresses, J.Rozema and J.A.C.Verkleij (eds.)
Dr.W.Junk Pub.

06. Dueck Th.A., 1990
Effect of ammonia and sulphur dioxide on the survival and growth of Calluna vulgaris (L.) Hull seedlings
Funct.Ecol., 4, 109-116

07. Dueck Th.A., Eerden L.J.van der and Berdowski J.J.M., 1990
Influence of ammonia and ammonium sulphate on heathland vegetation
In: Proc. 4th Int.Conf.Environm. Contam. (J.Barcelo, ed.) 30 - 32, Barcelona, Spain

08. Dueck Th.A., Dorel F.G., Horst ter R. and Eerden L.J.van der, 1990
The effect of ammonia, ammoniumsulphate and sulphur dioxide on the frost sensitivity of Scots pine (Pinus sylvestris L.)
Water, Air and Soil Pollut.
in press

09. Dueck Th.A., Eerden L.J.van der and Berdowski J.J.M., 1990
A risk analysis of SO_2 on heathland species in The Netherlands
Funct. Ecol. (submitted)

10. Dueck Th.A., Eerden L.J.van der, Beemsterboer B. and Elderson J., 1990
Nitrogen uptake and allocation by Calluna vulgaris (L.) Hull and Deschampsia flexuosa (L.) Trin., exposed to $^{15}NH_3$
Acta botanica
submitted

11. Dueck Th.A., Elderson J., 1991
Growth and reproduction of Arnica montana L., Viola canina L., in competition with Agrostis capillaris L. and exposed to ammonia and/or sulphur dioxide
in prep.

12. Pérez-Soba M., Stulen I. and Eerden L.J.van der, 1989
Effects of ammonia on the nitrogen metabolism of Pinus sylvestris
Proc.Int.Congr. on Forest Decline Research, Friedrichshafen, 2-9 Oct., 1989
in press

13. Pérez-Soba M., Eerden L.J.van der, Stulen I., 1990
Wet and dry deposition of ammonia to Scots pine
In: Proc.Int.Conf.Acidic Deposition; its Nature and Impacts, September 1990, Glasgow, United Kingdom, 274 CEP Consultants, Edinburgh

14. Prins A.H., Berdowski J.J.M. and Dueck Th.A., 1990
Influences of nitrogen deposition on gap forming in heathland vegetation; consequences for species composition
Acta Bot.Neerl.
submitted

15. Prins A.H. and Berdowski J.J.M., 1991
Effect of ammonium, aluminium and pH on dry heath herbs
in prep.

16. Termorshuizen A.J., Eerden L.J.van der, Dueck Th.A. and Berdowski J.J.M., 1989
The effects of SO_2 alone or in combination with NH_3 on mycorrhizal and non-mycorrhizal Pinus sylvestris seedlings
Neth.J.Pl.Path.
in prep.

17. Termorshuizen A.J., Eerden L.J.van der and Dueck L.J., 1989
The effects of SO_2 pollution on mycorrhizal and non-mycorrhizal seedlings of Pinus sylvestris
Agric., Ecos. and Environmn. 28, 513 - 518

18. Eerden L.J.van der, Lekkerkerk L.J., Smeulders S.M. and Jansen A.E., 1989

Effects of ozone and ammonia on Douglas fir (Pseudotsuga menziesii)
In: IUFRO Symp. P2.05 Interlaken, Switzerland, Oct. 1988, pp. 538-540

19. Eerden L.J.van der, Perez-Soba M. and Dueck Th.A., 1989
Responses of Pinus sylvestris to atmospheric ammonia
Proc.Int.Congr. on Forest Decline Research, Friedrichshafen 2-9 Oct., 1989
in press

20. Eerden L.J.van der, Dobben H.F.van, Dueck Th.A. and Berdowski J.J., 1990
Effects of atmospheric ammonia and ammonium on vegetation
In: Ammoniak und die Umwelt. KTBL/VDI, Braunschweig, Germany, 601 - 619

21. Eerden L.J.van der, Dueck Th.A., Berdowski J.J.M., Dobben M.F.van and Greven H., 1990
Influence of NH_3, SO_2 and $(NH_4)_2SO_4$ on heathland vegetation
Acta botanica Neerl., submitted
Abstract in: Proc.Int.Conf.Acidic Deposition; its Nature and Impacts, September 1990, Glasgow, United Kingdom, 85. CEP Consultants, Edinburgh, submitted

22. Eerden L.J.van der, Lekkerkerk L.J., Smeulders S.M. and Jansen A.E., 1990
Effects of ammonia and ammonium sulphate on Douglas fir (Pseudotsuga menziesii)
Atm.Env.
submitted

23. Eerden L.J.van der, 1991
The effect of ammonium sulphate on the growth and competition of Calluna vulgaris (L.) Hull, Deschampsia flexuosa (L.) Trin. and Molinia caerula (L.) Moench in mixed stands
in prep.

24. Eerden L.J.van der, 1991
Uptake and allocation of nitrogen by heathland species exposed to ammonia and ammonium sulphate
in prep.

Project no. **125**

Title: Effects of ammonia and ammonium sulphate on natural vegetation growing on dry nutrient poor soils

Projectleader: Ir.L.J.van der Eerden

Address: Research Institute for Plant Protection
P.O.Box 9060, 6700 GW WAGENINGEN

Tel. 08370-19151

see project 124

Project no. **126**

Title Quality assurance for the Dutch Priority Programme on Acidification; measurements of soil acidification

Projectleader Drs.J.C.T.Hollander

Address Organization for Applied Scientific Research TNO, Division of Technology for Society
P.O.Box 217, 2600 AE DELFT

Tel. 015-696900

01. Verhagen H.L.M. en Diederen H.S.M.A., 1991
Vergelijkingsmetingen van de analyse- en monsternemingsmethode van de vaste en de vloeibare fase van bodemmonsters
TNO-rapport

see project 120

Project no. **127**

Title Validation of methods and quality control for the Dutch Priority Programme on Acidification; air quality and deposition measurements

Projectleader Drs.J.C.Th.Hollander

Address Organization for Applied Scientific Research TNO, Division of Technology for Society
P.O.Box 217, 2600 AE DELFT

Tel. 015-696900

01. Hollander J.C.Th., 1990
Vergelijkbaarheid van resultaten van analysemethoden toegepast voor onderzoek van natte depositie in het verzuringsonderzoek. Resultaten tweede ringonderzoek
Additioneel Programma Verzuringsonderzoek, rapport 127-1 (Validation of methods and quality control for the Dutch Priority Programme on Acidification, air quality and desorption measurements
TNO report R 90/104

02. Verhagen H.L.M. and Diederen H.S.M.A., 1990
Methodevalidatie en kwaliteitscontrole ten behoeve van concentratiemetingen in de buitenlucht
Additioneel Programma Verzuringsonderzoek, rapport 127-2 (Validation of methods and quality control of concentration measurements in open air
TNO report R 90/401

Project no. **128**
Title Additional measurements for field experiments (ACIFORN)
Projectleader Dr.P.Wyers
Address Energy Research Foundation
P.O.Box 1, 1755 ZG PETTEN
Tel. 02246-4155

01. Keuken M.P., Wayers-Ypelaan A., Mols J.J., Otjes R.P. and Slanina J., 1989
The determination of ammonia in ambient air by an automated thermodenuder system
Atmospheric Environment 23, 2177 - 2185

02. Keuken M.P., Schoonebeek C.A.M., Wensveen A.van and Slanina J., 1988
Simultaneous sampling of NH_3, HNO_3, HCl, SO_2 and H_2O_2 in ambient air by a wet annular denuder system
Atmospheric Environment 22, 2541 - 2548

03. Slanina J., Arends B.G., Keuken M.P., Veltkamp A.C. and Wyers G.P.
Acidification research at ECN: Contribution to the second phase of the Dutch Priority Programme on Acidification
ECN report, Netherlands Energy Research Foundation
in press

Project no. **129**
Title Throughfall measurements
Projectleader Dr.P.Wyers
Address Energy Research Foundation
P.O.Box 1, 1755 ZG PETTEN
Tel. 02246-4251

01. Slanina J., Arends B.G., Keuken M.P., Veltkamp A.C. and Wyers G.P.
Acidification research at ECN: Contribution to the second phase of the Dutch Priority Programme on Acidification
ECN report, Netherlands Energy Research Foundation
in press

Project no. **130**
Title Field measurements of ammonia volatilization from grazed pastures at different N fertilizer levels
Projectleader Ir.D.J.den Boer
Address Netherlands Fertilizer Institute
Runderweg 6, 8219 PK LELYSTAD
Tel. 03200-22514

see project 67

Project no. **131**
Title Estimation of ammonia volatilization following application of liquid manure on grassland
Projectleader Drs.N.Vertregt
Address Centre for Agricultural Biological Research (CABO)
P.O.Box 14, 6700 AA WAGENINGEN
Tel. 08370-75700

01. Molen J.van der, Bussink D.W., Vertregt N., Faassen H.g.van, Boer D.J.den, 1989
Ammonia volatilization from arable and grassland soils
In: Nitrogen in organic wastes applied to soils (eds: A.A.Jens, Hansen, Kaj Henriksen), Academic Press
02. Vertregt N. and Selis H.E., 1990
Ammonia emission from cattle and pig slurry applied to grassland
CABO report 130, Dutch Priority Programme on Acidification, report 131-1

Project no. **132**
Title Emissions of ammonia after application of manure slurry
Projectleader Drs.H.G.van Faassen
Address Institute for Soil Fertility
P.O.Box 30003, 9750 RA HAREN (Gr.)
Tel. 050-337777

see project 65

Project no. **133**

Title Reduction of ammonia emissions from arable land and grassland by means of tillage techniques

Projectleader Ing.J.V.Klarenbeek

Address Institute for Agricultural Engineering
Mansholtlaan 10-12, 6708 PA WAGENINGEN

Tel. 08370-94556

01. Bussink D.W. and Klarenbeek J.V., 1990
Ammoniak en andere verliezen uit dunne mest
Praktijkonderzoek (3)3 pp. 13 - 16, Proefstation Rundveehouderij, Lelystad

02. Bruins M.A. and Huijsmans J.F.M., 1989
De reductie van ammoniakemissies uit varkensmest na de toediening op bouwland
Rapport 225, IMAG, Wageningen

03. Bruins M.A. and Hol J.M.G., 1990
De ammoniakemissie na aanwending van vaste varkensmest en varkensmengmest
IMAG-nota P-498, IMAG, Wageningen

04. Huijsmans J.M.G. and Klarenbeek J.V., 1988
Snel onderwerken van mest nuttig
Landbouwmechanisatie 8, pp. 28 - 29

05. Hol J.M.G. and Bruins M.A., 1990
Onderzoek naar de invloed van het tijdstip van uitrijden op de ammoniakemissie
IMAG-nota P-499, IMAG, Wageningen

06. Hol J.M.G., 1990
De ammoniakemissie na aanwending van slachtkuikenmest, droge- en natte kippenmest
IMAG-nota P-564, IMAG, Wageningen

07. Klarenbeek J.V. and Bruins M.A., 1990
Ammonia emissions during and after land spreading of animal slurries
In: Odour and ammonia emissions from livestock productions; eds. Nielsen V.C., Pain B.F. and Hartung J., CEC Brussels
in press

08. Pain B.F., Phillips V.R., Huijsmans J.F.M. and Klarenbeek J.V., 1991
Anglo-Dutch experiments on odour and ammonia emission following the spreading of piggery wastes on arable land
IMAG report, IMAG, Wageningen
in press

Project no. **150**
Title Interaction acid deposition soil
Projectleader Dr.P.Wyers
Address Energy Research Foundation
P.O.Box 1, 1755 ZG PETTEN
Tel. 02246-4251

01. Slanina J., Arends B.G., Keuken M.P., Veltkamp A.C. and Wyers G.P.
Acidification research at ECN: Contribution to the second phase of the Dutch Priority Programme on Acidification
ECN report, Netherlands Energy Research Foundation
in press

Project no. **190.1** (VROM project)
Title Measurement of dry deposition fluxes of NH_3 over coniferous forest
Projectleader Ing.J.H.Duyzer
Address Organization for Applied Scientific Research TNO, Division of Technology for Society
P.O.Box 217, 2600 AE DELFT
Tel. 015-696263

see project 120

Project no. **190.2** (VROM project)
Title Measurement of emission fluxes of NH_3 over pasture
Projectleader Ing.J.H.Duyzer
Address Organization for Applied Scienctific Research TNO, Division of Technology for Society
P.O.Box 217, 2600 AE DELFT
Tel. 015-696263

see project 120

Project no. **190.3** (VROM project)
Title Inputs of nitrogen compounds and photooxidants as compounds of the pollution climate in Europe
Projectleader Ing.J.H.Duyzer
Address Organization for Applied Scienctific Research TNO, Division of Technology for Society
P.O.Box 217, 2600 AE DELFT
Tel. 015-696263

see project 120

Project no. **191**
Title Deposition over forests and heathlands
Projectleader Dr.R.M.van Aalst
Address National Institute of Public Health and Environmental Protection
P.O.Box 1, 3720 BA BILTHOVEN
Tel. 030-742002

see project 120

Project no. **200**
Title Management Acidification Research
Projectleader Dr.T.Schneider/Ir.G.J.Heij
Address National Institute of Public Health and Environmental Protection
P.O.Box 1, 3720 BA BILTHOVEN
Tel. 030-742970

01. Schneider T., 1985
Voortgangsrapport verzuringsonderzoek
02. Schneider T. and Bresser A.H.M., 1986
Voortgangsverslag verzuringsonderzoek
RIVM report nr. 758475001
03. Schneider T. and Bresser A.H.M, 1986
Dutch priority programme on acidification; Programme and projects first phase
Report nr. 00-01
04. Schneider T. and Bresser A.H.M, 1986
Dutch priority programme on acidification; Programme and projects first phase
Report nr. 00-01 (revised)

05. Schneider T. and Bresser A.H.M, 1986
Dutch priority programme on acidification; Proceedings of the first symposium, December 1985
Report nr. 00-02

06. Schneider T. and Bresser A.H.M, 1987
Dutch priority programme on acidification; Proceedings of the second symposium, November 1986
Report nr. 00-03

07. Schneider T. and Bresser A.H.M, 1987
Verzuringsonderzoek eerste fase, tussentijdse evaluatie
Report nr. 00-04

08. Schneider T. and Bresser A.H.M, 1988
Kort verslag van het derde symposium verzuringsonderzoek 11-15 januari 1988
Report nr. 00-005

09. Schneider T. and Bresser A.H.M., 1988
Additioneel programma verzuringsonderzoek; evaluatierapport verzuring, 1988
Report nr. 00-06

10. Schneider T. and Bresser A.H.M., 1988
Summary report; Acidification research 1984 - 1988
Report nr. 00-06 (English translation)

11. Bresser A.H.M., 1988
Verslag van de derde bijeenkomst van coördinatoren verzuringsonderzoek MARC III
Report nr. 00-07

12. Schneider T., Bresser A.H.M. and Ruijter H.B.de, 1988
Financieel rapport
Report nr. 00-08

13. Schneider T. and Bresser A.H.M., 1989
Programme and projects 2nd phase
Report nr. 200-01

14. Schneider T. and Bresser A.H.M., 1989
Proceedings of the fourth symposium on acidification, January 23-27, 1989
Report nr. 200-02

15. Heij G.J. and Bresser A.H.M., 1989
Workshop report; Nitrogen - soil acidification and forest decline, May 22-24, 1989, The Netherlands
Report nr. 200-03

16. Schneider T. and Heij G.J., 1990
5e Symposium Verzuringsonderzoek, Bilthoven 5-9 maart 1990

Report nr. 200-04

17. Heij G.J. (ed.), 1990
Verslag NAPAP International Conference on Acidic Deposition: State of Science and technology, 11-16 Februari 1990
Report nr. 200-06

18. Voorburg J.H., Monteny G.J., Aalst R.M.van, Erisman J.W., Breemen N.van, Verstraten J.M., Vries W.de, Kros J., Posthumus A.C., Jansen A.E., Dobben H.F.van, Mohren G.M.J. and Olsthoorn T.N., 1990
Thematic Reports Dutch Priority Programme on Acidification, 1990
Report nr. 200-07

19. Wilson R.B., Nilsson J., Driscoll C.T., Freer Smith P., Ellenberg H., Fowler D., Hartung J. and Landmann G., 1990
Review Report Dutch Priority Programme on Acidification
Report nr. 200-08

<u>Project no.</u> **200.1**

<u>Title</u> A materials-flow model for dairy farms on sandy soils

<u>Projectleader</u> Prof.Dr.L.C.Zachariasse

<u>Address</u> Agricultural Economics Research Institute
Postbus 29703, 2502 LS DEN HAAG

<u>Tel.</u> 070-614161

01. Hoogervorst N.J.P., Dijk J., Veen M.Q.van der and Aarts H.F.M., 1990
Stofstromen op landbouwbedrijven: Scenario's voor melkveehouderijbedrijven in zandgebieden
In: Landbouw, milieu en ruimte; symposiumverslag; red.: Brouwer F.M. and Reinhard A.J., Den Haag Landbouw-Economisch Instituut, 1990, Mededeling 432, ISBN 90-5242-093-9

<u>Project no.</u> **200.2**

<u>Title</u> Research project on the relationship between micrometeorological deposition measurements and throughfall measurements in heathland

<u>Projectleader</u> Dr.G.W.Heil

<u>Address</u> University of Utrecht, Department of Plant Ecology
Lange Nieuwstraat 106, 3512 PN UTRECHT

<u>Tel.</u> 030-394515

see project 119

Project no. **201.1**
Title Review report: mycorrhiza aspects
Projectleader Dr.A.E.Jansen
Address Agricultural University, Department of Phytopathology
P.O.Box 8025, 6700 EE WAGENINGEN
Tel. 08370-83130

01. Jansen A.E. and Dighton J., 1990
Effects of air pollutants on ectomycorrhizas; A review
CEC Air Pollution Research Report nr. 30, Brussel October 1990
02. Dighton J. and A.E.Janssen
Atmospheric pollutants and ectomycorrhizas: more questions than answers? (abstract)
In: Proceedings American Conference on Mycorrhizas, September 1990 (in press)

Project no. **202.1**
Title Modeling of soil acidification in West German and Dutch forest ecosystems
Projectleader Prof.Dr.Ir.N.van Breemen
Address Agricultural University, Department of Soil Science and Geology
P.O.Box 37, 6700 AA WAGENINGEN
Tel. 08370-84145

01. Wesselink L.G., 1990
Modeling of soil acidification in Dutch and West-German forest ecosystems
Report project 202.1; Forschungszentrum Waldökosysteme, Göttingen, BRD and Department of Soil Science and Geology, Agricultural University Wageningen

Project no. **203**
Title CORRELACI (Correlative Evaluation of Aciforn data)
Projectleader Prof.Dr.Ir.N.van Breemen
Address Agricultural University, Department of Soil Science and Geology
P.O.Box 37, 6700 AA WAGENINGEN
Tel. 08370-84145
Projectleader Ir.P.Hofschreuder
Address Agricultural University, Department of Air Pollution
P.O.Box 8129, 6700 AA WAGENINGEN
Tel. 08370-82104

Projectleader Ir.J.van den Bos
Address Research Institute for Forestry and Urban Ecology "De Dorschkamp"
P.O.Box 23, 6700 AA WAGENINGEN
Tel. 08370-95246
Projectleader Dr.Ir.C.van den Akker
Address National Institute of Public Health and Environmental Protection
P.O.Box 1, 3720 BA BILTHOVEN
Tel. 030-743361
Projectleader Prof.Dr.J.M.Verstraten
Address University of Amsterdam, Laboratory of Physical Geography and Soil Science
Dapperstraat 115, 1093 BS AMSTERDAM
Tel. 020-5257451
Projectleader Dr.A.P.van Ulden
Address Royal Netherlands Meteorological Institute
P.O.Box 201, 3730 AE DE BILT
Tel. 030-206911

01. Swart W.A.J.M., Steingröver E., Jans W. and Evers P.W.
Carbon partitioning in a Douglas fir stand
Dorschkamp report
in prep.

02. Steingröver E., Swart W., Evers P.W., Vermetten A. and Meulen E.van der
Effects of ozone, PAR, temperature and VPD on photosynthesis
Dorschkamp report
in prep.

03. Maas M.P.van der, Evers P.W. and Grinsven J.J.M.
The effect of air pollution on the nutrient status of Douglas fir
Dorschkamp report
in prep.

04. Versluis A.H., Vermetten A.W.M., Bouten W., Maas M.P.van der and Hofschreuder P.
Canopy interactions in a Douglas fir forest: estimating dry deposition from data on canopy wetness, throughfall and wet deposition
Dorschkamp report
in prep.

05. Tiktak A., Bouten W., Jans W. and Olsthoorn A.F.M.
Temporal dynamics of shoot extension and fine root activity affected by traditional

stress factors
Dorschkamp report
in prep.

06. Bouten W., Bosveld F.C., Noppert F., Steingroever E. and Tiktak A.
Transpiration dynamics of a Douglas fir forest: evaluation of four measuring techniques
Dorschkamp report
in prep.

07. Bosveld F.C., Bouten W., Noppert F. and Steingröver E.
A transpiraton model for a Douglas fir forest
Dorschkamp report
in prep.

08. Noppert F., Bouten W., Bosveld F., Vermetten A., Tiktak A., Steingröver E. and Olsthoorn A.
Detection of effects of meteorology, soil moisture and air pollution on the transpiration dynamics of a Douglas fir stand; a stochastica approach
Dorschkamp report
in prep.